CAMBRIDGE STUDIES IN
ADVANCED MATHEMATICS 100

MARKOV PROCESSES, GAUSSIAN PROCESSES, AND LOCAL TIMES

Written by two of the foremost researchers in the field, this book studies the local times of Markov processes by employing isomorphism theorems that relate them to certain associated Gaussian processes. It builds to this material through self-contained but harmonized "mini-courses" on the relevant ingredients, which assume only knowledge of measure-theoretic probability. The streamlined selection of topics creates an easy entrance for students and experts in related fields.

The book starts by developing the fundamentals of Markov process theory and then of Gaussian process theory, including sample path properties. It then proceeds to more advanced results, bringing the reader to the heart of contemporary research. It presents the remarkable isomorphism theorems of Dynkin and Eisenbaum and then shows how they can be applied to obtain new properties of Markov processes by using well-established techniques in Gaussian process theory. This original, readable book will appeal to both researchers and advanced graduate students.

Cambridge Studies in Advanced Mathematics

All the titles listed below can be obtained from good booksellers or from Cambridge University Press. For a complete series listing, visit http://www.cambridge.org/us/mathematics

MARKOV PROCESSES, GAUSSIAN PROCESSES, AND LOCAL TIMES

MICHAEL B. MARCUS

City College and the CUNY Graduate Center

JAY ROSEN

College of Staten Island and the CUNY Graduate Center

CAMBRIDGE UNIVERSITY PRESS

Cambridge, New York, Melbourne, Madrid, Cape Town, Singapore, São Paulo

Cambridge University Press

32 Avenue of the Americas, New York, NY 10013-2473, USA

www.cambridge.org

Information on this title: www.cambridge.org/9780521863001

First published 2006

Printed in the United States of America

A catalog record for this publication is available from the British Library.

Library of Congress Cataloging in Publication Data

Marcus, Michael B.

Markov processes, Gaussian processes, and local times / Michael B. Marcus, Jay Rosen.

p. cm. – (Cambridge studies in advanced mathematics ; 100)

Includes bibliographical references and index.

ISBN-13: 978-0-521-86300-1 (hardback)

ISBN-10: 0-521-86300-7 (hardback)

1. Markov processes.   2. Gaussian processes.   3. Local times (Stochastic processes)

I. Rosen, Jay, 1948–   II. Title.   III. Series.

QA274.7.M35 2006

519.2′33–dc22   2006042511

ISBN-13   978-0-521-86300-1 hardback

ISBN-10   0-521-86300-7 hardback

To our wives

Jane Marcus

and

Sara Rosen

Contents

1

Introduction

We found it difficult to choose a title for this book. Clearly we are not covering the theory of Markov processes, Gaussian processes, and local times in one volume. A more descriptive title would have been "A Study of the Local Times of Strongly Symmetric Markov Processes Employing Isomorphisms That Relate Them to Certain Associated Gaussian Processes." The innovation here is that we can use the well-developed theory of Gaussian processes to obtain new results about local times.

Even with the more restricted title there is a lot of material to cover. Since we want this book to be accessible to advanced graduate students, we try to provided a self-contained development of the Markov process theory that we require. Next, since the crux of our approach is that we can use sophisticated results about the sample path properties of Gaussian processes to obtain similar sample path properties of the associated local times, we need to present this aspect of the theory of Gaussian processes. Furthermore, interesting questions about local times lead us to focus on some properties of Gaussian processes that are not usually featured in standard texts, such as processes with spectral densities or those that have infinitely divisible squares. Occasionally, as in the study of the p-variation of sample paths, we obtain new results about Gaussian processes.

Our third concern is to present the wonderful, mysterious isomorphism theorems that relate the local times of strongly symmetric Markov processes to associated mean zero Gaussian processes. Although some inkling of this idea appeared earlier in Brydges, Fröhlich and Spencer (1982) we think that credit for formulating it in an intriguing and usable format is due to E. B. Dynkin (1983), (1984). Subsequently, after our initial paper on this subject, Marcus and Rosen (1992d), in which we use Dynkin's Theorem, N. Eisenbaum (1995) found an unconditioned isomorphism that seems to be easier to use. After this Eisenbaum, Kaspi,

Marcus, Rosen and Shi (2000) found a third isomorphism theorem, which we refer to as the Generalized Second Ray–Knight Theorem, because it is a generalization of this important classical result.

Dynkin's and Eisenbaum's proofs contain a lot of difficult combinatorics, as does our proof of Dynkin's Theorem in Marcus and Rosen (1992d). Several years ago we found much simpler proofs of these theorems. Being able to present this material in a relatively simple way was our primary motivation for writing this book.

The classical Ray–Knight Theorems are isomorphisms that relate local times of Brownian motion and squares of independent Brownian motions. In the three isomorphism theorems we just referred to, these theorems are extended to give relationships between local times of strongly symmetric Markov processes and the squares of associated Gaussian processes. A Markov process with symmetric transition densities is strongly symmetric. Its associated Gaussian process is the mean zero Gaussian process with covariance equal to its 0-potential density. (If the Markov process, say X, does not have a 0-potential, one can consider $\widehat{X}$, the process X killed at the end of an independent exponential time with mean $1/\alpha$. The 0-potential density of $\widehat{X}$ is the α-potential density of X.)

As an example of how the isomorphism theorems are used and of the kinds of results we obtain, we mention that we show that there exists a jointly continuous version of the local times of a strongly symmetric Markov process if and only if the associated Gaussian process has a continuous version. We obtain this result as an equivalence, without obtaining conditions that imply that either process is continuous. However, conditions for the continuity of Gaussian processes are known, so we know them for the joint continuity of the local times.

M. Barlow and J. Hawkes obtained a sufficient condition for the joint continuity of the local times of Lévy processes in Barlow (1985) and Barlow and Hawkes (1985), which Barlow showed, in Barlow (1988), is also necessary. Gaussian processes do not enter into the proofs of their results. (Although they do point out that their conditions are also necessary and sufficient conditions for the continuity of related stationary Gaussian processes.) This stimulating work motivated us to look for a more direct link between Gaussian processes and local times and led us to Dynkin's isomorphism theorem.

We must point out that the work of Barlow and Hawkes just cited applies to all Lévy processes whereas the isomorphism theorem approach that we present applies only to symmetric Lévy processes. Nevertheless, our approach is not limited to Lévy processes and also opens up the possibility of using Gaussian process theory to obtain many other interesting properties of local times.

Another confession we must make is that we do not really understand the actual relationship between local times of strongly symmetric Markov processes and their associated Gaussian processes. That is, we have several functional equivalences between these disparate objects and can manipulate them to obtain many interesting results, but if one asks us, as is often the case during lectures, to give an intuitive description of how local times of Markov processes and Gaussian process are related, we must answer that we cannot. We leave this extremely interesting question to you. Nevertheless, there now exist interesting characterizations of the Gaussian processes that are associated with Markov processes. We say more about this in our discussion of the material in Chapter 13.

The isomorphism theorems can be applied to very general classes of Markov processes. In this book, with the exception of Chapter 13, we consider Borel right processes. To ease the reader into this degree of generality, and to give an idea of the direction in which we are going, in Chapter 2 we begin the discussion of Markov processes by focusing on Brownian motion. For Brownian motion these isomorphisms are old stuff but because, in the case of Brownian motion, the local times of Brownian motion are related to the squares of independent Brownian motion, one does not really leave the realm of Markov processes. That is, we think that in the classical Ray–Knight Theorems one can view Brownian motion as a Markov process, which it is, rather than as a Gaussian process, which it also is.

Chapters 2–4 develop the Markov process material we need for this book. Naturally, there is an emphasis on local times. There is also an emphasis on computing the potential density of strongly symmetric Markov processes, since it is through the potential densities that we associate the local times of strongly symmetric Markov processes with Gaussian processes. Even though Chapter 2 is restricted to Brownian motion, there is a lot of fundamental material required to construct the σ-algebras of the probability space that enables us to study local times. We do this in such a way that it also holds for the much more general Markov processes studied in Chapters 3 and 4. Therefore, although many aspects of Chapter 2 are repeated in greater generality in Chapters 3 and 4, the latter two chapters are not independent of Chapter 2.

In the beginning of Chapter 3 we study general Borel right processes with locally compact state spaces but soon restrict our attention to strongly symmetric Borel right processes with continuous potential densities. This restriction is tailored to the study of local times of Markov processes via their associated mean zero Gaussian processes. Also, even though this restriction may seem to be significant from the perspective of the general theory of Markov processes, it makes it easier to introduce the beautiful theory of Markov processes. We are able to obtain many deep and interesting results, especially about local times, relatively quickly and easily. We also consider h-transforms and generalizations of Kac's Theorem, both of which play a fundamental role in proving the isomorphism theorems and in applying them to the study of local times.

Chapter 4 deals with the construction of Markov processes. We first construct Feller processes and then use them to show the existence of Lévy processes. We also consider several of the finer properties of Borel right processes. Lastly, we construct a generalization of Borel right processes that we call local Borel right processes. These are needed in Chapter 13 to characterize associated Gaussian processes. This requires the introduction of Ray semigroups and Ray processes.

Chapters 5–7 are an exposition of sample path properties of Gaussian processes. Chapter 5 deals with structural properties of Gaussian processes and lays out the basic tools of Gaussian process theory. One of the most fundamental tools in this theory is the Borell, Sudakov–Tsirelson isoperimetric inequality. As far as we know this is stated without a complete proof in earlier books on Gaussian processes because the known proofs relied on the Brun–Minkowski inequality, which was deemed to be too far afield to include its proof. We give a new, analytical proof of the Borell, Sudakov–Tsirelson isoperimetric inequality due to M. Ledoux in Section 5.4.

Chapter 6 presents the work of R. M. Dudley, X. Fernique and M. Talagrand on necessary and sufficient conditions for continuity and boundedness of sample paths of Gaussian processes. This important work has been polished throughout the years in several texts, Ledoux and Talagrand (1991), Fernique (1997), and Dudley (1999), so we can give efficient proofs. Notably, we give a simpler proof of Talagrand's necessary condition for continuity involving majorizing measures, also due to Talagrand, than the one in Ledoux and Talagrand (1991). Our presentation in this chapter relies heavily on Fernique's excellent monograph, Fernique (1997).

Chapter 7 considers uniform and local moduli of continuity of Gaussian processes. We treat this question in general in Section 7.1. In most of the remaining sections in this chapter, we focus our attention on real-valued Gaussian processes with stationary increments, $\{G(t), t \in R^1\}$, for which the increments variance, $\sigma^2(t-s) := E(G(t) - G(s))^2$, is relatively smooth. This may appear old fashioned to the Gaussian purist but it is exactly these processes that are associated with real-valued Lévy processes. (And Lévy processes with values in R^n have local times only when $n = 1$.) Some results developed in this section and its applications in Section 9.5 have not been published elsewhere.

Chapters 2–7 develop the prerequisites for the book. Except for Section 3.7, the material at the end of Chapter 4 relating to local Borel right processes, and a few other items that are referenced in later chapters, they can be skipped by readers with a good background in the theory of Gaussian and Markov processes.

In Chapter 8 we prove the three main isomorphism theorems that we use. Even though we are pleased to be able to give simple proofs that avoid the difficult combinatorics of the original proofs of these theorems, in Section 8.3 we give the combinatoric proofs, both because they are interesting and because they may be useful later on.

Chapter 9 puts everything together to give sample path properties of local times. Some of the proofs are short, simply a reiteration of results that have been established in earlier chapters. At this point in the book we have given all the results in our first two joint papers on local times and isomorphism theorems (Marcus and Rosen, 1992a, 1992d). We think that we have filled in all the details and that many of the proofs are much simpler. We have also laid the foundation to obtain other interesting sample path properties of local times, which we present in Chapters 10–13.

In Chapter 10 we consider the p-variation of the local times of symmetric stable processes $1 < p \leq 2$ (this includes Brownian motion). To use our isomorphism theorem approach we first obtain results on the p-variation of fractional Brownian motion that generalize results of Dudley (1973) and Taylor (1972) that were obtained for Brownian motion. These are extended to the squares of fractional Brownian motion and then carried over to give results about the local times of symmetric stable processes.

Chapter 11 presents results of Bass, Eisenbaum and Shi (2000) on the range of the local times of symmetric stable processes as time goes to infinity and shows that the most visited site of such processes is transient. Our approach is different from theirs. We use an interesting bound for the behavior of stable processes in a neighborhood of the origin due to Molchan (1999), which itself is based on properties of the reproducing kernel Hilbert spaces of fractional Brownian motions.

In Chapter 12 we reexamine Ray's early isomorphism theorem for the h-transform of a transient regular symmetric diffusion, Ray (1963) and

give our own, simpler version. We also consider the Markov properties
of the local times of diffusions.

In Chapter 13, which is based on recent work of N. Eisenbaum and
H. Kaspi that appears in Eisenbaum (2003), Eisenbaum (2005), and
Eisenbaum and Kaspi (2006), we take up the problem of characterizing associated Gaussian processes. To obtain several equivalencies we
must generalize Borel right processes to what we call local Borel right
processes. In Theorem 13.3.1 we see that associated Gaussian processes
are just a little less general than the class of Gaussian processes that
have infinitely divisible squares. Gaussian processes with infinitely divisible squares are characterized in Griffiths (1984) and Bapat (1989).
We present their results in Section 13.2.

We began our joint research that led to this book over 19 years ago.
In the course of this time we received valuable help from R. Adler, M.
Barlow, H. Kaspi, E. B. Dynkin, P. Fitzsimmons, R. Getoor, E. Giné,
M. Talagrand, and J. Zinn. We express our thanks and gratitude to
them. We also acknowledge the help of P.-A. Meyer.

In the preparation of this book we received valuable assistance and
advice from O. Daviaud, S. Dhamoon, V. Dobric, N. Eisenbaum, S.
Evans, P. Fitzsimmons, C. Houdré, H. Kaspi, W. Li, and J. Rosinski.
We thank them also.

We are also grateful for the continued support of the National Science
Foundation and PSC–CUNY throughout the writing of this book.

1.1 Preliminaries

In this book $\mathcal{Z}$ denotes the integers both positive and negative and $\mathbb{N}$ or
sometimes $\mathcal{N}$ denotes the the positive integers including 0. R^1 denotes
the real line and R_+ the positive half line (including zero). $\overline{R}$ denotes
the extended real line $[-\infty, \infty]$. R^n denotes n-dimensional space and
$|\cdot|$ denotes Euclidean distance in R^n. We say that a real number a is
positive if $a \geq 0$. To specify that $a > 0$, we might say that it is strictly
positive. A similar convention is used for negative and strictly negative.

Measurable spaces: A measurable space is a pair $(\Omega, \mathcal{F})$, where Ω is a set
and $\mathcal{F}$ is a sigma-algebra of subsets of Ω. If Ω is a topological space, we
use $\mathcal{B}(\Omega)$ to denote the Borel σ-algebra of Ω. Bounded $\mathcal{B}(\Omega)$ measurable
functions on Ω are denoted by $\mathcal{B}_b(\Omega)$.

Let $t \in R_+$. A filtration of $\mathcal{F}$ is an increasing family of sub σ-algebras
$\mathcal{F}_t$ of $\mathcal{F}$, that is, for $0 \leq s < t < \infty$, $\mathcal{F}_s \subset \mathcal{F}_t \subset \mathcal{F}$ with $\mathcal{F} = \cup_{0 \leq t < \infty} \mathcal{F}_t$.

(Sometimes we describe this by saying that $\mathcal{F}$ is filtered.) To emphasize a specific filtration $\mathcal{F}_t$ of $\mathcal{F}$, we sometimes write $(\Omega, \mathcal{F}, \mathcal{F}_t)$.

Let $\mathcal{M}$ and $\mathcal{N}$ denote two σ-algebras of subsets of Ω. We use $\mathcal{M} \vee \mathcal{N}$ to denote the σ-algebra generated by $\mathcal{M} \cup \mathcal{N}$.

Probability spaces: A probability space is a triple $(\Omega, \mathcal{F}, P)$, where $(\Omega, \mathcal{F})$ is measurable space and P is a probability measure on Ω. A random variable, say X, is a measurable function on $(\Omega, \mathcal{F}, P)$. In general we let E denote the expectation operator on the probability space. When there are many random variables defined on $(\Omega, \mathcal{F}, P)$, say $Y, Z, \ldots$, we use E_Y to denote expectation with respect to Y. When dealing with a probability space, when it seems clear what we mean, we feel free to use E or even expressions like E_Y without defining them. As usual, we let ω denote the elements of Ω. As with E, we often use ω in this context without defining it.

When X is a random variable we call a number a a median of X if

$$P(X \le a) \ge \tfrac{1}{2} \qquad \text{and} \qquad P(X \ge a) \ge \tfrac{1}{2}. \tag{1.1}$$

Note that a is not necessarily unique.

A stochastic process X on $(\Omega, \mathcal{F}, P)$ is a family of measurable functions $\{X_t, t \in I\}$, where I is some index set. In this book, t usually represents "time" and we generally consider $\{X_t, t \in R_+\}$. $\sigma(X_r; r \le t)$ denotes the smallest σ-algebra for which $\{X_r; r \le t\}$ is measurable. Sometimes it is convenient to describe a stochastic process as a random variable on a function space, endowed with a suitable σ-algebra and probability measure.

In general, in this book, we reserve $(\Omega, \mathcal{F}, P)$ for a probability space. We generally use $(S, \mathcal{S}, \mu)$ to indicate more general measure spaces. Here μ is a positive (i.e., nonnegative) σ-finite measure.

Function spaces: Let f be a measurable function on $(S, \mathcal{S}, \mu)$. The $L^p(\mu)$ (or simply L^p), $1 \le p < \infty$, spaces are the families of functions f for which $\int_S |f(s)|^p \, d\mu(s) < \infty$ with

$$\|f\|_p := \left(\int_S |f(s)|^p \, d\mu(s) \right)^{1/p}. \tag{1.2}$$

Sometimes, when we need to be precise, we may write $\|f\|_{L^p(S)}$ instead of $\|f\|_p$. As usual we set

$$\|f\|_\infty = \sup_{s \in S} |f(s)|. \tag{1.3}$$

These definitions have analogs for sequence spaces. For $1 \le p < \infty$, ℓ_p

is the family of sequences $\{a_k\}_{k=0}^{\infty}$ of real or complex numbers such that $\sum_{k=0}^{\infty} |a_k|^p < \infty$. In this case, $\|a_k\|_p := \left(\sum_{k=0}^{\infty} |a_k|^p\right)^{1/p}$ and $\|a_k\|_{\infty} := \sup_{0 \le k < \infty} |a_k|$. We use ℓ_p^n to denote sequences in ℓ_p with n elements.

Let m be a measure on a topological space $(S, \mathcal{S})$. By an approximate identity or δ-function at y, with respect to m, we mean a family $\{f_{\epsilon,y} ; \epsilon > 0\}$ of positive continuous functions on S such that $\int f_{\epsilon,y}(x) \, dm(x) = 1$ and each $f_{\epsilon,y}$ is supported on a compact neighborhood K_{ϵ} of y with $K_{\epsilon} \downarrow \{y\}$ as $\epsilon \to 0$.

Let f and g be two real-valued functions on R^1. We say that f is asymptotic to g at zero and write $f \sim g$ if $\lim_{x \to 0} f(x)/g(x) = 1$. We say that f is comparable to g at zero and write $f \approx g$ if there exist constants $0 < C_1 \le C_2 < \infty$ such that $C_1 \le \liminf_{x \to 0} f(x)/g(x)$ and $\limsup_{x \to 0} f(x)/g(x) \le C_2$. We use essentially the same definitions at infinity.

Let f be a function on R^1. We use the notation $\lim_{y \uparrow\uparrow x} f(y)$ to be the limit of $f(y)$ as y increases to x, for all $y < x$, that is, the left-hand (or simply left) limit of f at x.

Metric spaces: Let (S, τ) be a locally compact metric or pseudo-metric space. A pseudo-metric has the same properties as a metric except that $\tau(s, t) = 0$ does not imply that $s = t$. Abstractly, one can turn a pseudo-metric into a metric by making the zeros of the pseudo-metric into an equivalence class, but in the study of stochastic processes pseudo-metrics are unavoidable. For example, suppose that $X = \{X(t), t \in [0, 1]\}$ is a real-valued stochastic process. In studying sample path properties of X it is natural to consider $(R^1, | \cdot |)$, a metric space. However, X may be completely determined by an L^2 metric, such as

$$d(s, t) := d_X(s, t) := (E(X(s) - X(t))^2)^{1/2} \tag{1.4}$$

(and an additional condition such as $E \, X^2(t) = 1$). Therefore, it is natural to also consider the space (R^1, d). This may be a pseudo-metric space since d need not be a metric on R^1.

If $A \subset S$, we set

$$\tau(s, A) := \inf_{u \in A} \tau(s, u). \tag{1.5}$$

We use $C(S)$ to denote the continuous functions on S, $C_b(S)$ to denote the bounded continuous functions on S, and $C_b^+(S)$ to denote the positive bounded continuous functions on S. We use $C_{\kappa}(S)$ to denote the continuous functions on S with compact support; $C_0(S)$ denotes the functions on S that go to 0 at infinity. Nevertheless, $C_0^{\infty}(S)$ denotes infinitely differentiable functions on S with compact support (whenever S

is a space for which this is defined). In all these cases we mean continuity with respect to the metric or pseudo-metric τ.

We say that a function is locally uniformly continuous on a measurable set in (S, τ) if it is uniformly continuous on all compact subsets of (S, τ). We say that a sequence of functions converges locally uniformly on (S, τ) if it converges uniformly on all compact subsets of (S, τ).

Separability: Let T be a separable metric space, and let $X = \{X(t), t \in T\}$ be a stochastic process on $(\Omega, \mathcal{F}, P)$ with values in $\overline{R}^n$. X is said to be separable if there is a countable set $D \subset T$ and a P-null set $\Lambda \subset \mathcal{F}$ such that, for any open set $U \subset T$ and closed set $A \subset \overline{R}^n$,

$$\{X(t) \in A, t \in D \cap U\}/\{X(t) \in A, t \in U\} \subset \Lambda. \tag{1.6}$$

If X is separable and $U \subset T$ is an open set and Λ is as above, then $\omega \notin \Lambda$ implies

$$\sup_{t \in D \cap U} |X(t, \omega)| = \sup_{t \in U} |X(t, \omega)| \tag{1.7}$$
$$\inf_{t \in D \cap U} |X(t, \omega)| = \inf_{t \in U} |X(t, \omega)|.$$

If T is a separable metric space, every stochastic process $X = \{X(t), t \in T\}$ with values in $\overline{R}^n$ has a separable version $\widetilde{X} = \{\widetilde{X}(t), t \in T\}$, that is, $P\left(\widetilde{X}(t) = X(t)\right) = 1$, for all $t \in T$, and $\widetilde{X}$ is separable for some D and Λ.

If X is stochastically continuous, that is, $\lim_{t \to t_0} P(|X(t) - X(t_0)| > \epsilon) = 0$, for every $\epsilon > 0$ and $t_0 \in T$, then any countable dense set $V \subset T$ serves as the set D in the separability condition (sometimes called the separability set). The P-null set Λ generally depends on the choice of V.

Fourier transform: We often give results with precise constants, so we need to describe what version of the Fourier transform we are using. Let $f \in L^2(R^1)$. Consistent with the standard definition of the characteristic function, the Fourier transform $\widehat{f}$ of f is defined by

$$\widehat{f}(\lambda) = \int_{-\infty}^{\infty} e^{i\lambda x} f(x) \, dx, \tag{1.8}$$

where the integral exists in the L^2 sense. The inverse Fourier transform is given by

$$f(x) = \frac{1}{2\pi} \int_{-\infty}^{\infty} e^{-i\lambda x} \widehat{f}(\lambda) \, d\lambda. \tag{1.9}$$

With this normalization, Parseval's Theorem is

$$\int_{-\infty}^{\infty} f(x)\overline{g(x)}\, dx = \frac{1}{2\pi} \int_{-\infty}^{\infty} \widehat{f}(\lambda)\overline{\widehat{g}(\lambda)}\, d\lambda. \qquad (1.10)$$

2

Brownian motion and Ray–Knight Theorems

In this book we develop relationships between the local times of strongly symmetric Markov processes and corresponding Gaussian processes. This was done for Brownian motion over 40 years ago in the famous Ray–Knight Theorems. In this chapter, which gives an overview of significant parts of the book, we discuss Brownian motion, its local times, and the Ray–Knight Theorems with an emphasis on those definitions and properties which we generalize to a much larger class of processes in subsequent chapters. Much of the material in this chapter is repeated in greater generality in subsequent chapters.

2.1 Brownian motion

A normal random variable with mean zero and variance t, denoted by $N(0, t)$, is a random variable with a distribution function that has density

$$p_t(x) = \frac{e^{-x^2/2t}}{\sqrt{2\pi t}} \qquad x \in R^1 \tag{2.1}$$

with respect to Lebesgue measure. (It is easy to check that a random variable with density $p_t(x)$ does have mean zero and variance t.) In anticipation of using p_t as the transition density of a Markov process, we sometimes use $p_t(x, y)$ to denote $p_t(y - x)$.

We give some important calculations involving p_t.

Lemma 2.1.1

(1) *The Fourier transform of $p_t(x)$ is*

$$\widehat{p}_t(\lambda) := \int_{-\infty}^{\infty} e^{i\lambda x} p_t(x)\, dx = e^{-t\lambda^2/2}. \qquad (2.2)$$

Equivalently, if ξ is $N(0, t)$

$$E(e^{i\lambda\xi}) = e^{-t\lambda^2/2}. \qquad (2.3)$$

(2) *If ξ is $N(0, t)$ and ζ is $N(0, s)$ and ξ and ζ are independent, then $\xi + \zeta$ is $N(0, s + t)$.*

(3)

$$\int_{-\infty}^{\infty} p_s(x, y) p_t(y, z)\, dy = p_{s+t}(x, z). \qquad (2.4)$$

This equation is called the Chapman–Kolmogorov equation.

(4) *For $\alpha > 0$,*

$$u^{\alpha}(x) := \int_{0}^{\infty} e^{-\alpha t} p_t(x)\, dt = \frac{e^{-\sqrt{2\alpha}|x|}}{\sqrt{2\alpha}}. \qquad (2.5)$$

(We see below that u^{α} is the α-potential density of standard Brownian motion).

Proof For (2.2), we write $i\lambda x - x^2/2t = -(x - i\lambda t)^2/2t - t\lambda^2/2$ so that

$$
\begin{aligned}
\widehat{p}_t(\lambda) &= e^{-t\lambda^2/2} \frac{1}{\sqrt{2\pi t}} \int_{-\infty}^{\infty} e^{-(x - i\lambda t)^2/2t}\, dx \qquad (2.6)\\
&= e^{-t\lambda^2/2} \frac{1}{\sqrt{2\pi t}} \int_{-\infty}^{\infty} e^{-x^2/2t}\, dx = e^{-t\lambda^2/2}.
\end{aligned}
$$

For the second equality note that $(2\pi t)^{-1/2} \int_{C_N} \exp(-z^2/(2t))\, dz = 0$, where C_N is the rectangle in the complex plane determined by $\{x \mid -N \le x \le N\}$ and $\{x - i\lambda t \mid -N \le x \le N\}$ (since $\exp(-z^2/(2t))$ is analytic), and then take the limit as N goes to infinity.

Equation (2.3) is simply a rewriting of (2.2). It immediately gives (2). Equation (2.4), for $z = 0$, follows from (2) and the fact that the density of $\xi + \zeta$, the sum of the independent random variables ξ and ζ, is given by the convolution of the densities of ξ and ζ. For general z we need only note that by a change of variables, $\int_{-\infty}^{\infty} p_s(x, y) p_t(y, z)\, dy = \int_{-\infty}^{\infty} p_s(x - z, y) p_t(y, 0)\, dy$.

For a slightly more direct proof of (2.4) consider $p_t(x, y) = p_t(y - x)$ as a function of y for some fixed x. The Fourier transform of $p_t(y - x)$

is $e^{i\lambda x}\widehat{p}_t(\lambda)$. Similarly, the Fourier transform of $p_t(z-y)$ is $e^{i\lambda z}\widehat{p}_t(\lambda)$. By Parseval's Theorem (1.10), the left-hand side of (2.4) is

$$\frac{1}{2\pi}\int_{-\infty}^{\infty} e^{i\lambda(x-z)}\widehat{p}_t(\lambda)\,\widehat{p}_s(\lambda)\,d\lambda \;=\; \frac{1}{2\pi}\int_{-\infty}^{\infty} e^{i\lambda(x-z)} e^{-(t+s)\lambda^2/2}\,d\lambda$$

$$= \; p_{s+t}(x,z) \qquad (2.7)$$

by (1.8), (1.9), and (2.2).

To prove (2.5) we note that by Fubini's Theorem and (2.2) the Fourier transform of $u^\alpha(x)$ is

$$\widehat{u^\alpha}(\lambda) = \int_0^\infty e^{-\alpha t}\widehat{p}_t(\lambda)\,dt = \frac{1}{\alpha+\lambda^2/2}. \qquad (2.8)$$

Taking the inverse Fourier transform we have

$$u^\alpha(x) = \frac{1}{2\pi}\int_{-\infty}^{\infty} e^{-i\lambda x}\frac{1}{\alpha+\lambda^2/2}\,d\lambda. \qquad (2.9)$$

Evaluating this integral in the complex plane using residues we get (2.5). (For $x \geq 0$, use the contour $(-\rho,\rho)\cup(\rho e^{-i\theta}, 0\leq\theta\leq\pi)$ in the clockwise direction and for $x < 0$ use the contour $(-\rho,\rho)\cup(\rho e^{i\theta},\pi\leq\theta\leq 2\pi)$ in the counterclockwise direction.) □

We define Brownian motion starting at 0 to be a stochastic process $W = \{W_t\,;\,t\in R_+\}$ that satisfies the following three properties:

(1) W has stationary and independent increments.
(2) $W_t \overset{law}{=} N(0,t)$, for all $t\geq 0$. (In particular $W_0 \equiv 0$.)
(3) $t\mapsto W_t$ is continuous.

Theorem 2.1.2 *The three conditions defining Brownian motion are consistent, that is, Brownian motion is well defined.*

Proof We construct a Brownian motion starting at 0. We first construct a probability $\widetilde{P}$ on R^{R_+}, the space of real-valued functions $\{f(t), t\in[0,\infty)\}$ equipped with the Borel product σ-algebra $\mathcal{B}(R^{R_+})$. Let X_t be the natural evaluation $X_t(f) = f(t)$. We first define $\widetilde{P}$ on sets of the form $\{X_{t_1}\in A_1,\ldots,X_{t_n}\in A_n\}$ for all Borel measurable sets $A_1,\ldots,A_n$ in R and $0 = t_0 < t_1 < \cdots < t_n$ by setting

$$\widetilde{P}(X_{t_1}\in A_1,\ldots,X_{t_n}\in A_n) = \int\prod_{i=1}^n p_{t_i-t_{i-1}}(z_{i-1},z_i)\prod_{i=1}^n 1_{A_i}(z_i)\,dz_i.$$

$$(2.10)$$

Here 1_{A_i} is the indicator function of A_i and we set $z_0 = 0$. That this construction is consistent follows from the Chapman–Kolmogorov equation

(2.4). The existence of $\widetilde{P}$ now follows from Kolmogorov's Construction Theorem.

It is obvious, by (2.10), that the random variable $(X_{t_1}, \ldots, X_{t_n})$ has probability density function $\prod_{i=1}^{n} p_{t_i - t_{i-1}}(z_{i-1}, z_i)$. Hence, for $0 = t_0 < t_1 < \cdots < t_n < v$ and measurable functions g and f,

$$E(g(X_{t_1}, \ldots, X_{t_n}) f(X_v - X_{t_n})) \tag{2.11}$$

$$= \int g(z_1, \ldots, z_n) f(y - z_{t_n})$$

$$\prod_{i=1}^{n} p_{t_i - t_{i-1}}(z_{i-1}, z_i) p_{v-t_n}(z_n, y) \prod_{i=1}^{n} dz_i \, dy$$

$$= \int g(z_1, \ldots, z_n) f(y) \prod_{i=1}^{n} p_{t_i - t_{i-1}}(z_{i-1}, z_i) p_{v-t_n}(y) \prod_{i=1}^{n} dz_i \, dy$$

$$= E(g(X_{t_1}, \ldots, X_{t_n})) \, E(f(X_{v-t_n})).$$

This shows that $X_v - X_{t_n}$ is independent of $X_{t_1}, \ldots, X_{t_n}$ and is equal in law to X_{v-t_n}. It follows from this that $\{X_t \, ; \, t \in R_+\}$ satisfies property (1). It is obvious that it satisfies property (2).

We now show that $\{X_t \, ; \, t \in R_+\}$ has a continuous version. To do this we note that by properties (1) and (2) and the fact that $N(0, t - r) \overset{law}{=} |t - r|^{1/2} N(0, 1)$,

$$\widetilde{E}\left(|X_t - X_r|^n\right) = |t - r|^{n/2} \widetilde{E}\left(|X_1|^n\right) \tag{2.12}$$

for all $0 \le r < t$. (It is also easy to check that $N(0, t)$ has moments of all orders.) Thus, by Kolmogorov's Theorem (Theorem 14.1.1), for any fixed $\epsilon > 0$ and some random variable $C(T)$, which is finite almost surely,

$$|X_t - X_r| \le C(T) |t - r|^{1/2 - \epsilon} \tag{2.13}$$

for all dyadic numbers $0 \le r < t \le T$.

We now define a version, $\widetilde{W} = \{\widetilde{W}_t \, ; \, t \in R_+\}$, of Brownian motion. Let D be the set of positive dyadic numbers. On the set $\cap_{N=1}^{\infty} \{C(N) < \infty\}$, we take $\widetilde{W}$ to be the continuous functions obtained by extending $\{X_t \, ; \, t \in D\}$ by continuity. On $(\cap_{N=1}^{\infty} \{C(N) < \infty\})^c$, which has probability 0, we take $\widetilde{W}$ to be identically 0. Obviously $\widetilde{W}$ satisfies property (3). Since $\{X_t \, ; \, t \in R_+\}$ satisfies properties (1) and (2), to see that $\widetilde{W}$ also satisfies properties (1) and (2) it suffices to note that for each t, $\widetilde{W}_t = X_t$ almost surely. This follows from (2.12). This shows that Brownian motion starting at zero is a well-defined stochastic process. $\square$

Remark 2.1.3 $\{\widetilde{W}_t \, ; \, t \in R_+\}$ is a version of Brownian motion. We present another version that has useful properties. Let $C[0, \infty)$ be the space of continuous real-valued functions $\{\omega(t), t \in [0, \infty)\}$ with the topology of uniform convergence on compact sets and let $\mathcal{B}(C)$ be the Borel σ-algebra of $C[0, \infty)$. Let $(\widetilde{P}, R^{R+}, \mathcal{B}(R^{R+}))$ denote the probability space of $\widetilde{W}$. The map

$$\widetilde{W} : (R^{R+}, \mathcal{B}(R^{R+})) \mapsto (C[0, \infty), \mathcal{B})$$

defined by $\widetilde{W}(f)(t) = \widetilde{W}_t(f)$ induces a map of the probability $\widetilde{P}$ on $(R^{R+}, \mathcal{B}(R^{R+}))$ to a probability P on $(C[0, \infty), \mathcal{B})$. Let $W_t := \omega(t)$. $\{W_t; t \in R_+\}$ is a Brownian motion on $(C[0, \infty), \mathcal{B})$. This version of Brownian motion, as a probability on the space of continuous functions, is called the canonical version of Brownian motion.

We mention two classical local limit laws for Brownian motion, $\{W_t; t \in R_+\}$ starting at 0. These laws are not difficult to prove, but we shall not do so at this point since their proofs are included in the more general Theorems 7.2.15 and 7.2.14; see also Example 7.4.13.

Khintchine's law of the iterated logarithm states that

$$\limsup_{\delta \to 0} \frac{W_{x+\delta} - W_x}{\sqrt{2\delta \log \log(1/\delta)}} = 1 \qquad \text{a.s.} \qquad (2.14)$$

Lévy's uniform modulus of continuity states that, for any $T > 0$,

$$\limsup_{\substack{|y-z| \to 0 \\ 0 \le y, z \le T}} \frac{|W_y - W_z|}{\sqrt{2|y-z| \log(1/|y-z|)}} = 1 \qquad \text{a.s.} \qquad (2.15)$$

It is remarkable that Khintchine's law also gives an iterated log law for the behavior of Brownian motion at infinity, namely,

$$\limsup_{t \to \infty} \frac{W_t}{\sqrt{2t \log \log t}} = 1 \qquad \text{a.s.} \qquad (2.16)$$

This is obtained by applying (2.14) to $\overline{W} := \{\overline{W}(t); t \in R_+\}$, for $\overline{W}(t) := tW(1/t)$ for $t \neq 0$ and $\overline{W}(0) = 0$, where $\{W(t); t \in R_+\}$ is a Brownian motion. The point here is that $\overline{W}$ is also a Brownian motion. We show this in Remark 2.1.6.

Remark 2.1.4 Let $\{W_t, t \in R_+\}$ be a Brownian motion starting at 0 on the probability space $(\Omega, \mathcal{F}^0, P)$, where $\mathcal{F}^0$ denotes the σ-algebra generated by W_s, $0 \le s < \infty$. (We use a superscript for $\mathcal{F}^0$, reserving $\mathcal{F}$ for the enlargement of $\mathcal{F}^0$ introduced in Section 2.3.) For each $x \in R^1$ we define a probability P^x on $\mathcal{F}^0$ by setting $P^x(F(W.)) = P(F(x + W.))$, for all measurable functions F on $\mathcal{F}_0$. $\{W_t; t \in R_+; P^x\}$ is called a Brownian motion starting at x.

Using the properties of Brownian motion one can easily show that, for $f_i \in \mathcal{B}_b(R)$, $i = 1, \ldots, n$,

$$
E^x \left(\prod_{i=1}^{n} f_i(W_{t_i}) \right) = \int p_{t_1}(x, z_1) \prod_{i=2}^{n} p_{t_i - t_{i-1}}(z_{i-1}, z_i) \prod_{i=1}^{n} f_i(z_i)\, dz_i
$$

$$
(2.17)
$$

for any $0 = t_0 < t_1 < \ldots < t_n$ (compare this with (2.10)).

For each $t > 0$ we define the transition operator $P_t : \mathcal{B}_b(R) \to \mathcal{B}_b(R)$ by

$$
P_t f(x) = E^x(f(W_t)) = \int p_t(x, z) f(z)\, dz \qquad (2.18)
$$

and take $P_0 = I$, where I denotes the identity operator.

The Chapman–Kolmogorov equation (2.4) shows that $\{P_t \,;\, t \in R_+\}$ is a semigroup of operators, that is, for all $f \in \mathcal{B}_b(R)$, $P_{t+s}f(\cdot) = P_s(P_t f)(\cdot)$. (This fact is often abbreviated by simply writing $P_s P_t = P_{s+t}$.)

For measurable sets $A \in \mathcal{B}(R)$, set $P_t(x, A) = P_t I_A(x)$, so that

$$
P_t(x, A) = \int_A p_t(x, z)\, dz. \qquad (2.19)
$$

We interpret $P_t(x, A)$ as the probability that a Brownian motion, starting at x at time zero, takes a value in A at time t. Because of (2.19) we call $p_t(x, z)$ the transition probability density function of Brownian motion.

For each $\alpha > 0$ we define the α-potential operator $U^\alpha : \mathcal{B}_b(R) \to \mathcal{B}_b(R)$ by

$$
U^\alpha f(x) = \int_0^\infty e^{-\alpha t} P_t f(x)\, dt = E^x \left(\int_0^\infty e^{-\alpha t} f(W_t)\, dt \right) \qquad \alpha > 0.
$$

$$
(2.20)
$$

The second equality uses Fubini's Theorem. To justify this we need to show that $\{W_t(\omega) \,;\, (t, \omega) \in R_+ \times \Omega\}$ is jointly $\mathcal{B}(R_+) \times \mathcal{F}^0$ measurable. This is easy to see because $W_t(\omega)$ can be written as a limit of $\mathcal{B}(R_+) \times \mathcal{F}^0$ measurable functions, that is,

$$
W_t(\omega) = \lim_{n \to \infty} \sum_{j=1}^{\infty} 1_{[j-1/n, j/n)}(t) W_{j/n}(\omega). \qquad (2.21)
$$

Let

$$
u^\alpha(x) = \int_0^\infty e^{-\alpha t} p_t(x)\, dt. \qquad (2.22)
$$

Using (2.18) we have

$$U^\alpha f(x) = \int u^\alpha(x, z) f(z) \, dz. \tag{2.23}$$

As in the case of P_t, for measurable sets $A \in \mathcal{B}(R)$, set $U^\alpha(x, A) = U^\alpha I_A(x)$. Because an equation similar to (2.19) holds in this case also, we call u^α the α-potential density of Brownian motion.

The numerical value of u^α is given in (2.5). In subsequent work we consider the 0-potential density of certain Markov processes. Since, for Brownian motion, $p_t(x) \sim 1/\sqrt{t}$ at infinity, (2.22) is infinite when $\alpha = 0$. Therefore, we say that the 0-potential density of Brownian motion is infinite.

Remark 2.1.5 For later use we note that Brownian motion is recurrent, that is, it takes on all values infinitely often, or, as is often said, it hits all points infinitely often. To see this note that $\{W(t), t \in R_+\}$ and $\{-W(t), t \in R_+\}$ have the same law. Therefore, by (2.16) we see that Brownian motion crosses the boundaries $(t \log \log t)^{1/2}$ and $-(t \log \log t)^{1/2}$ infinitely often, and, since it is continuous, it hits all points in between infinitely often.

Remark 2.1.6 Let $W = \{W(t); t \in R_+\}$ be a Brownian motion and consider $\overline{W} = \{\overline{W}(t); t \in R_+\}$, where $\overline{W}(t) = tW(1/t)$ for $t \neq 0$ and $\overline{W}(0) = 0$. For $t > s$ we write $W(t) = (W(t) - W(s)) + W(s)$. Then, using the fact that W has independent increments, we see that for all $s, t \in R_+$

$$EW(t)W(s) = s \wedge t = E\overline{W}(t)\overline{W}(s). \tag{2.24}$$

We show in Section 5.1 that W and $\overline{W}$ are both mean zero Gaussian processes, and, since mean zero Gaussian processes are determined uniquely by their covariances, they are equivalent as stochastic processes (see, in particular, Example 5.1.10). Here we give a direct proof of this equivalence, which does not require us to explicitly mention Gaussian processes.

Note that for any $0 < t_1 < \cdots < t_n$ and $a_1, \cdots, a_n \in R^1$,

$$\sum_{j=1}^n a_j W(t_j) = \sum_{j=1}^n \left(\sum_{k=j}^n a_k \right) (W(t_j) - W(t_{j-1})). \tag{2.25}$$

Therefore, using the independence of the increments of W, we see that

$$E \exp\left(i\lambda \sum_{j=1}^{n} a_j W(t_j)\right) = \exp\left(-\frac{\lambda^2}{2} \sum_{j=1}^{n} \left(\sum_{k=j}^{n} a_k\right)^2 (t_j - t_{j-1})\right).$$
(2.26)

This shows, by (2.3), that any linear combination of $W(t_j)$, $j = 1, \ldots, n$ is a normal random variable and necessarily

$$E \exp\left(i\lambda \sum_{j=1}^{n} a_j W(t_j)\right)$$
(2.27)

$$= \exp\left(-\frac{\lambda^2}{2} E\left(\left(\sum_{j=1}^{n} a_j W(t_j)\right)^2\right)\right)$$

$$= \exp\left(-\frac{\lambda^2}{2}\left(\sum_{j,k=1}^{n} a_j a_k E(W(t_j)W(t_k))\right)\right)$$

$$= \exp\left(-\frac{\lambda^2}{2}\left(\sum_{j,k=1}^{n} a_j a_k (t_j \wedge t_k)\right)\right).$$

Applying this to $\sum_{j=1}^{n} a_j \overline{W}(t_j) = \sum_{j=1}^{n} a_j t_j W(1/t_j)$ we obtain

$$E \exp\left(i\lambda \sum_{j=1}^{n} a_j \overline{W}(t_j)\right)$$
(2.28)

$$= \exp\left(-\frac{\lambda^2}{2}\left(\sum_{j,k=1}^{n} a_j t_j a_k t_k E(W(1/t_j)W(1/t_k))\right)\right)$$

$$= \exp\left(-\frac{\lambda^2}{2}\left(\sum_{j,k=1}^{n} a_j a_k E(\overline{W}(t_j)\overline{W}(t_k))\right)\right).$$

It follows from this and (2.24) that W and $\overline{W}$ have the same finite joint distributions. Consequently $\overline{W}$ satisfies the first two conditions in the definition of Brownian motion. The fact that $\overline{W}$ is continuous at any $t \neq 0$ is obvious. The continuity at $t = 0$ follows from (2.12) and (2.13) with $\overline{W}$ instead of X. Hence $\overline{W}$ is a Brownian motion.

Remark 2.1.7 Let $0 < t_1 < \cdots < t_n$ and let C denote the symmetric

$n \times n$ matrix with $C_{i,j} = t_j \wedge t_k$, $i, j = 1, \dots, k$. Since

$$\sum_{j,k=1}^{n} a_j a_k (t_j \wedge t_k) = E\left(\left(\sum_{j=1}^{n} a_j W(t_j)\right)^2\right) > 0$$

unless all $a_i = 0$, we have that C is strictly positive definite and hence invertible. In this case the distribution of $(B_{t_1}, \dots, B_{t_n})$ in R^n has density

$$\frac{1}{(2\pi)^{n/2}\sqrt{\det(C)}} e^{-(x, C^{-1}x)/2} \tag{2.29}$$

with respect to Lebesgue measure. To see this we compute characteristic functions. By a change of variables

$$\frac{1}{(2\pi)^{n/2}\sqrt{\det(C)}} \int e^{i\sum_{j=1}^{n} \lambda_j x_j} e^{-(x, C^{-1}x)/2} \prod_{j=1}^{n} dx_j \tag{2.30}$$

$$= \frac{1}{(2\pi)^{n/2}} \int e^{i(\lambda, C^{1/2}x)} e^{-(x,x)/2} \prod_{j=1}^{n} dx_j$$

$$= \frac{1}{(2\pi)^{n/2}} \int e^{i(C^{1/2}\lambda, x)} e^{-(x,x)/2} \prod_{j=1}^{n} dx_j$$

$$= \prod_{j=1}^{n} \frac{1}{(2\pi)^{1/2}} \int e^{i(C^{1/2}\lambda)_j x_j} e^{-x_j^2/2} dx_j$$

$$= e^{-\sum_{j=1}^{n}(C^{1/2}\lambda)_j^2/2} = e^{-(C^{1/2}\lambda, C^{1/2}\lambda)/2} = e^{-(\lambda, C\lambda)/2}.$$

Comparing this with (2.27) shows that $(B_{t_1}, \dots, B_{t_n})$ has probability density (2.29).

2.2 The Markov property

Let $\{W_t, t \in R_+\}$ be a Brownian motion. Let $\mathcal{F}_t^0$ denote the σ-algebra generated by $\{W_s, 0 \le s \le t\}$ so that $\mathcal{F}^0 = \cup_{t \ge 0} \mathcal{F}_t^0$.

Lemma 2.2.1 *For all $x \in R^1$*

$$E^x\left(f(W_{t+s}) \,|\, \mathcal{F}_t^0\right) = P_s f(W_t) \tag{2.31}$$

for all $s, t \ge 0$ and $f \in \mathcal{B}_b(R)$.

Since $P_s f(W_t)$ is W_t measurable, (2.31) implies that, for all $f \in \mathcal{B}_b(R)$,

$$E^x\left(f(W_{t+s}) \,|\, \mathcal{F}_t^0\right) = E^x\left(f(W_{t+s}) \,|\, W_t\right), \tag{2.32}$$

which says that the future of the process $\{W_t \,;\, t \geq 0\}$, given all its past values up to the present, say time t_0, only depends on its present value, $W(t_0)$. When (2.31) holds for any stochastic process $\{W_t; t \geq 0\}$, we say that $\{W_t; t \geq 0\}$ satisfies the simple Markov property.

Proof   We give two proofs of (2.31). In the first we use the fact that if Z is $\mathcal{A}$ measurable and Y is independent of $\mathcal{A}$, then

$$E(f(Z+Y)\,|\,\mathcal{A}) = E(f(z+Y))|_{z=Z}.$$

Writing $W_{t+s} = W_t + (W_{t+s} - W_t)$ we note that $Y := W_{t+s} - W_t$ is independent of $\mathcal{F}_t^0$ and for any $x \in R^1$, under the measure P^x, we have $Y \overset{law}{=} N(0, s)$. Therefore

$$
\begin{aligned}
E^x\left(f(W_{t+s})\,|\,\mathcal{F}_t^0\right) &= E^x\left(f(W_t+Y)\,|\,\mathcal{F}_t^0\right) \qquad (2.33)\\
&= E\left(f(z+Y)\right)|_{z=W_t}\\
&= E^{W_t}\left(f(W_s)\right) = P_s f(W_t).
\end{aligned}
$$

For the second proof we appeal to the definition of conditional expectation and show that, for all $\mathcal{F}_t^0$ measurable sets A,

$$\int_A f(W_{t+s})\,dP^x = \int_A P_s f(W_t)\,dP^x. \qquad (2.34)$$

Since $\mathcal{F}_t^0$ is generated by cylinder sets, that is, sets of the form $\cap_{i=1}^n \{a_i \leq W_{t_i} \leq b_i\}$, where $0 = t_0 < t_1 < \ldots < t_n \leq t$, it suffices to verify (2.34) for sets of this form. More generally we show that, for functions of the form $\prod_{i=1}^n f_i(W_{t_i})$ with $f_i \in \mathcal{B}_b(R)$, $i = 1, \ldots, n$ and $0 = t_0 < t_1 < \ldots < t_n \leq t$,

$$E^x\left(\prod_{i=1}^n f_i(W_{t_i})f(W_{t+s})\right) = E^x\left(\prod_{i=1}^n f_i(W_{t_i})P_s f(W_t)\right), \qquad (2.35)$$

thus completing the second proof.

By (2.17)

$$E^x\left(\prod_{i=1}^n f_i(W_{t_i})f(W_{t+s})\right) \qquad (2.36)$$

$$= \int p_{t_1}(x, z_1) \prod_{i=2}^n p_{t_i - t_{i-1}}(z_{i-1}, z_i) p_{t+s-t_n}(z_n, z)$$

$$\prod_{i=1}^n f_i(z_i)\,dz_i f(z)\,dz$$

$$= \int p_{t_1}(x, z_1) \prod_{i=2}^n p_{t_i - t_{i-1}}(z_{i-1}, z_i) \prod_{i=1}^n f_i(z_i) P_{t+s-t_n} f(z_n)$$

$$\prod_{i=1}^{n} dz_i.$$

Similarly, by (2.17) we have

$$E^x \left(\prod_{i=1}^{n} f_i(W_{t_i}) P_s f(W_t) \right) \tag{2.37}$$

$$= \int p_{t_1}(x, z_1) \prod_{i=2}^{n} p_{t_i - t_{i-1}}(z_{i-1}, z_i) p_{t-t_n}(z_n, z)$$

$$\prod_{i=1}^{n} f_i(z_i) \, dz_i P_s f(z) \, dz$$

$$= \int p_{t_1}(x, z_1) \prod_{i=2}^{n} p_{t_i - t_{i-1}}(z_{i-1}, z_i) \prod_{i=1}^{n} f_i(z_i) P_{t-t_n} P_s f(z_n) \prod_{i=1}^{n} dz_i.$$

It follows from the semigroup property of P_t that $P_{t-t_n} P_s = P_{t+s-t_n}$. Therefore (2.36) and (2.37) are equal, which is (2.35). $\qquad \square$

In the study of local times it is necessary to enlarge the σ-algebras $\mathcal{F}_t^0$. We do this in the next section. In preparation, we introduce the following definition, which generalizes (2.32). Let $\{\mathcal{G}_t; t \geq 0\}$ be an increasing family of σ-algebras with $\mathcal{F}_t^0 \subseteq \mathcal{G}_t$ for all $t \geq 0$. We say that the Brownian motion $\{W_t, t \in R_+\}$ is a simple Markov process with respect to $\{\mathcal{G}_t; t \geq 0\}$ if, for all $x \in R^1$,

$$E^x \left(f(W_{t+s}) \,|\, \mathcal{G}_t \right) = P_s f(W_t) \tag{2.38}$$

for all $s, t \geq 0$ and $f \in \mathcal{B}_b(R)$.

Let $\{W_t, t \in R_+\}$ be a Brownian motion on the probability space $(\Omega, \mathcal{F}^0)$. We assume that there exists a family $\{\theta_t, t \in R_+\}$ of shift operators for W, that is, operators $\theta_t : (\Omega, \mathcal{F}^0) \mapsto (\Omega, \mathcal{F}^0)$ with

$$\theta_t \circ \theta_s = \theta_{t+s} \qquad \text{and} \qquad W_t \circ \theta_s = W_{t+s} \qquad \forall s, t \geq 0. \tag{2.39}$$

For example, for the canonical version of Brownian motion, we can take the shift operators θ_t to be defined by

$$\theta_t(\omega)(s) = \omega(t+s) \qquad \forall s, t \geq 0. \tag{2.40}$$

In general, for any random variable Y on $(\Omega, \mathcal{F}^0)$, $Y \circ \theta_t := Y(\theta_t)$.

Lemma 2.2.2 *If $\{W_t, t \in R_+\}$ is a simple Markov process with respect to $\{\mathcal{G}_t; t \geq 0\}$, then*

$$E^x \left(Y \circ \theta_t \,|\, \mathcal{G}_t \right) = E^{W_t} (Y) \tag{2.41}$$

for all $\mathcal{F}^0$ measurable functions Y and all $x \in R^1$.

Proof It suffices to prove (2.41) for Y of the form $Y = \prod_{i=1}^{n} g_i(W_{t_i})$, with $0 < t_1 < \ldots < t_n$ and $g_i \in \mathcal{B}_b(S)$, $i = 1, \ldots, n$. In this case $Y \circ \theta_t = \prod_{i=1}^{n} g_i(W_{t+t_i})$. We then have

$$E^x \left(\prod_{i=1}^{n} g_i(W_{t+t_i}) \,|\, \mathcal{G}_{t+t_{n-1}} \right) = \prod_{i=1}^{n-1} g_i(W_{t+t_i}) E^x \left(g_n(W_{t+t_n}) \,|\, \mathcal{G}_{t+t_{n-1}} \right) \tag{2.42}$$

and, by (2.38),

$$\begin{aligned} E^x \left(g_n(W_{t+t_n}) \,|\, \mathcal{G}_{t+t_{n-1}} \right) &= P_{t_n - t_{n-1}} g_n(W_{t+t_{n-1}}) \tag{2.43} \\ &= \int p_{t_n - t_{n-1}}(W_{t+t_{n-1}}, z_n) g_n(z_n) \, dz_n \\ &:= h(W_{t+t_{n-1}}). \end{aligned}$$

Next we see that the conditional expectation of the left-hand side of (2.42) with respect to $\mathcal{G}_{t+t_{n-2}}$ is equal to

$$\prod_{i=1}^{n-2} g_i(W_{t+t_i}) \int p_{t_{n-1} - t_{n-2}}(W_{t+t_{n-2}}, z_{n-1}) g_{n-1}(z_{n-1}) h(z_{n-1}) \, dz_{n-1}$$

$$= \prod_{i=1}^{n-2} g_i(W_{t+t_i}) \int\int p_{t_{n-1} - t_{n-2}}(W_{t+t_{n-2}}, z_{n-1})$$

$$p_{t_n - t_{n-1}}(z_{n-1}, z_n) g_{n-1}(z_{n-1}) g_n(z_n) \, dz_{n-1} \, dz_n.$$

Continuing in this way see that the left-hand side of (2.41) is equal to

$$\int p_{t_1}(W_t, z_1) \prod_{i=2}^{n} p_{t_i - t_{i-1}}(z_{i-1}, z_i) \prod_{i=1}^{n} g_i(z_i) \, dz_i. \tag{2.44}$$

Clearly the right-hand side of (2.41) is also equal to this.

We now provide an alternative, more explicit proof for the important case in which $\mathcal{G}_t = \mathcal{F}_t^0$ for all $t \geq 0$. Thus we show that for Y as above

$$E^x \left(Y \circ \theta_t \,|\, \mathcal{F}_t^0 \right) = E^{W_t}(Y). \tag{2.45}$$

To prove (2.45) it is enough to show that

$$E^x \left(F \, Y \circ \theta_t \right) = E^x \left(F \, E^{W_t}(Y) \right) \tag{2.46}$$

for all F of the form $F = \prod_{j=1}^{m} f_j(W_{s_j})$, with $0 < s_1 < \ldots < s_m \leq t$ and $f_j \in \mathcal{B}_b(S)$, $j = 1, \ldots, m$.

By (2.17),

$$E^x \left(F \ Y \circ \theta_t \right) = E^x \left(\prod_{j=1}^{m} f_j(W_{s_j}) \prod_{i=1}^{n} g_i(W_{t+t_i}) \right) \tag{2.47}$$

$$= \int p_{s_1}(x, z_1) \prod_{j=2}^{m} p_{s_j - s_{j-1}}(z_{j-1}, z_j) p_{t+t_1 - s_m}(z_m, y_1)$$

$$\prod_{i=2}^{n} p_{t_i - t_{i-1}}(y_{i-1}, y_i) \prod_{j=1}^{m} f_j(z_j) \, dz_j \prod_{i=1}^{n} g_i(y_i) \, dy_i.$$

Similarly,

$$E^x \left(F \ E^{W_t}(Y) \right) = E^x \left(\prod_{j=1}^{m} f_j(W_{s_j}) E^{W_t}(Y) \right) \tag{2.48}$$

$$= \int p_{s_1}(x, z_1) \prod_{j=2}^{m} p_{s_j - s_{j-1}}(z_{j-1}, z_j) p_{t - s_m}(z_m, y)$$

$$\prod_{j=1}^{m} f_j(z_j) \, dz_j E^y(Y) \, dy$$

and

$$E^y(Y) = \int p_{t_1}(y, y_1) \prod_{i=2}^{n} p_{t_i - t_{i-1}}(y_{i-1}, y_i) \prod_{i=1}^{n} g_i(y_i) \, dy_i. \tag{2.49}$$

Since $p_{t+t_1 - s_m}(z_m, y_1) = \int p_{t - s_m}(z_m, y) p_{t_1}(y, y_1) \, dy$, we get (2.46). □

In the development of the theory of Markov processes it is crucial to extend the simple Markov property, which holds for fixed times, to certain random times. Let $(S, \mathcal{G}, \mathcal{G}_t, P)$ be a probability space, where $\mathcal{G}_t$ is an increasing sequence of σ-algebras and, as usual, $\mathcal{G} = \cup_{t \geq 0} \mathcal{G}_t$. A random variable T with values in $[0, \infty]$ is called a $\mathcal{G}_t$ stopping time if $\{T \leq t\}$ is $\mathcal{G}_t$ measurable for all $t \geq 0$.

Let T be a $\mathcal{G}_t$ stopping time and let

$$\mathcal{G}_T := \{A \subseteq S \mid A \cap \{T \leq t\} \in \mathcal{G}_t, \forall t\}.$$

It is easy to check that $\mathcal{G}_T$ is a σ-algebra. Note that if $A \in \mathcal{G}_T$, then

$$T(A)(\omega) := \begin{cases} T(\omega) & \text{if } \omega \in A \\ \infty & \text{if } \omega \notin A \end{cases} \tag{2.50}$$

is a $\mathcal{G}_t$ stopping time.

Set $\mathcal{G}_{t+} = \cap_{h>0} \mathcal{G}_{t+h}$. $\mathcal{G}_{t+}$ is an increasing sequence of σ-algebras. It is easy to check that T is a $\mathcal{G}_{t+}$ stopping time (that is, $\{T \leq t\}$ is $\mathcal{G}_{t+}$

measurable for all $t \geq 0$) if and only if $\{T < t\}$ is $\mathcal{G}_t$ measurable for all t. Similar to $\mathcal{G}_T$ we define $\mathcal{G}_{T+} = \{A \subseteq S \mid A \cap \{T \leq t\} \in \mathcal{G}_{t+}, \forall t\}$. One can check that $\mathcal{G}_{T+} = \{A \subseteq S \mid A \cap \{T < t\} \in \mathcal{G}_t, \forall t\}$.

Remark 2.2.3 We wrote the last three paragraphs on stopping times without specifically mentioning Brownian motion because they are relevant for general Markov processes. We did this so that in Chapter 3 we can discuss stopping times without defining them again. Several concepts that are developed in this chapter for the study of Brownian motion are presented in this way.

Let $W = \{W(t), t \in R_+\}$ be a Brownian motion on $(\Omega, \mathcal{F}^0, \mathcal{F}^0_t, P)$. An interesting class of stopping times is the first hitting times of certain subsets $A \subseteq R^1$, that is, the random times $T_A = \inf\{t > 0 \mid W_t \in A\}$.

Lemma 2.2.4 *Let $W = \{W(t), t \in R_+\}$ be a Brownian motion on $(\Omega, \mathcal{F}^0, \mathcal{F}^0_t, P)$. If $A \subseteq R^1$ is open, then T_A is an $\mathcal{F}^0_{t+}$ stopping time. If $A \subseteq R^1$ is closed, then T_A is an $\mathcal{F}^0_t$ stopping time.*

Proof Suppose that $A \subseteq R^1$ is open. Then $T_A < t$ if and only if $W_s \in A$ for some rational number $0 < s < t$. Therefore $\{T_A < t\} \in \mathcal{F}^0_t$.

Let $d(x, A) := \inf\{|x - y|, y \in A\}$. Since $d(x, A)$ is continuous, the sets $A_n = \{x \in R^1 \mid d(x, A) < 1/n\}$ are open, and if $A \subseteq R^1$ is closed, $A_n \downarrow A$. Using the facts that A is closed in the first equation and A_n is open in the last equation, we see that if $t > 0$,

$$
\begin{aligned}
\{T_A \leq t\} &= \{W_s \in A \text{ for some } 0 < s \leq t\} \qquad (2.51)\\
&= \bigcup_m \{W_s \in A \text{ for some } 1/m \leq s \leq t\}\\
&= \bigcup_m \bigcap_n \{W_s \in A_n \text{ for some } 1/m \leq s \leq t\}\\
&= \bigcup_m \bigcap_n \{W_s \in A_n \text{ for some rational } 1/m \leq s \leq t\}.
\end{aligned}
$$

We use the continuity of Brownian motion in the last line. Thus $\{T_A \leq t\} \in \mathcal{F}^0_t$ for $t > 0$. The case of $t = 0$ is immediate since, for a closed set A, $\{T_A = 0\} = \{W_0 \in A\} \in \mathcal{F}^0_0$. $\qquad \square$

The next lemma, which generalizes the simple Markov property of Brownian motion so that it holds for stopping times, is called the strong Markov property for Brownian motion.

Lemma 2.2.5 *If the Brownian motion $\{W_t, t \in R_+\}$ is a simple Markov process with respect to $\{\mathcal{G}_t; t \geq 0\}$, then for any $\mathcal{G}_{t+}$ stopping time T*

$$E^x\left(f(W_{T+s})1_{\{T<\infty\}} \mid \mathcal{G}_{T+}\right) = P_s f(W_T)1_{\{T<\infty\}} \qquad (2.52)$$

for all $s \geq 0$, $x \in R^1$, and bounded Borel measurable functions f.

Proof We need to show two things: first that $P_s f(W_T)1_{\{T<\infty\}}$ is $\mathcal{G}_{T+}$ measurable for each bounded Borel measurable function f and, second, that

$$E^x\left(1_A f(W_{T+s})1_{\{T<\infty\}}\right) = E^x\left(1_A P_s f(W_T)1_{\{T<\infty\}}\right) \qquad (2.53)$$

for each $A \in \mathcal{G}_{T+}$ and bounded Borel measurable function f. For both of these assertions it suffices to consider that f is also continuous.

Let $T_n = \dfrac{[2^n T] + 1}{2^n}$, that is, $T_n = j/2^n$ if and only if $(j-1)/2^n \leq T < j/2^n$ and check that, for each n, T_n is a $\mathcal{G}_t$ stopping time and $T_n \downarrow T$. Furthermore, $T_n < \infty$ if and only if $T < \infty$.

Using the continuity of $\{W_t, t \geq 0\}$ we have

$$W_T 1_{\{T<\infty\}} = \lim_{n\to\infty} \sum_{j=1}^{\infty} W_{(j-1)/2^n} 1_{\{T_n = j/2^n\}}.$$

Furthermore, it is easy to see that $W_{(j-1)/2^n} 1_{\{T_n = j/2^n\}}$ is $\mathcal{G}_{T+}$ measurable. Therefore the same is true for $W_T 1_{\{T<\infty\}}$ and consequently also for $P_s f(W_T)1_{\{T<\infty\}}$.

We proceed to verify (2.53). We claim that this equation holds with T replaced by T_n. Once this is established, (2.53) itself follows by taking the limit as n goes to infinity and using continuity. Replace T by T_n in the left-hand side of (2.53) and note that since $A \in \mathcal{G}_{T+}$ we have $A \cap \{T_n = j/2^n\} \in \mathcal{G}_{j/2^n}$. Then, using the simple Markov property, we see that

$$E^x\left(1_A f(W_{T_n+s})1_{\{T_n<\infty\}}\right) \qquad (2.54)$$

$$= \sum_{j=1}^{\infty} E^x\left(1_A 1_{\{T_n=j/2^n\}} f(W_{T_n+s})\right)$$

$$= \sum_{j=1}^{\infty} E^x\left(1_{A\cap\{T_n=j/2^n\}} f(W_{j/2^n+s})\right)$$

$$= \sum_{j=1}^{\infty} E^x\left(1_{A\cap\{T_n=j/2^n\}} P_s f(W_{j/2^n})\right)$$

$$= E^x\left(1_A 1_{\{T_n<\infty\}} P_s f(W_{T_n})\right).$$

This shows that (2.53) holds with T replaced by T_n. □

Remark 2.2.6 A similar proof shows that for any $\mathcal{G}_t$ stopping time T

$$E^x\left(f(W_{T+s})1_{\{T<\infty\}}\,|\,\mathcal{G}_T\right) = P_s f(W_T)1_{\{T<\infty\}} \qquad (2.55)$$

for all $s \geq 0$, $x \in R^1$ and bounded Borel measurable functions f.

In the same way we obtained (2.41) using (2.38), we use (2.52) to obtain the next lemma.

Lemma 2.2.7 *The strong Markov property (2.52) for the $\mathcal{G}_{t+}$ stopping time T can be extended to*

$$E^x\left(Y \circ \theta_T 1_{\{T<\infty\}}\,|\,\mathcal{G}_{T+}\right) = E^{W_T}(Y)\,1_{\{T<\infty\}} \qquad (2.56)$$

for all $\mathcal{F}^0$ measurable functions Y and $x \in R^1$.

Using (2.55), a proof similar to the proof of Lemma 2.2.7 shows that for the $\mathcal{G}_t$ stopping time T

$$E^x\left(Y \circ \theta_T 1_{\{T<\infty\}}\,|\,\mathcal{G}_T\right) = E^{W_T}(Y)\,1_{\{T<\infty\}} \qquad (2.57)$$

for all $\mathcal{F}^0$ measurable functions Y and $x \in R^1$.

Remark 2.2.8 Since a constant time is an $\mathcal{F}^0_t$ stopping time, it follows from (2.56) and (2.57) that, for any $t \geq 0$,

$$E^x\left(Y \circ \theta_t\,|\,\mathcal{F}^0_{t+}\right) = E^{W_t}(Y) = E^x\left(Y \circ \theta_t\,|\,\mathcal{F}^0_t\right) \qquad (2.58)$$

for all $\mathcal{F}^0$ measurable functions Y and $x \in R^1$. This in turn implies that

$$E^x\left(Z\,|\,\mathcal{F}^0_{t+}\right) = E^x\left(Z\,|\,\mathcal{F}^0_t\right) \qquad (2.59)$$

for all $\mathcal{F}^0$ measurable functions Z and $x \in R^1$. (To see this note that it suffices to verify it for Z of the form $Z = (Y \circ \theta_t)V$, where $V \in \mathcal{F}^0_t$, since such functions generate $\mathcal{F}^0$. For Z of this form (2.59) follows immediately from (2.58).)

It follows from (2.59) that $\mathcal{F}^0_{t+}$ and $\mathcal{F}^0_t$ differ merely by null sets. More precisely, let $A \in \mathcal{F}^0_{t+}$ and set $A' = \{\omega\,|\,E^x(1_A\,|\,\mathcal{F}^0_t) = 1\}$, which is clearly in $\mathcal{F}^0_t$. By (2.59) applied to $Z = 1_A$ we see that $1_{A'} = 1_A$, P^x almost surely, that is, $P^x(A\Delta A') = 0$.

Remark 2.2.8 leads to the next important lemma.

Lemma 2.2.9 (Blumenthal Zero–One Law for Brownian Motion) *Let $A \in \mathcal{F}^0_{0+}$. Then, for any $x \in R^1$, $P^x(A)$ is either zero or one.*

Proof Note that $\mathcal{F}_0^0 = \sigma(W_0)$, the σ-algebra generated by W_0. Thus any set $A' \in \mathcal{F}_0^0$ must be of the form $A' = W_0^{-1}(B)$ for some Borel set $B \subseteq R^1$. Therefore $P^x(A') = P^x(W_0 \in B) = 1_B(x)$, so that $P^x(A')$ is either zero or one. Applying the results stated in the second paragraph of Remark 2.2.8, in the case $t = 0$, we see that for $A \in \mathcal{F}_{0+}^0$, $P^x(A\Delta A') = 0$ for some $A' \in \mathcal{F}_0^0$. Thus $P^x(A)$ is either zero or one. $\qquad\square$

The following simple lemma is used in the next section.

Lemma 2.2.10 *Let $(S, \mathcal{G}, \mathcal{G}_t, P)$ be a probability space, where $\mathcal{G}_t$ is an increasing family of σ-algebras and let S and T be $\mathcal{G}_t$ stopping times. Then the sets*

$$\{S \leq T\}, \{S < T\}, \ and \ \{S = T\} \in \mathcal{G}_T. \tag{2.60}$$

Proof To see this simply use the facts that

$$\{S \leq T\} \cap \{T \leq t\} = \{S \leq t\} \cap \{T \leq t\} \cap \{S \wedge t \leq T \wedge t\} \in \mathcal{G}_t \tag{2.61}$$

and

$$\{S < T\} \cap \{T \leq t\} = \bigcup_{\substack{r < t \\ r \ \text{rational}}} \{S < r\} \cap \{r < T \leq t\} \in \mathcal{G}_t. \tag{2.62}$$

We use a stopping time argument to prove the following simple but useful result, known as the reflection principle.

Lemma 2.2.11 *Let $\{W_t, t \in R_+\}$ be a Brownian motion. Then*

$$P^0\left(\sup_{0 \leq s \leq t} W_s \geq x\right) = 2\, P^0\left(W_t \geq x\right). \tag{2.63}$$

Proof By Lemma 2.2.4, $T_x = \inf\{t > 0 \,|\, W_t = x\}$ is an $\mathcal{F}_t^0$ stopping time. For any $\alpha > 0$,

$$\int_0^\infty e^{-\alpha t} P^0\left(W_t \geq x\right) dt \tag{2.64}$$

$$= E^0\left(\int_0^\infty e^{-\alpha t} 1_{\{[x,\infty)\}}(W_t)\, dt\right)$$

$$= E^0\left(\int_{T_x}^\infty e^{-\alpha t} 1_{\{[x,\infty)\}}(W_t)\, dt\right)$$

$$= E^0\left(e^{-\alpha T_x}\left(\int_0^\infty e^{-\alpha t} 1_{\{[x,\infty)\}}(W_t)\right) \circ \theta_{T_x}\, dt\right).$$

Since for $\alpha > 0$, $e^{-\alpha T_x} = e^{-\alpha T_x}1_{\{T_x < \infty\}}$, and $W_{T_x} = x$, by the strong Markov property (2.57),

$$E^0\left(e^{-\alpha T_x}\left(\int_0^\infty e^{-\alpha t}1_{\{[x,\infty)\}}(W_t)\right)\circ\theta_{T_x}\,dt\right) \qquad (2.65)$$

$$= E^0\left(e^{-\alpha T_x}\right)E^x\left(\int_0^\infty e^{-\alpha t}1_{\{[x,\infty)\}}(W_t)\,dt\right)$$

and clearly

$$E^x\left(\int_0^\infty e^{-\alpha t}1_{\{[x,\infty)\}}(W_t)\,dt\right) \qquad (2.66)$$

$$= E^0\left(\int_0^\infty e^{-\alpha t}1_{\{[0,\infty)\}}(W_t)\,dt\right)$$

$$= \int_0^\infty e^{-\alpha t}P^0\left(W_t \geq 0\right)\,dt = \frac{1}{2\alpha}.$$

A standard integration by parts shows that

$$E^0\left(e^{-\alpha T_x}\right) = \int_0^\infty e^{-\alpha t}\,dP^0\left(T_x \leq t\right) = \alpha\int_0^\infty e^{-\alpha t}P^0\left(T_x \leq t\right)\,dt.$$
$$(2.67)$$

Combining (2.64)–(2.67) we have

$$\int_0^\infty e^{-\alpha t}P^0\left(W_t \geq x\right)\,dt = \frac{1}{2}\int_0^\infty e^{-\alpha t}P^0\left(T_x \leq t\right)\,dt. \qquad (2.68)$$

Since this holds for all $\alpha > 0$, and $P^0\left(W_t \geq x\right)$ and $P^0\left(T_x \leq t\right)$ are right continuous in t, we have

$$P^0\left(W_t \geq x\right) = \tfrac{1}{2}\,P^0\left(T_x \leq t\right). \qquad (2.69)$$

The lemma follows from the fact that

$$P^0\left(T_x \leq t\right) = P^0\left(\sup_{0 \leq s \leq t} W_s \geq x\right). \qquad (2.70)$$

$\square$

2.3 Standard augmentation

We initially defined Brownian motion on the probability space $(\Omega, \mathcal{F}^0, \mathcal{F}_t^0, P^x)$. However, as we pointed out on page 21, to continue with the introduction of local times in the next section, we must enlarge the σ-algebras $\mathcal{F}_t^0$, $t \geq 0$. We do this by using a construction referred to as the standard augmentation of $\mathcal{F}_t^0$ with respect to P^x, or just standard augmentation, when it is clear to which σ-algebras and probability measures we are referring. Heuristically, $\mathcal{F}_t^0$ is augmented with null sets that can depend on the whole trajectory of the path, not merely on the trajectory up to time t. An important task of this section is to show that the strong Markov property continues to hold on the standard augmentation of $\mathcal{F}_t^0$. (Note that the standard augmentation is sometimes referred to as the "usual augmentation.")

Let $\mathcal{B}$ denote the Borel σ-algebra on R^1 and let $\mathcal{M}$ denote the set of finite positive measures on $(R^1, \mathcal{B})$. For each $\mu \in \mathcal{M}$ and $A \in \mathcal{F}^0$, set

$$P^\mu(A) = \int_{-\infty}^{\infty} P^x(A)\, d\mu(x). \tag{2.71}$$

Let $\mathcal{F}^\mu$ be the P^μ completion of $\mathcal{F}^0$ and let $\mathcal{N}_\mu$ be the collection of null sets for $(\Omega, \mathcal{F}^\mu, P^\mu)$. Let $\mathcal{F}_t^\mu = \mathcal{F}_t^0 \vee \mathcal{N}_\mu$ and

$$\mathcal{F} = \cap_{\mu \in \mathcal{M}} \mathcal{F}^\mu \quad \text{and} \quad \mathcal{F}_t = \cap_{\mu \in \mathcal{M}} \mathcal{F}_t^\mu. \tag{2.72}$$

We note that

$$\mathcal{F}^\mu = \{ B \subseteq \Omega \,|\, B' \subseteq B \subseteq B'' \text{ with } B', B'' \in \mathcal{F}^0 \text{ and } P^\mu(B') = P^\mu(B'') \}. \tag{2.73}$$

The fact that the right-hand side of (2.73) is contained in $\mathcal{F}^\mu$ is immediate. The opposite inclusion follows by showing that the right-hand side of (2.73) is a σ-algebra. $\{\mathcal{F}_t, t \geq 0\}$ is the standard augmentation of $\mathcal{F}_t^0$ with respect to $\{P^x, x \in R^1\}$ or, more simply stated, the standard augmentation of $\mathcal{F}_t^0$ with respect to P^x. One can check that $\mathcal{F} = \cup_{t \geq 0} \mathcal{F}_t$. Since $\mathcal{F}$ and $\mathcal{F}_t$ are enlargements of the σ-algebras $\mathcal{F}^0$ and $\mathcal{F}_t^0$, respectively, it is clear that a Brownian motion $\{W_t, t \geq 0\}$ on $(\Omega, \mathcal{F}^0, \mathcal{F}_t^0, P^x)$ is also a Brownian motion on $(\Omega, \mathcal{F}, \mathcal{F}_t, P^x)$.

Note that for each $\mu \in \mathcal{M}$

$$E^\mu\left(f(W_{t+s}) \,|\, \mathcal{F}_t^\mu\right) = P_s f(W_t) \tag{2.74}$$

for all $s, t \geq 0$ and $f \in \mathcal{B}_b(R^1)$. This follows since the two measures $A \mapsto E^\mu(1_A f(W_{t+s}))$ and $A \mapsto E^\mu(1_A P_s f(W_t))$ agree on $\mathcal{F}_t^0$ by the simple Markov property, and also on $\mathcal{N}_\mu$, where they are both zero. Therefore they agree on $\mathcal{F}_t^\mu$, which gives (2.74).

It is also clear that in (2.74) we may replace $\mathcal{F}_t^\mu$ by $\mathcal{F}_t$, so that in particular $\{W_t, t \in R_+\}$ is a simple Markov process with respect to $\{\mathcal{F}_t; t \geq 0\}$. Hence, by Lemmas 2.2.5 and 2.2.7, for any $\mathcal{F}_{t+}$ stopping time T we have

$$E^x\left(Y \circ \theta_T 1_{\{T<\infty\}} \,|\, \mathcal{F}_{T+}\right) = E^{W_T}(Y)\, 1_{\{T<\infty\}} \tag{2.75}$$

for all $\mathcal{F}^0$ measurable functions Y. A simple integration then shows that

$$E^\mu\left(Y \circ \theta_T 1_{\{T<\infty\}} \,|\, \mathcal{F}_{T+}\right) = E^{W_T}(Y)\, 1_{\{T<\infty\}} \tag{2.76}$$

for all $\mathcal{F}^0$ measurable functions Y and $\mu \in \mathcal{M}$.

Lemma 2.3.1 $\mathcal{F}_{t+} = \mathcal{F}_t$ *for all* $t \geq 0$.

Proof   Just as we obtained (2.76) from (2.56), we can use (2.57) to see that for any $\mathcal{F}_t$ stopping time T we have

$$E^\mu \left(Y \circ \theta_T 1_{\{T < \infty\}} \,|\, \mathcal{F}_T \right) = E^{W_T} (Y) \, 1_{\{T < \infty\}} \tag{2.77}$$

for all $\mathcal{F}^0$ measurable functions Y. Applying (2.76) and (2.77) to the $\mathcal{F}_t$ stopping time $T = t$ we see that

$$E^\mu \left(Y \circ \theta_t \,|\, \mathcal{F}_{t+} \right) = E^\mu \left(Y \circ \theta_t \,|\, \mathcal{F}_t \right) \tag{2.78}$$

for all $\mathcal{F}^0$ measurable functions Y. By the argument in Remark 2.2.8, this implies that

$$E^\mu \left(Y \,|\, \mathcal{F}_{t+} \right) = E^\mu \left(Y \,|\, \mathcal{F}_t \right) \tag{2.79}$$

for all $\mathcal{F}^0$ measurable functions Y and hence for all $\mathcal{F}^\mu$ measurable functions Y.

Suppose that Y is $\mathcal{F}_{t+}$ measurable. Then, by (2.79), Y differs from an $\mathcal{F}_t$ measurable random variable by a set of μ measure zero. This shows that $\mathcal{F}_{t+} \subseteq \mathcal{F}_t^\mu$ for each $\mu \in \mathcal{M}$ and consequently $\mathcal{F}_{t+} = \mathcal{F}_t$.   □

Lemma 2.3.2 *Let* $\{W_t, t \geq 0\}$ *be a Brownian motion on* $(\Omega, \mathcal{F}, \mathcal{F}_t, P^x)$. *For any* $\mathcal{F}_t$ *stopping time* T

$$E^\mu \left(Y \circ \theta_T 1_{\{T < \infty\}} \,|\, \mathcal{F}_T \right) = E^{W_T} (Y) \, 1_{\{T < \infty\}} \tag{2.80}$$

for all $\mathcal{F}$ *measurable functions* Y *and* $\mu \in \mathcal{M}$.

Since $\mathcal{F}_{t+} = \mathcal{F}_t$ for all t, for any $\mathcal{F}_t$ stopping time T we have $\mathcal{F}_{T+} = \mathcal{F}_T$. Thus (2.80) is not more restrictive than (2.76) for $\mathcal{F}^0$ measurable functions Y. Of course, the significance of this lemma is that it holds for all $\mathcal{F}$ measurable functions Y.

Proof   Fix $\mu \in \mathcal{M}$. It suffices to obtain (2.80) for $Y = 1_B$, for $B \in \mathcal{F}$, that is, to show that for all $B \in \mathcal{F}$

$$P^{W_T} (B) \, 1_{\{T < \infty\}} \quad \text{is } \mathcal{F}_T^\mu \text{ measurable} \tag{2.81}$$

and

$$E^\mu \left(1_A 1_B \circ \theta_T 1_{\{T < \infty\}} \right) = E^\mu \left(1_A P^{W_T} (B) \, 1_{\{T < \infty\}} \right) \tag{2.82}$$

for each $A \in \mathcal{F}_T$.

Let $\nu \in \mathcal{M}$ be the measure defined by $\nu(C) = P^\mu(W_T \in C \,, T < \infty)$ for $C \in \mathcal{B}$. By (2.73) and the definition of $\mathcal{F}$ we can then find sets

$B', B'' \in \mathcal{F}^0$ with $B' \subseteq B \subseteq B''$ and $P^\nu(B') = P^\nu(B) = P^\nu(B'')$. By the definition of ν we have

$$\int f(x) \, d\nu(x) = E^\mu \left(f(W_T) 1_{\{T < \infty\}} \right). \tag{2.83}$$

Hence

$$P^\nu(B') = \int P^x(B') \, d\nu(x) = E^\mu \left(P^{W_T}(B') 1_{\{T < \infty\}} \right) \tag{2.84}$$

with a similar expression for $P^\nu(B'')$. Thus, since $P^\nu(B') = P^\nu(B'')$ and $B' \subseteq B \subseteq B''$, we see that

$$P^{W_T}(B') 1_{\{T < \infty\}} = P^{W_T}(B) 1_{\{T < \infty\}} = P^{W_T}(B'') 1_{\{T < \infty\}} \qquad P^\mu \text{ a.s.} \tag{2.85}$$

Using this in (2.76) with $Y = 1_{B'}$ we see that $P^{W_T}(B) 1_{\{T < \infty\}}$ is $\mathcal{F}_T^\mu$ measurable. This is (2.81).

Similarly, (2.76) with $Y = 1_{B'}$ and $Y = 1_{B''}$, combined with (2.85), show that

$$E^\mu(1_{B'} \circ \theta_T 1_{\{T < \infty\}}) = E^\mu(1_{B''} \circ \theta_T 1_{\{T < \infty\}}). \tag{2.86}$$

Since $B' \subseteq B \subseteq B''$, we see that $1_B \circ \theta_T 1_{\{T < \infty\}} = 1_{B'} \circ \theta_T 1_{\{T < \infty\}}$, P^μ almost surely. Therefore (2.82) follows by using (2.76), with $Y = 1_{B'}$, and (2.85).

We have thus established (2.80). $\qquad\qquad\qquad\qquad\qquad\qquad\qquad\square$

Lemma 2.3.3

$$P^\mu(A) = \int P^x(A) \, d\mu(x) \qquad \forall A \in \mathcal{F}. \tag{2.87}$$

Proof Equation (2.87) holds for $A \in \mathcal{F}^0$ by definition. To show that it continues to hold for $A \in \mathcal{F}$ we proceed as in the proof of Lemma 2.3.2. We find two sets $A', A'' \in \mathcal{F}^0$ with $A' \subseteq A \subseteq A''$ and $P^\mu(A') = P^\mu(A) = P^\mu(A'')$. Then, since $P^x(A) = P^x(A')$ almost surely with respect to μ, and since (2.87) holds with A replaced by $A' \in \mathcal{F}^0$, it also holds for A. $\square$

2.4 Brownian local time

In this section we establish the existence and continuity of a stochastic process $\{L_t^y, (t, y) \in R_+ \times R^1\}$ called the local time of Brownian motion. A primary objective of this book is to establish necessary and sufficient conditions for the continuity of local times of a wide class of Markov processes. For Brownian motion we can give a simple, direct proof.

The general result, which of course also applies to Brownian motion, is obtained by completely different methods.

Let $f(x)$ be a positive symmetric continuous function supported on $[-1, 1]$ with $\int f(x)\, dx = 1$. For any $\epsilon > 0$, set $f_\epsilon(x) = \frac{1}{\epsilon} f(x/\epsilon)$ and $f_{\epsilon,y}(x) = f_\epsilon(x - y)$. Thus, $f_{\epsilon,y}$ is an approximate identity with respect to Lebesgue measure. Let

$$L_t^{\epsilon,y} = \int_0^t f_{\epsilon,y}(W_s)\, ds, \tag{2.88}$$

where $\{W_s, s \geq 0\}$ is a Brownian motion on $(\Omega, \mathcal{F}, \mathcal{F}_t, P^x)$.

Lemma 2.4.1 *For any numbers $0 < T, M < \infty$, $L_t^{\epsilon,y}$ converges uniformly on $(t, y) \in [0, T] \times [-M, M]$ as $\epsilon \to 0$, almost surely and in $L^p(P^x)$ for all $p > 0$.*

Before proving this lemma we note some of its consequences. First, it allows us to define a jointly continuous local time for Brownian motion.

Theorem 2.4.2 *Let $\{W_t, t \geq 0\}$ be a Brownian motion on $(\Omega, \mathcal{F}, \mathcal{F}_t, P^x)$. There exists a set $\Omega' \subset \Omega$ with $P^x(\Omega') = 1$ such that, for $L_t^{\epsilon,y}$ in (2.88)*

$$L_t^y(\omega) := \begin{cases} \lim_{\epsilon \to 0} L_t^{\epsilon,y}(\omega) & \omega \in \Omega' \\ 0 & \omega \notin \Omega' \end{cases} \tag{2.89}$$

is continuous on $R_+ \times R^1$. L_t^y is called the local time of Brownian motion at y (up to time t).

The continuity of L_t^y on $R_+ \times R^1$ is often referred to by saying that it is "jointly continuous."

Proof Let Ω' denote the set on which $L_t^{\epsilon,y}$ converges locally uniformly on $(t, y) \in R_+ \times R^1$. By taking $T = M = n$, $n = 1, 2, \ldots$ in Lemma 2.4.1 we see that $P^x(\Omega') = 1$ for all $x \in R^1$. Note that for each $\epsilon > 0$, $f_{\epsilon,y}$ is simply a bounded continuous function with compact support. Consequently, one can use the Dominated Convergence Theorem to show that $\{L_t^{\epsilon,y}, (y, t) \in R_+ \times R^1\}$ is continuous for all $\epsilon > 0$. Therefore, the local uniform convergence of $\{L_t^{\epsilon,y} ; (t, y) \in R_+ \times R^1\}$ on Ω' implies that L_t^y is continuous on $R_+ \times R^1$. $\qquad\square$

A family $A = \{A_t ; t \in R_+\}$ of random variables on $(\Omega, \mathcal{F}, \mathcal{F}_t)$ is said to be a continuous additive functional (CAF) if it satisfies the following three properties:

(1) $t \mapsto A_t$ is almost surely continuous and nondecreasing with $A_0 = 0$;
(2) A_t is $\mathcal{F}_t$ measurable;

(3) $A_{t+s} = A_t + A_s \circ \theta_t$ for all $s, t \in R_+$ almost surely.

It is immediately obvious that L_t^y inherits from $L_t^{\epsilon,y}$ the property of being a continuous additive functional.

Note that the fact that $\mathcal{F}_t$ is the standard augmentation of $\mathcal{F}_t^0$ is crucial in showing that L_t^y satisfies property (2), since $\Omega' \notin \mathcal{F}_t^0$ (see (2.89)).

Since L_t^y is a continuous additive functional, it follows from property (1) that, for almost all $\omega \in \Omega$, $L_t^y(\omega)$ defines a positive measure $dL_t^y(\omega)$. It is easy to see from (2.89) that $dL_t^y(\omega)$ is supported on the set $\{t \mid W_t(\omega) = y\}$.

It also follows from the almost sure locally uniform convergence in (2.89) that for any continuous function g with compact support

$$\int g(y) L_t^y \, dy = \lim_{\epsilon \to 0} \int g(y) \left(\int_0^t f_{\epsilon,y}(W_s) \, ds \right) dy \qquad (2.90)$$

$$= \lim_{\epsilon \to 0} \int_0^t (g * f_\epsilon)(W_s) \, ds$$

$$= \int_0^t g(W_s) \, ds.$$

Therefore, for any bounded Borel set A,

$$\int_A L_t^y \, dy = \int_0^t 1_A(W_s) \, ds. \qquad (2.91)$$

In particular, let $A = [a, a + \epsilon]$. Using the continuity of L_t^y we have

$$L_t^a = \lim_{\epsilon \to 0} \frac{\text{measure } \{0 \le s \le t \,;\, a \le W_s \le a + \epsilon\}}{\epsilon}. \qquad (2.92)$$

This gives an intrinsic definition of L_t^a as the derivative of an occupation measure. It also shows that L_t^a is independent of the approximating sequence $f_{\epsilon,y}$ used in (2.88). This last fact also follows easily from the proof of Lemma 2.4.1 given below.

To further emphasize the fact that the local time is a density we note that (2.92) implies that, for each $a \in R^1$, $\lambda(\{t : W(t) = a\}) = 0$, where λ is Lebesgue measure.

Lemma 2.4.1 gives rise to three important formulas, which we give as a separate lemma.

Lemma 2.4.3

$$E^x(L_t^y) = \int_0^t p_s(x, y) \, ds. \qquad (2.93)$$

$$u^\alpha(x, y) = E^x \left(\int_0^\infty e^{-\alpha t} \, dL_t^y \right). \tag{2.94}$$

$$u^\alpha(x, y) = E^x \left(e^{-\alpha T_y} \right) u^\alpha(y, y). \tag{2.95}$$

Proof Since $L_t^{\epsilon, y}$ converges in $L^1(P^x)$, $E^x(L_t^y) = \lim_{\epsilon \to 0} E^x(L_t^{\epsilon, y})$. Using (2.88) we see that

$$
\begin{aligned}
\lim_{\epsilon \to 0} E^x(L_t^{\epsilon, y}) &= \lim_{\epsilon \to 0} \int\int_0^t f_{\epsilon, y}(z) p_s(x, z) \, dz \, ds \tag{2.96} \\
&= \int_0^t p_s(x, y) \, ds,
\end{aligned}
$$

which gives (2.93).

For (2.94) we note that

$$
\begin{aligned}
E^x \left(\int_0^\infty e^{-\alpha t} \, dL_t^y \right) &= \lim_{T \to \infty} E^x \left(\int_0^T e^{-\alpha t} \, dL_t^y \right) \tag{2.97} \\
&= \lim_{T \to \infty} E^x \left(e^{-\alpha T} L_T^y + \alpha \int_0^T e^{-\alpha t} L_t^y \, dt \right)
\end{aligned}
$$

by integration by parts. Using (2.93) and integration by parts again we see that the last line of (2.97) equals

$$
\begin{aligned}
\lim_{T \to \infty} e^{-\alpha T} E^x(L_T^y) &+ \alpha \int_0^T e^{-\alpha t} E^x(L_t^y) \, dt \tag{2.98} \\
&= \lim_{T \to \infty} e^{-\alpha T} \int_0^T p_s(x, y) \, ds + \alpha \int_0^T e^{-\alpha t} \left(\int_0^t p_s(x, y) \, ds \right) dt \\
&= u^\alpha(x, y).
\end{aligned}
$$

We now obtain (2.95). Since dL_t^y is supported on $\{t \,|\, W(t) = y\}$ almost surely, $L_t^y = 0$ for $t \in [0, T_y)$ almost surely. Therefore

$$
\begin{aligned}
u^\alpha(x, y) &= E^x \left(\int_0^\infty e^{-\alpha t} \, dL_t^y \right) \tag{2.99} \\
&= E^x \left(\int_{T_y}^\infty e^{-\alpha t} \, dL_t^y \right) \\
&= E^x \left(E^x \left(\int_{T_y}^\infty e^{-\alpha t} \, dL_t^y \,\middle|\, \mathcal{F}_{T_y} \right) \right).
\end{aligned}
$$

Note that it follows from Remark 2.1.5 that $T_y < \infty$ almost surely. Making the change of variables $t = s + T_y$ and then using the additivity property of local times, that is, that $L_{s+T_y}^y = L_{T_y}^y + L_s^y \circ \theta_{T_y}$, and the strong Markov property (2.80), we see that

$$
\begin{aligned}
E^x \left(\int_{T_y}^\infty e^{-\alpha t} \, dL_t^y \Big| \mathcal{F}_{T_y} \right) &= e^{-\alpha T_y} E^x \left(\int_0^\infty e^{-\alpha s} \, dL_s^y \circ \theta_{T_y} \Big| \mathcal{F}_{T_y} \right) \\
&= e^{-\alpha T_y} E^{W_{T_y}} \left(\int_0^\infty e^{-\alpha s} \, dL_s^y \right) \quad (2.100) \\
&= e^{-\alpha T_y} E^y \left(\int_0^\infty e^{-\alpha s} \, dL_s^y \right) \\
&= e^{-\alpha T_y} u^\alpha(y, y).
\end{aligned}
$$

(In the next to last line we use the fact that $W_{T_y} = y$ and (2.94).) Substituting this into the last line of (2.99), we get (2.95). $\qquad\square$

Proof of Lemma 2.4.1 We consider $L_t^{\epsilon,y}$ as a stochastic process indexed by $(t, y, \epsilon) \in R_+ \times R^1 \times (0, 1]$. Let $R_{T,M,1} = [0, T] \times [-M, M] \times (0, 1]$. We show that for all $0 < \gamma < 1$ and all integers n,

$$
E^x \left(\left(L_t^{\epsilon,y} - L_{t'}^{\epsilon',y'} \right)^{2n} \right) \leq C(T, M, n) |(t, y, \epsilon) - (t', y', \epsilon')|^{(1-\gamma)n} \quad (2.101)
$$

for all $(t, y, \epsilon), (t', y', \epsilon') \in R_{T,M,1}$. Given (2.101) it follows from Kolmogorov's Theorem (Theorem 14.1.1), with $h = 2n$, $d = 3$, and $r = n - \gamma n - 3$, that there exists a random variable $C_n(\omega)$ which is finite almost surely, such that

$$
|L_t^{\epsilon,y} - L_{t'}^{\epsilon',y'}| \leq C_n(\omega) |(t, y, \epsilon) - (t', y', \epsilon')|^{1/2 - (\gamma/2 + 3/(2n))} \quad (2.102)
$$

for all dyadic numbers $(t, y, \epsilon), (t', y', \epsilon') \in R_{T,M,1}$. However, since $L_t^{\epsilon,y}$ is continuous on $R_{T,M,1}$ almost surely, (2.102) holds for all $(t, y, \epsilon), (t', y', \epsilon') \in R_{T,M,1}$. Furthermore, since (2.102) holds for all n sufficiently large, we see that for any $\gamma' > 0$ we can find a random variable $C(\omega)$ which is finite almost surely, such that

$$
|L_t^{\epsilon,y} - L_{t'}^{\epsilon',y'}| \leq C(\omega) |(t, y, \epsilon) - (t', y', \epsilon')|^{1/2 - \gamma'}. \quad (2.103)
$$

Lemma 2.4.1 follows immediately from this.

To obtain (2.101) we use the triangle inequality to treat the variation in t, y, and ϵ separately. By (2.88), we have

$$
\begin{aligned}
E^x \left(\prod_{i=1}^{2n} L_t^{\epsilon_i, y_i} \right) &= \int_0^t \cdots \int_0^t E^x \left(\prod_{i=1}^{2n} f_{\epsilon_i, y_i}(W_{t_i}) \right) \prod_{i=1}^{2n} dt_i \quad (2.104) \\
&= \sum_\pi \int_{D_{2n,t}(\pi)} E^x \left(\prod_{i=1}^{2n} f_{\epsilon_i, y_i}(W_{t_{\pi_i}}) \right) \prod_{i=1}^{2n} dt_i,
\end{aligned}
$$

where $D_{2n,t}(\pi) = \{0 < t_{\pi_1} < \cdots < t_{\pi_{2n}} < t\}$ and the sum runs over all permutations π of $\{1, \ldots, 2n\}$. We decompose the integrals in this way so we can use (2.17), which requires that the t_i are increasing. (The reader should note well this technique, which is used repeatedly in this book.)

It now follows from (2.17), in which, following convention, we set $z_0 = x$ and $t_0 = 0$, that (2.104)

$$= \sum_{\pi} \int_{D_{2n,t}} \int \prod_{i=1}^{2n} p_{t_i - t_{i-1}}(z_{i-1}, z_i) \prod_{i=1}^{2n} f_{\epsilon_{\pi_i}, y_{\pi_i}}(z_i) \, dz_i \, dt_i \qquad (2.105)$$

$$= \sum_{\pi} \int_{D_{2n,t}} \int \prod_{i=1}^{2n} p_{t_i - t_{i-1}}(\epsilon_{\pi_{i-1}} z_{i-1} + y_{\pi_{i-1}}, \epsilon_{\pi_i} z_i + y_{\pi_i})$$

$$\prod_{i=1}^{2n} f(z_i) \, dz_i \, dt_i,$$

where we abbreviate $D_{2n,t} = D_{2n,t}(I)$, in which I denotes the identity partition, that is, $D_{2n,t}(I) = \{0 < t_1 < \ldots < t_{2n} < t\}$. We also take $\pi_0 = 0$, $\epsilon_0 = 1$, and $y_0 = 0$. Since, by (2.5), for any $r \leq T$

$$\int_0^r p_s(x, y) \, ds \leq e^T \int_0^\infty e^{-s} p_s(x, y) \, ds = e^T \frac{e^{-\sqrt{2}|x-y|}}{\sqrt{2}} \qquad (2.106)$$

we see that the last term of (2.105) is uniformly bounded for $t < T$. (Ultimately we set all $\epsilon_i = \epsilon$ and $y_i = y$. We introduce the subscripts to aid in following the next steps.)

We use the notation $\Delta_\delta h(z) = h(z+\delta) - h(z)$. When h is a function of several variables $(z_1, \ldots, z_{2n})$ we write Δ_{δ, z_i} for the operator Δ_δ applied to the variable z_i. Recall that $p_t(x, x') = p_t(x - x')$, so that the terms in p_t are actually functions of a single variable. Proceeding exactly as in (2.104), we have

$$E^x \left(\prod_{i=1}^{2n} \left\{ L_t^{\epsilon_i, y_i + \delta} - L_t^{\epsilon_i, y_i} \right\} \right) \qquad (2.107)$$

$$= \sum_{\pi} \int_{D_{2n,t}} \int \prod_{i=1}^{2n} p_{t_i - t_{i-1}}(z_{i-1}, z_i) \prod_{i=1}^{2n} \Delta_{\delta, y_{\pi_i}} f_{\epsilon_{\pi_i}, y_{\pi_i}}(z_i) \, dz_i \, dt_i$$

$$= \sum_{\pi} \prod_{i=1}^{2n} \Delta_{\delta, y_{\pi_i}} \int_{D_{2n,t}} \int \prod_{i=1}^{2n} p_{t_i - t_{i-1}}(\epsilon_{\pi_{i-1}} z_{i-1} + y_{\pi_{i-1}}, \epsilon_{\pi_i} z_i + y_{\pi_i})$$

$$\prod_{i=1}^{2n} f(z_i) \, dz_i \, dt_i.$$

Without loss of generality it suffices to find a bound for the summand in which π is the identity I, that is, to obtain a bound for

$$\prod_{i=1}^{2n} \Delta_{\delta, y_i} \int_{D_{2n,t}} \int \prod_{i=1}^{2n} p_{t_i - t_{i-1}}(\epsilon_{i-1} z_{i-1} + y_{i-1}, \epsilon_i z_i + y_i) \prod_{i=1}^{2n} f(z_i) \, dz_i \, dt_i.$$

(2.108)

It would be convenient to attach the operators Δ_{δ, y_i} to the individual p factors. We cannot do this because each y_i, except y_{2n}, occurs in two of the p factors. Instead we proceed as follows: We write (2.108) as the sum of the 2^{2n} terms obtained by expanding Δ_{δ, y_i} for each even i and using

$$\Delta_{\delta, y_i} fg = (\Delta_{\delta, y_i} f) g(y_i + \delta) + f \Delta_{\delta, y_i} g$$

(2.109)

for each odd i. Then we set $\epsilon_i = \epsilon$ and $y_i = y$ for all i. By doing this we get 2^{2n} terms of the form

$$\int_{D_{2n,t}} \int p_{t_1}^\sharp (y_1 + \epsilon z_1) \prod_{i=2}^{2n} p_{t_i - t_{i-1}}^\sharp (\epsilon(z_i - z_{i-1})) \prod_{i=1}^{2n} f(z_i) \, dz_i \, dt_i,$$

(2.110)

where $p^\sharp$ is either p, $p(\cdot + \delta)$, $\Delta_\delta p$, or $\Delta_{(-\delta)} p$. Furthermore, it follows from (2.109) that there are n terms of the form $\Delta_{\pm \delta} p$. (Note that the reason only the first factor in (2.110) contains a y_i, namely y_1, is that, since $p_t(x, x') = p_t(x - x')$, all the other terms in the y_i's drop out on setting $y_i = y$ for all i.)

To bound the absolute value of (2.110) we replace $\Delta_{\pm \delta} p$ by $|\Delta_{\pm \delta} p|$ and integrate successively with respect to the variables $t_{2n}, t_{2n-1}, \ldots, t_1$. The dt_i integral of a factor of the form $p_{t_i - t_{i-1}}(\epsilon(z_i - z_{i-1}))$ is bounded by (2.106). To bound the dt_i integral of a factor of the form $p_{t_i - t_{i-1}}(\epsilon(z_i - z_{i-1}) + \delta) - p_{t_i - t_{i-1}}(\epsilon(z_i - z_{i-1}))$, we note that

$$p_s(x) < p_s(y) \text{ for a single } s \Rightarrow p_s(x) < p_s(y) \text{ for all } s$$

(2.111)

so that

$$\int_0^r |p_s(x + \delta) - p_s(x)| \, ds$$

(2.112)

$$\leq e^T \int_0^\infty e^{-s} |p_s(x + \delta) - p_s(x)| \, ds$$

$$= e^T | \int_0^\infty e^{-s} (p_s(x + \delta) - p_s(x)) \, ds |$$

$$\leq c e^T |e^{-\sqrt{2}|x+\delta|} - e^{-\sqrt{2}|x|}| \leq c e^T \delta.$$

The same bound is obtained if δ is replaced by $-\delta$. After bounding all the dt_i integrals in this manner, the only terms in z_j that remain are in the integrals $\int f(z_j)\,dz_j = 1$. Thus we have shown that

$$E^x\left(\left(L_t^{\epsilon,y} - L_t^{\epsilon,y'}\right)^{2n}\right) \le C(T,M,n)|y - y'|^n. \tag{2.113}$$

Essentially the same analysis shows that

$$E^x\left(\left(L_t^{\epsilon,y} - L_t^{\epsilon',y}\right)^{2n}\right) \le C(T,M,n)|\epsilon - \epsilon'|^n \tag{2.114}$$

for all (t,y,ϵ), $(t',y',\epsilon') \in R_{T,M,1}$. (When carrying out the proof for $\{\epsilon_i\}$ rather than $\{y_i\}$ we get factors of the form $p_{t_i-t_{i-1}}(\epsilon_i(z_i - z_{i-1}))$, as in the proof for $\{y_i\}$, but the other n terms are of the form

$$p_{t_i-t_{i-1}}(\epsilon_i(z_i - z_{i-1}) + \delta(z_i - z_{i-1})) - p_{t_i-t_{i-1}}(\epsilon_i(z_i - z_{i-1}))$$

because all the ϵ_i are multiplied by $(z_i - z_{i-1})$. Nevertheless, we need only consider $|z_i| \le 1$ because f is supported on $[-1,1]$ (see (2.110)). Thus (2.112) applies in this case also.)

Finally we consider the variation in t. Let $t < t'$. Analogous to (2.104) and (2.105), but taking all the $\{\epsilon_i\}$ equal to ϵ and all the $\{y_i\}$ equal to y, we have

$$E^x\left(\{L_t^{\epsilon,y} - L_{t'}^{\epsilon,y}\}^{2n}\right) \tag{2.115}$$

$$= E^x\left(\prod_{i=1}^{2n}\int_t^{t'} f_{\epsilon,y}(W_{s_i})\,ds_i\right)$$

$$= (2n)!\int_{D_{2n,t'}} 1_{\{t\le t_i\le t',\forall i\}}\int \prod_{i=1}^{2n} p_{t_i-t_{i-1}}(\epsilon(z_i - z_{i-1}))$$

$$\prod_{i=1}^{2n} f(z_i)\,dz_i\,dt_i$$

$$\le (2n)!\int_{D_{2n,t'}} 1_{\{t\le t_i\le t',\forall i\}}\prod_{i=1}^{2n}\frac{1}{\sqrt{t_i - t_{i-1}}}\int \prod_{i=1}^{2n} f(z_i)\,dz_i\,dt_i.$$

Since $\int f(z_j)\,dz_j = 1$, the last line in (2.115) equals

$$(2n)!\int_{D_{2n,t'}} 1_{\{t\le t_i\le t',\forall i\}}\left(\prod_{i=1}^{2n}\frac{1}{\sqrt{t_i - t_{i-1}}}\right)dt_1\ldots dt_{2n}$$

$$\le (2n)!\,\|1_{\{D_{2n,t'}\}}\prod_{i=1}^{2n}\frac{1}{\sqrt{t_i - t_{i-1}}}\|_q\,\|1_{\{t\le t_i\le t',\forall i\}}\|_{q'},$$

where we use Hölder's inequality with $1/q + 1/q' = 1$ and $1 < q < 2$.

It is easy to see that

$$\left\|1_{\{D_{2n,t'}\}}\prod_{i=1}^{2n}\frac{1}{\sqrt{t_i-t_{i-1}}}\right\|_q \le C(T,n,q). \tag{2.116}$$

Also,

$$\begin{aligned}
\|1_{\{t\le t_i\le t',\forall i\}}\|_{q'} &= \left(\int_t^{t'}\cdots\int_t^{t'}\prod_{i=1}^{2n}dt_i\right)^{1/q'} \tag{2.117}\\
&= |t-t'|^{(1-1/q)2n}.
\end{aligned}$$

By taking q sufficiently close to 2 we see that, for all $\gamma > 0$,

$$E^x\left(\{L_t^{\epsilon,y}-L_{t'}^{\epsilon,y}\}^{2n}\right) \le C(T,n,q)|t-t'|^{(1-\gamma)n}. \tag{2.118}$$

The inequalities in (2.113), (2.114), and (2.118) yield (2.101), which is what we set out to show. $\qquad\square$

2.4.1 Inverse local time of Brownian motion

Let $\{L_t^x, (x,t)\in R^1\times R_+\}$ be the local times of Brownian motion. Let

$$\tau_z(s) := \inf\{t>0\,|\,L_t^z>s\}, \tag{2.119}$$

where $\inf\{\emptyset\}=\infty$. We refer to $\tau_z(\cdot)$ as the (right continuous) inverse local time (of Brownian motion) at z. It is easy to see that, for each s, $\tau_z(s)$ is a stopping time and

$$\tau_z(s+t) = \tau_z(s)+\tau_z(t)\circ\theta_{\tau_z(s)}. \tag{2.120}$$

We set $\tau(s)=\tau_0(s)$.

Lemma 2.4.4 *For each $s\in R^1$, $\tau(s)<\infty$ almost surely.*

Proof Since L_t^0 is continuous, it suffices to show that

$$L_\infty^0 := \lim_{t\to\infty}L_t^0 = \infty \quad\text{a.s.} \tag{2.121}$$

Define the sequence of stopping times $\mathcal{T}_0 = 0$ and $\mathcal{T}_{n+1} = \inf\{t > \mathcal{T}_n+1\,|\,W_t=0\}$, $n=0,1,\ldots$. It follows from Remark 2.1.5 that $\mathcal{T}_n<\infty$ almost surely. Therefore, using the monotonicity and additivity of L_t^0, we see that

$$L_\infty^0 \ge \sum_{n=0}^{\infty}\left(L_{\mathcal{T}_{n+1}}^0-L_{\mathcal{T}_n}^0\right) = \sum_{n=0}^{\infty}\left(L_1^0\circ\theta_{\mathcal{T}_n}\right). \tag{2.122}$$

Clearly $\{L_1^0 \circ \theta_{T_n}\}_{n=0}^{\infty}$ are independent and identically distributed. By the strong Markov property, Lemma 2.3.2, and (2.93), $E^0(L_1^0 \circ \theta_{T_n}) = E^0(L_1^0) > 0$. Thus the right-hand side of (2.122) is a sum of independent identically distributed random variables that are nonnegative and have a positive probability of being larger than ϵ for some $\epsilon > 0$. Consequently, (2.121) follows by the Borel–Cantelli Lemma.

Lemma 2.4.5 *Let W be a Brownian motion on $(\Omega, \mathcal{F}, P^0)$. The positive increasing stochastic process $\{\tau(s), \, s \in R_+\}$ has stationary and independent increments.*

Proof Let $F \in \mathcal{B}_b(R^n)$ and note that, by (2.120) and Lemma 2.3.2, for all $t > 0$

$$
\begin{aligned}
E^0 \left(F(\tau(s_1 + t) - \tau(t), \ldots, \tau(s_n + t) - \tau(t)) | \mathcal{F}_{\tau(t)} \right) \quad (2.123) \\
= E^0 \left(F(\tau(s_1), \ldots, \tau(s_n)) \circ \theta_{\tau(t)} | \mathcal{F}_{\tau(t)} \right) \\
= E^{W_{\tau(t)}} \left(F(\tau(s_1), \ldots, \tau(s_n)) \right) \\
= E^0 \left(F(\tau(s_1), \ldots, \tau(s_n)) \right)
\end{aligned}
$$

since $W_{\tau(t)} = 0$. Taking the expectation we get

$$
\begin{aligned}
E^0 \left(F(\tau(s_1 + t) - \tau(t), \ldots, \tau(s_n + t) - \tau(t)) \right) \quad (2.124) \\
= E^0 \left(F(\tau(s_1), \ldots, \tau(s_n)) \right),
\end{aligned}
$$

which completes the proof. $\qquad\square$

Lemma 2.4.6 *For any positive measurable function $f(t)$ and any $T \in [0, \infty]$ and $z \in R^1$ we have*

$$
\int_0^T f(t) \, dL_t^z = \int_0^{\infty} f(\tau_z(s)) 1_{\{\tau_z(s) < T\}} \, ds. \quad (2.125)
$$

If H_t is a positive continuous $\mathcal{F}_t$ measurable function, F a positive $\mathcal{F}$ measurable function, and T a stopping time (possibly $T \equiv \infty$), then

$$
E^x \left(\int_0^T H_t \, F \circ \theta_t \, dL_t^z \right) = E^z (F) E^x \left(\int_0^T H_t \, dL_t^z \right). \quad (2.126)
$$

Proof To prove (2.125) it suffices to show it for functions $f(t)$ of the form $f(t) = 1_{[0,r]}(t)$ for real numbers r. In this case (2.125) is simply

$$
L_{r \wedge T}^z = \mu\{s : \tau_z(s) \leq r, \, \tau_z(s) < T\}, \quad (2.127)
$$

where μ is Lebesgue measure. Note that

$$
L_t^z = \mu\{s : \tau_z(s) \leq t\} = \mu\{s : \tau_z(s) < t\}. \quad (2.128)
$$

Using this we get (2.127).

Using (2.125) we see that

$$E^x \left(\int_0^T H_t \, F \circ \theta_t \, dL_t^z \right) \tag{2.129}$$

$$= E^x \left(\int_0^\infty H_{\tau_z(s)} \, F \circ \theta_{\tau_z(s)} 1_{\{\tau_z(s) < T\}} \, ds \right)$$

$$= \int_0^\infty E^x \left(H_{\tau_z(s)} \, F \circ \theta_{\tau_z(s)} 1_{\{\tau_z(s) < T\}} \right) \, ds.$$

As we mentioned above, $\tau_z(s)$ is a stopping time. Therefore, by (2.60), $\{T \le \tau_z(s)\} \in \mathcal{F}_{\tau_z(s)}$. Taking complements we see that $\{\tau_z(s) < T\} \in \mathcal{F}_{\tau_z(s)}$.

Using the methods of the third paragraph in the proof of Lemma 2.2.5 and the fact that $\mathcal{F}_{\tau_z(s)+} = \mathcal{F}_{\tau_z(s)}$ (see the paragraph following the statement of Lemma 2.3.2), it follows that $H_{\tau_z(s)} 1_{\{\tau_z(s) < \infty\}}$ is $\mathcal{F}_{\tau_z(s)}$ measurable.

Hence, using the strong Markov property

$$E^x \left(H_{\tau_z(s)} \, F \circ \theta_{\tau_z(s)} 1_{\{\tau_z(s) < T\}} \right) \tag{2.130}$$

$$= E^x \left(H_{\tau_z(s)} \, 1_{\{\tau_z(s) < T\}} E^{W_{\tau_z(s)}} (F) \right)$$

$$= E^z (F) \, E^x \left(H_{\tau_z(s)} \, 1_{\{\tau_z(s) < T\}} \right)$$

since $W_{\tau_z(s)} = z$ on $\tau_z(s) < \infty$. Combining (2.129) and (2.130) we get

$$E^x \left(\int_0^T H_t \, F \circ \theta_t \, dL_t^z \right) = E^z (F) \int_0^\infty E^x \left(H_{\tau_z(s)} \, 1_{\{\tau_z(s) < T\}} \right) \, ds. \tag{2.131}$$

Using (2.125) again gives (2.126). □

Since $\tau(s) \ge 0$ for each $s \ge 0$, we can compute its Laplace transform.

Lemma 2.4.7

$$E^0(e^{-\alpha \tau(s)}) = e^{-s/u^\alpha(0,0)} = e^{-\sqrt{2\alpha} \, s}. \tag{2.132}$$

Proof Let $f(s) = E^0(e^{-\alpha \tau(s)})$. Note first that, by (2.120), the strong Markov property, and the fact that $W_{\tau(s)} = 0$, as in the preceding proof,

$$E^0(e^{-\alpha \tau(s+t)}) = E^0(e^{-\alpha \tau(s)} e^{-\alpha \tau(t) \circ \theta_{\tau(s)}}) \tag{2.133}$$

$$= E^0(e^{-\alpha \tau(s)}) \, E^0(e^{-\alpha \tau(t)}).$$

Thus $f(s+t) = f(s)f(t)$ and since $f(s)$ is right continuous we have $f(s) = e^{-h(\alpha)s}$ for some $h(\alpha) > 0$. To compute $h(\alpha)$ we use (2.125),

(2.94), and (2.5) to get

$$
\begin{aligned}
\frac{1}{h(\alpha)} &= \int_0^\infty f(s)\,ds = E^0\left(\int_0^\infty e^{-\alpha\tau(s)}\,ds\right) \qquad (2.134)\\
&= E^0\left(\int_0^\infty e^{-\alpha t}\,dL_t^0\right) = u^\alpha(0,0) = \frac{1}{\sqrt{2\alpha}}.
\end{aligned}
$$

Thus we obtain (2.132). $\qquad\square$

Remark 2.4.8 A canonical positive p-stable process $\{Z(t), t \in R_+\}$, $Z(0) = 0$, $0 < p < 1$, is defined by its Laplace transform,

$$
E^0 e^{-\lambda Z(t)} = e^{-t\lambda^p}. \qquad (2.135)
$$

By Lemma 2.4.7 we have

$$
E^0\left(e^{-\lambda\tau(s)/2}\right) = e^{-s\sqrt{\lambda}}. \qquad (2.136)
$$

Thus we see that $\tau(s)/2$ is a canonical positive $1/2$–stable process.

2.5 Terminal times

A stopping time T is called a terminal time if for every t

$$
T > t \Rightarrow T = t + T \circ \theta_t \qquad \text{a.s.} \qquad (2.137)
$$

It is easy to see that any first hitting time is a terminal time. Set

$$
u_T(x,y) = E^x\left(L_T^y\right), \qquad (2.138)
$$

which may be infinite. We evaluate $u_T(x,y)$ for Brownian motion when $T = T_0$, the first hitting time of 0. Since Brownian motion is continuous, it is obvious that $u_{T_0}(x,y) = 0$ unless x and y have the same sign.

Lemma 2.5.1 *Let $\{L_t^x, (x,t) \in R^1 \times R_+\}$ be the local times of Brownian motion and let u_{T_0} be given by (2.138). Then*

$$
u_{T_0}(x,y) = \begin{cases} 2\left(|x| \wedge |y|\right) & xy > 0 \\ 0 & xy \le 0. \end{cases} \qquad (2.139)
$$

Proof By (2.94),

$$
\begin{aligned}
u^\alpha(x,y) &= E^x\left(\int_0^\infty e^{-\alpha t}\,dL_t^y\right) \qquad (2.140)\\
&= E^x\left(\int_0^{T_0} e^{-\alpha t}\,dL_t^y\right) + E^x\left(\int_{T_0}^\infty e^{-\alpha t}\,dL_t^y\right).
\end{aligned}
$$

Using the arguments in (2.99), (2.100), and (2.95), we see that

$$
= E^x \left(\int_0^{T_0} e^{-\alpha t} \, dL_t^y \right) + E^x (e^{-\alpha T_0}) \, u^\alpha(0, y)
$$

$$
= E^x \left(\int_0^{T_0} e^{-\alpha t} \, dL_t^y \right) + \frac{u^\alpha(x, 0) \, u^\alpha(0, y)}{u^\alpha(0, 0)}.
$$

Therefore, by (2.5),

$$
E^x \left(\int_0^{T_0} e^{-\alpha t} \, dL_t^y \right) = u^\alpha(x, y) - \frac{u^\alpha(x, 0) u^\alpha(0, y)}{u^\alpha(0, 0)} \tag{2.141}
$$

$$
= \frac{e^{-\sqrt{2\alpha}|x-y|} - e^{-\sqrt{2\alpha}|x|} e^{-\sqrt{2\alpha}|y|}}{\sqrt{2\alpha}}.
$$

Taking the limit as $\alpha \to 0$ we obtain (2.139). $\qquad\square$

Let $A = \{A_t, t \in R_+\}$ be a continuous additive functional and let

$$
\tau_A(s) := \inf\{t > 0 \,|\, A_t > s\}, \tag{2.142}
$$

where $\inf\{\emptyset\} = \infty$. $\tau_A(s)$ is the right continuous inverse of A_t. It is easy to see that $\tau_A(s)$ is a stopping time but it is not a terminal time. However, we do have

$$
\tau_A(s) > t \Rightarrow \tau_A(s) = t + \tau_A(s - A_t) \circ \theta_t \qquad \text{a.s.} \tag{2.143}
$$

Lemma 2.5.2 *Let $A = \{A_t, t \in R_+\}$ be a continuous additive functional on $(\Omega, \mathcal{F}, \mathcal{F}_t, P^x)$ and let λ be an exponential random variable with mean $1/\alpha$ that is independent of $(\Omega, \mathcal{F}, \mathcal{F}_t, P^x)$. Then, for all $f \in \mathcal{B}_b(R^1)$,*

$$
E_\lambda \left(1_{\{\lambda > A_t\}} f(\tau_A(\lambda)) \right) = e^{-\alpha A_t} E_\lambda \left(f(t + \tau_A(\lambda) \circ \theta_t) \right) \qquad P^x \quad a.s., \tag{2.144}
$$

where E_λ denotes expectation with respect to λ.

Proof By (2.143), $\lambda > A_t$ implies that $\tau_A(\lambda) = t + \tau_A(\lambda - A_t) \circ \theta_t$. Therefore

$$
E_\lambda \left(1_{\{\lambda > A_t\}} f(\tau_A(\lambda)) \right) \tag{2.145}
$$

$$
= E_\lambda \left(1_{\{\lambda > A_t\}} f(t + \tau_A(\lambda - A_t) \circ \theta_t) \right)
$$

$$
= \alpha \int_{A_t}^\infty f(t + \tau_A(y - A_t) \circ \theta_t) e^{-\alpha y} \, dy
$$

$$
= e^{-\alpha A_t} \alpha \int_0^\infty f(t + \tau_A(y) \circ \theta_t) e^{-\alpha y} \, dy,
$$

which is the right-hand side of (2.144). $\qquad\square$

Note that even though we do not state it explicitly, we are really using the "memoryless" property of λ, that is, the fact that, conditioned on $\lambda > h$, $\lambda - h$ has the same law as λ.

Let $P_\lambda^x := P^x \times P_\lambda$. Similar to (2.138), we define

$$u_{\tau_A(\lambda)}(x, y) = E_\lambda^x \left(L_{\tau_A(\lambda)}^y \right). \qquad (2.146)$$

We note that

$$
\begin{aligned}
u_{\tau_A(\lambda)}(x, y) &= E_\lambda^x \left(\int_0^\infty 1_{\{\tau_A(\lambda) > t\}} \, dL_t^y \right) \qquad (2.147) \\
&= E_\lambda^x \left(\int_0^\infty 1_{\{\lambda > A_t\}} \, dL_t^y \right) \\
&= E^x \left(\int_0^\infty P_\lambda(\lambda > A_t) \, dL_t^y \right) \\
&= E^x \left(\int_0^\infty e^{-\alpha A_t} \, dL_t^y \right).
\end{aligned}
$$

We now give a simple formula that plays a crucial role in obtaining isomorphism theorems for local times of Brownian motion and, in later chapters, much more general classes of processes.

Theorem 2.5.3 (Kac's Moment Formula) *Let W be a Brownian motion and T a terminal time on $(\Omega, \mathcal{F}, \mathcal{F}_t, P^x)$. Let $\{L_t^y, (y, t) \in R^1 \times R_+\}$ be the local times of W. Then*

$$E^x \left(\prod_{i=1}^n L_T^{y_i} \right) = \sum_\pi u_T(x, y_{\pi_1}) \cdots u_T(y_{\pi_{n-1}}, y_{\pi_n}), \qquad (2.148)$$

where the sum runs over all permutations π of $\{1, \ldots, n\}$. In particular,

$$E^x \left((L_T^y)^n \right) = n! \, u_T(x, y) \, (u_T(y, y))^{n-1}. \qquad (2.149)$$

Let $A = \{A_t, t \in R_+\}$ be a CAF on $(\Omega, \mathcal{F}, \mathcal{F}_t)$ and λ an exponential random variable independent of $(\Omega, \mathcal{F}, \mathcal{F}_t, P^x)$. Then these equations are also valid with T replaced by $\tau_A(\lambda)$ and E^x replaced by E_λ^x.

If $u_T(x, y_i) = \infty$ for some y_i, both sides of (2.148) are infinite. Similarly when T is replaced by $\tau_A(\lambda)$ and E^x by E_λ^x.

Proof We first consider the case when T is a terminal time. Note that

$$E^x \left(\prod_{i=1}^n L_T^{y_i} \right) = E^x \left(\prod_{i=1}^n \int_0^T dL_{t_i}^{y_i} \right) \qquad (2.150)$$

$$= \sum_\pi E^x \left(\int_{\{0<t_1<\ldots<t_n<T\}} \prod_{i=1}^n dL_{t_i}^{y_{\pi_i}} \right)$$

(see (2.104)).

Let $y_{\pi_i} = z_i$, $i = 1, \ldots, n$. Note that

$$\int_{\{0<t_1<\ldots<t_n<T\}} \prod_{i=1}^n dL_{t_i}^{z_i} = \int_0^T \int_{t_1}^T \cdots \int_{t_{n-1}}^T \prod_{i=1}^n dL_{t_i}^{z_i}. \qquad (2.151)$$

Therefore, setting

$$F(t, T) = \int_{\{t<t_2<\ldots<t_n<T\}} \prod_{i=2}^n dL_{t_i}^{z_i}, \qquad (2.152)$$

we have

$$E^x \left(\int_{\{0<t_1<\ldots<t_n<T\}} \prod_{i=1}^n dL_{t_i}^{z_i} \right) = E^x \left(\int_0^T F(t, T) \, dL_t^{z_1} \right). \qquad (2.153)$$

Using the terminal time property (2.137) for the second equality and the additivity of L_t^y for the last equality, we can write

$$F(t, T) = \int_{\{t<t_2<\ldots<t_n<T\}} \prod_{i=2}^n dL_{t_i}^{z_i} \qquad (2.154)$$

$$= \int_{\{t<t_2<\ldots<t_n<t+T\circ\theta_t\}} \prod_{i=2}^n dL_{t_i}^{z_i}$$

$$= F(0, T) \circ \theta_t.$$

Hence

$$E^x \left(\int_0^T F(t, T) \, dL_t^{z_1} \right) = E^x \left(\int_0^T F(0, T) \circ \theta_t \, dL_t^{z_1} \right) ds. \qquad (2.155)$$

By (2.126) with $H_t \equiv 1$ we have

$$E^x \left(\int_0^T F(0, T) \circ \theta_t \, dL_t^{z_1} \right) = E^{z_1} \left(F(0, T) \right) E^x \left(L_T^{z_1} \right). \qquad (2.156)$$

(Since the right-hand side of (2.150) sums the terms in (2.151) over all permutations of $(y_1, \ldots, y_n)$, z_1 in (2.156) can be any of the y_i. Thus if $u_T(x, y_i) = \infty$ for some y_i, the left-hand side of (2.148) is infinite.)

Combining (2.153), (2.155), and (2.156), we have

$$E^x \left(\int_{\{0<t_1<\ldots<t_n<T\}} \prod_{i=1}^n dL_{t_i}^{z_i} \right) \qquad (2.157)$$

$$= u_T(x, z_1) E^{z_1} \left(\int_{\{0 < t_2 < \ldots < t_n < T\}} \prod_{i=2}^{n} dL_{t_i}^{z_i} \right).$$

Iterating this argument we get

$$E^x \left(\int_{\{0 < t_1 < \ldots < t_n < T\}} \prod_{i=1}^{n} dL_{t_i}^{z_i} \right) \tag{2.158}$$
$$= u_T(x, z_1) u_T(z_1, z_2) \cdots u_T(z_{n-1}, z_n),$$

which gives one of the terms in (2.148). Summing over π we get (2.148).

We now note that (2.150)–(2.153) also hold with T replaced by $\tau_A(\lambda)$ and E^x replaced by E_λ^x. Hence

$$E_\lambda^x \left(\int_{\{0 < t_1 < \ldots < t_n < \tau_A(\lambda)\}} \prod_{i=1}^{n} dL_{t_i}^{z_i} \right) \tag{2.159}$$
$$= E_\lambda^x \left(\int_0^{\tau_A(\lambda)} F(t, \tau_A(\lambda)) \, dL_t^{z_1} \right)$$
$$= E_\lambda^x \left(\int_0^{\infty} 1_{\{\lambda > A_t\}} F(t, \tau_A(\lambda)) \, dL_t^{z_1} \right),$$

where, as in (2.152),

$$F(t, \tau_A(\lambda)) = \int_{\{t < t_2 < \ldots < t_n < \tau_A(\lambda)\}} \prod_{i=2}^{n} dL_{t_i}^{z_i}. \tag{2.160}$$

Using Lemma 2.5.2 we see that

$$E_\lambda^x \left(\int_0^{\infty} 1_{\{\lambda > A_t\}} F(t, \tau_A(\lambda)) \, dL_t^{z_1} \right) \tag{2.161}$$
$$= E_\lambda^x \left(\int_0^{\infty} e^{-\alpha A_t} F(t, t + \tau_A(\lambda) \circ \theta_t) \, dL_t^{z_1} \right)$$
$$= E_\lambda^x \left(\int_0^{\infty} e^{-\alpha A_t} F(0, \tau_A(\lambda)) \circ \theta_t \, dL_t^{z_1} \right),$$

where the last equality uses the additivity of local times as in the last equality of (2.154).

By (2.126) with $T \equiv \infty$ we have

$$E_\lambda^x \left(\int_0^{\infty} e^{-\alpha A_t} F(0, \tau_A(\lambda)) \circ \theta_t \, dL_t^{z_1} \right) \tag{2.162}$$
$$= E_\lambda^{z_1} (F(0, \tau_A(\lambda))) E^x \left(\int_0^{\infty} e^{-\alpha A_t} \, dL_t^{z_1} \right)$$
$$= E_\lambda^{z_1} (F(0, \tau_A(\lambda))) u_{\tau_A(\lambda)}(x, z_1),$$

where, for the last equation, we use (2.147).

Combining (2.159)–(2.162) we see that

$$E_\lambda^x \left(\int_{\{0<t_1<...<t_n<\tau_A(\lambda)\}} \prod_{i=1}^n dL_{t_i}^{z_i} \right)$$
$$= u_{\tau_A(\lambda)}(x,z_1) E_\lambda^{z_1} \left(\int_{\{0<t_2<...<t_n<\tau_A(\lambda)\}} \prod_{i=2}^n dL_{t_i}^{z_i} \right).$$

Iterating this argument we get

$$E_\lambda^x \left(\int_{\{0<t_1<...<t_n<\tau_A(\lambda)\}} \prod_{i=1}^n dL_{t_i}^{z_i} \right) \tag{2.163}$$
$$= u_{\tau_A(\lambda)}(x,z_1) u_{\tau_A(\lambda)}(z_1,z_2) \cdots u_{\tau_A(\lambda)}(z_{n-1},z_n),$$

which gives one of the terms in (2.148) with T replaced by $\tau_A(\lambda)$ and E^x replaced by E_λ^x. Summing over π we get the other terms. □

We often suppress the λ in E_λ^x when the meaning is clear from the context.

When the CAF $A = \{A_t, t \in R_+\}$ is $L^0 = \{L_t^0, t \in R_+\}$, the local time of Brownian motion at 0, $\tau_{L^0}(s)$ is the same as $\tau(s)$, the inverse local time of Brownian motion at 0, defined in Subsection 2.4.1. We use the latter notation.

In order to apply Theorem 2.5.3 when $A = L^0$ we calculate $u_{\tau(\lambda)}$ in the next lemma.

Lemma 2.5.4 *Let W be a Brownian motion $(\Omega, \mathcal{F}, \mathcal{F}_t, P^x)$ and let $\{L_t^y, (y,t) \in R^1 \times R_+\}$ be the local times of W. Let λ be an exponential random variable with mean $1/\alpha$ which is independent of $(\Omega, \mathcal{F}, \mathcal{F}_t, P^x)$. Then*

$$u_{\tau(\lambda)}(x,y) = E_\lambda^x(L_{\tau(\lambda)}^y) = u_{T_0}(x,y) + \frac{1}{\alpha}, \tag{2.164}$$

where $u_{T_0}(x,y)$ is given in (2.139).

Proof Note that $\tau(\lambda) = T_0 + \tau(\lambda) \circ \theta_{T_0}$, so that by the strong Markov property, as in the proof of (2.95),

$$u^\beta(x,y) = E^x \left(\int_0^\infty e^{-\beta t} dL_t^y \right) \tag{2.165}$$
$$= E_\lambda^x \left(\int_0^{\tau(\lambda)} e^{-\beta t} dL_t^y \right) + E_\lambda^x \left(\int_{T_0+\tau(\lambda)\circ\theta_{T_0}}^\infty e^{-\beta t} dL_t^y \right)$$

$$= E_\lambda^x \left(\int_0^{\tau(\lambda)} e^{-\beta t}\, dL_t^y \right) + E^x(e^{-\beta T_0}) E_\lambda^0 \left(\int_{\tau(\lambda)}^\infty e^{-\beta t}\, dL_t^y \right)$$

$$= E_\lambda^x \left(\int_0^{\tau(\lambda)} e^{-\beta t}\, dL_t^y \right) + E^x(e^{-\beta T_0}) E_\lambda^0(e^{-\beta \tau(\lambda)}) u^\beta(0, y)$$

$$= E_\lambda^x \left(\int_0^{\tau(\lambda)} e^{-\beta t}\, dL_t^y \right) + \frac{u^\beta(x, 0) u^\beta(0, y)}{u^\beta(0, 0)} E_\lambda^0(e^{-\beta \tau(\lambda)}),$$

where for the last equality we again use (2.95). Consequently

$$u_{\tau(\lambda)}(x, y) = \lim_{\beta \to 0} E_\lambda^x \left(\int_0^{\tau(\lambda)} e^{-\beta t}\, dL_t^y \right) \tag{2.166}$$

$$= \lim_{\beta \to 0} \left\{ u^\beta(x, y) - \frac{u^\beta(x, 0) u^\beta(0, y)}{u^\beta(0, 0)} \right\}$$

$$+ \lim_{\beta \to 0} \frac{u^\beta(x, 0) u^\beta(0, y)}{u^\beta(0, 0)} (1 - E_\lambda^0(e^{-\beta \tau(\lambda)})).$$

It follows from (2.141) and (2.139) that

$$\lim_{\beta \to 0} \left\{ u^\beta(x, y) - \frac{u^\beta(x, 0) u^\beta(0, y)}{u^\beta(0, 0)} \right\} = u_{T_0}(x, y). \tag{2.167}$$

Also, by (2.5),

$$\lim_{\beta \to 0} \frac{u^\beta(x, 0) u^\beta(0, y)}{u^\beta(0, 0)} \left(1 - E_\lambda^0 \left(e^{-\beta \tau(\lambda)} \right) \right) \tag{2.168}$$

$$= \lim_{\beta \to 0} \frac{e^{-\sqrt{2\beta}(|x| + |y|)}}{\sqrt{2\beta}} \left(1 - E_\lambda^0 \left(e^{-\beta \tau(\lambda)} \right) \right)$$

$$= \lim_{\beta \to 0} \frac{1}{\sqrt{2\beta}} \left(1 - E_\lambda^0 \left(e^{-\beta \tau(\lambda)} \right) \right)$$

$$= \lim_{\beta \to 0} \frac{1}{\sqrt{2\beta}} \left(1 - \frac{\alpha}{\sqrt{2\beta} + \alpha} \right)$$

$$= \lim_{\beta \to 0} \frac{1}{\sqrt{2\beta}} \frac{\sqrt{2\beta}}{\sqrt{2\beta} + \alpha} = \frac{1}{\alpha},$$

where we use (2.132) for the third equality. $\qquad \square$

2.6 The First Ray–Knight Theorem

In addition to considering $\{L_t^x, (x, t) \in R^1 \times R_+\}$, the local times of Brownian motion, which is a stochastic process on $R^1 \times R_+$, we often can get interesting descriptions of the stochastic process on R^1, or subsets A of R^1, given by $\{L_T^x, x \in A\}$, where T is a stopping time. The classical First Ray–Knight Theorem is a relationship of this sort. It describes the behavior of the Brownian local times $\{L_{T_0}^r, r \in R_+\}$ in terms of squares of associated Gaussian processes, which, in this case, are two independent Brownian motions. This theorem, which was proved independently by Ray (1963) and Knight (1969), is the earliest example of such a result that we are aware of. We refer to results of this sort as isomorphism theorems.

To make matters transparent we initially state and prove this theorem in its simplest, restricted form (that is, we only consider a portion of the domain of L_{T_0}). Afterwards we prove the general theorem. There are many proofs of this theorem in the literature. We give a very simple proof which we generalize to a wider class of processes in Theorem 8.2.6.

Let B_t and $\bar{B}_t$ denote two independent Brownian motions starting at 0. By definition, the process $\{B_t^2 + \bar{B}_t^2 : t \in R_+\}$ is called a second order squared Bessel process (see Section 14.2).

Theorem 2.6.1 (First Ray–Knight Theorem, restricted form) Let $x > 0$. Then, under $P^x \times P_{B,\bar{B}}$,

$$\{L_{T_0}^r : r \in [0,x]\} \stackrel{law}{=} \{B_r^2 + \bar{B}_r^2 : r \in [0,x]\} \qquad (2.169)$$

on $C([0,x])$.

Equivalently, under P^x, between 0 and x, $L_{T_0}^r$ has the law of a second order squared Bessel process starting at 0.

The key to the proof of Theorem 2.6.1 is the following lemma which is a simple consequence of Kac's moment formula (Theorem 2.5.3).

Lemma 2.6.2 Let T be a terminal time with potential density $u_T(x,y)$ defined in (2.138). Let Σ be the matrix with elements $\Sigma_{i,j} = u_T(x_i, x_j)$, $i,j = 1,\ldots,n$. Let Λ be the matrix with elements $\{\Lambda\}_{i,j} = \lambda_i \delta_{i,j}$. For all $\lambda_1,\ldots,\lambda_n$ sufficiently small and $1 \le l \le n$,

$$E^{x_l} \exp\left(\sum_{i=1}^n \lambda_i L_T^{x_i}\right) = \frac{\det(I - \widehat{\Sigma}\Lambda)}{\det(I - \Sigma\Lambda)}, \qquad (2.170)$$

where

$$\widehat{\Sigma}_{i,j} = \Sigma_{i,j} - \Sigma_{l,j} \qquad i,j = 1,\ldots,n. \qquad (2.171)$$

These equations also hold when T is replaced by $\tau_A(\lambda)$, for any CAF, $A = \{A_t, t \in R_+\}$, and E^x is replaced by E_λ^x.

Proof By (2.148),

$$E^{x_l}\left(\left(\sum_{i=1}^{n}\lambda_i L_T^{x_i}\right)^k\right) \tag{2.172}$$

$$= k! \sum_{j_1,\ldots,j_k=1}^{n} u_T(x_l, x_{j_1})\lambda_{j_1} u_T(x_{j_1}, x_{j_2})\lambda_{j_2} u_T(x_{j_2}, x_{j_3})\cdots$$

$$u_T(x_{j_{k-2}}, x_{j_{k-1}})\lambda_{j_{k-1}} u_T(x_{j_{k-1}}, x_{j_k})\lambda_{j_k}$$

$$= k! \sum_{j_k=1}^{n}\{(\Sigma\Lambda)^k\}_{l,j_k}$$

for all k. Therefore

$$E^{x_l}\exp\left(\sum_{i=1}^{n}\lambda_i L_T^{x_i}\right) = \sum_{j=1}^{n}\{(I-\Sigma\Lambda)^{-1}\}_{l,j} = \{(I-\Sigma\Lambda)^{-1}\mathbf{1}^t\}_l, \tag{2.173}$$

where $\mathbf{1}^t$ denotes the transpose of the n-dimensional vector $(1,\ldots,1)$. Consequently

$$(I-\Sigma\Lambda)Y = \mathbf{1}^t, \tag{2.174}$$

where Y is the n-dimensional vector with components $E^{x_l}e^{\left(\sum_{i=1}^{n}\lambda_i L_T^{x_i}\right)}$, $l = 1,\ldots,n$. Therefore, by Cramer's Theorem,

$$E^{x_l}\exp\left(\sum_{i=1}^{n}\lambda_i L_T^{x_i}\right) = \frac{\det((I-\Sigma\Lambda)^{(l)})}{\det(I-\Sigma\Lambda)}, \tag{2.175}$$

where $(I-\Sigma\Lambda)^{(l)}$ is the matrix obtained by replacing the l-th column of $(I-\Sigma\Lambda)$ by a column with all of its elements equal to 1. Subtracting the l-th row of $(I-\Sigma\Lambda)^{(l)}$ from each of its other rows, we obtain a matrix A with its l-th column equal to $\delta_{j,l}$, $j = 1,\ldots,n$ and with entries $(I-\widehat{\Sigma}\Lambda)_{i,j}$ for $i,j = 1,\ldots,n$ for $i,j \neq l$. Note that

$$\det((I-\Sigma\Lambda)^{(l)}) = \det(A) \tag{2.176}$$

and expanding $\det(A)$ by the l-th column we see that

$$\det(A) = \det(B), \tag{2.177}$$

where B is the $(n-1)\times(n-1)$ matrix obtained by deleting the l-th column and l-th row from A. Note that $\{(I-\widehat{\Sigma}\Lambda)\}_{l,j} = \delta_{l,l}$, so that expanding $\det(I-\widehat{\Sigma}\Lambda)$ by the l-th row we see that

$$\det(I-\widehat{\Sigma}\Lambda) = \det(C), \tag{2.178}$$

where C is the $(n-1)\times(n-1)$ matrix obtained by deleting the l-th column and l-th row from $I - \widehat{\Sigma}\Lambda$. Considering the description of the matrix A in the sentence preceding (2.176), we see that $C = B$. Thus

$$\det((I - \Sigma\Lambda)^{(l)}) = \det((I - \widehat{\Sigma}\Lambda)). \qquad (2.179)$$

Using this in (2.175) we get (2.170). □

Proof of Theorem 2.6.1 To prove that two continuous stochastic processes are equal in law on $C([0, x])$ it suffices to show that they have the same finite-dimensional distributions. We do this by showing that they have the same finite-dimensional moment generating functions. Let $0 < x_1 < \cdots < x_n = x$. Since T_0 is a terminal time, we apply Lemma 2.6.2 with $T = T_0$ and $l = n$. By (2.139), $\Sigma_{i,j} = 2(x_i \wedge x_j)$, $i, j = 1, \ldots, n$. Since

$$\widehat{\Sigma}_{i,j} = \Sigma_{i,j} - \Sigma_{n,j} = 0 \qquad \text{for } j \le i \le n, \qquad (2.180)$$

we see that $(I - \widehat{\Sigma}\Lambda)$ is an upper triangular matrix with ones on the diagonal. Therefore

$$\det(I - \widehat{\Sigma}\Lambda) = 1. \qquad (2.181)$$

Theorem 2.6.1 now follows from Lemma 2.6.2 and the fact that

$$E\left(\exp\left(\sum_{i=1}^{n} \lambda_i B_{x_i}^2\right)\right) = \frac{1}{\sqrt{\det(I - \Sigma\Lambda)}}. \qquad (2.182)$$

This is a straightforward calculation. Note that $\Sigma = 2C$, where C is given on page 18. Using (2.29) we see that

$$Ee^{\sum_{i=1}^{n} \lambda_i B_{x_i}^2} = \frac{1}{(2\pi)^{n/2}(\det C)^{1/2}} \int e^{\sum_{i=1}^{n} \lambda_i y_i^2 - (y, C^{-1}y)/2} \, dy$$

$$= \frac{1}{(2\pi)^{n/2}(\det \Sigma)^{1/2}} \int e^{\sum_{i=1}^{n} \lambda_i y_i^2/2 - (y, \Sigma^{-1}y)/2} \, dy$$

$$= \frac{1}{(2\pi)^{n/2}(\det \Sigma)^{1/2}} \int e^{-(y, (\Sigma^{-1} - \Lambda)y)/2} \, dy.$$

$$= \left(\frac{\det(\Sigma^{-1} - \Lambda)^{-1}}{\det \Sigma}\right)^{1/2} = (\det(I - \Sigma\Lambda))^{-1/2}. \qquad (2.183)$$

□

In Theorem 2.6.1 we considered $L_{T_0}^r$ for $r \in [0, x]$. We now allow r to be in R_+. Note that $r < 0$ is not possible because the process is killed when it hits 0. On the other hand, it is clear that the following theorem is also valid for $r \in (-\infty, 0]$, with obvious modifications.

Theorem 2.6.3 (First Ray–Knight Theorem) *Let $x > 0$. Then,
under $P^x \times P_{B,\bar{B}}$,*

$$\{L^r_{T_0} + (B^2_{r-x} + \bar{B}^2_{r-x})1_{\{r \geq x\}} : r \in R_+\} \stackrel{law}{=} \{B^2_r + \bar{B}^2_r : r \in R_+\}$$
(2.184)

on $C(R_+)$.

*Equivalently, under P^x, $L^r_{T_0}$ between 0 and x has the law of a second
order squared Bessel process $Y = \{Y_r ; 0 \leq r \leq x\}$ with $Y_0 = 0$, and then
proceeds from x as a 0-th order squared Bessel process $Z = \{Z_r ; x \leq
r < \infty\}$ with $Z_x = Y_x$, where Z also has the property that, conditioned
on Y_x, it is independent of Y.*

Proof The proof is just a slight generalization of the previous proof.
Let $0 < x_1 < \cdots < x_l = x < \ldots < x_n$. It suffices to show that the mo-
ment generating functions of $\{L^{x_i}_{T_0} + B^2_{x_i-x_l}I_{\{x_i > x_l\}} + \bar{B}^2_{x_i-x_l}I_{\{x_i > x_l\}}, i \in
1, \ldots, n\}$ and $\{B^2_{x_i} + \bar{B}^2_{x_i}, i \in 1, \ldots, n\}$ are equal, for all $x_1, \ldots, x_n$ in
R_+, for all n. By Lemma 2.6.2,

$$E^{x_l} \exp \left(\sum_{i=1}^n \lambda_i L^{x_i}_{T_0} \right) = \frac{\det(I - \widehat{\Sigma}\Lambda)}{\det(I - \Sigma\Lambda)},$$
(2.185)

where

$$\widehat{\Sigma}_{i,j} = 2\left((x_i \wedge x_j) - (x_l \wedge x_j)\right) \qquad i,j = 1, \ldots, n.$$
(2.186)

Taking (2.182) into account we see that, to complete the proof, we need
only show that

$$E \left(\exp \left(\sum_{i=l+1}^n \lambda_i B^2_{x_i-x_l} \right) \right) = \frac{1}{\sqrt{\det(I - \widehat{\Sigma}\Lambda)}}.$$
(2.187)

It is easy to see that

$$\widehat{\Sigma}_{i,j} = 0 \qquad 1 \leq j \leq i \wedge l$$
(2.188)

and

$$\widehat{\Sigma}_{i,j} = 2((x_j - x_l) \wedge (x_i - x_l)) \qquad l < i \wedge j \leq n.$$
(2.189)

Thus

$$\det(I - \widehat{\Sigma}\Lambda) = \det(I - \overline{\Sigma}\Lambda),$$
(2.190)

where $\overline{\Sigma}$ is an $(n - l) \times (n - l)$ matrix with $\overline{\Sigma}_{i,j} = \widehat{\Sigma}_{i,j}$, $l < i \wedge j \leq n$.
Equation (2.187) follows from this and (2.182).

The equivalency is proved in Remark 14.2.3. $\square$

2.7 The Second Ray–Knight Theorem

Recall that $\tau(t) = \inf\{s \,|\, L_s^0 > t\}$ is the right continuous inverse of the local time at 0 of a standard Brownian motion. Heuristically, it is the amount of time it takes for the local time at zero to be equal to t and, by the continuity and support properties of L_s^0, $L_{\tau(t)}^0 = t$. The Second Ray–Knight Theorem describes the behavior of the Brownian local times $\{L_{\tau(t)}^r, r \in R_+\}$ in terms of squares of standard Brownian motion and Brownian motion starting from $\sqrt{t}$. As for the First Ray–Knight Theorem, there are many proofs of this theorem in the literature. We give a simple proof along the lines of our proof of the First Ray–Knight Theorem.

Theorem 2.7.1 (Second Ray–Knight Theorem) *Let $t > 0$. Then, under the measure $P^0 \times P_B$,*

$$\{L_{\tau(t)}^x + B_x^2; x \geq 0\} \overset{law}{=} \left\{\left(B_x + \sqrt{t}\right)^2; x \geq 0\right\} \tag{2.191}$$

on $C(R_+)$, where $\{B_x; x \geq 0\}$ is a real-valued Brownian motion starting at 0, independent of the original Brownian motion (that is, the Brownian motion with local times $L_{\tau(t)}^r$).

Equivalently, under P^0, $\{L_{\tau(t)}^r; r \in R_+\}$ has the law of a 0-th order squared Bessel process starting at t.

Proof Like the First Ray–Knight Theorem, this theorem is also a simple application of Lemma 2.6.2, this time applied to $\tau(\lambda)$. Let ξ be an exponential random variable with mean $1/\alpha$. By Lemma 2.5.4, $u_{\tau(\xi)}(x, y) = u_{T_0}(x, y) + 1/\alpha$, where $u_{T_0}(x, y) = 2(x \wedge y)$. Let $x_1 = 0$ and $x_2, \ldots, x_n > 0$. By Lemma 2.6.2,

$$E_\xi^{x_1} \exp\left(\sum_{i=1}^n \lambda_i L_{\tau(\xi)}^{x_i}\right) = \frac{\det(I - \widehat{\Sigma}\Lambda)}{\det(I - \Sigma\Lambda)}, \tag{2.192}$$

where

$$\Sigma_{i,j} = 2(x_i \wedge x_j) + 1/\alpha \qquad i, j = 1, \ldots, n \tag{2.193}$$

and

$$\widehat{\Sigma}_{i,j} = 2(x_i \wedge x_j) \qquad i, j = 1, \ldots, n \tag{2.194}$$

since $u_{T_0}(0, y) = 0$ for all $y \geq 0$.

By (2.182),

$$E\left(\exp\left(\sum_{i=1}^n \lambda_i B_{x_i}^2\right)\right) = \frac{1}{\sqrt{\det(I - \widehat{\Sigma}\Lambda)}}. \tag{2.195}$$

Let ρ be a $N(0,1)$ independent of $\{B_x, x \geq 0\}$. Using the fact that Brownian motion has stationary and independent increments, we can see that $\{B_x + (2\alpha)^{-1/2}\rho, x \geq 0\} \overset{law}{=} \{B_{x+(2\alpha)^{-1}}, x \geq 0\}$. Hence it follows from (2.195) that

$$
E\left(\exp\left(\sum_{i=1}^{n} \lambda_i \left(B_{x_i} + (2\alpha)^{-1/2}\rho\right)^2\right)\right) = \frac{1}{\sqrt{\det(I - \Sigma\Lambda)}}. \tag{2.196}
$$

Combining (2.192), (2.195), and (2.196), we see that under the measure $P^0 \times P_{B,\overline{B},\rho,\overline{\rho}}$

$$
\{L^x_{\tau(\xi)} + B^2_x + \overline{B}^2_x; x \geq 0\} \tag{2.197}
$$

$$
\overset{law}{=} \{(B_x + (2\alpha)^{-1/2}\rho)^2 + (\overline{B}_x + (2\alpha)^{-1/2}\overline{\rho})^2; x \geq 0\}.
$$

Let $\{B_x; x \geq 0\}$, $\{\overline{B}_x; x \geq 0\}$ be independent Brownian motions and let $\lambda_1, \ldots, \lambda_n$ be vectors in R^2. Comparison with (2.27) shows that

$$
E \exp\left(i \sum_{j=1}^{n} (\lambda_j, (B_{x_j}, \overline{B}_{x_j}))\right) = \exp\left(-\frac{1}{2}\left(\sum_{j,k=1}^{n} (\lambda_j, \lambda_k)(x_j \wedge x_k)\right)\right). \tag{2.198}
$$

It follows from this that if U is any orthogonal transformation in R^2, using the fact that $U^t = U^{-1}$, we have

$$
E \exp\left(i \sum_{j=1}^{n} (\lambda_j, U(B_{x_j}, \overline{B}_{x_j}))\right) \tag{2.199}
$$

$$
= E \exp\left(i \sum_{j=1}^{n} (U^{-1}\lambda_j, (B_{x_j}, \overline{B}_{x_j}))\right)
$$

$$
= \exp\left(-\frac{1}{2}\left(\sum_{j,k=1}^{n} (U^{-1}\lambda_j, U^{-1}\lambda_k)(x_j \wedge x_k)\right)\right)
$$

$$
= \exp\left(-\frac{1}{2}\left(\sum_{j,k=1}^{n} (\lambda_j, \lambda_k)(x_j \wedge x_k)\right)\right),
$$

and consequently

$$
\{U(B_x, \overline{B}_x); x \geq 0\} \overset{law}{=} \{(B_x, \overline{B}_x); x \geq 0\}. \tag{2.200}
$$

Let $\overline{\rho}$ be an independent copy of ρ, and let $U_{\rho,\overline{\rho}}$ be the orthogonal transformation in R^2 for which

$$
U_{\rho,\overline{\rho}}(\rho, \overline{\rho}) = (0, (\rho^2 + \overline{\rho}^2)^{1/2}). \tag{2.201}
$$

Using $|\cdot|$ to denote the Euclidean norm in R^2, it follows from (2.200) and the independence of $B, \overline{B}, \rho, \overline{\rho}$ that

$$\{|(B_x, \overline{B}_x) + (2\alpha)^{-1/2}(\rho, \overline{\rho})| \, ; \, x \geq 0\} \qquad (2.202)$$
$$= \{|U_{\rho, \overline{\rho}}(B_x, \overline{B}_x) + (2\alpha)^{-1/2}(0, (\rho^2 + \overline{\rho}^2)^{1/2})| \, ; \, x \geq 0\}$$
$$\stackrel{law}{=} \{|(B_x, \overline{B}_x) + (2\alpha)^{-1/2}(0, (\rho^2 + \overline{\rho}^2)^{1/2})| \, ; \, x \geq 0\}.$$

Note that $(2\alpha)^{-1}(\rho^2 + \overline{\rho}^2) \stackrel{law}{=} \xi$ (this follows easily using moment generating functions). Therefore, by (2.197) and (2.202), under the measure $P^0 \times P_\xi \times P_{B, \overline{B}}$,

$$\{L^x_{\tau(\xi)} + B^2_x + \overline{B}^2_x ; x \geq 0\} \stackrel{law}{=} \{\overline{B}^2_x + (B_x + \sqrt{\xi})^2 ; x \geq 0\}. \qquad (2.203)$$

By taking the moment generating function of each side of (2.203), it is easy to see that under the measure $P^0 \times P_\xi \times P_B$

$$\{L^x_{\tau(\xi)} + B^2_x ; x \geq 0\} \stackrel{law}{=} \{(B_x + \sqrt{\xi})^2 ; x \geq 0\}. \qquad (2.204)$$

This is actually the "Laplace transform" of (2.191). To see this note that by (2.204)

$$E^{x_1}_\xi \exp\left(\sum_{i=1}^n \lambda_i L^{x_i}_{\tau(\xi)}\right) E\left(\exp\left(\sum_{i=1}^n \lambda_i B^2_{x_i}\right)\right) \qquad (2.205)$$
$$= E\left(\exp\left(\sum_{i=1}^n \lambda_i \left(B_{x_i} + \sqrt{\xi}\right)^2\right)\right).$$

Since ξ is an exponential random variable with mean $1/\alpha$, we can write this as

$$E\left(\exp\left(\sum_{i=1}^n \lambda_i B^2_{x_i}\right)\right) \int_0^\infty E^{x_1} \exp\left(\sum_{i=1}^n \lambda_i L^{x_i}_{\tau(t)}\right) e^{-\alpha t} \, dt$$
$$= \int_0^\infty E\left(\exp\left(\sum_{i=1}^n \lambda_i \left(B_{x_i} + \sqrt{t}\right)^2\right)\right) e^{-\alpha t} \, dt. \qquad (2.206)$$

Therefore the moment generating functions of $\sum_{i=1}^n \lambda_i \left(L^{x_i}_{\tau(t)} + B^2_{x_i}\right)$ and $\sum_{i=1}^n \lambda_i \left(B_{x_i} + \sqrt{t}\right)^2$ are equal for all $(x_1, \ldots, x_n)$ in R_+ for all n for almost all t. However, since the first of these is increasing in t and the second is continuous in t, they are equal for all t. Thus we get (2.191). The equivalency follows from Theorem 14.2.2 as in Remark 14.2.3. $\square$

Remark 2.7.2 We state the First and Second Ray–Knight Theorems in terms of Bessel processes because this is how they often appear in the literature (see, e.g., Revuz and Yor (1991, Chapter XI, Section 2)). Also, describing the local times of Brownian motion as a squared Bessel process gives a stochastic integral representation of the local times which can be used to study them further. On the other hand, the isomorphism theorems that we presented first are in a form that easily extends to a much wider class of Markov processes than Brownian motion. The primary purpose of this book is to use the extensions to study the local times of these processes.

2.8 Ray's Theorem

The isomorphisms for the local time of Brownian motion in the preceding two sections are with respect to the spatial variable. The local time itself is evaluated in Theorem 2.6.3 at the first time the process hits zero and in Theorem 2.7.1 at the first time the local time at zero is equal to t. Ray's Theorem deals with the total accumulated local time of Brownian motion stopped (or, to be more precise, killed, as we explain in Section 3.5) at an independent exponential time. But it is more complicated than this. It actually is given for the local time of the h-transform of Brownian motion killed at an independent exponential time. We study h-transforms in Section 3.9 and obtain a simplified version of Ray's Theorem for transient regular diffusions in Chapter 12. We proceed to explain Ray's Theorem without justifying every statement.

Let λ be an independent exponential random variable with mean 2 and set $u_\lambda(x, y) = E_\lambda^x(L_\lambda^y)$. To compute this expectation we use Theorem 2.5.3 with $A_t = t$, so that $\tau_A(\lambda) = \lambda$. We then have

$$
\begin{aligned}
E_\lambda^x(L_\lambda^y) &= E_\lambda^x\left(\int_0^\lambda dL_t^y\right) \qquad\qquad (2.207)\\
&= E_\lambda^x\left(\int_0^\infty 1_{\{\lambda>t\}}\, dL_t^y\right)\\
&= E^x\left(\int_0^\infty P_\lambda(\lambda > t)\, dL_t^y\right)\\
&= E^x\left(\int_0^\infty e^{-t/2}\, dL_t^y\right) = e^{-|x-y|},
\end{aligned}
$$

where the last equality comes from (2.94) and (2.5).

Let $\{B_{r,\lambda}, r \in R_+\}$ denote Brownian motion killed at the end of an independent exponential time with mean 2. $\{B_{r,\lambda}, r \in R_+\}$ is a Markov process with 0-potential density $u_\lambda(x, y)$. The calculation in (2.207) should make this believable. We provide a complete proof of this fact in

Section 3.5. In the rest of this section we set

$$u(x,y) := u_{1/2}(x,y) = e^{-|x-y|}. \qquad (2.208)$$

Ray's Theorem for Brownian motion describes the total accumulated local time of the h-transform of $B_\lambda = \{B_{r,\lambda}, r \in R_+\}$, under the measure $P^{a,b}$, in terms of squares of associated Gaussian processes. Heuristically, the process with measure $P^{a,b}$ is B_λ, starting at a and conditioned to hit b and die at its last exit from b. $\{L^r_\infty, r \in R^1\}$ in Theorem 2.8.1 is the total accumulated local time of this process.

Let $h(x) = u(x,b)/u(b,b)$. The potential density of the h-transform process is

$$\frac{u(x,y)h(y)}{h(x)} = \frac{u(x,y)u(y,b)}{u(x,b)} \qquad (2.209)$$

(see (3.216)). This follows because the process with measure $P^{a,b}$ is a Markov process with transition density $e^{-t/2}p_t(x,y)h(y)/h(x)$ with respect to Lebesgue measure on R^1, where $p_t(x,y)$ is the transition probability density of Brownian motion.

The Gaussian process that enters into Ray's Theorem is the Ornstein–Uhlenbeck process, a mean zero stationary Gaussian process with covariance $u(x,y) = \exp(-|x-y|)$, the same function as the 0-potential density of the exponentially killed Brownian motion. In Section 5.1 we discuss how Gaussian processes are determined by their covariances. At this point we define the Ornstein–Uhlenbeck process as a rescaled Brownian motion $\{G_x = e^{-x}B_{e^{2x}}; \, x \in R^1\}$, where $\{B_x; \, x \geq 0\}$ is a Brownian motion starting at 0. For any $s \in R^1$ we set

$$G_{x,s} = G_x - e^{-|x-s|}G_s. \qquad (2.210)$$

This is the orthogonal complement of the projection of G_x onto G_s (see Remark 5.1.6).

Theorem 2.8.1 (Ray's Theorem) *Let $a < b$. Under the measure* $P^{a,b} \times P_{G,\bar{G}}$,

$$\{L^r_\infty + \left(\frac{G^2_{r,a}}{2} + \frac{\bar{G}^2_{r,a}}{2}\right)1_{\{r \leq a\}} + \left(\frac{G^2_{r,b}}{2} + \frac{\bar{G}^2_{r,b}}{2}\right)1_{\{r \geq b\}} : r \in R^1\}$$

$$\overset{law}{=} \{\frac{G^2_r}{2} + \frac{\bar{G}^2_r}{2} : r \in R^1\} \qquad (2.211)$$

on $C(R^1)$, where $\{G_x, \bar{G}_x; \, x \in R^1\}$ are independent Ornstein–Uhlenbeck processes, independent of the Brownian motion.

We call Theorem 2.8.1 Ray's Theorem because it describes the same process considered in Ray (1963). Ray's result actually looks quite different from (2.211). He describes L^r_∞ separately in the three regions $r \in (-\infty, a]$, $r \in [a, b]$, and $r \in [b, \infty)$ as follows:

Theorem 2.8.2 (Ray's Theorem, original version) *Let $B = \{B_t, t \geq 0\}$ denote a Brownian motion and let λ be an independent exponential random variable with mean $1/2$. Let $I = \inf_{t \leq \lambda} B_t$ and $S = \sup_{t \leq \lambda} B_t$, and let $\{G^{(i)}_x, x \in R^1\}$, $i = 1, \ldots, 4$ be four independent Ornstein–Uhlenbeck processes, independent of B. Then, under the measure $P^{a,b} \times P_{G^{(1)}, G^{(2)}, G^{(3)}, G^{(4)}}$,*

$$\{I, L^r_\infty ; r \leq a\} \stackrel{law}{=} \{I, 1_{\{r > I\}} \sum_{i=1}^4 \left(G^{(i)}_{r,I}\right)^2 /2 ; r \leq a\} \quad (2.212)$$

$$\{L^r_\infty ; a \leq r \leq b\} \stackrel{law}{=} \{\left(\left(G^{(1)}_r\right)^2 + \left(G^{(2)}_r\right)^2\right)/2 ; a \leq r \leq b\} \quad (2.213)$$

$$\{S, L^r_\infty ; r \geq b\} \stackrel{law}{=} \{S, 1_{\{r < S\}} \sum_{i=1}^4 \left(G^{(i)}_{r,S}\right)^2 /2 ; r \geq b\}, \quad (2.214)$$

where $\{I, L^r_\infty ; r \leq a\}$ is a random variable on $R^1 \times C((-\infty, a])$ and similarly for $\{S, L^r_\infty ; r \geq b\}$.

The proofs of Theorems 2.8.1 and 2.8.2 follow from more general results which are given in Chapter 12.

2.9 Applications of the Ray–Knight Theorems

The First and Second Ray–Knight Theorems and Ray's Theorem have been used to obtain interesting results about Brownian local times. We mention several of them here without proof. Some of them are contained in more general results that are proved in later chapters.

Joint continuity of Brownian local time. Knight (1969) used the First Ray–Knight Theorem (Theorem 2.6.3) to establish the joint continuity of Brownian local times $\{L^y_t ; (t, y) \in R_+ \times R^1\}$. Ito and McKean (1974) obtain this result via Ray's Theorem (Theorem 2.8.2). In Section 2.4 we obtained the joint continuity of Brownian local time by elementary methods. However, none of these methods can be extended to general Markov processes. In Chapter 9 we also use isomorphism theorems, but more general ones than those described in this chapter, to find necessary and sufficient conditions for strongly symmetric Markov processes to have jointly continuous local times.

Exact moduli of continuity for Brownian local time. Ray (1963) used the First Ray–Knight Theorem to show that for all $x \in R^1$

$$\limsup_{\delta \to 0} \frac{|L^{x+\delta}_{T_0} - L^x_{T_0}|}{\sqrt{\delta \log \log (1/\delta)}} = 2\sqrt{L^x_{T_0}} \qquad P^x \quad \text{a.s.} \qquad (2.215)$$

and

$$\limsup_{\substack{|y-z| \to 0 \\ 0 \le y,z \le x}} \frac{|L^y_{T_0} - L^z_{T_0}|}{\sqrt{|y-z|(1/\log |y-z|)}} = 2 \sup_{0 \le y \le x} \sqrt{L^y_{T_0}} \qquad P^x \quad \text{a.s.} \quad (2.216)$$

Note the similarity between these equations and (2.14) and (2.15). This is not a coincidence but an instance of a general principle given in Theorems 9.5.1 and 9.5.5.

Using Ray's Theorem, Ray also obtained (2.215) and (2.216) with T_0 replaced by an independent exponential time. Thus, by Fubini's Theorem, they also hold for some fixed time t. Then, using Brownian scaling, that is, the fact that $W_t \overset{law}{=} cW_{t/c^2}$ and consequently $L^y_t \overset{law}{=} cL^{y/c}_{t/c^2}$, it follows that (2.215) and (2.216) also hold with T_0 replaced by any fixed t.

Note that (2.215) is not very interesting when $x = 0$ since, in this case, $L^x_{T_0} = 0$. Indeed, using the First Ray–Knight Theorem (Theorem 2.6.3), it follows that

$$\limsup_{\delta \downarrow 0} \frac{L^\delta_{T_0}}{2\delta \log \log (1/\delta)} = 1 \qquad P^x \quad \text{a.s.} \qquad (2.217)$$

for $x > 0$.

We consider moduli of continuity of local times in Section 9.5.

Quadratic variation of Brownian local time. Using the Second Ray–Knight Theorem, Perkins (1982) showed that

$$\lim_{n \to \infty} \sum_{x_i \in \pi(n)} \left(L^{x_i}_t - L^{x_{i-1}}_t\right)^2 = 4 \int_0^a L^y_t \, dy \qquad (2.218)$$

in probability, for any sequence of partitions

$$\pi(n) = \{0 = x_0 < x_1 < \ldots < x_n = a\}.$$

We consider the p-variation of local times in Chapter 10. Quadratic variation is the case when $p = 2$.

Most visited site. Bass and Griffin (1985) used the Second Ray–Knight Theorem to study the most visited points for Brownian motion up to time t. Let

$$\mathcal{V}_t = \{x \in R^1 \mid L_t^x = \sup_y L_t^y\}.$$

Heuristically, the more time Brownian motion spends at a point in its range, the greater the local time at that point. Thus $\mathcal{V}$ can be thought of as the points at which the Brownian motion path spends the most time, up to time t. We refer to these points as the most visited sites of Brownian motion or as the favorite sites of Brownian motion. For specificity we consider

$$V_t = \inf_{x \in \mathcal{V}_t} |x| \tag{2.219}$$

the most visited point up to time t with the smallest absolute value. In Section 11.2 we show that, for any $\gamma > 3$,

$$\lim_{t \to \infty} \frac{(\log t)^\gamma}{t^{1/2}} V_t = \infty \qquad P^0 \quad \text{a.s.} \tag{2.220}$$

Following Bass and Griffin, we describe this phenomenon by saying that the most visited site of Brownian motion is transient. (Information on the rate of growth of $V(t)$ is also obtained in Bass and Griffin (1985).)

We also show in Section 11.2 that, for any $\gamma > 3$,

$$\lim_{t \to 0} \frac{|\log t|^\gamma}{t^{1/2}} V_t = \infty \qquad P^0 \quad \text{a.s.} \tag{2.221}$$

This shows, perhaps counterintuitively, that the most visited site by Brownian motion, as it starts from zero, stays away from zero.

Eisenbaum and Shi (1999) used the Second Ray–Knight Theorem to analyze the rarely visited points of Brownian motion, the set of points x such that L_t^x is greater than zero but is "small."

In Chapter 11 we study the most visited sites of symmetric β-stable processes, $1 < \beta \leq 2$.

Random walks. Hu and Shi (1998) used the First Ray–Knight Theorem for Brownian motion to derive almost sure properties of a simple random walk in a random environment and of Brox's diffusion, which is its continuous time analog.

The Edwards model. van der Hofstad, den Hollander and Konig (1997) used the Ray–Knight Theorems to derive a central limit theorem for the Edwards model of polymers.

2.10 Notes and references

Brownian motion is the pearl of probability theory. In this chapter we developed some of its properties that generalize to strongly symmetric Markov processes, our main concern in this book. It probably is not necessary for us to point out that there is much more to learn about Brownian motion. There are many excellent books on it. In particular, we mention Revuz and Yor (1991), Rogers and Williams (2000b), Port and Stone (1978), Chung (1982), and Knight (1981).

For more information on the standard augmentation, we refer the reader to Rogers and Williams (2000b), Chung (1982), and Blumenthal and Getoor (1968). Recall that the standard augmentation is also referred to as the usual augmentation.

Brownian local time was first studied in Lévy (1948). The joint continuity of Brownian local time is due to Trotter (1958). Our approach, which proves the existence and joint continuity of Brownian local time simultaneously, is based on ideas developed for studying intersection local times in Rosen (1986).

Kac's formula is discussed in great generality in Fitzsimmons and Pitman (1999). Lemma 2.6.2, which is a simple consequence of Kac's Lemma, is from Eisenbaum, Kaspi, Marcus, Rosen and Shi (2000), although this result was probably recognized earlier. Considered along with Lemma 5.2.1, it makes it evident that there is a relationship between local times and squares of Gaussian processes.

The Ray–Knight Theorems were proved independently by Ray (1963) and Knight (1969). For other proofs see Revuz and Yor (1991, Section XI.2).

Ray's Theorem first appeared in Ray (1963). It has been the subject of many investigations, reformulations, and new proofs. See, e.g., the work of Williams (1974), Sheppard (1985), Biane and Yor (1988), and Eisenbaum (1994).

3

Markov processes and local times

In this book we study the local times of Markov processes by means of isomorphism theorems that relate them to Gaussian processes. The natural class of Markov processes to consider are strongly symmetric Borel right processes with continuous potential densities. In this chapter we define these processes and study many of their properties.

In Chapter 2 we focus on Brownian motion to give an overview of some of our goals in this book. However, even for such a regular process as Brownian motion, there are many complexities in the development of Markov process theory. This is particularly evident in Section 2.3 on standard augmentation. Much of the material presented in Chapter 2 is presented in a manner that also makes it relevant to strongly symmetric Borel right processes. We use this material freely in this chapter, especially in the first two sections.

3.1 The Markov property

Let S be a locally compact space with a countable base. We use $\mathcal{B} = \mathcal{B}(S)$ to denote the Borel σ-algebra of S. A function $P = P(x, A)$ is called a sub-Markov kernel on $(S, \mathcal{B})$ if for each $x \in S$, $P(x, \cdot)$ is a positive measure on $(S, \mathcal{B})$ of total mass less than or equal to 1, and for each $A \in \mathcal{B}$, $P(\cdot, A)$ is a $\mathcal{B}$ measurable function on S. Unless otherwise indicated, a sub-Markov kernel will refer to a sub-Markov kernel on $(S, \mathcal{B})$. A sub-Markov kernel of total mass 1 is called a Markov kernel.

For a sub-Markov kernel P and $f \in \mathcal{B}_b(S)$, we set

$$Pf(x) = \int_S f(y) P(x, dy) \tag{3.1}$$

and note that P maps bounded Borel measurable functions into bounded Borel measurable functions. A family $\{P_t ; t \geq 0\}$ of sub-Markov kernels on $(S, \mathcal{B})$ is called a sub-Markov semigroup on $(S, \mathcal{B})$ if

$$P_{t+s}f(x) = P_t(P_s f)(x) \qquad (3.2)$$

for all $s, t \geq 0$, $x \in S$, and $f \in \mathcal{B}_b(S)$ (bounded Borel measurable functions S). Unless otherwise indicated, we always take $P_0 = I$. Semigroups such as $\{P_t \, ; \, t \geq 0\}$ that map $\mathcal{B}_b(S)$ to $\mathcal{B}_b(S)$ are referred to as Borel semigroups.

Imagine a particle starting at x at time $t = 0$. Intuitively, $P_t(x, A)$ can be thought of as the probability that the particle is in the set A at time t. The fact that we may have $P_t(x, S) < 1$ means that there is a chance that the particle has "disappeared" by time t. To deal with this mathematically we introduce a new state Δ, called the "cemetery state," and set $S_\Delta = S \cup \Delta$. If S is already compact, we consider Δ as an isolated point. If S is not compact, we consider Δ as the one-point compactification of S, the "point at infinity."

A family $\{P_t \, ; \, t \geq 0\}$ of sub-Markov kernels that is a sub-Markov semigroup can be extended to become a family of Markov kernels on S_Δ in a unique way. We define

$$P_t(x, \Delta) = 1 - P_t(x, S) \qquad (3.3)$$

for all $x \in S$ and

$$P_t(\Delta, \Delta) = 1. \qquad (3.4)$$

It is easily verified that the extended family $\{P_t \, ; \, t \geq 0\}$ forms a Markov semigroup on S_Δ. We use the convention that any function f defined on S, but not defined on Δ, is automatically extended to S_Δ by setting $f(\Delta) = 0$. Whenever we refer to a Markov semigroup $\{P_t \, ; \, t \geq 0\}$ on S_Δ we assume that it satisfies (3.3) and (3.4).

Let $\{X_t \, ; \, t \geq 0\}$ be a stochastic process on a probability space $(\Omega, \mathcal{G}, P)$ with values in $(S_\Delta, \mathcal{B}(S_\Delta))$. Let $\{\mathcal{G}_t \, ; \, t \geq 0\}$ be a filtration of $\mathcal{G}$ and $\{P_t \, ; \, t \geq 0\}$ a Markov semigroup on S_Δ. $\{X_t \, ; \, t \geq 0\}$ is called a simple Markov process with respect to $\{\mathcal{G}_t \, ; \, t \geq 0\}$ with transition semigroup $\{P_t \, ; \, t \geq 0\}$ if X_t is $\mathcal{G}_t$ measurable for each $t \geq 0$ and

$$E\left(f(X_{t+s}) \,|\, \mathcal{G}_t\right) = P_s f(X_t) \qquad (3.5)$$

for all $s, t \geq 0$ and $f \in \mathcal{B}_b(S_\Delta)$. We refer to S as the state space of $\{X_t \, ; \, t \geq 0\}$, even though, strictly speaking, X has values in S_Δ.

Since $P_s f(X_t)$ is X_t measurable, (3.5) implies that

$$E\left(f(X_{t+s}) \,|\, \mathcal{G}_t\right) = E\left(E\left(f(X_{t+s}) \,|\, \mathcal{G}_t\right) \,|\, X_t\right). \qquad (3.6)$$

Therefore, using a basic property of conditional expectation, we have

$$E\left(f(X_{t+s}) \mid \mathcal{G}_t\right) = E\left(f(X_{t+s}) \mid X_t\right). \tag{3.7}$$

This is a more familiar way to express the simple Markov property which says that the future values of the process $\{X_t \, ; \, t \geq 0\}$, given its past, depend only on its present values. Note that the semigroup property (3.2) is necessary for (3.5) to be consistent.

Remark 3.1.1 To check that $\{X_t \, ; \, t \geq 0\}$ is a simple Markov process, it suffices to verify (3.5) for $f \in C_b(S_\Delta)$. Furthermore, since (3.5) holds trivially for constant functions, it suffices to check it for $f \in C_0(S)$.

Repeated use of the Markov property (3.5) leads to

$$E\left(\prod_{i=1}^n f_i(X_{t_i})\right) = E\left(\int P_{t_1}(X_0, \, dz_1) \prod_{i=2}^n P_{t_i - t_{i-1}}(z_{i-1}, \, dz_i) \prod_{i=1}^n f_i(z_i)\right) \tag{3.8}$$

for any $0 < t_1 < \ldots < t_n$ and any $f_i \in \mathcal{B}_b(S_\Delta)$, $i = 1, \ldots, n$. This shows that the finite-dimensional distributions of the simple Markov process $\{X_t \, ; \, t \geq 0\}$ are determined by its transition semigroup $\{P_t \, ; \, t \geq 0\}$ and its initial distribution, the distribution of X_0.

Let $\{P_t \, ; \, t \geq 0\}$ be a Markov semigroup on S_Δ. The collection $X = (\Omega, \, \mathcal{G}, \, \mathcal{G}_t, \, X_t, \, \theta_t, \, P^x)$ is called a right continuous simple Markov process with transition semigroup $\{P_t \, ; \, t \geq 0\}$ if the following three conditions hold:

(1) $(\Omega, \, \mathcal{G}, \, \mathcal{G}_t)$ is a filtered measurable space, with $X_t : (\Omega, \, \mathcal{G}_t) \to (S_\Delta, \mathcal{B}(S_\Delta))$, and $\{X_t \, ; \, t \geq 0\}$ is right continuous, that is, $t \mapsto X_t(\omega)$ is a right continuous map from R_+ to S_Δ.
(2) $\{\theta_t \, ; \, t \geq 0\}$ is a collection of shift operators for X, that is, $\theta_t : \Omega \to \Omega$ with

$$\theta_t \circ \theta_s = \theta_{t+s} \quad \text{and} \quad X_t \circ \theta_s = X_{t+s} \quad \forall s, t \geq 0. \tag{3.9}$$

(3) For each $x \in S_\Delta$, $P^x(X_0 = x) = 1$, and $\{X_t \, ; \, t \geq 0\}$ on $(\Omega, \, \mathcal{G}, \, P^x)$ is a simple Markov process with respect to $\{\mathcal{G}_t \, ; \, t \geq 0\}$ with transition semigroup $\{P_t \, ; \, t \geq 0\}$.

Let X be a right continuous simple Markov process with transition semigroup $\{P_t \, ; \, t \geq 0\}$. It follows from (3.8) that

$$E^x\left(\prod_{i=1}^n f_i(X_{t_i})\right) = \int P_{t_1}(x, \, dz_1) \prod_{i=2}^n P_{t_i - t_{i-1}}(z_{i-1}, \, dz_i) \prod_{i=1}^n f_i(z_i) \tag{3.10}$$

for any $0 < t_1 < \ldots < t_n$ and any $f_i \in B_b(S_\Delta)$, $i = 1, \ldots, n$. In particular,

$$E^x(f(X_t)) = P_t f(x) = \int P_t(x, dz) f(z). \tag{3.11}$$

Let $A \in \mathcal{B}(S_\Delta)$. When $f = I_A$, (3.11) takes the form

$$P^x(X_t \in A) = P_t I_A(x) = P_t(x, A). \tag{3.12}$$

The process $\{X_s(\omega) ; (s, \omega) \in R_+ \times \Omega\}$ with values in $(S_\Delta, \mathcal{B}(S_\Delta))$ is $\mathcal{B}(R_+) \times \mathcal{G}$ measurable. This is easy to see since

$$X_s(\omega) = \lim_{n \to \infty} \sum_{j=1}^{\infty} 1_{[(j-1)/2^n, j/2^n)}(s) X_{j/2^n}(\omega), \tag{3.13}$$

which holds because of the right continuity of X_s. Consequently, $f(X_t)$ is $\mathcal{B}(R_+) \times \mathcal{G}$ measurable for any $f \in \mathcal{B}(S_\Delta)$. This allows us to define the α-potential operator

$$U^\alpha f(x) = E^x \left(\int_0^\infty e^{-\alpha t} f(X_t) \, dt \right) = \int_0^\infty e^{-\alpha t} P_t f(x) \, dt \tag{3.14}$$

for all $f \in B_b(S_\Delta)$, where the last equality follows from Fubini's Theorem. We also use the definition (3.14) for $\alpha = 0$ whenever the integral is defined, in particular when f is nonnegative, in which case $U^0 f(x)$ may be infinite.

Letting $f = I_A$ in (3.14) for $A \in \mathcal{B}(S_\Delta)$ and using (3.12) yields another measure on $\mathcal{B}(S_\Delta)$:

$$U^\alpha(x, A) := U^\alpha I_A(x) = \int_0^\infty e^{-\alpha t} P_t(x, A) \, dt. \tag{3.15}$$

One can think of $(\alpha e^{-\alpha t} P_t(x, dy))$ as a product probability measure and interpret $\alpha U^\alpha(x, A)$ as the probability that a particle, starting at x at time zero, is in the set A at the end of an independent exponential time with mean $1/\alpha$.

Lemma 3.1.2 *For any $\alpha > 0$ and $f \in C_b(S_\Delta)$,*

$$\lim_{\alpha \to \infty} \alpha U^\alpha f(x) = f(x) \qquad \forall \, x \in S_\Delta. \tag{3.16}$$

Proof Using (3.11) and the fact that X_t is right continuous, we see that $\lim_{t \to 0} P_t f(x) = f(x)$ for any $f \in C_b(S_\Delta)$. Therefore, by the Dominated Convergence Theorem,

$$\lim_{\alpha \to \infty} \alpha U^\alpha f(x) = \lim_{\alpha \to \infty} \alpha \int_0^\infty e^{-\alpha t} P_t f(x) \, dt \tag{3.17}$$

$$= \lim_{\alpha \to \infty} \int_0^\infty e^{-t} P_{t/\alpha} f(x) \, dt = f(x).$$

$\square$

The statement in the next lemma is called the resolvent equation.

Lemma 3.1.3 *For any $\alpha, \beta > 0$ and $f \in C_b(S_\Delta)$,*

$$
\begin{aligned}
U^\alpha f(x) - U^\beta f(x) &= (\beta - \alpha) U^\alpha U^\beta f(x). \qquad (3.18) \\
&= (\beta - \alpha) U^\beta U^\alpha f(x).
\end{aligned}
$$

This also holds for $\alpha = 0$ whenever $U^0 f(x)$ is defined.

Proof Consider first $\alpha, \beta > 0$. We have

$$U^\alpha f(x) - U^\beta f(x) \qquad (3.19)$$

$$= \int_0^\infty \left(e^{-\alpha s} - e^{-\beta s} \right) P_s f(x) \, ds$$

$$= (\beta - \alpha) \int_0^\infty \left(\int_0^s e^{-(\alpha - \beta) t} \, dt \right) e^{-\beta s} P_s f(x) \, ds$$

$$= (\beta - \alpha) \int_0^\infty e^{-(\alpha - \beta) t} \left(\int_t^\infty e^{-\beta s} P_s f(x) \, ds \right) dt$$

$$= (\beta - \alpha) \int_0^\infty e^{-\alpha t} \left(\int_0^\infty e^{-\beta s} P_{s+t} f(x) \, ds \right) dt$$

$$= (\beta - \alpha) \int_0^\infty e^{-\alpha t} \left(\int_0^\infty e^{-\beta s} P_t P_s f(x) \, ds \right) dt$$

$$= (\beta - \alpha) \int_0^\infty e^{-\alpha t} P_t \left(\int_0^\infty e^{-\beta s} P_s f(x) \, ds \right) dt$$

$$= (\beta - \alpha) U^\alpha U^\beta f(x).$$

To get the second equation in (3.18), simply interchange α and β in the first equation.

The case $\alpha = 0$ is proved separately for the positive and negative parts of f. We let α decrease to 0 in (3.18) and use the Monotone Convergence Theorem. $\square$

Let $X = (\Omega, \mathcal{G}, \mathcal{G}_t, X_t, \theta_t, P^x)$ be a right continuous simple Markov process with transition semigroup $\{P_t \, ; \, t \geq 0\}$. As in the discussion of standard augmentation in Section 2.3, let $\mathcal{M}$ denote the set of finite positive measures on $(S_\Delta, \mathcal{B}(S_\Delta))$. For each $\mu \in \mathcal{M}$ and $A \in \mathcal{G}$, set

$$P^\mu(A) = \int P^x(A) \, d\mu(x). \qquad (3.20)$$

Let $\mathcal{G}^\mu$ be the P^μ completion of $\mathcal{G}$ and let $\mathcal{N}_\mu$ be the collection of null sets for $(\Omega, \mathcal{G}^\mu, P^\mu)$. Let $\mathcal{G}_t^\mu = \mathcal{G}_t \vee \mathcal{N}_\mu$ and set

$$\overline{\mathcal{G}} = \cap_{\mu \in \mathcal{M}} \mathcal{G}^\mu \quad \text{and} \quad \overline{\mathcal{G}}_t = \cap_{\mu \in \mathcal{M}} \mathcal{G}_t^\mu. \tag{3.21}$$

We say that $\{\mathcal{G}_t \, ; \, t \geq 0\}$ is augmented if $\mathcal{G}_t = \overline{\mathcal{G}}_t$ for each t.

Suppose that $\{\mathcal{G}_t \, ; \, t \geq 0\}$ in the previous paragraph is not augmented. Nevertheless, we have that, for each $\mu \in \mathcal{M}$,

$$E^\mu \left(f(X_{t+s}) \, | \, \mathcal{G}_t^\mu \right) = P_s f(X_t) \tag{3.22}$$

for all $s, t \geq 0$ and $f \in \mathcal{B}_b(S_\Delta)$. To see this note that the two measures $A \mapsto E^\mu(1_A f(X_{t+s}))$ and $A \mapsto E^\mu(1_A P_s f(X_t))$ agree on $\mathcal{G}_t$ by the simple Markov property and they also agree on $\mathcal{N}_\mu$, where they are both 0. Therefore, they agree on $\mathcal{G}_t^\mu$, which is (3.22).

Clearly we can replace $\mathcal{G}_t^\mu$ by $\overline{\mathcal{G}}_t$ in (3.22). Hence, in dealing with a right continuous simple Markov process, we can always take $\{\mathcal{G}_t \, ; \, t \geq 0\}$ to be augmented.

Recall that $\{\mathcal{G}_t \, ; \, t \geq 0\}$ is said to be right continuous if $\mathcal{G}_t = \cap_{s>t} \mathcal{G}_s$ for each $t \geq 0$.

A collection $X = (\Omega, \, \mathcal{G}, \, \mathcal{G}_t, \, X_t, \, \theta_t, \, P^x)$ that satisfies the following three conditions is called a Borel right process with transition semigroup $\{P_t \, ; \, t \geq 0\}$:

(1) X is a right continuous simple Markov process with transition semigroup $\{P_t \, ; \, t \geq 0\}$.
(2) $U^\alpha f(X_t)$ is right continuous in t for all $\alpha > 0$ and all $f \in C_b(S_\Delta)$, the set of bounded continuous functions on S_Δ.
(3) $\{\mathcal{G}_t \, ; \, t \geq 0\}$ is augmented and right continuous.

The name "Borel right process" refers to the requirements of right continuity and the fact that $\{P_t \, ; \, t \geq 0\}$ is a Borel semigroup.

3.2 The strong Markov property

A right continuous simple Markov process $X = (\Omega, \, \mathcal{G}, \, \mathcal{G}_t, \, X_t, \, \theta_t, \, P^x)$ with transition semigroup $\{P_t \, ; \, t \geq 0\}$ is said to have the strong Markov property if for each $\mathcal{G}_{t+}$ stopping time T

$$E^x \left(f(X_{T+s}) 1_{\{T<\infty\}} \, | \, \mathcal{G}_{T+} \right) = P_s f(X_T) 1_{\{T<\infty\}} \tag{3.23}$$

for all $s \geq 0$, $x \in S_\Delta$ and all $f \in \mathcal{B}_b(S_\Delta)$.

The next theorem provides a criterion for determining when a simple

Markov process has the strong Markov property, which is tailored to our needs in this book.

Theorem 3.2.1 *Let* $X = (\Omega, \mathcal{G}, \mathcal{G}_t, X_t, \theta_t, P^x)$ *be a right continuous simple Markov process with transition semigroup* $\{P_t ; t \geq 0\}$. *If* $U^\alpha f(X_t)$ *is almost surely right continuous in* t *for all* $f \in C_b(S_\Delta)$, *for all* $\alpha > 0$, *the strong Markov property holds. In particular, a Borel right process has the strong Markov property.*

Proof As in the proof of Lemma 2.2.5, $P_s f(X_T) 1_{\{T < \infty\}}$ is $\mathcal{G}_{T+}$ measurable. Hence it suffices to show that

$$E^x \left(1_A f(X_{T+s}) 1_{\{T < \infty\}} \right) = E^x \left(1_A P_s f(X_T) 1_{\{T < \infty\}} \right) \qquad (3.24)$$

for each $A \in \mathcal{G}_{T+}$ and bounded continuous function f. As we point out after (3.11), since $\{X_t ; t \geq 0\}$ is right continuous in t, $P_t f(x) = E^x(f(X_t))$ is also right continuous in t for $f \in C_b(S_\Delta)$. Therefore we can obtain (3.24) by showing that their Laplace transforms are equal, that is, that

$$E^x \left(1_A 1_{\{T < \infty\}} \int_0^\infty e^{-\alpha s} f(X_{T+s}) \, ds \right) = E^x \left(1_A 1_{\{T < \infty\}} U^\alpha f(X_T) \right)$$
$$(3.25)$$

for each $\alpha > 0$.

Let $T_n = \dfrac{[2^n T] + 1}{2^n}$, as in the proof of Lemma 2.2.5, and recall that for each n, T_n is a $\mathcal{G}_t$ stopping time, $T_n \downarrow T$, and $T < \infty$ if and only if $T_n < \infty$, for all $n \geq 1$. We claim that (3.25) holds with T replaced by T_n. Once this is established, (3.25) itself follows by taking the limit as $n \to \infty$ and using right continuity.

Replace T by T_n in the left-hand side of (3.25) and note that since $A \in \mathcal{G}_{T+}$, we have $A \cap \{T_n = j/2^n\} \in \mathcal{G}_{j/2^n}$. Therefore, by the simple Markov property,

$$E^x \left(1_A 1_{\{T_n < \infty\}} \int_0^\infty e^{-\alpha s} f(X_{T_n+s}) \, ds \right) \qquad (3.26)$$

$$= \sum_{j=1}^\infty E^x \left(1_A 1_{\{T_n = j/2^n\}} \int_0^\infty e^{-\alpha s} f(X_{T_n+s}) \, ds \right)$$

$$= \sum_{j=1}^\infty \int_0^\infty e^{-\alpha s} E^x \left(1_{A \cap \{T_n = j/2^n\}} f(X_{j/2^n+s}) \right) \, ds$$

$$= \sum_{j=1}^\infty \int_0^\infty e^{-\alpha s} E^x \left(1_{A \cap \{T_n = j/2^n\}} P_s f(X_{j/2^n}) \right) \, ds$$

$$= \int_0^\infty e^{-\alpha s} E^x \left(1_A 1_{\{T_n < \infty\}} P_s f(X_{T_n}) \right) ds,$$

which gives the right-hand side of (3.25) with T replaced by T_n. $\qquad\square$

The strong Markov property (3.23) implies that for each $\mathcal{G}_{t+}$ stopping time T,

$$E^x \left(Y \circ \theta_T 1_{\{T < \infty\}} \,|\, \mathcal{G}_{T+} \right) = E^{X_T} \left(Y \right) 1_{\{T < \infty\}} \tag{3.27}$$

for all $x \in S$ and $\mathcal{F}^0$ measurable functions Y. (Of course, if $\{\mathcal{G}_t; t \geq 0\}$ is right continuous, $\mathcal{G}_{T+} = \mathcal{G}_T$.) To prove (3.27) it suffices to prove it for Y of the form $Y = \prod_{i=1}^n f_i(X_{t_i})$ with $0 < t_1 < \ldots < t_n$ and $f_i \in \mathcal{B}_b(S_\Delta)$, $i = 1, \ldots, n$. Then $Y \circ \theta_T = \prod_{i=1}^n f_i(X_{T+t_i})$ and (3.27) follows by repeated use of (3.23), precisely as in the proof of Lemma 2.2.2. Indeed, both sides of (3.27) are equal to

$$1_{\{T < \infty\}} \int P_{t_1}(X_T, dz_1) \prod_{i=2}^n P_{t_i - t_{i-1}}(z_{i-1}, dz_i) \prod_{i=1}^n f_i(z_i). \tag{3.28}$$

Note the similarity of (3.28) and (3.8).

Remark 3.2.2 Let $X = (\Omega, \mathcal{G}, \mathcal{G}_t, X_t, \theta_t, P^x)$ satisfy the first two conditions for a Borel right process. By Theorem 3.2.1, X has the strong Markov property (3.23). In particular, for the constant stopping time $T = t$ we have that for each $x \in S_\Delta$,

$$E^x \left(f(X_{t+s}) \,|\, \mathcal{G}_{t+} \right) = P_s f(X_t) \tag{3.29}$$

for all $s, t \geq 0$ and $f \in \mathcal{B}_b(S_\Delta)$. This shows that $X = (\Omega, \mathcal{G}, \mathcal{G}_{t+}, X_t, \theta_t, P^x)$ is a right continuous simple Markov process with transition semigroup $\{P_t \,;\, t \geq 0\}$. As mentioned near the end of Section 3.1, by replacing $\mathcal{G}_t$ with $\overline{\mathcal{G}}_t$, we can always assume that $\{\mathcal{G}_t \,;\, t \geq 0\}$ is augmented. It is easy to check that, in this case, $\{\mathcal{G}_{t+} \,;\, t \geq 0\}$ is also augmented. Since $\{\mathcal{G}_{t+}, t \geq 0\}$ is clearly right continuous, $(\Omega, \mathcal{G}, \mathcal{G}_{t+}, X_t, \theta_t, P^x)$ is a Borel right process.

Let $X = (\Omega, \mathcal{G}, \mathcal{G}_t, X_t, \theta_t, P^x)$ satisfy the first two conditions for a Borel right process. As in the study of Brownian motion, we set $\mathcal{F}_t^0 = \sigma(X_s; s \leq t)$ and $\mathcal{F}_t = \overline{\mathcal{F}}_t^0$, that is $\{\mathcal{F}_t, t \geq 0\}$ is the standard augmentation of $\{\mathcal{F}_t^0, t \geq 0\}$.

$\{\mathcal{F}_t^0, t \geq 0\}$ is the minimal σ-algebra with respect to which $\{X_s; s \leq t\}$ is measurable. Therefore, $\mathcal{F}_t^0 \subseteq \mathcal{G}_t$. Since X satisfies the first two conditions for a Borel right process, the same is true of $(\Omega, \mathcal{F}^0, \mathcal{F}_t^0, X_t, \theta_t, P^x)$ and, by the discussion near the end of Section 3.1, of $(\Omega, \mathcal{F}, \mathcal{F}_t, X_t, \theta_t, P^x)$.

As in the proof of Lemma 2.3.1, we can show that $\{\mathcal{F}_t\,;\,t \geq 0\}$ is right continuous. Thus, $(\Omega, \mathcal{F}, \mathcal{F}_t, X_t, \theta_t, P^x)$ is a Borel right process.

We summarize this discussion in the next lemma.

Lemma 3.2.3 *Let $X = (\Omega,\, \mathcal{G},\, \mathcal{G}_t,\, X_t,\, \theta_t,\, P^x)$ satisfy the first two conditions for a Borel right process. Let $\mathcal{F}_t^0 = \sigma(X_s;\, s \leq t)$ and let $\{\mathcal{F}_t, t \geq 0\}$ be the standard augmentation of $\{\mathcal{F}_t^0, t \geq 0\}$. Then $(\Omega,\, \mathcal{F},\, \mathcal{F}_t,\, X_t,\, \theta_t,\, P^x)$ is a Borel right process.*

As in the proof of Lemma 2.3.2, (3.27) extends to all $\mathcal{F}$ measurable functions Y.

Lemma 3.2.4 *Let $X = (\Omega,\, \mathcal{G},\, \mathcal{G}_t,\, X_t,\, \theta_t,\, P^x)$ be a Borel right process. Then, for each $\mathcal{G}_t$ stopping time T,*

$$E^x \left(Y \circ \theta_T 1_{\{T < \infty\}} \,|\, \mathcal{G}_T \right) = E^{X_T} \left(Y \right) 1_{\{T < \infty\}} \qquad (3.30)$$

for all $x \in S_\Delta$ and $\mathcal{F}$ measurable functions Y.

In particular, taking $\mathcal{G}_t = \mathcal{F}_t$ and using Lemma 3.2.3 we have that, for each $\mathcal{F}_t$ stopping time T,

$$E^x \left(Y \circ \theta_T 1_{\{T < \infty\}} \,|\, \mathcal{F}_T \right) = E^{X_T} \left(Y \right) 1_{\{T < \infty\}} \qquad (3.31)$$

for all $x \in S$ and $\mathcal{F}$ measurable functions Y.

Since the simple Markov property concerns the conditional expectation of a function of X, we can only obtain (3.27) and hence (3.30) for $\mathcal{F}$ measurable functions Y. On the other hand, because of later applications, we want to allow conditioning on a possibly larger σ-algebra. This is why we introduce the generic σ-field $\mathcal{G}$ in the definition of a Borel right process.

Following the arguments in Remark 2.2.8 and Lemma 2.2.9, we get

Lemma 3.2.5 (Blumenthal Zero–One Law) *Let $A \in \mathcal{F}_0$. Then for any $x \in S_\Delta$, $P^x(A)$ is either zero or one.*

We now discuss an important class of stopping times. For $A \subseteq S$ we define the first hitting time of A as

$$T_A = \inf\{s > 0 \,|\, X_s \in A\}. \qquad (3.32)$$

Theorem 3.2.6 *Let $X = (\Omega,\, \mathcal{G},\, \mathcal{G}_t,\, X_t,\, \theta_t,\, P^x)$ be a Borel right process. Then T_A is an $\mathcal{F}_t$ stopping time for every Borel set $A \subseteq S_\Delta$.*

Proof Since $\mathcal{F}_{t+} = \mathcal{F}_t$, it suffices to show that for each fixed $t > 0$,

$$\{T_A < t\} \in \mathcal{F}_t. \tag{3.33}$$

By the right continuity of X_s, we see that for any $s \leq t$,

$$X(s, \omega) = \lim_{n \to \infty} \sum_{j=1}^{2^n} 1_{[(j-1)t/2^n, jt/2^n)}(s) X(jt/2^n, \omega) + 1_{\{t\}}(s) X(t, \omega).$$

Hence $X : ([0, t] \times \Omega, \mathcal{B}([0, t]) \times \mathcal{F}_t) \mapsto (S_\Delta, \mathcal{B}(S_\Delta))$ and therefore we have $([0, t) \times \Omega) \cap X^{-1}(A) \in \mathcal{B}([0, t]) \times \mathcal{F}_t$. Let π denote the projection from $R^1 \times \Omega$ to Ω. Then, by the Projection Theorem (Theorem 14.3.1), for each $t \geq 0$, $\pi(\mathcal{B}([0, t]) \times \mathcal{F}_t) \subseteq \mathcal{F}_t$. Consequently,

$$\pi\{([0, t) \times \Omega) \cap X^{-1}(A)\} \in \mathcal{F}_t. \tag{3.34}$$

Set

$$D_A = \inf\{s \geq 0 \,|\, X_s \in A\}. \tag{3.35}$$

We claim that

$$\{D_A < t\} \in \mathcal{F}_t. \tag{3.36}$$

To see this, observe that

$$\begin{aligned}
\{D_A < t\} &= \{\omega \,|\, \inf\{s \geq 0 \,|\, X_s(\omega) \in A\} < t\} \tag{3.37} \\
&= \pi\{([0, t) \times \Omega) \cap X^{-1}(A)\}.
\end{aligned}$$

Therefore (3.36) follows from (3.34).

Next note that

$$\{T_A < t\} = \bigcup_{s>0} \{s + D_A \circ \theta_s < t\}, \tag{3.38}$$

so the proof of (3.33) reduces to showing that for each $s > 0$

$$\{s + D_A \circ \theta_s < t\} \in \mathcal{F}_t. \tag{3.39}$$

We observe that

$$\{s + D_A \circ \theta_s < t\} = \{D_A \circ \theta_s < t - s\} = \theta_s^{-1}\{D_A < t - s\}. \tag{3.40}$$

Therefore, by (3.36), to obtain (3.39) we need only show that

$$\theta_s^{-1} \mathcal{F}_{t-s} \subseteq \mathcal{F}_t. \tag{3.41}$$

Finally, considering how the augmented filtration is constructed, to obtain (3.41) it suffices to show that, for each $B \in \mathcal{F}_{t-s}$ and $\mu \in \mathcal{M}$, $\theta_s^{-1} B \in \mathcal{F}_t^\mu$.

Let $\nu \in \mathcal{M}$ be the measure defined by $\nu(C) = P^\mu(X_s \in C)$ for $C \in \mathcal{B}$.

By the definition of $\mathcal{F}_{t-s}$, for each $B \in \mathcal{F}_{t-s}$, we can find $B' \in \mathcal{F}^0_{t-s}$ with $P^\nu(B \triangle B') = 0$. Using the simple Markov property, as in (2.84), this is equivalent to $P^\mu(\theta_s^{-1}B \triangle \theta_s^{-1}B') = 0$. But clearly $\theta_s^{-1}\mathcal{F}^0_{t-s} \subseteq \mathcal{F}^0_t$, and therefore $\theta_s^{-1}B \in \mathcal{F}^\mu_t$. This implies (3.41), which completes the proof of Theorem 3.2.6. $\qquad\square$

Set $\zeta := \inf\{t > 0 \,|\, X_t = \Delta\}$. ζ is referred to, poetically, as the "death time" of X.

Lemma 3.2.7 *Let $X = (\Omega, \mathcal{G}, \mathcal{G}_t, X_t, \theta_t, P^x)$ be a Borel right process. Then, conditional on $\zeta < \infty$,*

$$X_t = \Delta \quad \text{for all } t \geq \zeta \qquad P^x \text{ a.s.} \tag{3.42}$$

Proof By the right continuity of X_t we have

$$P^x\left(X_{\zeta+t} = \Delta, \forall t \geq 0; \zeta < \infty\right) \tag{3.43}$$
$$= \lim_{n \to \infty} P^x\left(X_{\zeta+i/2^n} = \Delta, i = 0, 1, \ldots, n2^n; \zeta < \infty\right).$$

Since ζ is a stopping time, by the strong Markov property,

$$P^x\left(X_{\zeta+i/2^n} = \Delta, i = 0, 1, \ldots, n2^n; \zeta < \infty\right) \tag{3.44}$$
$$= E^x\left(I_{\{\zeta<\infty\}}P^\Delta\left(X_{i/2^n} = \Delta, i = 0, 1, \ldots, n2^n\right)\right)$$
$$= P^x\left(\zeta < \infty\right),$$

where we use the fact that

$$P^\Delta\left(X_{i/2^n} = \Delta, i = 0, 1, \ldots, n2^n\right) = 1, \tag{3.45}$$

which follows from (3.4) and (3.10). $\qquad\square$

Remark 3.2.8 By removing a set of measure 0 from Ω we can assume that (3.42) holds for every sample path, that is, that $X_t = \Delta$ on $\{t \geq \zeta\} \cap \{\zeta < \infty\}$.

Let Ω' be the space of right continuous S_Δ–valued functions $\{\omega'(t), t \in [0,\infty)\}$ such that $\omega'(t) = \Delta$ for all $\{t \geq \zeta'\} \cap \{\zeta' < \infty\}$, where $\zeta' = \inf\{s > 0 \,|\, \omega'(s) = \Delta\}$. It is sometimes useful to work with a Borel right process with the canonical sample space Ω'. Let X'_t be the natural evaluation $X'_t(\omega') = \omega'(t)$, and let $\mathcal{F}'^0$ and $\mathcal{F}'^0_t$ be the σ-algebras generated by $\{X'_s, s \in [0,\infty)\}$ and $\{X'_s, s \in [0,t]\}$, respectively. The map $\widetilde{X} : (\Omega, \mathcal{F}^0) \mapsto (\Omega', \mathcal{F}'^0)$ defined by $\widetilde{X}(\omega)(t) := X_t(\omega)$ induces a map of the probabilities P^x on $(\Omega, \mathcal{F}^0)$ to probabilities P'^x on $(\Omega', \mathcal{F}'^0)$.

Augmenting the σ-algebras $\mathcal{F}'^0$ and $\mathcal{F}'^0_t$, as in Remark 3.2.2, we obtain a Borel right process $X' = (\Omega', \mathcal{F}', \mathcal{F}'_t, X'_t, \theta_t, P'^x)$ with transition semigroup $\{P_t\,;\, t \geq 0\}$. X' has the property that $X'_t = \Delta$ for all $\{t \geq \zeta'\} \cap \{\zeta' < \infty\}$.

3.3 Strongly symmetric Borel right processes

Let $X = (\Omega, \mathcal{G}, \mathcal{G}_t, X_t, \theta_t, P^x)$ be a right continuous simple Markov process with state space $(S_\Delta, \mathcal{B}(S_\Delta))$. Let m be a σ-finite measure on $(S, \mathcal{B}(S))$. (Note that m is a measure on S, not S_Δ.) We say that X has α-potential densities with respect to m if, restricted to S,

$$U^\alpha(x, \cdot) \text{ is absolutely continuous with respect to } m, \quad \forall x \in S. \quad (3.46)$$

In this case we refer to m as a reference measure for X. Let $u^\alpha(x, y)$, $x, y \in S$, denote a density for $U^\alpha(x, \cdot)$ with respect to m. When $\alpha = 0$ we often write $u(x, y)$ instead of $u^0(x, y)$.

Lemma 3.3.1 *Let X be a right continuous simple Markov process. If X has α-potential densities with respect to m for some $\alpha \geq 0$, then X has β-potential densities with respect to m for all $\beta > 0$, and for any $x \in S$ and $\alpha, \beta > 0$*

$$u^\alpha(x, y) - u^\beta(x, y) = (\beta - \alpha) \int u^\beta(x, z) u^\alpha(z, y) \, dm(z) \quad (3.47)$$

for almost all $y \in S$, with respect to m. This also holds with $\beta = 0$ if $U^0(x, \cdot)$ is a σ-finite measure.

Furthermore,

$$U^\alpha f(x) = \int u^\alpha(x, z) f(z) \, dm(z) \quad (3.48)$$

for all $f \in \mathcal{B}(S)$, whenever either side exists.

Proof Suppose that (3.46) holds for some $\alpha \geq 0$. Let $A \in \mathcal{B}(S)$ have m measure zero. Then $U^\alpha I_A \equiv 0$ so $U^\beta U^\alpha I_A(x) = 0$. Therefore, by the resolvent equation (3.18), $U^\beta I_A(x) = 0$, that is, (3.46) holds with α replaced by β.

Note that since $U^\alpha(\cdot, A)$ is measurable with respect to $\mathcal{B}(S)$, we can choose $u^\alpha(x, y)$ to be measurable with respect to $\mathcal{B}(S) \times \mathcal{B}(S)$ (see Dellacherie and Meyer (1980, Theorem V.58)). For any set $A \in \mathcal{B}(S)$, (3.18) gives us

$$U^\alpha(x, A) - U^\beta(x, A) \quad (3.49)$$

$$= (\beta - \alpha) \int u^\beta(x, z) \int_A u^\alpha(z, y) \, dm(y) \, m(dz)$$

$$= (\beta - \alpha) \int_A \left(\int u^\beta(x, z) \, u^\alpha(z, y) \, dm(z) \right) m(dy),$$

which gives (3.47). The rest of the proof is obvious. $\qquad\square$

Theorem 3.3.2 (Kac's Moment Formula: II) *Let $X = (\Omega, \mathcal{G}, \mathcal{G}_t,$ $X_t, \theta_t, P^x)$ be a right continuous simple Markov process with 0-potential densities $u(x, y)$ with respect to some reference measure m. Then, for any $f_i \in \mathcal{B}_b^+(S)$, $i = 1, \ldots, n$,*

$$E^x \left(\prod_{i=1}^n \left(\int_0^\infty f_i(X_s) \, ds \right) \right) \tag{3.50}$$

$$= \sum_\pi \int u(x, y_1) \cdots u(y_{n-1}, y_n) \prod_{i=1}^n f_{\pi_i}(y_i) \, dm(y_i),$$

where the sum runs over all permutations π of $\{1, \ldots, n\}$. In particular,

$$E^x \left(\left(\int_0^\infty f(X_s) \, ds \right)^n \right) \tag{3.51}$$

$$= n! \int u(x, y_1) \cdots u(y_{n-1}, y_n) \prod_{i=1}^n f(y_i) \, dm(y_i).$$

Proof We have

$$E^x \left(\prod_{i=1}^n \left(\int_0^\infty f_i(X_s) \, ds \right) \right) \tag{3.52}$$

$$= \int_{R_+^n} E^x \left(\prod_{i=1}^n f_i(X_{s_i}) \right) \prod_{i=1}^n ds_i$$

$$= \sum_\pi \int_{\{0 \le s_1 \le \ldots \le s_n < \infty\}} E^x \left(\prod_{i=1}^n f_{\pi_i}(X_{s_i}) \right) \prod_{i=1}^n ds_i.$$

Using (3.10), the left-hand side of (3.52) can be written as

$$\sum_\pi \int \int_{\{0 \le s_1 \le \ldots \le s_n < \infty\}} P_{t_1}(x, dy_1) \prod_{i=2}^n P_{t_i - t_{i-1}}(y_{i-1}, dy_i) \tag{3.53}$$

$$\prod_{i=1}^n f_{\pi_i}(y_i) \prod_{i=1}^n ds_i$$

$$= \sum_\pi \int \int_{\{0 \le s_1 \le \ldots \le s_{n-1} < \infty\}} P_{t_1}(x, dy_1) \prod_{i=2}^{n-1} P_{t_i - t_{i-1}}(y_{i-1}, dy_i)$$

$$\prod_{i=1}^{n-1} f_{\pi_i}(y_i) \prod_{i=1}^{n-1} ds_i \int_0^\infty P_s(y_{n-1}, dy_n) f_{\pi_n}(y_n)\, ds.$$

By (3.14),

$$\int_0^\infty P_s(y_{n-1}, dy_n) f_{\pi_n}(y_n)\, ds = \int u(y_{n-1}, y_n) f_{\pi_n}(y_n)\, dm(y_n). \quad (3.54)$$

Iterating this procedure gives (3.50). Equation (3.51) is an obvious special case of (3.50). □

We say that X is symmetric with respect to m when its associated semigroup $\{P_t\,;\, t \geq 0\}$ is symmetric with respect to m, that is, if

$$\int g(x)\, P_t f(x)\, dm(x) = \int P_t g(x)\, f(x)\, dm(x) \quad (3.55)$$

for all nonnegative functions $f, g \in \mathcal{B}(S)$. By the Cauchy-Schwarz inequality,

$$|P_t f(x)|^2 = |E^x(f(X_t))|^2 \leq E^x(|f(X_t)|^2) = P_t |f|^2(x). \quad (3.56)$$

Therefore, by (3.55) and (3.56),

$$\int |P_t f(x)|^2\, dm(x) \leq \int P_t |f|^2(x)\, dm(x) \quad (3.57)$$

$$= \int |f(x)|^2\, P_t 1(x) dm(x) \leq \int |f(x)|^2\, dm(x),$$

where $1(x)$ is the constant function that takes the value 1. This shows that P_t can be extended to a contraction on $L^2(m)$. With this extension (3.55) holds for all $f, g \in L^2(m)$.

When (3.55) holds, we also have that

$$\int g(x)\, U^\alpha f(x)\, dm(x) = \int U^\alpha g(x)\, f(x)\, dm(x) \quad (3.58)$$

for all $\alpha > 0$. (Let $(\,\cdot\,,\,\cdot\,)$ be the canonical inner product on $L^2(m)$. Equations (3.55) and (3.58) are simply the statements that $(g, P_t f) = (P_t g, f)$ and $(g, U^\alpha f) = (U^\alpha g, f)$, respectively.)

We say that X is strongly symmetric with respect to m when X is symmetric with respect to m and has α-potential densities with respect to m for some (and consequently all) $\alpha > 0$. In this case, by (3.58),

$$u^\alpha(x, y) = u^\alpha(y, x) \quad \text{for } m \text{ a.e. } x, y \in S \text{ and } \alpha > 0. \quad (3.59)$$

Therefore, when $u^\alpha(x, y)$ is continuous on $S \times S$, (3.59) holds for all $x, y \in S$.

Lemma 3.3.3 *Let $X = (\Omega, \mathcal{G}, \mathcal{G}_t, X_t, \theta_t, P^x)$ be a strongly symmetric right continuous simple Markov process with semigroup $\{P_t; t \geq 0\}$ and continuous α-potential densities $u^\alpha(x, y)$, $\alpha > 0$, with respect to some reference measure m. Then $u^\alpha(x, y)$ is positive definite.*

This also holds for $\alpha = 0$ when $u(x, y)$ in continuous on S.

This elementary lemma plays a crucial role in in this book. It enables us to "associate" with the strongly symmetric Borel right process X a mean zero Gaussian process with covariance $u^\alpha(x, y)$ which we refer to as an associated Gaussian process (see Remark 5.1.7).

Proof Using the semigroup property and then symmetry, (3.55), we see that

$$\int \int u^\alpha(x, y) f(x) f(y) \, dm(x) \, dm(y) \tag{3.60}$$

$$= \int f(x) U^\alpha f(x) \, dm(x)$$

$$= \int_0^\infty \int f(x) e^{-\alpha t} P_t f(x) \, dm(x) \, dt$$

$$= \int_0^\infty \int f(x) e^{-\alpha t} P_{t/2} P_{t/2} f(x) \, dm(x) \, dt$$

$$= \int_0^\infty \int e^{-\alpha t} |P_{t/2} f(x)|^2 \, dm(x) \, dt \geq 0.$$

Let $f_{\epsilon,z}(x)$ be an approximate δ-function at z. For any $a_i \in R^1$ and $z_i \in S$, $i = 1, \ldots, n$, set $f(x) = \sum_{i=1}^n a_i f_{\epsilon,z_i}(x)$. Then, by (3.60),

$$\sum_{i,j=1}^n a_i a_j \int \int u^\alpha(x, y) f_{\epsilon,z_i}(x) f_{\epsilon,z_j}(y) \, dm(x) \, dm(y) \geq 0. \tag{3.61}$$

Taking the limit as ϵ goes to zero we get

$$\sum_{i,j=1}^n a_i a_j u^\alpha(z_i, z_j) \geq 0. \tag{3.62}$$

Thus $u^\alpha(x, y)$ is positive definite. $\qquad\square$

We say that the semigroup $\{P_t \, ; \, t \geq 0\}$ has regular transition densities with respect to m when there exists a family of nonnegative functions $\{p_t(x, y) \, ; \, (t, x, y) \in R_+ \times S \times S\}$ measurable with respect to $\mathcal{B}(R_+) \times \mathcal{B}(S) \times \mathcal{B}(S)$ such that

$$P_t f(x) = \int p_t(x, z) f(z) \, dm(z) \tag{3.63}$$

for all $t > 0$ and all bounded $\mathcal{B}(S)$ measurable functions f, and

$$p_{t+s}(x, y) = \int p_s(x, z) p_t(z, y) \, dm(z) \qquad (3.64)$$

for all $s, t > 0$. Note that the conditions on $p_t(x, y)$ apply only to $x, y \in S$; nothing is implied about Δ. Equation (3.64) is known as the Chapman–Kolmogorov equation, a special case of which is given in (2.4).

A set of functions $\{p_t(x, y) \, ; \, (t, x, y) \in R_+ \times S \times S\}$, measurable with respect to $\mathcal{B}(R_+) \times \mathcal{B}(S) \times \mathcal{B}(S)$, that satisfy (3.64) determines a semigroup, that is, (3.2) holds.

When regular transition densities exist it follows from (3.14) that the α-potential densities exist and are given by

$$u^\alpha(x, z) = \int_0^\infty e^{-\alpha t} p_t(x, z) \, dt. \qquad (3.65)$$

The next lemma presents the resolvent equations for potential densities.

Lemma 3.3.4 *Let* $X = (\Omega, \mathcal{G}, \mathcal{G}_t, X_t, \theta_t, P^x)$ *be a strongly symmetric right continuous simple Markov process with semigroup* $\{P_t; t \geq 0\}$, *which has regular transition densities with respect to some reference measure* m. *Then, for* $\alpha, \beta > 0$,

$$\begin{aligned}
u^\beta(x, y) &= u^\alpha(x, y) + (\alpha - \beta) \int u^\beta(x, z) u^\alpha(z, y) \, dm(z) \quad (3.66) \\
&= u^\alpha(x, y) + (\alpha - \beta) \int u^\alpha(x, z) u^\beta(z, y) \, dm(z)
\end{aligned}$$

for all $x, y \in S$.
This also holds for $\alpha = 0$ *when* $u(x, y) < \infty$ *for all* $x, y \in S$.

Proof We use (3.64) and (3.65) and mimic the proof of Lemma 3.1.3, with obvious modifications. $\square$

We say that the family $\{p_t(x, y) \, ; \, (t, x, y) \in R_+ \times S \times S\}$ of regular transition densities is symmetric when $p_t(x, y) = p_t(y, x)$ for all $(t, x, y) \in R_+ \times S \times S$. By (3.65), the symmetry of the transition densities $p_t(x, z)$ implies that of the potential densities $u^\alpha(x, z)$.

Remark 3.3.5 It follows from the above paragraph that a simple Markov process with a semigroup that has symmetric regular transition densities is strongly symmetric. In particular, this holds for a Borel right process. Conversely, Wittman (1986) showed that every strongly symmetric Borel right process has symmetric regular transition densities. We use this result from this point on and refer the reader to Wittman (1986) for the proof.

To get around the lack of proof of this result, one can often prove that a particular Borel right process that is being considered has symmetric regular transition densities. In fact, in many cases, one of the steps in the proof that a particular process is a strongly symmetric Borel right process is to show that it has symmetric regular transition densities (see Section 4.2).

Using the fact that a Borel right process has symmetric regular transition densities simplifies things considerably. Now we can simply define the α-potential densities by (3.65). Then the 0-potential density is also well defined as $\lim_{\alpha \to 0} u^\alpha(x, y)$, although it may be infinite. (Note that, by (3.65), $u^\alpha(x, y)$ is increasing as $\alpha \downarrow 0$.)

In the next lemma we show that when symmetric regular transition densities exist, it is not necessary to require that the α-potential densities $u^\alpha(x, y)$ are continuous to prove that they are positive definite.

Lemma 3.3.6 *Let $\{p_t(x, y) \, ; \, (t, x, y) \in R_+ \times S \times S\}$ be a family of symmetric regular transition densities with respect to some reference measure m. Then, for each $t > 0$, $p_t(x, y)$ is positive definite. Hence, whenever $u^\alpha(x, y)$ exists, for any $\alpha \geq 0$, it is positive definite.*

Proof Using the Chapman–Kolmogorov equation (3.64) and then symmetry we get

$$\sum_{i,j=1}^{n} a_i a_j p_t(x_i, x_j) = \sum_{i,j=1}^{n} a_i a_j \int p_{t/2}(x_i, z) p_{t/2}(z, x_j) \, dm(z)$$

$$= \int \Big| \sum_{i=1}^{n} a_i p_{t/2}(x_i, z) \Big|^2 dm(z) \geq 0. \quad (3.67)$$

Thus $p_t(x, y)$ is positive definite. Multiplying each side by $e^{-\alpha t}$ and integrating completes the proof. □

3.4 Continuous potential densities

Lemma 3.4.1 *Let $X = (\Omega, \mathcal{G}, \mathcal{G}_t, X_t, \theta_t, P^x)$ be a right continuous simple Markov process with α-potential densities $u^\alpha(x, y)$. If $u^\alpha(x, y)$ is* *continuous on $S \times S$ for some $\alpha > 0$, then $U^\alpha f(x)$ is continuous on S for any $f \in C_b(S_\Delta)$.*

Proof Since $u^\alpha(x, y)$ is continuous on $S \times S$, $U^\alpha f(x)$ is continuous on S for any continuous function f with compact support in S. Such functions are dense in $C_0(S_\Delta)$ in the uniform topology, where $C_0(S_\Delta)$ is the set of continuous functions on S_Δ that vanish at Δ. Since $U^\alpha 1(x) = 1/\alpha$ for all x, we see that if $f_n \to f$ uniformly, then $U^\alpha f_n(x) \to U^\alpha f(x)$ uniformly. It follows that $U^\alpha f(x)$ is continuous on S for any $f \in C_0(S_\Delta)$.

Let $f \in C_b(S_\Delta)$ and let f_Δ denote the constant function with value $f(\Delta)$. Then $f = f_\Delta + (f - f_\Delta)$ with $(f - f_\Delta) \in C_0(S_\Delta)$. The lemma now follows since $U^\alpha(f - f_\Delta)$ is continuous and $U^\alpha f_\Delta(x) = f(\Delta)/\alpha$. $\square$

Lemma 3.4.2 *Let $X = (\Omega, \mathcal{G}, \mathcal{G}_t, X_t, \theta_t, P^x)$ be a right continuous simple Markov process with $X_t = \Delta$ for all $t \geq \zeta$. If X has continuous α-potential densities on $S \times S$ for each $\alpha > 0$, then $(\Omega, \mathcal{F}, \mathcal{F}_t, X_t, \theta_t, P^x)$ is a Borel right process.*

Proof By Lemma 3.4.1, $U^\alpha f(X_t)$ is right continuous in $t < \zeta$ for any $f \in C_b(S_\Delta)$ and $\alpha > 0$. This also holds for all $t \geq \zeta$ by our assumption that $X_t = \Delta$ for all $t \geq \zeta$. This shows that X satisfies the first two conditions for a Borel right process. It then follows from Lemma 3.2.3 that $(\Omega, \mathcal{F}, \mathcal{F}_t, X_t, \theta_t, P^x)$ is a Borel right process. $\square$

We note the following remarkable properties of $u^\alpha(x, y)$ (to appreciate our use of the word "remarkable," see Remark 5.1.7).

Lemma 3.4.3 *Let X be a Borel right process with α-potential density $u^\alpha(x, y)$. Assume that $u^\alpha(x, y)$ is continuous on $S \times S$ for some $\alpha \geq 0$. Then*

$$u^\alpha(x, y) \leq u^\alpha(y, y) \quad \forall x, y \in S \tag{3.68}$$

and $u^\alpha(x, y)$ is continuous in y uniformly in $x \in S$.

Furthermore, for any compact $K \subseteq S$,

$$\sup_{x \in S;\, y, y' \in K} |u^\alpha(x, y) - u^\alpha(x, y')| \leq \sup_{x, y, y' \in K} |u^\alpha(x, y) - u^\alpha(x, y')|. \tag{3.69}$$

Proof Fix $y_0 \in S$ and some open neighborhood G of y_0 that has compact closure in S. If f is any bounded measurable function supported in G, then by (3.14) and the strong Markov property we have

$$\int u^\alpha(x, y) f(y) \, dm(y) \tag{3.70}$$

$$= E^x \left(\int_0^\infty e^{-\alpha s} f(X_s) \, ds \right)$$

$$= E^x \left(\int_{T_G}^\infty e^{-\alpha s} f(X_s) \, ds \, I_{\{T_G < \infty\}} \right)$$

$$= E^x \left(e^{-\alpha T_G} I_{\{T_G < \infty\}} \int_0^\infty e^{-\alpha s} f(X_{T_G + s}) \, ds \right)$$

$$= E^x \left(e^{-\alpha T_G} I_{\{T_G < \infty\}} E^{X_{T_G}} \left(\int_0^\infty e^{-\alpha s} f(X_s) \, ds \right) \right)$$

$$= E^x \left(e^{-\alpha T_G} I_{\{T_G < \infty\}} \int u^\alpha(X_{T_G}, y) \, f(y) \, dm(y) \right).$$

Let $\{f_n\}$ be a sequence of approximate identities converging to δ_{y_0}. Using (3.70) with $f = f_n$ and taking the limit as $n \to \infty$ we obtain

$$u^\alpha(x, y_0) = E^x \left(e^{-\alpha T_G} I_{\{T_G < \infty\}} u^\alpha(X_{T_G}, y_0) \right). \qquad (3.71)$$

Here we use the fact that $u^\alpha(x, y)$ is uniformly continuous on $\overline{G} \times \overline{G}$. Since (3.71) holds for all such neighborhoods G of y_0, we get (3.68).

It follows from (3.71) that, for $y, y' \in G$,

$$\begin{aligned}
|u^\alpha(x, y) - u^\alpha(x, y')| &\leq E^x \left(e^{-\alpha T_G} |u^\alpha(X_{T_G}, y) - u^\alpha(X_{T_G}, y')| \right) \\
&\leq \sup_{x \in \overline{G}} |u^\alpha(x, y) - u^\alpha(x, y')|, \qquad (3.72)
\end{aligned}$$

which gives

$$\sup_{x \in S;\, y, y' \in \overline{G}} |u^\alpha(x, y) - u^\alpha(x, y')| \leq \sup_{x, y, y' \in \overline{G}} |u^\alpha(x, y) - u^\alpha(x, y')|. \quad (3.73)$$

The fact that $u^\alpha(x, y)$ is continuous in y uniformly in $x \in S$ follows since by hypothesis $u^\alpha(x, y)$ is uniformly continuous on $\overline{G} \times \overline{G}$. Finally, (3.69) follows from (3.73) since we can find a sequence G_n of open sets with compact closure such that $K = \cap_n \overline{G}_n$. □

Remark 3.4.4 Let X be a strongly symmetric Borel right process with α-potential density $u^\alpha(x, y)$. An immediate consequence of Lemma 3.4.3 is that if $u^\alpha(x, y)$ is continuous on $S \times S$ for some $\alpha \geq 0$, then it is also continuous for any other $\alpha > 0$. This follows from the resolvent equation (3.66).

Remark 3.4.5 Under certain conditions that often arise in practice, we can establish the conclusion of the last remark without assuming the existence of regular transition densities. Thus, assume that X is a strongly symmetric Borel right process with $u^\alpha(x, y)$ continuous on

$S \times S$ for some $\alpha \geq 0$ and in addition that $U^\beta : C_b(S) \mapsto C_b(S)$ for some $\beta \neq \alpha$. It follows from (3.47) and Lemma 3.4.3 that for each fixed x we can choose the density $u^\beta(x, y)$ to be continuous in y, and this continuity is uniform in $x \in S$. With this choice, (3.47) holds for all y. Note once again from Lemma 3.4.3 that for fixed $y \in S$, $h_y(z) := u^\alpha(z, y)$ is bounded and continuous in z, and (3.47), which now holds for all y, can be written as $u^\alpha(x, y) - u^\beta(x, y) = (\beta - \alpha)U^\beta h_y(x)$. This shows that for each fixed y, $u^\beta(x, y)$ is continuous in x. Together with what we have just shown, this implies that $u^\beta(x, y)$ is continuous on $S \times S$, and then from (3.59) that it is symmetric.

We have a further consequence of the assumption that $u^\alpha(x, y)$ is continuous on $S \times S$:

Lemma 3.4.6 *Let X be a strongly symmetric Borel right process with α-potential density $u^\alpha(x, y)$. If $u^\alpha(x, y)$ is continuous on $S \times S$ for some $\alpha \geq 0$, then*

$$u^\alpha(y, y) > 0 \quad \forall\, y \in S. \tag{3.74}$$

Proof Fix $y_0 \in S$. If $u^\alpha(y_0, y_0) = 0$, then by (3.68) and symmetry we would have that $u^\alpha(y_0, x) = 0$, $\forall x \in S$. Now, since

$$U^\alpha(y_0, S) = \int_S u^\alpha(y_0, x)\, dm(x), \tag{3.75}$$

it follows that $U^\alpha(y_0, S) = 0$. This implies that $P^{y_0}(X_t \in S) = P_t(y_0, S) = 0$ for almost every t. This contradicts the fact that $P^{y_0}(X_0 = y_0) = 1$ and X_t is right continuous. □

3.5 Killing a process at an exponential time

Let $X = (\Omega, \mathcal{G}, \mathcal{G}_t, X_t, \theta_t, P^x)$ be a Borel right process with transition semigroup $\{P_t\,;\, t \geq 0\}$. For any $\alpha > 0$ we construct a Borel right process on $(S, \mathcal{B}(S))$ with transition semigroup $\{e^{-\alpha t} P_t\,;\, t \geq 0\}$. This new process is said to be obtained from the original process by killing it at an independent exponential time with mean $1/\alpha$.

The new process is useful for the following reason. Suppose that the original process is strongly symmetric with continuous α-potential density $u^\alpha(x, y)$ for some $\alpha > 0$ but that its 0-potential density is infinite. The exponentially killed process we construct is strongly symmetric with continuous 0-potential density (which is the α-potential density of the original process). We use this approach in the next section in the construction of local times.

Let $\widehat{\Omega} = \Omega \times R_+$ with elements $\widehat{\omega} = (\omega, s)$. Define the family of measures

$$\widehat{P}^x = P^x \times \alpha e^{-\alpha s}\, ds \qquad x \in S_\Delta \tag{3.76}$$

on $(\widehat{\Omega}, \mathcal{G} \times \mathcal{B}(R_+))$. Define $\lambda(\omega, s) = s$. λ is an exponential random variable with mean $1/\alpha$ under each of the measures $\widehat{P}^x$. Let

$$\widehat{X}_t(\omega, s) = \begin{cases} X_t(\omega) & t < s \\ \Delta & t \geq s. \end{cases} \tag{3.77}$$

Let $\widehat{\mathcal{G}} = \mathcal{G} \times \mathcal{B}(R_+)$. We define $\widehat{\mathcal{G}}_t$ to consist of all sets $\widehat{A} \in \widehat{\mathcal{G}}$ such that $\widehat{A} \cap (\Omega \times (t, \infty)) = A \times (t, \infty)$ for some $A \in \mathcal{G}_t$. It is easy to check that $\widehat{\mathcal{G}}_t$ is an increasing family of σ-algebras. We introduce shift operators on $\widehat{\Omega}$ given by $\widehat{\theta}_t(\omega, s) = (\theta_t \omega, (s - t) \vee 0)$ and verify that $\widehat{X}_r \circ \widehat{\theta}_t = \widehat{X}_{r+t}$.

Let $\widehat{P}_t f(x) := \widehat{E}^x(f(\widehat{X}_t))$ and note that, for $f \in \mathcal{B}_b(S)$,

$$\widehat{P}_t f(x) = \widehat{E}^x\left(f\left(\widehat{X}_t\right)\right) = \widehat{E}^x\left(f(X_t)1_{\{t<\lambda\}}\right) = e^{-\alpha t} P_t f(x). \tag{3.78}$$

Then when $\widehat{A} \in \widehat{\mathcal{G}}_t$, we have

$$\begin{aligned}
\widehat{E}^x\left(f(\widehat{X}_{r+t})1_{\widehat{A}}\right) &= \widehat{E}^x\left(f(X_{r+t})1_{\{r+t<\lambda\}}1_A\right) \tag{3.79} \\
&= e^{-\alpha(r+t)}E^x\left(f(X_{r+t})1_A\right) \\
&= e^{-\alpha(r+t)}E^x\left(P_r f(X_t)1_A\right) \\
&= e^{-\alpha t}E^x\left(\widehat{P}_r f(X_t)1_A\right) \\
&= \widehat{E}^x\left(\widehat{P}_r f(\widehat{X}_t)1_{\widehat{A}}\right).
\end{aligned}$$

Consequently,

$$\widehat{E}^x\left(f(\widehat{X}_{r+t})|\widehat{\mathcal{G}}_t\right) = \widehat{P}_r f(\widehat{X}_t). \tag{3.80}$$

Also, using (3.79) with $\widehat{A} = \widehat{\Omega}$ one can see that $\{\widehat{P}_t\,;\, t \geq 0\}$ is a semigroup. Thus we see that $\widehat{X}$ is a simple Markov process with respect to $\{\widehat{\mathcal{G}}_t\,;\, t \geq 0\}$ with transition semigroup $\{\widehat{P}_t\,;\, t \geq 0\}$.

It is clear that the map $t \mapsto \widehat{X}_t$ is right continuous. Also, by (3.78),

$$\widehat{U}^\beta f(x) := \int_0^\infty \widehat{P}_t f(x)\, dt = U^{\alpha+\beta} f(x). \tag{3.81}$$

Suppose that X is strongly symmetric with continuous α-potential density $u^\alpha(x, y)$ for some $\alpha > 0$. Then, by Remark 3.4.4 and Lemma 3.4.1, $\widehat{U}^\beta f(x)$ is continuous on S for each $f \in C_b(S_\Delta)$ and $\beta > 0$. Thus $\widehat{U}^\beta f(\widehat{X}_t)$ is right continuous in $t < \zeta$ for any $f \in C_b(S_\Delta)$ and $\beta > 0$. But this also holds for all $t \geq \zeta$ since $\widehat{X}_t = \Delta$ for all $t \geq \zeta$. Hence $\widehat{X}$ satisfies the second condition for Borel right processes.

We have now established that $\widehat{X}$ satisfies the first two conditions that define a Borel right process. Then, as discussed in Remark 3.2.2, we augment the σ-algebras $\widehat{\mathcal{G}}_t$ to obtain a Borel right process, which we again denote by $\widehat{X} = (\widehat{\Omega}, \widehat{\mathcal{G}}, \widehat{\mathcal{G}}_t, \widehat{X}_t, \widehat{\theta}_t, \widehat{P}^x)$.

3.6 Local times

Let $X = (\Omega, \mathcal{G}, \mathcal{G}_t, X_t, \theta_t, P^x)$ be a Borel right process. A family $A = \{A_t\,;\, t \geq 0\}$ of random variables on $(\Omega, \mathcal{G}, \mathcal{G}_t)$ is called a continuous additive functional (CAF) of X if

(1) $t \mapsto A_t$ is almost surely continuous and nondecreasing, with $A_0 = 0$ and $A_t = A_\zeta$, for all $t \geq \zeta$.

(2) A_t is $\mathcal{F}_t$ measurable.

(3) $A_{t+s} = A_t + A_s \circ \theta_t$ for all $s, t \in R_+$ a.s.

This is just a slight generalization of the definition of a CAF of Brownian motion on page 32 that incorporates the lifetime of X.

Let $\{A_t\,;\, t \geq 0\}$ be a continuous additive functional of X, and let

$$R_A(\omega) = \inf\{t \,|\, A_t(\omega) > 0\}. \tag{3.82}$$

We call $\{A_t\,;\, t \geq 0\}$ a local time of X at $y \in S$ if $P^y(R_A = 0) = 1$ and, for all $x \neq y$, $P^x(R_A = 0) = 0$.

In the next theorem we give general conditions for the existence of local times. However, we first assume that $\{A_t\,;\, t \geq 0\}$ is a local time of X at y and derive some important consequences. The first of these is that if X starts at y at time zero, it hits y infinitely often by time t' for all $t' > 0$, almost surely.

Lemma 3.6.1 *Let X be a Borel right process that has a local time at y. Then*

$$P^y(T_y = 0) = 1. \tag{3.83}$$

Proof Let $\{A_t\,;\, t \geq 0\}$ denote a local time of X at y. Since $P^y(R_A = 0) = 1$,

$$P^y(T_y > 0) = P^y(R_A = 0, T_y > 0). \tag{3.84}$$

If $T_y > 0$, then for some interval $(0, t_0]$ we have $X_t \neq y$ for all $0 < t \leq t_0$. Let $R_{A,r} = r + R_A \circ \theta_r$. Then, since $t \mapsto A_t$ is almost surely continuous and nondecreasing, with $A_0 = 0$ and $A_t > 0$ for $t > 0$ (since $\{A_t, t \geq 0\}$ is local time),

$$\{R_A = 0, T_y > 0\} \subseteq \bigcup_{\substack{r>0 \\ rational}} \{X_{R_{A,r}} \neq y\} \tag{3.85}$$

except, possibly, for a set of measure zero. Also, by definition, $R_A \circ \theta_{R_{A,r}} = 0$. Therefore,

$$P^y(R_A = 0, T_y > 0) \leq P^y\Big(\bigcup_{\substack{r>0 \\ rational}} \{R_A \circ \theta_{R_{A,r}} = 0, X_{R_{A,r}} \neq y\}\Big). \tag{3.86}$$

Clearly, for any $r > 0$,

$$P^y(R_A \circ \theta_{R_{A,r}} = 0, X_{R_{A,r}} \neq y) \tag{3.87}$$
$$= E^y\left(P^{X_{R_{A,r}}}(R_A = 0)I_{\{X_{R_{A,r}} \neq y\}}\right) = 0$$

since $P^x(R_A = 0) = 0$ when $x \neq y$. Putting these equations together gives (3.83). $\qquad\square$

Remark 3.6.2 By definition, a CAF $\{A_t \,; t \geq 0\}$ of a Markov process is nondecreasing and consequently determines a (random) measure dA_t on R^1. We now show that when $\{A_t \,; t \geq 0\}$ is a local time of a Borel right process X at y, dA_t is supported on $\mathcal{Z} := \{t : X_t = y\}$, almost surely. Set

$$\mathcal{I} = \{t : A_{t+\epsilon} - A_t > 0 \text{ for all } \epsilon > 0\}. \tag{3.88}$$

We first show that $\mathcal{I} \subseteq \mathcal{Z}$, almost surely. This is because, by the right continuity of X_t,

$$\{\mathcal{I} \cap \mathcal{Z}^c \neq \emptyset\} \subseteq \bigcup_{\substack{r>0 \\ rational}} \{X_{R_{A,r}} \neq y\}, \tag{3.89}$$

and, arguing as in (3.86) and (3.87), this has P^x measure zero for any $x \in S$. We now note that dA_t is supported on $\mathcal{J}$, where

$$\mathcal{J} = \{t : A_{t+\epsilon} - A_{t-\epsilon} > 0 \text{ for all } \epsilon > 0\}. \tag{3.90}$$

It is also easy to check that $\mathcal{J} - \mathcal{I}$ is countable. However, since $t \mapsto A_t$ is continuous, dA_t has no atoms. Therefore dA_t is supported on $\mathcal{I} \subseteq \mathcal{Z}$, almost surely.

We now show that when a Borel right process is strongly symmetric with continuous α-potential densities it has local times at all points in its state space.

Theorem 3.6.3 *Let* $X = (\Omega, \mathcal{G}, \mathcal{G}_t, X_t, \theta_t, P^x)$ *be a strongly symmetric Borel right process with continuous α-potential density $u^\alpha(x,y)$. For each $y \in S$ we can construct a local time of X at y, which we denote by $\{L_t^y ; t \geq 0\}$, such that, for each $x \in S$,*

$$E^x \left(\int_0^\infty e^{-\alpha t} \, dL_t^y \right) = u^\alpha(x,y). \tag{3.91}$$

Let $f_{\epsilon,y}$ be an approximate δ-function at y with respect to the reference measure m. There exists a sequence $\{\epsilon_n\}$ tending to zero, such that almost surely

$$L_t^y = \lim_{\epsilon_n \to 0} \int_0^t f_{\epsilon_n,y}(X_s) \, ds \tag{3.92}$$

uniformly for $t \in [0,T]$, where T is any finite time, which may be random.

Proof Consider first the case of $\alpha > 0$. Let λ be an exponential random variable with mean $1/\alpha$ that is independent of X. Let $\widehat{X} = (\widehat{\Omega}, \widehat{\mathcal{G}}, \widehat{\mathcal{G}}_t, \widehat{X}_t, \widehat{\theta}_t, \widehat{P}^x)$ be the Borel right process obtained by killing X at λ as described in Section 3.5. Note that the 0-potential density for $\widehat{X}$ is given by $u^\alpha(x,y)$. Kac's moment formula (3.50) shows that

$$I_{\epsilon,\epsilon'}(x) := \widehat{E}^x \left(\left(\int_0^\infty f_{\epsilon,y}(\widehat{X}_s) \, ds \right) \left(\int_0^\infty f_{\epsilon',y}(\widehat{X}_t) \, dt \right) \right) \tag{3.93}$$

$$= \int u^\alpha(x,z_1) u^\alpha(z_1,z_2) f_{\epsilon,y}(z_1) f_{\epsilon',y}(z_2) \, dm(z_1) \, dm(z_2)$$

$$+ \int u^\alpha(x,z_1) u^\alpha(z_1,z_2) f_{\epsilon',y}(z_1) f_{\epsilon,y}(z_2) \, dm(z_1) \, dm(z_2).$$

Since

$$\inf_{z,z' \in K_{\epsilon \vee \epsilon'}} u^\alpha(x,z) u^\alpha(z,z') \leq I_{\epsilon,\epsilon'}(x) \leq \sup_{z,z' \in K_{\epsilon \vee \epsilon'}} u^\alpha(x,z) u^\alpha(z,z') \tag{3.94}$$

and $u^\alpha(x,y)$ is continuous, we see that $\int_0^\infty f_{\epsilon,y}(\widehat{X}_s) \, ds$ converges in $L^2(\widehat{P}^x)$ as $\epsilon \to 0$. (K. is part of the definition of approximate δ-function on page 8.) In fact, using the uniform continuity of $u^\alpha(x,y)$ described in Lemma 3.4.3, we see that this convergence is uniform in $x \in S$.

Define the right continuous $\widehat{X}$-martingale

$$M_t^\epsilon = \widehat{E}^x \left(\int_0^\infty f_{\epsilon,y}(\widehat{X}_s) \, ds \big| \widehat{\mathcal{G}}_t \right) \tag{3.95}$$

$$= \int_0^t f_{\epsilon,y}(\widehat{X}_s) \, ds + \int u^\alpha(\widehat{X}_t, z) f_{\epsilon,y}(z) \, dm(z),$$

where for the last equation we use (3.50) again. Note that the right continuity of M_t^ϵ comes from the continuity of u^α. It follows from Doob's L_2 inequality (see, e.g., Rogers and Williams (2000b, II.70.2)) and the fact that $M_\infty^\epsilon = \int_0^\infty f_{\epsilon,y}(\widehat{X}_s)\, ds$ that

$$E^x\left(\sup_{t\geq 0}|M_t^\epsilon - M_t^{\epsilon'}|^2\right) \tag{3.96}$$

$$\leq 4E^x\left(\left|\int_0^\infty f_{\epsilon,y}(\widehat{X}_s)\, ds - \int_0^\infty f_{\epsilon',y}(\widehat{X}_s)\, ds\right|^2\right).$$

Therefore, M_t^ϵ converges in $L^2(\widehat{P}^x)$ uniformly in $t \in R_+$ and $x \in S$.

Using the bounds and uniform continuity of $u^\alpha(x,y)$ described in Lemma 3.4.3, we see that the last term in (3.95) also converges in $L^2(\widehat{P}^x)$ uniformly in $t \in R_+$ and $x \in S$. Consequently, we can find a sequence $\epsilon_n \to 0$ such that

$$\int_0^t f_{\epsilon_n,y}(\widehat{X}_s)\, ds = \int_0^{t\wedge\lambda} f_{\epsilon_n,y}(X_s)\, ds \tag{3.97}$$

converges uniformly in $t \in R_+$, $\widehat{P}^x$ almost surely for all $x \in S$.

Since $P(\lambda > v) = e^{-\alpha v} > 0$, it follows from Fubini's Theorem that, for this sequence $\{\epsilon_n\}$, the right-hand side of (3.92) converges uniformly for $t \in [0,v]$, P^x almost surely for all $x \in S$ and $v > 0$. Let

$$\widetilde{\Omega} = \left\{\omega\,\middle|\, \int_0^t f_{\epsilon_n,y}(X_s)\, ds \text{ converges locally uniformly in } t\right\}. \tag{3.98}$$

We have shown that $P^x(\widetilde{\Omega}) = 1$, for all $x \in S$. For $\omega \in \widetilde{\Omega}$ we define L_t^y by (3.92); otherwise set $L_t^y \equiv 0$. It is easy to verify that L_t^y is a continuous additive functional of X.

It follows from (3.93) that the sequence $\int_0^\infty f_{\epsilon_n,y}(\widehat{X}_s)\, ds$, $n = 1, 2, \ldots$ is uniformly integrable. Therefore

$$\widehat{E}^x(L_\lambda^y) = \lim_{n\to\infty} \widehat{E}^x\left(\int_0^\infty f_{\epsilon_n,y}(\widehat{X}_s)\, ds\right) = u^\alpha(x,y) \tag{3.99}$$

by Theorem 3.3.2. Also, clearly

$$\widehat{E}^x(L_\lambda^y) = \alpha E^x\left(\int_0^\infty e^{-\alpha t} L_t^y\, dt\right). \tag{3.100}$$

Note that if the integral in the right-hand side of (3.100) is finite, then $\limsup_{t\to\infty} e^{-\alpha t}L_t^y = 0$. Suppose to the contrary that there exists an infinite sequence $\{t_n\}$ for which $e^{-\alpha t_n}L_{t_n}^y \geq \epsilon$ for some $\epsilon > 0$. Then $e^{-\alpha t}L_t^y \geq \epsilon/2$ for $t \in [t_n, t_n + (1/\alpha)\log 2]$, which contradicts the fact that the integral is finite. Using this observation we see that (3.91) follows from (3.99), (3.100), and integration by parts.

To complete the proof we must show that $\{L_t^y \,;\, t \geq 0\}$ is a local time for X at y. To do this we need only to show that $R = R_{L^y}$ (see (3.82)) satisfies the conditions stated after (3.82). It follows immediately from (3.92) and the right continuity of X that $P^x(R = 0) = 0$ for all $x \neq y$. We now show that $P^y(R = 0) = 1$.

Suppose that $P^y(R = 0) = 0$, so that $P^x(R = 0) = 0$ for all $x \in S$. Since, for any $z \in S$,

$$P^z(L_t^y > 0) \leq P^z(R < t), \qquad (3.101)$$

we would have

$$\lim_{t \to 0} P^z(L_t^y > 0) = 0 \qquad \forall z \in S. \qquad (3.102)$$

This is not possible. To see this note that it follows from the definition of R that, for any x and $t > 0$,

$$P^x(R < \infty) = P^x\left(L_{R+t}^y > 0\,,\, R < \infty\right). \qquad (3.103)$$

It is easy to see that R is a stopping time. Therefore, using the additivity of L^y and the Markov property, we have

$$P^x(R < \infty) = E^x\left(P^{X_R}(L_t^y > 0)\, I_{\{R<\infty\}}\right). \qquad (3.104)$$

Using (3.102) in (3.103) gives us $P^x(R < \infty) = 0$ for all x, that is, that $L^y \equiv 0$ almost surely, which contradicts (3.91). Thus $P^y(R = 0) > 0$, and by the Blumenthal zero-one law (Lemma 3.2.5), $P^y(R = 0) = 1$.

When $\alpha = 0$, (3.91) is

$$E^x\left(L_\infty^y\right) = u^0(x, y). \qquad (3.105)$$

The proof of this theorem, in this case, is similar to the above, and in fact simpler, since we do not need to introduce the exponentially killed process. $\qquad \square$

Remark 3.6.4

(1) The version of $\{L_t^x \,;\, t \in R_+\}$ constructed in Theorem 3.6.3 is $\mathcal{F}^0$ measurable.

(2) It follows from the proof of Theorem 3.6.3 that, for any sequence $\{\epsilon_n'\}$ tending to zero, we can choose a subsequence $\{\epsilon_n\}$ tending to zero, such that (3.92) holds.

(3) It is easy to check that if $\{L_t^y \,;\, t \geq 0\}$ is a local time for $X = (\Omega, \mathcal{G}, \mathcal{G}_t, X_t, \theta_t, P^x)$ at $y \in S$, then $\{L_{t \wedge \lambda}^y \,;\, t \geq 0\}$ is a local time for $\widehat{X} = (\widehat{\Omega}, \widehat{\mathcal{G}}, \widehat{\mathcal{G}}_t, \widehat{X}_t, \widehat{\theta}_t, \widehat{P}^x)$, (see Section 3.5).

Theorem 3.6.5 *Let X be a strongly symmetric Borel right process and assume that its α-potential density, $u^\alpha(x,y)$, is finite for all $x, y \in S$. Let L_t^y be a local time of X at y, with*

$$E^x \left(\int_0^\infty e^{-\alpha t} \, dL_t^y \right) = u^\alpha(x,y). \tag{3.106}$$

Then

$$E^x \left(e^{-\alpha T_y} \right) = \frac{u^\alpha(x,y)}{u^\alpha(y,y)} \tag{3.107}$$

and for every t

$$E^x \left(L_t^y \right) = \int_0^t p_s(x,y) \, ds. \tag{3.108}$$

Furthermore, if $u(x,x)$ and $u(y,y)$ are finite,

$$P^x (T_y < \infty) = \frac{u(x,y)}{u(y,y)}. \tag{3.109}$$

Proof When $x = y$, (3.107) is just (3.83). When $x \neq y$, (3.107) follows from (3.106), the support property of dL_t^y, and the strong Markov property as follows:

$$
\begin{aligned}
u^\alpha(x,y) &= E^x \left(\int_0^\infty e^{-\alpha s} \, dL_s^y \right) \tag{3.110} \\
&= E^x \left(1_{\{T_y < \infty\}} \int_{T_y}^\infty e^{-\alpha s} \, dL_s^y \right) \\
&= E^x \left(e^{-\alpha T_y} 1_{\{T_y < \infty\}} \left(\int_0^\infty e^{-\alpha s} \, dL_s^y \right) \circ \theta_{T_y} \right) \\
&= E^x \left(e^{-\alpha T_y} 1_{\{T_y < \infty\}} E^{X_{T_y}} \left(\int_0^\infty e^{-\alpha s} \, dL_s^y \right) \right) \\
&= E^x \left(e^{-\alpha T_y} \right) u^\alpha(y,y)
\end{aligned}
$$

since $X_{T_y} = y$ on $\{T_y < \infty\}$.

We get (3.109) by taking the limit in (3.107) as α goes to zero.

To obtain (3.108) we first note that if λ is an independent exponential random variable with mean $1/\alpha$, then

$$
\begin{aligned}
\alpha \int_0^\infty e^{-\alpha t} L_t^y \, dt &= E_\lambda^x(L_\lambda^y) = E_\lambda^x \left(\int_0^\infty 1_{\{\lambda > t\}} \, dL_t^y \right) \tag{3.111} \\
&= E^x \left(\int_0^\infty e^{-\alpha t} \, dL_t^y \right) = u^\alpha(x,y).
\end{aligned}
$$

Therefore

$$\alpha \int_0^\infty e^{-\alpha t} E^x \left(L_t^y \right) dt \;=\; \int_0^\infty e^{-\alpha t} p_t(x,y) \, dt \tag{3.112}$$

$$= \; \alpha \int_0^\infty e^{-\alpha t} \left(\int_0^t p_s(x,y) \, ds \right) dt$$

by integration by parts. Here we use the fact that

$$\lim_{t \to \infty} e^{-\alpha t} \left(\int_0^t p_s(x,y) \, ds \right) \tag{3.113}$$

$$\leq \lim_{t \to \infty} e^{-\alpha t/2} \left(\int_0^t e^{-\alpha s/2} p_s(x,y) \, ds \right)$$

$$\leq \lim_{t \to \infty} e^{-\alpha t/2} u^{\alpha/2}(x,y) = 0.$$

Equation (3.112) holds as long as $u^\alpha(x,y)$ is finite. Since $u^\lambda(x,y)$ is decreasing in λ, (3.112) holds for all $\lambda \geq \alpha$. By the uniqueness property of Laplace transforms, this implies that (3.108) holds for almost every t. Since both sides of (3.108) are continuous in t, it follows that (3.108) holds for every t. $\qquad\qquad\square$

Remark 3.6.6 Fix some $x, y \in S$ with $P^x(T_y < \infty) > 0$. Equations (3.107) and (3.74) then imply that $u^\alpha(x,y) > 0$ and thus, by symmetry, that $u^\alpha(y,x) > 0$. This, in turn, implies that $P^y(T_x < \infty) > 0$.

Theorem 3.6.7 *If $\{A_t \,;\, t \geq 0\}$ is a local time of a Borel right process X at a point y and $K > 0$ is a constant, then $\{KA_t \,;\, t \geq 0\}$ is also a local time of X at y. Furthermore, if $\{A_t^1 \,;\, t \geq 0\}$ and $\{A_t^2 \,;\, t \geq 0\}$ are local times of X at y with*

$$E^y \left(\int_0^\infty e^{-\alpha t} \, dA_t^i \right) < \infty, \qquad i = 1, 2 \tag{3.114}$$

for some $\alpha > 0$, then

$$A_t^1 = C A_t^2 \qquad\qquad \forall \, t \geq 0 \quad a.s. \tag{3.115}$$

for some constant $C > 0$.

The proof of the first statement in this theorem is simple; one just checks that $\{KA_t \,;\, t \geq 0\}$ satisfies the definition of a local time. To prove the second statement we develop some material on the α-potential operator of a CAF. Let $A = \{A_t \,;\, t \geq 0\}$ be a continuous additive functional. We define its α-potential operator by

$$U_A^\alpha f(x) = E^x \left(\int_0^\infty e^{-\alpha t} f(X_t) \, dA_t \right) \tag{3.116}$$

for $f \in \mathcal{B}_b^+(S)$. When $f \equiv 1$ we write $u_A^\alpha(x) = U_A^\alpha 1(x)$. $u_A^\alpha(x)$ is called the α-potential of A. In particular, by (3.91),

$$u_{L^y}^\alpha(x) = u^\alpha(x, y). \tag{3.117}$$

If the α-potential of A is bounded, then in analogy with the resolvent equation, (3.18), for any $\alpha, \beta > 0$,

$$U_A^\alpha f(x) - U_A^\beta f(x) = (\beta - \alpha) U^\alpha U_A^\beta f(x). \tag{3.118}$$

The proof of (3.118) is similar to the proof of (3.18). We have

$$U_A^\alpha f(x) - U_A^\beta f(x) \tag{3.119}$$

$$= E^x \left(\int_0^\infty (e^{-\alpha s} - e^{-\beta s}) f(X_s) \, dA_s \right)$$

$$= (\beta - \alpha) E^x \left(\int_0^\infty \left(\int_0^s e^{-(\alpha - \beta)t} \, dt \right) e^{-\beta s} f(X_s) \, dA_s \right)$$

$$= (\beta - \alpha) E^x \left(\int_0^\infty e^{-(\alpha - \beta)t} \left(\int_t^\infty e^{-\beta s} f(X_s) \, dA_s \right) dt \right)$$

$$= (\beta - \alpha) E^x \left(\int_0^\infty e^{-\alpha t} \left(\int_0^\infty e^{-\beta s} f(X_s) \, dA_s \right) \circ \theta_t \, dt \right)$$

$$= (\beta - \alpha) E^x \left(\int_0^\infty e^{-\alpha t} E^{X_t} \left(\int_0^\infty e^{-\beta s} f(X_s) \, dA_s \right) dt \right)$$

$$= (\beta - \alpha) U^\alpha U_A^\beta f(x).$$

Lemma 3.6.8 *Let* $\{A_t^1 \, ; \, t \geq 0\}$ *and* $\{A_t^2 \, ; \, t \geq 0\}$ *be two continuous additive functionals. If* A_1, A_2 *have bounded* α-*potentials for some* $\alpha > 0$ *and* $U_{A_1}^\alpha f(x) = U_{A_2}^\alpha f(x)$ *for all bounded measurable functions* f *and* $x \in S$, *then* $A_t^1 = A_t^2$ *for all* $t \geq 0$ *almost surely.*

Proof It follows from (3.118) that $U_{A_1}^\alpha f(x) = U_{A_2}^\alpha f(x)$ for all $\alpha > 0$. Also, for any $i, j = 1, 2$ we have

$$E^x \left(\int_0^\infty \int_0^\infty e^{-\alpha(s+t)} \, dA_s^i \, dA_t^j \right) \tag{3.120}$$

$$= E^x \left(\int_0^\infty e^{-\alpha s} \int_s^\infty e^{-\alpha t} \, dA_t^j \, dA_s^i \right)$$

$$\quad + E^x \left(\int_0^\infty e^{-\alpha t} \int_t^\infty e^{-\alpha s} \, dA_s^i \, dA_t^j \right)$$

$$= E^x \left(\int_0^\infty e^{-2\alpha s} \left(\int_0^\infty e^{-\alpha t} \, dA_t^j \right) \circ \theta_s \, dA_s^i \right)$$

$$+ E^x \left(\int_0^\infty e^{-2\alpha t} \left(\int_0^\infty e^{-\alpha s} \, dA_s^i \right) \circ \theta_t \, dA_t^j \right).$$

Let $\tau_{A^i}(s)$ be as defined in (2.142). For any measurable function $f(t)$ we have the following analogue of (2.125):

$$\int_0^T f(t) \, dA_t^i = \int_0^\infty f(\tau_{A^i}(s)) 1_{\{\tau_{A^i}(s) < T\}} \, ds. \qquad (3.121)$$

This is proved in exactly the same way that (2.125) is proved.

Note that $\{\tau_{A^i}(s) < T\} \in \mathcal{F}_{\tau_{A^i}(s)}$. Using first (3.121) and then the strong Markov property we have

$$E^x \left(\int_0^\infty e^{-2\alpha s} \left(\int_0^\infty e^{-\alpha t} \, dA_t^j \right) \circ \theta_s \, dA_s^i \right) \qquad (3.122)$$

$$= \int_0^\infty E^x \left(e^{-2\alpha \tau_{A^i}(s)} \left(\int_0^\infty e^{-\alpha t} \, dA_t^j \right) \circ \theta_{\tau_{A^i}(s)} 1_{\{\tau_{A^i}(s) < \infty\}} \right) \, ds$$

$$= \int_0^\infty E^x \left(e^{-2\alpha \tau_{A^i}(s)} u_{A^j}^\alpha (X_{\tau_{A^i}(s)}) 1_{\{\tau_{A^i}(s) < \infty\}} \right) \, ds.$$

Interchanging the order of integration and applying (3.121) again, we see that this

$$= E^x \left(\int_0^\infty e^{-2\alpha s} u_{A^j}^\alpha (X_s) \, dA_s^i \right) = U_{A^i}^{2\alpha} u_{A^j}^\alpha (x). \qquad (3.123)$$

Thus, by (3.120),

$$E^x \left(\int_0^\infty \int_0^\infty e^{-\alpha(s+t)} \, dA_s^i \, dA_t^j \right) = U_{A^i}^{2\alpha} u_{A^j}^\alpha (x) + U_{A^j}^{2\alpha} u_{A^i}^\alpha (x). \qquad (3.124)$$

By hypothesis, the right-hand side of (3.124) is the same for all $i, j = 1, 2$. Therefore

$$E^x \left(\left(\int_0^\infty e^{-\alpha s} \, dA_s^1 - \int_0^\infty e^{-\alpha s} \, dA_s^2 \right)^2 \right) = 0, \qquad (3.125)$$

and consequently

$$\int_0^\infty e^{-\alpha s} \, dA_s^1 = \int_0^\infty e^{-\alpha s} \, dA_s^2 \qquad \forall \alpha > 0 \quad \text{a.s.} \qquad (3.126)$$

Since both A_s^1 and A_s^2 are continuous, the lemma follows (see, e.g., Feller (1971, Chapter XII)). $\qquad \square$

Proof of Theorem 3.6.7 Let $\{A_t \, ; \, t \geq 0\}$ be a local time of X at y. The support property of dA_t shows that

$$U_A^\alpha f(x) = E^x \left(\int_0^\infty e^{-\alpha t} f(X_t) \, dA_t \right) = f(y) u_A^\alpha(x). \tag{3.127}$$

Furthermore, for any $x \neq y$, the support property of dA_t also shows that

$$
\begin{aligned}
u_A^\alpha(x) &= E^x \left(\int_0^\infty e^{-\alpha s} \, dA_s \right) & (3.128) \\
&= E^x \left(1_{\{T_y < \infty\}} \int_{T_y}^\infty e^{-\alpha s} \, dA_s \right) \\
&= E^x \left(e^{-\alpha T_y} 1_{\{T_y < \infty\}} \left(\int_0^\infty e^{-\alpha s} \, dA_s \right) \circ \theta_{T_y} \right) \\
&= E^x \left(e^{-\alpha T_y} 1_{\{T_y < \infty\}} E^{X_{T_y}} \left(\int_0^\infty e^{-\alpha s} \, dA_s \right) \right) \\
&= E^x \left(e^{-\alpha T_y} \right) u_A^\alpha(y)
\end{aligned}
$$

since $X_{T_y} = y$ on $\{T_y < \infty\}$.

Let $\{A_t^1 \, ; \, t \geq 0\}$ and $\{A_t^2 \, ; \, t \geq 0\}$ be local times for X at y. Let

$$\widetilde{A}_t^i = \frac{A_t^i}{u_{A_i}^\alpha(y)} \qquad i = 1, 2. \tag{3.129}$$

It follows from (3.127) and (3.128) that $U_{\widetilde{A}_t^1} = U_{\widetilde{A}_t^2}$. Thus, by Lemma 3.6.8, $\widetilde{A}_t^1 = \widetilde{A}_t^2$. This gives (3.115). $\qquad\square$

Remark 3.6.9 Let X be a strongly symmetric Borel right process with local time at y given by $L^y = \{L_t^y ; t \geq 0\}$ as constructed in Theorem 3.6.3 and satisfying (3.91). By Theorem 3.6.7, $\overline{L}^y := \{f(y) L_t^y ; t \geq 0\}$ is also a local time of X at y, as long as $f(y) > 0$. In this case we have

$$E^x \left(\int_0^\infty e^{-\alpha t} \, d\overline{L}_t^y \right) = f(y) u^\alpha(x, y). \tag{3.130}$$

Therefore, when we speak of a local time for X we need to identify the left-hand side of (3.130). We refer to this as the normalization of the local time.

Note that by Lemma 3.6.8, if $\{L_t^y ; t \geq 0\}$ is a local time for X at y for which (3.91) holds and $\{\overline{L}_t^y ; t \geq 0\}$ is a local time for X at y for which (3.130) holds, then

$$\overline{L}_t^y = f(y) L_t^y \qquad \forall t \geq 0 \quad \text{a.s.} \tag{3.131}$$

3.6.1 Inverse local time

Let X be a strongly symmetric Borel right process with continuous α-potential density $u^\alpha(x, y)$. Let 0 denote a fixed element in S. By Theorem 3.6.3 we can construct a local time $\{L_t^0; t \geq 0\}$ for X at 0. Exactly as for Brownian motion in Subsection 2.4.1, we set

$$\tau(s) = \inf\{t > 0 \mid L_t^0 > s\} \tag{3.132}$$

with $\inf \emptyset = \infty$. We refer to $\tau(\cdot)$ as the (right continuous) inverse local time (of X), at 0. For each s, $\tau(s)$ is a stopping time and

$$\tau(s + t) = \tau(s) + \tau(t) \circ \theta_{\tau(s)}. \tag{3.133}$$

Lemma 3.6.10 *For $\alpha > 0$,*

$$E^0\big(e^{-\alpha\tau(s)}\big) = e^{-s/u^\alpha(0,0)}. \tag{3.134}$$

Proof The proof proceeds exactly like the proof of Lemma 2.4.7 once we note that, for $\alpha > 0$, $E^0(e^{-\alpha\tau(s)}) = E^0\big(e^{-\alpha\tau(s)}1_{\{\tau(s)<\infty\}}\big)$. This allows us to use the strong Markov property as in the proof of Lemma 2.4.7. $\qquad\square$

We now introduce the important concepts of transience and recurrence both for elements in the state space of X and for X itself. Let

$$\mathcal{L} := \sup\{s \mid X_s = 0\} \tag{3.135}$$

with $\sup \emptyset = 0$. If $\mathcal{L} = \infty$, P^0 almost surely, we say that 0 is recurrent (for X), while if $\mathcal{L} < \infty$, P^0 almost surely, we say that 0 is transient (for X).

Recall that $L_\infty^0 := \lim_{t\to\infty} L_t^0$.

Lemma 3.6.11

(1) *If $u(0,0) = \infty$, then 0 is recurrent for X (equivalently $\mathcal{L} = \infty$) and $L_\infty^0 = \infty$, P^0 almost surely.*

(2) *If $u(0,0) < \infty$, then 0 is transient for X (equivalently $\mathcal{L} < \infty$) and $L_\infty^0 < \infty$, P^x almost surely, for each $x \in S$.*

Proof Suppose that $u(0,0) = \infty$. Then, by (3.134), for each $s < \infty$ we have $\tau(s) < \infty$, P^0 almost surely. Now suppose that $\mathcal{L}(\omega) < \infty$. Since L_t^0 is continuous (and thus locally bounded) and grows only when $X_t = 0$, we would have $L_{\mathcal{L}}^0(\omega) < \infty$. Also, by the definition of inverse local time, $\tau(L_{\mathcal{L}}^0 + 1)(\omega) = \infty$. This contradiction proves that $\mathcal{L} = \infty$, P^0 almost surely. The fact that $L_\infty^0 = \infty$, P^0 almost surely then follows as in the proof of (2.121), where we now use the fact that $\mathcal{L} = \infty$, P^0 almost surely as a substitute for Remark 2.1.5.

Assume now that $u(0,0) < \infty$. Define the sequence of stopping times $\mathcal{T}_0 = 0$ and $\mathcal{T}_{n+1} = \inf\{t > \mathcal{T}_n + 1 \,|\, X_t = 0\}$, $n \geq 0$. Then

$$L_\infty^0 \geq \sum_{n=0}^{\infty} \left(L_{\mathcal{T}_n+1}^0 - L_{\mathcal{T}_n}^0\right) 1_{\{\mathcal{T}_n < \infty\}} = \sum_{n=0}^{\infty} \left(L_1^0 \circ \theta_{\mathcal{T}_n}\right) 1_{\{\mathcal{T}_n < \infty\}}. \quad (3.136)$$

By (3.91) and the strong Markov property,

$$u(0,0) \geq \sum_{n=0}^{\infty} E^0 \left(1_{\{\mathcal{T}_n < \infty\}} E^0(L_1^0)\right) = E^0(L_1^0) \sum_{n=0}^{\infty} P^0(\mathcal{T}_n < \infty). \quad (3.137)$$

By the definition of local time $L_1^0 > 0$, P^0 almost surely. Thus, obviously, $E^0(L_1^0) > 0$ and we have $\lim_{n \to \infty} P^0(\mathcal{T}_n < \infty) = 0$. Since the $\mathcal{T}_n$ are nondecreasing this implies that $P^0(\cap_{n=0}^{\infty}\{\mathcal{T}_n < \infty\}) = 0$, which in turn implies that $\mathcal{L} < \infty$, P^0 almost surely. Since $E^0(L_\infty^0) = u(0,0) < \infty$, we also get $L_\infty^0 < \infty$, P^0 almost surely. Thus we get (2) for $x = 0$. To prove that $\mathcal{L} < \infty$, P^x almost surely for a general $x \in S$, note that $\{\mathcal{L} > 0\} = \{T_0 < \infty\}$ and use the strong Markov property at T_0. To prove that $L_\infty^0 < \infty$, P^x almost surely for a general $x \in S$, note that $E^x(L_\infty^0) = u(x,0) \leq u(0,0) < \infty$ using (3.68).   □

Lemma 3.6.12 *If $P^x(T_y < \infty) = 1$ and $P^y(T_x < \infty) = 1$, then x and y are recurrent.*

Proof As in (3.136), we define the sequence of stopping times $\mathcal{T}_0 = 0$, $\mathcal{T}_1 = T_y + T_x \circ \theta_{T_y}$, and $\mathcal{T}_{n+1} = \mathcal{T}_n + \mathcal{T}_1 \circ \theta_{\mathcal{T}_n}$, $n = 1, 2, \ldots$. Using the monotonicity and additivity of L_t^x, we see that

$$L_\infty^x \geq \sum_{n=0}^{\infty} \left(L_{\mathcal{T}_{n+1}}^x - L_{\mathcal{T}_n}^x\right) = \sum_{n=0}^{\infty} \left(L_{\mathcal{T}_1}^x \circ \theta_{\mathcal{T}_n}\right). \quad (3.138)$$

By the hypothesis and the strong Markov property,

$$P^x(\mathcal{T}_1 < \infty) = P^x(T_y < \infty, T_x \circ \theta_{T_y} < \infty) = 1, \quad (3.139)$$

and consequently $\mathcal{T}_n < \infty$, P^x almost surely for all $n \geq 1$. Therefore, by the strong Markov property,

$$u(x,x) = E^x(L_\infty^x) \geq \sum_{n=0}^{\infty} E^x \left(L_{\mathcal{T}_1}^x \circ \theta_{\mathcal{T}_n}\right) = \sum_{n=0}^{\infty} E^x \left(L_{\mathcal{T}_1}^x\right). \quad (3.140)$$

Since $\mathcal{T}_1 > 0$, P^x almost surely, it follows from the definition of local time that $L_{\mathcal{T}_1}^x > 0$, P^x almost surely and therefore $E^x \left(L_{\mathcal{T}_1}^x\right) > 0$. Thus, by (3.140), we have that $u(x,x) = \infty$. It then follows from Lemma 3.6.11 that x is recurrent. The recurrence of y follows similarly. □

We say that the stochastic process X is recurrent (transient) if all the elements $x \in S$ are recurrent (transient).

The next lemma shows that if one element of the state space S is recurrent for X, then all the elements of S are recurrent for X, that is, that X is a recurrent process, and that X hits all the elements of S infinitely often.

Lemma 3.6.13 *If $u(0,0) = \infty$ and $P^x(T_0 < \infty) > 0$ for all $x \in S$, then $u(x,x) = \infty$ and $P^x(T_y < \infty) = 1$ for all $x, y \in S$, that is, X is a recurrent process.*

Proof Fix some $x \in S$. Since

$$\lim_{\alpha \to 0} E^x(e^{-\alpha T_0}) = P^x(T_0 < \infty), \tag{3.141}$$

we see from (3.107) that when $P^x(T_0 < \infty) > 0$ and $\lim_{\alpha \to 0} u^\alpha(0,0) = u(0,0) = \infty$, then $\lim_{\alpha \to 0} u^\alpha(x,0) = u(x,0) = \infty$. Since, by (3.68), $u^\alpha(x,0) \le u^\alpha(x,x)$, we see that $\lim_{\alpha \to 0} u^\alpha(x,x) = u(x,x) = \infty$.

We now show that $P^x(T_0 < \infty) = 1$. Since $P^x(T_0 < \infty) > 0$, we can find some $s < \infty$ for which $p := P^x(T_0 \le s) > 0$. Define the sequence of stopping times $T_0 = 0$ and $T_{n+1} = \inf\{t > T_n + s \mid X_t = x\}$, $n \ge 0$. Let $\mathcal{L}_x = \sup\{t \mid X_t = x\}$ with $\sup \emptyset = 0$. By Lemma 3.6.11 we have $\mathcal{L}_x = \infty$ almost surely, so that $T_n < \infty$ almost surely. By the strong Markov property,

$$
\begin{aligned}
P^x(T_0 = \infty) &\le P^x(T_0 > T_n) \tag{3.142} \\
&= E^x\left(\prod_{j=1}^n 1_{\{T_0 > T_j\}}\right) \\
&\le (1-p)^n.
\end{aligned}
$$

Letting $n \to \infty$, we see that $P^x(T_0 < \infty) = 1$.

By Remark 3.6.6 and the assumption that $P^x(T_0 < \infty) > 0$, we have $P^0(T_x < \infty) > 0$ and hence for any $y \in S$, $P^y(T_x < \infty) \ge P^y(T_x \circ \theta_{T_0} < \infty, T_0 < \infty) = P^y(T_0 < \infty)P^0(T_x < \infty) > 0$. Using Remark 3.6.6 again we see that $P^x(T_y < \infty) > 0$.

Finally we note that 0 is just a generic point in S. We could just as well have taken y. Since, by what we have shown already, $u(y,y) = \infty$ and $P^x(T_y < \infty) > 0$, our proof shows that $P^x(T_y < \infty) = 1$. □

Remark 3.6.14 When $\mathcal{L}(\omega) \neq 0$, $\mathcal{L}(\omega)$ is a left-limit point of zeros of $X_t(\omega)$. This is clear if $\mathcal{L}(\omega) = \infty$ almost surely. If $\mathcal{L}(\omega) \neq \infty$, then, by Lemma 3.6.11, $\mathcal{L}(\omega) < \infty$ almost surely.

Suppose that $\mathcal{L}(\omega) < \infty$ almost surely and $\mathcal{L}(\omega)$ is not a left-limit point of the zeros of $X_t(\omega)$. Then for some rational number r, $\mathcal{L}(\omega) = r + T_0 \circ \theta_r(\omega)$. Note that $T(r) := r + T_0 \circ \theta_r$ is a stopping time and $X_{T(r)} = 0$ on $T(r) < \infty$. Therefore, by the strong Markov property and (3.83), we have that, for any $x \in S$,

$$P^x(T(r) < \infty, \, T_0 \circ \theta_{T(r)} = \infty) = P^x(T(r) < \infty)P^0(T_0 = \infty) = 0. \tag{3.143}$$

Since the event

$$\{\omega \,|\, \mathcal{L}(\omega) < \infty, \text{ and } \mathcal{L}(\omega) \text{ is not a left-limit point of the zeros of } X_t(\omega)\}$$

is contained in $\bigcup_{r \in Q}\{T(r) < \infty\} \cap \{T_0 \circ \theta_T(r) = \infty\}$ where Q is the set of positive rational numbers, we have a contradiction.

Remark 3.6.15 We make the following observation for use in Theorem 13.1.2. Let $B = \{x_1, \ldots, x_p\} \in S$ be a fixed set and define

$$\sigma = \inf\{t \geq 0 \,|\, X_t \in B \cap \{X_0\}^c\}. \tag{3.144}$$

A proof similar to the proof of Lemma 3.6.12 shows that if $P^x(\sigma < \infty) = 1$ for all $x \in B$, then all $x \in B$ are recurrent.

Here are the details of the proof. We define the sequence of stopping times $T_0 = 0$, $T_1 = \sigma$, and $T_{n+1} = T_n + \sigma \circ \theta_{T_n}$, $n = 1, 2, \ldots$. Let $A_t = \sum_{x \in B} L_t^x$. Using the monotonicity and additivity of A_t, we see that

$$A_\infty \geq \sum_{n=0}^{\infty} (A_{T_{n+1}} - A_{T_n}) = \sum_{n=0}^{\infty} (A_\sigma \circ \theta_{T_n}). \tag{3.145}$$

It follows from the assumption that $P^x(\sigma < \infty) = 1$ for all $x \in B$ and the strong Markov property that $T_n < \infty$, P^x almost surely for all $x \in B$, for all $n \geq 1$. Hence, by (3.105) and the strong Markov property, for any $x \in B$,

$$\sum_{y \in B} u(x, y) = E^x(A_\infty) \geq \sum_{n=0}^{\infty} E^x(A_\sigma \circ \theta_{T_n}) = \sum_{n=0}^{\infty} E^x\left(E^{X_{T_n}}(A_\sigma)\right). \tag{3.146}$$

Since $\sigma > 0$, P^y almost surely for all $y \in B$, it follows from the definition of local time that $L_\sigma^y > 0$, P^y almost surely for all $y \in B$ and therefore

$r(y) := E^y(A_\sigma) > 0$ for all $y \in B$. Consequently,

$$E^{X_{T_n}}(A_\sigma) \geq \inf_{y \in B} r(y) > 0 \qquad P^x \text{ a.s.} \qquad (3.147)$$

Thus, by (3.146), we have $\sum_{y \in B} u(x, y) = \infty$, so that $u(x, y_0) = \infty$ for some $y_0 \in B$. By (3.68), $u(x, x) = \infty$, so that, by Lemma 3.6.11, x is recurrent P^x almost surely. $\qquad \square$

Analogous to Lemma 2.4.5, we have

Lemma 3.6.16 *If $u(0, 0) = \infty$, then $\{\tau(s); s \in R_+\}$ has stationary and independent increments under P^0.*

Proof The proof is the same as the proof for Brownian motion, except that we now use Lemma 3.2.4. $\qquad \square$

Remark 3.6.17 If $u(0, 0) < \infty$, a similar proof shows that, conditional on the event $\tau(s) < \infty$, $\tau(t) \circ \theta_{\tau(s)}$ is independent of $\mathcal{F}_{\tau(s)}$ and has the same law, under P^0, as $\tau(t)$.

Let

$$\tau^-(s) = \inf\{t > 0 \mid L_t^0 \geq s\} \qquad (3.148)$$

with $\inf \emptyset = \infty$. We refer to $\tau^-(\cdot)$ as the left continuous inverse local time (of X), at 0. For each s, $\tau^-(s)$ is a stopping time.

Lemma 3.6.18 *For any $t > 0$,*

$$\tau^-(t) = \tau(t) \qquad P^0 \quad \text{a.s.} \qquad (3.149)$$

Proof Clearly $\tau^-(t) \leq \tau(t)$. Since

$$\{\tau^-(t) < \tau(t)\} = \{\tau^-(t) < \infty, \ R_{L^0} \circ \theta_{\tau^-(t)} > 0\}$$

and $\tau^-(t)$ is a stopping time, by the strong Markov property,

$$P^0(\tau^-(t) < \tau(t)) = E^0\left(I_{\{\tau^-(t) < \infty\}} P^{X_{\tau^-(t)}}(R_{L^0} > 0)\right). \qquad (3.150)$$

It is clear from the definition that on $\{\tau^-(t) < \infty\}$ we have that $\tau^-(t)$ is in the support of dL_s^0. Therefore, by Remark 3.6.2, $X_{\tau^-(t)} = 0$ almost surely. The proof follows since, by the definition of local time, $P^0(R_{L^0} > 0) = 0$. $\qquad \square$

Lemma 3.6.19

$$\tau^-(L_\infty^0) = \mathcal{L} \qquad P^0 \quad \text{a.s.} \qquad (3.151)$$

Proof　Since L_s^0 cannot increase after time $\mathcal{L}$, we have $\tau^-(L_\infty^0) \leq \mathcal{L}$. Also,

$$\{\tau^-(L_\infty^0) < \mathcal{L}\} \subseteq \bigcup_{r \text{ rational}} \{\tau^-(L_\infty^0) < r < \mathcal{L}\} \tag{3.152}$$

$$\subseteq \bigcup_{r \text{ rational}} \{T_0 \circ \theta_r < \infty, \, R_{L^0} \circ \theta_{T_0 \circ \theta_r} > 0\}.$$

By the strong Markov property and the facts that $X_{T_0 \circ \theta_r} = 0$ on $\{T_0 \circ \theta_r < \infty\}$ and $P^0(R_{L^0} > 0) = 0$, we have

$$P^0 \left(T_0 \circ \theta_r < \infty, \, R_{L^0} \circ \theta_{T_0 \circ \theta_r} > 0\right) \tag{3.153}$$

$$= E^0 \left(I_{\{T_0 \circ \theta_r < \infty\}} P^{X_{T_0 \circ \theta_r}}(R_{L^0} > 0)\right)$$

$$= E^0 \left(I_{\{T_0 \circ \theta_r < \infty\}} P^0(R_{L^0} > 0)\right) = 0.$$

Therefore $P\left(\tau^-(L_\infty^0) < \mathcal{L}\right) = 0$. $\qquad\square$

3.7 Jointly continuous local times

Let $X = (\Omega, \mathcal{G}, \mathcal{G}_t, X_t, \theta_t, P^x)$ be a strongly symmetric Borel right process with continuous α-potential density $u^\alpha(x, y)$. In Section 3.6 we construct a local time L_t^y for each $y \in S$. Now we consider the stochastic process $L = \{L_t^y, (t, y) \in R_+ \times S\}$. We ask whether we can obtain a version of L that is continuous or, as is often stated, "jointly continuous." This is a classical and important problem in the theory of Markov processes. One of the main results in this book is a necessary and sufficient condition for a strongly symmetric Borel right process to have jointly continuous local times.

When we say that a stochastic process $\bar{L} = \{\bar{L}_t^y, (t, y) \in R_+ \times S\}$ is a version of the local time of a Markov process X, we mean more than the traditional statement that one stochastic process is a version of the other. Besides this we also require that the version is itself a local time for X, that is, for each $y \in S$, $\bar{L}^y$ is a local time for X at y, as defined in Section 3.6. To be more specific, suppose that $L = \{L_t^y, (t, y) \in R_+ \times S\}$ is a local time for X. When we say that we can find a version of the local time that is jointly continuous on $T \times S$, where $T \subset R_+$, we mean that we can find a stochastic process $\bar{L} = \{\bar{L}_t^y, (t, y) \in R_+ \times S\}$ that is continuous on $T \times S$ for all $x \in S$ and which satisfies, for each $x, y \in S$,

$$\bar{L}_t^y = L_t^y \quad \forall t \in R_+ \quad \text{a.s. } P^x \tag{3.154}$$

Following convention, we often say that a Markov process has a continuous local time when we mean that we can find a continuous version for the local time.

In Theorem 3.7.3 we give conditions that imply that X has a jointly continuous local time, but first we note a simple but useful consequence of joint continuity.

Theorem 3.7.1 (Occupation Density Formula) *If $L = \{L_t^y, (t,y) \in R_+ \times S\}$ is jointly continuous, then, for any $f \in C_\kappa(S)$ and $t \geq 0$,*

$$\int_0^t f(X_s)\,ds = \int f(y)L_t^y\,dm(y) \qquad a.s. \qquad (3.155)$$

Proof Let

$$A_t = \int f(y)L_t^y\,dm(y). \qquad (3.156)$$

It is easy to see that $A = \{A_t; t \geq 0\}$ is a CAF, as defined on page 83. It is clear that A satisfies conditions (1) and (2). To check that it satisfies condition (3), additivity, note that for each $y \in S$,

$$L_{t+s}^y = L_t^y + L_s^y \circ \theta_t \qquad \text{for all } s,t \in R_+ \text{ a.s.} \qquad (3.157)$$

Clearly (3.157) also holds on a countable dense set of S almost surely, and since L is jointly continuous, it holds for all $y \in S$ almost surely. It follows from this that condition (3), in the definition of a CAF is also satisfied.

Let $g \in \mathcal{B}_b(S)$. It follows from Remark 3.6.2 and (3.91) that

$$\begin{aligned}
U_A^\alpha g(x) &= E^x \left(\int \int_0^\infty e^{-\alpha t} g(X_t)\,dL_t^y\,f(y)m(dy) \right) \qquad (3.158) \\
&= E^x \left(\int \int_0^\infty e^{-\alpha t} g(y)\,dL_t^y\,f(y)m(dy) \right) \\
&= \int u^\alpha(x,y)g(y)f(y)\,dm(y).
\end{aligned}$$

Clearly $B_t = \int_0^t f(X_s)\,ds$ is a CAF and

$$\begin{aligned}
U_B^\alpha g(x) &= E^x \left(\int \int_0^\infty e^{-\alpha t} g(X_t)f(X_t)dt \right) \qquad (3.159) \\
&= \int u^\alpha(x,y)g(y)f(y)\,dm(y).
\end{aligned}$$

Thus $U_A^\alpha g = U_B^\alpha g$ for all $g \in \mathcal{B}_b(S)$. Equation (3.155) now follows from Lemma 3.6.8. $\qquad \square$

Let L_t^y be a local time for X at y. Since L_t^y is increasing in t, $L_\infty^y = \lim_{t\to\infty} L_t^y$ exists. We often refer to the process $L_\infty := \{L_\infty^y, y \in S\}$ as the total accumulated local time of X. The isomorphism theorems that we develop, which relate local times to Gaussian processes, usually are obtained (at least initially) for L_∞, and so we require that $L_\infty^y < \infty$ for some, or all, $y \in S$. Of course sometimes it is. When it is not we consider X killed at some stopping time and study the total accumulated local time of the killed process. For example, if L_∞^y is infinite, we could consider $\widehat{L}_t^y := L_{t \wedge \lambda}$, the local time for the killed process $\widehat{X}$, which was introduced in Section 3.5, since $\widehat{L}_\infty^y$ is finite almost surely (see Remark 3.6.4 (3)).

In this section we reduce the problem of showing that $L = \{L_t^y, (t, y) \in R_+ \times S\}$ is jointly continuous to showing that $\{L_\infty^y, y \in S\}$ is continuous. Note that since S is a locally compact space with a countable base, we can always find a metric ρ compatible with the topology of S and consider (S, ρ) as a locally compact separable metric space.

Lemma 3.7.2 *Let $X = (\Omega, \mathcal{G}, \mathcal{G}_t, X_t, \theta_t, P^x)$ be a strongly symmetric Borel right process with continuous 0-potential density $u(x, y)$ and state space (S, ρ), where S is a locally compact separable metric space. Let $L = \{L_t^y, (t, y) \in R_+ \times S\}$ be local times for X. Suppose there exists a $p > 1$ such that, for any compact set $K \subseteq S$,*

$$E^x \sup_{y \in D \cap K} (L_\infty^y)^p < \infty, \tag{3.160}$$

where D is countable dense subset of S. Then

$$\left(E^x \sup_{t \geq 0} \sup_{\substack{\rho(y,z) \leq \delta \\ y,z \in D \cap K}} |L_t^y - L_t^z|^p \right)^{1/p} \tag{3.161}$$

$$\leq \frac{p}{p-1} \left(E^x \sup_{\substack{\rho(y,z) \leq \delta \\ y,z \in D \cap K}} |L_\infty^y - L_\infty^z|^p \right)^{1/p}$$

$$+ \sup_{\substack{\rho(y,z) \leq \delta \\ x,y,z \in D \cap K}} |u(x,y) - u(x,z)|.$$

Proof   Consider the martingale

$$A_t^y = E^x(L_\infty^y \mid \mathcal{F}_t) \tag{3.162}$$

and note that

$$L_\infty^y = L_t^y + L_\infty^y \circ \theta_t.$$

Therefore,

$$A_t^y = L_t^y + E^x(L_\infty^y \circ \theta_t \mid \mathcal{F}_t) = L_t^y + E^{X_t}(L_\infty^y)1_{\{t < \zeta\}}, \tag{3.163}$$

where we use the Markov property. Since $E^x(L_\infty^y) = u(x, y)$, we have

$$A_t^y = L_t^y + u(X_t, y)1_{\{t<\zeta\}}. \tag{3.164}$$

Let F be a finite subset of $D \cap K$. Using (3.160) along with the fact that $u(X_t, y) \le u(y, y)$, we see that $E^x \sup_{t\ge0} \sup_{x\in F} A_t^x < \infty$ and

$$\left(E^x \sup_{\substack{t\ge0 \\ y,z\in F}} \sup_{\rho(y,z)\le\delta} |L_t^y - L_t^z|^p\right)^{1/p} \tag{3.165}$$

$$\le \left(E^x \sup_{\substack{t\ge0 \\ y,z\in F}} \sup_{\rho(y,z)\le\delta} |A_t^y - A_t^z|^p\right)^{1/p}$$

$$+ \left(E^x \sup_{\substack{t\ge0 \\ y,z\in F}} \sup_{\rho(y,z)\le\delta} |u(X_t, y) - u(X_t, z)|^p 1_{\{t<\zeta\}}\right)^{1/p}.$$

It follows from Lemma 3.4.3 and the continuity of u that the last term in (3.165) is dominated by the last term in (3.162). Furthermore, since A_t^y is right continuous, we have that

$$\sup_{\substack{\rho(y,z)\le\delta \\ y,z\in F}} A_t^y - A_t^z = \sup_{\substack{\rho(y,z)\le\delta \\ y,z\in F}} |A_t^y - A_t^z| \tag{3.166}$$

is a right continuous nonnegative submartingale. Therefore, by Doob's L_p inequality (see, e.g., Rogers and Williams (2000b, II.70.2)) and the fact that $A_\infty^\cdot = L_\infty^\cdot$, we have

$$\left(E^x \sup_{\substack{t\ge0 \\ y,z\in F}} \sup_{\rho(y,z)\le\delta} |A_t^y - A_t^z|^p\right)^{1/p} \le \frac{p}{p-1}\left(E^x \sup_{\substack{\rho(y,z)\le\delta \\ y,z\in F}} |L_\infty^y - L_\infty^z|^p\right)^{1/p}, \tag{3.167}$$

which when substituted in (3.165) gives (3.162). □

Theorem 3.7.3 *Suppose that, in addition to the hypotheses of Lemma 3.7.2,*

$$\lim_{\delta\to0} E^x \sup_{\substack{\rho(y,z)\le\delta \\ y,z\in D\cap K}} |L_\infty^y - L_\infty^z|^p = 0. \tag{3.168}$$

Then we can find a version of L that is continuous on $R_+ \times S$.

Proof Fix $T < \infty$ and a compact set $K \subseteq S$. We first show that

$$E^x \sup_{\substack{|s-t| \leq h(\delta) \\ s,t \in [0,T]}} \sup_{\substack{\rho(y,z) \leq \delta \\ y,z \in D \cap K}} |L_s^y - L_t^z|^p \leq H(\delta), \qquad (3.169)$$

where both $h(\delta)$ and $H(\delta)$ decrease to zero as δ decreases to zero. Note that it follows from Lemma 3.7.2, (3.168), and the uniform continuity of $u(x,y)$ on the compact set $K \times K$ that

$$E^x \sup_{s \geq 0} \sup_{\substack{\rho(y,z) \leq 2\delta \\ y,z \in D \cap K}} |L_s^y - L_s^z|^p \leq H'(\delta), \qquad (3.170)$$

where $H'(\delta)$ decreases to zero as δ decreases to zero.

Let $B(v_j, \delta)$, $j = 1, \ldots, N(\delta)$ be a family of closed balls of radius δ in the metric ρ, centered at $v_j \in D \cap K$ such that $D \cap K \subseteq \cup_{j=1}^{N(\delta)} B(v_j, \delta)$. Note that $\rho(y,z) \leq \delta$ implies that both y and z are contained in $B(v_j, 2\delta)$ for at least one $j \in [1, N(\delta)]$. For each $j \in [1, N(\delta)]$,

$$\sup_{\substack{\rho(y,z) \leq \delta \\ y,z \in B(v_j, 2\delta)}} |L_s^y - L_t^z| \leq \sup_{y \in B(v_j, 2\delta)} |L_s^y - L_s^{v_j}| \qquad (3.171)$$

$$+ \sup_{z \in B(v_j, 2\delta)} |L_t^{v_j} - L_t^z| + |L_s^{v_j} - L_t^{v_j}|.$$

Consequently,

$$\sup_{\substack{\rho(y,z) \leq \delta \\ y,z \in D \cap K}} |L_s^y - L_t^z|^p = \sup_{j \in [1, N(\delta)]} \sup_{\substack{\rho(y,z) \leq \delta \\ y,z \in B(v_j, 2\delta) \cap D \cap K}} |L_s^y - L_t^z|^p \qquad (3.172)$$

$$\leq C_p \left(\sup_{s \geq 0} \sup_{\substack{\rho(y,z) \leq 2\delta \\ y,z \in D \cap K}} |L_s^y - L_s^z|^p + \sup_{j \in [1, N(\delta)]} |L_s^{v_j} - L_t^{v_j}|^p \right).$$

We use (3.172) to obtain (3.169). The first term in the last line of (3.172) is controlled by (3.170). To control the second term in the last line of (3.172), we note that for each v_j, $L_t^{v_j}$ is uniformly continuous on $[0, T]$. Therefore, it follows from (3.160) and the Dominated Convergence Theorem that

$$\lim_{\delta' \to 0} E^x \sup_{\substack{|s-t| \leq \delta' \\ s,t \in [0,T]}} |L_s^{v_j} - L_t^{v_j}|^p = 0. \qquad (3.173)$$

Thus we can find a function $h(\delta)$ that goes to zero as δ goes to zero, such that

$$E^x \sup_{\substack{j \in [1, N(\delta)] \\ }} \sup_{\substack{|s-t| \le h(\delta) \\ s,t \in [0,T]}} |L_s^{v_j} - L_t^{v_j}|^p \le \sum_{j=1}^{N(\delta)} E^x \sup_{\substack{|s-t| \le h(\delta) \\ s,t \in [0,T]}} |L_s^{v_j} - L_t^{v_j}|^p \le H'(\delta).$$

(3.174)

Thus we have (3.169).

Let K_n be a sequence of compact subsets of S such that $S = \cup_{n=1}^{\infty} K_n$ and let $s > 0$. It follows from (3.169) that L_t^y is uniformly continuous on $[0, s] \times (K_n \cap D)$, P^x almost surely.

Let

$$\widetilde{\Omega} = \{\omega \mid L_t^y(\omega) \text{ is locally uniformly continuous on } R_+ \times D\}$$

(3.175)

and note that

$$\widetilde{\Omega}^c = \bigcup_{\substack{s \in \mathcal{R} \\ 1 \le n \le \infty}} \{\omega \mid L_t^y(\omega) \text{ is not uniformly continuous on } [0, s] \times (K_n \cap D)\},$$

(3.176)

where $\mathcal{R}$ denotes the rational numbers. By the remark in the previous paragraph, $P^x(\widetilde{\Omega}^c) = 0$. Thus we have

$$P^x(\widetilde{\Omega}) = 1 \qquad \forall x \in S.$$

(3.177)

We now construct a stochastic process $\widetilde{L} = \{\widetilde{L}_t^y, (t, y) \in R_+ \times S\}$ that is continuous and is a version of L. For $\omega \in \widetilde{\Omega}$, let $\{\widetilde{L}_t^y(\omega), (t, y) \in R_+ \times S\}$ be the continuous extension of $\{L_t^y(\omega), (t, y) \in R_+ \times D\}$ to $R_+ \times S$, and for $\omega \in \widetilde{\Omega}^c$ set

$$\widetilde{L}_t^y \equiv 0 \qquad \forall t, y \in R_+ \times S.$$

(3.178)

$\{\widetilde{L}_t^y, (t, y) \in R_+ \times S\}$ is a well-defined stochastic process that, clearly, is jointly continuous on $R_+ \times S$. We now show that $\widetilde{L}$ satisfies (3.154).

We have been working with a countable dense set D. We could just as well have used $D \cup \{y\}$ and thus obtained (3.177) with D replaced by $D \cup \{y\}$ in the definition of $\widetilde{\Omega}$. Therefore we can assume that $\{y\} \in D$.

Let $\{y_i\}_{i=1}^{\infty}$, with all $y_i \in D$, be such that $\lim_{i \to \infty} y_i = y$. We then have

$$\lim_{i \to \infty} L_t^{y_i} = L_t^y \qquad \text{locally uniformly on } R_+, \quad P^x \text{ a.s.}$$

By the definition of $\widetilde{L}$ we also have

$$\lim_{i \to \infty} L_t^{y_i} = \widetilde{L}_t^y \qquad \text{locally uniformly on } R_+, \quad P^x \text{ a.s.}$$

This shows that

$$\widetilde{L}_t^y = L_t^y \qquad \forall t \quad P^x \text{ a.s.,} \tag{3.179}$$

which is (3.154).   □

At first glance it might appear as though we require that X has a 0-potential density to obtain conditions for X to have a jointly continuous local time. This is not the case, as we see from the following corollary of Lemma 3.7.2 and Theorem 3.7.3.

Corollary 3.7.4 *Let* $X = (\Omega, \mathcal{G}, \mathcal{G}_t, X_t, \theta_t, P^x)$ *be a strongly symmetric Borel right process with continuous* α-*potential density* $u^\alpha(x,y)$ *and state space* (S, ρ), *where* S *is a locally compact separable metric space. Let* $L = \{L_t^y, (t,y) \in R_+ \times S\}$ *be local times for* X. *Suppose there exists a* $p > 1$ *such that, for all compact sets* $K \subseteq S$

$$E_\lambda^x \sup_{y \in D \cap K} (L_\lambda^y)^p < \infty, \tag{3.180}$$

where D *is a countable dense subset of* S *and* λ *is an exponential random variable independent of* X. *Then*

$$\left(E_\lambda^x \sup_{t \geq 0} \sup_{\substack{\rho(y,z) \leq \delta \\ y,z \in D \cap K}} |L_{t \wedge \lambda}^y - L_{t \wedge \lambda}^z|^p \right)^{1/p} \tag{3.181}$$

$$\leq \frac{p}{p-1} \left(E_\lambda^x \sup_{\substack{\rho(y,z) \leq \delta \\ y,z \in D \cap K}} |L_\lambda^y - L_\lambda^z|^p \right)^{1/p} + \sup_{\substack{\rho(y,z) \leq \delta \\ x,y,z \in D \cap K}} |u(x,y) - u(x,z)|.$$

If, in addition,

$$\lim_{\delta \to 0} E_\lambda^x \sup_{\substack{\rho(y,z) \leq \delta \\ y,z \in D \cap K}} |L_\lambda^y - L_\lambda^z|^p = 0, \tag{3.182}$$

we can find a version of L *that is continuous on* $R_+ \times S$.

Proof Let $\widehat{X} = (\widehat{\Omega}, \widehat{\mathcal{G}}, \widehat{\mathcal{G}}_t, \widehat{X}_t, \widehat{\theta}_t, \widehat{P}^x)$ be the strongly symmetric Borel right process obtained by killing X at an independent exponential time λ with mean $1/\alpha$, as described in Section 3.5. The 0-potential density of $\widehat{X}$ is the α-potential density of X. Thus we have a Borel right process $\widehat{X}$ with state space S and continuous 0-potential density $u^\alpha(x,y)$. As we pointed out in Remark 3.6.4 (3) when L_t^y is a local time for X, $\widehat{L}_t^y := L_{t \wedge \lambda}^y$ is a local time for $\widehat{X}$.

It follows from Lemma 3.7.2 and Theorem 3.7.3 that $\widehat{L}_t^y$ is locally uniformly continuous on $R_+ \times D$, almost surely with respect to $\mu \times P^x$, where μ is the probability measure of λ and D is a countable dense subset of S. Equivalently, L_t^y is locally uniformly continuous on $[0, \lambda] \times D$, almost surely with respect to $\mu \times P^x$. Using Fubini's Theorem we can now see that L^y_t is locally uniformly continuous on $[0, q_i] \times D$ for all $q_i \in Q$, where Q is a countable dense subset of R_+. As in Theorem 3.7.3, we can extend L^y_t so that it is jointly continuous on $R_+ \times D$ almost surely with respect to P^x. The rest of the proof, showing that the extended L^y_t is a local time for X, is the same as in Theorem 3.7.3. $\square$

Remark 3.7.5 Suppose that we only assume (3.182) for a single compact set $K \subseteq S$. The proof of Corollary 3.7.4 shows that we can find a version of L that is continuous on $R_+ \times K$. This is easy to check.

3.8 Calculating u_{T_0} and $u_{\tau(\lambda)}$

Let $X = (\Omega, \mathcal{G}, \mathcal{G}_t, X_t, \theta_t, P^x)$ be a strongly symmetric Borel right process with continuous α-potential density $u^\alpha(x, y)$. In this section we generalize the concepts introduced for Brownian motion in Section 2.5. We consider terminal times, as defined in the first paragraph of Section 2.5, and, for a CAF $A = \{A_t; t \geq 0\}$ on $(\Omega, \mathcal{F}_t, \mathcal{F})$, the inverse times τ_A as defined in the paragraph containing (2.143). We also consider u_T and $u_{\tau_A(\lambda)}$ as defined in (2.138) and (2.146). Note that Lemma 2.5.2 continues to hold.

A particularly important example of a continuous additive functional is the local time itself. As a generalization of (3.132) we set $\tau_x(s) = \inf\{t > 0 \,|\, L^x_t > s\}$, where $\inf\{\emptyset\} = \infty$. (We generally write $\tau(s)$ for $\tau_0(s)$.) Note that, just as in (3.133),

$$\tau_x(s + t) = \tau_x(s) + \tau_x(t) \circ \theta_{\tau_x(s)}. \tag{3.183}$$

Recall that T_0 denotes the first hitting time of 0, where we use 0 to indicate some fixed point in S. It is obvious that T_0 is a terminal time. In the next two lemmas we obtain explicit descriptions of the potential densities u_{T_0} and $u_{\tau(\lambda)}$.

Lemma 3.8.1

$$u_{T_0}(x, y) = \lim_{\alpha \to 0} \left\{ u^\alpha(x, y) - \frac{u^\alpha(x, 0)u^\alpha(0, y)}{u^\alpha(0, 0)} \right\}, \tag{3.184}$$

where the limit always exists. If $P^x(T_0 < \infty) > 0$ for all $x \in S$, then the limit in (3.184) is finite and positive definite.

Note that $u_{T_0}(x, 0) = 0$ for all $x \in S$.

Proof We first note that, as in (3.128),

$$
\begin{aligned}
E^x \left(\int_{T_0}^{\infty} e^{-\alpha t} dL_t^y \right) &= E^x \left(1_{\{T_0 < \infty\}} \int_{T_0}^{\infty} e^{-\alpha t} dL_t^y \right) \qquad (3.185) \\
&= E^x \left(e^{-\alpha T_0} 1_{\{T_0 < \infty\}} \left(\int_0^{\infty} e^{-\alpha t} dL_t^y \right) \circ \theta_{T_0} \right) \\
&= E^x \left(e^{-\alpha T_0} 1_{\{T_0 < \infty\}} E^{X_{T_0}} \left(\int_0^{\infty} e^{-\alpha t} dL_t^y \right) \right) \\
&= E^x \left(e^{-\alpha T_0} \right) u^{\alpha}(0, y)
\end{aligned}
$$

since $X_{T_0} = 0$ on $T_0 < \infty$. We use (3.185) and (3.107) precisely as in (2.140) to obtain

$$
E^x \left(\int_0^{T_0} e^{-\alpha t} dL_t^y \right) = u^{\alpha}(x, y) - \frac{u^{\alpha}(x, 0) u^{\alpha}(0, y)}{u^{\alpha}(0, 0)}. \qquad (3.186)
$$

Consequently, by the Monotone Convergence Theorem,

$$
u_{T_0}(x, y) = \lim_{\alpha \to 0} \left\{ u^{\alpha}(x, y) - \frac{u^{\alpha}(x, 0) u^{\alpha}(0, y)}{u^{\alpha}(0, 0)} \right\}. \qquad (3.187)
$$

We now show that if $P^x(T_0 < \infty) > 0$ for all $x \in S$, then the limit in (3.187) is finite. Using (3.183) and the terminal time property, (2.137) for T_0, we have

$$
\begin{aligned}
P^x(L_{T_0}^x > s + t) &= P^x(\tau_x(s + t) < T_0) \qquad (3.188) \\
&= P^x(\tau_x(s) < T_0, \, \tau_x(t) \circ \theta_{\tau_x(s)} < T_0 \circ \theta_{\tau_x(s)}).
\end{aligned}
$$

Since $X_{\tau_x(s)} = x$ on $\{\tau_x(s) < \infty\}$, the strong Markov property shows us that

$$
\begin{aligned}
P^x(L_{T_0}^x > s + t) &= P^x(\tau_x(s) < T_0) P^x(\tau_x(t) < T_0) \qquad (3.189) \\
&= P^x(L_{T_0}^x > s) P^x(L_{T_0}^x > t).
\end{aligned}
$$

It follows from (3.189) that if there exists an s for which $P^x(L_{T_0}^x > s) < 1$, then $L_{T_0}^x$ is finite almost surely and is an exponential random variable under P^x with mean $u_{T_0}(x, x) < \infty$. On the other hand, $L_{T_0}^x$ cannot be infinite almost surely. To see this we note that $u^{\alpha}(x, x) = E^x \left(\int_0^{\infty} e^{-\alpha t} dL_t^x \right)$ is finite and L_t^x is increasing in t. Therefore

$$
e^{-\alpha T_0} L_{T_0}^x \leq \int_0^{\infty} e^{-\alpha t} dL_t^x < \infty \qquad P^x \quad \text{a.s.} \qquad (3.190)
$$

By assumption $P^x(T_0 < \infty) > 0$, and when $T_0 < \infty$ we see by (3.190) that $L_{T_0}^x < \infty$. Thus we see that $u_{T_0}(x, x) < \infty$ for all $x \in S$.

We know that $u^\alpha(x,y)$ is positive definite by Lemma 3.3.3, and therefore, by Remark 5.1.6, so is $u^\alpha_{(0)}(x,y) := u^\alpha(x,y) - \dfrac{u^\alpha(x,0)u^\alpha(0,y)}{u^\alpha(0,0)}$.

Taking the limit as α goes to zero we see that $u_{T_0}(x,y)$ is positive definite. Since $u_{T_0}(x,x) < \infty$ for all $x \in S$, the finiteness of $u_{T_0}(x,y)$ follows by the Cauchy–Schwarz inequality; see (5.27).   □

Remark 3.8.2 In the course of the proof of Lemma 3.8.1 we show that when $P^x(T_0 < \infty) > 0$, $L^x_{T_0}$ is an exponential random variable, under P^x, with mean $u_{T_0}(x,x) < \infty$.

Remark 3.8.3 It follows from (3.184) that if X has continuous 0-potential densities $u(x,y)$, then

$$u_{T_0}(x,y) = u(x,y) - \frac{u(x,0)u(0,y)}{u(0,0)}. \tag{3.191}$$

In this case it is obvious that $u_{T_0}(x,y)$ is continuous and it follows from Remark 5.1.6 that it is positive definite.

Lemma 3.8.4 *Assume that $P^x(T_0 < \infty) > 0$ for all $x \in S$. Then*

$$u_{\tau(\lambda)}(x,y) = u_{T_0}(x,y) + \frac{1}{\alpha + 1/u(0,0)} P^x(T_0 < \infty)P^y(T_0 < \infty), \tag{3.192}$$

where $1/\alpha = E(\lambda)$ and $1/u(0,0) = 0$ if $u(0,0) = \infty$. In fact, when $u(0,0) = \infty$ we have $P^x(T_0 < \infty) = 1$ for all $x \in S$, so that

$$u_{\tau(\lambda)}(x,y) = u_{T_0}(x,y) + \frac{1}{\alpha}. \tag{3.193}$$

Proof We first note that

$$E^x_\lambda\left(\int_{T_0+\tau(\lambda)\circ\theta_{T_0}}^\infty e^{-\beta t}\,dL^y_t\right) \tag{3.194}$$

$$= E^x_\lambda\left(1_{\{T_0<\infty\}}\int_{T_0+\tau(\lambda)\circ\theta_{T_0}}^\infty e^{-\beta t}\,dL^y_t\right)$$

$$= E^x_\lambda\left(e^{-\beta T_0}1_{\{T_0<\infty\}}\left(\int_{\tau(\lambda)}^\infty e^{-\beta t}\,dL^y_t\right)\circ\theta_{T_0}\right)$$

$$= E^x\left(e^{-\beta T_0}1_{\{T_0<\infty\}}E^{X_{T_0}}_\lambda\left(\int_{\tau(\lambda)}^\infty e^{-\beta t}\,dL^y_t\right)\right)$$

$$= E^x\left(e^{-\beta T_0}\right)E^0_\lambda\left(\int_{\tau(\lambda)}^\infty e^{-\beta t}\,dL^y_t\right)$$

since $X_{T_0} = 0$ on $T_0 < \infty$. Similarly,

$$E_\lambda^0 \left(\int_{\tau(\lambda)}^\infty e^{-\beta t} \, dL_t^y \right) \tag{3.195}$$

$$= E_\lambda^x \left(1_{\{\tau(\lambda) < \infty\}} \int_{\tau(\lambda)}^\infty e^{-\beta t} \, dL_t^y \right)$$

$$= E_\lambda^x \left(e^{-\beta \tau(\lambda)} 1_{\{\tau(\lambda) < \infty\}} \left(\int_0^\infty e^{-\beta t} \, dL_t^y \right) \circ \theta_{\tau(\lambda)} \right)$$

$$= E_\lambda^x \left(e^{-\beta \tau(\lambda)} 1_{\{\tau(\lambda) < \infty\}} E^{X_{\tau(\lambda)}} \left(\int_0^\infty e^{-\beta t} \, dL_t^y \right) \right)$$

$$= E_\lambda^x \left(e^{-\beta \tau(\lambda)} \right) E^0 \left(\int_0^\infty e^{-\beta t} \, dL_t^y \right)$$

since $X_{\tau(\lambda)} = 0$ on $\tau(\lambda) < \infty$. Combining (3.194) and (3.195), we see that

$$E_\lambda^x \left(\int_{T_0 + \tau(\lambda) \circ \theta_{T_0}}^\infty e^{-\beta t} \, dL_t^y \right) = E^x \left(e^{-\beta T_0} \right) E_\lambda^x \left(e^{-\beta \tau(\lambda)} \right) u^\beta(0, y). \tag{3.196}$$

Using this and proceeding exactly as in (2.165)–(2.166), we see that

$$u_{\tau(\lambda)}(x, y) = \lim_{\beta \to 0} \left\{ u^\beta(x, y) - \frac{u^\beta(x, 0) u^\beta(0, y)}{u^\beta(0, 0)} \right\} \tag{3.197}$$

$$+ \lim_{\beta \to 0} \frac{u^\beta(x, 0) u^\beta(0, y)}{u^\beta(0, 0)} (1 - E_\lambda^0(e^{-\beta \tau(\lambda)})).$$

By (3.184),

$$\lim_{\beta \to 0} \left\{ u^\beta(x, y) - \frac{u^\beta(x, 0) u^\beta(0, y)}{u^\beta(0, 0)} \right\} = u_{T_0}(x, y). \tag{3.198}$$

Also, by (3.107),

$$\lim_{\beta \to 0} \frac{u^\beta(x, 0) u^\beta(0, y)}{u^\beta(0, 0)} (1 - E_\lambda^0(e^{-\beta \tau(\lambda)})) \tag{3.199}$$

$$= \lim_{\beta \to 0} E^x(e^{-\beta T_0}) E^y(e^{-\beta T_0}) u^\beta(0, 0)(1 - E_\lambda^0(e^{-\beta \tau(\lambda)}))$$

$$= P^x(T_0 < \infty) P^y(T_0 < \infty) \lim_{\beta \to 0} u^\beta(0, 0)(1 - E_\lambda^0(e^{-\beta \tau(\lambda)})),$$

and, by (3.134),

$$u^\beta(0, 0)(1 - E_\lambda^0(e^{-\beta \tau(\lambda)})) \tag{3.200}$$

$$= u^\beta(0, 0)(1 - E_\lambda(e^{-\lambda / u^\beta(0,0)}))$$

$$= u^\beta(0, 0)(1 - \frac{\alpha}{\alpha + 1/u^\beta(0, 0)}) = \frac{1}{\alpha + 1/u^\beta(0, 0)}.$$

Thus we get (3.192).

The latter statement of the lemma follows from Lemma 3.6.13. □

We end this section with an extension of (3.68) (which is given by (3.201) with $T = \infty$).

Lemma 3.8.5 *Let X be a strongly symmetric Borel right process with state space S and continuous 0-potential densities $u(x,y)$. Let T be a terminal time for X. Then*

$$u_T(x,y) \leq u_T(y,y) \quad \forall\, x, y \in S. \tag{3.201}$$

Proof Using the additivity and support properties of L_t^y and the fact that T is a terminal time, we have

$$L_T^y = L_{T_y + T \circ \theta_{T_y}}^y = L_T^y \circ \theta_{T_y} \quad \text{on } T_y < T. \tag{3.202}$$

Therefore, by the strong Markov property,

$$
\begin{aligned}
u_T(x,y) = E^x(L_T^y) &= E^x(L_T^y \circ \theta_{T_y}, T_y < T) \tag{3.203} \\
&= E^x(T_y < T) E^y(L_T^y) \leq E^y(L_T^y) = u_T(y,y).
\end{aligned}
$$

□

3.9 The *h*-transform

Let $X = (\Omega, \mathcal{F}, \mathcal{F}_t, X_t, \theta_t, P^x)$ be a strongly symmetric Borel right process with transition semigroup $\{P_t \,;\, t \geq 0\}$ and continuous 0-potential density $u(x,y)$. Let 0 denote some fixed element in S and assume that $h(x) := P^x(T_0 < \infty) > 0$ for all $x \in S$. By (3.109),

$$h(x) = \frac{u(x,0)}{u(0,0)}. \tag{3.204}$$

In this section we construct another Borel right process, called the *h*-transform of X, that is, essentially, X conditioned to hit 0 and die at its last exit from 0.

Let $\mathcal{L}$ be as defined in (3.135). By Remark 3.2.8 we assume that Ω is the space of right continuous S_Δ-valued functions $\{\omega(t), t \in [0, \infty)\}$ such that $\omega(t) = \Delta$ for all $t \geq \zeta(\omega)$. Furthermore, by Lemma 3.6.11 and Remark 3.6.14 and using the same procedure as in Remark 3.2.8, we assume that $\mathcal{L}(\omega) < \infty$ for all $\omega \in \Omega$ and that $\mathcal{L}(\omega)$ is a left limit point of zeros of $\omega(t)$ on $\{\mathcal{L}(\omega) > 0\}$. (Recall that we have $X_t(\omega) = \omega(t)$.)

We define the killing operator $k_{\mathcal{L}} : \Omega \mapsto \Omega$ as

$$k_{\mathcal{L}}(\omega)(s) = \begin{cases} \omega(s) & s < \mathcal{L}(\omega) \\ \Delta & s \geq \mathcal{L}(\omega). \end{cases} \tag{3.205}$$

Let $\Omega_h = k_{\mathcal{L}}(\Omega)$. We then have

$$\Omega_h = \{\omega \in \Omega \,|\, \zeta(\omega) = \mathcal{L}(\omega)\}. \tag{3.206}$$

Note that $\{\mathcal{L} > 0\} = \{T_0 < \infty\}$ and, more generally, $\{\mathcal{L} > s\} = \theta_s\{T_0 < \infty\}$. Therefore $\mathcal{L}$ is $\mathcal{F}$ measurable. Furthermore, for any $f \in \mathcal{B}(S_\Delta)$,

$$f(X_t) \circ k_{\mathcal{L}} := f(k_{\mathcal{L}}(X(t)) = f(X_t)1_{\{t<\mathcal{L}\}} + f(\Delta)1_{\{t\geq\mathcal{L}\}}, \tag{3.207}$$

which implies that $f(X_t) \circ k_{\mathcal{L}}$ is $\mathcal{F}$ measurable. Thus, for any $a < b$, we have $k_{\mathcal{L}}^{-1}(\{a \leq f(X_t) \leq b\}) = (f(X_t) \circ k_{\mathcal{L}})^{-1}[a,b] \in \mathcal{F}$. Since sets of the form $\{a \leq f(X_t) \leq b\}$ generate $\mathcal{F}^0$, it follows that $k_{\mathcal{L}}^{-1} : \mathcal{F}^0 \mapsto \mathcal{F}$.

Since $k_{\mathcal{L}}^{-1} : \mathcal{F}^0 \mapsto \mathcal{F}$, we can define the probability measures $\{P^{x/h}; x \in S\}$ on $(\Omega_h, \mathcal{F}^0)$ by setting

$$P^{x/h}(A) = \frac{1}{h(x)} P^x(k_{\mathcal{L}}^{-1}(A) \cap \{\mathcal{L} > 0\}) \qquad A \in \mathcal{F}^0. \tag{3.208}$$

We then have

$$E^{x/h}(F) = \frac{1}{h(x)} E^x \left(F \circ k_{\mathcal{L}} \, 1_{\{\mathcal{L}>0\}} \right) \tag{3.209}$$

for any positive $\mathcal{F}^0$ measurable function F. This follows from (3.208) by first considering F of the form $F = 1_A$ with $A \in \mathcal{F}^0$.

Consider $E^{x/h}(F)$ for $F = f_1(X_{t_1}) \cdots f_n(X_{t_n})$ with $t_1 < \cdots < t_n$, where $f_i \in \mathcal{B}(S)$, $i = 1, \ldots, n$, and recall our convention that functions on S are extended to S_Δ by setting $f(\Delta) = 0$. Thus

$$F \circ k_{\mathcal{L}} = F 1_{\{\mathcal{L}>t_n\}} = F \, 1_{\{\mathcal{L}>0\}} \circ \theta_{t_n}. \tag{3.210}$$

It now follows from (3.209), (3.210), and the Markov property for X that

$$\begin{aligned} E^{x/h}(F) &= \frac{1}{h(x)} E^x \left(F 1_{\{\mathcal{L}>t_n\}} \right) \tag{3.211} \\[1ex] &= \frac{1}{h(x)} E^x \left(F \, P^{X_{t_n}}(\mathcal{L}>0) \right) \\[1ex] &= \frac{1}{h(x)} E^x (F \, h(X_{t_n})). \end{aligned}$$

Using the abbreviation $F_{n-1} = f_1(X_{t_1}) \cdots f_{n-1}(X_{t_{n-1}})$ in (3.211), we have

$$E^{x/h}(F) \quad = \quad \frac{1}{h(x)} E^x(Fh(X_{t_n})) \tag{3.212}$$

$$= \quad \frac{1}{h(x)} E^x(F_{n-1} f_n(X_{t_n}) h(X_{t_n}))$$

$$= \quad \frac{1}{h(x)} E^x(F_{n-1} E^{X_{t_{n-1}}} \{ f_n(X_{t_n-t_{n-1}}) h(X_{t_n-t_{n-1}}) \})$$

$$= \quad \frac{1}{h(x)} E^x(F_{n-1} h(X_{t_{n-1}}) E^{X_{t_{n-1}}/h} \{ f_n(X_{t_n-t_{n-1}}) \})$$

$$= \quad E^{x/h}(F_{n-1} E^{X_{t_{n-1}}/h} \{ f_n(X_{t_n-t_{n-1}}) \}).$$

Using (3.212) for the first equality and (3.211) for the second, we see that, for any $t_{n-1} < t_n$ and $f_n \in \mathcal{B}(S)$,

$$E^{x/h}(f_n(X_{t_n}) \,|\, \mathcal{F}^0_{t_{n-1}}) \tag{3.213}$$

$$= E^{X_{t_{n-1}}/h} \left(f_n(X_{t_n-t_{n-1}}) \right)$$

$$= \frac{1}{h(X_{t_{n-1}})} E^{X_{t_{n-1}}} \left(f_n(X_{t_n-t_{n-1}}) h(X_{t_n-t_{n-1}}) \right)$$

$$= \frac{1}{h(X_{t_{n-1}})} P_{t_n-t_{n-1}} fh(X_{t_{n-1}}).$$

Let

$$Q_t f(x) := \frac{1}{h(x)} P_t fh(x). \tag{3.214}$$

Using this, (3.213), and Remark 3.1.1, it is easy to verify the following lemma.

Lemma 3.9.1 $\overline{X} = (\Omega_h, \mathcal{F}^0, \mathcal{F}^0_t, X_t, \theta_t, P^{x/h})$ *is a right continuous simple Markov process with transition semigroup* $\{Q_t \,;\, t \geq 0\}$.

Let $\overline{U}^\alpha$ denote the α-potential of of $\overline{X}$. Then, for any $f \in \mathcal{B}_b(S)$,

$$\overline{U}^\alpha f(x) \quad = \quad \int_0^\infty e^{-\alpha t} Q_t f(x) \, dt \tag{3.215}$$

$$= \quad \frac{1}{h(x)} \int_0^\infty e^{-\alpha t} P_t fh(x) \, dt$$

$$= \quad \frac{1}{h(x)} U^\alpha fh(x)$$

$$= \quad \frac{1}{h(x)} \int u^\alpha(x, y) h(y) f(y) \, dm(y).$$

Thus

$$\overline{X} \text{ has } \alpha\text{-potential density } \frac{1}{h(x)} u^\alpha(x,y) h(y). \qquad (3.216)$$

Let $\mathcal{F}^h, \mathcal{F}_t^h$ denote the standard augmentation of $\mathcal{F}^0, \mathcal{F}_t^0$ under $\{P^{x/h} ; x \in S\}$. Statement (3.216) together with the continuity and strict positivity of h show that $\overline{X}$ has α-potential densities that are continuous on $S \times S$ for each α. It then follows from Lemmas 3.9.1 and 3.4.2 that $\overline{X}$ is a Borel right process. As usual, let m denote the reference measure of the original Borel right process X. Considering (3.216), it is clear that $\overline{X}$ is not strongly symmetric with respect to m. Nevertheless, with a different reference measure, $\overline{X}$ is a strongly symmetric Borel right process. We refer to this process as $\widetilde{X}$.

Theorem 3.9.2 $\widetilde{X} = (\Omega_h, \mathcal{F}^h, \mathcal{F}_t^h, X_t, \theta_t, P^{x/h})$ *is a Borel right process that is strongly symmetric with respect to the measure* $\widetilde{m}(dy) := h^2(y)\, dm(y)$ *and has α-potential densities*

$$\widetilde{u}^\alpha(x,y) := \frac{u^\alpha(x,y)}{h(x)h(y)} \qquad (3.217)$$

with respect to the $\widetilde{m}(dy)$.

$\widetilde{X}$ *is called the h-transform of X.*

Proof We just showed that $\widetilde{X}$ is a Borel right process. It follows from (3.216) that $\widetilde{u}^\alpha(x,y)$ is the potential density of $\widetilde{X}$ with respect to $\widetilde{m}$. It is easy to see that $\widetilde{U}^\alpha$, the potential with density $\widetilde{u}^\alpha$, is symmetric with respect to $\widetilde{m}$. $\qquad \square$

Remark 3.9.3 $\overline{X}$ and $\widetilde{X}$ are the same Borel right process. They simply have different α-potentials with respect to different reference measures. It is also customary to refer to $\overline{X}$ as the h-transform process.

Since $\widetilde{X}$ is a strongly symmetric Borel right process, by Theorem 3.6.3 we know that it has a local time $\widetilde{L} = \{\widetilde{L}_t^y, (y,t) \in S \times R_+\}$ that satisfies

$$E^x \left(\int_0^\infty e^{-\alpha t}\, d\widetilde{L}_t^y \right) = \widetilde{u}^\alpha(x,y). \qquad (3.218)$$

By Remark 3.6.9,

$$\overline{L}_t^y := h^2(y)\widetilde{L}_t^y \qquad \forall\, t \in R_+ \qquad (3.219)$$

is also a local time for $\widetilde{X}$ or, equivalently, $\overline{X}$. Furthermore,

$$E^x \left(\int_0^\infty e^{-\alpha t}\, d\overline{L}_t^y \right) = h^2(y)\widetilde{u}^\alpha(x,y) = \frac{1}{h(x)} u^\alpha(x,y) h(y) \qquad (3.220)$$

is the α-potential density of $\overline{X}$.

Remark 3.9.4 The role of the killing operator in the definition of the probability measures $P^{x/h}$ on $(\Omega_h, \mathcal{F}^0)$ in (3.208) justifies our interpretation of $\widetilde{X}$ as the paths of X conditioned to hit 0 and die on their last exit from 0. For this reason $P^{x/h}$ is often written as $P^{x,0}$. We use both notations in this book.

Let $L = \{L_t^x \,;\, (x,t) \in S \times R_+\}$ be the local times of a strongly symmetric Borel right process X and $\widetilde{L} = \{\widetilde{L}_t^x \,;\, (x,t) \in S \times R_+\}$ the local times of the corresponding h-transform process $\widetilde{X}$. We consider the versions of L and $\widetilde{L}$ satisfying

$$E^y \left(\int_0^\infty e^{-\alpha t} dL_t^x \right) = u^\alpha(y,x) \qquad \forall \, \alpha > 0 \qquad (3.221)$$

and

$$E^y \left(\int_0^\infty e^{-\alpha t} d\widetilde{L}_t^x \right) = \widetilde{u}^\alpha(y,x) \qquad \forall \, \alpha > 0. \qquad (3.222)$$

These exist by Theorem 3.6.3.

In the next lemma we compare $\widetilde{L}$ under $P^{0/h}$ with L under P^0.

Lemma 3.9.5 *For any countable subset $D \subseteq S$,*

$$\{h^2(x)\widetilde{L}_t^x \,;\, (x,t) \in D \times R_+ \,,\, P^{0/h}\} \overset{law}{=} \{L_t^x \circ k_{\mathcal{L}} \,;\, (x,t) \in D \times R_+ \,,\, P^0\}. \qquad (3.223)$$

Proof Let $f_{\epsilon,x}(y)$ be an approximate δ-function at x with respect to the reference measure m. By Theorem 3.6.3 there exists a sequence $\{\epsilon_n\}$ tending to zero, such that, for

$$\widetilde{\Omega}_x := \{\omega \in \Omega \,|\, \int_0^t f_{\epsilon_n,x}(X_s) \, ds \text{ converges locally uniformly in } t\}, \qquad (3.224)$$

we have $P^y(\widetilde{\Omega}_x) = 1$ for all $y \in S$ and $\{L_t^x \,;\, t \in R_+\}$ defined by

$$L_t^x = \begin{cases} \lim_{\epsilon_n \to 0} \int_0^t f_{\epsilon_n,x}(X_s) \, ds & \omega \in \widetilde{\Omega}_x \\ 0 & \omega \notin \widetilde{\Omega}_x \end{cases} \qquad (3.225)$$

is a version of the local time of X at x satisfying (3.221). Clearly, $\{L_t^x \,;\, t \in R_+\}$ is $\mathcal{F}^0$ measurable. Using (3.209), we see that for any $x_1, \ldots, x_n \in S$, $t_1, \ldots, t_n \in R_+^1$, and $\lambda_1, \ldots, \lambda_n \in R^1$,

$$E^{0/h} \left(e^{i \sum_{j=1}^n \lambda_j L_{t_j}^{x_j}} \right) = E^0 \left(e^{i \sum_{j=1}^n \lambda_j L_{t_j}^{x_j} \circ k_{\mathcal{L}}} \right). \qquad (3.226)$$

Using the continuity of $h(y)$ and the fact that $h(y) > 0$, we see that $h^{-2}(x)f_{\epsilon,x}(y)$ is an approximate δ-function at x with respect to the reference measure $\widetilde{m}$. Exactly as above, by Theorem 3.6.3, there exists a sequence $\{\epsilon_n\}$ tending to zero, such that, for

$$\widetilde{\Omega}_{h,x} := \tag{3.227}$$

$$\left\{ \omega \in \Omega_h \mid \int_0^t h^{-2}(x)f_{\epsilon_n,x}(X_s)\,ds \text{ converges locally uniformly in } t \right\},$$

we have $P^{y/h}(\widetilde{\Omega}_{h,x}) = 1$ for all $y \in S$ and $\{\widetilde{L}_t^x \, ; \, t \in R_+\}$ defined by

$$\widetilde{L}_t^x = \begin{cases} \lim_{\epsilon_n \to 0} \int_0^t h^{-2}(x)f_{\epsilon_n,x}(X_s)\,ds & \omega \in \widetilde{\Omega}_{h,x} \\ 0 & \omega \notin \widetilde{\Omega}_{h,x} \end{cases} \tag{3.228}$$

is a version of the local time of $\widetilde{X}$ at x satisfying (3.222). By Remark 3.6.4 (2) we can choose the same sequence $\{\epsilon_n\}$ in (3.224) and (3.227). It follows that

$$\widetilde{\Omega}_{h,x} = \Omega_h \cap \widetilde{\Omega}_x \tag{3.229}$$

and that

$$L_t^x = h^2(x)\widetilde{L}_t^x \quad \text{on} \quad \widetilde{\Omega}_{h,x}. \tag{3.230}$$

Since $P^{0/h}(\widetilde{\Omega}_{h,x}) = 1$, the lemma follows from (3.230) and (3.226). $\square$

Recall that $\tau(s) = \inf\{t > 0 \mid L_t^0 > s\}$ and $\tau^-(s) = \inf\{t > 0 \mid L_t^0 \geq s\}$. Similarly we set $\widetilde{\tau}(s) = \inf\{t > 0 \mid \widetilde{L}_t^0 > s\}$ (in all of these expressions, $\inf\{\emptyset\} = \infty$).

Lemma 3.9.6 *For any countable subset $D \subseteq S$ and $t > 0$,*

$$\left\{ h^2(x)\widetilde{L}_{\widetilde{\tau}(t)}^x \, ; \, x \in D, \, P^{0/h} \right\} \overset{law}{=} \left\{ L_{\tau^-(t \wedge L_\infty^0)}^x \, ; \, x \in D, \, P^0 \right\}. \tag{3.231}$$

Proof We continue the notation of the previous lemma. Note that

$$\int_0^{t \wedge \mathcal{L}} f_{\epsilon_n,0}(X_s)\,ds = \int_0^t 1_{[0,\mathcal{L}]}(s)f_{\epsilon_n,0}(X_s)\,ds \tag{3.232}$$

$$= \int_0^t f_{\epsilon_n,0}(X_s \circ k_\mathcal{L})\,ds,$$

where, according to our convention, $f_{\epsilon_n,0}(\Delta) = 0$ for all n, so that the last two integrands agree except (possibly) for $s = \mathcal{L}$

Using (3.232) and (3.225) we see that $L_{t \wedge \mathcal{L}}^0 = L_t^0 \circ k_\mathcal{L}$, for all t P^0 almost surely. Also, by Remark 3.6.2 we have $L_t^0 = L_{t \wedge \mathcal{L}}^0$ for all t, P^0 almost surely. Therefore,

$$L_t^0 = L_t^0 \circ k_\mathcal{L}, \qquad \forall t \geq 0 \qquad P^0 \quad \text{a.s.} \tag{3.233}$$

It follows from (3.223) and the fact that $h(0) = 1$ that

$$\left\{ h^2(x)\widetilde{L}_t^x \,;\, (x,t) \in D \times R_+ \,,\, \widetilde{L}_s^0 \,;\, s \in R_+ \,,\, P^{0/h} \right\} \tag{3.234}$$

$$\overset{law}{=} \left\{ L_t^x \circ k_{\mathcal{L}} \,;\, (x,t) \in D \times R_+ \,,\, L_s^0 \circ k_{\mathcal{L}} \,;\, s \in R_+ \,,\, P^0 \right\}.$$

Therefore, using (3.233) we see that for any countable subset $D \subseteq S$

$$\left\{ h^2(x)\widetilde{L}_t^x \,;\, (x,t) \in D \times R_+ \,,\, \widetilde{L}_s^0 \,;\, s \in R_+ \,,\, P^{0/h} \right\} \tag{3.235}$$

$$\overset{law}{=} \left\{ L_t^x \circ k_{\mathcal{L}} \,;\, (x,t) \in D \times R_+ \,,\, L_s^0 \,;\, s \in R_+ \,,\, P^0 \right\}.$$

This, together with $\{\tau(s) < r\} = \{L_r^0 > s\}, \{\widetilde{\tau}(s) < r\} = \{\widetilde{L}_r^0 > s\}$ and the right continuity of $\tau(s), \widetilde{\tau}(s)$ shows that

$$\left\{ h^2(x)\widetilde{L}_t^x \,;\, (x,t) \in D \times R_+ \,,\, \widetilde{\tau}(s) \,;\, s \in R_+ \,,\, P^{0/h} \right\} \tag{3.236}$$

$$\overset{law}{=} \left\{ L_t^x \circ k_{\mathcal{L}} \,;\, (x,t) \in D \times R_+ \,,\, \tau(s) \,;\, s \in R_+ \,,\, P^0 \right\}.$$

Consequently, for any $t > 0$,

$$\left\{ h^2(x)\widetilde{L}_{\widetilde{\tau}(t)}^x \,;\, x \in D \,,\, P^{0/h} \right\} \overset{law}{=} \left\{ L_{\tau(t)}^x \circ k_{\mathcal{L}} \,;\, x \in D \,,\, P^0 \right\}. \tag{3.237}$$

It only remains to show that, for any $x \in S$,

$$L_{\tau(t)}^x \circ k_{\mathcal{L}} = L_{\tau^-(t \wedge L_\infty^0)}^x, \qquad P^0 \quad \text{a.s.} \tag{3.238}$$

As in the beginning of the proof, we have $L_{\tau(t) \wedge \mathcal{L}}^x = L_{\tau(t)}^x \circ k_{\mathcal{L}}$, P^0 almost surely, so we need only show that

$$\tau(t) \wedge \mathcal{L} = \tau^-(t \wedge L_\infty^0) \qquad P^0 \quad \text{a.s.} \tag{3.239}$$

This follows from Lemmas 3.6.18 and 3.6.19 and the fact that $\tau^-(t)$ is monotonically increasing. $\qquad\square$

Remark 3.9.7 By (3.232) with $t = \infty$, $L_\infty^x \circ k_{\mathcal{L}} = L_{\mathcal{L}}^x$ and by (3.151), $L_{\mathcal{L}}^x = L_{\tau^-(L_\infty^0)}^x$. Thus we have

$$L_\infty^x \circ k_{\mathcal{L}} = L_{\tau^-(L_\infty^0)}^x \qquad P^0 \quad \text{a.s.} \tag{3.240}$$

3.10 Moments and moment generating functions of local times

In Section 3.8 we define terminal times and inverse times, that is, stopping times of the form $\tau_A(\lambda)$, where $A = \{A_t; t \geq 0\}$ is a CAF and λ is an exponential random variable independent of A. In Theorem 2.5.3 we give Kac's Moment Formula for Brownian motion, which holds for terminal times and inverse times. Clearly, the proof of Theorem 2.5.3 does not involve any specific properties of Brownian motion. As is easily seen, that proof suffices to prove the next, much more general result.

Theorem 3.10.1 (Kac's Moment Formula: III) *Let $X = (\Omega, \mathcal{G}, \mathcal{G}_t, X_t, \theta_t, P^x)$ be a strongly symmetric Borel right process and assume that its α-potential density, $u^\alpha(x, y)$, is finite for all $x, y \in S$. Let $\{L_t^y, (t, y) \in R_+ \times S\}$ be a local time of X with*

$$E^x \left(\int_0^\infty e^{-\alpha t} \, dL_t^y \right) = u^\alpha(x, y). \tag{3.241}$$

Then, if T is a terminal time for X,

$$E^x \left(\prod_{i=1}^n L_T^{y_i} \right) = \sum_\pi u_T(x, y_{\pi_1}) \cdots u_T(y_{\pi_{n-1}}, y_{\pi_n}), \tag{3.242}$$

where the sum runs over all permutations π of $\{1, \ldots, n\}$. In particular,

$$E^x \left((L_T^{y_i})^n \right) = n! u_T(x, y) \left(u_T(y, y) \right)^{n-1}. \tag{3.243}$$

Let $A = \{A_t, t \geq 0\}$ be a CAF on $(\Omega, \mathcal{G}, \mathcal{G}_t)$ and λ an exponential random variable independent of $(\Omega, \mathcal{G}, \mathcal{G}_t, P^x)$. These equations are also valid for inverse times, that is, with T replaced by $\tau_A(\lambda)$ and E^x replaced by E^x_λ.

Using Theorem 3.10.1, the proof of Lemma 2.6.2 can be immediately extended to give the following important relationship for the moment generating function of local times evaluated at terminal times and inverse times.

Lemma 3.10.2 *Let $X = (\Omega, \mathcal{G}, \mathcal{G}_t, X_t, \theta_t, P^x)$ be a strongly symmetric Borel right process with continuous α-potential density, and let T be a terminal time with potential density $u_T(x, y)$. Let Σ be the matrix with elements $\Sigma_{i,j} = u_T(x_i, x_j)$, $i, j = 1, \ldots, n$. Let Λ be the matrix with elements $\{\Lambda\}_{i,j} = \lambda_i \delta_{i,j}$. For all $\lambda_1, \ldots, \lambda_n$ sufficiently small and $1 \leq l \leq n$,*

$$E^{x_l} \exp \left(\sum_{i=1}^n \lambda_i L_T^{x_i} \right) = \frac{\det(I - \widehat{\Sigma}\Lambda)}{\det(I - \Sigma\Lambda)}, \tag{3.244}$$

where

$$\widehat{\Sigma}_{j,k} = (\Sigma_{j,k} - \Sigma_{l,k}) \qquad j, k = 1, \ldots, n. \tag{3.245}$$

These equations also hold when T is replaced by $\tau_A(\lambda)$, for any CAF, $A = \{A_t, t \in R_+\}$, and E^x is replaced by E^x_λ.

Remark 3.10.3 We note for future reference that the proof of Lemma 3.10.2 gives the following analog of (2.173):

$$E^{x_l} \exp\left(\sum_{i=1}^{n} \lambda_i L_T^{x_i}\right) = \sum_{j=1}^{n} \{(I - \Sigma\Lambda)^{-1}\}_{l,j} = \{(I - \Sigma\Lambda)^{-1}\mathbf{1}^t\}_l, \quad (3.246)$$

where $\mathbf{1}^t$ denotes the transpose of the n-dimensional vector $(1, \ldots, 1)$.

The next lemma gives analogous results for the h-transform process described in Section 3.9.

Lemma 3.10.4 *Let $X = (\Omega, \mathcal{G}, \mathcal{G}_t, X_t, \theta_t, P^x)$ be a strongly symmetric Borel right process with continuous 0-potential density $u(x, y)$. Let 0 denote a fixed element of S. Assume that $h(x) > 0$ for all $x \in S$ (see (3.204)). Let $\widetilde{X} = (\Omega_h, \mathcal{F}^h, \mathcal{F}_t^h, X_t, \theta_t, P^{x,0})$ denote the h-transform of X, with 0-potential density $\widetilde{u}(x, y)$ (see (3.217)). Let $\overline{L} = \{\overline{L}_t^y ; (y, t) \in S \times R_+\}$ denote the local time for $\widetilde{X}$ normalized so that*

$$E^{x,0}\left(\overline{L}_\infty^y\right) = h^2(y)\widetilde{u}(x, y) = \frac{u(x, y)h(y)}{h(x)}. \quad (3.247)$$

Then

$$E^{x,0}\left(\prod_{i=1}^{n} \overline{L}_\infty^{y_i}\right) = \frac{1}{h(x)} \sum_{\pi} u(x, y_{\pi_1}) \cdots u(y_{\pi_{n-1}}, y_{\pi_n})h(y_{\pi_n}), \quad (3.248)$$

where the sum runs over all permutations π of $\{1, \ldots, n\}$.

Let Σ be the matrix with elements $\Sigma_{i,j} = u(x_i, x_j)$, $i, j = 1, \ldots, n$. Let Λ be the matrix with elements $\{\Lambda\}_{i,j} = \lambda_i \delta_{i,j}$. For all $\lambda_1, \ldots, \lambda_n$ sufficiently small and $1 \leq l \leq n$,

$$E^{x_l,0} \exp\left(\sum_{i=1}^{n} \lambda_i \overline{L}_\infty^{x_i}\right) = \frac{\det(I - \widehat{\Sigma}\Lambda)}{\det(I - \Sigma\Lambda)}, \quad (3.249)$$

where

$$\widehat{\Sigma}_{j,k} = \left(\Sigma_{j,k} - \frac{h(x_j)\Sigma_{l,k}}{h(x_l)}\right) \quad j, k = 1, \ldots, n. \quad (3.250)$$

Proof Let m be the reference measure for the original process X. We show in Theorem 3.9.2 that the h-transform process $\widetilde{X}$ is not strongly symmetric with respect to m but that it is a strongly symmetric Borel right process with respect to the reference measure $\widetilde{m}(dy) = h^2(y)\, dm(y)$ and has 0-potential density

$$\widetilde{u}(x, y) = \frac{u(x, y)}{h(x)h(y)}. \quad (3.251)$$

Let $\widetilde{L} = \{\widetilde{L}_t^x \, ; \, (x,t) \in S \times R_+\}$ be the local time for the h-transform process $\widetilde{X}$ normalized so that $E^{x,0}(\widetilde{L}_t^y) = \widetilde{u}(x,y)$. Then, by Theorem 3.10.1,

$$E^{x,0}\left(\prod_{i=1}^n \widetilde{L}_\infty^{y_i}\right) = \sum_\pi \widetilde{u}(x, y_{\pi_1}) \cdots \widetilde{u}(y_{\pi_{n-1}}, y_{\pi_n}), \qquad (3.252)$$

where the sum runs over all permutations π of $\{1, \ldots, n\}$. (Obviously, $T = \infty$ is a terminal time.) By Remark 3.9.3,

$$h^{-2}(x_i)\overline{L}_t^{x_i} = \widetilde{L}_t^{x_i}. \qquad (3.253)$$

Substituting this in (3.252) gives (3.248).

In the same way as the first equation in (2.173) follows from (2.148) we see that

$$E^{x_l,0}\exp\left(\sum_{i=1}^n \lambda_i \overline{L}_\infty^{x_i}\right) = \frac{1}{h(x_l)}\sum_{j=1}^n \{(I - \Sigma\Lambda)^{-1}\}_{l,j}\, h(x_j). \qquad (3.254)$$

Let H be the matrix with elements $H_{j,k} = h(x_j)\delta_{j,k}$. Let Y be the $1 \times n$ vector with elements $E^{x_l,0}\exp\left(\sum_{i=1}^n \lambda_i \overline{L}_\infty^{x_i}\right)$ and h the $1 \times n$ vector with elements $h(x_i)$, $i = 1, \ldots, n$. It follows from (3.254) that

$$(I - \Sigma\Lambda)HY = h. \qquad (3.255)$$

Consequently, by Cramer's Theorem,

$$h(x_l)E^{x_l,0}\exp\left(\sum_{i=1}^n \lambda_i L_\infty^{x_i}\right) = (HY)_l = \frac{\det((I - \Sigma\Lambda)^{(l,h)})}{\det((I - \Sigma\Lambda))}, \qquad (3.256)$$

where $(I - \Sigma\Lambda)^{(l,h)}$ is the matrix obtained by replacing the l-th column of $(I - \Sigma\Lambda)$ by h. Thus

$$E^{x_l,0}\exp\left(\sum_{i=1}^n \lambda_i L_\infty^{x_i}\right) = \frac{\det((I - \Sigma\Lambda)^{(l,\widehat{h})})}{\det((I - \Sigma\Lambda))}, \qquad (3.257)$$

where $\widehat{h} = \dfrac{1}{h(x_l)}h$. Note that $\widehat{h}_l = 1$.

Let B be the matrix obtained by subtracting the $h(x_j)/h(x_l)$ times the l-th row of $(I - \Sigma\Lambda)^{(l,\widehat{h})}$ from the j-th row for each $j \neq l$. We see that

$$\begin{aligned} B_{j,l} &= \delta_{l,l} & (3.258) \\ B_{j,k} &= (I - \widehat{\Sigma}\Lambda)_{j,k} & j,k \neq l. \end{aligned}$$

Thus

$$\det((I - \Sigma\Lambda)^{(l, \widehat{h})}) = \det B = M_{l,l}, \qquad (3.259)$$

where $M_{l,l}$ is the (l, l)-th minor of $(I - \widehat{\Sigma}\Lambda)$. Since by (3.250), $\widehat{\Sigma}_{l,k} = 0$ for all k, we also have $\det(I - \widehat{\Sigma}\Lambda) = M_{l,l}$. Thus we get (3.249) from (3.257). □

Example 3.10.5 We can write (3.243) as

$$E^x \left((L_T^{y_i})^n\right) = \frac{u_T(x, y)}{u_T(y, y)} n! \left(u_T(y, y)\right)^n, \qquad (3.260)$$

from which it is easy to see that, for $\lambda < 1/u_T(y, y)$,

$$E^x \exp\left(\lambda L_T^y\right) = \left(1 - \frac{u_T(x, y)}{u_T(y, y)}\right) + \frac{u_T(x, y)}{u_T(y, y)} \frac{1}{1 - \lambda u_T(y, y)}. \qquad (3.261)$$

Thus, under P^x, we can write

$$L_T^y = \xi_1 \xi_2, \qquad (3.262)$$

where ξ_1 and ξ_2 are independent random variables, $\xi_1 = 1$ with probability $u_T(x, y)/u_T(y, y)$ and $\xi_1 = 0$ otherwise, and ξ_2 is an exponential random variable with mean $u_T(y, y)$. The same result holds for $L_{T_A(\lambda)}^y$ under $\widehat{P}^x$ (see (3.76)), if u_T is replaced by $u_{\tau(\lambda)}$. Similarly, by (3.252), the same result holds for $\widetilde{L}_\infty^x$ under $P^{x,0}$, if u_T is replaced by $\widetilde{u}$ in (3.251).

3.11 Notes and references

The standard references on right processes are Sharpe (1988), Getoor (1975), Dellacherie and Meyer (1987) and Dellacherie, Maisonneuve and Meyer (1992). Other classics on the theory of Markov processes are Rogers and Williams (2000b), Chung (1982), and Blumenthal and Getoor (1968). Borel right processes on locally compact Hausdorff spaces are the simplest right processes.

Our restriction to strongly symmetric processes is due to the fact that our main tools in the study of local times are isomorphism theorems that relate the local times of a Markov process to an associated Gaussian process. The association is that the covariance of the Gaussian process, which is necessarily symmetric, is the same as the potential density of the Markov process. Strongly symmetric Markov processes are treated in Fukushima, Oshima and Takeda (1994). Le Jan (1988) has an isomorphism theorem for nonsymmetric Markov processes that involves complex-valued measures. We have not been able to use his results to study local times of nonsymmetric processes.

Beginning in Section 3.4 we assume that the potential densities are continuous. This is a very strong condition, which allows us to simplify many arguments concerning Markov processes. The justification for this assumption is that we are interested in processes with jointly continuous local times, and, as explained in Section 4.6, if a strongly symmetric Borel right process with finite potential densities has a jointly continuous local time, then the potential densities are continuous.

We mention in Section 4.4 that a strongly symmetric Borel right process with continuous potential densities is in fact a Hunt process. We could have considered only Hunt processes from the onset, but instead we chose to work with Borel right processes, both to clarify the generality of our results and because we never explicitly use quasi left continuity.

There is an extensive literature on the continuity of local times of Markov processes, beginning with the celebrated result of Trotter (1958) on the joint continuity of the local time of Brownian motion and the results of McKean (1962) and Ray (1963), which give the exact uniform modulus of continuity for the local times of Brownian motion. Boylan (1964) found a sufficient condition for the joint continuity of the local time for a wide class of Markov processes that inspired considerable efforts to obtain necessary and sufficient conditions. A variation of Boylan's result is given in Blumenthal and Getoor (1968, Section 3, Chapter V) following an approach of Meyer (1966). Further improvements are given in Getoor and Kesten (1972) and Millar and Tran (1974). After a long hiatus, M. Barlow and J. Hawkes obtained necessary and sufficient conditions for the joint continuity of local times of Lévy processes. Their work: Barlow (1985), Hawkes (1985), Barlow and Hawkes (1985), and Barlow (1988) inspired our own, as we say in Chapter 1.

Our treatment of local times in Section 3.6 is simplified by the assumption that the Markov processes are strongly symmetric Borel right processes with continuous potential densities. Theorem 3.7.3 is the core of our argument that gives a sufficient condition for the joint continuity local times, a condition that we also show is necessary in Chapter 9. Gaussian processes are not mentioned in this theorem. They come in later in the isomorphism theorems in Chapter 8. Theorem 3.7.3 combines several theorems in our paper Marcus and Rosen (1992d) , but with considerable simplifications due the simpler, more efficient isomorphism theorems that are now available to us.

Kac's formula is discussed in great generality in Fitzsimmons and Pitman (1999).

4

Constructing Markov processes

So far in this book, we have simply assumed that we are given a strongly symmetric Borel right process with continuous α-potential densities $u^\alpha(x, y)$, $\alpha > 0$ and also $u(x, y)$ when the 0-potential exists. In general, constructing such processes is not trivial. However, given additional conditions on transition semigroups or potentials, we can construct special classes of Borel right processes. In this chapter we show how to construct Feller and Lévy processes. (For references to the general question of establishing the existence of Borel right processes, see Section 3.11.) In Sections 4.7–4.8, we show how to construct certain strongly symmetric right continuous processes with continuous α-potential densities that generalize the notion of Borel right processes and are used in Chapter 13.

In Sections 4.4–4.5 we present certain material, on quasi left continuity and killing at a terminal time, which is of interest in its own right and is needed for Sections 4.7–4.10. In Section 4.6 we tie up a loose end by showing that if a strongly symmetric Borel right process has a jointly continuous local time, then the potential densities $\{u^\alpha(x, y), (x, y) \in S \times S\}$ are continuous.

In Section 4.10 we present an extension theorem of general interest which is needed for Chapter 13.

4.1 Feller processes

A Feller process is a Borel right process with transition semigroup $\{P_t; t \geq 0\}$ such that, for each $t \geq 0$, $P_t : C_0(S) \mapsto C_0(S)$. Such a semigroup is called a Feller semigroup. We consider $C_0(S)$ as a Banach space in the uniform or sup norm, that is, $\|f\| = \sup_{x \in S} |f(x)|$. (Sometimes, to avoid confusion, we use $\| \cdot \|_\infty$ to denote the sup norm.)

A family $\{P_t; t \geq 0\}$ of bounded linear operators on $C_0(S)$ (that is, $P_t : C_0(S) \to C_0(S)$) is called a strongly continuous contraction semigroup if

(1) $P_t P_s = P_{t+s}$, $\forall s, t \geq 0$
(2) $\|P_t f\| \leq \|f\|$, $\forall f \in C_0(S)$ and $\forall t \geq 0$
(3) $\lim_{t \to 0} \|P_t f - f\| = 0$, $\forall f \in C_0(S)$.

If, in addition, for each $t \geq 0$, P_t is a positive operator, that is, $P_t f \geq 0$ when $f \geq 0$, then each P_t can be associated with a sub-Markov kernel on S, which we also denote by P_t. $\{P_t; t \geq 0\}$ is clearly a Borel semigroup.

Theorem 4.1.1 *Let S be a locally compact space with a countable base, and let $\{P_t; t \geq 0\}$ be a strongly continuous contraction semigroup of positive linear operators on $C_0(S)$. Then we can construct a Feller process X with transition semigroup $\{P_t; t \geq 0\}$.*

Proof As we did in Section 3.1, we extend $\{P_t; t \geq 0\}$ to be a Borel semigroup of Markov kernels on the compact set S_Δ. Since any function $f \in C_b(S_\Delta)$ can be written as the sum of a constant plus a function in $C_b(S_\Delta)$ that is zero at Δ, it is easy to check that the extended $\{P_t; t \geq 0\}$ is a strongly continuous contraction semigroup on $C_b(S_\Delta)$ (recall that S_Δ is compact, so that $C_0(S_\Delta) = C_b(S_\Delta)$).

We now construct a Borel right process with transition semigroup $\{P_t; t \geq 0\}$. Let $x \in S_\Delta$. We first construct a probability $\widetilde{P}^x$ on $S_\Delta^{R_+}$, the space of S_Δ-valued functions $\{f(t), t \in [0, \infty)\}$ equipped with the Borel product σ-algebra $\mathcal{B}(S_\Delta^{R_+})$. Let X_t be the natural evaluation $X_t(f) = f(t)$. We define $\widetilde{P}^x$ on sets of the form $\{X_{t_1} \in A_1, \ldots, X_{t_n} \in A_n\}$ for all Borel measurable sets $A_1, \ldots, A_n$ in S_Δ and $0 = t_0 < t_1 < \cdots < t_n$ by setting

$$\widetilde{P}^x(X_{t_1} \in A_1, \ldots, X_{t_n} \in A_n) = \int \prod_{i=1}^n I_{A_1}(z_i) \prod_{i=1}^n P_{t_i - t_{i-1}}(z_{i-1}, dz_i),$$
(4.1)

where $z_0 = x$. Here I_{A_i} is the indicator function of A_i. It follows from the semigroup property of $\{P_t; t \geq 0\}$ that this construction is consistent. Therefore, by the Kolmogorov Construction Theorem, we can extend $\widetilde{P}^x$ to $S_\Delta^{R_+}$.

Let $\mathcal{F}_t^0 = \sigma(X_s; s \leq t)$. As in the second proof of (2.31), it follows from (4.1) that

$$\widetilde{E}\left(f(X_{t+s}) \mid \mathcal{F}_t^0\right) = P_s f(X_t)$$
(4.2)

for all $s, t \geq 0$ and $f \in \mathcal{B}_b(S_\Delta)$. Thus, $\{X_t; t \geq 0\}$ is a simple Markov process with respect to the filtration $\mathcal{F}_t^0 = \sigma(X_s; s \leq t)$.

As usual we set

$$U^\lambda = \int_0^\infty e^{-\lambda t} P_t \, dt \quad \forall \lambda > 0. \tag{4.3}$$

Using the fact that the $\{P_t; t \geq 0\}$ are contractions on $C_b(S_\Delta)$, we also have $U^\lambda : C_b(S_\Delta) \to C_b(S_\Delta)$ and $\|\lambda U^\lambda\| \leq 1$ for each $\lambda > 0$. Furthermore, the fact that $\lim_{t \to 0} \|P_t f - f\| = 0$ for $f \in C_b(S_\Delta)$ and

$$\lambda U^\lambda f(x) = \lambda \int_0^\infty e^{-\lambda t} P_t f(x) \, dt = \int_0^\infty e^{-t} P_{t/\lambda} f(x) \, dt \tag{4.4}$$

implies that

$$\lim_{\lambda \to \infty} \|\lambda U^\lambda f - f\| = 0 \quad \forall f \in C_b(S_\Delta). \tag{4.5}$$

We now show that $\{X_t; t \geq 0\}$ has a right continuous modification. We begin by noting that for any $\alpha > 0$ and $f \in C_b^+(S_\Delta)$, $e^{-\alpha t} U^\alpha f(X_t)$ is a $\mathcal{F}_t^0$ supermartingale for any $\widetilde{P}^x$. To see this note that by the simple Markov property, for any $s < t$,

$$\widetilde{E}^x \left(e^{-\alpha t} U^\alpha f(X_t) \mid \mathcal{F}_s^0 \right) = e^{-\alpha t} P_{t-s} U^\alpha f(X_s), \tag{4.6}$$

and it is easily checked using (4.3) that $P_{t-s} U^\alpha f \leq e^{\alpha(t-s)} U^\alpha f$.

Since S_Δ is compact we can find a sequence of functions $\mathcal{F} = \{f_n\}$ with each $f_n \in C_b^+(S_\Delta)$ that separate points of S_Δ, that is, for each pair x, y in S_Δ with $x \neq y$ there exists an $f_n \in \mathcal{F}$ such that $f_n(x) \neq f_n(y)$. Since $\lim_{\alpha \to \infty} \alpha U^\alpha f_n = f_n$, we see that the countable collection $\mathcal{G} := \{U^\alpha f_n; \alpha, n = 1, 2, \ldots\}$ also separates points. Let Q be a countable dense set in R_+. By the convergence theorem for supermartingales, Revuz and Yor (1991, Chapter II, Theorem 2.5), $g(X_t)$ has right-hand limits along Q for all $g \in \mathcal{G}$, almost surely.

We now show that almost surely $t \mapsto X_t$ has right-hand limits in S_Δ along Q. Since S_Δ is compact, it suffices to show that whenever $\{s_n\}$ and $\{s_n'\}$ are sequences in Q such that $s_n \downarrow t$, $s_n' \downarrow t$, $\lim_{n \to \infty} X_{s_n} = y$, and $\lim_{n \to \infty} X_{s_n'} = y'$, then $y = y'$. However, for any $g \in \mathcal{G}$, the continuity of g implies that $\lim_{n \to \infty} g(X_{s_n}) = g(y)$ and $\lim_{n \to \infty} g(X_{s_n'}) = g(y')$. Furthermore, since $g(X_t)$ has right-hand limits along Q, we must have $g(y) = g(y')$. Since $\mathcal{G}$ separates points, we see that $y = y'$.

For all ω for which the following limit exists for all t, set $\widetilde{X}_t(\omega) = \lim_{s \downarrow t, s \in Q} X_s$, and for those ω for which it does not exist for all t, set $\widetilde{X}_\cdot(\omega) = \Delta$. Note that $\{\widetilde{X}_t; t \geq 0\}$ is right continuous. We claim that for each t, $\widetilde{X}_t = X_t$ almost surely. To see this note that, for any $f, g \in C_b(S_\Delta)$,

$$
\begin{aligned}
\widetilde{E}^x\left(f(X_t)g(\widetilde{X}_t)\right) &= \lim_{s\downarrow t, s\in Q} \widetilde{E}^x\left(f(X_t)g(X_s)\right) \qquad (4.7)\\
&= \lim_{s\downarrow t, s\in Q} \widetilde{E}^x\left(f(X_t)P_{s-t}g(X_t)\right)\\
&= \widetilde{E}^x\left(f(X_t)g(X_t)\right)
\end{aligned}
$$

since $\lim_{s\downarrow t} P_{s-t}g = g$ uniformly. Since this holds for all $f, g \in C_b(S_\Delta)$, the Monotone Class Theorem (see, e.g., Rogers and Williams (2000b, II.3.1)) shows that

$$
\widetilde{E}^x\left(h(X_t, \widetilde{X}_t)\right) = \widetilde{E}^x\left(h(X_t, X_t)\right), \qquad (4.8)
$$

for all bounded Borel measurable functions h on $S_\Delta \times S_\Delta$. In particular this holds for $h(y, z) = I_{\{(y,z)\,|\,y\neq z\}}$. This shows that for each t, $\widetilde{X}_t = X_t$ almost surely.

Let Ω' be the space of right continuous S_Δ-valued functions $\{\omega'(t), t \in [0, \infty)\}$. Let X_t' be the natural evaluation $X_t'(\omega') = \omega'(t)$, and let $\mathcal{F}'^0$ and $\mathcal{F}_t'^0$ be the σ-algebras generated by $\{X_s', s \in [0, \infty)\}$ and $\{X_s', s \in [0, t]\}$, respectively. The map $\widetilde{X} : (S_\Delta^{R+}, \mathcal{F}^0) \mapsto (\Omega', \mathcal{F}'^0)$ defined for $\omega \in S_\Delta^{R+}$ by $\widetilde{X}(\omega)(t) = \widetilde{X}_t(\omega)$ induces a map of the probabilities $\widetilde{P}^x$ on $(S_\Delta^{R+}, \mathcal{F}^0)$ to probabilities P'^x on $(\Omega', \mathcal{F}'^0)$. It is easy to check that $X' = (\Omega', \mathcal{F}'^0, \mathcal{F}_t'^0, X_t', \theta_t, P'^x)$ is a right continuous simple Markov process with transition semigroup $\{P_t; t \geq 0\}$.

Since $U^\alpha : C_b(S_\Delta) \mapsto C_b(S_\Delta)$ for every $\alpha > 0$, it is clear that X' satisfies the second condition for a Borel right process. It then follows from Lemma 3.2.3 that we can find a Borel right process with transition semigroup $\{P_t; t \geq 0\}$. $\qquad \square$

Remark 4.1.2

(1) In Theorem 4.1.1 we construct a Feller process with state space S_Δ. Assume that the original family $\{P_t; t \geq 0\}$ is a Markov semigroup on S. It follows immediately that, for each $x \in S$, $P^x(X_t \in S, \forall t \in Q) = 1$, which implies that $P^x(\zeta < \infty) = 0$ (see Lemma 3.2.7). Thus, if the process starts in S, almost surely it never reaches Δ. Therefore, we can drop the cemetery state Δ so that the corresponding Feller process has state space S.

(2) The paths of the Feller process $\{X_t, t \geq 0\}$ constructed in Theorem 4.1.1 have left limits in S_Δ almost surely. This follows from the proof of Theorem 4.1.1. As in (4.6), we see that, for

any $\alpha > 0$ and $f \in C_b^+(S_\Delta)$, $e^{-\alpha t}U^\alpha f(X_t)$ is a right continuous supermartingale. It is well known that a right continuous supermartingale has left limits for all $t \in [0, \infty)$ almost surely; see Revuz and Yor (1991, Chapter II, Theorem 2.8). Let $\mathcal{G}$ be as defined in the paragraph following (4.6). Following the argument of that paragraph, we can show that, almost surely, $t \mapsto X_t$ has left-hand limits.

As a simple consequence of Theorem 4.1.1 and the Hille–Yosida Theorem (Theorem 14.4.1) we can give conditions for the construction of Feller processes in terms of potential operators. Let $X = (\Omega, \mathcal{G}, \mathcal{G}_t, X_t, \theta_t, P^x)$ be a Borel right process in a locally compact space S with continuous α-potential densities $u^\alpha(x, y)$, $\alpha > 0$. Recall the definition given in (3.14) of the potential operators U^α on $C_b(S)$. It is clear that U^α is a positive operator. Also, we show in Lemma 3.4.1 that $U^\alpha : C_0(S) \mapsto C_b(S)$, and, using the fact that $\|P_t f\| \leq \|f\|$ for each $t \geq 0$, it is easy to see that we have the contraction property

$$\|\alpha U^\alpha f\| \leq \|f\| \qquad \forall f \in C_0(S) \tag{4.9}$$

(indeed, this holds for all $f \in C_b(S)$). In Lemma 3.1.3 we show that the family $\{U^\alpha; \alpha > 0\}$ satisfies the resolvent equation

$$U^\lambda - U^\mu = (\mu - \lambda)U^\lambda U^\mu \qquad \forall \lambda, \mu > 0. \tag{4.10}$$

A family $\{U^\lambda; \lambda > 0\}$ of operators on a Banach space $(B, \|\| \cdot \|\|)$ that satisfies (4.9) and (4.10) is called a contraction resolvent on B (we use the notation $\|\| \cdot \|\|$ to emphasize that we are referring to an arbitrary norm, not necessarily the sup norm).

Note that for general Borel right processes with continuous α-potential densities, we do not have $U^\alpha : C_0(S) \mapsto C_0(S)$. However, by (3.16), we do have

$$\lim_{\lambda \to \infty} \lambda U^\lambda f(x) = f(x) \qquad \forall f \in C_0(S) \text{ and } x \in S. \tag{4.11}$$

Theorem 4.1.3 *Let S be a locally compact space with a countable base, and let $\{U^\lambda; \lambda > 0\}$ be a family of positive linear operators on $C_0(S)$ that is a contraction resolvent on $C_0(S)$ and in addition satisfies (4.11). Then we can construct a Feller process X with potential operators $\{U^\lambda; \lambda > 0\}$.*

Proof By Theorems 14.4.2 and 14.4.1, we can find a strongly continuous contraction semigroup $\{P_t; t \geq 0\}$ on $C_0(S)$ with

$$U^\lambda = \int_0^\infty e^{-\lambda t} P_t \, dt \qquad \forall \lambda > 0. \tag{4.12}$$

The construction of P_t in the proof of Theorem 14.4.1 and the fact that $\{U^\lambda; \lambda > 0\}$ are positive operators implies that each P_t is a positive operator on $C_0(S)$. The theorem now follows from Theorem 4.1.1. $\square$

Corollary 4.1.4 *The following are equivalent:*

(1) X is a Feller process.

(2) X is a Borel right process with potentials U^α with the property that $U^\alpha : C_0(S) \mapsto C_0(S)$ for all $\alpha > 0$.

(3) X is a Borel right process with transition semigroup $\{P_t; t \geq 0\}$, which is a strongly continuous contraction semigroup on $C_0(S)$.

Proof Let $\{P_t; t \geq 0\}$ be the transition semigroup for X. To show that (1) implies (2) we use the fact that $\{P_t; t \geq 0\}$ are contractions from $C_0(S) \mapsto C_0(S)$ along with the Dominated Convergence Theorem in (4.12) to see that $U^\alpha : C_0(S) \mapsto C_0(S)$ for all $\alpha > 0$.

To show that (2) implies (3), we first note that X has a semigroup, say $\{P_t; t \geq 0\}$, associated with it, and the potential operators $\{U_\alpha, \alpha > 0\}$ are related to $\{P_t; t \geq 0\}$ by (4.12). Also, as we show in the discussion preceding Theorem 4.1.3, $\{U^\lambda; \lambda > 0\}$ is a contraction resolvent on $C_0(S)$ satisfying (4.11). Therefore, it follows from the proof of Theorem 4.1.3 that there exists a strongly continuous contraction semigroup, say $\{P'_t; t \geq 0\}$, such that $P'_t : C_0(S) \to C_0(S)$, and $\{U_\alpha, \alpha > 0\}$ are also related to $\{P'_t; t \geq 0\}$ by (4.12) (with P_t replaced by P'_t). It follows from the uniqueness of the Laplace transform that the semigroups $\{P_t; t \geq 0\}$ and $\{P'_t; t \geq 0\}$ are the same. Thus (2) implies (3).

That (3) implies (1) follows from the definition of Feller processes. $\square$

Remark 4.1.5 Let X be a Borel right process with continuous α-potential densities. As mentioned in the discussion preceding Theorem 4.1.3, we show in Lemma 3.4.1 that $U^\alpha : C_0(S) \mapsto C_b(S)$. If S is compact, $C_b(S) = C_0(S)$. Hence it follows from Corollary 4.1.4 that a Borel right process with continuous α-potential densities and compact state space is a Feller process.

Theorem 4.1.3 gives conditions for a contraction resolvent on $C_0(S)$ to be the potential operators of a Feller process. We next present a theorem of Hunt that provides conditions for a bounded operator on $C_0(S)$ to be the 0-potential operator of a Feller process.

Let $G : C_0(S) \mapsto C_0(S)$ be a positive operator. We say that G satisfies the positive maximum principle on $C_0(S)$ if, for all $h \in C_0(S)$ for which $\sup_x Gh(x) > 0$,

$$\sup_{\{x:h(x)>0\}} Gh(x) \le a \Rightarrow \sup_{x \in S} Gh(x) \le a. \qquad (4.13)$$

Theorem 4.1.6 (Hunt's Theorem) *Let S be a locally compact space with a countable base and let $G : C_0(S) \mapsto C_0(S)$ be a positive, bounded linear operator. The following are equivalent:*

(1) G is the 0-potential operator of a Feller process on S.

(2) G satisfies the positive maximum principle on $C_0(S)$, and the image of $C_0(S)$ under G is dense in $C_0(S)$ in the uniform norm.

Proof $(1) \Rightarrow (2)$ Assume that there exists a Feller process X on S with potential operators $\{U^\lambda, \lambda \ge 0\}$ such that $U^0 = G$. Let $f \in C_0(S)$. It follows from Corollary 4.1.4 that the transition semigroup of a Feller process is a strongly continuous contraction semigroup on $C_0(S)$. Consequently, we see by (4.5) that

$$\lim_{\alpha \to \infty} \|\alpha U^\alpha f - f\|_\infty = 0 \qquad \forall f \in C_0(S). \qquad (4.14)$$

Thus it suffices to show that any function of the form $U^\alpha f$ is in the image of $C_0(S)$ under $U^0 = G$. This follows from the resolvent equation, Lemma 3.1.3, which can be written as

$$U^\alpha f(x) = U^0 f(x) - \alpha U^0 U^\alpha f(x) = U^0 \left(f - \alpha U^\alpha f \right)(x). \qquad (4.15)$$

We now show that G satisfies the positive maximum principle. Let $h \in C_0(S)$, with $\sup_x Gh(x) > 0$, and assume that for some $a > 0$,

$$\sup_{\{x:h(x)>0\}} Gh(x) \le a. \qquad (4.16)$$

Note that since G is a positive operator, the condition $\sup_x Gh(x) > 0$ ensures that the set $A := \{x : h(x) > 0\} \ne \emptyset$. Let $H_A f(x) := E^x(f(X_{T_A}); T_A < \infty)$ and let $x \in A^c$. Using the strong Markov property we see that

$$
\begin{aligned}
Gh(x) &= E^x \left(\int_0^\infty h(X_t)\, dt \right) \qquad\qquad (4.17) \\
&= E^x \left(\int_0^{T_A} h(X_t)\, dt \right) + E^x \left(\int_{T_A}^\infty h(X_t)\, dt \right) \\
&= E^x \left(\int_0^{T_A} h(X_t)\, dt \right) + H_A Gh(x).
\end{aligned}
$$

The integral in the last line of (4.17) is less than or equal to zero because $h(X_t) \leq 0$ for $t \in [0, T_A)$. Therefore,

$$Gh(x) \leq H_A Gh(x) \qquad \forall x \in A^c. \tag{4.18}$$

We now note that

$$H_A Gh(x) = E^x(Gh(X_{T_A}); T_A < \infty) \leq a. \tag{4.19}$$

To see this we first observe that $Gh(y) \leq a$ for all $y \in \overline{A}$ by (4.16) and the continuity of Gh, and then use the fact that $X_{T_A} \in \overline{A}$ when $T_A < \infty$, by the right continuity of X. Using (4.19) and (4.18), we see that

$$Gh(x) \leq a \qquad \forall x \in A^c. \tag{4.20}$$

This shows that G satisfies the positive maximum principle.

$(2) \Rightarrow (1)$ In Lemma 4.1.9, below, we construct a family $\{U^\lambda, \lambda \geq 0\}$ of positive linear operators on $C_0(S) \mapsto C_0(S)$, with $U^0 = G$, satisfying

$$\|\lambda U^\lambda\| \leq 1 \qquad \lambda > 0, \tag{4.21}$$

where $\|U^\lambda\|$ denotes the operator norm of U^λ as an operator on $C_0(S)$, that is, $\|U^\lambda\| = \sup_{\{f \in C_0(S), \|f\|_\infty \leq 1\}} \|U^\lambda f\|_\infty$, and

$$U^r - U^s = (s - r)U^s U^r \qquad \forall r, s \geq 0. \tag{4.22}$$

We refer to such a family $\{U^\lambda, \lambda \geq 0\}$ as a contraction resolvent. This is the natural extension of the concept of contraction resolvent, introduced after (4.10), to the index set $\lambda \geq 0$.

Taking $r = 0$ in (4.22) and using (4.21), we see that for any $f \in C_0(S)$,

$$\lim_{s \to \infty} sU^s Gf = Gf. \tag{4.23}$$

Using (4.21) and the hypothesis that the image of $C_0(S)$ under G is dense in $C_0(S)$ in the uniform norm, it follows from (4.23) that, for any $f \in C_0(S)$,

$$\lim_{s \to \infty} sU^s f = f. \tag{4.24}$$

The theorem now follows by Theorem 4.1.3. $\qquad \square$

To motivate the rather abstract construction of $\{U^\lambda, \lambda \geq 0\}$, we note that (4.22) shows that

$$U^p(1 + pG) = G. \tag{4.25}$$

Thus, if $(1 + pG)$ is invertible, we would have

$$U^p = (1 + pG)^{-1} G \tag{4.26}$$

for $p > 0$. This suggests that when we are given only G and want to define $\{U^\lambda, \lambda \geq 0\}$ with $U^0 = G$, we should use the right-hand side of (4.26). This is what is done in (4.38).

The next two lemmas are used in the proof of Lemma 4.1.9.

Lemma 4.1.7 *Let S be a locally compact space with a countable base and let $G : C_0(S) \mapsto C_0(S)$ be a positive, bounded linear operator that satisfies the positive maximum principle on $C_0(S)$. Then $(I + pG)$ is one–one.*

Proof It follows from the hypothesis that G satisfies the positive maximum principle on $C_0(S)$ that for any $p > 0$ and $f \in C_0(S)$,

$$\|(I + pG)f\|_\infty \geq \|pGf\|_\infty. \tag{4.27}$$

To see this, note first that for any $f \in C_0(S)$,

$$\sup_{\{x:f(x)\geq 0\}} Gf(x) \leq \sup_{\{x:f(x)\geq 0\}} \left(\frac{1}{p}I + G\right)f(x) \tag{4.28}$$

$$\leq \left\|\left(\frac{1}{p}I + G\right)f\right\|_\infty.$$

Therefore, since G satisfies the positive maximum principle on $C_0(S)$,

$$\sup_{x \in S} Gf(x) \leq \left\|\left(\frac{1}{p}I + G\right)f\right\|_\infty. \tag{4.29}$$

Using (4.28) with f replaced by $-f$ we see that (4.29) also holds with f replaced by $-f$. Since

$$\|Gf\|_\infty = \max\left(\sup_x Gf(x), \sup_x G(-f)(x)\right), \tag{4.30}$$

we get (4.27).

If $\|(I + pG)f\|_\infty = 0$, by (4.27) we also have $\|pGf\|_\infty = 0$. Consequently, $f = (I + pG)f - pGf \equiv 0$. Thus $(I + pG)$ is one–one. $\square$

Lemma 4.1.8 *Under the hypotheses of Lemma 4.1.7, let W be a bounded operator that commutes with G and satisfies*

$$G - W = pGW \quad \text{or, equivalently,} \quad (I + pG)W = G. \tag{4.31}$$

Then W is a positive operator and

$$\|W\| \leq 1/p. \tag{4.32}$$

Proof By straightforward manipulations (4.31) implies that

$$(I + pG)(I - pW) = I \quad \text{and} \quad G(I - pW) = W. \tag{4.33}$$

Since G satisfies the positive maximum principle on $C_0(S)$, it follows from the first line of (4.28) that

$$\sup_x Gf(x) \le \sup_x \left(\frac{1}{p}I + G\right) f(x). \tag{4.34}$$

Using this with $f = (I - pW)g$ and (4.33), we see that

$$\sup_x pWg(x) \le \sup_x g(x). \tag{4.35}$$

Therefore, if $g \le 0$, the same is true for Wg. This shows that W is a positive operator. We then have $\|pWg\|_\infty \le \|pW|g|\|_\infty \le \|g\|_\infty$, so that (4.32) holds. $\qquad \square$

We say that a positive linear operator H is symmetric with respect to m if

$$\int g(x) \, Hf(x) \, dm(x) = \int Hg(x) \, f(x) \, dm(x) \tag{4.36}$$

for all nonnegative functions $f, g \in \mathcal{B}(S)$, whenever both sides exist.

Lemma 4.1.9 *Let S be a locally compact space with a countable base, and let $G : C_0(S) \mapsto C_0(S)$ be a positive, bounded linear operator that satisfies the positive maximum principle on $C_0(S)$. Then there is a unique contraction resolvent $\{U^\lambda, \lambda \ge 0\}$ on $C_0(S)$, with $U^0 = G$. Furthermore, when G is symmetric, so is $\{U^\lambda, \lambda \ge 0\}$.*

Proof We construct the operators $\{U^\lambda, \lambda \ge 0\}$. For any $\epsilon < 1/\|G\|$, $I + \epsilon G$ is invertible. In fact,

$$(I + \epsilon G)^{-1} = \sum_{j=0}^\infty (-1)^j \epsilon^j G^j. \tag{4.37}$$

To see this, note that the sum on the right-hand side converges in the operator norm so that when multiplied by $I + \epsilon G$ the sum converges to the identity.

It follows from (4.37) that G and all the operators $(I + \epsilon G)^{-1}$, for $\epsilon < 1/\|G\|$, commute with each other. For $p < 1/\|G\|$, set

$$U^p = (I + pG)^{-1}G = G(I + pG)^{-1}. \tag{4.38}$$

It follows from this that for all $p < 1/\|G\|$,

$$G - U^p = pGU^p = pU^pG. \tag{4.39}$$

Therefore, U^p has the properties of the operator W in Lemma 4.1.8. In particular, U^p is a positive linear operator with

$$\|U^p\| \le 1/p. \tag{4.40}$$

We extend the definition of U^p to all $p \ge 0$ by induction. Assume that for some integer k and all $p < k/\|G\|$ we have constructed operators U^p satisfying (4.39) and (4.40). Fix a $p < k/\|G\|$. We repeat the procedure in the last paragraph starting from (4.37) but with G replaced by U^p. For any $\epsilon < 1/\|U^p\|$, $I + \epsilon U^p$ is invertible and all the operators G, U^p and $(I + \epsilon U^p)^{-1}$ commute with each other. For $\epsilon < 1/\|U^p\|$ we mimic (4.38) and set

$$U^{p,\epsilon} = (I + \epsilon U^p)^{-1}U^p = U^p(I + \epsilon U^p)^{-1}. \tag{4.41}$$

Thus

$$U^p - U^{p,\epsilon} = \epsilon U^p U^{p,\epsilon} = \epsilon U^{p,\epsilon} U^p, \tag{4.42}$$

so that by (4.39)

$$pG(U^p - U^{p,\epsilon}) = \epsilon pG U^p U^{p,\epsilon} = \epsilon(G - U^p)U^{p,\epsilon} \tag{4.43}$$

or, equivalently,

$$pGU^p + \epsilon U^p U^{p,\epsilon} = (p + \epsilon)GU^{p,\epsilon}. \tag{4.44}$$

Using (4.39) and (4.42) again we find that

$$G - U^{p,\epsilon} = (p + \epsilon)GU^{p,\epsilon} \tag{4.45}$$

or, equivalently, that

$$(I + (p + \epsilon)G)U^{p,\epsilon} = G. \tag{4.46}$$

Note that $U^{p,\epsilon}$ and G commute, so that $U^{p,\epsilon}$ also has the properties of the operator W in Lemma 4.1.8. Therefore, in particular, $U^{p,\epsilon}$ is a positive linear operator with

$$\|U^{p,\epsilon}\| \le \frac{1}{p + \epsilon}. \tag{4.47}$$

How large can $p+\epsilon$ be? We can take any $p < k/\|G\|$ and any $\epsilon < 1/\|U^p\|$. Since $\|U^p\| \le 1/p$, this implies that we can take any $\epsilon < p$. Thus we can choose ϵ and p to achieve any number $p + \epsilon < 2k$.

We now define $U^{p+\epsilon} = U^{p,\epsilon}$. However, we must check for consistency since $U^{p+\epsilon}$ already exists for $p + \epsilon < k/\|G\|$. In this case, by (4.39), we have

$$(I + (p + \epsilon)G)U^{p+\epsilon} = G, \tag{4.48}$$

and since $I + (p + \epsilon)G$ is one–one, we see by (4.46) and (4.48) that $U^{p+\epsilon} = U^{p,\epsilon}$. Thus, we have constructed positive operators U^p satisfying (4.39) and (4.40) for all $p < 2k/\|G\|$. We can proceed in this manner to define positive operators U^p satisfying (4.39) and (4.40) for all $p \geq 0$.

By repeatedly using (4.39), which now holds for all $p \geq 0$, we see that for any $r, s \geq 0$,

$$
\begin{aligned}
(I + rG)(U^r - U^s) &= G - (I + rG)U^s \\
&= (s - r)GU^s \\
&= (I + rG)(s - r)U^r U^s.
\end{aligned}
\tag{4.49}
$$

Since, by Lemma 4.1.7, $I + rG$ is one–one, we obtain (4.22).

Uniqueness follows from (4.48) and the fact that $I + (p+\epsilon)G$ is one–one. And, lastly, it follows from (4.37) and (4.41) that when G is symmetric, so is $\{U^\lambda, \lambda \geq 0\}$. $\qquad\square$

Remark 4.1.10 If one reexamines the proof of Lemma 4.1.9, one sees that the only property of $C_0(S)$ that is used is the fact that $C_0(S)$ is a Banach space of functions on S. An almost identical proof shows that if $G : C_b(S) \mapsto C_b(S)$ is a positive, bounded linear operator that satisfies the positive maximum principle on $C_b(S)$, then we can construct a contraction resolvent $\{U^\lambda; \lambda \geq 0\}$ on $C_b(S)$ with $U^0 = G$. Furthermore, when G is symmetric, so is $\{U^\lambda, \lambda \geq 0\}$.

Remark 4.1.11 Let X and X' be strongly symmetric Borel right processes on compact state spaces S and S', respectively, with continuous 0-potential densities $u(x, y)$ and $u'(x, y)$. Assume that $S' \subseteq S$ and $u'(x, y) = u(x, y)$ on $S' \times S'$. Let

$$
\widetilde{X}_t = X_{\tau_t},
\tag{4.50}
$$

where $\tau.$ is the right continuous inverse of

$$
A_t = \int_0^t 1_{S'}(X_s)\, ds.
\tag{4.51}
$$

Note that $\{\widetilde{X}_t\,;\, t \geq 0\}$ is right continuous with values in S'. We claim that

$$
\{\widetilde{X}_t\,;\, t \geq 0\} \overset{law}{=} \{X'_t\,;\, t \geq 0\}.
\tag{4.52}
$$

To see this, first note that it follows from the fact that $\tau_{t+s} = \tau_t + \tau_s \circ \theta_{\tau_t}$ for any s, t and the strong Markov property for X_t that for any

$x \in S'$

$$E^x \left(h(\widetilde{X}_{t+s}) \,|\, \mathcal{F}_{\tau_t} \right) = E^x \left(h(X_{\tau_{t+s}}) \,|\, \mathcal{F}_{\tau_t} \right) \tag{4.53}$$

$$= E^x \left(h(X_{\tau_s}) \circ \theta_{\tau_t} \,|\, \mathcal{F}_{\tau_t} \right)$$

$$= E^{X_{\tau_t}} \left(h(X_{\tau_s}) \right) = E^{\widetilde{X}_t} \left(h(\widetilde{X}_s) \right)$$

for any $h \in \mathcal{B}(S')$. Let

$$\widetilde{P}_t h(x) := E^x \left(h(\widetilde{X}_t) \right). \tag{4.54}$$

Taking the expectation E^x of the first and last equations in (4.53) we see that $\{\widetilde{P}_t\,;\, t \geq 0\}$ is a Borel semigroup on $\mathcal{B}(S')$. Thus, $\widetilde{X} = (\Omega, \mathcal{F}_{\tau_\infty}, \mathcal{F}_{\tau_t}, \widetilde{X}_t, \theta_{\tau_t})$ is a right continuous simple Markov process with transition semigroup $\{\widetilde{P}_t\,;\, t \geq 0\}$. Let $\widetilde{U}^0$ denote the 0-potential operator for $\widetilde{X}$.

Let $f \in C(S')$ and $x \in S'$. As in the proof of (2.125) (with $T = \infty$):

$$\widetilde{U}^0 f(x) = E^x \left(\int_0^\infty f(X_{\tau_t})\,dt \right) \tag{4.55}$$

$$= E^x \left(\int_0^\infty f(X_s)\,dA_s \right)$$

$$= E^x \left(\int_0^\infty f(X_s) 1_{S'}(X_s)\,ds \right) \tag{4.56}$$

$$= U^0 f(x) = \int u(x,y) f(y)\,dy.$$

We see that $\widetilde{X}$ is a right continuous simple Markov process in S' with a symmetric, continuous 0-potential density, $\{u(x,y)\,;\, x, y \in S'\}$. This shows that $\widetilde{X}$ and X' have the same 0-potential operators. We now use Lemma 4.1.9 to extend these to potential operators on S'. These are the same by the uniqueness property in Lemma 4.1.9.

It follows by (3.15) and the uniqueness of the Laplace transform that $\widetilde{X}$ and X' have that same transition semigroup. Therefore we have (4.52).

Remark 4.1.12 Not every Feller process has a finite 0-potential operator, and even when it does, its 0-potential operator may not be a bounded operator on $C_0(S)$. Brownian motion is an example of the first assertion. An example of the second is given by Brownian motion killed the first time it hits 0. In that case,

$$U^0 f(x) = 2 \int_0^\infty (x \wedge y) f(y)\,dy. \tag{4.57}$$

On the other hand, every Feller process has a 1-potential operator that is bounded as an operator on $C_0(S)$.

The next corollary deals with an important special case of this phenomenon.

Corollary 4.1.13 *Let S be a locally compact space with a countable base and let $G : C_0(S) \mapsto C_0(S)$ be a positive linear operator with $G1(x) = 1$ for all $x \in S$. The following are equivalent:*

(1) G is the 1-potential operator of a Feller process on S.

(2) G satisfies the positive maximum principle on $C_0(S)$ and the image of $C_0(S)$ under G is dense in $C_0(S)$ in the uniform norm.

Proof Since $G : C_0(S) \mapsto C_0(S)$ is a positive linear operator with $G1(x) = 1$ for all $x \in S$, it follows that for any $f \in C_0(S)$,

$$Gf(x) \le G\|f\|_\infty(x) = \|f\|_\infty. \tag{4.58}$$

This also holds with f replaced by $-f$, so we have $\|G\| \le 1$.

Since the 1-potential operator of a Feller process X on S is the 0-potential operator of the Feller process obtained by killing X at an independent exponential time with mean one, it follows from Theorem 4.1.6 that (1) implies (2).

To see that (2) implies (1), we first use Theorem 4.1.6 to see that G is the 0-potential operator of a Feller process X on S. The resolvent equation (3.18) together with the fact that $G1(x) = 1$ for all $x \in S$ shows that

$$U^\alpha 1(x) = (1 + \alpha)^{-1} \tag{4.59}$$

for all $x \in S$ and all $\alpha \ge 0$.

Let $\{P_t; t \ge 0\}$ be the strongly continuous contraction semigroup of positive linear operators on $C_0(S)$ associated with the Feller process X. We define the linear operators $\{\overline{P}_t = e^t P_t; t \ge 0\}$. It is clear that $\{\overline{P}_t; t \ge 0\}$ satisfies conditions (1) and (3) in the definition of a strongly continuous contraction semigroup of positive linear operators on $C_0(S)$ on page 122. We show that it also satisfies condition (2). For any $\alpha \ge 0$,

$$\int_0^\infty e^{-(1+\alpha)t}\overline{P}_t 1(x)\, dt = \int_0^\infty e^{-\alpha t} P_t 1(x)\, dt = U^\alpha 1(x) = (1 + \alpha)^{-1}. \tag{4.60}$$

By (3.15) and the uniqueness of the Laplace transform, this implies that for each $x \in S$,

$$\overline{P}_t 1(x) = 1 \qquad \text{for almost all } t. \tag{4.61}$$

Assume that for some x and t we have $\overline{P}_t 1(x) > 1$. Then we could find some $f \in C_0(S)$ with $0 \leq f \leq 1$ and $\overline{P}_t f(x) > 1$. However, by (4.61) we can find a sequence $t_n \downarrow t$ with $\overline{P}_{t_n} 1(x) = 1$ and consequently, $\overline{P}_{t_n} f(x) \leq 1$ for all n. This implies, by condition (3) on page 122, that $\overline{P}_t f(x) \leq 1$. This contradiction shows that condition (2) on page 122 is satisfied. Thus we see that $\{\overline{P}_t; t \geq 0\}$ is a strongly continuous contraction semigroup of positive linear operators on $C_0(S)$.

By Theorem 4.1.1 we can construct a Feller process $\overline{X}$ with transition semigroup $\{\overline{P}_t; t \geq 0\}$. If $\{\overline{U}^\lambda; \lambda > 0\}$ are the potential operators for $\overline{X}$, we have

$$\overline{U}^1 f(x) = \int_0^\infty e^{-t} \overline{P}_t f(x)\, dt = \int_0^\infty P_t f(x)\, dt = U^0 f(x) = Gf(x). \tag{4.62}$$

□

4.2 Lévy processes

Symmetric Lévy processes in R^1 give us a rich class of examples of Borel right processes for which we can give explicit examples of many general results about local times.

We define an R^n-valued Lévy process, starting at 0, to be a stochastic process $X = \{X_t\,; t \in R_+\}$ that satisfies the following two properties:

(1) X has stationary and independent increments.
(2) $t \mapsto X_t$ is right continuous and has limits from the left.

Let $\phi_t(\lambda)$ denote the characteristic function of X_t. It is easy to see that for all u, v rational, $\phi_{u+v}(\lambda) = \phi_u(\lambda)\phi_v(\lambda)$. It then follows that

$$Ee^{i(\lambda \cdot X_t)} = e^{-t\psi(\lambda)} \qquad \forall t \geq 0 \tag{4.63}$$

for some continuous function ψ with $\mathrm{Re}\ \psi \geq 0$. The function ψ is called the Lévy exponent of X.

A stochastic process $Z = \{Z(t); t \in S\}$ is said to be infinitely divisible, if for each n there exists a stochastic process $Z_n = \{Z_n(t); t \in S\}$ such that

$$Z \stackrel{law}{=} \sum_{i=1}^n Z_{n,i}, \tag{4.64}$$

where $Z_{n,i}$, $i = 1, \ldots, n$ are independent copies of Z_n. It is easy to see that X is infinitely divisible and for each n the independent random processes for which (4.64) holds are those determined by (4.63) with characteristic function $\exp(-t\psi(\lambda)/n)$. Obviously, X_1 is an infinitely divisible random variable.

Infinitely divisible random variables are characterized by the Lévy–Khintchine Theorem, which states that $\psi(\lambda)$ is of the form

$$\psi(\lambda) = i(a \cdot \lambda) + Q(\lambda) + \int \left(1 - e^{i(\lambda \cdot x)} + i(\lambda \cdot x)1_{|x|<1}\right) \nu(dx), \quad (4.65)$$

where $a \in R^n$, $Q(\lambda)$ is a positive semidefinite quadratic form and ν is a measure on $R^n - \{0\}$ satisfying

$$\int (1 \wedge |x|^2)\, \nu(dx) < \infty. \quad (4.66)$$

In the next theorem, starting with the Lévy–Khintchine Theorem, we use the material in Section 4.1 to prove that for all functions ψ satisfying (4.65)–(4.66) there exists a Lévy process with Lévy exponent $\psi(\lambda)$.

Theorem 4.2.1 *Let $\psi(\lambda)$, $\lambda \in R^n$ be a function that satisfies (4.65)–(4.66). Then there exists a Lévy process with Lévy exponent $\psi(\lambda)$.*

Proof The Lévy–Khintchine Theorem implies that for each $t \geq 0$, there exists an R^n-valued random variable Z_t with characteristic function $\exp(-t\psi(\lambda))$. Let μ_t be the probability measure on R^n induced by Z_t. It follows that $\widehat{\mu}_t(\lambda) = \exp(-t\psi(\lambda))$ and $\widehat{\mu}_t\widehat{\mu}_s = \widehat{\mu}_{t+s}$. Consequently, $\mu_t * \mu_s = \mu_{s+t}$ for all $s, t \geq 0$. Set

$$P_t f(x) = \int_{R^n} f(x+y)\, d\mu_t(y) \quad f \in C_b(R^n). \quad (4.67)$$

P_t is a positive operator on $C_b(R^n)$ satisfying $\|P_t f\| \leq \|f\|$ for each $t \geq 0$, where $\|\cdot\|$ is the uniform norm.

It also follows from the fact that $\mu_t * \mu_s = \mu_{s+t}$ for all $s, t \geq 0$ that $\{P_t;\ t \geq 0\}$ is a Markov semigroup of operators on $C_b(R^n)$. In addition, using the fact that $|e^{-t\psi(\lambda)}| \leq 1$, we see that for $f \in C_0^\infty(R^n)$,

$$P_t f(x) = \frac{1}{2\pi} \int_{R^n} e^{-i\lambda x}\, e^{-t\psi(\lambda)}\, \widehat{f}(\lambda)\, d\lambda. \quad (4.68)$$

It follows from the Riemann-Lebesgue Lemma that $P_t : C_0^\infty(R^n) \mapsto C_0(R^n)$. Using the fact that $\|P_t f\| \leq \|f\|$, we then see that $P_t : C_0(R^n) \mapsto C_0(R^n)$.

Furthermore, for $f \in C_0^\infty(R^n)$, we see that

$$\|P_t f - f\| \leq \frac{1}{2\pi} \int_{R^n} |e^{-t\psi(\lambda)} - 1|\, |\widehat{f}(\lambda)|\, d\lambda, \quad (4.69)$$

so that by the Dominated Convergence Theorem $\lim_{t \to 0} \|P_t f - f\| = 0$

for $f \in C_0^\infty(R^n)$. As above, using the fact that $\|P_t f\| \le \|f\|$ we then see that $\lim_{t \to 0} \|P_t f - f\| = 0$ for all $f \in C_0(R^n)$. Thus, $\{P_t; t \ge 0\}$ is a strongly continuous contraction semigroup of positive linear operators on $C_0(R^n)$.

We can now use Theorem 4.1.1 and Remark 4.1.2 (1) to construct a Feller process $\{X_t, t \ge 0\}$ with transition semigroup $\{P_t; t \ge 0\}$ and state space R^n. It is easy to check that this Feller process is a Lévy process. Condition (2) on page 135 follows from Remark 4.1.2 (2). Also, from the proof of Theorem 4.1.1, we see that for $0 = t_0 < t_1 < \ldots < t_n$, $(X_{t_1}, \ldots, X_{t_n})$ has probability measure $\prod_{j=1}^n P_{t_j - t_{j-1}}(z_{j-1}, dz_j)$, where $z_0 = 0$. Note that, by definition, for all $f \in C_b(R^n)$,

$$\int f(a+y) P_t(x, dy) = \int f(a+y+x) \mu_t(dy).$$

Therefore, for any functions $f, g \in C_b(R^n)$,

$$E(g(X_{t_1}, \ldots, X_{t_{n-1}}) f(X_{t_n} - X_{t_{n-1}})) \qquad (4.70)$$

$$= \int g(z_1, \ldots, z_{n-1}) f(z_n - z_{n-1}) \prod_{j=1}^n P_{t_j - t_{j-1}}(z_{j-1}, dz_j)$$

$$= \int g(z_1, \ldots, z_{n-1}) \left(\int f(z_n - z_{n-1}) P_{t_n - t_{n-1}}(z_{n-1}, dz_n) \right)$$

$$\prod_{j=1}^{n-1} P_{t_j - t_{j-1}}(z_{j-1}, dz_j)$$

$$= \int g(z_1, \ldots, z_{n-1}) \left(\int f(z_n) \mu_{t_n - t_{n-1}}(dz_n) \right)$$

$$\prod_{j=1}^{n-1} P_{t_j - t_{j-1}}(z_{j-1}, dz_j)$$

$$= E(g(X_{t_1}, \ldots, X_{t_{n-1}})) E(f(X_{t_n - t_{n-1}})).$$

As in (2.11), in the proof of the existence of Brownian motion, this shows that $\{X_t, t \ge 0\}$ satisfies condition (1) for Lévy processes on page 135. $\square$

We now restrict our attention to the symmetric Lévy processes that satisfy the additional condition

$$\frac{1}{1 + \psi(\lambda)} \in L^1(d\lambda). \qquad (4.71)$$

(We point out later in this section why this condition is fundamental in this book.) The symmetry of X implies that $\psi(\lambda)$ is a positive real-valued even function. Using this in (4.65), one can easily show that $\psi(\lambda) = O(|\lambda|^2)$ as $|\lambda| \to \infty$. Thus (4.71) can only hold on R^1. In this case (4.65) takes the form

$$\psi(\lambda) = C\lambda^2 + 2 \int_0^\infty (1 - \cos \lambda x)\, \nu(dx) \qquad (4.72)$$

for some constant C and measure ν on $(0, \infty)$ such that

$$\int_0^\infty (1 \wedge x^2)\, \nu(dx) < \infty. \qquad (4.73)$$

Lévy processes with Lévy exponents ψ that satisfy (4.71) have continuous transition probability density functions. To see this note first that condition (4.71) implies that $\exp(-t\psi(\lambda)) \in L^1(d\lambda)$ for each $t > 0$ (since for each $t > 0$ there exists a constant C such that $\exp(-tx) \leq C/(1+x)$ for all $x \geq 0$). Hence, for each $t > 0$,

$$p_t(x) = \frac{1}{2\pi} \int_{-\infty}^\infty e^{-i\lambda x}\, e^{-t\psi(\lambda)}\, d\lambda \qquad (4.74)$$

is continuous. It is easily checked that $\{p_t(x, y) := p_t(y - x), t > 0\}$ are regular transition densities with respect to Lebesgue measure.

Let $\{X_t; t \in R_+\}$ be a symmetric Lévy process. Substituting (4.72) in (4.63) we see that we can write X_t as the sum of two independent stochastic processes, one of which is a constant times standard Brownian motion and the other which has characteristic function $\exp(-t\widetilde{\psi}(\lambda))$, where

$$\widetilde{\psi}(\lambda) = 2 \int_0^\infty (1 - \cos \lambda x)\, \nu(dx). \qquad (4.75)$$

The conditions on the Lévy measure ν in (4.73) impose restraints on the growth of $\widetilde{\psi}$.

Lemma 4.2.2

$$\widetilde{\psi}(\lambda) = o(\lambda^2) \quad as \quad \lambda \to \infty \qquad (4.76)$$

$$\lim_{\lambda \to 0} \frac{\widetilde{\psi}(\lambda)}{\lambda^2} \geq (0.7) \int_0^\infty (x^2 \wedge 1)\, \nu(dx). \qquad (4.77)$$

$$\lim_{\lambda \to \infty} \widetilde{\psi}(\lambda) = \infty. \qquad (4.78)$$

Proof By (4.75) we have for $\lambda > 1$

$$\widetilde{\psi}(\lambda) = 4 \int_0^\infty \sin^2 \frac{\lambda x}{2} \nu(dx) \qquad (4.79)$$

$$\leq \lambda^2 \int_0^{1/\lambda} x^2 \nu(dx) + 4\nu[1/\lambda, \infty).$$

The first term in the last line of (4.79) is obviously $o(\lambda^2)$ as $\lambda \to \infty$. Therefore, it remains to show that $\lim_{\lambda \to \infty} \nu[1/\lambda, \infty) = o(\lambda^2)$ or, equivalently, that $\lim_{\epsilon \to 0} \epsilon^2 \nu[\epsilon, \infty) = 0$. For any $\delta > 0$ and $\epsilon < \delta$,

$$\epsilon^2 \nu[\epsilon, \delta) \leq \int_\epsilon^\delta x^2 \nu(dx) \leq \int_0^\delta x^2 \nu(dx). \qquad (4.80)$$

Therefore, since $\lim_{\epsilon \to 0} \epsilon^2 \nu[\delta, \infty) = 0$, we see that

$$\lim_{\epsilon \to 0} \epsilon^2 \nu[\epsilon, \infty) \leq \int_0^\delta x^2 \nu(dx). \qquad (4.81)$$

Thus $\lim_{\epsilon \to 0} \epsilon^2 \nu[\epsilon, \infty) = 0$ and we get (4.76). Equation (4.77) is elementary. Since $\sin x > (0.84)x$ for $x \in [0, 1]$,

$$\widetilde{\psi}(\lambda) > (0.84)^2 \lambda^2 \int_0^{2/\lambda} x^2 \, \nu(dx), \qquad (4.82)$$

which gives (4.77).

Equation (4.78) is an immediate consequence of the Riemann–Lebesgue Lemma since X_t has a probability density function (see (4.74)), and so $e^{-t\psi(\lambda)}$ is the Fourier transform of a function in L^1. □

It follows from (4.76) that the class of Lévy processes with characteristic function $\exp(-t\,\widetilde{\psi}(\lambda))$ does not include Brownian motion. Therefore, Lévy processes for which $C = 0$ in (4.72) are often referred to as Lévy processes without a Gaussian component.

Remark 4.2.3 Although it is not needed in this book, it is interesting to note that Lévy processes without a Gaussian component are pure jump processes. When (4.71) holds the paths of these processes are not of bounded variation on bounded intervals of time, although the sum of the squares of their jumps on a bounded interval is finite. A necessary and sufficient condition for the paths to be of bounded variation, on bounded intervals of time, is that

$$\int_0^\infty (1 \wedge x) \, \nu(dx) < \infty. \qquad (4.83)$$

It follows from (4.74) that the α-potential density of X is given by

$$u^\alpha(x, y) = \frac{1}{\pi} \int_0^\infty \frac{\cos \lambda(x - y)}{\alpha + \psi(\lambda)} \, d\lambda. \tag{4.84}$$

It follows from (4.71) that $u^\alpha(x, y)$ is continuous for all $x, y \in R^1$. We also write $u^\alpha(y - x) = u^\alpha(x, y)$. Here we see the significance of condition (4.71). It is necessary and sufficient for the existence of a continuous α-potential density for X, for $\alpha > 0$. Obviously, $u(0, 0)$ exists if and only if $(\psi(\lambda))^{-1} \in L^1(d\lambda)$.

Theorem 4.2.4 *Suppose that 0 is recurrent for X (equivalently, $u(0) = \infty$). Then*

$$P^x(T_0 < \infty) = 1 \qquad \forall x \in R^1 \tag{4.85}$$

and

$$u_{T_0}(x, y) = \phi(x) + \phi(y) - \phi(x - y), \tag{4.86}$$

where

$$\phi(x) := \frac{1}{\pi} \int_0^\infty \frac{1 - \cos \lambda x}{\psi(\lambda)} \, d\lambda \qquad x \in R^1. \tag{4.87}$$

Proof We first note that $\phi(x)$ is bounded and continuous. This follows by breaking the integral into two pieces. The integral over $[1, \infty]$ is controlled by (4.71), and the integral over $[0, 1]$ is controlled by (4.77).

Note that

$$\begin{aligned} u^\alpha(0) - u^\alpha(x) &= \frac{1}{\pi} \int_0^\infty \frac{1 - \cos \lambda x}{\alpha + \psi(\lambda)} \, d\lambda \tag{4.88} \\ &:= \phi_\alpha(x). \end{aligned}$$

Since $\phi(x)$ is bounded and $u(0) = \infty$ (see Lemma 3.6.11), we see that

$$\lim_{\alpha \to 0} \frac{u^\alpha(x)}{u^\alpha(0)} = 1. \tag{4.89}$$

Equation (4.85) now follows from (3.107).

It follows from (4.88) that

$$\begin{aligned} \lim_{\alpha \to 0} (u^\alpha(0) - u^\alpha(x)) &= \frac{1}{\pi} \int_0^\infty \frac{1 - \cos(\lambda x)}{\psi(\lambda)} \, d\lambda \tag{4.90} \\ &= \phi(x). \end{aligned}$$

Using the Markov property at time T_0 shows that

$$u^\alpha(x - y) = E^x \left(\int_0^\infty e^{-\alpha t} \, dL_t^y \right) \tag{4.91}$$

$$= E^x \left(\int_0^{T_0} e^{-\alpha t} \, dL_t^y \right) + E^x \left(\int_{T_0}^{\infty} e^{-\alpha t} \, dL_t^y \right)$$

$$= u_{T_0}^{\alpha}(x, y) + E^x \left(e^{-\alpha T_0} \right) u^{\alpha}(y).$$

It follows from (4.91) that

$$u_{T_0}^{\alpha}(x, y) - u_{T_0}^{\alpha}(x, 0) \qquad (4.92)$$
$$= u^{\alpha}(x - y) - u^{\alpha}(x) - \{u^{\alpha}(y) - u^{\alpha}(0)\} E^x(e^{-\alpha T_0})$$
$$= \phi_{\alpha}(x) + \phi_{\alpha}(y) E^x(e^{-\alpha T_0}) - \phi_{\alpha}(x - y).$$

Using (4.85) we see that $\lim_{\alpha \to 0} E^x(e^{-\alpha T_0}) = 1$. Also, clearly, $u_{T_0}^{\alpha}(x, y)$ and $u_{T_0}^{\alpha}(x, 0)$ tend to $u_{T_0}(x, y)$ and $u_{T_0}(x, 0)$ as α decreases to zero. Therefore we can take the limits in (4.92) and use (4.90) and the fact that $u_{T_0}(x, 0) = 0$ (see Lemma 3.8.1) to obtain (4.86). $\qquad \square$

An infinitely divisible stochastic process Z is called stable if Z_n in (4.64) is equal to $c_n Z$ for some $c_n > 0$. It follows from (4.63) that for this to happen we must have $\psi(c_n \lambda) = \psi(\lambda)/n$. It is easy to see that this is the case when

$$\psi(\lambda) = |\lambda|^p \qquad 0 < p \le 2 \qquad (4.93)$$

and $c_n = n^{-1/p}$ (in fact, except for multiplication by a constant, this is the only way this can happen; see, e.g., Feller (1971, XVII.4)). We call a Lévy process with characteristic exponent ψ given by (4.93) a canonical stable or canonical p-stable process. It is obvious from (4.72) that $\psi(\lambda) = \lambda^2$ is possible. To obtain $\psi(\lambda) = |\lambda|^p$ for $0 < p < 2$, take $\nu(dx) = (\pi C_{p+1} x^{p+1})^{-1} \, dx$ and $C = 0$ in (4.72), where

$$C_p = \frac{2}{\pi} \int_0^{\infty} \frac{1 - \cos s}{s^p} \, ds. \qquad (4.94)$$

Example 4.2.5 We define the canonical symmetric stable process on R^1, of index β, to be a Lévy process with Lévy exponent $\psi(\lambda) = |\lambda|^{\beta}$. Then, for $1 < \beta \le 2$, by Theorem 4.2.4,

$$\phi(x) = \frac{C_{\beta}}{2} |x|^{\beta - 1} \qquad (4.95)$$

and

$$u_{T_0}(x, y) = C_{\beta} \left(|x|^{\beta - 1} + |y|^{\beta - 1} - |x - y|^{\beta - 1} \right). \qquad (4.96)$$

Note that condition (4.71) requires that $\beta > 1$ (C_{β} is given by (4.94)) with $\beta = p$.

Remark 4.2.6 Note that the symmetric stable process of index 2 is standard Brownian motion multiplied by $\sqrt{2}$. Because of this, when we give results for stable processes, those for 2-stable processes differ, by the effects of the square root, from known results for standard Brownian motion.

In order to compare results we obtain with others in the literature, it is useful to have numerical values for C_p. We can do this easily for C_2. For standard Brownian motion (4.87) is

$$\phi(x) = \frac{1}{\pi} \int_0^\infty \frac{1 - \cos \lambda x}{\lambda^2/2} \, d\lambda \qquad x \in R^1. \tag{4.97}$$

By a change of variables this is equivalent to $\phi\left(x/\sqrt{2}\right) = \sqrt{2}C_2|x|$. Therefore, by Theorem 4.2.4 for Brownian motion, $\phi(x) = C_2|x|$ and

$$
\begin{aligned}
u_{T_0}(x, y) &= C_2 \left(|x| + |y| - |x - y|\right) \tag{4.98} \\
&= 2\, C_2 \left(|x| \wedge |y|\right) 1_{[xy \geq 0]}.
\end{aligned}
$$

It follows from Lemma 2.5.1 that $C_2 = 1$.

We state the results for $1 < p < 2$ and give references. By Ibragamov and Linnik (1971, 2.6.32, page 88)

$$C_p = -\frac{2}{\pi} \cos\left(\frac{\pi}{2}(p-1)\right) \Gamma(1-p). \tag{4.99}$$

Since

$$\Gamma(p)\Gamma(1-p) = \frac{\pi}{\sin \pi p} \tag{4.100}$$

(see, e.g., Ahlfors (1966, (30), page 198)) and

$$\cos\left(\frac{\pi}{2}(p-1)\right) = \sin\left(\frac{\pi}{2}p\right) \tag{4.101}$$

we can also write

$$C_p = \frac{2}{\Gamma(p) \sin\left(\frac{\pi}{2}(p-1)\right)}. \tag{4.102}$$

Condition (4.71) implies that the symmetric Lévy process X has a continuous α-potential density for all $\alpha > 0$ and hence, by Theorem 3.6.3, that X has local times. In fact, (4.71) is a necessary condition for (3.83), so it is fundamental for us (see Bertoin (1996, Theorem 19, II.5)).

4.2.1 Lévy processes on the torus

We obtain a large and interesting class of symmetric Lévy processes taking values in a torus, T^1. One aspect of these processes that interests us is that their associated Gaussian processes are random Fourier series with normal coefficients. These Gaussian processes are relatively easy to work with.

Let X be a symmetric Lévy process on R^1 with characteristic function given by (4.63) and transition probability density $p_t(x,0)=p_t(x)$. Let $\pi : R^1 \mapsto T^1$ denote the natural projection, $\pi(x) = x(\mathrm{mod}\ 2\pi)$, onto T^1. Consider

$$Y_t = \pi(X_t). \tag{4.103}$$

It is easy to see that $Y = \{Y_t,\, t \in R_+\}$ satisfies conditions (1) and (2) in the definition of a Lévy process on page 135 (here we extend the definition of a Lévy process to stochastic processes $X = \{X_t,\, t \in R_+\}$ with values in an Abelian group). Also, it is easy to check that the transition probability density function of Y is given by

$$q_t(x,y) = 2\pi \sum_{j=-\infty}^{\infty} p_t(x - y + 2\pi j) \qquad \forall\, x, y \in T^1 \tag{4.104}$$

with respect to the normalized Lebesgue measure $dx/(2\pi)$ on T^1. As usual we write $q_t(x,y) = q_t(x - y)$. It is clear from (4.104) that q_t is symmetric.

We have

$$
\begin{aligned}
Ee^{ijY_t} &= \frac{1}{2\pi} \int_0^{2\pi} e^{ijx} q_t(x)\, dx \tag{4.105} \\
&= \int_0^{2\pi} e^{ijx} \sum_{j=-\infty}^{\infty} p_t(x + 2\pi j)\, dx \\
&= \int_{-\infty}^{\infty} e^{ijx} p_t(x)\, dx = e^{-t\psi(j)}.
\end{aligned}
$$

The distribution of Y on T^1 is determined by its characteristic sequence $\{\exp(-t\psi(j))\}$. Since $q_t(x)$ is the density function for Y_t,

$$q_t(x) = \sum_{j=-\infty}^{\infty} e^{ijx} e^{-t\psi(j)}, \tag{4.106}$$

and the α-potential of Y is given by

$$\tilde{u}^\alpha(x) = \sum_{j=-\infty}^{\infty} \frac{e^{ijx}}{\alpha + \psi(j)} \tag{4.107}$$

$$= \frac{1}{\alpha} + 2 \sum_{j=1}^{\infty} \frac{\cos jx}{\alpha + \psi(j)}$$

for all $\alpha > 0$. Note that $\widetilde{u}(0) = \lim_{\alpha \to 0} \widetilde{u}^{\alpha}(0) = \infty$. Thus every point in T^1 is recurrent for Y, which should not be a surprise.

As in Theorem 4.2.4,

$$\widetilde{u}_{T_0}(x, y) = \widetilde{\phi}(x) + \widetilde{\phi}(y) - \widetilde{\phi}(x - y), \qquad (4.108)$$

where

$$\widetilde{\phi}(x) = 2 \sum_{j=1}^{\infty} \frac{1 - \cos jx}{\psi(j)}. \qquad (4.109)$$

4.3 Diffusions

Let $I \subset R^1$ be an interval that can be infinite. For the purpose of studying local times of diffusions we define a diffusion to be a Borel right process with state space I that has continuous paths. A diffusion is called regular and without traps if

$$P^x(T_y < \infty) > 0 \qquad \forall x, y \in I. \qquad (4.110)$$

Let X be a transient regular diffusion without traps on I. It is known that we can always find a positive σ-finite measure m on the state space I, called the speed measure, so that the 0-potential density $u(x, y)$ of X, with respect to m, is symmetric and continuous. Furthermore, there exist two continuous positive functions p and q with p strictly increasing and q strictly decreasing such that for all $x, y \in I$

$$u(x, y) = \begin{cases} p(x)q(y) & x \leq y \\ p(y)q(x) & y < x. \end{cases} \qquad (4.111)$$

See Ray (1963, (1.4), (1.6)) (see also Rogers and Williams (2000a, Theorem V, 50.7) for an exponentially killed diffusion).

The reader unfamiliar with these facts can simply take them as assumptions. However, in the next lemma we show how to derive the representation in (4.111), for particular potential densities that we are interested in, from the facts that X has continuous paths and continuous symmetric potential densities.

Lemma 4.3.1 *Let X be a diffusion in I with continuous symmetric α-potential densities, $\alpha > 0$, and with $P^x(T_y < \infty) > 0$ for all $x, y \in I$.*

Then, for some continuous functions p and q on I and $x, y \in I$, we have

$$u^\alpha(x, y) = \begin{cases} p(x)q(y) & x \leq y \\ p(y)q(x) & y < x. \end{cases} \tag{4.112}$$

Furthermore, p is strictly increasing and q is strictly decreasing.
   This is also true for $\alpha = 0$ when $u(x, y)$ exists.

Proof By (3.91), $u^\alpha(x, y) = E^x \left(\int_0^\infty e^{-\alpha t} \, dL_t^y \right)$. Making use of the fact that X has continuous paths, we see that for $x < v < y$

$$\begin{aligned} u^\alpha(x, y) &= E^x \left(\int_{T_v}^\infty e^{-\alpha t} \, dL_t^y \right) \tag{4.113} \\ &= E^x \left(e^{-\alpha T_v} E^{X_{T_v}} \left(\int_0^\infty e^{-\alpha t} \, dL_t^y \right) \right) \\ &= E^x \left(e^{-\alpha T_v} \right) u^\alpha(v, y) = \frac{u^\alpha(x, v) u^\alpha(v, y)}{u^\alpha(v, v)}. \end{aligned}$$

Here we also use (3.107). By Remark 3.6.6 and the assumption that $P^x(T_y < \infty) > 0$, we have $u^\alpha(x, y) > 0$ for all $x, y \in I$. Consequently, the following functions are well defined:

$$G(x) = \begin{cases} u^\alpha(x, x_0) & x \leq x_0 \\ \dfrac{u^\alpha(x, x) u^\alpha(x_0, x_0)}{u^\alpha(x, x_0)} & x > x_0 \end{cases}$$

and

$$H(x) = \begin{cases} \dfrac{u^\alpha(x, x)}{u^\alpha(x, x_0)} & x \leq x_0 \\ \dfrac{u^\alpha(x, x_0)}{u^\alpha(x_0, x_0)} & x > x_0. \end{cases}$$

Using (4.113) one can check that

$$u^\alpha(x, y) = G(x \wedge y) H(x \vee y). \tag{4.114}$$

This shows that $u^\alpha(x, y)$ can be represented as in (4.112).
   Furthermore, by the continuity of X, $E^x(e^{-T_v}) < 1$. Therefore, by (4.113), $u^\alpha(x, y)$ is strictly less than $u^\alpha(v, y)$. This shows that p is strictly increasing. A similar argument using (4.113) and the fact that

$$\frac{u^\alpha(v, y)}{u^\alpha(v, v)} = \frac{u^\alpha(y, v)}{u^\alpha(v, v)} = E^y(e^{-\alpha T_v}) \tag{4.115}$$

shows that q is strictly decreasing.

It is clear that everything goes through with $\alpha = 0$ when the 0-potential density of X exists. □

Lemma 4.3.2 *Let X be a recurrent diffusion in R^1 with continuous α-potential densities. Assume that $P^x(T_y < \infty) = 1$ for all $x, y \in R^1$. Then $u_{T_0}(x, x)$ is strictly increasing on R_+ and*

$$u_{T_0}(x, y) = \begin{cases} u_{T_0}(x, x) \wedge u_{T_0}(y, y) & xy \geq 0 \\ 0 & xy < 0. \end{cases} \tag{4.116}$$

Proof Let $y > x > 0$ and recall that $u_{T_0}(x, y) = E^x(L^y_{T_0})$. By the support properties of L^y_t and the strong Markov property,

$$u_{T_0}(x, y) = E^x\left(1_{\{T_y < T_0\}} L^y_{T_0} \circ \theta_{T_y}\right) = P^x(T_y < T_0) u_{T_0}(y, y) \tag{4.117}$$

and, similarly,

$$u_{T_0}(y, x) = P^y(T_x < T_0) u_{T_0}(x, x). \tag{4.118}$$

Furthermore, using continuity of the paths of X and the strong Markov property, we have

$$P^y(T_0 < \infty) = P^y(T_x < T_0, T_0 \circ \theta_{T_x} < \infty) = P^y(T_x < T_0) P^x(T_0 < \infty), \tag{4.119}$$

so by the assumption that $P^x(T_y < \infty) = 1$ for all $x, y \in R^1$ we see that $P^y(T_x < T_0) = 1$. Using this in (4.118) we see that

$$u_{T_0}(y, x) = u_{T_0}(x, x). \tag{4.120}$$

Let $\overline{\tau}_0 = 0$ and for $i = 1, 2, \ldots$. Define

$$\tau_i = \inf\{t \geq \overline{\tau}_{i-1} : X_t = x\}$$
$$\overline{\tau}_i = \inf\{t \geq \tau_i : X_t = y\}.$$

Using continuity of the paths of X, the fact that T_0 is a terminal time, and the strong Markov property we have

$$P^y(T_0 < \infty) = \sum_{i=1}^{\infty} P^y(\tau_i < T_0 < \overline{\tau}_i) \tag{4.121}$$

$$\leq \sum_{i=1}^{\infty} P^y(T_0 \circ \theta_{T_{\tau_i}} < T_y \circ \theta_{T_{\tau_i}}, \tau_i < \infty)$$

$$= \sum_{i=1}^{\infty} P^x(T_0 < T_y),$$

since $P^y(T_x < \infty) = 1$. Since, by assumption, $P^y(T_0 < \infty) = 1$, we must have $P^x(T_0 < T_y) > 0$. Furthermore, since

$$\begin{aligned} P^x(T_0 < \infty) &= P^x(T_0 < T_y) + P^x(T_y < T_0, T_0 \circ \theta_{T_y} < \infty) \\ &= P^x(T_0 < T_y) + P^x(T_y < T_0), \end{aligned}$$

it follows that $P^x(T_y < T_0) < 1$. Using this in (4.117) we see that

$$u_{T_0}(x, y) < u_{T_0}(y, y). \tag{4.122}$$

Finally, we note that by Lemma 3.8.1, $u_{T_0}(x, y) = u_{T_0}(y, x)$. Combining this with (4.120) and (4.122) we get (4.116) when $x, y \geq 0$. The extension to all x, y is obvious. $\qquad\square$

4.4 Left limits and quasi left continuity

We continue to explore the sample path properties of strongly symmetric Borel right processes with continuous α-potential densities. We first show that X has left-hand limits.

Theorem 4.4.1 *Let* $X = (\Omega, \mathcal{G}, \mathcal{G}_t, X_t, \theta_t, P^x)$ *be a strongly symmetric Borel right process. Then* $X_{s^-} := \lim_{r \uparrow\uparrow s} X_r$ *exists in* S *for all* $s \in (0, \zeta)$.

Proof Let $D \subseteq R_+$ denote the positive rational numbers . Since X_t is right continuous, it suffices to show that $\lim_{r \uparrow\uparrow s, r \in D} X_r$ exists in S for every $s \in (0, \zeta)$. Fix $t \in D$. Let $\{P_t; t \geq 0\}$ denote the transition semigroup and m denote the reference measure for X. Then, using the fact that X is symmetric with respect to m, we have that, for any $0 < s_1 < s_2 < \ldots < s_n \in D \cap (0, t)$ and $f_i \in \mathcal{B}_b(S)$, $i = 1, \ldots, n$,

$$E^m \left(\prod_{i=1}^n f_i(X_{t-s_i}) I_{\{t < \zeta\}} \right) \tag{4.123}$$

$$= \int P_{t-s_n}(x, dy_n) f_n(y_n) P_{s_n-s_{n-1}}(y_n, dy_{n-1}) f_{n-1}(y_{n-1}) \cdots$$

$$\cdots P_{s_2-s_1}(y_2, dy_1) f_1(y_1) P_{s_1}(y_1, dz) 1_S(z) \, dm(x)$$

$$= \int 1_S(x) P_{t-s_n}(f_n P_{s_n-s_{n-1}}(f_{n-1} \cdots P_{s_2-s_1}(f_1 P_{s_1} 1_S)) \cdots)(x) \, dm(x).$$

On page 75 we express the symmetry condition in inner product notation. Using this notation we note that for $f, g, h \in L^2(dm)$, $(h, P_t fg) =$

$(P_t h, fg) = (g, f P_t h)$. Using this we see that the last line of (4.123) equals

$$\int 1_S(x) P_{s_1}(f_1 P_{s_2-s_1} \cdots (f_{n-1} P_{s_n-s_{n-1}}(f_n P_{t-s_n} 1_S)) \cdots)(x)\, dm(x)$$

$$= E^m \left(\prod_{i=1}^{n} f_i(X_{s_i}) I_{\{t<\zeta\}} \right). \tag{4.124}$$

Hence, under $P^m(\cdot \,|\, I_{\{t<\zeta\}})$, $\{X_s\,;\, s \in D \cap (0,t)\}$ has the same law as $\{X_{t-s}\,;\, s \in D \cap (0,t)\}$. Because X is right continuous, $\{X_{t-s}\,;\, s \in D \cap (0,t)\}$ has left-hand limits in S, and therefore so does $\{X_s\,;\, s \in D \cap (0,t)\}$, $P^m(\cdot \,|\, I_{\{t<\zeta\}})$ almost surely. Since this is true for all $t \in D$, $P^m(\Gamma) = 0$, where

$$\Gamma := \left\{ \lim_{r \uparrow\uparrow s,\, r \in D} X_r \text{ fails to exist in } S \text{ for some } s \in (0,\zeta) \right\}. \tag{4.125}$$

We now show that $P^x(\Gamma) = 0$ for all $x \in S$. Let $f(x) = P^x(\Gamma)$ so that $\int f(x)\, dm(x) = 0$. Since, by hypothesis, X has α-potential densities, we have $U^\alpha f(x) = 0$ for all $x \in S$ and $\alpha > 0$.

Assume that $\Gamma \in \mathcal{F}$. Note that $\theta_t^{-1}\Gamma = \{\theta_t \omega \in \Gamma\}$. Therefore, by the Markov property, for any $t > 0$

$$\begin{aligned} P^x(\theta_t^{-1}\Gamma) &= E^x(1_\Gamma \circ \theta_t) \tag{4.126} \\ &= E^x\left(E^x(1_\Gamma \circ \theta_t | \mathcal{G}_t)\right) \\ &= E^x\left(E^{X_t}(1_\Gamma)\right) = P_t f(x) \end{aligned}$$

(see (3.27) and (3.11)). Since

$$\theta_t^{-1}\Gamma = \left\{ \lim_{r \uparrow\uparrow s,\, r \in D} X_r \text{ fails to exist in } S \text{ for some } s \in (t,\zeta) \right\}, \tag{4.127}$$

we see that $P_t f(x) \uparrow f(x)$ as $t \to 0$. Therefore, by the Monotone Convergence Theorem,

$$\begin{aligned} 0 = \lim_{\alpha \to \infty} \alpha U^\alpha f(x) &= \lim_{\alpha \to \infty} \alpha \int_0^\infty e^{-\alpha t} P_t f(x)\, dt \tag{4.128} \\ &= \lim_{\alpha \to \infty} \int_0^\infty e^{-t} P_{t/\alpha} f(x)\, dt = f(x), \end{aligned}$$

which completes the proof of this theorem.

It remains to show that $\Gamma \in \mathcal{F}$. We prove this first when $S = R_+$. In this case, we denote Δ, the one-point compactification of R_+ by ∞ (as usual). Note that $\sup_{R_+} x = \infty$. We write $\overline{R}_+ = R_+ \cup \{\infty\}$. With these definitions set

$$\overline{Z}_t = \limsup_{r \uparrow\uparrow t,\, r \in D} X_r \qquad \underline{Z}_t = \liminf_{r \uparrow\uparrow t,\, r \in D} X_r \qquad t \in (0,\infty). \tag{4.129}$$

Suppose that the map $(t, \omega) \mapsto \overline{Z}_t(\omega)$ from $(0, \infty) \times \Omega$ to $\overline{R}_+$ is measurable from $\mathcal{B}((0, \infty)) \times \mathcal{F}$ to $\mathcal{B}(\overline{R}_+)$, and similarly for $\underline{Z}_t$. Then Γ is the projection on Ω of the $\mathcal{B}((0, \infty)) \times \mathcal{F}$ measurable set

$$\left\{ (t, \omega) \,|\, \overline{Z}_t(\omega) \neq \underline{Z}_t(\omega), \ X_t(\omega) \in R_+ \right\} \bigcup \left\{ (t, \omega) \,|\, \overline{Z}_t(\omega) = \infty, \right.$$

$$\left. X_t(\omega) \in R_+ \right\}$$

and hence is in $\mathcal{F}$ by the Projection Theorem (Theorem 14.3.1).

To prove that $(t, \omega) \mapsto \overline{Z}_t(\omega)$ is measurable we note first that

$$\overline{Z}_t = \lim_n Z_t^n, \tag{4.130}$$

where

$$Z_t^n = \sum_{k=0}^{\infty} 1_{(k2^{-n}, (k+1)2^{-n}]}(t) \sup_{\{k2^{-n} < r < t, \, r \in D\}} X_r =: \sum_{k=0}^{\infty} Y_t^{k,n} \tag{4.131}$$

with $Y_t^{k,n} = 1_{(k2^{-n}, (k+1)2^{-n}]}(t) \sup_{\{k2^{-n} < r < t, \, r \in D\}} X_r$. Fix k and n and note that $Y_t^{k,n}$ is left continuous. Therefore

$$Y_t^{k,n} = \lim_m \sum_{j=0}^{\infty} 1_{(j2^{-m}, (j+1)2^{-m}]}(t) Y_{j2^{-m}}^{k,n}. \tag{4.132}$$

Thus $Y_t^{k,n}$ is measurable. Then Z_t^n is measurable, and consequently so is $\overline{Z}_t$. A similar analysis applies to $(t, \omega) \mapsto \underline{Z}_t(\omega)$. This proves the theorem when $S = R_+$.

Finally, let S be a general locally compact Hausdorff space and let $d(x, y)$ be a metric compatible with the topology of S. Set $d(x, \Delta) = \infty$ for each $x \in S$. Let $x_1, x_2, \ldots$ be a countable dense set in S. Then

$$\Gamma = \bigcup_{n=1}^{\infty} \left\{ \lim_{r \uparrow \uparrow s, \, r \in D} d(x_n, X_r) \text{ fails to exist in } R_+ \text{ for some } s \in (0, \zeta) \right\}. \tag{4.133}$$

Repeating the argument given when $S = R_+$ applied to $d(x_n, X_r)$, we see that each set in the union in (4.133) is in $\mathcal{F}$. Thus $\Gamma \in \mathcal{F}$. This completes the proof of Theorem 4.4.1. $\qquad \square$

We say that X is quasi left continuous on $[0, \zeta)$ if, for any increasing sequence $\{T_n\}$ of $\mathcal{G}_t$ stopping times with limit T, we have $X_{T_n} \to X_T$ on $\{T < \zeta\}$ almost surely.

Theorem 4.4.2 *Let $X = (\Omega, \mathcal{G}, \mathcal{G}_t, X_t, \theta_t, P^x)$ be a strongly symmetric Borel right process with continuous α-potential density $u^{\alpha}(x, y)$. Then X_t is quasi left continuous on $[0, \zeta)$.*

Proof Let $\{T_n\}$ be an increasing sequence of stopping times with limit T and define $L = \lim_n X_{T_n}$, which exists in S when $T < \zeta$ by Theorem 4.4.1. Let $f, g \in C_b(S)$. Using the fact that $L \in \mathcal{F}_T$ and the strong Markov property, we have that for any $\alpha > 0$,

$$E^x \left(f(L) U^\alpha g(X_T) I_{\{T < \zeta\}} \right) \tag{4.134}$$

$$= E^x \left(f(L) I_{\{T < \zeta\}} E^{X_T} \left(\int_0^\infty e^{-\alpha t} g(X_s) \, ds \right) \right)$$

$$= E^x \left(f(L) I_{\{T < \zeta\}} \int_0^\infty e^{-\alpha t} g(X_{T+s}) \, ds \right)$$

$$= E^x \left(f(L) I_{\{T < \zeta\}} e^{\alpha T} \int_T^\infty e^{-\alpha t} g(X_s) \, ds \right)$$

$$= \lim_{n \to \infty} E^x \left(f(X_{T_n}) I_{\{T_n < \zeta\}} e^{\alpha T_n} \int_{T_n}^\infty e^{-\alpha t} g(X_s) \, ds \right),$$

by the Dominated Convergence Theorem. Here we use the fact that

$$\left| f(X_{T_n}) I_{\{T_n < \zeta\}} e^{\alpha T_n} \int_{T_n}^\infty e^{-\alpha t} g(X_s) \, ds \right|$$

$$\leq \|f\|_\infty \|g\|_\infty e^{\alpha T_n} \int_{T_n}^\infty e^{-\alpha t} \, ds = \frac{1}{\alpha} \|f\|_\infty \|g\|_\infty$$

is uniformly bounded.

Using the same procedure involving the strong Markov property we have that for any $\alpha > 0$

$$E^x \left(f(L) U^\alpha g(X_T) I_{\{T < \zeta\}} \right) = \lim_{n \to \infty} E^x \left(f(X_{T_n}) U^\alpha g(X_{T_n}) I_{\{T_n < \zeta\}} \right). \tag{4.135}$$

By Lemma 3.4.1, $U^\alpha g$ is bounded and continuous on S and therefore $\lim_n U^\alpha g(X_{T_n}) = U^\alpha g(L)$ on $T < \zeta$. Hence, by (4.135) we get

$$E^x \left(f(L) U^\alpha g(X_T) I_{\{T < \zeta\}} \right) = E^x \left(f(L) U^\alpha g(L) I_{\{T < \zeta\}} \right). \tag{4.136}$$

By (3.16) and the Dominated Convergence Theorem, this implies that

$$E^x \left(f(L) g(X_T) I_{\{T < \zeta\}} \right) = E^x \left(f(L) g(L) I_{\{T < \zeta\}} \right). \tag{4.137}$$

The Monotone Class Theorem (see, e.g., Rogers and Williams (2000b, II.3.1)) then shows that

$$E^x \left(h(L, X_T) I_{\{T < \zeta\}} \right) = E^x \left(h(L, L) I_{\{T < \zeta\}} \right) \tag{4.138}$$

for all bounded measurable functions h, in particular for $h(x, y) = 1_{\{x=y\}}$, which yields $P^x \left(X_T = L \, | \, T < \zeta \right) = 1$. $\qquad \square$

A Borel right process is said to be a standard Markov process if it is quasi left continuous on $[0, \zeta)$. Thus we see that a strongly symmetric Borel right process with continuous α-potential density is a standard Markov process.

In Theorem 3.2.6 we use the Projection Theorem to show that, for a Borel right process, the first hitting time $T_A = \inf\{s > 0 \,|\, X_s \in A\}$ is a stopping time for every Borel set A. In this book all our results are for standard Markov processes, and we only consider sets A that are open or closed. In this case we can give a direct proof of Theorem 3.2.6 that does not use the Projection Theorem.

Theorem 4.4.3 *Let $X = (\Omega, \mathcal{G}, \mathcal{G}_t, X_t, \theta_t, P^x)$ be a standard Markov process. Then T_A is an $\mathcal{F}_t$ stopping time for every open or closed set $A \subseteq S$.*

Proof The proof for A open is exactly the same as for Brownian motion. We note that

$$\{T_A < t\} = \bigcup_{0 < r < t,\, r \in D} \{X_r \in A\} \in \mathcal{F}_t \tag{4.139}$$

by the right continuity of X.

Now let A be a closed set. Let $A_n \supseteq A$ be a decreasing sequence of open sets such that $A_n \supseteq \overline{A_{n+1}}$ and $A = \cap_n A_n$. Note that

$$A = \cap_n \overline{A_n}. \tag{4.140}$$

Recall the definition $D_B = \inf\{s \geq 0 \,|\, X_s \in B\}$ in (3.35). Clearly $D_{A_n} \leq D_A$ is increasing in n. Set

$$D = \lim D_{A_n} \leq D_A. \tag{4.141}$$

By right continuity we have $X_{D_{A_n}} \in \overline{A_n}$.

Note that D_{A_n} is a stopping time. This can be shown by using (4.139) but with the union taken over $\{0 \leq r < t, r \in D\}$. Since $\{D \leq t\} = \cap_n \{D_{A_n} \leq t\}$, we see that D is also a stopping time. Therefore, since X is quasi left continuous on $[0, \zeta)$, we have

$$X_D = \lim_n X_{D_{A_n}} \in \cap_n \overline{A_n} = A \qquad \text{on } D < \zeta. \tag{4.142}$$

Hence $D \geq D_A$ on $\{D < \zeta\}$, so that by (4.141) we have $D = D_A$ on $\{D < \zeta\}$ and a fortiori on $\{D_A < \zeta\}$. Now, from the definitions, we have that if $D_A \geq \zeta$, then $D_A = \infty$, so that in fact $D = D_A$ on $D_A < \infty$. Thus D_A is an $\mathcal{F}_t$ stopping time. The proof is completed as it was in the proof of Theorem 3.2.6, starting at (3.38). □

A Hunt process is a Borel right process that is quasi left continuous on $[0, \infty)$, that is, for any increasing sequence $\{T_n\}$ of $\mathcal{G}_t$ stopping times with limit T we have $X_{T_n} \to X_T$ on $\{T < \infty\}$ almost surely. It can be shown that a strongly symmetric Borel right process with continuous α-potential density is a Hunt process. The proof of this would take us too far afield, so we refer the interested reader to Theorem 3.8 in our paper Marcus and Rosen (1992d).

4.5 Killing at a terminal time

Let $X = (\Omega, \mathcal{G}, \mathcal{G}_t, X_t, \theta_t, P^x)$ be a strongly symmetric Borel right process with transition semigroup $\{P_t \,;\, t \geq 0\}$ and continuous α-potential density $u^\alpha(x, y)$. Let T be a terminal time. In this section we consider the process obtained by killing X at T.

For $f \in C(S)$, set

$$\widetilde{P}_t f(x) = E^x(f(X_t) 1_{\{t < T\}}). \tag{4.143}$$

Using the terminal time property we show that $\{\widetilde{P}_t \,;\, t \geq 0\}$ is a semigroup,

$$
\begin{aligned}
\widetilde{P}_{s+t} f(x) &= E^x(f(X_{s+t}) 1_{\{s+t < T\}}) & (4.144)\\
&= E^x\left((f(X_s) 1_{\{s < T\}}) \circ \theta_t \, 1_{\{t < T\}}\right)\\
&= E^x(\widetilde{P}_s f(X_t) 1_{\{t < T\}})\\
&= \widetilde{P}_t(\widetilde{P}_s f)(x).
\end{aligned}
$$

Let

$$\widetilde{X}_t(\omega) = \begin{cases} X_t(\omega) & \text{if } t < T \\ \Delta & \text{otherwise.} \end{cases} \tag{4.145}$$

Clearly, $t \mapsto \widetilde{X}_t$ is right continuous.

Let S be a $\mathcal{G}_t$ stopping time and let $H \in \mathcal{G}_S$. Noting that $\{S < T\} \in \mathcal{G}_S$ and using the strong Markov property of X, we see that

$$
\begin{aligned}
E^x\left(f(\widetilde{X}_{S+t}) 1_H\right) &= E^x(f(X_{S+t}) 1_{\{S+t < T\}} 1_H) & (4.146)\\
&= E^x\left((f(X_t) 1_{\{t < T\}}) \circ \theta_S 1_{\{S < T\}} 1_H\right)\\
&= E^x(\widetilde{P}_t f(X_S) 1_{\{S < T\}} 1_H)\\
&= E^x(\widetilde{P}_t f(\widetilde{X}_S) 1_H).
\end{aligned}
$$

Set

$$\widetilde{\theta}_t(\omega) = \begin{cases} \theta_t(\omega) & \text{if } t < T(\omega) \\ \Delta & \text{otherwise.} \end{cases} \tag{4.147}$$

We see that $\widetilde{X} = (\Omega, \mathcal{G}_t, \mathcal{G}, \widetilde{X}_t, \widetilde{\theta}_t, \widetilde{P}^x)$ satisfies all the conditions in the definition of a right continuous simple Markov process with transition semigroup $\{\widetilde{P}_t \, ; \, t \geq 0\}$ on S_Δ, with the possible exception of the condition that $\widetilde{P}^x(\widetilde{X}_0 = x) = 1$ for all $x \in S_\Delta$.

We now restrict our attention to the case when $T = T_B$, the first hitting time of a set B. If, for example, B is open, then it is clear that we do not have $\widetilde{P}^x(\widetilde{X}_0 = x) = 1$ for $x \in B$, since in fact $\widetilde{X}_0 = \Delta$, $\widetilde{P}^x$ a.s. We remedy this defect by restricting the state space. We begin by showing that the semigroup $\{\widetilde{P}_t \, ; \, t \geq 0\}$ is symmetric for certain sets B.

Lemma 4.5.1 *Let* $\{\widetilde{P}_t ; t \geq 0\}$ *be the semigroup defined in (4.143) with* $T = T_B$, *the first hitting time of the set* B. *If* B *is open,*

$$\int g(x)\widetilde{P}_t f(x)\, dm(x) = \int f(x)\widetilde{P}_t g(x)\, dm(x) \qquad (4.148)$$

for all $t \geq 0$ *and all* $f, g \in \mathcal{B}_b(S)$. *This also holds if* B *is closed and* $P^x(T_B = 0) = 1$ *for all* $x \in B$.

Proof Let B be open. By the right continuity of X and the symmetry of P_t, we have

$$\int g(x)\widetilde{P}_t f(x)\, dm(x) \qquad (4.149)$$

$$= \int g(x)E^x(f(X_t)1_{\{t<T_B\}})\, dm(x)$$

$$= \lim_{n\to\infty} \int g(x)E^x(f(X_t)\prod_{j=0}^{n} 1_{\{S\cap B^c\}}(X_{jt/n}))\, dm(x)$$

$$= \lim_{n\to\infty} \int g(x_0)\prod_{i=1}^{n} P_{t/n}(x_{i-1},\, dx_i)f(x_n)\prod_{j=0}^{n} 1_{\{S\cap B^c\}}(x_j)\, dm(x_0)$$

$$= \lim_{n\to\infty} \int g(x)1_{S\cap B^c}(x)$$
$$P_{t/n}1_{S\cap B^c}(P_{t/n}1_{S\cap B^c}\cdots(P_{t/n}1_{S\cap B^c}f))\cdots)(x)\, dm(x)$$

$$= \lim_{n\to\infty} \int f(x)1_{S\cap B^c}(x)$$
$$P_{t/n}1_{S\cap B^c}(P_{t/n}1_{S\cap B^c}\cdots(P_{t/n}1_{S\cap B^c}g))\cdots)(x)\, dm(x)$$

$$= \int f(x)\widetilde{P}_t g(x)\, dm(x).$$

Now let B be closed. Let $B_n \supseteq B$ be a decreasing sequence of open sets such that $B_n \supseteq \overline{B_{n+1}}$ and $B = \cap_n B_n$. Let D_B be as in the proof of Theorem 4.4.3. We show, in the proof of Theorem 4.4.3 that $D_{B_n} \uparrow D$, for some stopping time D, with $D \leq D_B$ and $D = D_B$ on $\{D < \zeta\}$. It is easy to see that $D_{B_n} = T_{B_n}$ for open sets B_n. The condition that $P^x(T_B = 0) = 1$ for all $x \in B$ implies that $D_B = T_B$ as well. Therefore, $T_{B_n} \uparrow D$ with $D \leq T_B$ and $D = T_B$ on $\{D < \zeta\}$. Using this we see that

$$E^x(f(X_t)1_{\{t < T_{B_n}\}}) \to E^x(f(X_t)1_{\{t < D\}}) = E^x(f(X_t)1_{\{t < T_B\}}) \quad (4.150)$$

for all $x \in S$. Consequently, (4.148) holds when $T = T_B$ since it holds when $T = T_{B_n}$, because the sets B_n are open. $\qquad \square$

Theorem 4.5.2 *Let $X = (\Omega, \mathcal{G}, \mathcal{G}_t, X_t, \theta_t, P^x)$ be a strongly symmetric Borel right process with continuous α-potential density $u^\alpha(x, y)$. Let B be an open set. Let $\widetilde{X} = (\Omega, \mathcal{G}_t, \mathcal{G}, \widetilde{X}_t, \widetilde{\theta}_t, \widetilde{P}^x)$ with semigroup $\{\widetilde{P}_t; t \geq 0\}$ be as defined in (4.143) with $T = T_B$ and state space $\widetilde{S} = \overline{B}^c$. Then $\widetilde{X}$ is a strongly symmetric Borel right process with continuous α-potential density $\widetilde{u}^\alpha(x, y)$.*

This also holds when B is a closed set with the additional property that $P^x(T_B = 0) = 1$ for all $x \in B$.

Proof Using the Markov property at the stopping time T_B, we see that for any bounded continuous function f we have

$$\int u^\alpha(x, y) f(y) \, dm(y) \qquad (4.151)$$

$$= E^x \left(\int_0^\infty e^{-\alpha t} f(X_t) \, dt \right)$$

$$= E^x \left(\int_0^{T_B} e^{-\alpha t} f(X_t) \, dt \right) + E^x \left(\int_{T_B}^\infty e^{-\alpha t} f(X_t) \, dt \right)$$

$$= \widetilde{U}^\alpha f(x) + E^x \left(e^{-\alpha T_B} E^{X_{T_B}} \left(\int_0^\infty e^{-\alpha t} f(X_t) \, dt \right) \right)$$

$$= \widetilde{U}^\alpha f(x) + E^x \left(e^{-\alpha T_B} \int u^\alpha(X_{T_B}, y) f(y) \, dm(y) \right).$$

To say that $\widetilde{X}$ has an α-potential density means that we can find a function $\widetilde{u}^\alpha(x, y)$ such that $\widetilde{U}^\alpha f(x) = \int \widetilde{u}^\alpha(x, y) f(y) \, dy$. Thus we see from (4.151) that $\widetilde{X}$ has the α-potential density

$$\widetilde{u}^\alpha(x, y) = u^\alpha(x, y) - E^x \left(e^{-\alpha T_B} u^\alpha(X_{T_B}, y) \right) \qquad x, y \in S. \quad (4.152)$$

By Lemma 3.4.3 we see that $\widetilde{u}^\alpha(x, y)$ is continuous in y uniformly in x. By Lemma 4.5.1 we have

$$\widetilde{u}^\alpha(x, y) = \widetilde{u}^\alpha(y, x) \qquad m \text{ a.e. } x, y \in S \qquad (4.153)$$

and thus by (4.152)

$$E^x \left(e^{-\alpha T_B} u^\alpha (X_{T_B}, y) \right) = E^y \left(e^{-\alpha T_B} u^\alpha (X_{T_B}, x) \right), \quad m \text{ a.e. } x, y \in S. \tag{4.154}$$

We now show that this holds for all $x, y \in S$, and consequently (4.153) holds for all $x, y \in S$.

Let $h(x, y) = E^x \left(e^{-\alpha T_B} u^\alpha (X_{T_B}, y) \right)$. By the strong Markov property

$$E^x \left(h(X_t, y) \right) = E^x \left(e^{-\alpha T_B \circ \theta_t} u^\alpha (X_{t + T_B \circ \theta_t}, y) \right). \tag{4.155}$$

Now let $\bar{X} = (\bar{\Omega}, \bar{\mathcal{G}}, \bar{\mathcal{G}}_t, \bar{X}_t, \bar{\theta}_t, \bar{P}^x)$ be an independent copy of X. We have

$$\bar{E}^y \left(E^x \left(h(X_t, \bar{X}_s) \right) \right) = \bar{E}^y \left(E^x \left(e^{-\alpha T_B \circ \theta_t} u^\alpha \left(X_{t + T_B \circ \theta_t}, \bar{X}_s \right) \right) \right). \tag{4.156}$$

Since $t + T_B \circ \theta_t \downarrow T_B$ as $t \to 0$, using the right continuity of X and the continuity of $u^\alpha(x, y)$, we have

$$\lim_{s, t \to 0} \bar{E}^y \left(E^x \left(h(X_t, \bar{X}_s) \right) \right) = h(x, y). \tag{4.157}$$

Fix $x, y \in S$. Let $K \subset S$ be a compact set containing x, y. Let $f : K \times K \to [0, 1]$ be a continuous symmetric function with $f(x, y) = f(y, x) = 1$. Then clearly (4.157) implies that

$$\lim_{s, t \to 0} \bar{E}^y \left(E^x \left(h(X_t, \bar{X}_s) f(X_t, \bar{X}_s) \right) \right) = h(x, y). \tag{4.158}$$

Note that by (3.68) and the fact that $f(x', y')$ is bounded and compactly supported we have that $h(x', y') f(x', y')$ is bounded.

Let $g \in \mathcal{B}_b(S)$. Then it follows from the basic definitions in Section 3.1 that

$$\int g(y') U^\beta(y, dy') = \int_0^\infty e^{-\beta t} E^y (g(X_t)) \, dt. \tag{4.159}$$

Using this, we see that

$$\lim_{\beta, \beta' \to \infty} \beta \beta' \int \int h(x', y') f(x', y') U^\beta(x, dx') U^{\beta'}(y, dy') \tag{4.160}$$

$$= \lim_{\beta, \beta' \to \infty} \beta \beta' \int_0^\infty \int_0^\infty e^{-\beta t} e^{-\beta' s} \bar{E}^y$$
$$\left(E^x \left(h(X_t, \bar{X}_s) f(X_t, \bar{X}_s) \right) \right) ds \, dt$$

$$= \lim_{\beta, \beta' \to \infty} \int_0^\infty \int_0^\infty e^{-t} e^{-s} \bar{E}^y$$
$$\left(E^x \left(h(X_{t/\beta}, \bar{X}_{s/\beta'}) f(X_{t/\beta}, \bar{X}_{s/\beta'}) \right) \right) ds \, dt$$

$$= h(x, y)$$

by (4.158). It follows from (4.154) that $h(x', y') = h(y', x')$ for m almost all $x', y' \in S$. Therefore, since $f(x', y')$ is symmetric and the measures $U^\beta(x, dx')$ and $U^{\beta'}(y, dy')$ are absolutely continuous with respect to m, we see that

$$\int \int h(x', y') f(x', y') U^\beta(x, dx') U^{\beta'}(y, dy') \qquad (4.161)$$
$$= \int \int h(y', x') f(y', x') U^\beta(x, dx') U^{\beta'}(y, dy').$$

Applying the sequence of equalities in (4.160) to this last expression, we see that $h(x, y)$ is symmetric, which in turn implies, as we pointed out above, that $\tilde{u}^\alpha(x, y)$ is symmetric. Finally, since $\tilde{u}^\alpha(x, y)$ is continuous in each variable uniformly in the other, we see that $\tilde{u}^\alpha(x, y)$ is also jointly continuous.

We now note that since $\bar{B}^c$ is open, it is clear that for each $x \in \bar{B}^c$ we have $P^x(T_B = 0) = 0$, so that $\tilde{P}^x(\tilde{X}_0 = x) = P^x(X_0 = x) = 1$. We next show that $P^x(T_{\bar{B}} = T_B) = 1$ for each $x \in \bar{B}^c$, that is, that $\tilde{X}$, starting in $\tilde{S} = \bar{B}^c$, does not leave $\tilde{S}$ before being killed. This is a tautology when B is closed. When B is open, $\tilde{u}^\alpha(x, y) = 0$ for any $x \in B$; hence by continuity for any $x \in \partial B$ and therefore for such x we have

$$E^x\left(\int_0^{T_B} e^{-\alpha t} 1_S(X_t)\, dt\right) = \int_S \tilde{u}^\alpha(x, y)\, dm(y) = 0. \qquad (4.162)$$

This implies that $P^x(T_B > 0) = 0$ for any $x \in \partial B$. Hence, by the strong Markov property, for any $x \in S$,

$$P^x(T_B > T_{\partial B}) = E^x\left(P^{X_{\partial B}}(T_B > 0)\right) = 0 \qquad (4.163)$$

and therefore $P^x(T_{\bar{B}} = T_B) = 1$.

We have obtained enough information to establish that $\tilde{X}$ restricted to $\tilde{S} = \bar{B}^c$ is a right continuous simple Markov process with continuous α-potential densities. The theorem now follows from Lemma 3.4.2. $\square$

Remark 4.5.3 Let X and X' be strongly symmetric Borel right processes on compact state spaces S and S', respectively, with continuous 0-potential densities $u^0(x, y)$ and $u'^0(x, y)$. Assume that $S' \subseteq S$ and $u'^0(x, y) = u^0(x, y)$ on $S' \times S'$. Let $B \subseteq S$ be an open set with compact closure contained in S'. Let $\tilde{X}$ be the process obtained in Theorem 4.5.2 by killing X at $T_{\bar{B}^c}$ and $\tilde{X}'$ be the process obtained by killing X' at $T_{\bar{B}^c}$. Then $\tilde{X}$ and $\tilde{X}'$ have the same distribution. This follows from Remark 4.1.11. Using the notation of Remark 4.1.11, we see that $\{X_{\tau_t}\,;\, t \geq 0\}$

has the same distribution as $\{X'_t ; t \geq 0\}$. It is clear that for $t < T_{\bar{B}^c}$, $A_t = t$, so that $\tau_t = t$.

Example 4.5.4 Let 0 be a point in the state space S. When $T = T_0$ we can describe the potential densities $\widetilde{u}^\alpha(x, y)$ more explicitly than in (4.152). Recall from (3.83) that $P^0(T_0 = 0) = 1$, so the condition in Theorem 4.5.2 is satisfied. Using the fact that $X_{T_0} = 0$, it follows from (4.152) that

$$\widetilde{u}^\alpha(x, y) = u^\alpha(x, y) - E^x \left(e^{-\alpha T_0} \right) u^\alpha(0, y). \tag{4.164}$$

Therefore, by (3.107), we have

$$\widetilde{u}^\alpha(x, y) = u^\alpha(x, y) - \frac{u^\alpha(x, 0) u^\alpha(0, y)}{u^\alpha(0, 0)}. \tag{4.165}$$

If we consider $u^\alpha(x, y)$ as the 0-potential density of X (in the notation of Theorem 4.5.2), killed at the end of an independent exponential time with mean $1/\alpha$, then, by Remark 3.8.3, the right-hand side of (4.165) is equal to $u^\alpha_{T_0}(x, y)$. It is not surprising that $\widetilde{u}^\alpha(x, y) = u^\alpha_{T_0}(x, y)$.

Example 4.5.5 By Theorem 4.5.2, Brownian motion killed at T_0 is a strongly symmetric Borel right process. By (4.165) and (2.5), its α-potential density for $\alpha > 0$ is

$$\widetilde{u}^\alpha(x, y) = \frac{e^{-\sqrt{2\alpha}|y-x|}}{\sqrt{2\alpha}} - \frac{e^{-\sqrt{2\alpha}|x|} e^{-\sqrt{2\alpha}|y|}}{\sqrt{2\alpha}}. \tag{4.166}$$

Because Brownian motion has continuous paths, the killed process naturally breaks up into two processes, one with state space $(0, \infty)$ and the other with state space $(-\infty, 0)$. To be specific we deal with the process with state space $(0, \infty)$. In this case, using (2.5) once more, we see that

$$\begin{aligned} \widetilde{u}^\alpha(x, y) &= \frac{e^{-\sqrt{2\alpha}|y-x|}}{\sqrt{2\alpha}} - \frac{e^{-\sqrt{2\alpha}|x+y|}}{\sqrt{2\alpha}} \\ &= \int_0^\infty e^{-\alpha t} \left(p_t(y - x) - p_t(y + x) \right) dt, \end{aligned} \tag{4.167}$$

where $p_t(y - x)$ is the transition density for Brownian motion. By the uniqueness of the Laplace transform, we see that the Borel right process $\widetilde{X}$ with state space $(0, \infty)$, obtained by killing Brownian motion the first time it hits 0, has transition densities

$$\widetilde{p}_t(x, y) = p_t(y - x) - p_t(y + x) \tag{4.168}$$

with respect to Lebesgue measure. Note that $(0, \infty)$ is locally compact.

A one-point compactification of $(0, \infty)$ can be obtained by first embedding it in the compact set $[0, \infty]$ and then considering 0 and ∞ as a single point. By considering the limits of $\tilde{u}^{\alpha}(x, y)$, both as $x \to \infty$ and $x \to 0$, we see from (4.167) and Corollary 4.1.4 (2) that $\tilde{X}$ is a Feller process.

We note that the simple relation

$$\int_0^{\infty} p_t(x - y) \, dy + \int_0^{\infty} p_t(x + y) \, dy = 1 \qquad (4.169)$$

together with (4.168), for standard Brownian motion B_t, show that

$$\begin{aligned}
P^x(T_0 \le t) &= 1 - P^x(T_0 > t) & (4.170) \\
&= 1 - \int_0^{\infty} p_t(y - x) \, dy + \int_0^{\infty} p_t(y + x) \, dy \\
&= 2 \int_0^{\infty} p_t(y + x) \, dy = 2P^0(B_t \ge x).
\end{aligned}$$

This gives an alternative derivation of the reflection principle, Lemma 2.2.11.

4.5.1 Brownian motion killed at T_0 and the three-dimensional Bessel process

The general class of Bessel processes is defined in Section 14.2. In (4.185) we obtain an explicit relationship between Brownian motion killed at T_0 and the three-dimensional Bessel process.

Let $W^{(i)}$, $i = 1 \ldots, 3$ be independent standard Brownian motions and set $W_t = (W_t^{(1)}, W_t^{(2)}, W_t^{(3)})$. We refer to $\{W_t, t \ge 0\}$ as three-dimensional Brownian motion. $\{|W_t|, t \ge 0\}$ is a three-dimensional Bessel process.

Let

$$q_t(x, y) = \frac{1}{x} \left(p_t(y - x) - p_t(y + x) \right) y \qquad x, y > 0. \qquad (4.171)$$

Note that

$$\int_0^{\infty} q_t(x, y) \, dy = 1. \qquad (4.172)$$

This follows since

$$\int_0^{\infty} y \left(p_t(y - x) - p_t(y + x) \right) dy \qquad (4.173)$$

$$= \int_{-x}^{\infty} (s + x) p_t(s) \, ds - \int_x^{\infty} (s - x) p_t(s) \, ds$$

$$= x \left(\int_{-x}^{\infty} p_t(s) \, ds + \int_{x}^{\infty} p_t(s) \, ds \right)$$
$$= x.$$

It is easy to check that $q_t(x, y)$ satisfies the Chapman–Kolmogorov equation (3.64) with respect to Lebesgue measure on the state space $S = (0, \infty)$. We now show that $q_t(x, y)$ are the transition densities of $\{|W_t|, t \geq 0\}$. A priori, $\{|W_t|, t \geq 0\}$ has state space $[0, \infty)$, but the next lemma shows that we may omit $\{0\}$ and take the state space to be $S = (0, \infty)$.

Lemma 4.5.6 *Let $\{W_t, t \geq 0\}$ be three-dimensional Brownian motion. Then $P^x(T_0 < \infty) = 0$ for $x \neq 0$.*

In other words, starting from any point away from the origin, the three-dimensional Brownian motion will never hit the origin, almost surely.

Proof Fix $0 < a < b < \infty$. We claim that

$$\{W_t = 0 \text{ for some } a \leq t < b\} \tag{4.174}$$

$$\subseteq \bigcup_{m \geq 1} \bigcap_{n \geq m} \bigcup_{an \leq k \leq bn} \bigcap_{i \leq 3} \left\{ |W_{k/n}^{(i)}| \leq \sqrt{\frac{3}{n} \log n} \right\},$$

except, possibly, for a set of P^x measure 0, where $x > 0$.

To see this we first note that $\{W_t = 0 \text{ for some } a \leq t < b\} = \{T_0 \circ \theta_a < b\}$. Therefore, by the strong Markov property at $T_0 \circ \theta_a$ and Khintchine's law of the iterated logarithm, (2.14), applied to each component of $W_t = (W_t^{(1)}, W_t^{(2)}, W_t^{(3)})$, we have

$$P^x \left(T_0 \circ \theta_a < b, \bigcap_{i \leq 3} \left\{ \limsup_{\delta \to 0} \frac{W_{T_0 \circ \theta_a + \delta}^{(i)}}{\sqrt{2\delta \log \log(1/\delta)}} = 1 \right\} \right)$$
$$= P^x \left(T_0 \circ \theta_a < b \right). \tag{4.175}$$

Hence

$$\{W_t = 0 \text{ for some } a \leq t < b\} \tag{4.176}$$

$$\subseteq \{T_0 \circ \theta_a < b\} \bigcap_{i \leq 3} \left\{ \limsup_{\delta \to 0} \frac{W_{T_0 \circ \theta_a + \delta}^{(i)}}{\sqrt{2\delta \log \log(1/\delta)}} = 1 \right\},$$

except, possibly, for a set of P^x measure 0, from which (4.174) follows. Using (4.174) we see that

$$P^x \left(W_t = 0 \text{ for some } a \leq t < b \right) \tag{4.177}$$

$$\leq P^x \left(\bigcup_{m \geq 1} \bigcap_{n \geq m} \bigcup_{an \leq k \leq bn} \bigcap_{i \leq 3} \left\{ |W^{(i)}_{k/n}| \leq \sqrt{\frac{3}{n} \log n} \right\} \right)$$

$$\leq \liminf_{n \to \infty} P^x \left(\bigcup_{an \leq k \leq bn} \bigcap_{i \leq 3} \left\{ |W^{(i)}_{k/n}| \leq \sqrt{\frac{3}{n} \log n} \right\} \right)$$

$$\leq \liminf_{n \to \infty} \sum_{k=an}^{bn} \left(P^x \left(|W^{(1)}_1| \leq \sqrt{\frac{3}{k} \log n} \right) \right)^3$$

$$\leq \lim_{n \to \infty} C(a,b) \frac{(\log n)^{3/2}}{n^{1/2}} = 0,$$

where $C(a,b)$ is a constant depending only on a and b. The proof is completed by noting that that for any sequences $\{a_n\} \downarrow 0$ and $\{b_n\} \uparrow \infty$,

$$\{W_t = 0 \text{ for some } t > 0\}$$
$$= \bigcup_n \{W_t = 0, \text{ for some } a_n \leq t < b_n\} \qquad (4.178)$$

$\square$

It follows from Lemma 4.5.6 that, starting from $x \neq 0$, the value of $f(|W_t|)$ is almost surely well defined for $f \in \mathcal{B}_b(S)$, even though $0 \notin S$. For notational convenience we extend $f \in \mathcal{B}_b(S)$ by setting $f(0) = 0$. Then, for $x \neq 0$, $f_i \in \mathcal{B}_b(S)$, $i = 1, \ldots, n$, and $0 < t_1 < \cdots < t_n$,

$$E^x \left(\prod_{i=1}^n f_i(|W_{t_i}|) \right) = \int_{R^{3n}} \bar{p}_{t_1}(x, z_1) \prod_{i=2}^n \bar{p}_{t_i - t_{i-1}}(z_{i-1}, z_i) \prod_{i=1}^n f_i(|z_i|) \, d^3 z_i,$$
$$(4.179)$$

where $\bar{p}_t(x, y)$ are the transition probabilities of three-dimensional Brownian motion. We use the law of cosines to write

$$\bar{p}_t(x, y) = \frac{1}{(2\pi t)^{3/2}} e^{-|y-x|^2/2t} = \frac{e^{-|x|^2/2t} e^{-|y|^2/2t}}{(2\pi t)^{3/2}} e^{-|y||x| \cos \phi/t} \quad (4.180)$$

(ϕ is the angle between x and y). Using polar coordinates and integrating out the angular coordinates, we see that

$$\int \bar{p}_t(x, y) f(|y|) \, d^3 y \qquad\qquad\qquad\qquad\qquad (4.181)$$

$$= 2\pi \frac{e^{-|x|^2/2t}}{(2\pi t)^{3/2}} \int_0^\infty \left(\int_0^\pi e^{-r|x| \cos \phi/t} \sin \phi \, d\phi \right) e^{-r^2/2t} r^2 f(r) \, dr$$

$$= 2\pi \frac{e^{-|x|^2/2t}}{(2\pi t)^{3/2}} \int_0^\infty \frac{t}{r|x|} \left(e^{r|x|/t} - e^{-r|x|/t} \right) e^{-r^2/2t} r^2 f(r) \, dr$$

$$= \int_S q_t(|x|, r) f(r) \, dr,$$

where q_t is given in (4.171). Thus we can write (4.179) as

$$E^x \left(\prod_{i=1}^{n} f_i(|W_{t_i}|) \right) = \int_{S^n} q_{t_1}(|x|, z_1) \prod_{i=2}^{n} q_{t_i - t_{i-1}}(z_{i-1}, z_i) \prod_{i=1}^{n} f_i(z_i) \, dz_i.$$
(4.182)

Note that E^x depends only on $|x|$. Therefore, when dealing with $\{|W_t|, t \geq 0\}$ we use the notation P^y with $y \in S$ to denote any of the measures $P^{y'}$ with $y' \in R^3$ and $|y'| = y$.

Writing $\{Y_t, t \geq 0\}$ for $\{|W_t|, t \geq 0\}$ under the family of measures P^x, $x \in S$, we see that $\{Y_t, t \geq 0\}$ has state space S and we can write (4.182) as

$$E^x \left(\prod_{i=1}^{n} f_i(Y_{t_i}) \right) = \int_{S^n} q_{t_1}(x, z_1) \prod_{i=2}^{n} q_{t_i - t_{i-1}}(z_{i-1}, z_i) \prod_{i=1}^{n} f_i(z_i) \, dz_i$$
(4.183)

for $f_i \in \mathcal{B}_b(S)$, $i = 1, \ldots, n$ and $0 < t_1 < \cdots < t_n$. Thus we see that the $q_t(x, y)$ are the transition densities of $\{|W_t|, t \geq 0\}$ under the family of measures P^x, $x \in S$.

Let $\mathcal{F}_t^0 = \sigma(Y_s; s \leq t)$. As in the second proof of Lemma 2.2.1, it follows from (4.183) that

$$E \left(f(Y_{t+s}) \, | \, \mathcal{F}_t^0 \right) = Q_s f(X_t)$$
(4.184)

for all $s, t \geq 0$ and $f \in \mathcal{B}_b(S)$, where $Q_s f(x) = \int q_s(x, y) f(y) \, dy$. Thus, $\{Y_t; t \geq 0\}$ is a continuous simple Markov process with respect to the filtration $\mathcal{F}_t^0 = \sigma(Y_s; s \leq t)$.

It is clear from from (4.166), (4.167), and (4.171) that $\{Y_t; t \geq 0\}$ has continuous α-potential densities for $\alpha > 0$. Therefore, we see by Lemma 3.4.2 that, after augmentation, $\{Y_t; t \geq 0\}$ is a Borel right process. In what follows, when we refer to the three-dimensional Bessel process we mean the Borel right process $\{Y_t; t \geq 0\}$.

We now obtain an explicit connection between Brownian motion killed at T_0 and the three-dimensional Bessel process. Let $\widetilde{X}$, with state space $S = (0, \infty)$, denote the Borel right process obtained by killing Brownian motion the first time it hits 0 (see Example 4.5.5). We see from (4.168) and (4.171) that the transition probability density functions of $\widetilde{X}_t$ and Y_t, the three-dimensional Bessel process, are closely related. Therefore, using (4.183) and (4.172), we see that for $f_i \in \mathcal{B}_b(S)$, $i = 1, \ldots, n$, and $0 < t_1 < \cdots < t_n$,

$$E^x \left(\prod_{i=1}^{n} f_i(Y_{t_i}) \right) = \widetilde{E}^x \left(\frac{\widetilde{X}_{t_n}}{x} \prod_{i=1}^{n} f_i(\widetilde{X}_{t_i}) \right)$$
(4.185)

$$= \tilde{E}^x \left(\frac{\tilde{X}_t}{x} \prod_{i=1}^{n} f_i(\tilde{X}_{t_i}) \right)$$

for any $t \geq t_n$. Note that in the last equality we use the fact that $\tilde{X}_t$ is a martingale, which follows from the calculation in (4.173). Equation (4.185) is used in Subsection 8.1.1.

Remark 4.5.7 Recall from (2.139) that $\tilde{u}(x, y) = 2(x \wedge y)$. Although ∞ is not in $S = (0, \infty)$, we have formally that $\tilde{u}(x, \infty) = 2x$. Comparing (4.185) with (3.211) we can say, heuristically, that the three-dimensional Bessel process has the law of Brownian motion killed at T_0, conditioned to die at ∞.

4.6 Continuous local times and potential densities

In all the results in this book that prove the joint continuity of local times, we assume that the potential densities $\{u^\alpha(x, y), (x, y) \in S \times S\}$ are continuous. The next theorem shows that we lose nothing by this assumption.

Theorem 4.6.1 *Let $X = (\Omega, \mathcal{G}, \mathcal{G}_t, X_t, \theta_t, P^x)$ be a strongly symmetric Borel right process and assume that its α-potential density, $u^\alpha(x, y)$, is finite for all $x, y \in S$. Let $L_t^y = \{L_t^y, (t, y) \in R_+ \times S\}$ be a local time of X with*

$$E^x \left(\int_0^\infty e^{-\alpha t} \, dL_t^y \right) = u^\alpha(x, y). \tag{4.186}$$

If $\{L_t^y, y \in S\}$ is continuous for all $t \in R_+$ almost surely, then $\{u^\alpha(x, y), (x, y) \in S \times S\}$ is continuous.

Proof We show first that for compact sets $K \subset S$,

$$\sup_{y \in K} u^\alpha(y, y) < \infty. \tag{4.187}$$

Suppose that (4.187) does not hold. Then we can find a sequence $\{y_n\}_{n=1}^\infty$ in K and a $y \in K$ such that $\lim_{n \to \infty} y_n = y$ and

$$\lim_{n \to \infty} u^\alpha(y_n, y_n) = \infty. \tag{4.188}$$

It follows from the definition of local time that for any real number $T > 0$, $P^y(L_T^y > 0) = 1$. Let G_n be a decreasing sequence of open sets such that $y_n \in G_n$ and $\cap_{n=1}^\infty G_n = y$. The continuity of L_T^y implies that

$$\{L_T^y > 0\} = \bigcup_{n=1}^\infty \{L_T^z > 0, \forall z \in G_n\}. \qquad (4.189)$$

Hence there exists an n_0 such that for all $n \geq n_0$

$$P^y(L_T^z > 0, \forall z \in G_n) \geq \tfrac{1}{2}. \qquad (4.190)$$

Let T_n denote the first hitting time of y_n by (X, P^y). It follows that

$$P^y(T_n \leq T, \forall n \geq n_0) \geq \tfrac{1}{2}, \qquad (4.191)$$

which implies that

$$E^y\left(e^{-\alpha T_n}\right) \geq \tfrac{1}{2} e^{-\alpha T}. \qquad (4.192)$$

However, by (3.107),

$$E^y\left(e^{-\alpha T_n}\right) = \frac{u^\alpha(y, y_n)}{u^\alpha(y_n, y_n)} \qquad (4.193)$$

and for any z

$$\frac{u^\alpha(y, z)}{u^\alpha(y, y)} = E^z\left(e^{-\alpha T_y}\right) \leq 1. \qquad (4.194)$$

Hence

$$E^y\left(e^{-\alpha T_n}\right) = \frac{u^\alpha(y, y_n)}{u^\alpha(y_n, y_n)} \leq \frac{u^\alpha(y, y)}{u^\alpha(y_n y_n)}. \qquad (4.195)$$

Therefore, by (4.188) and (4.195), the left-hand side of (4.192) goes to zero as n approaches infinity. This contradiction establishes (4.187).

Now let λ be an exponential random variable with mean α that is independent of X and let μ be the probability measure of λ. It follows from Theorem 3.10.1 along with (4.194) and (4.187) that the random variables $\{L_\lambda^y, y \in K\}$ are uniformly bounded in $L^p(P^x \times \mu)$ for all $1 \leq p < \infty$. Therefore, both $\{L_\lambda^y, y \in K\}$ and $\{L_\lambda^y L_\lambda^z, y, z \in K\}$ are uniformly integrable. Using this and the continuity of L_t^y for all $t \in R_+$, we see that both

$$y \to E^x(L_\lambda^y) = u^\alpha(x, y) \qquad (4.196)$$

and

$$(y, z) \to E^x(L_\lambda^y L_\lambda^z) = (u^\alpha(x, y) + u^\alpha(x, z))u^\alpha(y, z) \qquad (4.197)$$

are continuous.

To see that $\{u^\alpha(x,y), (x,y) \in K \times K\}$ is continuous, let $(y_n, z_n) \to (y_0, z_0)$, where all $\{y_n\}$, $\{z_n\}$, y_0, and z_0 are contained in K. We see from (4.196) and (4.197) that

$$u^\alpha(y_0, y_n) \to u^\alpha(y_0, y_0), \qquad u^\alpha(y_0, z_n) \to u^\alpha(y_0, z_0), \qquad (4.198)$$

and

$$(u^\alpha(y_0, y_n) + u^\alpha(y_0, z_n))u^\alpha(y_n, z_n) \to (u^\alpha(y_0, y_0) + u^\alpha(y_0, z_0))u^\alpha(y_0, z_0). \tag{4.199}$$

Recall that by (3.74), $u^\alpha(y_0, y_0) > 0$. Therefore, it follows from (4.198) and (4.199) that $u^\alpha(y_n, z_n) \to u^\alpha(y_0, z_0)$. Thus, $\{u^\alpha(x,y), (x,y) \in K \times K\}$ is continuous. Since this holds for all compact sets $K \subset S$, the theorem is proved. $\qquad \square$

4.7 Constructing Ray semigroups and Ray processes

Until Chapter 13, Borel right processes seem to be sufficiently general for all the results we obtain using Gaussian process theory to study local times. However, in the important Theorems 13.1.2 and 13.3.1, Borel right process are too limited to enable us to establish the equivalencies we seek. For this we introduce local Borel right processes in Section 4.8. In this section we develop prerequisite material that is also interesting and important in its own right.

Let S be a locally compact space with a countable base. Let $\{U^\lambda; \lambda \geq 0\}$ be a contraction resolvent on $C_b(S)$, that is, $\{U^\lambda; \lambda \geq 0\}$ are positive bounded linear operators on $C_b(S)$, and similar to (4.9) and (4.10)

$$\|\alpha U^\alpha f\| \leq \|f\| \quad \forall f \in C_b(S) \text{ and } \alpha > 0 \tag{4.200}$$

(clearly, this also holds for $\alpha = 0$), and

$$U^\lambda - U^\mu = (\mu - \lambda)U^\lambda U^\mu \quad \forall \lambda, \mu \geq 0. \tag{4.201}$$

A function $f \in \mathcal{B}_b^+(S)$ is called a supermedian function (with respect to $\{U^\lambda; \lambda \geq 0\}$), or just supermedian, if

$$\alpha U^\alpha f(x) \leq f(x) \quad \forall x \in S \text{ and } \alpha > 0. \tag{4.202}$$

Let $\mathcal{M}^+$ denote the set of supermedian functions. Set $\mathcal{M} := \mathcal{M}^+ - \mathcal{M}^+$ and $\mathcal{H} := \mathcal{M}^+ \cap C_b(S) - \mathcal{M}^+ \cap C_b(S)$. If $f \in \mathcal{M}^+$ and $\lambda > \mu$, it follows from the resolvent equation (4.201) that

$$\lambda U^\lambda f(x) - \mu U^\mu f(x) \tag{4.203}$$
$$= (\lambda - \mu)U^\lambda f(x) + \mu(U^\lambda f(x) - U^\mu f(x))$$

$$= (\lambda - \mu)U^\lambda f(x) - \mu(\lambda - \mu)U^\mu U^\lambda f(x)$$
$$= (\lambda - \mu)U^\lambda (f(x) - \mu U^\mu f(x)) \geq 0.$$

It follows from this that for $f \in \mathcal{M}^+$

$$\lambda U^\lambda f(x) \uparrow \qquad \text{as} \quad \lambda \to \infty. \tag{4.204}$$

Consequently,

$$\widehat{f}(x) := \lim_{\lambda \to \infty} \lambda U^\lambda f(x) \tag{4.205}$$

exists. Furthermore, we see from (4.202) that

$$\widehat{f}(x) \leq f(x) \quad \forall f \in \mathcal{M}^+. \tag{4.206}$$

Using (4.205), (4.200), and then (4.201), we see that

$$
\begin{aligned}
\mu U^\mu \widehat{f}(x) &= \mu U^\mu \lim_{\lambda \to \infty} \lambda U^\lambda f(x) \tag{4.207} \\
&= \lim_{\lambda \to \infty} \lambda U^\lambda \mu U^\mu f(x) \\
&= \lim_{\lambda \to \infty} \frac{\lambda \mu}{\lambda - \mu}(U^\mu f(x) - U^\lambda f(x)) \\
&= \mu U^\mu f(x) \leq \widehat{f}(x),
\end{aligned}
$$

so that $\widehat{f} \in \mathcal{M}^+$. $\widehat{f}$ is called the excessive regularization of f.

Theorem 4.7.1 *Let S be a locally compact space with a countable base, and let $\{U^\lambda;\ \lambda \geq 0\}$ be a contraction resolvent on $C_b(S)$. Assume that $\mathcal{H} \cap C_0(S)$ is dense in $C_0(S)$ in the uniform norm. Then we can construct a sub-Markov semigroup $\{P_t;\ t \geq 0\}$ on $(S, \mathcal{B})$, without the additional requirement that $P_0 = I$, such that for all $f \in C_b(S)$, $P_t f(x)$ is right continuous in $t \in R_+$ and*

$$U^\lambda f(x) = \int_0^\infty e^{-\lambda t} P_t f(x)\, dt. \tag{4.208}$$

The sub-Markov semigroup $\{P_t;\ t \geq 0\}$ constructed in Theorem 4.7.1 is called a Ray semigroup.

We break the proof of Theorem 4.7.1 into a series of lemmas, Lemmas 4.7.2–4.7.5, that give the details of the construction. The reader should note that each of these lemmas employs the notation, hypotheses, and results of the previous ones. In all of these lemmas we assume that S is a locally compact space with a countable base and that there exists a contraction resolvent $\{U^\lambda;\ \lambda \geq 0\}$ on $C_b(S)$. Moreover, we assume that $\mathcal{H} \cap C_0(S)$ is dense in $C_0(S)$ in the uniform norm.

Lemma 4.7.2 *For each $f \in \mathcal{M}^+$ and $x \in S$ we can find a positive function of $t \in R_+$, denoted by $P_t(x, f)$, that is decreasing, right continuous in t, and satisfies*

$$U^\lambda f(x) = \int_0^\infty e^{-\lambda t} P_t(x, f) \, dt. \qquad (4.209)$$

Proof of Lemma 4.7.2 Note that by the resolvent equation (4.201), for any $f \in \mathcal{B}_b(S)$ and $\lambda > 0$,

$$\frac{\partial}{\partial \lambda} U^\lambda f(x) = \lim_{\epsilon \to 0} \frac{U^{\lambda+\epsilon} f(x) - U^\lambda f(x)}{\epsilon} \qquad (4.210)$$
$$= -\lim_{\epsilon \to 0} U^{\lambda+\epsilon} U^\lambda f(x) = -(U^\lambda)^2 f(x),$$

where for the last equality we use (4.201) and (4.200). Then by induction we have

$$\frac{\partial^k}{\partial \lambda^k} U^\lambda f(x) = k!(-1)^k (U^\lambda)^{k+1} f(x) \quad \forall \lambda > 0. \qquad (4.211)$$

Hence for any $f \in \mathcal{B}_b^+(S)$ and $x \in S$, $U^\lambda f(x)$ is completely monotone in λ, and consequently it is a Laplace transform (see, e.g., Feller (1971, XIII.4)). That is, there exists a positive measure $\mu_{f,x}$ on R_+ such that

$$U^\lambda f(x) = \int_0^\infty e^{-\lambda t} \, d\mu_{f,x}(t) \quad \forall \lambda > 0. \qquad (4.212)$$

Using the differentiation formula $D^k(fg) = \sum_{j=0}^k \binom{k}{j} D^j f D^k g$ and (4.211) we have

$$\frac{\partial^k}{\partial \lambda^k} (\lambda U^\lambda) f(x) = k \frac{\partial^{k-1}}{\partial \lambda^{k-1}} U^\lambda f(x) + \lambda \frac{\partial^k}{\partial \lambda^k} U^\lambda f(x) \qquad (4.213)$$
$$= k!(-1)^{k-1} (U^\lambda)^k (I - \lambda U^\lambda) f(x).$$

Let $f \in \mathcal{M}^+$. By (4.204)–(4.205) we see that $\widehat{f}(x) - \lambda U^\lambda f(x) \geq 0$. Furthermore, for any $k \geq 1$ it follows from (4.213) and (4.202) that

$$(-1)^k \frac{\partial^k}{\partial \lambda^k} (\widehat{f}(x) - \lambda U^\lambda f(x)) = (-1)^{k-1} \frac{\partial^k}{\partial \lambda^k} (\lambda U^\lambda) f(x) \geq 0.$$

Thus $\widehat{f}(x) - \lambda U^\lambda f(x)$ is completely monotone in λ, and consequently for some positive measure $\rho_{f,x}$ on R_+ we have

$$\widehat{f}(x) - \lambda U^\lambda f(x) = \int_0^\infty e^{-\lambda t} \, d\rho_{f,x}(t) \quad \lambda > 0. \qquad (4.214)$$

It follows from the resolvent equation (4.201) that $U^\lambda f(x) \leq U^0 f(x) < \infty$, so that $\lim_{\lambda \to 0} \lambda U^\lambda f(x) = 0$. Therefore, taking the limit in (4.214) as $\lambda \to 0$ we see that $\rho_{f,x}$ has total mass $\widehat{f}(x) \leq f(x) < \infty$. Taking the limit as $\lambda \to \infty$ and using (4.205) shows us that $\rho_{f,x}(\{0\}) = 0$. Consequently,

$$
\begin{aligned}
\int_0^\infty e^{-\lambda t} \rho_{f,x}((t, \infty)) \, dt &= \int_0^\infty e^{-\lambda t} \left(\int_{(t,\infty)} d\rho_{f,x}(s) \right) dt \qquad (4.215) \\
&= \int_0^\infty \left(\int_{[0,s)} e^{-\lambda t} \, dt \right) d\rho_{f,x}(s) \\
&= \int_0^\infty \frac{1 - e^{-\lambda s}}{\lambda} \, d\rho_{f,x}(s) \\
&= \frac{1}{\lambda} \left(\widehat{f}(x) - [\widehat{f}(x) - \lambda U^\lambda f(x)] \right) = U^\lambda f(x),
\end{aligned}
$$

where for the last line we use (4.214). Set $P_t(x, f) = \rho_{f,x}((t, \infty))$. It is easy to see that $P_t(x, f)$ is decreasing and right continuous in t. □

Lemma 4.7.3 *For x and t fixed consider $P_t(x, f)$ as a functional on $\mathcal{M}^+$. We can extend it to be a positive bounded linear functional on $\mathcal{M}$, such that for fixed $x \in S$ and $f \in \mathcal{M}$, $P_t(x, f)$ is right continuous in t and for all $f \in \mathcal{M}$*

$$
U^\lambda f(x) = \int_0^\infty e^{-\lambda t} P_t(x, f) \, dt \qquad \forall \, x \in S. \qquad (4.216)
$$

Proof of Lemma 4.7.3 We show in Lemma 4.7.2 that $t \mapsto P_t(x, f)$ is decreasing and right continuous in t for $f \in \mathcal{M}^+$. Therefore, using (4.205) and (4.209), we have

$$
\begin{aligned}
\widehat{f}(x) &= \lim_{\lambda \to \infty} \lambda U^\lambda f(x) \qquad (4.217) \\
&= \lim_{\lambda \to \infty} \lambda \int_0^\infty e^{-\lambda t} P_t(x, f) \, dt \\
&= \lim_{\lambda \to \infty} \int_0^\infty e^{-t} P_{t/\lambda}(x, f) \, dt = P_0(x, f).
\end{aligned}
$$

Thus, by (4.206),

$$
P_0(x, f) = \widehat{f}(x) \le f(x) \qquad \forall f \in \mathcal{M}^+. \qquad (4.218)
$$

In particular, taking $f \equiv 1 \in \mathcal{M}^+$ and again using the fact that $t \mapsto P_t(x, 1)$ is decreasing, we have

$$
P_t(x, 1) \le 1. \qquad (4.219)
$$

These arguments also show that $P_t(x, f)$ is bounded.

Now let $g, h \in \mathcal{M}^+$. It follows from (4.209) that

$$U^\lambda(h - g)(x) = \int_0^\infty e^{-\lambda t} \left(P_t(x, h) - P_t(x, g) \right) dt. \qquad (4.220)$$

This has two consequences. First, if $h - g = h' - g'$ with $h, g, h', g' \in \mathcal{M}^+$, then it follows from (4.220), using the uniqueness property of Laplace transforms and the right continuity of $t \mapsto P_t(x, \cdot)$, that $P_t(x, h) - P_t(x, g) = P_t(x, h') - P_t(x, g')$. Thus we can obtain a consistent extension of $P_t(x, f)$ to $f \in \mathcal{M}$ by defining

$$P_t(x, f) = P_t(x, h) - P_t(x, g) \qquad (4.221)$$

whenever $f = h - g$ with $g, h \in \mathcal{M}^+$. With this definition, (4.216) holds for all $f \in \mathcal{M}$. Using this, the uniqueness property of Laplace transforms, and the right continuity of $t \mapsto P_t(x, \cdot)$, it is easy to see that $P_t(x, \cdot)$ is linear. The second consequence of (4.220) comes from the fact that if $f \in \mathcal{M}$ with $f = h - g \ge 0$ and $g, h \in \mathcal{M}^+$, then by (4.212) we have that $U^\lambda f(x)$ is the Laplace transform of a positive measure. Using, once again, the uniqueness property of Laplace transforms, (4.221), and the right continuity of $t \mapsto P_t(x, \cdot)$, we see that $P_t(x, f) \ge 0$. $\qquad \square$

Lemma 4.7.4 *For each $t \in R_+$ we can extend the linear functionals $f \mapsto P_t(x, f)$, $f \in \mathcal{M}$, to become a sub-Markov kernel P_t on $(S, \mathcal{B})$ such that for all $f \in C_b(S)$ and $x \in S$, $t \mapsto P_t f(x)$, defined in (3.1), is right continuous in t and*

$$U^\lambda f(x) = \int_0^\infty e^{-\lambda t} P_t f(x) \, dt. \qquad (4.222)$$

Proof of Lemma 4.7.4 To begin, fix x and t. By the previous lemma, $f \mapsto P_t(x, f)$ is a positive bounded linear functional on $\mathcal{M}$. By the assumption that any function in $C_0(S)$ is the limit in the uniform norm of functions in $\mathcal{H} \cap C_0(S)$, we can extend $f \mapsto P_t(x, f)$ to be a positive bounded linear functional on $C_0(S)$. By the Riesz Representation Theorem, there is a positive finite Borel measure $\widetilde{P}_t(x, \cdot)$ so that for all $f \in C_0(S)$

$$P_t(x, f) = \int f(y) \widetilde{P}_t(x, dy). \qquad (4.223)$$

Therefore we can extend $P_t(x, \cdot)$ so that it is a positive finite Borel measure on $(S, \mathcal{B})$. Obviously $P_t(x, dy) = \widetilde{P}_t(x, dy)$.

Since, by Lemma 4.7.3, $P_t(x, f)$ is right continuous in t for $f \in \mathcal{M}$, the extension $t \mapsto P_t(x, f)$ is right continuous and (4.216) or, equivalently,

(4.222) continues to hold for all $f \in C_0(S)$. By (4.219), $P_t(x, \cdot)$ is a subprobability measure, so that

$$P_t f(x) := \int f(y) P_t(x, dy) \qquad (4.224)$$

extends naturally to $f \in C_b(S)$.

We now show that $t \mapsto P_t f(x)$ is right continuous. Without loss of generality we can assume that $f \geq 0$. Fix t_0 and $\epsilon > 0$ and let K be a compact set with $P_{t_0}(x, K^c) \leq \epsilon$. Let $0 \leq g_K \leq 1$ be a continuous compactly supported function that is 1 on K and set $h_K = 1 - g_K$. Clearly $P_{t_0} h_K(x) \leq \epsilon$. By (4.224), $P_t h_K(x) = P_t 1 - P_t g_K(x)$, and since $1 \in \mathcal{M}^+ \cap C_b(S)$, we have by Lemma 4.7.3 that $t \to P_t h_K(x)$ is right continuous. Consequently, for some $\delta > 0$, we have $P_t h_K(x) \leq 2\epsilon$ for $t \in [t_0, t_0 + \delta]$. For these values of t,

$$|P_{t_0} f(x) - P_t f(x)| \leq |P_{t_0}(f(x) g_K(x)) - P_t(f(x) g_K(x))| + 3\epsilon \|f\|_\infty. \qquad (4.225)$$

The right continuity of $t \mapsto P_t f(x)$ at t_0 now follows from that of $t \mapsto P_{t_0}(f(x) g_K(x))$, which holds since $f g_K \in C_0(S)$. Therefore $t \mapsto P_t f(x)$ is right continuous for all $f \in C_b(S)$. Using the Dominated Convergence Theorem we can further extend (4.222) from all $f \in C_0(S)$ to all $f \in C_b(S)$.

It remains to show that for all $A \in \mathcal{B}$, $x \mapsto P_t(x, A)$ is Borel measurable. A simple monotone class argument shows that to prove this, it suffices to show that for all $f \in C_b(S)$, $x \mapsto P_t f(x)$ is Borel measurable. By hypothesis, $U^\lambda f \in C_b(S)$ for all $f \in C_b(S)$ and $\lambda \geq 0$. It follows that $\int_0^\infty \psi(t) P_t f(x) \, dt$ is Borel measurable for any $\psi \in C_0(R_+)$. This last remark follows from the fact that the linear span of $\{e^{-\lambda t} ; \lambda \geq 0\}$ is dense in $C_0(R_+)$, since supposing otherwise we obtain a contradiction: The Hahn–Banach Theorem and the Riesz Representation Theorem give a finite nonzero Borel measure whose Laplace transform is identically zero.

To conclude the proof, for any fixed $s \geq 0$, we take $\psi_{\epsilon,s}(\cdot) \in C_0(R_+)$ to be an approximate identity supported on $[s, s + \epsilon]$. Then, using the right continuity of $P_t f(x)$ in t, $P_s f(x) = \lim_{\epsilon \to 0} \int_0^\infty \psi_{\epsilon,s}(t) P_t f(x) dt$ is Borel measurable in x for each $s \geq 0$ and $f \in C_b(S)$. $\qquad \square$

Lemma 4.7.5 *The family of sub-Markov kernels $\{P_t ; t \geq 0\}$ is a sub-Markov semigroup on $(S, \mathcal{B})$.*

Proof of Lemma 4.7.5: We only need to verify the semigroup property, and for this it suffices to show that

$$P_s P_t f(x) = P_{s+t} f(x) \tag{4.226}$$

for all $s, t \geq 0$ and $f \in C_b(S)$. By Lemma 4.7.4, $t \mapsto P_t f(x)$ is right continuous and $P_s(x, \cdot)$ is a bounded measure. Using these observations and the Dominated Convergence Theorem, it follows that $t \mapsto P_s P_t f(x)$ is right continuous, and similarly for $t \mapsto P_{s+t} f(x)$. Thus, by the uniqueness property of Laplace transforms, to prove (4.226), it suffices to show that the Laplace transforms of $P_s P_t f(x)$ and $P_{s+t} f(x)$ are equal or, equivalently, that

$$P_s U^\lambda f(x) = \int_0^\infty e^{-\lambda t} P_{s+t} f(x) \, dt \tag{4.227}$$

for all $\lambda > 0$.

By hypothesis, $U^\lambda f \in C_b(S)$, so that by the Dominated Convergence Theorem again, both sides of (4.227) are right continuous in s. Therefore, taking Laplace transforms again, we see that to prove (4.226), it suffices to show that

$$U^\mu U^\lambda f(x) = \int_0^\infty \int_0^\infty e^{-\mu s} e^{-\lambda t} P_{s+t} f(x) \, ds \, dt \tag{4.228}$$

for all $\mu, \lambda > 0$. By the first five lines of (3.19) we see that

$$U^\mu f(x) - U^\lambda f(x) = (\lambda - \mu) \int_0^\infty \int_0^\infty e^{-\mu s} e^{-\lambda t} P_{s+t} f(x) \, ds \, dt, \tag{4.229}$$

so that (4.228) follows from the resolvent equation (4.201) when $\lambda \neq \mu$. By continuity, (4.228) also holds when $\lambda = \mu$. $\square$

Proof of Theorem 4.7.1 This follows from Lemmas 4.7.2–4.7.5. $\square$

Let $\{P_t \, ; \, t \geq 0\}$ be a Ray semigroup on $(S, \mathcal{B})$. Let δ_x denote the unit mass at x and set

$$\mathcal{N} = \{x \in S \, ; \, P_0(x, \, dy) \neq \delta_x(dy)\}. \tag{4.230}$$

$\mathcal{N}$ is called the set of branch points for $\{P_t \, ; \, t \geq 0\}$. Heuristically, one may think of a branch point in the following way: If $\{X_t \, ; \, t \geq 0\}$ is a process with semigroup $\{P_t \, ; \, t \geq 0\}$, then a branch point is a point x that is never visited by the path X but may appear as a left limit point of the path, that is, we may have $\lim_{s \uparrow t} X_s = x$ for some t, even though $X_t \neq x$. The path "branches" just before hitting x. This is clarified further by Lemma 4.7.9.

We set $D = S - \mathcal{N}$. D is the set of points in S that are not branch points. We note from the definition of $\mathcal{N}$ that

$$P_0(x, \{x\}) = P_0(x, S) = 1 \qquad \forall x \in D. \qquad (4.231)$$

Lemma 4.7.6 *Under the hypotheses of Theorem 4.7.1, for all $x \in S$ and for all $t \geq 0$, the measures $P_t(x, \cdot)$ are carried by D, that is, for all measurable sets $A \subset S$,*

$$P_t(x, A \cap D) = P_t(x, A). \qquad (4.232)$$

Proof Let f_n be a sequence of functions in $\mathcal{H} \cap C_0(S)$ that is dense in $C_0(S)$ in the uniform norm. Each f_n can be written as a difference $f_n = f_{n,1} - f_{n,2}$ of two functions in $\mathcal{M}^+ \cap C_b(S)$. Let $\{g_n\}$ be the set of functions $\{f_{n,1}, f_{n,2}, \ n = 1, \ldots\}$. Since $P_0 g_n(x) = g_n(x)$ for all n implies that $P_0 f(x) = f(x)$ for all $f \in C_0(S)$, we have

$$D = \cap_n \{x \in S \mid P_0 g_n(x) = g_n(x)\}.$$

Consequently,

$$\mathcal{N} = \cup_n \{x \in S \mid |g_n(x) - P_0 g_n(x)| > 0\}. \qquad (4.233)$$

Then, for any $t \geq 0$ and $y \in S$,

$$P_t(y, \mathcal{N}) \leq \sum_n P_t(y, \{x \in S \mid |g_n(x) - P_0 g_n(x)| > 0\}). \qquad (4.234)$$

Using (4.218) we have that for any $\epsilon > 0$,

$$P_t(y, \{x \in S \mid |g_n(x) - P_0 g_n(x)| > \epsilon\}) \qquad (4.235)$$

$$\leq \frac{1}{\epsilon} \int |g_n(x) - P_0 g_n(x)| \, P_t(y, dx)$$

$$= \frac{1}{\epsilon} \int (g_n(x) - P_0 g_n(x)) \, P_t(y, dx)$$

$$\leq \frac{P_t g_n(y) - P_t P_0 g_n(y)}{\epsilon} = 0$$

since by the semigroup property, $P_t P_0 = P_t$. Therefore, $P_t(y, \mathcal{N}) = 0$ and we get (4.232). □

On page 63 we show how to extend a sub-Markov semigroup with state space S so that it becomes a Markov semigroup, by introducing a cemetery state. We repeat this process here for Ray semigroups because there are some additional issues we need to consider. Let S be a locally compact space with a countable base and let $\{P_t \, ; \, t \geq 0\}$ be a Ray semigroup on $(S, \mathcal{B})$. Recall the definition of $S_\Delta = S \cup \Delta$ as the one-point compactification of S, with Δ the "point at infinity." Recall also that $\{P_t\,;\,t \geq 0\}$ can be extended to become a Markov semigroup on S_Δ, which for clarity we denote in this section as $\{\overline{P}_t\,;\,t \geq 0\}$, in a unique way by setting

$$\overline{P}_t(x, \Delta) = 1 - P_t(x, S) \tag{4.236}$$

for all $x \in S$ and

$$\overline{P}_t(\Delta, \Delta) = 1. \tag{4.237}$$

Let $\{\overline{U}^\lambda\,;\,\lambda \geq 0\}$ be the corresponding potentials. If f is a function defined on S, we let $\overline{f}$ be the extension of f to S_Δ obtained by setting $\overline{f}(\Delta) = 0$. Note that for functions f defined on S

$$\overline{f} \in C_b(S_\Delta) \iff f \in C_0(S). \tag{4.238}$$

If follows from the above definitions that

$$\overline{P}_t \overline{f}(x) = \overline{P_t f}(x) \qquad \overline{U}^\lambda \overline{f}(x) = \overline{U^\lambda f}(x), \tag{4.239}$$

so that if $\overline{\mathcal{M}}^+$ denotes the supermedian functions for $\{\overline{U}^\lambda\,;\,\lambda \geq 0\}$,

$$\overline{f} \in \overline{\mathcal{M}}^+ \iff f \in \mathcal{M}^+. \tag{4.240}$$

Note that in general $\overline{U}^\lambda \overline{f}(x)$ is not continuous at Δ, even for compactly supported f.

Lemma 4.7.7 *Assume all the hypotheses of Theorem 4.7.1. Set $D_\Delta = D \cup \Delta$. Then D_Δ is the set of nonbranch points for the Markov semigroup $\{\overline{P}_t\,;\,t \geq 0\}$, and for all $x \in S_\Delta$ and $t \geq 0$, $\overline{P}_t(x, \cdot)$ is carried by D_Δ.*

Furthermore, let $\{g_n,\ n = 1, 2, \ldots\}$ be the sequence of functions in $\mathcal{M}^+ \cap C_b(S)$ described in the proof of Lemma 4.7.6. We have

$$D_\Delta = \cap_n \{x \in S_\Delta \,|\, \overline{P}_0 \overline{g}_n(x) = \overline{g}_n(x)\}. \tag{4.241}$$

Proof   It is clear that if $x \in D$ is not a branch point for $\{P_t\,;\,t \geq 0\}$, it is not a branch point for $\{\overline{P}_t\,;\,t \geq 0\}$, and by (4.237), Δ is not a branch point for $\{\overline{P}_t\,;\,t \geq 0\}$. The remainder of the first paragraph follows immediately from Lemma 4.7.6.

The result in the second paragraph follows from what is proved in the first paragraph and the proof of Lemma 4.7.6. $\square$

Let $\{P_t\,;\,t \geq 0\}$ be a Ray semigroup on $(S, \mathcal{B})$. A Ray process is a collection $X = (\Omega, \mathcal{G}, \mathcal{G}_t, X_t, \theta_t, P^x)$ that satisfies all the properties of a right continuous simple Markov process (see page 64) with semigroup

$\{P_t\,;\,t \geq 0\}$, with the exception of the requirement that $P^x(X_0 = x) = 1$, which holds only for $x \in D$.

Theorem 4.7.8 *Let S be a locally compact space with a countable base, and let $\{P_t\,;\,t \geq 0\}$ be a Ray semigroup on $(S, \mathcal{B})$ with $\{U^\lambda;\,\lambda \geq 0\}$ the associated contraction resolvent on $C_b(S)$. Assume that $\mathcal{H} \cap C_0(S)$ is dense in $C_0(S)$ in the uniform norm. Then we can construct a Ray process $X = (\Omega, \mathcal{F}^0, \mathcal{F}_t^0, X_t, \theta_t, P^x)$ with state space S_Δ and semigroup $\{P_t\,;\,t \geq 0\}$ such that X has left limits in S_Δ.*

Proof   The proof of this theorem has much in common with the proof of Theorem 4.1.1.

Let $x \in S_\Delta$. We first construct a probability $\widetilde{P}^x$ on $S_\Delta^{R^+}$, the space of S_Δ-valued functions $\{f(t), t \in [0, \infty)\}$ equipped with the Borel product σ-algebra $\mathcal{B}(S_\Delta^{R^+})$. Let X_t be the natural evaluation $X_t(f) = f(t)$. We define $\widetilde{P}^x$ on sets of the form $\{X_0 \in A_0, X_{t_1} \in A_1, \ldots, X_{t_n} \in A_n\}$ for all Borel measurable sets $A_0, A_1, \ldots, A_n$ in S_Δ and $0 = t_0 < t_1 < \cdots < t_n$ by setting

$$\widetilde{P}^x(X_0 \in A_0, X_{t_1} \in A_1, \ldots, X_{t_n} \in A_n) \qquad (4.242)$$
$$= \int \prod_{i=0}^{n} I_{A_i}(z_i) \overline{P}_0(x, dz_0) \prod_{i=1}^{n} \overline{P}_{t_i - t_{i-1}}(z_{i-1}, dz_i).$$

Here I_{A_i} is the indicator function of A_i and $\{\overline{P}_t, t \geq 0\}$ is the extension of $\{P_t, t \geq 0\}$ to S_Δ defined in (4.236). It follows from the semigroup property of $\{\overline{P}_t;\,t \geq 0\}$ that this construction is consistent. Therefore, by the Kolmogorov Construction Theorem, we can extend $\widetilde{P}^x$ to $S_\Delta^{R^+}$.

Let $\mathcal{F}_t^0 = \sigma(X_s;\,s \leq t)$. As in the second proof of (2.31), it follows from (4.242) that

$$\widetilde{E}\left(f(X_{t+s}) \,|\, \mathcal{F}_t^0\right) = \overline{P}_s f(X_t) \qquad (4.243)$$

for all $s, t \geq 0$ and $f \in \mathcal{B}_b(S_\Delta)$. Thus, $\{X_t;\,t \geq 0\}$ is a simple Markov process with respect to the filtration $\mathcal{F}_t^0 = \sigma(X_s;\,s \leq t)$.

We now show that $\{X_t;\,t \geq 0\}$ has a modification that is right continuous with left limits. We begin by noting that, for any $f \in \mathcal{M}^+ \cap C_b(S)$, it follows from (4.218) that $P_0(x, f) = \widehat{f}(x) \leq f(x)$, and, using the fact that $t \mapsto P_t(x, f)$ is decreasing, we have $P_t(x, f) \leq f(x)$. It then follows from (4.239) and (4.243) that for any $f \in \mathcal{M}^+ \cap C_b(S)$, $\overline{f}(X_t)$ is a $\mathcal{F}_t^0$ supermartingale, for any $\widetilde{P}^x$.

Let Q be a countable dense set in R^+. By the convergence theorem for supermartingales, Revuz and Yor (1991, Chapter II, Theorem 2.5), $\overline{f}(X_t)$ has right- and left-hand limits along Q, almost surely. Let $\mathcal{G} =$

$\{g_n,\ n = 1, 2, \ldots\}$ be the sequence of functions in $\mathcal{M}^+ \cap C_b(S)$ described in the proof of Lemma 4.7.6. Since, by hypothesis, any function in $C_0(S)$ can be approximated in the uniform norm by the difference of two functions in $\mathcal{G}$, we have that $\overline{f}(X_t)$ has right- and left-hand limits along Q for all $f \in C_0(S)$, almost surely. Since $\{\overline{f} \mid f \in C_0(S)\}$ is a collection of continuous functions that separate points in S_Δ, we see from the argument in the seventh paragraph of the proof of Theorem 4.1.1 that, almost surely, $t \mapsto X_t$ has right-hand limits in S_Δ along Q. The same argument shows that, almost surely, $t \mapsto X_t$ has left-hand limits in S_Δ along Q; see also Remark 4.1.2 (2).

For all ω for which the following limit exists for all t, set $\widetilde{X_t}(\omega) = \lim_{s \downarrow t, s \in Q} X_s$, and for those ω for which it does not exist set $\widetilde{X}.(\omega) = \Delta$. Note that $\{\widetilde{X_t};\ t \geq 0\}$ is right continuous with left-hand limits. We claim that for each t, $\widetilde{X_t} = X_t$ almost surely. To see this note that for any $f, g \in \mathcal{M}^+ \cap C_b(S)$,

$$\widetilde{E}^x\left(\overline{f}(X_t)\overline{g}(\widetilde{X_t})\right) = \lim_{s \downarrow t, s \in Q} \widetilde{E}^x\left(\overline{f}(X_t)\overline{g}(X_s)\right) \qquad (4.244)$$

$$= \lim_{s \downarrow t, s \in Q} \widetilde{E}^x\left(\overline{f}(X_t)\overline{P}_{s-t}\overline{g}(X_t)\right)$$

$$= \widetilde{E}^x\left(\overline{f}(X_t)\overline{P}_0\overline{g}(X_t)\right),$$

since $t \mapsto P_t g$ is bounded and right continuous. It follows from the semigroup property $\overline{P}_t\overline{P}_0 = \overline{P}_t$ that $\widetilde{E}^x\left(\overline{P}_0\overline{g}(\widetilde{X_t})\right) = \widetilde{E}^x\left(\overline{g}(\widetilde{X_t})\right)$. In addition, by (4.218) we have that $\overline{P}_0(x, \overline{g}) \leq \overline{g}(x)$. Therefore

$$\overline{P}_0\overline{g}(X_t) = \overline{g}(X_t) \qquad \text{a.s.} \qquad (4.245)$$

Using this, it follows from (4.244) that

$$\widetilde{E}^x\left(\overline{f}(X_t)\overline{g}(\widetilde{X_t})\right) = \widetilde{E}^x\left(\overline{f}(X_t)\overline{g}(X_t)\right), \qquad (4.246)$$

and by our assumption that $\mathcal{H} \cap C_0(S)$ is dense in $C_0(S)$ we see that (4.246) continues to hold for all $f, g \in C_0(S)$.

The same proof works if the functions $\overline{f}$ and $\overline{g}$ are replaced by functions that are constant on S_Δ. Consequently, (4.246) holds for all functions $f, g \in C_b(S_\Delta)$. As in the eighth paragraph of the proof of Theorem 4.1.1, this shows that for each t, $\widetilde{X_t} = X_t$ almost surely.

The rest of the theorem follows as in the ninth paragraph of the proof of Theorem 4.1.1. □

Given a set Ω and a filtration $\mathcal{F}_t^0$, we define the optional σ-algebra $\mathcal{O}$ to be the σ-algebra of subsets of $R^+ \times \Omega$ generated by all real-valued processes Y_t that are right continuous with left limits and are adapted to the filtration $\mathcal{F}_t^0$. A set $A \in \mathcal{O}$ is called an optional set, or simply optional. A process Z_t is said to be optional if the function $(t, \omega) \mapsto Z_t(\omega)$ is $\mathcal{O}$ measurable.

Lemma 4.7.9 *Let* $X = (\Omega, \mathcal{F}^0, \mathcal{F}_t^0, X_t, \theta_t, P^x)$ *be the Ray process constructed in Theorem 4.7.8. Then, almost surely,* $X_t \in D_\Delta$ *for all* $t \geq 0$.

Proof   As in Lemma 4.7.7, let g_n be the sequence of functions in $\mathcal{M}^+ \cap C_b(S)$ for which

$$D_\Delta = \cap_n \{x \in S_\Delta \,|\, \overline{P}_0 \bar{g}_n(x) = \bar{g}_n(x)\}. \tag{4.247}$$

It suffices to show that for each n, $\overline{P}_0 \bar{g}_n(X_t) = \bar{g}_n(X_t)$ for all t, almost surely. Since this equality is trivial if $X_t = \Delta$, we need only show that $\overline{P}_0 \bar{g}_n(X_t) 1_S(X_t) = \bar{g}_n(X_t) 1_S(X_t)$ for all t, almost surely. Let K_m be a sequence of compact sets with $\cup_m K_m = S$, such that each K_n is the closure of its interior and $K_n \subset \text{int } K_{n+1}$. Let f_m be a continuous function that is supported on K_m and equal to 1 on K_{m-1}. It suffices to show that for all n, m,

$$P^x \left(\overline{P}_0 \left(\bar{g}_n(X_t) f_m(X_t) \right) = \bar{g}_n(X_t) f_m(X_t), \text{ for all } t \right) = 1. \tag{4.248}$$

Note that $\bar{g}_n f_m \in C_0(S)$ by the support property of f_m. This implies that the process $\{\bar{g}_n(X_t) f_m(X_t), t \geq 0\}$ is optional. The same reasoning shows that $\{\overline{U}^\lambda \bar{g}_n(X_t) f_m(X_t), t \geq 0\}$ is optional. Since, by (4.217), $\overline{P}_0 \bar{g}_n(x) = \lim_{\lambda \to \infty} \lambda \overline{U}^\lambda \bar{g}_n(x)$, we have that $\{\overline{P}_0 \bar{g}_n(X_t) f_m(X_t), t \geq 0\}$ is optional. Let

$$A = \{(t, \omega) \,|\, \overline{P}_0 \left(\bar{g}_n(X_t) f_m(X_t) \right) \neq \bar{g}_n(X_t) f_m(X_t)\}. \tag{4.249}$$

A is an optional set. Therefore, by the Optional Section Theorem (Dellacherie and Meyer (1978, Theorem IV.84), Rogers and Williams (2000a, Theorem VI.5.1)), if (4.248) does not hold, we can find an $\mathcal{F}_t^0$ stopping time $T_{n,m}$ with $P^x(T_{n,m} < \infty) > 0$, such that $(T_{n,m}(\omega), \omega) \in A$ on $\{T_{n,m} < \infty\}$. Set $X_\infty = \Delta$. Since $\overline{P}_0 \bar{g}_n(x) \leq \bar{g}_n(x)$ by (4.218), we see that if (4.248) does not hold,

$$P^x \left(X_{T_{n,m}} \in K_m \text{ and } \overline{P}_0 \bar{g}_n(X_{T_{n,m}}) < \bar{g}_n(X_{T_{n,m}}) \right) > 0. \tag{4.250}$$

Using once more the fact that $\overline{P}_0 \bar{g}_n(x) \leq \bar{g}_n(x)$, it is clear that in order to prove (4.248), it suffices to show that

$$E^x \left(\overline{P}_0 \bar{g}_n(X_{T_{n,m}}) \right) = E^x \left(\bar{g}_n(X_{T_{n,m}}) \right). \tag{4.251}$$

Let $T_{n,m,k} = \dfrac{[2^k T_{n,m}] + 1}{2^k}$, as in the proof of Lemma 2.2.5, and recall that for each k, $T_{n,m,k}$ is an $\mathcal{F}_t^0$ stopping time, $T_{n,m,k} \downarrow T_{n,m}$, and $T_{n,m} < \infty$ if and only if $T_{n,m,k} < \infty$, for all $k \geq 1$. Since, by (4.245), $\overline{P}_0 \overline{g}_n(X_t) = \overline{g}_n(X_t)$ almost surely for each fixed t, and $T_{n,m,k}$ is rational valued, we have

$$\overline{P}_0 \overline{g}_n(X_{T_{n,m,k}}) = \overline{g}_n(X_{T_{n,m,k}}), \quad \text{a.s.} \quad (4.252)$$

Recall that $X_{T_{n,m}} \in K_m$. Thus $\overline{g}_n(X_t)$ is right continuous at $t = T_{n,m}$, so that

$$E^x\left(\overline{g}_n(X_{T_{n,m}})\right) = \lim_{k \to \infty} E^x\left(\overline{g}_n(X_{T_{n,m,k}})\right). \quad (4.253)$$

Therefore, to establish (4.251) it suffices to show that

$$E^x\left(\overline{P}_0 \overline{g}_n(X_{T_{n,m}})\right) = \lim_{k \to \infty} E^x\left(\overline{P}_0 \overline{g}_n(X_{T_{n,m,k}})\right). \quad (4.254)$$

Note that by (4.217) and (4.204), $\overline{P}_0 \overline{g}_n(X_t) = \lim_{\lambda \to \infty} \lambda \overline{U}^\lambda \overline{g}_n(X_t)$ for all t, where the limit is an increasing limit. Since $\overline{U}^\lambda \overline{g}_n$ is continuous on S, it follows, as in (4.253), that

$$E^x\left(\overline{U}^\lambda \overline{g}_n(X_{T_{n,m}})\right) = \lim_{k \to \infty} E^x\left(\overline{U}^\lambda \overline{g}_n(X_{T_{n,m,k}})\right). \quad (4.255)$$

Since

$$U^\lambda g_n - \mu U^\mu \left(U^\lambda g_n\right) = U^\lambda \left(g_n - \mu U^\mu g_n\right) \geq 0$$

for all $\mu > 0$, we see that $U^\lambda g_n \in \mathcal{M}^+ \cap C_b(S)$. Consequently, it follows from the fourth paragraph of the proof of Theorem 4.7.8 that $\overline{U}^\lambda \overline{g}_n(X_t)$ is a supermartingle. Therefore, since $T_{n,m,k}$ is rational valued, the limit in (4.255) is also an increasing limit. Using the interchangeability of increasing limits we see that

$$
\begin{aligned}
E^x(\overline{P}_0 \overline{g}_n(X_{T_{n,m}})) &= \lim_{\lambda \to \infty} \lambda E^x(\overline{U}^\lambda \overline{g}_n(X_{T_{n,m}})) \quad (4.256) \\
&= \lim_{\lambda \to \infty} \lambda \lim_{k \to \infty} E^x(\overline{U}^\lambda \overline{g}_n(X_{T_{n,m,k}})) \\
&= \lim_{k \to \infty} \lim_{\lambda \to \infty} \lambda E^x(\overline{U}^\lambda \overline{g}_n(X_{T_{n,m,k}})) \\
&= \lim_{k \to \infty} E^x(\overline{P}_0 \overline{g}_n(X_{T_{n,m,k}})),
\end{aligned}
$$

which gives (4.254). $\qquad \square$

We say that a function $u(x, y)$ on $S \times S$ is strongly continuous, if $u(x, y)$ is continuous on $S \times S$, bounded in x for each fixed $y \in S$, and continuous in x uniformly in $y \in S$ (to motivate this definition, recall Lemma 3.4.3, which states that a strongly symmetric Borel right process with continuous potential densities has strongly continuous potential densities).

Theorem 4.7.10 *Let S be a locally compact space with a countable base, and let $\{U^\lambda; \lambda \geq 0\}$ be a symmetric contraction resolvent on $C_b(S)$. Assume that U^0 has a symmetric strongly continuous density $u(x, y)$ with $u(y, y) > 0$ for all $y \in S$. Assume also that $\mathcal{H} \cap C_0(S)$ is dense in $C_0(S)$ in the uniform norm. Then we can construct a strongly symmetric right continuous simple Markov process $X = (\Omega, \mathcal{F}^0, \mathcal{F}^0_t, X_t, \theta_t, P^x)$ in S with potential operators $\{U^\lambda; \lambda \geq 0\}$.*

Proof Under these hypotheses, Theorem 4.7.1 gives a Ray semigroup $\{P_t; t \geq 0\}$ with potentials $\{U^\lambda; \lambda \geq 0\}$. Then, by Theorem 4.7.8 and Lemma 4.7.9, we can construct a Ray process $X = (\Omega, \mathcal{F}^0, \mathcal{F}^0_t, X_t, \theta_t, P^x)$ with state space S_Δ such that $X_t \in D_\Delta$ for all $t \geq 0$ almost surely. We now use the assumption that $u(x, y)$ is strongly continuous to show that $\mathcal{N} = \emptyset$. This implies that $P^x(X_0) = 1$ for all $x \in S_\Delta$ and completes the proof of this theorem.

Let $y \in \mathcal{N}$. Let $f_{\epsilon, y}$ be an approximate δ-function at y with respect to some reference measure m. Following the proof of Theorem 3.6.3, we see that there exists a sequence $\{\epsilon_n\}$ tending to zero, such that almost surely

$$A^y_t := \lim_{\epsilon_n \to 0} \int_0^t f_{\epsilon_n, y}(X_s) \, ds \qquad (4.257)$$

exists and the convergence is uniform in t. Furthermore, $t \mapsto A^y_t$ is almost surely increasing and continuous, and for each $x \in S$

$$E^x(A^y_\infty) = u(x, y). \qquad (4.258)$$

(Note that in the proof of Theorem 3.6.3 we initially use the α-potential for $\alpha > 0$. Consequently, we get uniform convergence in (3.92) only on finite intervals. As we note at the end of the proof Theorem 3.6.3, when the 0-potential exists, we can dispense with the exponentially killed process. This gives us uniform convergence in (4.257) for all $t \in R_+$.)

Clearly dA^y_t is supported on

$$\mathcal{J} = \{t : A^y_{t+\epsilon} - A^y_{t-\epsilon} > 0 \text{ for all } \epsilon > 0\}. \qquad (4.259)$$

It follows from the definition of A^y_t in (4.257), and the fact that $y \in \mathcal{N}$ and $X_s \in D_\Delta$ for all $s \geq 0$, that

$$\mathcal{J} \subseteq \{t \mid X_{t-} = y \text{ and } X_t \neq X_{t-}\}. \qquad (4.260)$$

Since X is right continuous, the set of discontinuities, which by (4.260) includes $\mathscr{J}$, must be countable almost surely. However, since $t \mapsto A_t^y$ is continuous, dA_t^y has no atoms, so $A_t^y \equiv 0$ for all t. This implies, by (4.258), that $u(x, y) = 0$ for all $x \in S$. The hypothesis that $u(y, y) > 0$ for all $y \in S$ then implies that $\mathscr{N} = \emptyset$. $\square$

4.8 Local Borel right processes

The strongly symmetric right continuous simple Markov process X constructed in Theorem 4.7.10 is not necessarily a Borel right process. To see this, let $\zeta := \inf\{t > 0 \mid X_t = \Delta\}$ and

$$\overline{\zeta} := \inf\{t > 0 \mid X_s = \Delta, \forall s \geq t\}. \tag{4.261}$$

By Lemma 3.2.7, if X is a Borel right process, $\overline{\zeta} = \zeta$.

To see that this need not hold for the strongly symmetric right continuous simple Markov process X constructed in Theorem 4.7.10, let $\{U^\lambda; \lambda \geq 0\}$ be the symmetric contraction resolvent on $C_b(S)$ with λ-potential densities

$$u^\lambda(x, y) = \frac{e^{-\sqrt{2(1+\lambda)}|x-y|}}{\sqrt{2(1+\lambda)}} \tag{4.262}$$

with respect to Lebesgue measure on the locally compact space $S = R^1 - \{0\} = (-\infty, 0) \cup (0, \infty)$. $u^\lambda(x, y)$ is the λ-potential density with respect to Lebesgue measure of Brownian motion on R^1 that is killed at the end of an independent exponential time with mean 1. However, in the state space S we remove 0 from R^1. The one-point compactification for S can be viewed as embedding S in $[-\infty, \infty]$ and identifying the three points $\{-\infty, \infty, 0\}$ as the single point Δ. By the uniqueness property of Laplace transforms it is immediate that the semigroup corresponding to $u^\lambda(x, y)$ is the semigroup $\{P_t \,;\, t \geq 0\}$ for exponentially killed Brownian motion on R^1 restricted to S.

The construction in Theorem 4.7.8 proceeds by first using the semigroup $\{P_t \,;\, t \geq 0\}$ together with the Kolmogorov Extension Theorem to obtain a process X_t indexed by t in a countable dense set $D \subset R_+^1$, and then extending it by taking limits from the right. The process $\{X_t, t \in D\}$ has the same distribution in S as exponentially killed Brownian motion B_t, but since $P^x(B_t = 0) = 0$ for fixed t and D is countable, we see that, almost surely, exponentially killed Brownian motion does not hit 0 for all $t \in D$. It follows that the process constructed in Theorem 4.7.10 on S_Δ, for $u^\lambda(x, y)$ as given in (4.262), is obtained from exponentially killed Brownian motion by projecting R_Δ^1 onto S_Δ, which simply identifies 0 and Δ. Since, with positive probability, exponentially killed Brownian motion hits 0 often before it is killed, we see that $\zeta < \bar{\zeta}$, with positive probability.

Therefore, in general, the processes constructed in Theorem 4.7.10 are not Borel right processes. Nevertheless, they are used in Theorem 4.8.4 to construct a class of Markov processes that have most of the nice properties of Borel right processes. We call these processes local Borel right process.

A collection $X = (\Omega, \mathcal{G}, \mathcal{G}_t, X_t, \theta_t, P^x)$ that satisfies the following three conditions is called a local Borel right process with state space S and transition semigroup $\{P_t \,;\, t \geq 0\}$:

(1) X is a right continuous simple Markov process with transition semigroup $\{P_t \,;\, t \geq 0\}$.

(2) $U^\alpha f(X_t)$ is right continuous at all t such that $X_t \in S$, for all $\alpha > 0$ and all $f \in C_b(S_\Delta)$.

(3) $\{\mathcal{G}_t \,;\, t \geq 0\}$ is augmented and right continuous.

We emphasize that condition (1) requires that $t \mapsto X_t(\omega)$ is a right continuous map from R_+ to S_Δ. In fact, the definition of a local Borel right process differs from that of a Borel right process only in condition (2). For a Borel right process we must have that $U^\alpha f(X_t)$ is right continuous at all t such that $X_t \in S_\Delta$, which of course is all $t \in R_+$.

To clarify this distinction we show that for the process considered at the beginning of this section, $U^\alpha f(X_t)$ is not right continuous for all $f \in C_b(S_\Delta)$ at values of t for which $X_t = \Delta$. Let $f \in C_0^+(S)$ have support in $[1, 2]$ and note that

$$U^\alpha f(x) = (\sqrt{2(1+\lambda)})^{-1} \int e^{-\sqrt{2(1+\lambda)}|x-y|} f(y)\, dy, \qquad (4.263)$$

so that

$$\lim_{x \to 0} U^\alpha f(x) = (\sqrt{2(1+\lambda)})^{-1} \int e^{-\sqrt{2(1+\lambda)}|y|} f(y)\, dy > 0. \qquad (4.264)$$

If $\lim_{t \downarrow t_0} Y_t = 0$ for exponentially killed Brownian motion Y_t, then for X_t, its projection in S, we have $\lim_{t \downarrow t_0} X_t = \Delta$. Therefore, by (4.264),

$$\lim_{t \downarrow t_0} U^\alpha f(X_t) = \lim_{x \to 0} U^\alpha f(x) > 0, \qquad (4.265)$$

but $U^\alpha f(X_{t_0}) = U^\alpha f(\Delta) = 0$. On the other hand, since X_t is right continuous and $U^\alpha f \in C_b(S)$, $\lim_{t \downarrow t_0} U^\alpha f(X_t) = U^\alpha f(X_{t_0})$ for all t_0 for which $X_{t_0} \in S$.

Before presenting Theorem 4.7.10 we collect some facts about local Borel right processes.

Lemma 4.8.1 *If S is compact, then any local Borel right process with state space S is a Borel right process.*

Proof Let $X_t = \Delta$. When S is compact, recall that Δ is added as an isolated point. Since by assumption X_s is right continuous, we must have $X_s = \Delta$ for all $t \leq s \leq s + \epsilon$ for some $\epsilon > 0$, and the result follows. $\square$

Lemma 4.8.2 *Let X be a local Borel right process with locally compact state space S. If, for some $\alpha > 0$ and $f \in C_b^+(S)$, we have that $U^\alpha f \in C_0(S)$ and is strictly positive on S, then X is a Borel right process.*

Proof Recall that we show in the fifth paragraph of the proof of Theorem 4.1.1 that $e^{-\alpha t} U^\alpha f(X_t)$ is a supermartingale. Furthermore, since $U^\alpha f \in C_0(S)$ and the paths of X are right continuous, we have that $e^{-\alpha t} U^\alpha f(X_t)$ is right continuous and, by hypothesis, $U^\alpha f(X_t)$ is 0 only when $X_t = \Delta$. It then follows from a standard result about right continuous supermartingales that $\bar{\zeta} = \zeta$ (see Dellacherie and Meyer (1980, Theorem VI.17)).

$\square$

The proof of Lemma 3.2.7 shows that if X is right continuous and satisfies the strong Markov property, then $\bar{\zeta} = \zeta$. Hence, the example considered at the beginning of this section, and consequently local Borel right processes in general, do not satisfy the strong Markov property. The next theorem provides a substitute for the strong Markov property that is sufficient for our needs.

Lemma 4.8.3 *Let $X = (\Omega, \mathcal{G}, \mathcal{G}_t, X_t, \theta_t, P^x)$ be a right continuous simple Markov process. Assume that $U^\alpha f(X_t)$ is almost surely right continuous at all t such that $X_t \in S$, for all $f \in C_b(S_\Delta)$ and all $\alpha > 0$. If T is a $\mathcal{G}_{t+}$ stopping time such that $X_T \in S$ on $T < \infty$, almost surely, then*

$$E^x \left(f(X_{T+s}) 1_{\{T < \infty\}} \,|\, \mathcal{G}_{T+} \right) = P_s f(X_T) 1_{\{T < \infty\}} \qquad (4.266)$$

for all $s \geq 0$, $x \in S$, and all $f \in \mathcal{B}_b(S_\Delta)$.
Furthermore,

$$E^x \left(f(X_{t+s}) \,|\, \mathcal{G}_{t+} \right) = P_s f(X_t) \qquad (4.267)$$

for all $s, t \geq 0$, $x \in S$, and all $f \in \mathcal{B}_b(S_\Delta)$.

Proof The proof of (4.266) follows exactly as in the proof of Theorem 3.2.1. Equation (4.267) does not follow immediately from that proof since we may have $X_t = \Delta$ with positive probability. However, by the simple Markov property and (4.237), $P^x(X_r = \Delta \text{ and } X_s \neq \Delta) = 0$ for any $r < s$ and $x \in S_\Delta$. Therefore, if t is fixed and D is a countable dense set in R_+,

$$P^x(X_t = \Delta \text{ and } X_s \neq \Delta \text{ for some } s > t, s \in D) = 0.$$

Consequently, using the right continuity of the paths of X, we have

$$P^x(X_t = \Delta \text{ and } X_s \neq \Delta \text{ for some } s > t) = 0. \tag{4.268}$$

It is clear from this that, for any fixed t, if $X_t \in \Delta$, then $U^\alpha f(X_s)$ is right continuous at t, almost surely. Since by hypothesis $U^\alpha f(X_s)$ is almost surely right continuous at all t such that $X_t \in S$ for all $f \in C_b(S_\Delta)$ and all $\alpha > 0$, we see that for any fixed t, $U^\alpha f(X_s)$ is almost surely right continuous at t for all $f \in C_b(S_\Delta)$ and all $\alpha > 0$. Using this, the proof of (4.267) follows exactly as in the proof of Theorem 3.2.1 (see also Remark 3.2.2). $\qquad\square$

Theorem 4.8.4 *Let S be a locally compact space with a countable base and let $\{U^\lambda; \lambda \geq 0\}$ be a symmetric contraction resolvent on $C_b(S)$. Assume that U^0 has a symmetric strongly continuous density $u(x,y)$ with $u(y,y) > 0$ for all $y \in S$. Assume also that $\mathcal{H} \cap C_0(S)$ is dense in $C_0(S)$ in the uniform norm. Then we can construct a strongly symmetric local Borel right process $X = (\Omega, \mathcal{F}, \mathcal{F}_t, X_t, \theta_t, P^x)$ with state space S and potential operators $\{U^\lambda; \lambda \geq 0\}$.*

Proof Let $X = (\Omega, \mathcal{F}^0, \mathcal{F}^0_t, X_t, \theta_t, P^x)$ be the strongly symmetric right continuous simple Markov process constructed in Theorem 4.7.10. It follows from the fact that $U^\alpha f \in C_b(S)$ for all $f \in C_b(S_\Delta)$ and all $\alpha > 0$ that condition (2) for a local Borel right process is satisfied. Let $\{\mathcal{F}_t, t \geq 0\}$ be the standard augmentation of $\{\mathcal{F}^0_t, t \geq 0\}$. As in the proof of (2.75), it follows from (4.267) that

$$E^x(Y \circ \theta_t \mid \mathcal{F}_{t+}) = E^{W_t}(Y) \tag{4.269}$$

for all $\mathcal{F}^0$ measurable functions Y. It then follows from the proof of Lemma 2.3.1 that $\mathcal{F}_{t+} = \mathcal{F}_t$ for all $t \geq 0$. Using this, the theorem follows as in the proof of Lemma 3.2.3. $\qquad\square$

As in the proof of Lemma 2.3.2, (4.266) extends to all $\mathcal{F}$ measurable functions Y.

Lemma 4.8.5 *Let* $X = (\Omega, \mathcal{F}, \mathcal{F}_t, X_t, \theta_t, P^x)$ *be a local Borel right process. If* T *is an* $\mathcal{F}_t$ *stopping time such that* $X_T \in S$ *on* $T < \infty$, *almost surely, then*

$$E^x \left(Y \circ \theta_T 1_{\{T < \infty\}} \,|\, \mathcal{F}_T \right) = E^{X_T} (Y) \, 1_{\{T < \infty\}} \qquad (4.270)$$

for all $x \in S$ *and* $\mathcal{F}$ *measurable functions* Y.

We call the result of this lemma the local strong Markov property.

4.9 Supermedian functions

Let S be a locally compact space with a countable base. Supermedian functions are functions in $\mathcal{B}_b^+(S)$ that satisfy (4.202), where $\{U^\lambda; \lambda \geq 0\}$ is a contraction resolvent. We use $\mathcal{M}^+$ to denote these functions and $\mathcal{M}$ to denote $\mathcal{M}^+ - \mathcal{M}^+$. We now give some properties of $\mathcal{M}^+$ and $\mathcal{M}$ for use in the next section.

Lemma 4.9.1 *Let* $\{f_n\}$ *be a sequence of functions in* $\mathcal{M}^+$. *Then*

(1) $f_n \uparrow f \Longrightarrow f \in \mathcal{M}^+$.
(2) $\inf_n f_n \in \mathcal{M}^+$.
(3) $\liminf_n f_n \in \mathcal{M}^+$.
(4) $\mathcal{M}$ *is closed under* $\wedge$ *and* $\vee$.
(5) $\mathcal{M}$ *contains the constant functions.*

Proof Property (1) follows from the Monotone Convergence Theorem and (2) from the definition of supermedian functions. Property (3) follows from Fatou's Lemma. Closure in (4) under $\wedge$ follows from (2) and the fact that

$$
\begin{aligned}
(f - g) \wedge (u - v) &= [(f + v) - (g + v)] \wedge [(u + g) - (v + g)] \\
&= [(f + v) \wedge (u + g)] - (v + g).
\end{aligned}
$$

Since both $(f + v) \wedge (u + g)$ and $v + g$ are in $\mathcal{M}^+$, we see that $(f - g) \wedge (u - v) \in \mathcal{M}$. Writing $(f - g) \vee (u - v) = -[(g - f) \wedge (v - u)]$, the same proof gives closure under $\vee$. Property (5) follows from (4.200). $\square$

A function $f \in \mathcal{M}^+$ is said to be excessive if

$$\lambda U^\lambda f(x) \uparrow f(x) \quad \text{as } \lambda \to \infty \quad \forall x \in S. \qquad (4.271)$$

Lemma 4.9.2 *Assume that* U^0 *has a symmetric strongly continuous density* $u(x, y)$. *Then*

(1) $u_v(x) := u(x, v)$ *is excessive for each* $v \in S$.

(2)

$$U^0\rho(x) := \int u(x,v)\,d\rho(v) \tag{4.272}$$

is excessive for any finite positive measure ρ on S. Such functions are called potentials.

(3) All excessive functions are the increasing limits of potentials.

Proof To obtain (1), let $f \in C_b^+(S)$ and let $\alpha < \beta$. It follows from the resolvent equation (4.201) that

$$\alpha U^\alpha U^0 f(x) = U^0 f(x) - U^\alpha f(x) \tag{4.273}$$

$$\beta U^\beta U^0 f(x) = U^0 f(x) - U^\beta f(x) \tag{4.274}$$

and

$$U^\alpha f(x) - U^\beta f(x) = (\beta - \alpha)U^\alpha U^\beta f(x) \geq 0. \tag{4.275}$$

Combining these equations, we see that

$$\alpha U^\alpha U^0 f(x) \leq \beta U^\beta U^0 f(x). \tag{4.276}$$

Since this is true for all $f \in C_b^+(S)$, it follows from the strong continuity of u that

$$\alpha U^\alpha u_v(x) \leq \beta U^\beta u_v(x). \tag{4.277}$$

Consequently, $h(v) := \lim_{\alpha \to \infty} \alpha U^\alpha u_v(x)$ exists and, by the strong continuity of u, $h \in C_b^+(S)$.

Let $f \in C_b^+(S)$ with compact support. By the Monotone Convergence Theorem, (4.273), and (4.200),

$$
\begin{aligned}
\int h(v)f(v)\,dm(v) &= \lim_{\alpha \to \infty} \int \alpha U^\alpha u_v(x)f(v)\,dm(v) \quad (4.278)\\
&= \lim_{\alpha \to \infty} \alpha U^\alpha U^0 f(x)\\
&= U^0 f(x) = \int u_v(x)f(v)\,dm(v).
\end{aligned}
$$

Since this is true for all $f \in C_b^+(S)$ with compact support, it follows from the continuity of h and u_v that $h = u_v$. This proves (1).

(2) follows from (1).

To obtain (3), we note that when $f \in \mathcal{M}^+$ is excessive, $0 \leq f - \lambda U^\lambda f \downarrow 0$. Using the resolvent equation, we then have $\lambda U^0(f - \lambda U^\lambda f) = \lambda U^\lambda f \uparrow f$ pointwise as $\lambda \to \infty$. $\qquad \square$

4.10 An extension theorem for local Borel right processes

All the results obtained about strongly symmetric Borel right process on S make perfect sense when S is finite (although much of the machinery we develop is not needed to handle this simple case). Also, since every finite set is compact, by Lemma 4.8.1 for such sets, there is no difference between Borel right processes and local Borel right processes. We now present a theorem that allows us to reduce the study of certain properties of strongly symmetric local Borel right processes, to processes with finite state spaces. This is used in Chapter 13. Whenever we consider a process on a finite state space we take the reference measure to be the measure that assigns mass 1 to each point in the state space. Such a measure is called a counting measure.

Theorem 4.10.1 (Extension Theorem) *Let S be a locally compact space with a countable base. Let $\Gamma(x, y)$ be a continuous function on $S \times S$. Assume that for every finite subset $S' \subseteq S$, $\Gamma(x, y)$ is the 0-potential density of a strongly symmetric transient Borel right process X' on S'. Then $\Gamma(x, y)$ is the 0-potential density of a strongly symmetric transient local Borel right process X on S.*

In particular, if S is compact, then $\Gamma(x, y)$ is the 0-potential density of a strongly symmetric transient Borel right process X on S.

Proof We can always find a probability measure μ on S that gives strictly positive mass to each open set of S. By Lemma 3.4.6 applied to all finite subsets $S' \subseteq S$, $\Gamma(y, y) > 0$ for all $y \in S$. Set $dm(y) = \dfrac{1}{\Gamma(y, y)} d\mu(y)$. By Lemma 3.4.3 again applied to all finite subsets $S' \subseteq S$, $\Gamma(x, y) \leq \Gamma(y, y)$ for all $x \in S$. We define the operator

$$Hf(x) = \int_S \Gamma(x, y) f(y) dm(y), \qquad f \in C_b(S). \tag{4.279}$$

We have

$$\begin{aligned}
|Hf(x)| &\leq \|f\|_\infty \int_S \Gamma(x, y) dm(y) \tag{4.280} \\
&= \|f\|_\infty \int_S \frac{\Gamma(x, y)}{\Gamma(y, y)} d\mu(y) \leq \|f\|_\infty.
\end{aligned}$$

Hence H is a positive bounded linear operator on $f \in C_b(S)$.

We show that H is the 0-potential operator of a local Borel right process in S. It then follows from (4.279) that H has 0-potential density $\Gamma(x, y)$ with respect to the measure m.

To begin, we show that H satisfies the positive maximum principle on $C_b(S)$. Considering the form of H in (4.279), it is easy to see that it suffices to show that for each compact set $K \subseteq S$, the restriction of H to K, which we denote by H_K, satisfies the positive maximum principle on $C(K)$.

For each n write K as a finite disjoint union $\cup_{j=1}^{m_n} K_{n,j}$ of neighborhoods $K_{n,j}$ with diameter less than or equal to $1/n$. For each j choose a point $x_{n,j} \in K_{n,j}$. Define the operators

$$H_n h(x_{n,i}) = \sum_{j=1}^{m_n} \Gamma(x_{n,i}, x_{n,j}) h(x_{n,j}) m(K_{n,j}). \qquad (4.281)$$

We claim that for each n the operator H_n satisfies the positive maximum principle on the set $E_n = \{x_{n,j}, 1 \le j \le m_n\}$. It is easy to check that this claim follows once we show that

$$\overline{H}_n h(x_{n,i}) = \sum_{j=1}^{m_n} \Gamma(x_{n,i}, x_{n,j}) h(x_{n,j}) \qquad (4.282)$$

satisfies the positive maximum principle on E_n. To prove this last assertion note that, by hypothesis, $\overline{H}_n$ is the 0-potential operator of a strongly symmetric transient Borel right process on E_n. Obviously, since E_n is finite, this process is also a Feller process. Therefore, by Theorem 4.1.6, $\overline{H}_n$ satisfies the positive maximum principle on E_n.

We now show that this implies that H_K satisfies the positive maximum principle on $C(K)$. Suppose that it does not. Then there exists some $a > 0$ and a continuous function h on K with $\sup_{x \in K} H_K h(x) > 0$ for which

$$\sup_{\{x:h(x)>0\}} H_K h(x) \le a \qquad (4.283)$$

but for some $x_* \in K$, with $h(x_*) \le 0$, and some $b > 0$

$$H_K h(x_*) > a + b. \qquad (4.284)$$

By replacing $h(x)$ by $h(x) - \epsilon$ for $\epsilon > 0$ sufficiently small, we may assume that $h(x_*) < 0$. Since h is continuous, for all sufficiently large n we can find some $x_{**} \in E_n$ for which $h(x_{**}) \le 0$. Using the continuity of Γ and h on the compact set K it follows from (4.283) and (4.284) that for all sufficiently large n,

$$\sup_{\{x \in E_n : h(x)>0\}} H_n h(x) \le a + b/2 \qquad (4.285)$$

and

$$H_n h(x_{**}) > a + b/2. \qquad (4.286)$$

This contradicts the assumption that H_n satisfies the positive maximum principle on E_n. Therefore, H_K satisfies the positive maximum principle on $C(K)$.

By Remark 4.1.10 we can construct a symmetric contraction resolvent $\{U^\lambda; \lambda \geq 0\}$ on $C_b(S)$ with $U^0 = H$.

By Lemma 3.4.3 applied to all finite subsets $S' \subseteq S$, we see that $u(x,y) = \Gamma(x,y)$ is strongly continuous, and by Lemma 3.4.6 applied to all finite subsets $S' \subseteq S$, we see that $u(y,y) > 0$ for all $y \in S$. Set $u_v(x) = u(x,v)$. It follows from Lemma 4.9.2 (1) that $u_v(x)$ is excessive for each $v \in S$ and thus is a supermedian function on S for each $v \in S$.

Let $x, y \in S$. By hypothesis, $u(x,y) = \Gamma(x,y)$ is the 0-potential density of a strongly symmetric transient Borel right process on the two point set $S' = \{x,y\}$. Therefore, it follows from (3.109) and Lemma 3.6.12 that

$$u(x,y) < u(x,x) \vee u(y,y) \qquad \forall\, x, y \in S. \tag{4.287}$$

If $u(x,y) < u(x,x)$, then $u_x(x) \neq u_x(y)$, whereas if $u(x,y) < u(y,y)$, then $u_y(x) \neq u_y(y)$. This shows that the collection $\mathcal{M}^+ \cap C_b(S)$ of continuous bounded supermedian functions separates points of S. It follows from Lemma 4.9.1 (4) and (5) that $\mathcal{M} \cap C_b(S)$ is a vector lattice that contains the constants.

Suppose S is compact. Then it follows from Theorem 14.5.1 that $\mathcal{M} \cap C_b(S)$ is dense in $(C_b(S), \|\cdot\|_\infty)$. Therefore, when S is compact this theorem follows from Theorem 4.8.4.

We now consider the general case in which S is a locally compact space with a countable base. We show below that for any open set $B \subseteq S$ with compact closure and any $x \in B$, we can find a function $v_{B,x} \in \mathcal{H} \cap C_0(S)$ that is supported in $\overline{B}$ and $v_{B,x}(x) > 0$. Because S is locally compact, this shows that $\mathcal{H} \cap C_0(S)$ separates points of S. Since $\mathcal{H}$ is a vector lattice that also contains the constants, it follows from Theorem 14.5.2 that $\mathcal{H} \cap C_0(S)$ is dense in $(C_0(S), \|\cdot\|_\infty)$. We can now use Theorem 4.8.4 to complete the proof of this theorem.

We now construct the functions $v_{B,x}$. Let $\{K_n\}$ be an increasing sequence of compact subsets of S with $\cup_{n=1}^\infty K_n = S$, such that each K_n is the closure of its interior and $K_n \subseteq \operatorname{int} K_{n+1}$. For $f \in C(K_n)$ and $x \in K_n$, we define the operator

$$V_n f(x) = \int_{K_n} u^0(x,y) f(y) dm(y). \tag{4.288}$$

Since we have proved this theorem for compact state spaces, we know that V_n is the 0-potential of a strongly symmetric Borel right process $X_t^{(n)}$ in K_n with continuous potential densities.

By Theorem 4.5.2, for any open set $B \subseteq \text{int } K_n$, the process $\widetilde{X}_t^{(n)}$, obtained by killing $X_t^{(n)}$ at $T_{\overline{B}^c}^{(n)}$, is a strongly symmetric Borel right process in B with continuous potential densities, which we denote by $\widetilde{u}^{(n),\alpha}(x,y)$. By (4.152),

$$\widetilde{u}^{(n),0}(x,y) = u^0(x,y) - E^x\left(u^0\left(X_{T_{\overline{B}^c}^{(n)}}, y\right) 1_{T_{\overline{B}^c}^{(n)} < \infty}\right) \qquad \forall x, y \in K_n.$$
(4.289)

It is clear that the right-hand side of (4.289) is zero for $x \in \overline{B}^c$. Consequently, by symmetry, it is zero for all for $x, y \in \overline{B}^c$.

Fix $x \in B$. Let

$$v_x^{(n)}(y) = \begin{cases} \widetilde{u}^{(n),0}(x,y) & y \in K_n \\ 0 & y \in K_n^c. \end{cases}$$

$v_x^{(n)}(y)$ is clearly a continuous function on S with support in $\overline{B}$. Furthermore, since $\widetilde{u}^{(n),0}(x,y)$ is the 0-potential density of a strongly symmetric Borel right process, it follows from Lemma 3.4.6 that $v_x^{(n)}(x) > 0$ for $x \in B$. Consequently, $v_x^{(n)}(y)$ is not identically zero. Note that

$$v_x^{(n)}(y) = u^0(x,y) 1_{\{K_n\}}(y)$$
(4.290)
$$- E^x\left(u^0\left(X_{T_{\overline{B}^c}^{(n)}}, y\right) 1_{\{K_n\}}(y) 1_{T_{\overline{B}^c}^{(n)} < \infty}\right).$$

Let n_0 be such that that $\overline{B} \subseteq \text{int } K_{n_0}$. It follows from Remark 4.5.3 that $v_x^{(n)}$ is independent of n for $n \geq n_0$. We denote this function by $v_{B,x}(y)$.

It remains to show that $\{v_{B,x}(y), y \in S\} \in \mathcal{M}$. For $n \geq n_0$ we have, by (4.290),

$$v_{B,x}(y) = u^0(x,y) 1_{\{K_n\}}(y)$$
(4.291)
$$- E^x\left(u^0\left(X_{T_{\overline{B}^c}^{(n)}}, y\right) 1_{\{K_n\}}(y) 1_{T_{\overline{B}^c}^{(n)} < \infty}\right).$$

So it also holds if we take the limit as $n \to \infty$, providing that the limit exists. Clearly $u^0(x,y) 1_{\{K_n\}}(y) \uparrow u^0(x,y) = u_x^0(y)$, and by Lemma 4.9.2 (1), $u_x^0(y) \in \mathcal{M}^+$. Now set

$$h_n(y) = E^x\left(u^0\left(X_{T_{\overline{B}^c}^{(n)}}, y\right) 1_{\{K_n\}}(y) 1_{T_{\overline{B}^c}^{(n)} < \infty}\right).$$
(4.292)

We see from (4.291) that $h_n(y)$ increases in n to some continuous, non-negative limit that we denote by $h_\infty(y)$. We conclude the proof by showing that $h_\infty \in \mathcal{M}^+$.

Consider the contraction resolvent $\{U^{(n),\lambda}; \lambda \geq 0\}$ on $C_b(K_n)$ with $U^{(n),0} = V_n$. It follows from Lemma 4.9.2 (1) that $u_v^0(y)1_{\{K_n\}}(y)$ is excessive for this contraction resolvent. Therefore, using Lemma 4.9.2 (2) and (4.292), we see that $h_\infty(y)$ is excessive for this contraction resolvent.

It now follows from Lemma 4.9.2 (3) that there is a sequence of non-negative functions $\{g_{n,k}\}$, $k = 1,\ldots$ on K_n with $V_n g_{n,k} \uparrow h_\infty$ or, equivalently, that $U^0(g_{n,k}1_{\{K_n\}}) \uparrow h_\infty$ pointwise on K_n. Let

$$w_n(y) := \liminf_{k \to \infty} U^0(g_{n,k}1_{\{K_n\}})(y) \qquad y \in S. \qquad (4.293)$$

By Lemma 4.9.2 (2), $U^0(g_{n,k}1_{\{K_n\}}) \in \mathcal{M}^+$. Consequently, by Lemma 4.9.1 (3), $w_n \in \mathcal{M}^+$. On the other hand, $w_n = h_\infty$ on K_n. This implies that $h_\infty = \liminf_{n \to \infty} w_n$ on S. Using Lemma 4.9.1 (3) again, we see that $h_\infty \in \mathcal{M}^+$. $\qquad \square$

4.11 Notes and references

We refer the reader to Bertoin (1996), Kallenberg (2001), Khoshnevisan (2002), and Stroock (1993) for many different interesting proofs of the existence of Lévy processes.

The material in Sections 4.4–4.5 is adapted from the classic references on Markov processes mentioned in Section 3.11. The content of Section 4.6, which comes from our paper Marcus and Rosen (1992d) is due to P. Fitzsimmons.

Section 4.7 is adapted from Getoor (1975) and Dellacherie and Meyer (1987), both of which are based on Ray (1963). All these sources, however, only consider compact state spaces. In their treatment, a state space that is not compact is compactified by a procedure known as the Ray–Knight compactification. This compactification changes the topology and can substantially enlarge the state space. For the applications of the results in Sections 4.7 that we present in Chapter 13, it is necessary to deal directly with locally compact spaces.

The basic idea for the extension theorem of Section 4.10 is due to P. Fitzsimmons.

5

Basic properties of Gaussian processes

5.1 Definitions and some simple properties

A real-valued random variable X is a Gaussian random variable if it has characteristic function

$$Ee^{i\lambda X} = \exp\left(im\lambda - \frac{\sigma^2\lambda^2}{2}\right) \tag{5.1}$$

for some real numbers m and σ. It follows from (5.1), by differentiating with respect to λ and then setting $\lambda = 0$, that

$$E(X) = m \quad \text{and} \quad \text{Var}(X) = \sigma^2. \tag{5.2}$$

An R^n-valued random variable ξ is a Gaussian random variable if (y, ξ) is a real-valued Gaussian random variable for each $y \in R^n$. By the above, this is equivalent to saying that it has characteristic function

$$\phi_\xi(y) \stackrel{def}{=} Ee^{i(y,\xi)} = \exp\left(iE((y,\xi)) - \frac{\text{Var}((y,\xi))}{2}\right) \tag{5.3}$$

for each $y \in R^n$. Let $m = (m_1, \ldots, m_n)$. Setting

$$E\xi_j = m_j \quad \text{and} \quad E(\xi_j - m_j)(\xi_k - m_k) = \Sigma_{j,k}, \tag{5.4}$$

we see that (5.3) can be written as

$$\phi_\xi(y) = \exp\left(imy^t - \frac{y\Sigma y^t}{2}\right), \tag{5.5}$$

where $\Sigma = \{\Sigma_{j,k}\}_{j,k=1}^n$ is a symmetric $n \times n$ matrix with real components. (We also consider a vector in R^n as a $1 \times n$ matrix. Thus we can express (m, y) as my^t). We refer to m as the mean vector, or simply the mean, of ξ and Σ as the covariance matrix of ξ. We note that the distribution of an R^n-valued Gaussian random variable is completely determined by its mean vector and covariance matrix. The rank of Σ equals the dimension of the subspace of R^n, which supports the distribution of ξ.

A remarkable property of Gaussian random variables is that uncorrelated Gaussian random variables are independent.

Lemma 5.1.1 *Let ξ be an R^n-valued Gaussian random variable with mean vector m and assume that*

$$E(\xi_j - m_j)(\xi_k - m_k) = 0 \qquad j \neq k. \tag{5.6}$$

Then $\xi_1, \ldots, \xi_n$ are independent.

Proof In this case

$$\phi_\xi(y) = \prod_{j=1}^{n} \exp\left(im_j y_j - \frac{\Sigma_{j,j} y_j^2}{2}\right). \tag{5.7}$$

$\square$

A matrix $A = \{A_{j,k}\}_{j,k=1}^{n}$ with real components is positive definite if $\sum_{j,k=1}^{n} a_j a_k A_{j,k} \geq 0$ for all real numbers $a_1, \ldots, a_n$. We say that the positive definite matrix A is strictly positive definite if $\sum_{j,k=1}^{n} a_j a_k A_{j,k} = 0$ implies that all the a_i are zero.

A function $\Sigma(s,t)$ on $T \times T$ is a positive definite function if, for all $n \geq 1$ and $t_1, \ldots, t_n$ in T, the $n \times n$ matrix Σ, defined by $\Sigma_{j,k} = \Sigma(t_j, t_k)$ is a positive definite matrix. A function $\{\phi(u), u \in T\}$ is positive definite if the function $\Sigma(s,t) := \phi(s - t)$ is positive definite. Both of these definitions hold similarly for strict positive definiteness.

The covariance matrix Σ in (5.5) is positive definite. This is easy to see, since

$$\sum_{j,k=1}^{n} a_j a_k \Sigma_{j,k} = E\left(\sum_{j=1}^{n} a_j(\xi_j - m_j)\right)^2. \tag{5.8}$$

Let ξ be an R^n-valued Gaussian random variable with mean m and covariance matrix Σ and let U be a $p \times n$ matrix. Then $\eta = (U\xi^t)^t$ is an R^p-valued Gaussian random variable with mean $(Um^t)^t$ and covariance matrix $U\Sigma U^t$. This follows from (5.5) since

$$\begin{aligned}
\phi_\eta(z) &= Ee^{i\eta z^t} = Ee^{i\xi(zU)^t} = \phi_\xi(zU) \tag{5.9}\\
&= \exp\left(im(zU)^t - \frac{zU\Sigma(zU)^t}{2}\right)\\
&= \exp\left(i(Um^t)^t z^t - \frac{zU\Sigma U^t z^t}{2}\right).
\end{aligned}$$

When Σ is strictly positive definite its rank is equal to n.

Lemma 5.1.2 *Let ξ be an R^n-valued Gaussian random variable with mean m and strictly positive definite covariance matrix Σ. Then the probability distribution of ξ is absolutely continuous with respect to Lebesgue measure on R^n and has probability density function*

$$f(x) = \frac{1}{(2\pi)^{n/2}(\det \Sigma)^{1/2}} \exp\left(-\frac{(x-m)\Sigma^{-1}(x-m)^t}{2}\right), \qquad (5.10)$$

where $\det \Sigma$ denotes the determinant of Σ and $x \in R^n$.

We also use the notation $|\Sigma|$ for $\det \Sigma$.

Proof Since the characteristic function is unique, it suffices to show that

$$\frac{1}{(2\pi)^{n/2}(\det \Sigma)^{1/2}} \int e^{i(y,x)} \exp\left(-\frac{(x-m)\Sigma^{-1}(x-m)^t}{2}\right) dx = \phi_\xi(y),$$
$$(5.11)$$

where $\phi_\xi(y)$ is given in (5.5). Since Σ is a strictly positive definite symmetric matrix, it has a symmetric square root $\Sigma^{1/2}$. Using this we see that

$$\int e^{i(y,x)} \exp\left(-\frac{(x-m)\Sigma^{-1}(x-m)^t}{2}\right) dx \qquad (5.12)$$

$$= e^{i(y,m)} \int e^{i(y,x)} \exp\left(-\frac{x\Sigma^{-1}x^t}{2}\right) dx$$

$$= (\det \Sigma)^{1/2} e^{i(y,m)} \int e^{i(y\Sigma^{1/2},x)} \exp\left(-\frac{xx^t}{2}\right) dx.$$

Let η be an R^n-valued Gaussian random variable with mean zero and covariance matrix I. Then

$$\frac{1}{(2\pi)^{n/2}} \int e^{i(y\Sigma^{1/2},x)} \exp\left(-\frac{xx^t}{2}\right) dx = \phi_\eta(y\Sigma^{1/2}) \qquad (5.13)$$

$$= \exp\left(-\frac{y\Sigma y^t}{2}\right),$$

where, for the second equality we use (5.5). Using (5.13) to substitute for the integral in (5.12), we get (5.11). $\qquad \square$

Let ξ be an R^n-valued Gaussian random variable with mean zero and covariance matrix Σ. Let $a \in R^n$. Then, by (5.10), in dimension one

$$P\left(\alpha \leq (a,\xi) \leq \beta\right) = \frac{1}{(2\pi\sigma^2)^{1/2}} \int_\alpha^\beta e^{-u^2/2\sigma^2}\, du, \qquad (5.14)$$

where $\sigma^2 = E((a, \xi))^2 = a\Sigma a^t$. Using this we easily see that

$$Ee^{(a,\xi)} = \exp\left(\frac{a\Sigma a^t}{2}\right). \tag{5.15}$$

We use the standard notation

$$\Phi(x) = \frac{1}{\sqrt{2\pi}} \int_{-\infty}^{x} e^{-u^2/2}\, du \tag{5.16}$$

$$\phi(x) = \frac{d}{dx}\Phi(x) = \frac{1}{\sqrt{2\pi}}e^{-x^2/2}. \tag{5.17}$$

Lemma 5.1.3 *Let ξ be a real-valued Gaussian random variable with mean zero with variance σ^2. Then, for all $a > 0$,*

$$P(|\xi| > a) \le \exp\left(-\frac{a^2}{2\sigma^2}\right), \tag{5.18}$$

and when $a/\sigma \ge 1$,

$$(\sigma/a)\phi(a/\sigma) \le P(|\xi| > a) \le 2\,(\sigma/a)\phi(a/\sigma). \tag{5.19}$$

Proof Dividing $|\xi|$ by σ we see that it suffices to verify (5.18) when $\sigma = 1$. In this case,

$$P(|\xi| > a) = \frac{2}{\sqrt{2\pi}} \int_{a}^{\infty} e^{-u^2/2}\, du \tag{5.20}$$

$$\le \frac{2}{a\sqrt{2\pi}} \int_{a}^{\infty} u e^{-u^2/2}\, du = \frac{2}{a\sqrt{2\pi}}e^{-a^2/2}.$$

This gives us (5.18) for $a \ge 2/\sqrt{2\pi}$. Also, taking derivatives, we see that $\exp(-a^2/2) - P(|\xi| > a)$ is increasing for $a < 2/\sqrt{2\pi}$, and since this difference is zero when $a = 0$ we get (5.18) for $a < 2/\sqrt{2\pi}$.

To obtain the right-hand side of (5.19) we note that

$$P(|\xi| > a) = \frac{2}{\sqrt{2\pi}} \int_{a/\sigma}^{\infty} e^{-u^2/2}\, du \le \frac{2\sigma}{\sqrt{2\pi}a} \int_{a/\sigma}^{\infty} u e^{-u^2/2}\, du. \tag{5.21}$$

The left-hand side of (5.19) follows from Lemma 14.8.1. □

A sequence of independent Gaussian random variables with mean zero and variance one is called a standard normal sequence. Sometimes the word "normal" is a synonym for Gaussian. Sometimes we use the notation $N(\mu, \sigma^2)$ to indicate a normal random variable with mean μ and variance σ^2.

The following simple limit theorem plays an important role in this book.

Theorem 5.1.4 *Let $\{\xi_n\}$ be a standard normal sequence. Then*

$$\limsup_{n\to\infty} \frac{\xi_n}{(2\log n)^{1/2}} = 1 \qquad a.s. \tag{5.22}$$

Proof For $\epsilon > 0$, let $a_n = P\left(\xi_n > ((2+\epsilon)\log n)^{1/2}\right)$ and $b_n = P(\xi_n > (2\log n)^{1/2})$ and observe that $\sum a_n < \infty$ and $\sum b_n = \infty$. The result follows by the Borel–Cantelli Lemmas. $\qquad\square$

Consider the function

$$\mathcal{U}(x) \overset{def}{=} (\phi \circ \Phi^{-1})(x) \qquad x \in [0,1] \tag{5.23}$$

with $\mathcal{U}(0) = \mathcal{U}(1) = 0$. We note the following important relationship.

Lemma 5.1.5 *For $x \in (0,1)$*

$$\mathcal{U}\mathcal{U}''(x) \overset{def}{=} \mathcal{U}(x)\mathcal{U}''(x) = -1. \tag{5.24}$$

Proof This follows simply, using the relations $\phi'(x) = -x\phi(x)$ and $(\Phi^{-1}(x))' = 1/\mathcal{U}(x)$. $\qquad\square$

We use γ_n, or simply γ, to denote the canonical Gaussian measure on R^n. This is the probability distribution on R^n that is induced by the standard normal sequence with n terms. Clearly the probability density function of γ is $(2\pi)^{-n/2} \exp(-xx^t/2)$, $x \in R^n$.

A real-valued stochastic process $\{X(t), t \in T\}$ (T is some index set) is a Gaussian process if its finite-dimensional distributions are Gaussian. It is characterized by its mean function m and its covariance kernel Σ, which are given by

$$m(t) = EX(t) \quad \text{and} \quad \Sigma(s,t) = E(X(t) - m(t))(X(s) - m(s)). \tag{5.25}$$

It follows from (5.8) that $\Sigma(s,t)$ is a positive definite function on $T \times T$.

Conversely, and this is the route we generally take, given a real-valued function $m(t)$ on T and a positive definite function $\Sigma(s,t)$ on $T \times T$, we obtain a real-valued Gaussian process $\{X(t), t \in T\}$, with mean function m and covariance kernel Σ. This is because m, Σ, and (5.5) allow us to obtain a consistent family of finite-dimensional Gaussian distributions on the finite subsets of T.

When T is the positive integers we write the Gaussian process $\{X(t), t \in T\}$ as $\{\xi_j\}_{j=1}^{\infty}$, or simply $\{\xi_j\}$, and refer to it as a Gaussian sequence. It follows from (5.14) that a real-valued Gaussian random variable has moments of all orders. Thus, given a Gaussian process $\{X(t), t \in T\}$, we can define a natural L^2 metric on $T \times T$ by

$$
\begin{aligned}
d(s,t) \quad &:= \quad d_X(s,t) = \left(E(X(s) - X(t))^2\right)^{1/2} \qquad (5.26) \\
&= \quad (\Sigma(s,s) + \Sigma(t,t) - 2\,\Sigma(s,t))^{1/2} .
\end{aligned}
$$

Remark 5.1.6 The fact that there is a one–one correspondence between mean zero Gaussian processes and positive definite functions allows us to very easily obtain many important properties of positive definite functions. Let $\Sigma(s,t)$, $s,t \in T$ be a positive definite function and let $\{X(t), t \in T\}$ be a mean zero Gaussian process with covariance Σ. Then, because $\Sigma(s,t) = EX(s)X(t)$, it follows from the Schwarz inequality that

$$
\Sigma(s,t) \le (\Sigma(s,s)\Sigma(t,t))^{1/2} . \qquad (5.27)
$$

Let 0 be a point in T. Then

$$
\Sigma(s,t) - \frac{\Sigma(s,0)\,\Sigma(t,0)}{\Sigma(0,0)} \qquad (5.28)
$$

is also a positive definite function on $T \times T$. This is because it is the covariance of

$$
\eta(t) = X(t) - \frac{\Sigma(t,0)}{\Sigma(0,0)} X(0). \qquad (5.29)
$$

(It follows from the remarks in the paragraph containing (5.9) that η is a Gaussian process.)

Let (Ω, P) denote the probability space of a Gaussian process, $\{X(t), t \in T\}$. This process exists in $L^2(P)$ with the inner product given by $(X(s), X(t)) = EX(s)X(t)$, $s,t \in T$. Note that $\eta(t)$ is the projection of $X(t)$, onto the orthogonal complement of $X(0)$ with respect to $L^2(P)$.

Remark 5.1.7 We show in Lemma 3.3.3 that the α-potential density of a strongly symmetric Borel right process with continuous α-potential density is positive definite. Therefore, to every such process we can associate a mean zero Gaussian process with covariance equal to its α-potential density. We call these Gaussian processes associated processes. Not all Gaussian processes are associated processes. In fact, associated processes have some stronger properties than Gaussian processes in general. Let $\{X(t), t \in T\}$ be an associated process. In contrast with (5.27), it follows from Lemma 3.4.3 that

$$
EX(s)X(t) \le EX^2(s) \wedge EX^2(t) \qquad \forall\, s,t \in T. \qquad (5.30)
$$

Also, by Lemma 3.4.6, $EX^2(t) > 0$ for all $t \in T$. In Chapter 13 we give several properties that characterize when a Gaussian process is an associated process.

5.1.1 Gaussian Markov processes

We consider Gaussian processes $\{X(t), t \in R_+\}$ that are also Markov processes.

Lemma 5.1.8 *Let p and q be positive functions on $T \subset R^1$ with p/q strictly increasing. Set*

$$\Sigma(s,t) = \begin{cases} p(s)q(t) & s \leq t \\ p(t)q(s) & t < s \end{cases} \tag{5.31}$$

and assume that p and q are such that $\Sigma(s,t) > 0$ for all $s, t \in T$. Then $\Sigma(s,t)$ is a strictly positive definite function on $T \times T$.

Proof Let $t_1 < \cdots < t_n \in T$ and set $p_j = p(t_j)$ and $q_k = q(t_k)$. Using (5.31), we see that the matrix $D_n := \{\Sigma(t_j, t_k)\}_{j,k=1}^n$ is given by

$$\begin{pmatrix} p_1q_1 & p_1q_2 & \cdots & p_1q_{n-1} & p_1q_n \\ p_1q_2 & p_2q_2 & \cdots & p_2q_{n-1} & p_2q_n \\ \vdots & \vdots & \ddots & \vdots & \vdots \\ p_1q_n & p_2q_n & \cdots & p_{n-1}q_n & p_nq_n \end{pmatrix}.$$

Multiply the $(n-1)$-st column of D_n by $-q_n/q_{n-1}$ and add it to the last column to see that

$$\det D_n = (\det D_{n-1}) \left(p_nq_n - \frac{p_{n-1}q_n^2}{q_{n-1}} \right). \tag{5.32}$$

Since p/q is strictly increasing, the last term in (5.32) is greater than zero. Since $\det D_1 > 0$, it follows by induction that $\det D_j > 0$ for all $1 \leq j \leq n$. By Lemma 14.9.2, this implies that the matrix D_n is strictly positive definite. $\qquad\square$

Lemma 5.1.9 *Let $I \in R^1$ be an interval, open or closed, and let $X = \{X(t), t \in I\}$ be a mean zero Gaussian process with continuous strictly positive definite covariance Σ. Then X is a Gaussian Markov process, that is, for all increasing sequences $t_1, \ldots, t_n \in R_+$, for all n,*

$$E(X(t_n)|X(t_{n-1})) = E(X(t_n)|X(t_{n-1}), \ldots, X(t_1)) \tag{5.33}$$

if and only if Σ can be expressed as in (5.31).

Proof We write

$$X(t_n) = \left(X(t_n) - \frac{\Sigma(t_n, t_{n-1})}{\Sigma(t_{n-1}, t_{n-1})} X(t_{n-1}) \right) + \frac{\Sigma(t_n, t_{n-1})}{\Sigma(t_{n-1}, t_{n-1})} X(t_{n-1})$$

$$:= \widetilde{X}_n + \widetilde{Y}_{n-1}. \tag{5.34}$$

Suppose Σ can be expressed as in (5.31). Then $E\widetilde{X}_n X(t_j) = 0$, $1 \leq j \leq n - 1$. Thus $\widetilde{X}_n$ is independent of $X(t_j)$, $1 \leq j \leq n - 1$. Therefore, obtaining (5.33) reduces to showing that

$$E(\widetilde{Y}_{n-1} | X(t_{n-1})) = E(\widetilde{Y}_{n-1} | X(t_{n-1}), \ldots, X(t_1)), \tag{5.35}$$

which is trivially true.

Now suppose that X is a Gaussian Markov process. Let $t_1 < t_2 < t_3$ and consider the Gaussian vector $X(t_1), X(t_2), X(t_3)$, which we also denote by X_1, X_2, X_3. Set $(X_i, X_j) = E(X_i X_j)$, $i, j = 1, 2, 3$. Let $\widetilde{X}_2 = X_2 - \frac{(X_1, X_2)}{(X_1, X_1)} X_1$. We write

$$X_3 = \left[X_3 - \frac{(X_3, \widetilde{X}_2)}{(\widetilde{X}_2, \widetilde{X}_2)} \widetilde{X}_2 - \frac{(X_3, X_1)}{(X_1, X_1)} X_1 \right] + \frac{(X_3, \widetilde{X}_2)}{(\widetilde{X}_2, \widetilde{X}_2)} \widetilde{X}_2 + \frac{(X_3, X_1)}{(X_1, X_1)} X_1. \tag{5.36}$$

The term in the square brackets is independent of X_1 and X_2. Therefore,

$$E(X_3 | X_2, X_1) = \frac{(X_3, \widetilde{X}_2)}{(\widetilde{X}_2, \widetilde{X}_2)} X_2 + \left[\frac{(X_3, X_1)}{(X_1, X_1)} - \frac{(X_3, \widetilde{X}_2)}{(\widetilde{X}_2, \widetilde{X}_2)} \frac{(X_1, X_2)}{(X_1, X_1)} \right] X_1. \tag{5.37}$$

In order for (5.33) to hold, the term in the square bracket must be zero, that is, we must have

$$(X_3, X_1)(\widetilde{X}_2, \widetilde{X}_2) = (X_3, \widetilde{X}_2)(X_1, X_2). \tag{5.38}$$

Since $(\widetilde{X}_2, \widetilde{X}_2) = (X_2, X_2) - \frac{(X_1, X_2)^2}{(X_1, X_1)}$ and $(\widetilde{X}_2, X_3) = (X_2, X_3) - \frac{(X_1, X_2)(X_1, X_3)}{(X_1, X_1)}$, we see that (5.38) holds if and only if

$$(X_1, X_3) = \frac{(X_1, X_2)(X_2, X_3)}{(X_2, X_2)}. \tag{5.39}$$

Consequently, when X is a Gaussian Markov process, for $u < s < t$,

$$\Sigma(u, t) = \frac{\Sigma(u, s)\Sigma(s, t)}{\Sigma(s, s)}. \tag{5.40}$$

We now show that (5.40) implies that, for any $t_0 \in I$,

$$\Sigma(t_0, t) > 0 \qquad \forall t \in I. \tag{5.41}$$

Suppose this is false and $\Sigma(t_0, t) = 0$ for some $t > t_0$. Let $s = \inf\{v > t_0 | \Sigma(t_0, v) = 0\}$. Clearly, $s > t_0$ because, by hypothesis, Σ is continuous and $\Sigma(t_0, t_0) > 0$. Using (5.40) we see that, for $t_0 < s' < s$,

$$0 = \Sigma(t_0, s) = \frac{\Sigma(t_0, s')\Sigma(s', s)}{\Sigma(s', s')}. \tag{5.42}$$

By our choice of s, $\Sigma(t_0, s') > 0$. Therefore, $\Sigma(s', s) = 0$ for all $t_0 < s' < s$. Since the covariance is assumed to be continuous, this implies that $\Sigma(s, s) = 0$, which is a contradiction. Therefore, $\Sigma(t_0, t) > 0$ for all $t \geq t_0$. A similar proof gives this result for $t \leq t_0$.

The covariance Σ satisfies (5.40) and (5.41). These are precisely the properties satisfied by u^α in Lemma 4.3.1 that are used to show that u^α can be represented as in (4.112). This shows that Σ can be represented as in (5.31). Since $\Sigma(s, t)$ is strictly positive definite, it follows from (5.32) that p/q is strictly increasing. $\qquad\square$

Example 5.1.10 Let $p(s) = s$ and $q(t) = 1$ in (5.31), where $T = R_+ - \{0\}$. This gives rise to a Gaussian Markov process $\widetilde{W} := \{\widetilde{W}(t), t \in R_+ - \{0\}\}$. We extend it to R_+ by setting $\widetilde{W}(0) = 0$ and denote the extended process by $W = \{W(t), t \in R_+\}$. By Lemma 5.1.1, it is easy to see that W has stationary independent increments and $W(t) - W(s)$ is $N(0, |t - s|)$. Thus by (2.12) and the argument following it, W has a continuous version. Thus, in a certain sense, W is a Brownian motion. We say in a certain sense because, in the definition of Brownian motion in Section 2.1, we require that all the paths of Brownian motion are continuous. In general, when we consider Gaussian processes we only require that a continuous version exists. Thus Brownian motion, as a Markov process, is considered differently from Brownian motion as a Gaussian process. This does not cause us any difficulties.

Consider $\overline{W} := \{\overline{W}(t) ; t \in R_+\}$, for $\overline{W}(t) := tW(1/t)$ for $t \neq 0$ and $\overline{W}(0) = 0$. Since a mean zero Gaussian process is determined by its covariance and since $E\overline{W}(t)\overline{W}(s) = s \wedge t$, $\overline{W}$ is also a Brownian motion (see page 15).

Remark 5.1.11 Let $I \in R^1$ be an interval, open or closed, and let $X = \{X(t), t \in I\}$ be a mean zero Gaussian Markov process with continuous strictly positive definite covariance Σ. Then X is a simple modification of time changed Brownian motion. This follows from Lemma 5.1.9 because, for any functions p and q satisfying the conditions of Lemma 5.1.8 on I with $\Sigma(s, t)$ continuous, $\{X(t), t \in I\}$ and $\{q(t)W(p(t)/q(t)) ; t \in I\}$, where W is Brownian motion, are both mean zero Gaussian processes with the same covariance. Furthermore, since Brownian motion is continuous, X has a version with continuous sample paths.

5.2 Moment generating functions

Lemma 5.2.1 *Let $\zeta = (\zeta_1, \ldots, \zeta_n)$ be a mean zero, n-dimensional Gaussian random variable with covariance matrix Σ. Assume that Σ is invertible. Let $\lambda = (\lambda_1, \ldots, \lambda_n)$ be an n-dimensional vector and Λ an $n \times n$ diagonal matrix with λ_j as its j-th diagonal entry. Let $u = (u_1, \ldots, u_n)$ be an n-dimensional vector. For all λ_i, $i = 1 \ldots, n$ with $\lambda_i < \epsilon$ for some $\epsilon > 0$ sufficiently small, $(\Sigma^{-1} - \Lambda)$ is invertible and*

$$E \exp \left(\sum_{i=1}^{n} \lambda_i (\zeta_i + u_i)^2 / 2 \right) \tag{5.43}$$

$$= \frac{1}{(\det(I - \Sigma \Lambda))^{1/2}} \exp \left(\frac{u \Lambda u^t}{2} + \frac{(u \Lambda \widetilde{\Sigma} \Lambda u^t)}{2} \right),$$

where

$$\widetilde{\Sigma} \overset{def}{=} (\Sigma^{-1} - \Lambda)^{-1} = (I - \Sigma \Lambda)^{-1} \Sigma \tag{5.44}$$

and $u = (u_1, \ldots, u_n)$. Equivalently,

$$E \exp \left(\sum_{i=1}^{n} u_i \lambda_i \zeta_i + \lambda_i \frac{\zeta_i^2}{2} \right) \tag{5.45}$$

$$= \frac{1}{(\det(I - \Sigma \Lambda))^{1/2}} \exp \left(\frac{u \Lambda \widetilde{\Sigma} \Lambda u^t}{2} \right).$$

Proof We prove (5.45), which immediately gives (5.43). Using (5.10) twice we see that

$$E \exp \left(\sum_{j=1}^{n} u_j \lambda_j \zeta_j + \lambda_j \frac{\zeta_j^2}{2} \right) \tag{5.46}$$

$$= \frac{1}{(2\pi)^{n/2} (\det \Sigma)^{1/2}} \int \exp \left(u \Lambda \zeta^t - \frac{\zeta \left(\Sigma^{-1} - \Lambda \right) \zeta^t}{2} \right) d\zeta$$

$$= \frac{(\det \widetilde{\Sigma})^{1/2}}{(\det \Sigma)^{1/2}} \widetilde{E} e^{u \Lambda \xi^t} = \frac{1}{(\det(I - \Sigma \Lambda))^{1/2}} \widetilde{E} e^{u \Lambda \xi^t},$$

where ξ is an n-dimensional Gaussian random variable with mean zero and covariance matrix $\widetilde{\Sigma}$ and $\widetilde{E}$ is expectation with respect to the probability measure of ξ. Equation (5.45) now follows from (5.15). $\qquad\square$

We have the following immediate corollary of Lemma 5.2.1.

Corollary 5.2.2 *Let* $\eta = \{\eta_x; x \in S\}$ *be a mean zero Gaussian process and* f_x *a real-valued function on* S. *It follows from Lemma 5.2.1 that for* $a^2 + b^2 = c^2 + d^2$,

$$\{(\eta_x + f_x a)^2 + (\widetilde{\eta}_x + f_x b)^2; x \in S\} \qquad (5.47)$$
$$\overset{law}{=} \{(\eta_x + f_x c)^2 + (\widetilde{\eta}_x + f_x d)^2; x \in S\},$$

where $\widetilde{\eta}$ *is an independent copy of* η.

Lemma 5.2.3 *Let* (ζ_1, ζ_2) *be an* R^2-*valued Gaussian random variable with mean zero. Then, for all* $s \neq 0$,

$$\frac{E(\zeta_1 \exp(s\zeta_2))}{sE(\exp(s\zeta_2))} = E(\zeta_1 \zeta_2). \qquad (5.48)$$

Proof $t\zeta_1 + s\zeta_2$ is a mean zero Gaussian random variable with variance $t^2 E(\zeta_1^2) + 2ts E(\zeta_1 \zeta_2) + s^2 E(\zeta_2^2)$. Therefore,

$$E(\exp(t\zeta_1 + s\zeta_2)) \qquad (5.49)$$
$$= \exp\left(t^2 E(\zeta_1^2)/2 + ts E(\zeta_1 \zeta_2) + s^2 E(\zeta_2^2)/2\right).$$

Differentiating this with respect to t and then setting $t = 0$, we get

$$E(\zeta_1 \exp(s\zeta_2)) = sE(\zeta_1 \zeta_2) \exp\left(s^2 E(\zeta_2^2)/2\right), \qquad (5.50)$$

which gives (5.48). $\qquad\square$

Here is an alternative proof of Lemma 5.2.3. We write

$$\zeta_1 = \left(\zeta_1 - \frac{E(\zeta_1 \zeta_2)}{E(\zeta_2^2)} \zeta_2\right) + \frac{E(\zeta_1 \zeta_2)}{E(\zeta_2^2)} \zeta_2 \qquad (5.51)$$

$$:= \zeta_2^\perp + \frac{E(\zeta_1 \zeta_2)}{E(\zeta_2^2)} \zeta_2. \qquad (5.52)$$

Observe that $\zeta_2^\perp$ and ζ_2 are orthogonal and hence independent. Since $\zeta_2^\perp$ has mean zero, we have

$$E(\zeta_1 \exp(s\zeta_2)) = \frac{E(\zeta_1 \zeta_2)}{E(\zeta_2^2)} E(\zeta_2 \exp(s\zeta_2)). \qquad (5.53)$$

By (5.15) we see that

$$
\begin{aligned}
E(\zeta_2 \exp(s\zeta_2)) &= \frac{d}{ds} E(\exp(s\zeta_2)) && (5.54) \\
&= \frac{d}{ds} \exp\left(\frac{s^2 E\left(\zeta_2^2\right)}{2}\right) \\
&= s E\left(\zeta_2^2\right) E\left(\exp\left(s\zeta_2\right)\right).
\end{aligned}
$$

Using this in (5.53) we get (5.48). $\qquad\square$

Remark 5.2.4 We define a probability measure $\widetilde{P}$ on R^n, in terms of its expectation operator $\widetilde{E}$, by

$$
\widetilde{E}(g(\zeta_1, \ldots, \zeta_n)) = \frac{E\left(g(\zeta_1, \ldots, \zeta_n) \exp\left(\sum_{i=1}^{n} \lambda_i \zeta_i^2 / 2\right)\right)}{E\left(\exp\left(\sum_{i=1}^{n} \lambda_i \zeta_i^2 / 2\right)\right)} \tag{5.55}
$$

for all measurable functions g on R^n. Under $\widetilde{P}$, $\zeta = (\zeta_1, \ldots, \zeta_n)$ is a mean zero, n-dimensional Gaussian random variable with covariance matrix $\widetilde{\Sigma}$ given in (5.44). To see this, take $g(\zeta_1, \ldots, \zeta_n) = \exp(i(u, \zeta))$ in (5.55). Using (5.46) with u_j replaced by iu_j, $j = 1, \ldots, n$, we get

$$
E\left(\exp(i(u, \zeta)) \exp\left(\sum_{j=1}^{n} \lambda_j \zeta_j^2 / 2\right)\right) = \frac{1}{(\det(I - \Sigma\,\Lambda))^{1/2}} E e^{iu\xi^t}, \tag{5.56}
$$

where ξ is a mean zero normal random variable with variance $\widetilde{\Sigma}$. By (5.5), $E e^{iu\xi^t} = \exp(-(u\widetilde{\Sigma}u^t)/2)$ and by (5.44), $E(\exp(\sum_{j=1}^{n} \lambda_j \zeta_j^2 / 2)) = (\det(I - \Sigma\,\Lambda))^{-1}$. Thus we see that when $g(\zeta_1, \ldots, \zeta_n) = \exp(i(u, \zeta))$, the right-hand side of (5.55) is equal to $\exp(-(u\widetilde{\Sigma}u^t)/2)$, that is, we have

$$
\widetilde{E} \exp(i(u, \zeta)) = \exp(-(u\widetilde{\Sigma}u^t)/2). \tag{5.57}
$$

Corollary 5.2.5 *Let $\zeta = (\zeta_1, \ldots, \zeta_n)$ be an R^n-valued Gaussian random variable. We have*

$$
\frac{E\left(\zeta_j \zeta_k \exp\left(\sum_{i=1}^{n} \lambda_i \zeta_i^2 / 2\right)\right)}{E\left(\exp\left(\sum_{i=1}^{n} \lambda_i \zeta_i^2 / 2\right)\right)} = \{\widetilde{\Sigma}\}_{j,k}, \tag{5.58}
$$

and for all $s \neq 0$

$$
\frac{E\left(\zeta_1 \exp\left(\sum_{i=1}^{n} \lambda_i (\zeta_i + s)^2 / 2\right)\right)}{s E \exp\left(\sum_{i=1}^{n} \lambda_i (\zeta_i + s)^2 / 2\right)} = \{\widetilde{\Sigma}\Lambda\mathbf{1}^t\}_1, \tag{5.59}
$$

where $\mathbf{1}$ denotes a vector with all its components equal to 1.

Proof Equation (5.58) follows immediately from Remark 5.2.4. To prove (5.59) we expand the squares $(\zeta_i + s)^2$ and cancel the terms in s^2 to see that the left-hand side of (5.59) is equal to

$$\frac{E\left(\zeta_1 \exp\left(s\sum_{i=1}^{n}\lambda_i\zeta_i\right)\exp\left(\sum_{i=1}^{n}\lambda_i\zeta_i^2/2\right)\right)}{sE\left(\exp\left(s\sum_{i=1}^{n}\lambda_i\zeta_i\right)\exp\left(\sum_{i=1}^{n}\lambda_i\zeta_i^2/2\right)\right)} = \frac{\widetilde{E}\left(\zeta_1\exp\left(s\sum_{i=1}^{n}\lambda_i\zeta_i\right)\right)}{s\widetilde{E}\left(\exp\left(s\sum_{i=1}^{n}\lambda_i\zeta_i\right)\right)}.$$
$$(5.60)$$

Using Lemma 5.2.3, we see that this last term is equal to

$$\widetilde{E}\left(\zeta_1\left(\sum_{i=1}^{n}\lambda_i\zeta_i\right)\right) = \sum_{i=1}^{n}\lambda_i\{\widetilde{\Sigma}\}_{1,i} = \{\widetilde{\Sigma}\Lambda 1^t\}_1. \qquad (5.61)$$

$\square$

We end this section with an important moment identity for products of Gaussian random variables.

Lemma 5.2.6 *Let $\{g_i\}_{i=1}^{k}$ be an R^k-valued Gaussian random variable with mean zero. Then, when k is even,*

$$E\left(\prod_{i=1}^{k}g_i\right) = \sum_{D_1\cup\ldots\cup D_{k/2}=\{1,\ldots,k\}}\prod_{i=1}^{k/2}\mathrm{cov}(D_i), \qquad (5.62)$$

where the sum is over all pairings $(D_1,\ldots,D_{k/2})$ of $\{1,\ldots,k\}$, that is, over all partitions of $\{1,\ldots,k\}$ into disjoint sets each containing two elements, and where

$$\mathrm{cov}(\{i,j\}) := \mathrm{cov}(g_i,g_j) := E(g_ig_j). \qquad (5.63)$$

When k is odd, the left-hand side of (5.62) equals zero.

Proof By (5.15),

$$E\exp\left(\sum_{i=1}^{k}\lambda_ig_i\right) = \exp\left(\frac{1}{2}E\left(\sum_{i=1}^{k}\lambda_ig_i\right)^2\right). \qquad (5.64)$$

Clearly

$$\frac{\partial}{\partial\lambda_1}\cdots\frac{\partial}{\partial\lambda_k}E\exp\left(\sum_{i=1}^{k}\lambda_ig_i\right)\Bigg|_{\lambda_1=\cdots=\lambda_k=0} = E\left(\prod_{i=1}^{k}g_i\right). \qquad (5.65)$$

Also,

$$\frac{\partial}{\partial\lambda_1}\cdots\frac{\partial}{\partial\lambda_k}\exp\left(\frac{1}{2}E\left(\sum_{i=1}^{k}\lambda_ig_i\right)^2\right) \qquad (5.66)$$

$$= \sum_{n=0}^{\infty} \frac{1}{2^n n!} \frac{\partial}{\partial \lambda_1} \cdots \frac{\partial}{\partial \lambda_k} \left(\sum_{i=1}^{k} \sum_{j=1}^{k} \lambda_i \lambda_j \mathrm{cov}(g_i, g_j) \right)^n.$$

It is easy to see that

$$\frac{\partial}{\partial \lambda_1} \cdots \frac{\partial}{\partial \lambda_k} \left(\sum_{i=1}^{k} \sum_{j=1}^{k} \lambda_i \lambda_j \mathrm{cov}(g_i, g_j) \right)^n \bigg|_{\lambda_1 = \ldots = \lambda_k = 0} \tag{5.67}$$

is zero when $n \neq k/2$. Thus, in particular, when k is odd the left-hand side of (5.62) is equal to zero.

When $n = k/2$, (5.67) is not zero only for those terms in

$$\left(\sum_{i=1}^{k} \sum_{j=1}^{k} \lambda_i \lambda_j \mathrm{cov}(g_i, g_j) \right)^{k/2} \tag{5.68}$$

in which each λ_i, $i = 1, \ldots, k$ appears only once. The terms with this property form a pairing, say $(D_1, \ldots, D_{k/2})$, of $\{1, \ldots, k\}$. For this pairing (4.36) is equal to $\prod_{i=1}^{k/2} \mathrm{cov}(D_i)$. Finally, it is easy to see that there are $2^{k/2}(k/2)!$ terms in (5.67), corresponding to each pairing of $\{1, \ldots, k\}$. This establishes (5.62). $\qquad\square$

5.2.1 *Exponential random variables*

There are some simple relationships between exponential and normal random variables that play an important role in this book. An exponential random variable is a positive random variable λ with density $\exp(-x/\gamma)\gamma^{-1}$ on R_+ for some $\gamma > 0$. Equivalently, $P(\lambda > u) = \exp(-u/\gamma)$ for all $u \geq 0$. The importance of exponential random variables in the theory of Markov processes is due to the fact that an exponential variable is "memoryless," that is,

$$P(\lambda > t + s | \lambda > s) = P(\lambda > t). \tag{5.69}$$

It is elementary to check this as well as the facts that $E\lambda = \gamma$ and the moment generating function of λ is

$$Ee^{v\lambda} = \frac{1}{1 - \gamma v}. \tag{5.70}$$

Let ξ be a normal random variable with mean zero and variance γ, and let ξ' be an independent copy of ξ. It follows from Lemma 5.2.1 that

$$\xi^2 + (\xi')^2 \stackrel{law}{=} 2\lambda. \tag{5.71}$$

The following immediate consequence of Corollary 5.2.2 is used in the proofs of isomorphism theorems for local times:

Corollary 5.2.7 *Let $\eta = \{\eta_x; x \in S\}$ be a mean zero Gaussian process and η' be an independent copy of η. Let ξ, ξ', and λ be as above and be independent of η and η'. Then*

$$\{(\eta_x + \xi)^2 + (\eta'_x + \xi')^2; x \in S\} \overset{law}{=} \{\eta_x^2 + (\eta'_x + \sqrt{2\lambda})^2; x \in S\}. \quad (5.72)$$

Remark 5.2.8 We record another fact for use in the proofs of isomorphism theorems. Let λ and ρ be two independent exponential random variables with $E\lambda = \gamma$ and $E\rho = \bar{\gamma}$. Then $\lambda \wedge \rho$ is an exponential random variable with mean

$$E(\lambda \wedge \rho) = \frac{1}{\gamma^{-1} + \bar{\gamma}^{-1}}. \quad (5.73)$$

To prove this note that $P(\lambda \wedge \rho > u) = P(\lambda > u)P(\rho > u)$.

Finally we note another relationship that associates Gaussian random variables with exponential random variables, the identity

$$E|\eta|^p = \frac{2^{p/2}}{\sqrt{\pi}} \Gamma\left(\frac{p+1}{2}\right), \quad (5.74)$$

where Γ is the gamma function, $\Gamma(t) = \int_0^\infty x^{t-1} e^{-x}\, dx$. This is obtained by writing out the integral for $E|\eta|^p$ and making the change of variables $y = \sqrt{2v}$.

5.3 Zero–one laws and the oscillation function

A Gaussian process with a continuous covariance can be expressed as an infinite series of independent continuous functions. This representation allows us to easily obtain several important zero–one laws for the process. Since everything in this section applies to a larger class of processes than Gaussian processes, we give the results for this larger class of processes.

Let (T, d) be a separable metric space and let Γ be a positive definite function on $T \times T$. We sometimes refer to Γ as a covariance kernel.

Theorem 5.3.1 (Reproducing kernel Hilbert space) *Let (T, d) be a separable metric space and let Γ be a continuous covariance kernel on $T \times T$. Then there exists a separable Hilbert space $H(\Gamma)$ of continuous real-valued functions on T such that*

$$\Gamma(t, \cdot) \in H(\Gamma) \qquad t \in T \quad (5.75)$$

$$(f(\cdot), \Gamma(t, \cdot)) = f(t) \qquad f \in H(\Gamma) \quad t \in T, \tag{5.76}$$

where $(\cdot, \cdot)$ *denotes the inner product on* $H(\Gamma)$.

If H_1 *is a separable Hilbert space of continuous real-valued functions on* T *such that (5.75) and (5.76) hold with* $H(\Gamma)$ *replaced by* H_1 *and the inner product by the inner product on* H_1, *then* $H_1 = H(\Gamma)$ *as Hilbert spaces.*

$H(\Gamma)$ is called the reproducing kernel Hilbert space of the covariance kernel Γ.

Proof Let

$$S = \left\{ \sum_{j=1}^{n} a_j \Gamma(t_j, \cdot), a_1, \ldots, a_n \in R^1, t_1, \ldots, t_n \in T, n \geq 1 \right\}. \tag{5.77}$$

On S we define the bilinear form

$$\left(\sum_{j=1}^{n} a_j \Gamma(t_j, \cdot), \sum_{k=1}^{m} b_k \Gamma(t_k, \cdot) \right) = \sum_{j=1}^{n} \sum_{k=1}^{m} a_j b_k \Gamma(t_j, t_k). \tag{5.78}$$

Note that if $f(t) = \sum_{j=1}^{n} a_j \Gamma(t_j, t)$,

$$f(t) = (f(\cdot), \Gamma(t, \cdot)). \tag{5.79}$$

If $f \in S$, then $(f, f) \geq 0$, because Γ is positive definite. Also, $(f, f) = 0$ implies that $f \equiv 0$ since

$$|f(t)|^2 = |(f, \Gamma(t, \cdot))|^2 \leq (f, f)(\Gamma(t, \cdot), \Gamma(t, \cdot)) = 0. \tag{5.80}$$

Thus we see that (5.78) defines an inner product on S.

Let $\{f_n\}$ be a sequence of functions in S. Then

$$|f_n(t) - f_m(t)|^2 = |(f_n(\cdot) - f_m(\cdot), \Gamma(t, \cdot))|^2 \leq \|f_n - f_m\|^2 \Gamma(t, t), \tag{5.81}$$

where $\|f\|^2 := (f, f)$ for $f \in S$ and $\|\Gamma(t, \cdot)\|^2 = \Gamma(t, t)$ by (5.78). This shows that if $\{f_n\}$ is a Cauchy sequence with respect to the inner product norm, then it is also a Cauchy sequence pointwise on T. We close S in the inner product norm and identify the limits that are not already in S with the pointwise limits in T. $H(\Gamma)$ is this closure of S. It is a Hilbert space of real-valued functions.

Let $\{t_n\}$ be a dense subset of (T, d). Since T is separable and the covariance kernel Γ is continuous,

$$S_1 = \left\{ \sum_{j=1}^{k} a_j \Gamma(s_j, \cdot), a_1, \ldots, a_k \text{ rational}, s_1, \ldots, s_k \in \{t_n\}, k \geq 1 \right\}$$

$$\tag{5.82}$$

is dense in $H(\Gamma)$. The reproducing property (5.79) immediately extends to $H(\Gamma)$. Furthermore,

$$|f(s)-f(t)| = |(f(\,\cdot\,),\Gamma(s,\,\cdot\,)-\Gamma(t,\,\cdot\,))| \le \|f\|\|\Gamma(s,\,\cdot\,)-\Gamma(t,\,\cdot\,)\|, \quad (5.83)$$

which goes to zero as $d(s,t) \to 0$, since Γ is continuous. Thus $H(\Gamma)$ consists of continuous functions. This completes the proof of the statements in the first paragraph of this theorem.

For the second part of the theorem note that (5.75) and (5.76) with $H(\Gamma)$ replaced by H_1 and the inner product by the inner product on H_1, let us call it $(\,\cdot\,,\,\cdot\,)_1$, show that the set S in (5.77) is contained in H_1 and $(\,\cdot\,,\,\cdot\,)_1$ agrees with the reproducing kernel Hilbert space norm $(\,\cdot\,,\,\cdot\,)$ on S. Therefore, $H(\Gamma) \subset H_1$ and the two inner products agree on $H(\Gamma)$. Let $f \in H_1$ and suppose that $f \notin H(\Gamma)$. Then f can be written uniquely as $f = f_1 + f_2$, where $f_1 \in H(\Gamma) \cap H_1$, $f_2 \in H_1$ and $(g, f_2)_1 = 0$ for every $g \in H(\Gamma)$. In particular $(\Gamma(t,\,\cdot\,), f_2)_1 = f_2(t) = 0$ for all $t \in T$. Hence $f_2 \equiv 0$, so $f = f_1 \in H(\Gamma)$, which is a contradiction. Therefore $H_1 = H(\Gamma)$. $\qquad\qquad\qquad\qquad\qquad\qquad\qquad\qquad\qquad\qquad\qquad\square$

Theorem 5.3.2 *Let* $X = \{X(t), t \in T\}$, T *a separable metric space, be a stochastic process on* $(\Omega, \mathcal{F}, P)$ *with* $EX^2(t) < \infty$, *for all* $t \in T$. *Let* $EX(t) = m(t)$ *and assume that* $\Gamma(s,t) = EX(s)X(t)$ *is continuous on* $T \times T$. *Then*

$$X(t) = \sum_{j=1}^{\infty} \phi_j(t)Y_j + m(t), \quad (5.84)$$

where $\{\phi_j\}$ *are continuous functions on* T, $\{Y_j\}$ *is an orthonormal set in* $L^2(\Omega, \mathcal{F}, P)$, *and convergence and equality in (5.84) are in* $L^2(\Omega, \mathcal{F}, P)$.

Proof Let $H(\Gamma)$ be the reproducing kernel Hilbert space of Γ and let $\widetilde{X}(t) = X(t) - m(t)$. We establish an isomorphism between $H(\Gamma)$ and a subspace of $L^2(\Omega, \mathcal{F}, P)$ containing $\widetilde{X} = \{\widetilde{X}(t), t \in T\}$. Let

$$\mathcal{L}_2(\widetilde{X}) = \text{Closure of} \left\{ \sum_{j=1}^{n} a_j \widetilde{X}(t_j), a_1, \ldots, a_n \text{ real}, t_1, \ldots, t_n \in T, n \ge 1 \right\} \quad (5.85)$$

in $L^2(\Omega, \mathcal{F}, P)$. Consider the map

$$\Theta_P \left(\sum_{j=1}^{n} a_j \Gamma(t_j, \,\cdot\,) \right) = \sum_{j=1}^{n} a_j \widetilde{X}(t_j) \quad (5.86)$$

from S, given in (5.77), into $\mathcal{L}_2(\widetilde{X})$. Θ_P is linear, one–one, and norm preserving. It extends to all of $H(\Gamma)$ with range equal to $\mathcal{L}_2(\widetilde{X})$. Furthermore, each element in $\mathcal{L}_2(\widetilde{X})$ has mean zero.

Let $\{\phi_j\}$ be a complete orthonormal set in $H(\Gamma)$ and set $Y_j = \Theta_P(\phi_j)$. Then $\{Y_j\}_{j=1}^{\infty}$ is a complete orthonormal set in $\mathcal{L}_2(\widetilde{X})$, $EY_j = 0$, and

$$X(t) = \sum_{j=1}^{\infty} E(X(t)Y_j)Y_j + m(t), \tag{5.87}$$

where the convergence is in $L^2(\Omega, \mathcal{F}, P)$. By the isometry Θ_P,

$$E(X(t)Y_j) = E(\widetilde{X}(t)Y_j) = (\Gamma(t, \,\cdot\,), \phi_j(\,\cdot\,)) = \phi_j(t), \tag{5.88}$$

where $(\,\cdot\,, \,\cdot\,)$ is the inner product in $H(\Gamma)$ and the last equality is the reproducing property (5.76). Equations (5.87) and (5.88) give (5.84).$\square$

Remark 5.3.3 One should be careful when interpreting the equalities in (5.84) and (5.87). The left-hand side, $X(t)$, is a random variable that takes a unique value at each $\omega \in \Omega$. The right-hand side represents an equivalence class in $L^2(\Omega, \mathcal{F}, P)$. The equalities simply mean that $X(t)$ belongs to this equivalence class.

When the stochastic process in Theorem 5.3.2 is a Gaussian process one can be much more explicit about the series representation.

Corollary 5.3.4 *Let* $X = \{X(t), t \in T\}$, T *a separable metric space, be a Gaussian process with* $EX(t) = m(t)$. *Then* $X(t)$ *has a version given by*

$$X'(t) = \sum_{j=1}^{\infty} \phi_j(t)\xi_j + m(t), \tag{5.89}$$

where $\{\phi_j\}$ *are continuous functions on* T, $\{\xi_j\}$ *is a standard normal sequence, and convergence in (5.89) is in* $L^2(\Omega, \mathcal{F}, P)$. *If* X *has a continuous version and* T *is compact, the series in (5.89) converges uniformly on* T *almost surely.*

The series in (5.89) is called the Karhunen–Loéve expansion of X.

Proof Since X is a Gaussian process, $\sum_{j=1}^{n} a_j \widetilde{X}(t_j)$ in the right-hand side of (5.86) is a Gaussian random variable. Since all random variables in $\mathcal{L}_2(\widetilde{X})$ are limits in $L^2(\Omega, \mathcal{F}, P)$ of Gaussian random variables, each Y_j in the previous proof is a Gaussian random variable. Since $\{Y_j\}$ is also an orthonormal sequence, they must be independent. We label them $\{\xi_j\}$ in this corollary. Thus (5.89) is just a statement of Theorem 5.3.2. The variance of $X(t)$ is $\sum_{j=1}^{\infty} \phi_j^2(t)$. Thus this sum must converge.

It then follows from the three-series theorem that the sum in (5.89) converges almost surely for each fixed $t \in T$. Therefore, by Corollary 14.6.4, the series in (5.89) converges uniformly almost surely on T. □

We have the following simple application of Corollary 5.3.4, which we use in Chapter 9.

Lemma 5.3.5 *Let $X = \{X(t), t \in K\}$, K a compact separable metric space, be a mean zero Gaussian process with continuous sample paths. Then, for all $\epsilon > 0$, we have*

$$P\left(\sup_{t \in K} |X(t)| \leq \epsilon\right) > 0. \tag{5.90}$$

Proof We represent X by its Karhunen–Loéve expansion

$$X(t) = \sum_{j=1}^{\infty} \xi_j \phi_j(t) \qquad t \in K \tag{5.91}$$

as in Corollary 5.3.4. Since X is continuous, this series converges uniformly almost surely on K. Hence, given any $\epsilon > 0$, we can find an $N(\epsilon)$ such that

$$P\left(\sup_{t \in K} \left| \sum_{j=N(\epsilon)+1}^{\infty} \xi_j \phi_j(t) \right| < \epsilon/2\right) \geq 1/2. \tag{5.92}$$

Using independence, we see that

$$P\left(\sup_{t \in K} \left| \sum_{j=1}^{\infty} \xi_j \phi_j(t) \right| \leq \epsilon\right) \tag{5.93}$$

$$\geq P\left(\sup_{t \in K} \left| \sum_{j=1}^{N(\epsilon)} \xi_j \phi_j(t) \right| \leq \epsilon/2\right) P\left(\sup_{t \in K} \left| \sum_{j=N(\epsilon)+1}^{\infty} \xi_j \phi_j(t) \right| < \epsilon/2\right)$$

$$\geq \frac{1}{2} P\left(\sup_{t \in K} \left| \sum_{j=1}^{N(\epsilon)} \xi_j \phi_j(t) \right| \leq \epsilon/2\right).$$

It is easy to see that this last probability is strictly positive since the ϕ_j are bounded on K and the $\{\xi_j\}_{j=1}^{N(\epsilon)}$ are simply a finite collection of independent normal random variables with mean zero and variance one. Thus we get (5.90). □

The sample paths of a Gaussian process have interesting zero–one properties that follow because the Gaussian process can be represented by a series as in (5.89). Since these properties are shared by all processes with such a representation, we present the next few results in this greater generality.

Let (T, d) be a compact metric space. A stochastic process $X = \{X(t), t \in T\}$ is said to be of class $\mathcal{S}$, if there exist real-valued continuous functions $\{\phi_j\}$ on T and independent symmetric real-valued random variables $\{\xi_j\}$ such that

$$X(t) = \sum_{j=1}^{\infty} \phi_j(t)\xi_j \qquad t \in T, \tag{5.94}$$

where the series converges almost surely for each fixed $t \in T$.

In the rest of this section we take T to be a compact metric space.

Let $Z = \{Z(t); t \in T\}$ be a real (or complex)-valued stochastic process, where (T, d) is a separable metric space. Let $t_0 \in T$. $Z(t_0)$ is a real (or complex)-valued random variable and hence is finite almost surely. We say that the process Z has a bounded discontinuity at t_0 if

$$0 < \lim_{\delta \to 0} \sup_{t \in B_d(t_0, \delta)} |Z(t) - Z(t_0)| < \infty, \tag{5.95}$$

where $B_d(t_0, \delta)$ is a closed ball of radius δ at t_0 in the metric d.

We say that the process Z has an unbounded discontinuity at t_0 if

$$\lim_{\delta \to 0} \sup_{t \in B_d(t_0, \delta)} |Z(t) - Z(t_0)| = \infty. \tag{5.96}$$

Corollary 5.3.6 *Let $X = \{X(t), t \in T\}$ be a separable process of class $\mathcal{S}$. Let $t_0 \in T$. The following events have probability zero or one:*

(1) X is continuous at t_0;
(2) X has a bounded discontinuity at t_0;
(3) X is unbounded at t_0;
(4) X is continuous on T;
(5) X has a bounded discontinuity on T;
(6) X is unbounded on T.

Proof Let D be the separability set of T. It suffices to prove this corollary for $\{X(t), t \in D\}$ (see page 9). There exists a P-null set Λ such that, for all $\omega \notin \Lambda$,

$$X(t, \omega) = \sum_{j=1}^{\infty} \phi_j(t)\xi_j(\omega) \qquad t \in D, \tag{5.97}$$

where the series converges as a series of numbers. Also, since X is a process of class $\mathcal{S}$, $\{\xi_j\}$ are independent random variables and $\{\phi_j\}$ are continuous functions.

For $t \in D$, $\omega \notin \Lambda$ let

$$X_n(t,\omega) = \sum_{j=n+1}^{\infty} \phi_j(t)\xi_j(\omega). \qquad (5.98)$$

Note that $S_n := \{\sum_{j=1}^{n} \phi_j(t)\xi_j(\omega), t \in T\}$ is a continuous function on T for each ω and $n \geq 1$. Consequently, the events in (1) to (6) are in $\sigma(\{X_n(t), t \in D\}) \subset \sigma(\{\xi_j, j > n\})$, for all n. The fact that these events have probability zero or one follows from the Kolmogorov zero–one law. $\square$

With a little more effort we can obtain a more precise description of the discontinuities of processes in class $\mathcal{S}$. Let (T, d) be a separable metric space and f an extended real-valued function on T. Let W_f be the oscillation function of f, that is,

$$W_f(t) = \lim_{\epsilon \to 0} \sup_{u,v \in B_d(t,\epsilon)} |f(u) - f(v)|, \qquad (5.99)$$

where $B_d(t, \epsilon)$ is a closed ball of radius ϵ in (T, d). We also define

$$M_f(t) = \lim_{\epsilon \to 0} \sup_{u \in B_d(t,\epsilon)} f(u) \quad \text{and} \quad m_f(t) = \lim_{\epsilon \to 0} \inf_{u \in B_d(t,\epsilon)} f(u) \qquad (5.100)$$

so that

$$W_f(t) = M_f(t) - m_f(t). \qquad (5.101)$$

We take $\infty - \infty = 0$ and $(-\infty) - (-\infty) = 0$. Note that $W_f(t) = 0$ if and only if f is continuous at t.

Theorem 5.3.7 *Let $X = \{X(t), t \in T\}$ be a separable process of class $\mathcal{S}$ defined on the probability space $(\Omega, \mathcal{F}, P)$. There exists an extended real-valued, upper semicontinuous function α on T, called the oscillation function of the process X, such that*

$$P(W_X(t,\omega) = \alpha(t), t \in T) = 1, \qquad (5.102)$$

where $W_X(t,\omega) := W_{X(\cdot,\omega)}(t)$. Furthermore, for all $t \in T$,

$$P(M_X(t,\omega) = X(t,\omega) + \alpha(t)/2, m_X(t,\omega) = X(t,\omega) - \alpha(t)/2) = 1. \qquad (5.103)$$

Proof Let Λ be the P-null set in the first paragraph of the proof of

Corollary 5.3.6. For a closed subset F of T, define

$$W_X(F,\omega) = \lim_{n\to\infty} \lim_{k\to\infty} \sup_{\substack{d(s,t)\leq 1/k \\ s,t\in F_n}} |X(t,\omega) - X(s,\omega)|, \qquad (5.104)$$

where $F_n = \{u \in T, d(u,F) \leq 1/n\}$. Because X is separable, we can choose Λ so that the supremum in (5.104) and in the definition of $W_X(t,\omega)$ is achieved over $s,t \in D$, $\omega \notin \Lambda$. It then follows that $W_X(t,\cdot)$ and $W_X(F,\cdot)$ are random variables.

Let $X(t,\omega)$ and $X_n(t,\omega)$ be as given in (5.97) and (5.98). For any set $F \subset T$ and $\omega \notin \Lambda$, $W_X(F,\omega) = W_{X_n}(F,\omega)$. Furthermore, $W_{X_n}(F,\omega) \in \sigma(\{X_n(t), t \in D\}) \subset \sigma(\{\xi_j, j > n\})$. Therefore, by the Kolmogorov zero–one law,

$$P(W_X(F,w) = \alpha(F)) = 1 \qquad (5.105)$$

for some number $\alpha(F)$.

It is simple to check that for each $\omega \in \Omega$,

$$W_X(F,\omega) = W_X(t,\omega) \qquad \text{when } F = \{t\} \qquad (5.106)$$

and

$$\lim_{m\to\infty} W_X(F_m,\omega) = W_X(F,\omega). \qquad (5.107)$$

Let $\{O_n, n \geq 1\}$ be a countable basis for the separable metric space T. Let J_n be the closure of O_n and set $J = \{J_n, n \geq 1\}$. By (5.105),

$$P(W_X(J_n,\omega) = \alpha(J_n), n \geq 1) = 1. \qquad (5.108)$$

By (5.108) there exists a set $\Omega_0 \subset \Omega$, with $P(\Omega_0) = 1$ such that $W_X(J_n,\omega) = \alpha(J_n)$ for all $\omega \in \Omega_0$. Choose any $t \in T$. There exists a nested sequence $\{J_{n_k}, n_k \geq 1\}$ of sets in J such that $\cap_k J_{n_k} = \{t\}$. Therefore, for this t and $\omega \in \Omega_0$, by (5.107), $W_X(J_{n_k},\omega) \downarrow W_X(\{t\},\omega)$. Also, since $W_X(J_{n_k},\omega) = \alpha(J_{n_k})$, $\alpha(J_{n_k})$ decreases, as $k \to \infty$, to some number β, which is independent of ω. Consequently, $W_X(\{t\},\omega) = \beta$. It follows from (5.106) and (5.105) that $\beta = \alpha(\{t\})$. Thus we get (5.102) where $\alpha(t) = \alpha(\{t\})$, $t \in T$.

Recall that, by definition, a function f on (T,d) is upper semicontinuous if $\limsup_{s\to t} f(s) = f(t)$ (see, e.g., McShane and Botts (1959, page 74)). Since, obviously, $\limsup_{s\to t} f(s) \geq f(t)$, one can prove that a function is upper semicontinuous by showing that $\limsup_{s\to t} f(s) \leq f(t)$. Thus, in general, the oscillation function of an extended real-valued function on a separable metric space is an upper semicontinuous function since $\limsup_{s\to t} W_f(s) \leq W_f(t)$.

To verify (5.103) note that

$$\lim_{s \to t} \sup_{s \in D}(X(s) - X(t)) = \lim_{s \to t} \sup_{s \in D}(X_n(s) - X_n(t)). \tag{5.109}$$

Therefore, by Kolmogorov's zero–one law,

$$\lim_{s \to t} \sup_{s \in D}(X(s) - X(t)) = M_X(t) - X(t) = \gamma(t) \qquad \text{a.s.} \tag{5.110}$$

for some constant $\gamma(t)$. Since X is symmetric, we also have

$$\lim_{s \to t} \sup_{s \in D}(-X(s) + X(t)) = \gamma(t) \qquad \text{a.s.}, \tag{5.111}$$

which implies that

$$\lim_{s \to t} \inf_{s \in D}(X(s) - X(t)) = m_X(t) - X(t) = -\gamma(t) \qquad \text{a.s.} \tag{5.112}$$

It follows from (5.110) and (5.112) that $\alpha(t) = M_X(t) - m_X(t) = 2\gamma(t)$, which together with (5.110) and (5.112) again gives (5.103). $\qquad\square$

Corollary 5.3.8 *Let $X = \{X(t), t \in T\}$ be a separable process of class $\mathcal{S}$. Then*

(1) X has continuous paths almost surely if and only if it is continuous at each fixed $t \in T$ almost surely, that is, if and only if

$$P\left(\lim_{s \to t} X(s) = X(t)\right) = 1 \qquad \text{for each} \quad t \in T. \tag{5.113}$$

(2) The paths of X are either almost surely continuous on T or almost surely discontinuous on T.

(3) Let K be a compact subset of T. If X is unbounded on K with positive probability, there exists a $t_0 \in T$ such that X is unbounded almost surely at t_0, that is $\alpha(t_0) = \infty$, where α is the oscillation function of X.

Proof Statements (1) and (2) are immediate consequences of Theorem 5.3.7. We prove statement (3). Suppose that X is unbounded on K with probability greater than zero. Then, by Corollary 5.3.6, X is unbounded on K almost surely. For each j let $\{\epsilon_j\}_{j=1}^{\infty}$ be a decreasing sequence of positive numbers satisfying $\lim_{j \to \infty} \epsilon_j = 0$. Let $\{B(z_{k,j}, \epsilon_j)\}_{k=1}^{N_j}$ be a cover of K. If

$$\sup_{z \in B(z_{k,j}, \epsilon_j)} |X(z)| < \infty \tag{5.114}$$

on a set of positive measure, then, by Corollary 5.3.6, it is finite almost surely. Hence, if X is unbounded almost surely on K, there exists a $z_{k,j} \in K$ such that

$$\sup_{z \in B(z_{k,j}, \epsilon_j)} |X(z)| = \infty \qquad \text{a.s.} \tag{5.115}$$

Note that for each $z \in K$, $X(z)$ is just a mean zero random variable with finite variance. Thus (5.115) implies that

$$\sup_{y,z \in B(z_{k,j}, \epsilon_j)} |X(y) - X(z)| = \infty \qquad \text{a.s.} \tag{5.116}$$

Since K is compact, there exists a subsequence $\{z_{k(i),j}\}_{i=1}^{\infty}$ of $\{z_{k,j}\}_{k=1}^{\infty}$ such that $\lim_{i \to \infty} z_{k(i),j} = z_0$ for some $z_0 \in K$. It is easy to see that

$$\sup_{y,z \in B(z_0, \epsilon)} |X(y) - X(z)| = \infty \quad \text{a.s.} \quad \forall \epsilon > 0. \tag{5.117}$$

$\square$

All mean zero Gaussian processes are in class $\mathcal{S}$, so their sample paths have the properties described in the previous corollary. Note that, generally speaking, independent increment processes, say on $[0,1]$, are continuous almost surely for t fixed but not continuous on the whole interval. Lévy processes that have only a finite number of jumps on $[0,1]$ are continuous with positive probability that is less than 1 on all subintervals of $[0,1]$ (this is because the jump times are uniformly distributed on $[0,1]$; see Sato (1999, Theorem 21.3)).

In the next theorem we give conditions on the nature of the discontinuities of processes of class $\mathcal{S}$.

Theorem 5.3.9 *Let* $X = \{X(t), t \in T\}$ *be a stochastically continuous (see page 9) separable process of class $\mathcal{S}$ with oscillation function α.*

(1) *Suppose that $\alpha(t) \geq a > 0$ on a dense subset S of an open set $I \subset T$. Then*

$$P(M_X(t) = \infty, m_X(t) = -\infty, t \in I) = 1. \tag{5.118}$$

(2) *The set $\{t \in T : a \leq \alpha(t) < \infty\}$ is nowhere dense (i.e., its closure contains no open ball of T).*

Proof Since X is stochastically continuous, we can assume that the separability set D is such that $D \cap I = S$. For fixed $t \in I$, set $F_n(t) = \{u \in T : d(t, u) < 1/n\}$. Then, by (5.103), almost surely

$$\begin{aligned} M_X(t, w) &= \limsup_{\substack{s \to t \\ s \in T}} X(s, \omega) \tag{5.119} \\ &= \lim_{n \to \infty} \sup_{s \in D \cap F_n(t)} X(s, \omega). \end{aligned}$$

Let $s \in D \cap F_n(t)$ and suppose that $u \in D \cap F_j(t)$ for some $j \geq n$. Then $u \in D \cap F_{n/2}(t)$. Therefore

$$\lim_{n \to \infty} \sup_{s \in D \cap F_n(t)} X(s, \omega) \geq \lim_{n \to \infty} \sup_{s \in D \cap F_n(t)} \left(\lim_{j \to \infty} \sup_{u \in D \cap F_j(s)} X(u, \omega) \right)$$

$$= \lim_{n \to \infty} \sup_{s \in D \cap F_n(t)} (X(s, \omega) + \alpha(s)/2).$$

Since $\alpha(s) \geq a$ on $D \cap I$, it follows from (5.119) that

$$M_X(t, \omega) \geq M_X(t, \omega) + a/2. \tag{5.120}$$

Therefore, since $a > 0$,

$$P(M_X(t, \omega) = \infty, t \in D \cap I) = 1. \tag{5.121}$$

Since D is dense in I we see that, for ω in a set of probability one, $M_X(t, \omega) = \infty$ for all $t \in T$. The result for $m_X(t, \omega)$ follows similarly.

To obtain (2) we note that, because α is upper semicontinuous, the sets $T_a := \{t : \alpha(t) \geq a\}$ are closed. Suppose that there exists an $0 < a < \infty$ such that $T_a - T_\infty$ contains a dense subset S of I. Then by part (1) of this theorem $S \subset T_\infty$, contradicting the supposition. Therefore, for each $a > 0$, $T_a - T_\infty$ is nowhere dense. □

We now present the Belyaev dichotomy, which says that the sample paths of processes of class $\mathcal{S}$ with stationary increments are either continuous or completely irregular.

Theorem 5.3.10 (Belyaev) *Let $X = \{X(t), t \in T\}$, where T is an interval of R^n, be a stochastically continuous separable process of class $\mathcal{S}$ with stationary increments. Then either X has continuous paths almost surely on all open subsets of T or it is unbounded almost surely on all open subsets of T.*

Proof Since X has stationary increments, its oscillation function takes the same value on all open subsets of T. Therefore, by Theorem 5.3.9, the oscillation function can only be zero or infinity. □

One might ask whether further restrictions can be imposed on the oscillation function of processes of class $\mathcal{S}$. Essentially the answer is no, as can be seen from the following theorem (a proof can be found in Ito and Nisio (1968b) or Jain and Kallianpur (1972)).

Theorem 5.3.11 *Let $T = [0, 1]^n$. Let α be an extended real-valued upper semicontinuous function on T such that, for all $a > 0$, the set $\{t \in T : a \leq \alpha < \infty\}$ is nowhere dense in T. There exists a mean* zero Gaussian process on T with continuous covariance and oscillation function α.

5.4 Concentration inequalities

One of the most important results in the theory of Gaussian processes that is used repeatedly in this book is a consequence of an isoperimetric inequality that states that for all measurable sets in R^n with the same canonical Gaussian measure γ, half-spaces achieve the minimal surface measure with respect to γ.

Let A be a Borel set in R^n. Define $A_r = \{x + rB, x \in A\}$, where B is the open unit ball in R^n. The surface measure of A is defined as

$$\gamma_S(\partial A) = \liminf_{r \to 0} \frac{1}{r} \left(\gamma(A_r) - \gamma(A) \right). \tag{5.122}$$

Let H be the half-space $\{x \in R^n : (x, u) < a\}$ for some $u \in R^n$ with $|u| = 1$ and $a \in [-\infty, \infty]$. Let $A \subset R^n$ be such that $\gamma(A) = \gamma(H)$. The isoperimetric inequality just referred to is

$$\gamma_S(\partial A) \geq \gamma_S(\partial H). \tag{5.123}$$

It is convenient to express (5.123) only in terms of A. Since $\gamma(H) = \Phi(a)$ and $\gamma_S(\partial H) = \phi(a)$, we can rewrite (5.123) as

$$\gamma_S(\partial A) \geq \phi(a) = \phi \circ \Phi^{-1}(\gamma(A)) \tag{5.124}$$

with equality when $A = H$.

It is more useful for us to have an integrated version of this isoperimetric inequality, which we now state.

Theorem 5.4.1 (Borell, Sudakov–Tsirelson) *Let A be a measurable subset of R^n such that*

$$\gamma(A) = \Phi(a) \qquad -\infty \leq a \leq \infty. \tag{5.125}$$

Then, for $r \geq 0$,

$$\gamma(A + rB) \geq \Phi(a + r), \tag{5.126}$$

where B is the open unit ball in R^n.

Proof Let $\mathcal{C}$ denote the sets in R^n that are finite unions of open balls. We prove below that (5.124) holds for all $C \in \mathcal{C}$. We now show that this implies (5.126) for all measurable sets $A \subseteq R^n$.

To begin we first show that if (5.124) holds for $C \in \mathcal{C}$ and $\gamma(C) = \gamma(H) = \Phi(a)$, then

$$\gamma(C_r) \geq \gamma(H_r) = \Phi(a + r), \tag{5.127}$$

which is (5.126) for $C \in \mathcal{C}$. Note that for $C \in \mathcal{C}$, the lim inf in (5.122) is actually a limit, so that $\dfrac{d}{dr}\gamma(C_r) = \gamma_S(\partial C_r)$. Consider the function $v(r) = \Phi^{-1}(\gamma(C_r))$, $r \geq 0$. It follows from (5.124) that for $C_r \in \mathcal{C}$,

$$v'(r) = \frac{\gamma_S(\partial C_r)}{\phi \circ \Phi^{-1}(\gamma(C_r))} \geq 1. \tag{5.128}$$

Therefore, $\int_0^r v'(u)\, du \geq r$ and consequently $v(r) \geq v(0) + r$. Since $v(0) = a$, we get (5.127) for sets in $\mathcal{C}$.

Suppose that $A \subset R^n$ is open. Then we can find an increasing sequence of sets C_n in $\mathcal{C}$ such that $C_n \subset A$ and $\gamma(C_n)$ increases to $\gamma(A)$. Applying (5.127) to each C_n and taking the limit, we see that (5.127) holds for open sets.

Now let A be a measurable set in R^n. Let $\rho > 0$. Then A_ρ is an open set and $\gamma(A_{\rho+r}) = \gamma((A_\rho)_r) \geq \Phi(a_\rho + r)$, where a_ρ is such that $\gamma(A_\rho) = \Phi(a_\rho)$. Taking the limit as ρ goes to zero we get $\gamma(A_r) \geq \Phi(\tilde{a} + r)$, where $\tilde{a}$ is such that $\Phi(\tilde{a}) = \lim_{\rho \to 0} \gamma(A_\rho) \geq \gamma(A)$. Thus $\tilde{a} \geq a$ and we get (5.127) in the general case.

We show below that for sufficiently smooth functions $f : R^n \longrightarrow [0, 1]$,

$$\mathcal{U}\left(\int f\, d\gamma\right) \leq \int \sqrt{\mathcal{U}^2(f) + |\bigtriangledown f|^2}\, d\gamma, \tag{5.129}$$

where $|\bigtriangledown f|$ denotes the Euclidean length of the gradient of f (see (5.23)). We now show that (5.129) implies that (5.124) holds for all $A \in \mathcal{C}$.

Suppose that the set A in (5.124) is in $\mathcal{C}$. This implies that $\gamma(\partial A) = 0$. Suppose that (5.129) holds for all sufficiently smooth functions with values in $[0, 1]$; then it holds for all Lipschitz functions with values in $[0, 1]$, that is, functions $f : R^n \to R^1$ with $|f(x) - f(y)| \leq M|x - y|$, for some constant M (see Dudley (1989, Theorem 11.2.4)). In particular it holds for the functions

$$f_r(x) \overset{def}{=} \left(1 - \frac{d(x, A)}{r}\right)^+ \qquad r > 0, \tag{5.130}$$

where d is the Euclidean distance on R^n (see also (1.5)).

Let $\bar{A}$ be the closure of A. If $x \in \bar{A}$, $f_r(x) = 1$ for all $r > 0$. If $x \in \bar{A}^c$, $\lim_{r \to 0} f_r(x) = 0$. Thus $\lim_{r \to 0} f_r(x) = I_{\bar{A}}(x)$ and $\lim_{r \to 0} \mathcal{U}(f_r(x)) = 0$, since $\mathcal{U}(0) = \mathcal{U}(1) = 0$. Moreover, $|\bigtriangledown f_r| = 0$ on A and on the complement of the closure of A_r and $|\bigtriangledown f_r| \leq 1/r$ everywhere.

We now use (5.129) with $f = f_r$ and take the limit inferior as $r \to 0$. Since $\gamma(A) = \gamma(\bar{A})$ and $\gamma(\partial A_r) = 0$ for all r, we get

$$\mathcal{U}(\gamma(A)) \leq \liminf_{r \to 0} \int \mathcal{U}(f_r) \, d\gamma + \liminf_{r \to 0} \int |\nabla f_r| \, d\gamma \quad (5.131)$$

$$= \liminf_{r \to 0} \int |\nabla f_r| \, d\gamma$$

$$\leq \liminf_{r \to 0} \frac{1}{r} \left(\gamma(A_r) - \gamma(A)\right) = \gamma_S(\partial A),$$

which is (5.124).

Since (5.129) implies (5.124), which implies the statement of this theorem, we complete the proof of this theorem by verifying (5.129). To do this we use the Ornstein–Uhlenbeck or Hermite semigroup with invariant measure γ. For $f \in L^1(\gamma)$ set

$$P_t f(x) = \int_{R^n} f(e^{-t} x + (1 - e^{-2t})^{1/2} y) \, d\gamma(y) \qquad x \in R^n, \quad t > 0. \tag{5.132}$$

We note the following properties of $\{P_t, t \geq 0\}$.

Lemma 5.4.2

(1) *The operators P_t are contractions on the function spaces $L^p(\gamma)$, for all $p \geq 1$.*

(2) *For all sufficiently smooth integrable functions f and g and every $t > 0$,*

$$\int f P_t g \, d\gamma = \int g P_t f \, d\gamma. \tag{5.133}$$

(3) *$\{P_t, t \geq 0\}$ is a semigroup of operators, that is, $P_s \circ P_t = P_{s+t}$. P_0 is the identity and for any $f \in L^1(\gamma)$, $\lim_{t \to \infty} P_t f = \int f \, d\gamma$.*

Proof   For any $0 \leq \alpha \leq 1$, set $\widetilde{\alpha} = (1 - \alpha^2)^{1/2}$. Then let $\alpha = e^{-t}$ and Y be a standard normal random variable. We can write

$$P_t f(x) = E(f(\alpha x + \widetilde{\alpha} Y)).$$

Let X be a standard normal random variable independent of Y and note that $Z = \alpha X + \widetilde{\alpha} Y$ is also a standard normal random variable. We have

$$\|P_t f\|_p = (E_X |E_Y(f(\alpha X + \widetilde{\alpha} Y))|^p)^{1/p}$$

$$\leq (E_X E_Y |f(\alpha X + \widetilde{\alpha} Y)|^p)^{1/p}$$

$$= (E|f(Z)|^p)^{1/p} = \|f\|_p$$

which is (1) (in the last line we use the conditional Hölder's inequality).

To obtain (2) we note that

$$\int g P_t f \, d\gamma = E g(X) f(\alpha X + \widetilde{\alpha} Y) = E g(X) f(Z).$$

Here (X, Z) is a Gaussian random variable, $EX^2 = EZ^2 = 1$, and $E(XZ) = \alpha$. Clearly then $Eg(X)f(Z) = Eg(Z)f(X)$, which is (2).

When $\beta = e^{-s}$ we have

$$P_t \circ P_s f(x) = E(P_s f(\alpha x + \widetilde{\alpha} Y)) = E(f(\beta \alpha x + \beta \widetilde{\alpha} Y + \widetilde{\beta} X)).$$

Since $\beta \widetilde{\alpha} Y + \widetilde{\beta} X$ is normal with variance $1 - (\alpha \beta)^2$, we see that $P_s \circ P_t = P_{s+t}$. The rest of (3) is immediate. $\qquad \square$

Proof of Theorem 5.4.1 continued Let L be the infinitesimal operator for the semigroup $\{P_t, t \geq 0\}$. That is, L satisfies

$$\frac{d}{dt} P_t f = P_t L f = L P_t f \qquad (5.134)$$

for all sufficiently smooth functions $f \in R^n$. One can check that

$$L f(x) = \triangle f(x) - (x, \bigtriangledown f(x)). \qquad (5.135)$$

It follows, by repeatedly integrating by parts on each component of R^n, that

$$-\int f(x)(Lg(x)) \, d\gamma(x) = \int (\bigtriangledown f(x), \bigtriangledown g(x)) \, d\gamma(x). \qquad (5.136)$$

Let $0 \leq f \leq 1$ be a smooth function on R^n. We assume that $0 < P_t f < 1$. This is the case unless f is equal to one or zero on a set of γ measure one, in which case (5.129) is satisfied. To verify (5.129) when $0 < P_t f < 1$, it suffices to show that the function

$$F(t) = \int \sqrt{\mathcal{U}^2(P_t f) + |\bigtriangledown P_t f|^2} \, d\gamma \qquad (5.137)$$

is nonincreasing in $t \geq 0$, since if this is true, then $F(\infty) \leq F(0)$, which implies (5.129) by property (3) of $\{P_t, t \geq 0\}$.

Showing that the derivative of F is less than or equal to zero is a straightforward, albeit tedious, calculation. However, since it is critical in the proof of this theorem, we go through the details. Using (5.134), we see that

$$\frac{dF}{dt} = \int \frac{\mathcal{U}\mathcal{U}'(P_t f) L P_t f + (\bigtriangledown(P_t f), \bigtriangledown(L P_t f))}{\sqrt{\mathcal{U}^2(P_t f) + |\bigtriangledown P_t f|^2}} \, d\gamma. \qquad (5.138)$$

To simplify the notation we set $P_t f = h$ and $\mathcal{U}^2(h) + |\bigtriangledown h|^2 = K(h)$. In this notation (5.138) becomes

$$\frac{dF}{dt} = \int \frac{\mathcal{U}\mathcal{U}'(h)Lh + (\nabla h, \nabla(Lh))}{\sqrt{K(h)}} \, d\gamma. \tag{5.139}$$

Let $h_i = \dfrac{\partial h}{\partial x_i}$. Note that

$$(\nabla h, \nabla(Lh)) = \sum_{i=1}^{n} h_i Lh_i - |\nabla h|^2 \tag{5.140}$$

and

$$\nabla K(h) = 2\mathcal{U}\mathcal{U}'(h) \nabla h + \nabla |\nabla h|^2. \tag{5.141}$$

Also, by (5.136) and (5.140),

$$\begin{aligned}
\frac{dF}{dt} &= -\int \left(\nabla \frac{\mathcal{U}\mathcal{U}'(h)}{\sqrt{K(h)}}, \nabla h \right) d\gamma - \int \frac{|\nabla h|^2}{\sqrt{K(h)}} \, d\gamma \tag{5.142} \\
&\quad - \sum_{i=1}^{n} \int \left(\nabla \frac{h_i}{\sqrt{K(h)}}, \nabla h_i \right) d\gamma \\
&= I + II + III.
\end{aligned}$$

By Lemma 5.1.5,

$$I = -\int \frac{(\mathcal{U}')^2 - 1}{\sqrt{K(h)}} |\nabla h|^2 \, d\gamma + \frac{1}{2} \int \frac{\mathcal{U}\mathcal{U}'(h)}{K^{3/2}(h)} (\nabla K(h), \nabla h) \, d\gamma \tag{5.143}$$

and by and (5.141),

$$I + II \tag{5.144}$$

$$\begin{aligned}
&= -\int \frac{(\mathcal{U}'(h))^2}{K^{3/2}(h)} \left(\mathcal{U}^2(h) + |\nabla h|^2 \right) |\nabla h|^2 \, d\gamma \\
&\quad + \int \frac{(\mathcal{U}\mathcal{U}'(h))^2}{K^{3/2}(h)} |\nabla h|^2 \, d\gamma + \frac{1}{2} \int \frac{\mathcal{U}\mathcal{U}'(h)}{K^{3/2}(h)} \left(\nabla |\nabla h|^2, \nabla h \right) d\gamma \\
&= -\int \frac{(\mathcal{U}'(h))^2}{K^{3/2}(h)} |\nabla h|^4 \, d\gamma + \frac{1}{2} \int \frac{\mathcal{U}\mathcal{U}'(h)}{K^{3/2}(h)} \left(\nabla |\nabla h|^2, \nabla h \right) d\gamma \\
&= I' + II'.
\end{aligned}$$

Also,

$$\begin{aligned}
III &= -\sum_{i=1}^{n} \int \frac{1}{K^{3/2}(h)} \left(K(h)|\nabla h_i|^2 - \frac{1}{2} h_i (\nabla K(h), \nabla h_i) \right) d\gamma \\
&= -\sum_{i=1}^{n} \int \frac{1}{K^{3/2}(h)} \left(\mathcal{U}^2(h)|\nabla h_i|^2 + |\nabla h|^2 |\nabla h_i|^2 \right)
\end{aligned}$$

$$-h_i \mathcal{U}\mathcal{U}'(h)(\nabla h, \nabla h_i) - \frac{1}{2} h_i \left(\nabla | \nabla h|^2, \nabla h_i \right) \Big) \, d\gamma$$

$$= \quad IV + V + VI + VII,$$

where

$$IV = -\sum_{i=1}^{n} \int \frac{\mathcal{U}^2(h)| \nabla h_i|^2}{K^{3/2}(h)} \, d\gamma \qquad (5.145)$$

and similarly for V, VI, and VII.

Note that $\sum_{i=1}^{n} h_i \nabla h_i = \frac{1}{2} \nabla | \nabla h|^2$, so that $VI = II'$. Therefore,

$$I' + II' + IV + VI = -\sum_{i,j=1}^{n} \int \frac{(h_i h_j \mathcal{U}'(h) - h_{i,j} \mathcal{U}(h))^2}{K^{3/2}(h)} \, d\gamma \le 0, \quad (5.146)$$

where $h_{i,j} := \dfrac{\partial^2 h}{\partial x_i \partial x_j}$. Also,

$$\sum_{i}^{n} | \nabla h|^2 | \nabla h_i|^2 = \sum_{j=1}^{n} \left(\sum_{i>k} \left(h_i^2 h_{j,k}^2 + h_k^2 h_{j,i}^2 \right) + \sum_{i=1}^{n} h_i^2 h_{j,i}^2 \right) \quad (5.147)$$

and

$$\frac{1}{2} \sum_{i}^{n} h_i \left(\nabla | \nabla h|^2, \nabla h_i \right) = \frac{1}{4} \left(\nabla | \nabla h|^2, \nabla | \nabla h|^2 \right) \qquad (5.148)$$

$$= \sum_{j=1}^{n} \left(\sum_{i>k} 2 h_i h_k h_{j,i} h_{j,k} + \sum_{i=1}^{n} h_i^2 h_{j,i}^2 \right).$$

This shows that $V + VII \le 0$. Thus $\dfrac{dF}{dt} \le 0$. $\qquad \square$

The next theorem, which is a consequence of Theorem 5.4.1, plays a critical role in this book.

Theorem 5.4.3 *Let $X = \{X(z), z \in T\}$ be a real-valued mean zero Gaussian process where T is a countable set. Let a be a median of $\sup_{z \in T} X(z)$ and let*

$$\sigma := \sup_{z \in T} (EX^2(z))^{1/2} < \infty. \qquad (5.149)$$

Then, for all $t > 0$, we have

$$P \left(\sup_{z \in T} X(z) > a - \sigma t \right) \ge \Phi(t), \qquad (5.150)$$

$$P \left(\sup_{z \in T} X(z) < a + \sigma t \right) \ge \Phi(t), \qquad (5.151)$$

and

$$P\left(\left|\sup_{z\in T} X(z) - a\right| \geq \sigma t\right) \leq 2(1 - \Phi(t)). \tag{5.152}$$

Furthermore, the median a is unique.

Proof To begin, we take T to be a finite set and assume that the covariance matrix, say R, of $\{X(z), z \in T\}$ is strictly positive definite. This implies that R is invertible and we can write $R = PDP^t$, where P is an orthogonal matrix and D is a diagonal matrix with strictly positive entries. Let $n = \mathrm{card}\{T\}$ and let ρ_n denote the measure induced by $\{X(z), z \in T\}$ on R^n. For $C \subset R^n$ we have

$$\rho_n(C) = \int_C \exp\left(-\frac{xR^{-1}x^t}{2}\right) \frac{dx}{(2\pi)^{n/2}|R|^{1/2}}. \tag{5.153}$$

Under the change of variables $x = uD^{1/2}P^t$, we get

$$\rho_n(C) = \int_{CPD^{-1/2}} \exp\left(-\frac{u^t u}{2}\right) \frac{du}{(2\pi)^{n/2}} \tag{5.154}$$

so that

$$\rho_n(C) = \gamma_n(CPD^{-1/2}). \tag{5.155}$$

Let $f : R^n \to R^1$ be given by $f(x) = \sup_k x_k$, where $x = (x_1, \ldots, x_n)$. Let $C_b = \{f \leq b\}$. We see, by (5.153), that $\rho_n(C_b)$ is strictly increasing. This shows that the median of f is unique. Let a be the median of f, that is, $\rho_n(\{f \leq a\}) = 1/2$, and set $A = \{f \leq a\}$. By (5.155), since $\rho_n(A) = 1/2$, $\gamma_n(APD^{-1/2}) = \Phi(0)$. Therefore, by Theorem 5.4.1,

$$\gamma_n(APD^{-1/2} + tB) \geq \Phi(t), \tag{5.156}$$

where B is the unit ball in R^n, and by (5.155)

$$\rho_n(A + tBD^{1/2}P^t) \geq \Phi(t). \tag{5.157}$$

We show that

$$A + tBD^{1/2}P^t \subset \{f \leq a + \sigma t\}, \tag{5.158}$$

where

$$\sigma = \sup_{z\in T}(EX^2(z))^{1/2} = \sup_k R_{k,k}^{1/2}. \tag{5.159}$$

This along with (5.157) gives (5.151).

To get (5.158) let $x = \alpha + t\beta D^{1/2}P^t$, where $\alpha \in A$ and $\beta \in B$. Then

$$\sup_k x_k \leq a + t\sup_k |(\beta D^{1/2}P^t)_k|. \tag{5.160}$$

Note that

$$(\beta D^{1/2} P^t)_k = \sum_{j=1}^{n} \beta_j (D^{1/2} P^t)_{j,k}. \tag{5.161}$$

Therefore, by the Schwarz inequality and the fact that $|\beta| \leq 1$

$$|(\beta D^{1/2} P^t)_k| \leq \left(\sum_{j=1}^{n} (D^{1/2} P^t)_{j,k}^2 \right)^{1/2}. \tag{5.162}$$

Also,

$$\sum_{j=1}^{n} (D^{1/2} P^t)_{j,k}^2 \tag{5.163}$$

$$= \sum_{j=1}^{n} (D^{1/2} P^t)_{j,k} (D^{1/2} P^t)_{j,k} = \sum_{j=1}^{n} (D^{1/2} P^t)_{k,j}^t (D^{1/2} P^t)_{j,k}$$

$$= ((D^{1/2} P^t)^t (D^{1/2} P^t))_{k,k} = (PDP^t)_{k,k} = R_{kk} \leq \sigma^2.$$

Combining the inequalities in this paragraph, we get (5.158) and hence (5.151).

To obtain (5.150) consider the set $A' = \{f \geq a\}$. Since $\rho_n(A') = 1/2$, we have, as in (5.157),

$$\rho_n(A' + tBD^{1/2} P^t) \geq \Phi(t). \tag{5.164}$$

Also, we have

$$(A' + tBD^{1/2} P^t) \subset \{f \geq a - \sigma t\} \tag{5.165}$$

since, if $\alpha \in A'$, $\beta \in B$ and $x = \alpha - t\beta D^{1/2} P^t$, then

$$\sup_k x_k \geq a - t \sup_k |(\beta D^{1/2} P^t)_k| \geq a - \sigma t \tag{5.166}$$

by (5.162) and (5.163). The inequality in (5.150) now follows from (5.164) and (5.165).

We now remove the condition that the covariance matrix of X is strictly positive definite. By Corollary 5.3.4 there is a version of X that can be represented by $X(z) = \sum_{j=1}^{m} \phi_j(z)\xi_j$, $t \in T$, where $\{\xi_j\}$ is a standard normal sequence. (Clearly, if the covariance matrix of X is not strictly positive definite, m is strictly less than the cardinality of T.) Note that

$$P\left(\sup_{z \in T} X(z) \leq b \right) = \int_{C_b} \exp\left(-\frac{xx^t}{2} \right) \frac{dx}{(2\pi)^{n/2} |R|^{1/2}}, \tag{5.167}$$

where $C_b = \{x : \sup_{z \in T} \sum_{j=1}^{m} \phi_j(z) x_j \leq b\}$. It follows from this that the distribution function of $\sup_{z \in T} X(z)$ is continuous and strictly increasing. In particular, $\sup_{z \in T} X(z)$ has a unique median, which we denote by a.

Let $\{g_i\}_{i=1}^n$ be a standard normal sequence and set $X_\epsilon = \{X(z_1) + \epsilon g_1, \ldots, X(z_n) + \epsilon g_n\}$. It is easy to check that the covariance matrix of X_ϵ is strictly positive definite. Let a_ϵ be the median of $\sup_{t \in T} X_\epsilon(t)$ and $\sigma_\epsilon = \sup_{t \in T}(EX_\epsilon^2(t))^{1/2}$. By (5.150),

$$P\left(\sup_{z \in T} X_\epsilon(z) > a_\epsilon - \sigma_\epsilon t\right) \geq \Phi(t). \tag{5.168}$$

It is also easy to check that $\sup_{z \in T} X_\epsilon(z)$ converges to $\sup_{z \in T} X(z)$ almost surely as ϵ goes to zero. Clearly $\lim_{\epsilon \to 0} \sigma_\epsilon = \sigma$, and since the distribution function of $\sup_{z \in T} X(z)$ is continuous and strictly increasing, $\lim_{\epsilon \to 0} a_\epsilon = a$. Therefore we can take the limit in (5.168) to get (5.150) for X without the requirement that the covariance matrix of X is strictly positive definite. A similar argument gives (5.151) for finite sets T.

Now let T be a countable index set. Choose an increasing sequence of finite sets $\{T_n\}$, $T_n \subset T$, such that $\lim_{n \to \infty} T_n = T$. Let $a_n = \text{median} \sup_{z \in T_n} X(z)$ and $\sigma_n = \sup_{z \in T_n}(EX^2(z))^{1/2}$. Note that both $\{a_n\}$ and $\{\sigma_n\}$ are increasing in n. Let $\alpha = \lim_{n \to \infty} a_n$ and $\sigma = \lim_{n \to \infty} \sigma_n$. We always have $a_n < 2\sigma_n \leq 2\sigma$. Since we assume that σ is finite, we see that α is also finite. By (5.151), for any n,

$$\Phi(t) \leq P\left(\sup_{z \in T_n} X(z) < \alpha + \sigma t\right). \tag{5.169}$$

Since the sets $E_n = \{\sup_{z \in T_n} X(z) < \alpha + \sigma t\}$ are decreasing and $\cap_n E_n = \{\sup_{z \in T} X(z) \leq \alpha + \sigma t\}$, we can take the limit in (5.169) as $n \to \infty$ to get

$$\Phi(t) \leq P\left(\sup_{z \in T} X(z) \leq \alpha + \sigma t\right). \tag{5.170}$$

Let $t > 0$ and let $\{t_n\}$ be a sequence of positive real numbers, $t_n < t$, that increases to t. Then, by (5.170),

$$\Phi(t_n) \leq P\left(\sup_{z \in T} X(z) < \alpha + \sigma t\right). \tag{5.171}$$

Taking the limit as $n \to \infty$, we see that

$$\Phi(t) \leq P\left(\sup_{z \in T} X(z) < \alpha + \sigma t\right). \tag{5.172}$$

We also note that by (5.170),

$$\frac{1}{2} \leq P\left(\sup_{z \in T} X(z) \leq \alpha\right). \tag{5.173}$$

By (5.150), for any n,

$$\Phi(t) \leq P\left(\sup_{z \in T_n} X(z) > a_n - \sigma t\right) \tag{5.174}$$

$$\leq P\left(\sup_{z \in T} X(z) > a_n - \sigma t\right).$$

Since the sets $F_n = \{\sup_{z \in T} X(z) > a_n - \sigma t\}$ are decreasing and $\cap_n F_n = \{\sup_{z \in T} X(z) \geq \alpha - \sigma t\}$, we can take the limit in (5.174) as n→ ∞ to get

$$\Phi(t) \leq P\left(\sup_{z \in T} X(z) \geq \alpha - \sigma t\right). \tag{5.175}$$

As above, let $t > 0$ and let $\{t_n\}$ be a sequence of positive real numbers, $t_n < t$, that increases to t. Then, by (5.175),

$$\Phi(t_n) \leq P\left(\sup_{z \in T} X(z) > \alpha - \sigma t\right) \tag{5.176}$$

and taking the limit as $n \to \infty$ we see that

$$\Phi(t) \leq P\left(\sup_{z \in T} X(z) > \alpha - \sigma t\right). \tag{5.177}$$

Letting t decrease to zero, we obtain

$$\frac{1}{2} \leq P\left(\sup_{z \in T} X(z) \geq \alpha\right). \tag{5.178}$$

We see from (5.173) and (5.178) that α is a median of $\sup_{z \in T} X(z)$. Thus we get (5.150) and (5.151) from (5.177) and (5.172). The inequality in (5.152) follows from (5.150) and (5.151).

To see that the median is unique, note that it follows from (5.151) that for $t > 0$

$$P\left(\sup_{z \in T} X(z) - \operatorname{median} \sup_{z \in T} X(z) \geq \sigma t\right) \leq \Psi(t), \tag{5.179}$$

where $\Psi(t) = 1 - \Phi(t)$. Suppose that $\operatorname{median} \sup_{z \in T} X(z)$ is equal to both a_1 and a_2 with $a_2 > a_1$. Let $t = (a_2 - a_1)$. We then have

$$\frac{1}{2} \leq P\left(\sup_{z \in T} X(z) \geq a_2\right) = P\left(\sup_{z \in T} X(z) \geq a_1 + t\right) \leq \Psi(t/\sigma^2) < \frac{1}{2}. \tag{5.180}$$

Therefore $a_1 = a_2$. $\qquad\square$

Remark 5.4.4 It is useful to replace X by $|X|$ in (5.152). Using (5.150) and (5.151) we get

$$P\left(\left|\sup_{z\in T}|X(z)| - a\right| \geq \sigma t\right) \leq 3(1 - \Phi(t)). \qquad (5.181)$$

To see this note that, by symmetry, (5.151) also holds with X replaced by $-X$. Therefore

$$P\left(\sup_{z\in T}|X(z)| < a + \sigma t\right) \qquad (5.182)$$

$$= P\left(\left\{\sup_{z\in T} X(z) < a + \sigma t\right\} \cap \left\{\sup_{z\in T} -X(z) < a + \sigma t\right\}\right)$$

$$\geq 2\Phi(t) - 1.$$

Using (5.182) and the fact that (5.150) holds with X replaced by $|X|$, we get (5.181).

Statements like (5.152) and (5.181) are called concentration inequalities because they show that $\sup_{z\in T} X(z)$ and $\sup_{z\in T}|X(z)|$ are concentrated at the median of $\sup_{z\in T} X(z)$ when this median is large relative to σ.

It is often useful to replace a in Theorem 5.4.3 by $E\sup_{z\in T} X(z)$. The median of a random variable Y is less than or equal to $2EY$, but in general we can say nothing about bounding EY by the median of Y. However, for Gaussian processes, using Theorem 5.4.3, we can show that the mean and median of $\sup_{z\in T} X(z)$ can be very close.

Corollary 5.4.5 *Under the hypotheses and notation of Theorem 5.4.3,*

$$\left|a - E\sup_{z\in T} X(z)\right| \leq \frac{\sigma}{\sqrt{2\pi}}. \qquad (5.183)$$

Also,

$$\sigma \leq 2\,\text{median}\left(\sup_{z\in T}|X(z)|\right). \qquad (5.184)$$

Proof It follows from (5.150) that

$$P\left(\frac{a - \sup_{z\in T} X(z)}{\sigma} \geq t\right) \leq \frac{1}{\sqrt{2\pi}}\int_t^\infty e^{-u^2/2}\,du. \qquad (5.185)$$

Since, for any random variable Y, $EY \leq \int_0^\infty P(Y \geq u)\,du$, we see that

$$E\left(\frac{a - \sup_{z\in T} X(z)}{\sigma}\right) \leq \frac{1}{\sqrt{2\pi}}\int_0^\infty\int_t^\infty e^{-u^2/2}\,du\,dt = \frac{1}{\sqrt{2\pi}}. \qquad (5.186)$$

Therefore

$$a - E \sup_{z \in T} X(z) \leq \frac{\sigma}{\sqrt{2\pi}}. \tag{5.187}$$

Using (5.151) and essentially the same argument we get $E \sup_{z \in T} X(z) - a \leq \sigma/\sqrt{2\pi}$ and hence (5.183).

For a fixed $z \in T$,

$$(EX^2(z))^{1/2} \leq 2 \text{ median } |X(z)| \leq 2 \text{ median } \sup_{z \in T} |X(z)|.$$

This implies (5.184). (Note that for the first inequality we use a table of the cumulative normal distribution.) □

Theorem 5.4.3 enables us to make some very strong statements about integrability properties of the supremum of Gaussian processes. These immediately extend to Banach space–valued Gaussian processes and we express them in this way. Let $(B, \| \cdot \|)$ be a Banach space with the following property: There exists a countable subset D of the unit ball of B^*, the dual of B, such that $\|x\| = \sup_{f \in D} |f(x)|$ for all $x \in B$. X is a B-valued Gaussian random variable if $f(X)$ is measurable for all $f \in D$, and every finite linear combination $\sum_i \alpha_i f_i(X)$, $\alpha_i \in R^1$, $f_i \in D$, is a Gaussian random variable. X has mean zero if $E(f(X)) = 0$ for all $f \in D$.

Corollary 5.4.6 *Let X be a $(B, \| \cdot \|)$-valued Gaussian random variable with mean zero, and let*

$$\widetilde{\sigma} := \sup_{f \in D} (E(f(X))^2)^{1/2}. \tag{5.188}$$

Then

$$\lim_{t \to \infty} \frac{1}{t^2} \log P(\|X\| \geq t) = -\frac{1}{2\widetilde{\sigma}^2} \tag{5.189}$$

and, equivalently,

$$E \exp\left(\frac{1}{2\alpha^2}\|X\|^2\right) < \infty \qquad \text{if and only if } \alpha > \widetilde{\sigma}. \tag{5.190}$$

Proof By hypothesis on the Banach space $(B, \| \cdot \|)$, there exists a countable set D in the unit ball of B^* such that $\|X\| = \sup_{f \in D} |X(f)|$. It follows from (5.181) that

$$\lim_{t \to \infty} \frac{1}{t^2} \log P\left(\sup_{f \in D} |X(f)| \geq t\right) \leq -\frac{1}{2\widetilde{\sigma}^2}. \tag{5.191}$$

This gives us the upper bound in (5.189). The lower bound is trivial since $P(\sup_{f \in D} |X(f)| > t) \geq P(|X(f)| > t)$ for all $f \in D$.

Using (5.189), we get sufficiency in (5.190). The reverse implication follows since

$$
\begin{aligned}
E \exp\left(\frac{1}{2\alpha^2} \|X\|^2\right) &\geq \sup_{f \in D} E \exp\left(\frac{1}{2\alpha^2} X^2(f)\right) \\
&= \sup_{f \in D} \left(1 - \frac{\sigma_f^2}{\alpha^2}\right)^{-1/2},
\end{aligned}
$$

where $\sigma^2(f) = EX^2(f)$. This is finite only when $\alpha^2 > \widetilde{\sigma}^2$. $\qquad\square$

For a $(B, \|\cdot\|)$-valued random variable X, let $\|X\|_p := (E\|X\|^p)^{1/p}$. In the next corollary of Theorem 5.4.3 we show that all the moments of the norm of a mean zero $(B, \|\cdot\|)$-valued Gaussian random variable are equivalent.

Corollary 5.4.7 *Let X be a $(B, \|\cdot\|)$-valued Gaussian random variable with mean zero. Then all the moments of $\|X\|$ exist. Furthermore, the moments are equivalent, that is, for all $p, q > 0$ there exist constants $K_{p,q}$ such that*

$$
\|X\|_p \leq K_{p,q} \|X\|_q. \tag{5.192}
$$

In particular, $K_{p,2} \leq K\sqrt{p}$ for $p \geq 1$, where K is an absolute constant.

Proof The fact that $\|X\|$ has all its moments follows immediately from (5.189). To continue, as in the proof of Corollary 5.4.6, we obtain (5.192) with $\|X\|$ replaced by $\sup_{f \in D} |X(f)|$. Let $a = \text{med}\,(\sup_{f \in D} |X(f)|)$. By (5.181), for all $p > 0$,

$$
\begin{aligned}
E \left| \sup_{f \in D} |X(f)| - a \right|^p &= \int_0^\infty P\left(\left| \sup_{f \in D} |X(f)| - a \right| > t\right) dt^p \\
&\leq \frac{3}{\sqrt{2\pi}} \sigma^p \int_0^\infty \int_t^\infty \exp(-u^2/2)\, du\, dt^p \\
&= \frac{3}{\sqrt{2\pi}} \sigma^p \int_0^\infty t^p \exp(-t^2/2)\, dt \\
&\leq (K\sqrt{p}\sigma)^p \tag{5.193}
\end{aligned}
$$

for some absolute constant K. Therefore,

$$
\left(E \sup_{f \in D} |X(f)|^p \right)^{1/p} \leq a + K\sqrt{p}\sigma. \tag{5.194}
$$

For a real-valued random variable $|\xi|$, median $|\xi| \leq (2E|\xi|^q)^{1/q}$ for all

$q > 0$. Thus $a \le (2E \sup_{f \in D} |X(f)|^q)^{1/q}$. Also, by (5.184), $\sigma \le 2a$. Using these observations in (5.194) we get (5.192) and the comment following it. $\qquad\square$

5.5 Comparison theorems

Gaussian processes are determined by their covariance functions. In this section we show the rather remarkable fact that, generally speaking, the smoother the covariance function, the better behaved are the sample paths of the processes.

Lemma 5.5.1 (Slepian) *Let ξ and ζ be mean zero R^n-valued Gaussian random variables such that*

$$E\xi_j^2 = E\zeta_j^2 \qquad \forall\, 1 \le j \le n \tag{5.195}$$

and

$$E\xi_j \xi_k \le E\zeta_j \zeta_k \qquad \forall\, 1 \le j, k \le n. \tag{5.196}$$

Then, for any real numbers $\lambda_1, \ldots, \lambda_n$,

$$P\left(\cup_{j=1}^n \{\xi_j > \lambda_j\}\right) \ge P\left(\cup_{j=1}^n \{\zeta_j > \lambda_j\}\right). \tag{5.197}$$

Proof Let η be mean zero R^n-valued Gaussian random variable with a strictly positive definite covariance matrix $\Gamma = \{\Gamma_{j,k}\}$. Taking the inverse Fourier transform of the characteristic function of η, we can express the probability density function of η by

$$g(z, \Gamma) = \frac{1}{(2\pi)^n} \int \exp\left(-i(z, x) - \frac{(\Gamma x, x)}{2}\right) dx, \tag{5.198}$$

where $x \in R^n$. Differentiating the right-hand side of (5.198), one sees that

$$\frac{\partial g(z, \Gamma)}{\partial \Gamma_{j,k}} = \frac{\partial^2 g(z, \Gamma)}{\partial z_j \partial z_k} \qquad \forall\, j, k = 1, \ldots, n \qquad j \ne k. \tag{5.199}$$

Let

$$Q(\eta, \Gamma) = P\left(\cap_{j=1}^n \{\eta_j \le \lambda_j\}\right) = \int_{-\infty}^{\lambda_1} \cdots \int_{-\infty}^{\lambda_n} g(z, \Gamma)\, dz. \tag{5.200}$$

For $1 \le j < k \le n$, we see by (5.199) that

$$\frac{\partial Q(\eta, \Gamma)}{\partial \Gamma_{j,k}} = \int_{-\infty}^{\lambda_1} \cdots \int_{-\infty}^{\lambda_n} \frac{\partial^2}{\partial z_j \partial z_k} g(z, \Gamma)\, dz. \tag{5.201}$$

The right-hand side of (5.201) is an $(n-2)$-fold integral of the function g, with the arguments z_j and z_k replaced by λ_j and λ_k and the domain of integration of the remaining variables the same as before. This shows us that

$$\partial Q(\eta, \Gamma)/\partial \Gamma_{j,k} > 0. \tag{5.202}$$

Let $\Sigma = \{\Sigma_{j,k}\}$ and $Z = \{Z_{j,k}\}$ be the covariance matrices of ξ and ζ, respectively, and assume that both matrices are strictly positive definite. For $0 \le \theta \le 1$ set $\Gamma_{j,k}(\theta) = \theta \Sigma_{j,k} + (1-\theta) Z_{j,k}$ and note that $\Gamma(\theta) = \{\Gamma_{j,k}(\theta)\}$ is strictly positive definite for all $0 \le \theta \le 1$. Let $q(\theta) = 1 - Q(\eta, \Gamma(\theta))$. Using (5.195), (5.196), and (5.202), we see that

$$
\begin{aligned}
\frac{dq(\theta)}{d\theta} &= -\sum_{j,k=1}^{n} \frac{\partial Q(\eta, \Gamma(\theta))}{\partial \Gamma_{j,k}} \frac{d\Gamma_{j,k}(\theta)}{d\theta} \\
&= -\sum_{j,k=1}^{n} \frac{\partial Q(\eta, \Gamma(\theta))}{\partial \Gamma_{j,k}} (\Sigma_{j,k} - Z_{j,k}) \\
&= -\sum_{j,k=1, j \ne k}^{n} \frac{\partial Q(\eta, \Gamma(\theta))}{\partial \Gamma_{j,k}} (\Sigma_{j,k} - Z_{j,k}) \ge 0.
\end{aligned}
\tag{5.203}
$$

Since q is increasing, we get

$$q(1) = P\left(\cup_{j=1}^{n}\{\xi_j > \lambda_j\}\right) \ge q(0) = P\left(\cup_{j=1}^{n}\{\zeta_j > \lambda_j\}\right). \tag{5.204}$$

This proves the lemma when both Σ and Z are strictly positive definite. Suppose this assumption is not satisfied. Let ϕ be a standard normal sequence with n components independent of ξ and ζ. Consider $\xi' = \xi + \epsilon\phi$ and $\zeta' = \zeta + \epsilon\phi$. The hypotheses of this lemma hold for these Gaussian random variables and their covariance matrices are strictly positive definite for all $\epsilon > 0$. Therefore the lemma holds for ξ' and ζ'. By considering their characteristic functions it is easy to see that ξ' and ζ' converge in distribution to ξ and ζ as $\epsilon \to 0$. Thus the lemma holds as stated. $\square$

The following simple corollary of Lemma 5.5.1 is fundamental and used often in this book.

Corollary 5.5.2 *Let $\{Y_j\}$ be a Gaussian sequence such that*

$$EY_j^2 = 1 \quad \text{and} \quad EY_j Y_k \le \delta \quad j \ne k. \tag{5.205}$$

Then

$$\limsup_{j \to \infty} \frac{Y_j}{(2\log j)^{1/2}} \ge (1-\delta)^{1/2} \quad a.s. \tag{5.206}$$

Proof Let $\{\xi_j\}$ be a standard normal sequence and set

$$Z_j = \delta^{1/2}\xi_1 + (1-\delta)^{1/2}\xi_j \qquad j = 2, \ldots \tag{5.207}$$

Clearly

$$EY_j Y_k \leq E Z_j Z_k \qquad j \neq k. \tag{5.208}$$

Therefore, by Lemma 5.5.1 and the Monotone Convergence Theorem,

$$P\left(\cup_{j=k}^\infty \{Y_j/(2\log j)^{1/2} \geq (1-\delta)^{1/2}\}\right) \tag{5.209}$$

$$\geq P\left(\cup_{j=k}^\infty \{Z_j/(2\log j)^{1/2} \geq (1-\delta)^{1/2}\}\right).$$

By Theorem 5.1.4, this last term is equal to one. Using the Monotone Convergence Theorem again we get (5.206). $\qquad\square$

Lemma 5.5.3 *Let ξ and ζ be mean zero R^n-valued Gaussian random variables such that*

$$E(\zeta_j - \zeta_k)^2 \leq E(\xi_j - \xi_k)^2 \qquad \forall\, 0 \leq j, k \leq n. \tag{5.210}$$

Then

$$E \sup_j \zeta_j \leq E \sup_j \xi_j \tag{5.211}$$

and

$$E \sup_{j,k} |\zeta_j - \zeta_k| \leq E \sup_{j,k} |\xi_j - \xi_k|. \tag{5.212}$$

A much simpler proof gives this result with the factor 2 on the right-hand sides of (5.211) and (5.212) (see Chapter 15, Section 3 in Kahane (1985)). However, we need this degree of sharpness in the proof of Theorem 6.3.4.

Proof Note that (5.211) implies (5.212) because

$$E \sup_{j,k} |\zeta_j - \zeta_k| = E \sup_{j,k} (\zeta_j - \zeta_k) = E(\sup_j \zeta_j + \sup_k (-\zeta_k)) = 2E \sup_j \zeta_j, \tag{5.213}$$

where the last step uses the fact that ζ is symmetric. Therefore it suffices to prove (5.211).

The proof follows along the lines of Lemma 5.5.1 but is more complicated. We use the notation of Lemma 5.5.1. As we show in Lemma 5.5.1, it is enough to consider that Σ and Z are strictly positive definite. Also, without loss of generality, we can assume that ξ and ζ are independent. Consider $\xi(\theta) = \theta^{1/2}\xi + (1-\theta)^{1/2}\zeta$, $0 \leq \theta \leq 1$. The covariance matrix of $\xi(\theta)$ is $\Gamma(\theta)$, which is strictly positive definite for all $0 \leq \theta \leq 1$. We obtain (5.211) by showing that

$$h(\theta) = E \sup_j \xi_j(\theta) \tag{5.214}$$

is increasing in θ. The probability density function of $\xi(\theta)$ is given by

$$g(z, \Gamma(\theta)) = \frac{1}{(2\pi)^n} \int \exp\left(i(z, x) - \frac{(\Gamma(\theta)x, x)}{2}\right) dx. \tag{5.215}$$

Also,

$$\frac{dh(\theta)}{d\theta} = \int_{R^n} \max(z_1, \ldots, z_n) \frac{d(g(z, \Gamma(\theta)))}{d\theta} dz. \tag{5.216}$$

It follows from (5.215) that

$$\frac{dg(z, \Gamma(\theta))}{d\theta} = -\frac{1}{(2\pi)^n} \int \left(\frac{1}{2} \sum_{j,k=1}^n x_j x_k \frac{d\Gamma_{j,k}(\theta)}{d\theta}\right) \tag{5.217}$$

$$\exp\left(i(z, x) - \frac{(\Gamma(\theta)x, x)}{2}\right) dx$$

and

$$\frac{\partial^2 g(z, \Gamma(\theta))}{\partial z_j \partial z_k} = -\frac{1}{(2\pi)^n} \int x_j x_k \exp\left(i(z, x) - \frac{(\Gamma(\theta)x, x)}{2}\right) dx. \tag{5.218}$$

Using (5.217) and (5.218), we see that

$$\frac{1}{2} \sum_{j,k=1}^n \frac{d\Gamma_{j,k}(\theta)}{d\theta} \frac{\partial^2 g(z, \Gamma(\theta))}{\partial z_j \partial z_k} = \frac{dg(z, \Gamma(\theta))}{d\theta}. \tag{5.219}$$

Using this in (5.216) gives

$$\frac{dh(\theta)}{d\theta} = \frac{1}{2} \sum_{j,k=1}^n \frac{d\Gamma_{j,k}(\theta)}{d\theta} \int_{R^n} \max(z_1, \ldots, z_n) \frac{\partial^2 g(z, \Gamma(\theta))}{\partial z_j \partial z_k} dz. \tag{5.220}$$

We now show that the right-hand side of (5.220) is greater than or equal to zero. To this end let us consider

$$I_{1,2} := \int_{R^n} \max(z_1, \ldots, z_n) \frac{\partial^2 g(z, \Gamma(\theta))}{\partial z_1 \partial z_2} dz. \tag{5.221}$$

Let $u_1 = \max(z_2, \ldots, z_n)$. We can write

$$I_{1,2} := \int_{R^{n-1}} \left(\int_{u_1}^{\infty} z_1 \frac{d}{dz_1}\left(\frac{\partial g(z, \Gamma(\theta))}{\partial z_2}\right) dz_1 \right. \tag{5.222}$$

$$\left. + \int_{-\infty}^{u_1} u_1 \frac{d}{dz_1}\left(\frac{\partial g(z, \Gamma(\theta))}{\partial z_2}\right) dz_1 \right) dz_2 \cdots dz_n.$$

It is easy to do the integration with respect to z_1, using integration by parts on the first integral, to get

$$I_{1,2} = -\int_{R^{n-1}} \left(\int_{u_1}^{\infty} \frac{\partial g(z, \Gamma(\theta))}{\partial z_2} \, dz_1 \right) dz_2 \cdots dz_n. \qquad (5.223)$$

Interchanging the order of integration between z_1 and z_2 and setting $u_{1,2} = \max(z_3, \ldots, z_n)$, we get

$$\begin{aligned}
I_{1,2} &= -\int_{R^{n-2}} \left(\int_{u_{1,2}}^{\infty} \int_{-\infty}^{z_1} \frac{\partial g(z, \Gamma(\theta))}{\partial z_2} \, dz_2 \, dz_1 \right) dz_3 \cdots dz_n. \quad (5.224) \\
&= -\int_{R^{n-2}} \left(\int_{u_{1,2}}^{\infty} g((z_1, z_1, \ldots, z_n), \Gamma(\theta)) \, dz_1 \right) dz_3 \cdots dz_n.
\end{aligned}$$

For $2 < k \leq n$ we define $I_{1,k}$ as we defined $I_{1,2}$ in (5.221) but with 2 replaced by k. Following the argument above we get

$$I_{1,k} = -\int_{R^{n-2}} \left(\int_{u_{1,k}}^{\infty} g((z_1, z_2, \ldots, z_{k-1}, z_1, z_{k+1}, \ldots, z_n), \Gamma(\theta)) \, dz_1 \right)$$
$$dz_2 \cdots dz_{k-1}, dz_{k+1} dz_n,$$

where $u_{1,k} = \max(z_2, \ldots, z_{k-1}, z_{k+1}, \ldots, z_n)$. Since g is a probability density function, it is clear that $I_{1,k} < 0$ for all $k = 2, \ldots, n$.

Next, consider

$$\begin{aligned}
I_{1,1} &:= \int_{R^n} \max(z_1, \ldots, z_n) \frac{\partial^2 g(z, \Gamma(\theta))}{\partial z_1^2} \, dz. \qquad (5.225) \\
&= -\int_{R^{n-1}} \left(\int_{u_1}^{\infty} \frac{\partial g(z, \Gamma(\theta))}{\partial z_1} \, dz_1 \right) dz_2 \cdots dz_n \\
&= \int_{R^{n-1}} g((u_1, z_2, \ldots, z_n), \Gamma(\theta)) \, dz_2 \cdots dz_n,
\end{aligned}$$

where, for the second equality, we follow the computations in (5.222) and (5.223).

Divide R^{n-1} into $n - 1$ disjoint sets $B_k = \{z \in R^{n-1}, z_k > u_{1,k}\}$. Thus

$$I_{1,1} = \sum_{k=2}^{n} \int_{B_k} g((z_k, z_2, \ldots, z_n) \, dz_2 \cdots dz_n \qquad (5.226)$$

since $u_1 = z_k$ on B_k. The integral over B_2 is

$$\int_{R^{n-2}} \left(\int_{u_{1,2}}^{\infty} g((z_2, z_2, \ldots, z_n), \Gamma(\theta)) \, dz_2 \right) dz_3 \cdots dz_n = -I_{1,2}. \quad (5.227)$$

The integrals over the other sets B_k are obtained similarly and we see that

$$I_{1,1} = -\sum_{k=2}^{n} I_{1,k}. \qquad (5.228)$$

Define $I_{j,j}$ and $I_{j,k}$ as in (5.221) and (5.225). Rearranging the components of ξ and ζ, we obtain a result similar to (5.228) for $I_{j,j}$ and $I_{j,k}$ for all j and $k \neq j$. That is,

$$I_{j,j} = -\sum_{1 \leq k \leq n, k \neq j} I_{j,k} \qquad \forall\, j = 1, \ldots, n. \qquad (5.229)$$

Also, as in the case $j = 1$, $I_{j,k} < 0$ for all $j \neq k$.

Using this in (5.220) we see that

$$\begin{aligned}
\frac{dh(\theta)}{d\theta} &= \frac{1}{2} \sum_{1 \leq j,k \leq n, j \neq k} \frac{d\Gamma_{j,k}(\theta)}{d\theta} I_{j,k} + \sum_{j=1}^{n} \frac{d\Gamma_{j,j}(\theta)}{d\theta} I_{j,j} \qquad (5.230) \\
&= \frac{1}{2} \sum_{1 \leq j,k \leq n, j \neq k} \left(\frac{d\Gamma_{j,k}(\theta)}{d\theta} - \frac{d\Gamma_{j,j}(\theta)}{d\theta} \right) I_{j,k} \\
&= -\frac{1}{2} \sum_{1 \leq j,k \leq n, j > k} \frac{d}{d\theta} \Big(\Gamma_{j,j}(\theta) + \Gamma_{k,k}(\theta) - 2\Gamma_{j,k}(\theta) \Big) I_{j,k}.
\end{aligned}$$

Note that

$$\begin{aligned}
\Gamma_{j,j}(\theta) + \Gamma_{k,k}(\theta) - 2\Gamma_{j,k}(\theta) &= E(\xi_j(\theta) - \xi_k(\theta))^2 \qquad (5.231) \\
&= \theta E(\xi_j - \xi_k)^2 + (1-\theta) E(\zeta_j - \zeta_k)^2.
\end{aligned}$$

The derivative of $\Gamma_{j,j}(\theta) + \Gamma_{k,k}(\theta) - 2\Gamma_{j,k}(\theta)$ is greater than or equal to zero by (5.210). Since $I_{j,k}$, $j \neq k$, is less than zero, we see that $dh(\theta)/d\theta \geq 0$, which proves this lemma. $\qquad \square$

It is worthwhile to give a simple equality that is used often in this chapter.

Lemma 5.5.4 *Let (ξ_1, ξ_2) be an R^2-valued mean zero normal random variable with $(E(\xi_1 - \xi_2)^2)^{1/2} = D$. Then*

$$E(\xi_1 \vee \xi_2) = \frac{D}{\sqrt{2\pi}}. \qquad (5.232)$$

Proof Let a and b be real numbers. Then $a \vee b = \frac{1}{2}((a+b) + |a-b|)$. Using this and the fact that ξ_1 and ξ_2 have mean zero, we see that

$$E(\xi_1 \vee \xi_2) = \tfrac{1}{2} E(|\xi_1 - \xi_2|) = \tfrac{1}{2} DE(|\xi|), \qquad (5.233)$$

where ξ is $N(0,1)$. This is (5.232).

Lemma 5.5.5 *Let* $\xi_1, \ldots, \xi_n$ *be a standard normal sequence. Then, for all* $n \geq 1$,

$$E \sup_{1 \leq j \leq n} \xi_j \geq \tfrac{1}{12}(\log n)^{1/2}. \tag{5.234}$$

Proof Since $\log 1 = 0$ and, by (5.232), $E(\xi_1 \vee \xi_2) = 1/\sqrt{\pi}$, we need only consider $n \geq 3$. Recall that, for any real-valued function f, $f^+ = f \vee 0$ and $f^- = |f \wedge 0|$, so that $f = f^+ - f^-$. Let $M_n = \sup_{1 \leq j \leq n} \xi_n$. Clearly,

$$EM_n = EM_n^+ + E(-M_n^-). \tag{5.235}$$

We show below that $P(M_n > (\log n)^{1/2}) > 0.18$. This implies that $EM_n^+ \geq (0.18)(\log n)^{1/2}$. Also, note that $(-M_n^-)$ is increasing as n increases. Therefore, for $n \geq 3$, $E(-M_n^-) \geq E(-M_3^-)$ and

$$
\begin{aligned}
E(-M_3^-) &= E(-M_3^- | M_3 < 0)(1/8) \geq E(-\xi_1^- | M_3 < 0)(1/8) \\
&= E(-\xi_1^- | \xi_1 < 0)(1/8) = -(2/\pi)^{1/2}(1/8),
\end{aligned}
$$

where in the next to last equality we use the fact that $\{M_3 < 0\} = \cap_{i=1}^{3}\{\xi_i < 0\}$. Using these observations in (5.235) we get (5.234).

To see that $P(M_n > (\log n)^{1/2}) > 0.18$, we note the elementary equality $P(M_n > b) = 1 - (1 - P(\xi_1 > b))^n$. Furthermore, by the right-hand side of (5.19) and symmetry, we see that for $n \geq 3$ and $\alpha = 1/(8\pi)^{1/2}$,

$$P\left(\xi_1 > (\log n)^{1/2}\right) \geq \frac{1}{(8\pi n \log n)^{1/2}} > \frac{\alpha}{n}. \tag{5.236}$$

Since $(1 - \frac{\alpha}{n})^n < e^{-\alpha}$, we get the desired estimate for M_n. $\qquad \square$

Let (T, d) be a metric space. A set $S \subset T$ is said to be u-distinguishable if, for all $x, y \in S$, $x \neq y$, $d(x, y) > u$. Whenever we consider a Gaussian process $X = \{X_t, t \in T\}$, unless otherwise stated, we take the metric $d(s, t) = (E(X(s) - X(t))^2)^{1/2}$. We sometimes refer to this as the standard L_2 metric.

Lemma 5.5.6 (Sudakov) *Let* $X = \{X_t, t \in T\}$ *be a mean zero Gaussian process. Let* $S \subset T$ *be finite and* u-*distinguishable. Then*

$$E \sup_{s \in S} X(s) \geq \frac{1}{17} u (\log \#S)^{1/2}, \tag{5.237}$$

where $\#S$ *is the cardinality of the set* S. *If* $E \sup_{t \in T} X(t) < \infty$ *and* $S \subset T$ *is* u-*distinguishable, then*

$$\#S \leq \exp\left(\frac{17 E \sup_{t \in T} X(t)}{u}\right)^2. \tag{5.238}$$

Proof Let $\{s_1, \ldots, s_n\}$ denote the points of S. Let $\xi = (\xi_1, \ldots, \xi_n)$ be a standard normal sequence. Consider the Gaussian process $\{Y(s), s \in S\}$, where $Y(s_i) = (u/\sqrt{2})\xi_i$, $i \in [1, n]$. By hypothesis,

$$E(Y(s_i) - Y(s_j))^2 \leq E(X(s_i) - X(s_j))^2 \qquad \forall \, 1 \leq i, j \leq n. \quad (5.239)$$

Therefore, by Lemmas 5.5.3 and 5.5.5,

$$E \sup_{s \in S} X(s) \geq E \sup_{s \in S} Y(s) = (u/\sqrt{2})E \sup_{1 \leq i \leq n} \xi_i \geq u(\log n)^{1/2}/17,$$
$$(5.240)$$

which is (5.237). The inequality in (5.238) follows from (5.237) when $\#S$ is finite. Moreover, we see from (5.237) that when $E \sup_{t \in T} X(t) < \infty$, $\#S$ must be finite. Hence we get (5.238) as stated. $\qquad \square$

We have the following sharpening of Lemma 5.5.6. It seems to be an obvious and minor improvement over Lemma 5.5.6. However, it is exactly what is needed in Talagrand's elegant proof of Theorem 6.3.4.

Lemma 5.5.7 (Talagrand) *Let $X = \{X_t, t \in T\}$, where T is countable, be a mean zero Gaussian process. Let $S \subset T$ be $4u$-distinguishable, $u > 0$. Then*

$$E \sup_{t \in T} X(t) \geq \frac{1}{12} u(\log \#S)^{1/2} + \inf_{s \in S} E \left(\sup_{t \in B(s,u)} X(t) \right), \quad (5.241)$$

where $B(s, u) = \{t \in T \mid d(s, t) \leq u\}$.

Proof We can assume that $E \sup_{t \in T} X(t) < \infty$. This implies, by Lemma 5.5.6, that $S = \{s_k, k \in [1, M]\}$ is a finite set, where, of course, $M = \#S$.

Let $\{X_k, k \in [1, M]\}$ be independent copies of X and $\{g_k, k \in [1, M]\}$ be a standard normal sequence independent of $\{X_k, k \in [1, M]\}$. Let $T' = \cup_{k \in [1, M]} B(s_k, u)$. We now define a Gaussian process on T' as follows:

$$Y(t) = X_k(t) - X_k(s_k) + ug_k \qquad t \in B(s_k, u). \quad (5.242)$$

This is possible because the balls $B(s_k, u)$ are disjoint.

Let $t, t' \in B(s_k, u)$. Then $Y(t) - Y(t') \stackrel{law}{=} X(t) - X(t')$, so $d_Y(t, t') = d_X(t, t')$. If t, t' are in disjoint balls, $d_Y^2(t, t') = E(X_k(t) - X_k(s_k))^2 + E(X_{k'}(t') - X_{k'}(s_{k'}))^2 + 2u^2$ for some $k \neq k'$. Therefore $d_Y(t, t') \leq 2u$. On the other hand, since t, t' are in disjoint balls, $d_X(t, t') \geq 2u$. Thus we see that $d_Y(s, t) \leq d_X(s, t)$ for all $s, t \in T'$. It follows from Lemma 5.5.3 that

$$E \sup_{t \in T} X(t) \geq E \sup_{t \in T'} X(t) \geq E \sup_{t \in T'} Y(t). \quad (5.243)$$

We proceed to find a lower bound for (5.243):

$$\sup_{t \in T'} Y(t) = \sup_{k \in [1,M]} \sup_{t \in B(s_k,u)} Y(t) \tag{5.244}$$

$$= \sup_{k \in [1,M]} \left(u g_k + \left(\sup_{t \in B(s_k,u)} X_k(t) \right) - X_k(s_k) \right).$$

Recall that $\{X_k, k \in [1, M]\}$ and $\{g_k, k \in [1, M]\}$ are independent. First we hold $\{g_k, k \in [1, M]\}$ fixed and take the expectation with respect to $\{X_k, k \in [1, M]\}$ to get

$$E_{\{X_k\}} \sup_{t \in T'} Y(t) \geq \sup_{k \in [1,M]} \left(u g_k + E \sup_{t \in B(s_k,u)} X_k(t) \right) \tag{5.245}$$

$$\geq \sup_{k \in [1,M]} u g_k + \inf_{k \in [1,M]} E \sup_{t \in B(s_k,u)} X_k(t).$$

Taking the expectation with respect to $\{g_k, k \in [1, M]\}$, we get

$$E \sup_{t \in T'} Y(t) \geq E \sup_{k \in [1,M]} u g_k + \inf_{k \in [1,M]} E \sup_{t \in B(s_k,u)} X_k(t). \tag{5.246}$$

The inequality in (5.241) now follows from (5.243) and Lemma 5.5.5. $\Box$

5.6 Processes with stationary increments

A stochastic process $\{Y(t), t \in T\}$ is called a stationary process if, for all $t_1, \ldots, t_n$ in T, for all $n \geq 1$, and all $\tau \in T$, $(Y(t_1), \ldots, Y(t)) \overset{law}{=} (Y(t_1 + \tau), \ldots, Y(t_n + \tau))$. A stochastic process $\{Z(t), t \in T\}$ is called a process with stationary increments if $(Z(t_1) - Z(s_1), \ldots, Z(t_n) - Z(s_n)) \overset{law}{=} (Z(t_1 + \tau) - Z(s_1 + \tau), \ldots, Z(t_n + \tau) - Z(s_n + \tau))$ for all $t_1, \ldots, t_n$ and $s_1, \ldots, s_n$ in T, for all $n \geq 1$, and all $\tau \in T$.

These definitions do not make sense unless T is an additive semigroup. For our purposes it suffices to take $T = R^d$ or $T = [0, 1]^d$. R^d is a locally compact Abelian group, and, when dealing with Gaussian processes that can be expressed as random Fourier series, we view $[0, 1]^d$ as R^d/Z^d, which is a compact Abelian group. Most of the results we give about Gaussian processes on R^d and $[0, 1]^d$, when considered to be compact or locally compact Abelian groups, are also valid on general compact or locally compact Abelian groups.

Since a Gaussian process is determined by its finite-dimensional distributions, a necessary and sufficient condition for a Gaussian process to be stationary is that its covariance $\Sigma(s, t)$ is a function of $s - t$. Thus we may talk about a stationary Gaussian process $Y = \{Y(t), t \in T\}$ with covariance $\phi(s)$, meaning that $EY(t)Y(t+s) = \phi(s)$. Note that ϕ must be an even positive definite function, with $\phi(0) > 0$ (or else $Y(t) \equiv 0$ almost surely, for all $t \in T$).

A theorem of Bochner states that every continuous, even, positive definite function ϕ on R^d is the cosine transform of a finite positive symmetric measure on R^d (see page 184 in Donoghue (1969)). Therefore, each stationary Gaussian process on R^d has a covariance ϕ that can be represented as

$$\phi(u) = \int_{R^d} \cos(2\pi \langle \lambda, u \rangle)\, \mu(d\lambda), \qquad (5.247)$$

where μ is a finite positive symmetric measure on R^d.

We now consider the representation of characteristic functions of Gaussian processes with stationary increments. To motivate this, let $Z(t) = Y(t) - Y(0)$ for Y as given above, and set $Z = \{Z(t), t \in T\}$. Clearly, Z is a Gaussian process with stationary increments and obviously $Z(0) = 0$ almost surely. Because of this $Z(t) - Z(s)$ is equal in law to $Z(t-s)$ and

$$\psi(t-s) := EZ^2(t-s) = 2 \int_{R^d} (1 - \cos(2\pi \langle \lambda, t-s \rangle))\, \mu(d\lambda). \quad (5.248)$$

We note that

$$EZ(t)Z(s) = \frac{1}{2} \left(\psi(s) + \psi(t) - \psi(s-t) \right), \qquad (5.249)$$

which is positive definite since it is the covariance of Z.

We extend the class of functions ψ in (5.248) to integrals with respect to a larger family of measures on R^d, that is, we consider

$$\psi(u) = 2 \int_{R^d} (1 - \cos 2\pi \langle \lambda, u \rangle)\, \nu(d\lambda), \qquad (5.250)$$

where ν is a symmetric positive measure on R^d satisfying

$$\int_{R^d} (1 \wedge |\lambda|^2)\, \nu(d\lambda) < \infty, \qquad (5.251)$$

the same condition as for a symmetric Lévy measure. It is easy to see that $\psi(s) + \psi(t) - \psi(s-t)$ is still positive definite since

$$1 - \cos 2\pi \langle \lambda, s \rangle - \cos 2\pi \langle \lambda, t \rangle + \cos 2\pi \langle \lambda, s-t \rangle \qquad (5.252)$$
$$= \mathrm{Re}(e^{i2\pi \langle \lambda, s \rangle} - 1)(e^{-i2\pi \langle \lambda, t \rangle} - 1).$$

Thus ψ as given in (5.250) determines a mean zero Gaussian process with stationary increments with covariance given by the right-hand side of (5.249).

Let $\{G(t), t \in R_+\}$ be a mean zero Gaussian process with stationary increments and $G(0) \equiv 0$. We can set $G(t) = \xi t + G_1(t)$, where ξ is a normal random variable with mean zero and G_1 and ξ are independent ($\xi \equiv 0$ is possible). It follows from Chapter 11, Section 11 in Doob (1953) that the covariance of $G_1(t)$ is given by (5.250) for some measure ν satisfying (5.251).

Consider a real-valued stationary Gaussian processes on R^1. Using symmetry, we can set $F(\lambda) = \mu(\{0\}) + 2\mu((0, \lambda])$ and integrate over $[0, \infty)$ in (5.247). When $f(\lambda) = d/d\lambda(F(\lambda))$ exists on $[0, \infty)$ we can write (5.247) as

$$\phi(u) = \int_0^\infty \cos(2\pi\lambda u) f(\lambda)\, d\lambda. \qquad (5.253)$$

In this case we can represent a real-valued stationary Gaussian process $\widetilde{Y} = \{\widetilde{Y}_t, t \in R^1\}$ with covariance $\phi(u)$ by

$$\widetilde{Y}(t) = \int_0^\infty \cos(2\pi\lambda t) f^{1/2}(\lambda)\, dB(\lambda) + \int_0^\infty \sin(2\pi\lambda t) f^{1/2}(\lambda)\, dB'(\lambda), \qquad (5.254)$$

where B and B' are independent Brownian motions on $[0, \infty)$. Also, by (5.253),

$$
\begin{aligned}
E\left(\widetilde{Y}(s) - \widetilde{Y}(t)\right)^2 &= 2\int_0^\infty (1 - \cos 2\pi\lambda(s - t)) f(\lambda)\, d\lambda \\
&= 4\int_0^\infty (\sin^2 \pi\lambda(s - t)) f(\lambda)\, d\lambda. \qquad (5.255)
\end{aligned}
$$

F is called the spectral distribution and f the spectral density of the stationary process $\widetilde{Y}$.

For Gaussian processes with stationary increments, using symmetry, we set $H(\lambda) = 2\nu([\lambda, \infty))$ and integrate over $[0, \infty)$ in (5.250). When $h(\lambda) = -d/d\lambda(H(\lambda))$ exists on $[0, \infty)$, we can write (5.250) as

$$\psi(u) = \int_0^\infty (1 - \cos 2\pi\lambda u)\, h(\lambda)\, d\lambda. \qquad (5.256)$$

In this case we can represent a real-valued Gaussian process $\{\widetilde{Z}(t), t \in R^1\}$ with stationary increments and covariance $\psi(s) + \psi(t) - \psi(s - t)$ by

$$
\begin{aligned}
\widetilde{Z}(t) = 2\Bigg(&\int_0^\infty (1 - \cos 2\pi\lambda t)\, h^{1/2}(\lambda)\, dB(\lambda) \qquad (5.257) \\
&+ \int_0^\infty \sin(2\pi\lambda t) h^{1/2}(\lambda)\, dB'(\lambda)\Bigg),
\end{aligned}
$$

where B and B' are independent Brownian motions on $[0, \infty)$. Analogous to (5.255),

$$E\left(\widetilde{Z}(s) - \widetilde{Z}(t)\right)^2 = 4 \int_0^\infty \sin^2(\pi\lambda(s-t))h(\lambda)\, d\lambda. \qquad (5.258)$$

We stress (5.254) and (5.257) because Gaussian processes associated with Lévy processes that have local times can be expressed in these ways. Although it is not common to do so, we shall also refer to ν as the spectral distribution and h as the spectral density of $\widetilde{Z}$, the process with stationary increments.

Consider the characterization of the characteristic function of a stationary Gaussian process given in (5.247) and assume that μ is supported on Z^d with $\mu(\{k\}) = a_k^2$, for $k \in Z^d$. In this case its characteristic function is

$$\chi(u) = \sum_{k \in Z^d} a_k^2 \cos 2\pi(u, k) \qquad u \in [0,1]^d \qquad (5.259)$$

(since $\chi(u)$ is periodic it suffices to consider it on $[0,1]^d$). We can represent a real-valued stationary Gaussian process $\{\widetilde{W}_t, t \in [0,1]^d\}$ with covariance $\chi(u)$ by the random Fourier series

$$\widetilde{W}(t) = \sum_{k \in Z^d} a_k \left(g_k \cos 2\pi(t, k) + g_k' \sin 2\pi(t, k)\right) \qquad t \in [0,1]^d, \quad (5.260)$$

where $\{g_k\}_{k \in Z^d}$ and $\{g_k'\}_{k \in Z^d}$ are independent standard normal sequences. Series like this for $d = 1$ and 2 are associated Gaussian processes for Lévy processes on the torus. In this context we are actually considering the processes on R^d/Z^d rather than on $[0,1]^d$.

The next lemma is a technical result due to Fernique, which shows that, given a Gaussian process with stationary increments, we can find a Gaussian random Fourier series that is close to it in a very important way. This may appear as a very specialized observation, but it greatly simplifies the proof of a fundamental result of Fernique, Theorem 6.2.2 in this book, that the finiteness of Dudley's metric entropy integral is a necessary condition for the continuity and boundedness of a Gaussian process with stationary increments.

Lemma 5.6.1 (Fernique) *Let $Z = \{Z(t), t \in [0,1]^d\}$ be a Gaussian process with stationary increments and covariance given by (5.249) where*

$$\psi(t-s) := EZ^2(t-s) = 2 \int_{R^d} (1 - \cos(2\pi(\lambda, t-s)))\, \nu(d\lambda) \qquad (5.261)$$

for a positive symmetric measure ν satisfying (5.251). We can find a Gaussian random Fourier series $W = \{W(t), t \in [0,1]^d\}$ such that

$$\left| E \sup_{t \in [0,1]^d} Z(t) - E \sup_{t \in [0,1]^d} W(t) \right| \leq C_d \left(\int_{R^d} (1 \wedge |\lambda|^2) \, \nu(d\lambda) \right)^{1/2},$$

$$(5.262)$$

where C_d is a constant depending only on the dimension d and ν is the measure in (5.250).

Proof The covariance of Z is given in terms of ψ, which is an integral with respect to $\nu(d\lambda)$, $\lambda \in R^d$, where $\lambda = (\lambda_1 \ldots, \lambda_d)$. We partition R^d as follows: Let $\|\lambda\| = \sup_{1 \leq k \leq d} |\lambda_k|$. For all $n \in Z^d$, where $n = (n_1, \ldots, n_d)$, we set

$$A_n = \{\lambda \in R^d : (2n_k - 1) \leq \lambda_k < (2n_k + 1), \forall \, k = [1, d]\}. \quad (5.263)$$

Note that $\|\lambda\| \leq 1$ on A_0, where $0 = (0, \ldots, 0)$, and $\|\lambda\| \geq 1$ on A_n, $n \neq 0$.

For $t \in R^d$ we define

$$W(t) = \frac{1}{2} \sum_{n \in Z^d / \{0\}} \nu^{1/2}(A_n) \left(g_n \cos 2\pi(2n, t) + g'_n \sin 2\pi(2n, t) \right),$$

$$(5.264)$$

where $\{g_n\}_{n \in Z^d}$ and $\{g'_n\}_{n \in Z^d}$ are independent standard normal sequences independent of the Gaussian process Z. We define another Gaussian process $U = \{U(t), t \in R^d\}$, which is independent of both Z and W by

$$U(t) = \quad (5.265)$$

$$\left(\int_{R^d} (1 \wedge |\lambda|^2) \, \nu(d\lambda) \right)^{1/2} \left(2\pi \, d^{1/2} \sum_{k=1}^{d} g''_k t_k + (2\pi)^{1/2} \sum_{k=1}^{d} B_k(t_k) \right),$$

where $t = (t_1, \ldots, t_d)$; $g'' = (g''_1, \ldots, g''_d)$ is a standard normal sequence and $B_1, \ldots, B_d$ are independent Brownian motions independent of g''. Using the trivial inequality $\sup_{t \in [0,1]^d} \sum_{k=1}^{d} f_k(t_k) \leq \sum_{k=1}^{d} \sup_{s \in [0,1]} f_k(s)$, we see that

$$E \sup_{t \in [0,1]^d} U(t) = C_d \left(\int_{R^d} (1 \wedge |\lambda|^2) \, \nu(d\lambda) \right)^{1/2}. \quad (5.266)$$

Using the inequalities $1 - \cos \theta \leq \theta^2 / 2$ and $|\cos \alpha - \cos \beta| \leq |\alpha - \beta|$, we see that

$$\lambda \in A_0 \implies (1 - \cos 2\pi(\lambda, s - t)) \leq 2\pi^2 \|\lambda\|^2 \left(\sum_{k=1}^{d} |s_k - t_k| \right)^2$$

and

$$\lambda \in A_n, n \neq 0 \Longrightarrow |\cos 2\pi(\lambda, s-t) - \cos 2\pi(2n, s-t)| \leq 2\pi \sum_{k=1}^{d} |s_k - t_k|.$$

Using these two statements and the Schwarz inequality, we see that

$$\left| d_Z^2(s,t) - d_W^2(s,t) \right| \tag{5.267}$$

$$\leq \int_{A_0} (1 - \cos 2\pi(\lambda, s-t))\, \nu(d\lambda)$$

$$+ \sum_{n \neq 0} \left(\int_{A_n} |\cos 2\pi(\lambda, s-t) - \cos 2\pi(2n, s-t)|\, \nu(d\lambda) \right)$$

$$\leq 2\pi^2 \left(\int_{A_0} \|\lambda\|^2 \nu(d\lambda) \right) \left(\sum_{k=1}^{d} |s_k - t_k| \right)^2$$

$$+ 2\pi\nu(R^d \setminus A_0) \sum_{k=1}^{d} |s_k - t_k|$$

$$\leq d_U^2(s,t).$$

It now follows from Lemma 5.5.3 and (5.266) that

$$E \sup_{t \in [0,1]^d} Z(t) \leq E \sup_{t \in [0,1]^d} W(t) + C_d \left(\int_{R^d} (1 \wedge |\lambda|^2)\, \nu(d\lambda) \right)^{1/2} \tag{5.268}$$

and

$$E \sup_{t \in [0,1]^d} W(t) \leq E \sup_{t \in [0,1]^d} Z(t) + C_d \left(\int_{R^d} (1 \wedge |\lambda|^2)\, \nu(d\lambda) \right)^{1/2}, \tag{5.269}$$

which give us (5.262). $\qquad\square$

5.7 Notes and references

In this survey of aspects of the theory of Gaussian processes, we present the material that we need to obtain results about the local times of symmetric Markov processes. Basically this comes down to sample path properties of Gaussian processes such as continuity and limit laws. We have in mind a reader who wants proofs of the fundamental results but who is also eager to move on, to see how these results are used in the study of local times. To this end we have tried to put the most important results in the statements of theorems, corollaries, and occasional remarks, so that they serve as an outline of the basic results. In this way, the reader who wishes to can skip some of the proofs on a first reading. In our attempt to streamline the presentation we have omitted many interesting sidelights that exhibit the depth and beauty of this subject. Indeed, this survey is not a substitute for the profound elaborations of this subject in *Probability in Banach Spaces* by Ledoux and Talagrand (1991) and *Fonctions Aléatoires Gaussiennes Vecteurs Aléatoires Gaussiennes* by Fernique (1997).

In this presentation we rely most heavily on the two books mentioned above and the long article "Continuity of subgaussian process" by Jain and Marcus (1978) and *Uniform Central Limit Theorems* by Dudley (1999). Other exposés that we know of that the reader might want to consult are Lifshits (1995) and Adler (1991). Also, the Saint Flour notes by Fernique (1975) are useful, since they present the subject in its early days when the proofs were less elegant but perhaps easier on a first reading. The same is true of Jain and Marcus (1978) and some of the expository material in *Random Fourier Series with Applications to Harmonic Analysis* by Marcus and Pisier (1981), which considers stationary Gaussian processes.

Many people have contributed to the material presented in this chapter. Since this is only a survey, we cannot possibly give all the credit that is deserved. The book by Ledoux and Talagrand (1991) has a detailed and we think accurate history of the development of the theory of Gaussian processes in its chapter notes. Dudley (1999), who, like us, surveys some aspects of Gaussian process theory relevant to his presentation, has good historical notes on the material he treats in his notes on Chapter 2.

Details about the moment generating function of Gaussian processes that we present in Section 5.2 are not usually considered in books on Gaussian processes. We use them in the proofs of the isomorphism theorems in Chapter 8 that relate Gaussian processes to local times. The material in Section 5.3 is taken from Jain and Marcus (1978). It depends on the results of Jain and Kallianpur (1972) and Ito and Nisio (1968b). Theorem 5.3.10 is given in Belyaev (1961).

In Marcus and Rosen (1992d) we refer to Theorem 5.4.1 as Borell's inequality; we learned it from Borell (1975). More recently, the attribution of this result has broadened. We quote Ledoux (1998) whom we consider an expert on this topic. "The Gaussian isoperimetric inequality has been established in 1974 independently by Borell (1975) and Sudakov and Tsirelson (1978) on the basis of the isoperimetric inequality on the sphere and a limiting procedure known as Poincaré's lemma. A proof using Gaussian symmetrizations was developed by Ehrhard (1983)." A complete proof of this fundamentally important result is not included in earlier books on Gaussian processes because the known proofs depended on the Brun–Minkowski inequality. Fortunately, Ledoux (1998) has given a relatively simple and straightforward proof of this result based in part on a paper by Bobkov (1996) and his joint paper, Bakry and Ledoux (1996). We present Ledoux's proof here.

The argument about the uniqueness of the median in Theorem 5.4.3 is taken from Section 1.1 in Ledoux and Talagrand (1991). Corollary 5.4.6 is presented as a simple consequence of Theorem 5.4.3, but it predated Theorem 5.4.3. It was originally proved, without the best constant, independently by Landau and Shepp (1971) and Fernique (1970), and as given, independently by Marcus and Shepp (1972) and Fernique (1971). For a simple direct proof of Corollary 5.4.6 based on Fernique (1970) and Marcus and Shepp (1972), see Chapter II, Theorem 4.8 in Jain and Marcus (1978). Corollary 5.4.6 is very useful, but there are several critical points in this book where we must use Theorem 5.4.3.

Lemma 5.5.1 is the famous result of Slepian (1962). It seems to be a very special property of Gaussian processes. The ideas behind it have been studied intensively, and it has been generalized in many directions. We refer the reader to Chapter 3, Section 3 in Ledoux and Talagrand (1991) for details and further references. See also Li and Shao (2002) for a reverse Slepian-type lemma.

Lemma 5.5.3 is due to Fernique (1974). A slightly weaker result with an easier proof is due to Marcus and Shepp (1972) (see Corollary 3.14, Ledoux and Talagrand (1991)). Our treatment of Lemmas 5.5.1 and 5.5.3 is taken from Jain and Marcus (1978). Lemma 5.5.6 is by Sudakov (1973), and Lemma 5.5.7 is by Talagrand (1992).

It is not necessary for F to be differentiable to get an expression like (5.254). When it is not we can write

$$\widetilde{Y}(t) = \int_0^\infty (\cos 2\pi\lambda t)\, dB(F(\lambda)) + \int_0^\infty (\sin 2\pi\lambda t)\, dB'(F(\lambda)) \quad (5.270)$$

and similarly in (5.257). We stress (5.254) and (5.257) because Gaussian processes associated with Lévy processes killed at different stopping times are of this form. Lemma 5.6.1 is taken from 3.2.9 Lemme in Fernique (1997).

6

Continuity and boundedness of Gaussian processes

According to legend, Kolmogorov proposed the problem of finding necessary and sufficient conditions for the continuity of the sample paths of a Gaussian process, in terms of its covariance function. This problem is well posed since a mean zero Gaussian process is determined by its covariance. One can tackle this question using Kolmogorov's continuity theorem, which is given in Section 14.1, and get results that in some cases are pretty sharp. In fact, refinements of Kolmogorov's approach actually give necessary and sufficient conditions for continuity when the covariance function is sufficiently regular. Nevertheless, to fully describe the criteria for the continuity and boundedness of Gaussian processes, we must abandon this approach totally. Here is the main point. Suppose we have a Gaussian process $X = \{X(t), t \in T\}$, where T has a natural topology. For example, suppose $T = [0, 1]^d$, where, of course, the natural topology is the one induced by Euclidean distance. In the classical approach to studying continuity, one considers $X(s) - X(t)$ when $|s - t|$ is small. However, the Gaussian random variable $X(s) - X(t)$ is small when

$$d_X(s, t) := (E(X(s) - X(t))^2)^{1/2} \qquad (6.1)$$

is small. Thus, one should prove the continuity of X on the metric or pseudometric space (T, d_X), not on $(T, | \cdot |)$. When X is continuous on (T, d_X) and d_X is continuous on $(T, | \cdot |)$, X is also continuous on $(T, | \cdot |)$.

Note that on a pseudometric space $d_X(s, t) = 0$ does not necessarily imply that $s = t$. The continuity results we prove are with respect to pseudometric spaces.

Many people have worked on the problem of finding necessary and sufficient conditions for the continuity of Gaussian processes. Nevertheless, it seems appropriate to mention that the major results on this problem were obtained by R. M. Dudley, X. Fernique, and M. Talagrand.

6.1 Sufficient conditions in terms of metric entropy

We obtain a sufficient condition for the continuity and boundedness of a Gaussian process $X = \{X(t), t \in T\}$ on the metric or pseudometric space (T, d_X), where $d_X(s, t) = (E(X(s) - X(t))^2)^{1/2}$. These conditions are given in terms of the metric entropy of T with respect to d_X.

Let (T, d) be a separable metric or pseudometric space with the topology induced by d. We use $B_d(t, u)$ to denote a closed ball of radius u in T centered at $t \in T$. The diameter of T is the radius of the smallest ball with center in T that contains T. Often, when it is clear which metric is being used, we abbreviate $B_d(t, u)$ by $B(t, u)$.

The following functions help describe the structure of T:

(1) $N(u) = N(T, d, u)$ is the minimum number of closed balls of radius u in the metric d with centers in T that cover T. The family $\{N(u) : u > 0\}$ is called the metric entropy of T.

(2) $D(u) = D(T, d, u)$ is the maximum number of closed disjoint balls of radius u in the metric d with centers in T. The family $\{D(u) : u > 0\}$ is called the packing numbers of T.

(3) $M(u) = M(T, d, u)$ is the maximum number of points of T such that each distinct pair of these points is u-distinguishable.

Lemma 6.1.1 *For each $u > 0$,*

$$N(2u) \leq D(u) \leq M(u) \leq N(u/2). \tag{6.2}$$

Proof   We consider the three inequalities in Lemma 6.1.1 in order.

(1) Let $v_1, \ldots, v_{D(u)}$ be the centers of the balls in $D(u)$. Then $T \subset \cup_{k=1}^{D(u)} B_d(v_k, 2u)$. To prove this, suppose the contrary, that there exists a $w \in T$ that is not in $\cup_{k=1}^{D(u)} B_d(v_k, 2u)$. Then $B_d(w, u)$ is disjoint from all the balls in $D(u)$, thus contradicting the definition of $D(u)$. This follows because, for any $k \in [1, D(u)]$, $d(w, v_k) > 2u$, whereas for any $s \in B_d(w, u)$, $d(s, w) \leq u$. Thus $d(s, v_k) \geq d(w, v_k) - d(s, w) > u$. This shows that the $D(u)$ balls $\cup_{k=1}^{D(u)} B_d(v_k, 2u)$ cover T and therefore $D(u) \geq N(2u)$, since $N(2u)$ is the cardinality of the smallest such covering.

(2) This is much simpler since the centers of the balls in $D(u)$ are u-distinguishable.

(3) Let $v_1, \ldots, v_{M(u)}$ be the u-distinguishable points in $M(u)$. Let $r_1, \ldots, r_{N(u/2)}$ be the centers of the balls in $N(u/2)$. Each ball $B_d(r_k, u/2)$

can contain only one of the points $v_1, \ldots, v_{M(u)}$. We see from this that $N(u/2) \geq M(u)$. $\square$

We now come to the famous theorem of R. M. Dudley, the first major result in the modern theory of Gaussian processes.

Theorem 6.1.2 (Dudley) *Let $X = \{X(t), t \in T\}$ be a mean zero Gaussian process and assume that*

$$\int_0^D (\log N(T, d_X, u))^{1/2} \, du < \infty, \tag{6.3}$$

where D denotes the diameter of T with respect to d_X. Then there exists a version X' of X with bounded uniformly continuous sample paths on (T, d_X) such that

$$E \sup_{t \in T} X'(t) \leq 16\sqrt{2} \int_0^{D/2} (\log N(T, d_X, u))^{1/2} \, du \tag{6.4}$$

and

$$E \left(\sup_{\substack{s, t \in T \\ d_X(s,t) \leq \epsilon}} X'(t) - X'(s) \right) \leq 99 \int_0^\epsilon (\log N(T, d_X, u))^{1/2} \, du. \tag{6.5}$$

Proof If T contains only one point, then both sides of (6.4) are equal to zero. Thus we may assume that T contains at least two points. In this case, $N(T, d_X, u) \geq 2$ for $u \in (0, D)$.

Let $S \subset T$ be finite and consider $\{X(s), s \in S\}$, along with the pseudometric space (S, d_X). For $s \in S$, let

$$F_n(s) = \sup_{t \in B(s, D2^{-n}) \cap S} X(t) - X(s). \tag{6.6}$$

Clearly, $\sup_{t \in B(s, D2^{-n}) \cap S} (E(X(t) - X(s))^2)^{1/2} \leq D2^{-n}$. Let $N_{n+1} = N(B(s, D2^{-n}), d_X, D2^{-n-1})$ and let S_k, $k = 1, \ldots, N_{n+1}$, be a covering of $B(s, D2^{-n})$ by balls of radius $D2^{-n-1}$. Let s_k denote the center of S_k. Set

$$F_{n+1}(s, s_k) = \sup_{t \in B(s_k, D2^{-n-1}) \cap S} X(t) - X(s). \tag{6.7}$$

Clearly

$$E F_n(s) \leq E \sup_{k \leq N_{n+1}} F_{n+1}(s, s_k). \tag{6.8}$$

Note that

$$E F_n(s) = E \sup_{t \in B(s, D2^{-n}) \cap S} X(t) \tag{6.9}$$

and similarly for $F_{n+1}(s, s_k)$. Also, note that

$$\sup_{t \in B(s_k, D2^{-n-1}) \cap S} (E(X(t) - X(s))^2)^{1/2} \leq 2D2^{-n}. \tag{6.10}$$

By (6.10) and (5.151) we see that

$$P\left(F_{n+1}(s, s_k) - \text{med } F_{n+1}(s, s_k) \geq 4D2^{-n-1}u\right) \leq \frac{1}{\sqrt{2\pi}} \int_u^\infty e^{-s^2/2} \, ds, \tag{6.11}$$

where med indicates the median. Consequently,

$$P\left(\sup_{k \leq N_{n+1}} (F_{n+1}(s, s_k) - \text{med } F_{n+1}(s, s_k)) \geq 4D2^{-n-1}u\right) \tag{6.12}$$

$$\leq 1 \wedge \left(N_{n+1} \frac{1}{\sqrt{2\pi}} \int_u^\infty e^{-s^2/2} \, ds\right).$$

Integrating as in the proof of Corollary 5.4.5, we get

$$E \sup_{k \leq N_{n+1}} (F_{n+1}(s, s_k) - \text{med } F_{n+1}(s, s_k)) \tag{6.13}$$

$$\leq 4D2^{-n-1} \left(\int_0^\infty \left(1 \wedge N_{n+1} \frac{1}{\sqrt{2\pi}} \int_u^\infty e^{-s^2/2} \, ds\right) du\right)$$

$$\leq 4D2^{-n-1} \left(\int_0^{(2 \log N_{n+1})^{1/2}} 1 \, du \right.$$

$$\left. + \int_{(2 \log N_{n+1})^{1/2}}^\infty \left(N_{n+1} \frac{1}{\sqrt{2\pi}} \int_u^\infty e^{-s^2/2} \, ds\right) du\right)$$

$$\leq 4D2^{-n-1} \left((2 \log N_{n+1})^{1/2} + N_{n+1} \frac{1}{\sqrt{2\pi}} \int_{(2 \log N_{n+1})^{1/2}}^\infty s e^{-s^2/2} \, ds\right)$$

$$\leq 4D2^{-n-1}((2 \log N_{n+1})^{1/2} + (2\pi)^{-1/2}).$$

By Corollary 5.4.5, med $F_{n+1}(s, s_k) \leq EF_{n+1}(s, s_k) + 4D2^{-n-1}/\sqrt{2\pi}$. Using this and (6.8), we see that

$$EF_n(s) \leq E \sup_{k \leq N_{n+1}} F_{n+1}(s, s_k) \tag{6.14}$$

$$\leq \sup_{k \leq N_{n+1}} EF_{n+1}(s, s_k) + 4D2^{-n-1} \left((2 \log N_{n+1})^{1/2} + \sqrt{\frac{2}{\pi}}\right).$$

Finally, using (6.9) and (6.14) we get

$$E \sup_{t \in B(s, D2^{-n}) \cap S} X(t) \tag{6.15}$$

$$\leq \sup_{k \leq N_{n+1}} E \sup_{t \in B(s_k, D2^{-n-1}) \cap S} X(t) + 8D2^{-n-1}(2 \log N_{n+1})^{1/2}.$$

Let $K(S, d_X, u) = \sup_{s \in S} N(B(s, 2u), d_X, u)$ and set

$$E_n = \sup_{s \in S} E \sup_{t \in B(s, D2^{-n}) \cap S} X(t). \tag{6.16}$$

It follows from (6.15) that

$$E_n \leq E_{n+1} + 8D2^{-n-1}(2 \log K(S, d_X, D2^{-n-1}))^{1/2} \qquad \forall n \geq 0. \tag{6.17}$$

Since S is finite, E_n is equal to zero for n sufficiently large. Thus we can sum the terms in (6.17) to obtain

$$E \sup_{t \in S} X(t) = E_0 \leq 8 \sum_{n=0}^{\infty} D2^{-n-1}(2 \log K(S, d_X, D2^{-n-1}))^{1/2}. \tag{6.18}$$

Using the fact that $K(S, d_X, D2^{-n-1}) \leq N(S, d_X, D2^{-n-1})$, we can dominate the sum in (6.18) by an integral and obtain (6.4) for X, under the assumption that S is finite. Using the Monotone Convergence Theorem we see that (6.4) continues to hold for X when S is a countable dense subset, say S^*, of T. Therefore, since X is separable, we can find a version X' of X for which (6.4) holds as stated.

For all $s \in S^*$ and $\epsilon > 0$, we can apply (6.4) to the process $\{X'(t), t \in B(s, \epsilon)\}$. Using the fact that $N(B(s, \epsilon), d_X, u) \leq N(T, d_X, u)$, we see that

$$E \sup_{\substack{t \in S^* \\ d_X(s,t) \leq \epsilon}} X'(t) - X'(s) \leq 16\sqrt{2} \int_0^{\epsilon} (\log N(T, d_X, u))^{1/2} \, du$$

$$\leq 32\sqrt{2} \int_0^{\epsilon/2} (\log N(T, d_X, u))^{1/2} \, du. \tag{6.19}$$

It follows from (6.19) and Chebyshev's inequality that X' is continuous at s almost surely.

We now obtain (6.5). Assume, to begin with, that $S \subset S^*$ is finite. Set $H_\epsilon = \{(t, t') : t, t' \in S, d_X(t, t') \leq \epsilon\}$. Let S_k, $k = 1, \ldots, N(S, d_X, \eta)$, be a covering of S by balls of radius η and let s_k denote the center of S_k. Note that

$$G_k := \{(u, v), u \in B(s_k, \eta), v \in B(s_k, \eta + \epsilon)\} \quad k = 1, \ldots, N(S, d_X, \eta) \tag{6.20}$$

is a covering of H_ϵ. Therefore

$$E \left(\sup_{\substack{t, t' \in S \\ d_X(t,t') \leq \epsilon}} X'(t) - X'(t') \right) \tag{6.21}$$

$$\leq E \left(\sup_{k \leq N(S, d_X, \eta)} \sup_{\substack{t, t' \in S, d_X(s_k, t) \leq \eta \\ d_X(s_k, t') \leq \epsilon + \eta}} X'(t) - X'(t') \right).$$

Following the argument used in (6.13)–(6.15), we see that the right-hand side of (6.21) is less than or equal to

$$\sup_{k \leq N(\eta)} E \Big(\sup_{\substack{t,t' \in S, d_X(s_k,t) \leq \eta \\ d_X(s_k,t') \leq \epsilon+\eta}} X'(t) - X'(t') \Big) + 2(2\eta + \epsilon)(2 \log N(S, d_X, \eta))^{1/2}.$$

$$(6.22)$$

Furthermore, the expectation in (6.22) is less than or equal to

$$E \Big(\sup_{\substack{t \in S \\ d_X(s_k,t) \leq \eta}} X'(t) - X'(s_k) \Big) + E \Big(\sup_{\substack{t' \in S \\ d_X(s_k,t') \leq \epsilon+\eta}} X'(t') - X'(s_k) \Big). \quad (6.23)$$

Using (6.23) in (6.21) and (6.22) and taking $\epsilon = \eta$, we get

$$E \Big(\sup_{\substack{t,t' \in S \\ d_X(t,t') \leq \epsilon}} X'(t) - X'(t') \Big) \ \leq \ 2 \sup_{s \in S} E \Big(\sup_{\substack{t \in S \\ d_X(s,t) \leq 2\epsilon}} X'(t) - X'(s) \Big)$$

$$+ 6\epsilon(2 \log N(S, d_X, \epsilon))^{1/2}. \quad (6.24)$$

The inequality in (6.5) now follows from (6.19) and (6.24) when S is a finite set. As above, it also holds on S^*. Thus we get (6.5) as stated.□

The condition in (6.3) is also necessary for boundedness and continuity when the Gaussian process has stationary increments. We take this up in Section 6.2. Here we give an example that shows that (6.3) is not necessary in general.

Example 6.1.3 Let $T = \{\{t_k\}_{k=4}^\infty \cup 0\}$, where $t_4 = 1$ and $t_k \downarrow 0$. Let $X(t_k) := \xi_k/(2 \log k \log \log k)^{1/2}$, where $\{\xi_k\}_{k=4}^\infty$ is a standard normal sequence and $X(0) = 0$. It follows easily from the Borel–Cantelli Lemma that $\{X(t), t \in T\}$ is continuous on (T, d_X) (we only need to check this at 0). Note that

$$d_X(t_j, t_k) = \Big(\frac{1}{2 \log j \log \log j} + \frac{1}{2 \log k \log \log k} \Big)^{1/2}. \quad (6.25)$$

Let $u_k = 1/(\log k \log \log k)^{1/2}$. It follows from (6.25) that $B_{d_X}(t_k, u_k)$ contains the points $\{t_j\}_{j \geq k}$ and 0. Therefore, $N(T, d_X, u_k) = k - 3$. This implies that the metric entropy integral in (6.3) is infinite since it is essentially equivalent to

$$\sum_k (u_k - u_{k+1})(\log N(T, d_X, u_{k+1}))^{1/2} \approx \sum_k \frac{(\log k)^{1/2}}{k(\log k)^{3/2}(\log \log k)^{1/2}}.$$

$$(6.26)$$

To make this an example in which X has continuous sample paths on $[0, 1]$, simply define $X(t) = \sum_{k=4}^\infty \xi_k \phi_k(t)$, where $\phi_k(t)$ is a positive continuous function supported on $[(t_{k+1} + t_k)/2, (t_k + t_{k-1})/2]$, which is zero at the endpoints of this interval and takes as its largest value $1/(2 \log k \log \log k)^{1/2}$, at t_k.

Another drawback of the metric entropy approach is that it does not distinguish between bounded and continuous processes, since when (6.3) holds we have

$$\lim_{\epsilon \to 0} \int_0^\epsilon (\log N(T, d_X, u))^{1/2} \, du = 0, \tag{6.27}$$

so we also get (6.5). To appreciate the relevance of this comment see Theorem 6.3.1 and Example 6.3.7 .

Here is an example that shows that, for certain metrics, Dudley's metric entropy integral is not large. It is used in the proof of Theorem 6.2.2.

Example 6.1.4 Consider the metric space $([0,1]^k, d)$, where $d(s,t) \leq K|s-t|^\alpha$, $s, t \in [0,1]^k$, K is a constant, and $0 < \alpha \leq 1$. Note that $d(s,t) \leq \epsilon$ when $|s-t| \leq (\epsilon/K)^{1/\alpha}$, so that

$$B_e(s, (\epsilon/K)^{1/\alpha}) \subseteq B_d(s, \epsilon), \tag{6.28}$$

where $B_e(x, r)$ is the ball of radius r in the Euclidean metric centered at x. A closed ball in R^k of radius u in the Euclidean metric covers a cube with sides $2u/\sqrt{k}$. Therefore we can cover $[0,1]^k$ with $(\sqrt{k}/2u)^k$ balls of radius u in the Euclidean metric. Consequently, taking $u = (\epsilon/K)^{1/\alpha}$ and using (6.28), we see that

$$N_d([0,1]^k, \epsilon) \leq \left(\frac{k^{\alpha/2} K}{2^\alpha \epsilon} \right)^{k/\alpha} := \left(\frac{K'}{\epsilon} \right)^{k/\alpha}. \tag{6.29}$$

Note that (6.29) implies that the diameter of $[0,1]^k$ in the metric d is bounded by K'. We have

$$\int_0^{K'} (\log N_d([0,1]^k, \epsilon))^{1/2} \, d\epsilon \leq \left(\frac{k}{\alpha} \right)^{1/2} \int_0^{K'} \left(\log \frac{K'}{\epsilon} \right)^{1/2} d\epsilon$$

$$= \frac{k^{(\alpha+1)/2}}{\alpha^{1/2} 2^\alpha} K \int_0^1 \left(\log \frac{1}{u} \right)^{1/2} du.$$

Here is an important example that satisfies the criteria of the previous paragraph. Let $Y = \{Y(t), t \in [0,1]^k\}$ be a Gaussian process with stationary increments and covariance given by (5.249) for ψ as given in

(5.250). Assume that ν has support in $[-T,T]^k$. For $s,t \in [-T,T]^k$, we have

$$
\begin{aligned}
\left(E(Y(s) - Y(t))^2\right)^{1/2} &= \left(\int_{[-T,T]^k} (1 - \cos 2\pi(\lambda, s - t))\, \nu(d\lambda)\right)^{1/2} \\
&\leq \left(C \int_{[-T,T]^k} |\lambda|^2 \nu(d\lambda)\right)^{1/2} |s - t| \quad (6.30)
\end{aligned}
$$

for some constant C independent of k.

6.2 Necessary conditions in terms of metric entropy

The principle result in this section is that (6.3) is also necessary for the continuity and boundedness of Gaussian processes with stationary increments. Before we get to this we give two necessary conditions that are valid for all Gaussian processes. Note that (6.3) and the monotonicity of $u \to (\log N(T, d_X, u))^{1/2}$ imply that

$$
\lim_{u \to 0} u \, (\log N(T, d_X, u))^{1/2} = 0 \qquad (6.31)
$$

but, obviously, is not implied by it. We show in the next theorem that (6.31) is a necessary condition for the continuity of the Gaussian process X.

Theorem 6.2.1 *Let $X = \{X_t, t \in T\}$ be a mean zero Gaussian process with bounded sample paths. Then $E \sup_{t \in T} X(t) < \infty$, and for all $u > 0$,*

$$
E \sup_{t \in T} X(t) \geq \frac{1}{17} u (\log M(T, d_X, u))^{1/2}. \qquad (6.32)
$$

If X is also uniformly continuous on T, then

$$
\lim_{u \to 0} u \, (\log M(T, d_X, u))^{1/2} = 0. \qquad (6.33)
$$

Note that by Lemma 6.1.1, (6.33) and (6.31) are equivalent.

Proof The fact that $E \sup_{t \in T} |X(t)| < \infty$ is proved in the paragraph containing (5.191). The inequality in (6.32) is simply a restatement of (5.237). Now suppose that X is uniformly continuous on T. Since $EX(t) = 0$, we have

$$
\begin{aligned}
E \sup_{s \in B(t,\epsilon)} X(s) &= E\left(\sup_{s \in B(t,\epsilon)} (X(s) - X(t)) + X(t)\right) \quad (6.34) \\
&\leq E \sup_{s \in B(t,\epsilon)} |X(s) - X(t)|
\end{aligned}
$$

$$\le \quad E \sup_{\substack{d(s,t)\le\epsilon \\ s,t\in T}} |X(s) - X(t)|.$$

It follows from the Dominated Convergence Theorem that this last term goes to zero as ϵ goes to zero. Consequently, for any $h > 0$ we can find an $\epsilon > 0$ such that, for all $t \in T$, $E\sup_{s\in B(t,\epsilon)} X(s) \le h/34$. It follows from (6.32) that for all $t \in T$ and $u > 0$,

$$\frac{h}{2} \ge u(\log M(B(t,\epsilon), d_X, u))^{1/2}. \tag{6.35}$$

For the ϵ for which (6.35) holds, let $\{B(s_k,\epsilon)\}_{k=1}^{N(T,\epsilon)}$ be a cover of T. Clearly, for all $u > 0$,

$$M(T,u) \le \sum_{k=1}^{N(T,\epsilon)} M(B(s_k,\epsilon), u) \le N(T,\epsilon) \sup_{t\in T} M(B(t,\epsilon), u). \tag{6.36}$$

Therefore, for all $u > 0$,

$$u(\log M(T,u))^{1/2} \le u(\log N(T,\epsilon))^{1/2} + u\sup_{t\in T}(\log M(B(t,\epsilon), u))^{1/2}. \tag{6.37}$$

By (6.35), the last term in (6.37) is less than or equal to $h/2$ for all $u > 0$. Thus, for $u \le h/(2(\log N(T,\epsilon))^{1/2})$ (we assume that $\log N(T,\epsilon) > 0$), the left-hand side of (6.37) is less than or equal to h. Since this holds for all h, we get (6.33). $\qquad\square$

A second milestone in the theory of Gaussian processes is Fernique's Theorem, which shows that Dudley's metric entropy condition is necessary for Gaussian processes with stationary increments.

Theorem 6.2.2 (Fernique) *Let $Z = \{Z(t), t \in [0,1]^d\}$ be a Gaussian process with stationary increments and covariance given by (5.249) for ψ as given in (5.250). Then*

$$\int_0^D \left(\log N([0,1]^d, d_Z, u)\right)^{1/2} du \;\le\; (10\cdot 4^5)E \sup_{t\in[0,1]^d} Z(t) \tag{6.38}$$

$$+ C_d \left(\int_{R^d} (1\wedge |\lambda|^2)\,\nu(d\lambda)\right)^{1/2},$$

where C_d is a constant depending only on the dimension d and ν is the measure in (5.250).

Proof To begin, let X be a random Fourier series on $[0,1]^d$, that is, a process of the form of (5.260). In this case, $d_X(s,t)$ is a function of $|s - t|$. Given $s, t \in [0,1]^d$, $s - t$ may or may not be in $[0,1]^d$. To make sure it is, we define addition modulo $[0,1]^d$. That is, we consider $[0,1]^d$ as a compact Abelian group. Therefore, since $d_X(s,t)$ is translation invariant, the number

$$L_n = N(B(t, D4^{-n}), d_X, D4^{-n-1}) \qquad (6.39)$$

is independent of t.

Let $T = [0,1]^d$. Let $q = 1/4$ and for $n \geq 0$ set $u_n = q^{n+3}D$. For each $s \in T$ consider the subset $T_n(s) = B(s, q^{n-1}D)$ and a set $\mathcal{S}_n$ of $4u_n$-distinguishable points in $B(s, q^n D)$. Note that if $s' \in \mathcal{S}_n$, then $B(s', u_n) \subset T_n(s)$. It now follows from Lemma 5.5.7 and the fact that $\mathcal{S}_n \subset T$, that

$$E \sup_{t \in T_n(s)} X(t) \qquad (6.40)$$

$$\geq \inf_{s' \in \mathcal{S}_n} E \left(\sup_{t \in B(s', u_n)} X(t) \right) + \tfrac{1}{12} u_n (\log M(B(s, q^n D), d_X, 4u_n)))^{1/2}.$$

(In Lemma 5.5.7, the index set of the stochastic process is countable, but it is easy to see that (5.241) holds in this case as well.)

Since X is stationary, $E(\sup_{t \in B(s', u_n)} X(t))$ is independent of s'. Thus (6.40) can be written more simply as

$$E \sup_{t \in B(0, q^{n-1} D)} X(t) \qquad (6.41)$$

$$\geq E \left(\sup_{t \in B(0, q^{n+3} D)} X(t) \right) + \tfrac{1}{12} u_n (\log M(B(0, q^n D), d_X, 4u_n)))^{1/2}.$$

Note that by Lemma 6.1.1, $M(B(0, q^n D), d_X, 4u_n) \geq L_n$. Define $E_n = E(\sup_{t \in B(0, q^{n-1} D)} X(t))$. We can write (6.41) as

$$E_n \geq E_{n+4} + \tfrac{1}{12} q^{n+3} D (\log L_n)^{1/2} \qquad \forall n \geq 0. \qquad (6.42)$$

Iterating this over $n = 1, \ldots, K$ and recognizing that $T_0 = T$, we see that

$$4E \sup_{t \in T} X(t) \geq \sum_{n=0}^{4} E_n \geq \frac{1}{12 \cdot 4^3} \sum_{n=0}^{K} q^n D (\log L_n)^{1/2} \qquad (6.43)$$

for all K.

We next note that $L_n N(T, d_X, Dq^n) \geq N(T, d_X, Dq^{n+1})$. This is easy to see since a covering of T with balls of radius Dq^{n+1} can be achieved by covering each of the $N(T, d_X, Dq^n)$ balls in the covering of T by balls of radius Dq^n with balls of radius Dq^{n+1}. By definition this can be done with L_n balls. Also note that

$$\forall y \geq x \geq 1 \qquad \left(\log \frac{y}{x}\right)^{1/2} \geq (\log y)^{1/2} - (\log x)^{1/2}. \qquad (6.44)$$

Therefore,

$$\sum_{n=0}^{\infty} q^n D(\log L_n)^{1/2} \qquad (6.45)$$

$$\geq \sum_{n=0}^{\infty} q^n D\left((\log N(T, d_X, Dq^{n+1}))^{1/2} - (\log N(T, d_X, Dq^n))^{1/2}\right)$$

$$= \sum_{n=0}^{\infty} \left(q^n D - q^{n+1}D\right)(\log N(T, d_X, Dq^{n+1}))^{1/2}$$

$$\geq \sum_{n=0}^{\infty} \int_{q^{n+1}D}^{q^n D} (\log N(T, d_X, u))^{1/2}\, du = \int_0^D (\log N(T, d_X, u))^{1/2}\, du.$$

Thus we get

$$(3 \cdot 4^5) E \sup_{t \in [0,1]^d} X(t) \geq \int_0^D (\log N([0,1]^d, d_X, u))^{1/2}\, du. \qquad (6.46)$$

Let $Z = \{Z(t), t \in [0,1]^d\}$ be a Gaussian process with stationary increments. We know that the covariance of Z can be expressed as in (5.249) for ψ as given in (5.250). Let $Z_N = \{Z_N(t), t \in [0,1]^d\}$ be the Gaussian process with stationary increments but with ν in (5.250) replaced by ν_N, where $\nu_N(A) = \nu(A \cap [-N, N]^d)$, for all measurable sets $A \subset R^d$. Let $d_{Z_N}(s,t) = (E(Z_N(s) - Z_N(t))^2)^{1/2}$. It follows from (6.30) in Example 6.1.4 and Theorem 6.1.2 that $\int_0^D (\log N(T, d_{Z_N}, u))^{1/2}\, du < \infty$ and $E \sup_{t \in T} Z_N(t) < \infty$.

It follows from Lemma 5.6.1 and (6.46) that there exists a random Fourier series $\{W(t), t \in [0,1]^d\}$ such that

$$E \sup_{t \in [0,1]^d} Z_N(t) + C_d \left(\int_{R^d} (1 \wedge |\lambda|^2)\, \nu(d\lambda)\right)^{1/2} \qquad (6.47)$$

$$\geq E \sup_{t \in [0,1]^d} W(t) \geq \frac{1}{3 \cdot 4^5} \int_0^D \left(\log N([0,1]^d, d_W, u)\right)^{1/2}\, du.$$

Furthermore, by (5.267),

$$d_{Z_N}^2(s,t) \leq d_W^2(s,t) + d_U^2(s,t), \qquad (6.48)$$

where U is given in (5.265) with ν replaced by ν_N.

It follows from (6.48) that

$$N\left(T, d_{Z_N}, 2\sqrt{2}\delta\right) \leq N(T, d_W, \delta)N(T, d_U, \delta) \qquad \forall \delta > 0. \qquad (6.49)$$

To see why this holds, let w_j, $j = 1, \ldots, N(T, d_W, \delta)$ denote the centers of the $N(T, d_W, \delta)$ balls that cover T and let u_k, $k = 1, \ldots, N(T, d_U, \delta)$ denote the centers of the $N(T, d_U, \delta)$ balls that cover T. Let $s_{j,k}$ denote some point in $B(w_j, \delta) \cap B(u_k, \delta)$. Then, for any $s \in B(w_j, \delta) \cap B(u_k, \delta)$, by (6.48),

$$d_{Z_N}^2(s_{j,k}, s) \leq d_W^2(s_{j,k}, s) + d_U^2(s_{j,k}, s) \qquad (6.50)$$

and $d_W(s_{j,k}, s) \leq d_W(s_{j,k}, w_j) + d_W(w_j, s) \leq 2\delta$, and similarly for $d_U(s_{j,k}, s)$. Thus we get (6.49).

It follows from (6.49) that

$$\int_0^D (\log N(T, d_{Z_N}, u))^{1/2} \, du \qquad (6.51)$$

$$\leq 2\sqrt{2} \left(\int_0^D (\log N(T, d_W, u))^{1/2} \, du + \int_0^D (\log N(T, d_U, u))^{1/2} \, du \right).$$

Also, it is easy to see that

$$d_U(s,t) \leq C_d \left(\int_{R^d} (1 \wedge |\lambda|^2) \, \nu(d\lambda) \right)^{1/2} |t - s|^{1/2} \qquad (6.52)$$

(the constants C_d are not necessarily the same at each stage). Thus, by Example 6.1.4,

$$\int_0^D (\log N(T, d_U, u))^{1/2} \, du \leq C_d \left(\int_{R^d} (1 \wedge |\lambda|^2) \, \nu(d\lambda) \right)^{1/2}. \qquad (6.53)$$

Using (6.47), (6.51), and (6.53), we see that

$$\int_0^D \left(\log N([0,1]^d, d_{Z_N}, u) \right)^{1/2} \, du \ \leq \ (10 \cdot 4^5) E \sup_{t \in [0,1]^d} Z_N(t) \qquad (6.54)$$

$$+ C_d \left(\int_{R^d} (1 \wedge |\lambda|^2) \, \nu(d\lambda) \right)^{1/2}.$$

It is easy to see that $d_{Z_{N+1}}(s,t) \geq d_{Z_N}(s,t)$ for all $s, t \in [0,1]^d$. Consequently, both terms in (6.54) that depend on N are increasing in N (to see this for $E \sup_{t \in [0,1]^d} Z_N(t)$, we use Lemma 5.5.3). Taking the limit we get (6.38). $\qquad \square$

Remark 6.2.3 We single out two special cases of Theorem 6.2.2. If the process Z is a random Fourier series on $[0,1]^d$, we showed in (6.46) that

$$(3 \cdot 4^5)E \sup_{t \in [0,1]^d} Z(t) \geq \int_0^D (\log N([0,1]^d, d_Z, u))^{1/2} \, du. \qquad (6.55)$$

This same inequality is valid for a Gaussian process on any compact Abelian group.

If the process Z is a stationary Gaussian process on R^d, $\{Z(t) - Z(0), t \in R^d\}$ is a Gaussian process with stationary increments and the measure ν in (6.38) is finite. In fact, in this case, $EZ^2(t) = \nu(R^d)$ for all $t \in [0,1]^d$. Since

$$E \sup_{t \in [0,1]^d} (Z(t) - Z(0)) = E \sup_{t \in [0,1]^d} Z(t) \leq E \sup_{t \in [0,1]^d} |Z(t)| \qquad (6.56)$$

and $E \sup_{t \in [0,1]^d} |Z(t)| \geq (3/4)(EZ^2(0))^{1/2}$, we get

$$E \sup_{t \in [0,1]^d} |Z(t)| \geq C_d \int_0^D (\log N([0,1]^d, d_Z, u))^{1/2} \, du \qquad (6.57)$$

for some constant C_d depending only on d. Note that explicit values for the constants C_d can be determined both here and in (6.38).

6.3 Necessary and sufficient conditions in terms of majorizing measures

In Example 6.1.3 we show that the metric entropy condition (6.3) is not a necessary condition for a Gaussian process to be continuous. In this section we give sufficient conditions for the continuity and boundedness of Gaussian processes that are also necessary. We show, in Example 6.3.7, that these conditions give the continuity and boundedness of the process considered in Example 6.1.3.

Given a probability measure μ on a pseudometric space (T, d_X), we are often interested in $\mu(B_{d_X}(t, u))$. Sometimes we suppress the d_X when it is clear what metric we are referring to.

The following is the best result so far on the continuity and boundedness of Gaussian processes.

Theorem 6.3.1 *Let $X = \{X(t), t \in T\}$ be a mean zero Gaussian process, where (T, d_X) is a separable metric or pseudometric space with finite diameter D. Suppose there exists a probability measure μ on T*

such that

$$\sup_{t \in T} \int_0^D \left(\log \frac{1}{\mu(B(t, u))} \right)^{1/2} du < \infty. \qquad (6.58)$$

Then there exists a version $X' = \{X'(t), t \in T\}$ of X such that

$$E \sup_{t \in T} X'(t) \leq 1056 \sup_{t \in T} \int_0^D \left(\log \frac{1}{\mu(B(t, u))} \right)^{1/2} du. \qquad (6.59)$$

If

$$\lim_{\epsilon \to 0} \sup_{t \in T} \int_0^\epsilon \left(\log \frac{1}{\mu(B(t, u))} \right)^{1/2} du = 0, \qquad (6.60)$$

then X' is uniformly continuous on T almost surely and

$$E \sup_{\substack{s, t \in T \\ d_X(s,t) \leq \delta}} |X'(s) - X'(t)| \leq 1056 \sup_{s \in T} \int_0^\delta \left(\log \frac{1}{\mu(B(s, u))} \right)^{1/2} du. \qquad (6.61)$$

It is customary to call μ a majorizing measure because it leads to an upper bound for $E \sup_{t \in T} X(t)$, that is (6.59) holds.

Theorem 6.3.1 is a consequence of Theorem 6.3.3. The next lemma is used in the proof of Theorem 6.3.3, but it is also quite interesting in itself.

Lemma 6.3.2 *Let $\{X(t), t \in T\}$ be a mean zero Gaussian process on the probability space $(\Omega, \mathcal{F}, P)$, with $EX^2(t) \leq 1$, for all $t \in T$. Let μ be a probability measure on T. Then there exists a random variable Z on $(\Omega, \mathcal{F}, P)$, with $EZ < 5/2$, such that, for all measurable functions f on T for which, $0 < \int |f(t)| \, \mu(dt) < \infty$*

$$\int |X(\omega, t) f(t)| \, \mu(dt) \qquad (6.62)$$

$$\leq 3Z(\omega) \int |f(t)| \left(\log \left(1 + \frac{|f(t)|}{\int |f(t)| \, \mu(dt)} \right) \right)^{1/2} \mu(dt).$$

Proof For all $\beta < 1/2$, $\sup_{t \in T} E \exp(\beta X^2(t)) < \infty$. This implies, by Fubini's Theorem, that for all $\beta < 1/2$, the set $\{\omega : \exp(\beta X^2(\omega, \cdot)) \in L^1(T, \mu)\}$ has measure one. Let

$$Z(\omega) = \inf \left\{ \alpha > 0 : \int \left(\exp \left(\frac{X^2(\omega, t)}{\alpha^2} \right) - 1 \right) \mu(dt) \leq 1 \right\}. \qquad (6.63)$$

Since, by the Dominated Convergence Theorem

$$\lim_{\alpha \to \infty} \int \left(\exp \left(\frac{X^2(\omega, t)}{\alpha^2} \right) - 1 \right) \mu(dt) = 0 \qquad \text{a.s.} \qquad (6.64)$$

$Z(\omega)$ is a well-defined random variable.

It follows from Young's inequality, that for all $x, y \geq 0$ and $\alpha, \beta > 0$,

$$xy \leq \alpha\beta \left(\exp \left(\frac{x^2}{\alpha^2} \right) - 1 \right) + \alpha y \left(\log \left(1 + \frac{y}{\beta} \right) \right)^{1/2}. \qquad (6.65)$$

It is simple to verify this directly. Dividing both sides of (6.65) by $\alpha\beta$ we see that to obtain (6.65) it suffices to verify it for $\alpha = \beta = 1$. Next, setting $y = \exp(u^2) - 1$, we see that it suffices to show that, for all $x, u \geq 0$,

$$\delta(x, u) = (x - u)\left(\exp(u^2) - 1\right) - \left(\exp(x^2) - 1\right) \leq 0. \qquad (6.66)$$

This is obvious for $x \leq u$. It is also easy when $x \in [u, u+1]$ since, in this case,

$$(x - u)\left(\exp(u^2) - 1\right) \leq \left(\exp(u^2) - 1\right) \leq \left(\exp(x^2) - 1\right). \qquad (6.67)$$

For $x \geq u + 1$, note that $\dfrac{\partial}{\partial x}\delta(x, u) \leq 0$, which implies that $\delta(x, u) \leq \delta(u + 1, u)$, which we have just shown is less than or equal to zero.

To simplify the notation we suppress ω in the rest of this proof. Apply (6.65) with $x = |X(t)|$, $y = |f(t)|$, and $\beta = \int |f(t)| \mu(dt)$ and then integrate with respect to $\mu(dt)$ and set $\alpha = Z$ to get

$$\int |X(t)f(t)| \mu(dt) \qquad (6.68)$$

$$\leq \alpha \int |f(t)| \mu(dt) \int \left(\exp \left(\frac{X^2(t)}{\alpha^2} \right) - 1 \right) \mu(dt)$$

$$+ \alpha \int |f(t)| \left(\log \left(1 + \frac{|f(t)|}{\int |f(t)| \mu(dt)} \right) \right)^{1/2} \mu(dt).$$

$$= Z \left(\int |f(t)| \mu(dt) \right.$$

$$\left. + \int |f(t)| \left(\log \left(1 + \frac{|f(t)|}{\int |f(t)| \mu(dt)} \right) \right)^{1/2} \mu(dt) \right).$$

Let $F(u) = u \left(\log \left(1 + u / \int |f(t)| \mu(dt) \right) \right)^{1/2}$ and note that

$$F \left(\int |f(t)| \mu(dt) \right) = (\log 2)^{1/2} \int |f(t)| \mu(dt). \qquad (6.69)$$

Also note that the function $u \left(\log \left(1 + u\right)\right)^{1/2}$ or, equivalently, the function $u(\log(1+u/C))^{1/2}$ is convex and increasing for all $C > 0$. Therefore, by Jensen's inequality and (6.69),

$$(\log 2)^{1/2} \int |f(t)| \, \mu(dt) \tag{6.70}$$

$$\leq \int F(|f(t)|) \, \mu(dt)$$

$$= \int |f(t)| \left(\log \left(1 + \frac{|f(t)|}{\int |f(t)| \, \mu(dt)}\right)\right)^{1/2} \mu(dt).$$

Combining (6.70) and (6.68), we get (6.62).

We now get the estimate for EZ. By Chebyshev's inequality followed by Hölder's inequality, for all $u > 0$ and $p \geq 1$,

$$P(Z > u) \quad \leq \quad P\left(\int \exp\left(\frac{X^2(t)}{u^2}\right) \mu(dt) \geq 2\right)$$

$$\leq \quad 2^{-p} E\left(\int \exp\left(\frac{X^2(t)}{u^2}\right) \mu(dt)\right)^p$$

$$\leq \quad 2^{-p} E \int \exp\left(\frac{pX^2(t)}{u^2}\right) \mu(dt).$$

Using Lemma 5.2.1 and the fact that $EX^2(t) \leq 1$, we see that $P(Z > u) \leq 2^{-p}(1 - 2p/u^2)^{-1/2}$. For $u > (2 + 1/\log 2)^{1/2}$, this last term is minimized by $p = u^2/2 - 1/(2 \log 2)$. Thus we see that for all $u > (2 + 1/\log 2)^{1/2}$, $P(Z > u) \leq (e \log 2)^{1/2} u 2^{-u^2/2}$. Consequently $EZ < 5/2$. $\square$

The next, rather wonderful, theorem immediately proves Theorem 6.3.1 and also gives a modulus of continuity for all continuous Gaussian processes $\{X(t), t \in T\}$ on the pseudometric space (T, d_X), as long as (T, d_X) has a finite diameter.

Theorem 6.3.3 *Let $\{X(t), t \in T\}$ be a mean zero Gaussian process with probability space $(\Omega, \mathcal{F}, P)$, where (T, d_X) is a separable metric space with finite diameter D. For all probability measures μ on T there exists a version X' of X and a positive random variable Z' on $(\Omega, \mathcal{F}, P)$, with $EZ' \leq 526$, such that, for all $\omega \in \Omega$ and $s, t \in T$,*

$$X'(\omega, s) - X'(\omega, t) \tag{6.71}$$

$$\leq Z'(\omega) \left(\int_0^{d(s,t)} \left(\left(\log \frac{1}{\mu(B(s,u))}\right)^{1/2} + \left(\log \frac{1}{\mu(B(t,u))}\right)^{1/2}\right) du\right),$$

where $d(s,t) = d_X(s,t)$, and also

$$\left| X'(\omega, t) - \int_T X'(\omega, t)\,\mu(dt) \right| \tag{6.72}$$

$$\leq Z'(\omega) \left(2D + \int_0^D \left(\log \frac{1}{\mu(B(t, u))} \right)^{1/2} du \right).$$

Proof It suffices to prove this theorem for those elements $s, t \in T$ for which the integrals on the right-hand sides of (6.71) and (6.72) are finite.

Without loss of generality we assume that T contains at least two points. Let

$$\mu_k(t) = \mu(B(t, D2^{-k})) \qquad \rho_k(t, \cdot) = \frac{I_{B(t, D2^{-k})}(\cdot)}{\mu_k(t)}$$

and

$$M_k(t) = \int_T \rho_k(t, u) X(u)\,\mu(du).$$

Note that for all $t \in T$, $B(t, D) = T$, so that $M_0(t) = \int_T X(u)\,\mu(du)$ does not depend on t. Note also that

$$E|X(t) - M_k(t)| \leq \frac{1}{\mu_k(t)} \int_{B(t, D2^{-k})} d(u, t)\,\mu(du) \leq D2^{-k}. \tag{6.73}$$

This shows, by the Borel–Cantelli Lemma, that for all $t \in T$, $\lim_{k \to \infty} M_k(t) = X(t)$ almost surely.

We consider the Gaussian process $\{Y(u, v), u, v \in T \times T\}$ defined by

$$Y(u, v) = \begin{cases} \dfrac{X(u) - X(v)}{d(u, v)} & d(u, v) \neq 0 \\ \\ 0 & \text{otherwise.} \end{cases}$$

Note that for all $n \geq 1$,

$$M_n(t) - M_{n-1}(t) \tag{6.74}$$

$$= \int_T \rho_n(t, u) X(u)\,\mu(du) - \int_T \rho_{n-1}(t, v) X(v)\,\mu(dv)$$

$$= \int_{T \times T} \rho_n(t, u) \rho_{n-1}(t, v)(X(u) - X(v))\,\mu(du)\mu(dv).$$

Therefore

$$|M_n(t) - M_{n-1}(t)| \tag{6.75}$$

$$\leq \int_{T \times T} |Y(u, v)| d(u, v) \rho_n(t, u) \rho_{n-1}(t, v)\,\mu(du)\mu(dv).$$

$$\le 3D2^{-n} \int_{T \times T} |Y(u,v)| \rho_n(t,u) \rho_{n-1}(t,v)\, \mu(du)\mu(dv),$$

where, for the last inequality, we use the simple fact that $d(u,v) \le 3D2^{-n}$, since $u \in B(t, D2^{-n})$ and $v \in B(t, D2^{-(n-1)})$.

We now apply Lemma 6.3.2 to the last line of (6.75), where the process is $\{Y(u,v), (u,v) \in T \times T\}$ and the measure is the product measure $\mu \times \mu$. Using the facts that $\int_{T \times T} \rho_n(t,u) \rho_{n-1}(t,v)\, \mu(du)\mu(dv) = 1$ and $\rho_n(t,u)\rho_{n-1}(t,v) \le (\mu_n(t)\mu_{n-1}(t))^{-1}$ we get

$$|M_n(\omega,t) - M_{n-1}(\omega,t)| \tag{6.76}$$

$$\le Z(\omega) 9 D2^{-n} \int (\log (1 + \rho_n(t,u)\rho_{n-1}(t,v)))^{1/2}$$

$$\rho_n(t,u)\rho_{n-1}(t,v)\, \mu(du)\mu(dv)$$

$$\le Z(\omega) 9 D2^{-n} \left(\log \left(1 + \frac{1}{\mu_n(t)\mu_{n-1}(t)} \right) \right)^{1/2},$$

where $EZ < 5/2$. Since $(1 + ab) \le (1 + a)(1 + b)$ for all $a, b \ge 0$, this implies that

$$|M_n(\omega,t) - M_{n-1}(\omega,t)| \le Z(\omega) 9\sqrt{2} D2^{-n} \left(\log \left(1 + \frac{1}{\mu_n(t)} \right) \right)^{1/2}, \tag{6.77}$$

and for all $m \ge n$

$$|M_m(\omega,t) - M_{n-1}(\omega,t)| \tag{6.78}$$

$$\le Z(\omega) 18\sqrt{2} \int_0^{D/2^n} \left(\log \left(1 + \frac{1}{\mu(B(t,u))} \right) \right)^{1/2} du.$$

This shows us that there exists a set $\Omega_1 \subset \Omega$ with $P(\Omega_1) = 1$ on which $M_n(\omega,t)$ converges to a limit, say $X'(\omega,t)$, for all $t \in T$. For $\omega \ne \Omega_1$, set $X'(\omega,t) = 0$. Since, as we stated above, $M_n(\omega,t)$ converges to $X(\omega,t)$, for each $t \in T$, $\{X'(t), t \in T\}$ is a version of $\{X(t), t \in T\}$.

It follows from (6.78) that for all $k \ge 1$,

$$|M_k(\omega,t) - M_0(\omega,t)| \tag{6.79}$$

$$\le Z(\omega) 18\sqrt{2} \int_0^{D/2} \left(\log \left(1 + \frac{1}{\mu(B(t,u))} \right) \right)^{1/2} du.$$

So, taking the limit as $k \to \infty$, we get

$$|X'(\omega,t) - M_0(\omega,t)| \le Z(\omega) 18\sqrt{2} \int_0^{D/2} \left(\log \left(1 + \frac{1}{\mu(B(t,u))} \right) \right)^{1/2} du. \tag{6.80}$$

To proceed, for any pair $s, t \in T$, choose k such that $D/2^{k+1} < d(s,t) \leq D/2^k$ and, for $\omega \in \Omega_1$, consider

$$|X'(s) - X'(t)| \leq |M_k(s) - M_k(t)| + |X'(s) - M_k(s)| + |X'(t) - M_k(t)| \tag{6.81}$$

in which we suppress the ω. We use (6.78) again, taking the limit as $m \to \infty$, to see that

$$|X'(s) - M_k(s)| \leq Z18\sqrt{2} \int_0^{D/2^{k+1}} \left(\log \left(1 + \frac{1}{\mu(B(s, u))} \right) \right)^{1/2} du, \tag{6.82}$$

and similarly with s replaced by t.

Similar to (6.75) and (6.76), we have

$$
\begin{aligned}
|M_k(s) - M_k(t)| &\leq 3D2^{-k} \int_{T \times T} |Y(u, v)| \rho_k(s, u) \rho_k(t, v) \, \mu(du)\mu(dv) \\
&\leq Z9D2^{-k} \left(\log \left(1 + \frac{1}{\mu_k(s)\mu_k(t)} \right) \right)^{1/2}.
\end{aligned}
$$

Consequently,

$$|M_k(s) - M_k(t)| \tag{6.83}$$

$$
\leq Z18 \int_0^{D/2^{k+1}} \left(\log \left(1 + \frac{1}{\mu(B(s, u))} \right) \right. \\
\left. + \log \left(1 + \frac{1}{\mu(B(t, u))} \right) \right)^{1/2} du.
$$

It now follows from (6.81) and (6.82) applied to both s and t that

$$|X'(s) - X'(t)| \tag{6.84}$$

$$
\leq Z54 \int_0^{D/2^{k+1}} \left(\log \left(1 + \frac{1}{\mu(B(s, u))} \right) \right. \\
\left. + \log \left(1 + \frac{1}{\mu(B(t, u))} \right) \right)^{1/2} du
$$

$$
\leq Z54\sqrt{2} \int_0^{D/2^{k+1}} \left(\log \left(1 + \frac{1}{\mu(B(s, u)) \wedge \mu(B(t, u))} \right) \right)^{1/2} du
$$

$$
\leq Z108\sqrt{2} \int_0^{D/2^{k+2}} \left(\log \left(1 + \frac{1}{\mu(B(s, u)) \wedge \mu(B(t, u))} \right) \right)^{1/2} du,
$$

where, for the second inequality, we replace each measure by the infimum of both of them and in the last line we use the fact that, for positive nonincreasing functions f on $[0, a]$,

$$\int_0^a f(u)\, du \le 2 \int_0^{a/2} f(u)\, du.$$

Recall that $d(s, t) > D/2^{k+1}$. This implies that the balls $B(s, u)$ and $B(t, u)$, in the last line of (6.83), are disjoint and, consequently, that $\mu(B(s, u)) \wedge \mu(B(t, u)) \le 1/2$. Note that

$$\log(1 + x) \le (\log 3)(\log 2)^{-1} \log x \qquad x \ge 2 \qquad (6.85)$$

since they are equal when $x = 2$ and $\log x$ grows more rapidly than $\log(1 + x)$ for $x \ge 2$. Therefore

$$\log\left(1 + \frac{1}{\mu(B(s, u)) \wedge \mu(B(t, u))}\right) \qquad\qquad (6.86)$$
$$\le \frac{\log 3}{\log 2} \log\left(\frac{1}{\mu(B(s, u)) \wedge \mu(B(t, u))}\right)$$
$$\le \frac{\log 3}{\log 2}\left(\log\left(\frac{1}{\mu(B(s, u))}\right) + \log\left(\frac{1}{\mu(B(t, u))}\right)\right).$$

We now use this last inequality in (6.84) to get (6.71) with $Z' = 108\sqrt{2} \left(\frac{\log 3}{\log 2}\right)^{1/2} Z$, so that $EZ' \le 295\sqrt{2} \left(\frac{\log 3}{\log 2}\right)^{1/2}$.

Using (6.85) along with (6.80) when $\mu(B(t, u)) \le 1/2$ and making allowance for the fact that $\mu(B(t, u))$ may be larger than $1/2$, we get (6.72). $\qquad\square$

Proof of Theorem 6.3.1 It follows from (6.71) that

$$E \sup_{s \in T} X'(\omega, s) \le EX'(\omega, t) + 1056 \sup_{s \in T} \int_0^D \left(\log \frac{1}{\mu(B(s, u))}\right)^{1/2} du. \qquad (6.87)$$

The inequality in (6.59) follows because $EX'(\omega, t) = 0$.

To show that X' is uniformly continuous, we use (6.71) to see that

$$\sup_{\substack{s, t \in T \\ d(s, t) \le \delta}} |X'(\omega, s) - X'(\omega, t)| \le 2Z'(\omega) \sup_{s \in T} \int_0^\delta \left(\log \frac{1}{\mu(B(s, u))}\right)^{1/2} du. \qquad (6.88)$$

It now follows from (6.60) that the limit of the left-hand side of (6.88) goes to zero as δ goes to zero. Clearly (6.61) follows immediately from (6.88). $\qquad\square$

The shining stars in theory of Gaussian processes are the next two theorems by M. Talagrand. The first gives a necessary condition for the boundedness of Gaussian processes.

Theorem 6.3.4 (Talagrand) *Using the notation of Theorem 6.3.1, let* $\{X(t), t \in T\}$ *be a Gaussian process with bounded sample paths. Then there exists a probability measure μ on T such that*

$$\sup_{t \in T} \int_0^\infty \left(\log \frac{1}{\mu(B(t, u))} \right)^{1/2} du \le CE \sup_{t \in T} X(t), \quad (6.89)$$

where C=835.

Proof It suffices to take T to be countable and to suppose that the diameter of T is greater than 0. Under these stipulations and the fact that the theorem remains unchanged if we replace d_X by Dd_X, we can suppose that the diameter of T is equal to 1.

The crux of this proof is the construction of a nested sequence $\mathcal{P}_n$ of finite partitions of T, and for each n, functions $\tau_n : \mathcal{P}_n \to T$ and functions $m_n : \mathcal{P}_n \to \mathcal{N}$ with the following properties, in which $\rho = 1/4$:

(1) For all $n \ge 0$ and all elements $C \in \mathcal{P}_n$, $C \subset B(\tau_n(C), \rho^n)$.
(2) For all $n \ge 1$, every set $C \in \mathcal{P}_{n-1}$, and any two sets $D \ne D'$ in $\mathcal{P}_n$ that are also in C, $m_n(D) \ne m_n(D')$, $\tau_n(D)$ and $\tau_n(D')$ are in C, and $d(\tau_n(D), \tau_n(D')) > \rho^n$.

With regard to (2), note that because $\mathcal{P}_n$ is a partition of T, all elements of $\mathcal{P}_n$ are disjoint, and, to elaborate further, we show below that the function m_n counts the elements of $\mathcal{P}_n$ that cover C.

The partitions $\mathcal{P}_n$ are defined recursively. $\mathcal{P}_0$ is the single set T itself, $m_0(T) = 1$, and $\tau_0(T) = a$, where a can be chosen to be any point in T. Suppose now that $\mathcal{P}_j$, $j = 0, \ldots, n$ has been chosen. Let C be an element of $\mathcal{P}_n$. C is decomposed into the union of disjoint sets that are part of the partition $\mathcal{P}_{n+1}$ in the following way: Let $A_0 = \phi$ and $W_0 = C$. Choose a point $t_1 \in C$ such that

$$\mathcal{E}(B(t_1, \rho^{n+2}) \cap W_0) \ge \sup_{t \in W_0} \mathcal{E}(B(t, \rho^{n+2}) \cap W_0) - \rho^{n+2}\mathcal{E}(T), \quad (6.90)$$

where we use the notation $\mathcal{E}(U) = E\left(\sup_{t \in U} X(t)\right)$ for any $U \subset T$.

We now consider the larger ball centered at t_1, $B(t_1, \rho^{n+1})$. If this ball covers C, we are done. The element in $\mathcal{P}_{n+1}$ that covers C is C itself, and we set $\tau_{n+1}(C) = t_1$ and $m_{n+1}(C) = 1$. If $B(t_1, \rho^{n+1})$ does not cover C, we consider $C \setminus B(t_1, \rho^{n+1})$ and repeat the procedure.

To be more precise, for $r \ge 1$, let $A_r = \cup_{i \le r} B(t_i, \rho^{n+1})$. If C is not included in A_r, we choose the point t_{r+1} in $W_r = C \setminus A_r$ such that

$$\mathcal{E}(B(t_{r+1}, \rho^{n+2}) \cap W_r) \geq \sup_{t \in W_r} \mathcal{E}(B(t, \rho^{n+2}) \cap W_r) - \rho^{n+2}\mathcal{E}(T). \quad (6.91)$$

Note that $d(t_i, t_j) > \rho^{n+1}$. Let $M(C)$ be the smallest integer k for which $C \subset A_k$. We stop the construction at the k-th stage. We know that $M(C)$ is finite because, for all

$$i, j = 1, \ldots, k \quad \text{with} \quad i \neq j, \quad d(t_i, t_j) > \rho^{n+1}, \quad (6.92)$$

and thus $M(C) \leq \exp\left((17\mathcal{E}(T)^2)/\rho^{2(n+1)}\right)$ by Lemma 5.5.6.

For each $r \in [1, M(C)]$ let $D_r := D_r(C) = B(t_r, \rho^{n+1}) \cap W_{r-1}$. We construct $\mathcal{P}_{n+1}$ by taking $\{D_r(C), r \in [1, M(C)]\}$ as the elements in $\mathcal{P}_{n+1}$ that cover C. We define $\tau_{n+1}(D_r) = t_r$ and $m_{n+1}(D_r) = r$, $r = 1, \ldots, M(C)$. We do this for each element of $\mathcal{P}_n$. This completes the description of the recursive construction. We see that (1) and (2) hold.

Given this construction, the rest of the proof is pretty straightforward. Let $s \in T$ and $C_n(s)$ be the element of $\mathcal{P}_n$ that contains s. We now show that

$$\sum_{n=0}^{\infty} \rho^n (\log m_n(C_n(s)))^{1/2} \leq 104\mathcal{E}(T). \quad (6.93)$$

Set $C = C_n(s)$. The element $C_{n+1}(s)$ of $\mathcal{P}_{n+1}$ is a subset of C; indeed, it is one of the sets $\{D_r(C), r \in [1, M(C)]\}$ that are the elements $\mathcal{P}_{n+1}$ that cover C. Let us suppose that it is $D_m := D_m(C)$, that is, $D_m(C) = C_{n+1}(s)$. Set

$$m = m_{n+1}(C_{n+1}(s)) \quad \text{and} \quad t_m = \tau(C_{n+1}(s)). \quad (6.94)$$

Consider the family of sets $\{D_r, r \in [1, m]\}$. Since they are contained in C and the points $\{\tau(D_r), r \in [1, m]\}$ are ρ^{n+1}–distinguishable, it follows from Lemma 5.5.7 that

$$\mathcal{E}(C) \geq \tfrac{1}{12}\rho^{n+2}(\log m)^{1/2} + \inf_{i \leq m} \mathcal{E}(B(t_i, \rho^{n+2}) \cap W_{i-1}) \quad (6.95)$$

($W.$ is as defined above (6.91).)

Notice also that by (6.91), for each $\{i \in [1, m]\}$,

$$\rho^{n+2}\mathcal{E}(T) + \mathcal{E}(B(t_i, \rho^{n+2}) \cap W_{i-1}) \quad (6.96)$$
$$\geq \mathcal{E}(B(t_m, \rho^{n+2}) \cap W_{i-1})$$
$$\geq \mathcal{E}(B(t_m, \rho^{n+2}) \cap W_{m-1})$$

since $t_m \in W_{i-1}$. Combining (6.95) and (6.96), we get

$$\mathcal{E}(C_n(s)) = \mathcal{E}(C) \geq \tfrac{1}{12}\rho^{n+2}(\log m_{n+1}(C_{n+1}(s)))^{1/2} \tag{6.97}$$
$$+ \mathcal{E}(B(t_m, \rho^{n+2}) \cap W_{m-1}) - \rho^{n+2}\mathcal{E}(T).$$

Furthermore, since $C_{n+2}(s) \subset C_{n+1}(s)$, $\tau(C_{n+2}(s)) \in C_{n+1}(s) = D_m$. Since t_m is maximal in the sense of (6.91), we see that

$$\mathcal{E}(B(t_m, \rho^{n+2}) \cap W_{m-1}) \geq \mathcal{E}(B(\tau(C_{n+2}(s)), \rho^{n+2}) \cap W_{m-1}) - \rho^{n+2}\mathcal{E}(T). \tag{6.98}$$

It is clear that $B(\tau(C_{n+2}(s)), \rho^{n+2}) \cap W_{m-1}$ contains $C_{n+2}(s)$. Using this in (6.98) along with (6.97), we see that

$$\mathcal{E}(C_n(s)) \geq \tfrac{1}{12}\rho^{n+2}(\log m_{n+1}(C_{n+1}(s)))^{1/2} + \mathcal{E}(C_{n+2}(s)) - 2\rho^{n+2}\mathcal{E}(T). \tag{6.99}$$

For a fixed integer N, keeping in mind that $\rho = 1/4$, we get

$$\sum_{n=0}^{N} \rho^{n+2}(\log m_{n+1}(C_{n+1}(s)))^{1/2} \tag{6.100}$$

$$\leq 12 \left(\mathcal{E}(C_0(s)) + \mathcal{E}(C_1(s)) + 2\,\mathcal{E}(T) \sum_{n=0}^{\infty} \rho^{n+2} \right)$$

$$\leq 26\,\mathcal{E}(T).$$

Thus $\sum_{n=1}^{\infty} \rho^n(\log m_n(C_n(s)))^{1/2} \leq (26/\rho)\mathcal{E}(T)$, which gives us (6.93), since $C_0(s) = T$ and $m_0(T) = 1$.

We now define the probability measure μ in the left-hand side of (6.89). For each $n \in \mathcal{N}$, let $Q(n) = \{\tau_n(C), C \in \mathcal{P}_n\}$. Recall that $Q(0) = \{a\}$. We set $\mu_0(a) = 1/2$. We continue to define μ_n, $n \geq 1$, recursively. Each μ_n is supported on $Q(n)$ and for each $t \in Q(n)$

$$\mu_n(t) = \frac{\mu_{n-1}(\tau_{n-1}(C_{n-1}(t)))}{2m_n^2(C_n(t))} \left(\sum \frac{1}{m_n^2(C_n(u))} \right)^{-1}, \tag{6.101}$$

where the sum is taken over $\{u \in Q(n) : \tau_{n-1}(C_{n-1}(u)) = \tau_{n-1}(C_{n-1}(t))\}$. We set $\mu = \sum_{n=0}^{\infty} \sum_{t \in Q(n)} \mu_n(t)\delta_t$.

For each $s \in T$, let $\phi_n(s) = \tau_n(C_n(s))$ and $m_n(s) = m_n(C_n(s))$. It is obvious that for each $v \in Q(n-1)$,

$$\sum\{\mu_n(t) : t \in Q(n), \phi_{n-1}(t) = v\} = \mu_{n-1}(v)/2. \tag{6.102}$$

Therefore $A_n = \sum\{\mu_n(t) : t \in Q(n)\} = \tfrac{1}{2}\sum\{\mu_{n-1}(v) : v \in Q(n-1)\}$. That is, $A_n = A_{n-1}/2$. Since $A_0 = \mu(a) = 1/2$, we see that $A_n = 2^{-(n+1)}$. Thus μ is a probability measure.

For each $u \in T$ and $n \geq 0$, we have $d(u, \phi_n(u)) < \rho^n$. Thus $\phi_n(u) \in B(u, \rho^n)$ and consequently

$$\mu(B(u, \rho^n)) \geq \mu_n(\phi_n(u)) \geq 3\pi^{-2}\mu_{n-1}(\phi_{n-1}(u))/m_n^2(u). \qquad (6.103)$$

For the second inequality we use (6.101) and the facts that $\phi_{n-1}(\phi_n(u)) = \phi_{n-1}(u)$ and $\sum_{m=1}^{\infty} 1/m^2 = \pi^2/6$. Iterating and noting that, $\mu_0(\phi_0(u)) = 1/2$ for all u, we see that, for all $u \in T$ and all $n \geq 0$,

$$\mu(B(u, \rho^n)) \geq \left(\frac{3}{\pi^2}\right)^n \frac{1}{2\, m_n^2(u)\, m_{n-1}^2(u) \cdots m_1^2(u)}. \qquad (6.104)$$

Since the diameter of $T = 1$, we see that

$$\begin{aligned}
I(u, \mu) &:= \int_0^{\infty} \left(\log \frac{1}{\mu(B(u, r))}\right)^{1/2} dr & (6.105) \\
&\leq \sum_{j=1}^{\infty} \int_{\rho^j}^{\rho^{j-1}} \left(\log \frac{1}{\mu(B(u, r))}\right)^{1/2} dr \\
&\leq \sum_{j=1}^{\infty} \rho^{j-1} \left(\log \frac{1}{\mu(B(u, \rho^j))}\right)^{1/2}.
\end{aligned}$$

Using (6.104), we see that

$$\log \frac{1}{\mu(B(u, \rho^j))} \leq j \log \frac{\pi^2}{3} + \log 2 + \sum_{i=1}^{j} 2 \log m_i(u). \qquad (6.106)$$

Consequently,

$$\begin{aligned}
I(u, \mu) &\leq 4 \sum_{j=1}^{\infty} \rho^j \left(\left(j \log \frac{\pi^2}{3}\right)^{1/2} + (\log 2)^{1/2} + \sum_{i=1}^{j} (2 \log m_i(u))^{1/2} \right) \\
&\leq \frac{16}{9} \left(\log \frac{\pi^2}{3}\right)^{1/2} + \frac{4}{3} (\log 2)^{1/2} & (6.107) \\
&\qquad + \frac{16\sqrt{2}}{3} \sum_{i=1}^{\infty} \rho^i (\log m_i(u))^{1/2}.
\end{aligned}$$

By (6.93), the last term in the second line of (6.107) is bounded by $785\,\mathcal{E}(T)$. The sum of the first two terms in the second line of (6.107) is bounded by 3.06. Now, since the diameter of T is not less than b for any $b < 1$, there are two points in T that are 3/4-distinguishable. Therefore, by Lemma 5.5.6, $50\,\mathcal{E}(T) > 3.06$. Thus we obtain (6.89). $\qquad \square$

We now obtain a modification of Theorem 6.3.4 that gives a necessary condition for the continuity of Gaussian processes.

Theorem 6.3.5 (Talagrand) *Using the notation of Theorem 6.3.1, let $\{X(t), t \in T\}$ be a Gaussian process with bounded uniformly continuous sample paths. Then there exists a probability measure μ on T such that*

$$\sup_{t \in T} \int_0^\infty \left(\log \frac{1}{\mu(B(t,u))} \right)^{1/2} du \leq CE \sup_{t \in T} X(t) \tag{6.108}$$

and

$$\lim_{\delta \to 0} \sup_{t \in T} \int_0^\delta \left(\log \frac{1}{\mu(B(t,u))} \right)^{1/2} du = 0. \tag{6.109}$$

Proof Let D be the diameter of T. By (6.33) and the sentence immediately following it, for all $n \geq 0$ we can find a family of balls $\{B(s, D2^{-n}), s \in S_n\}$ such that $\cup_{s \in S_n} B(s, D2^{-n}) = T$ and

$$\lim_{n \to \infty} 2^{-n} \left(\log \#S_n \right)^{1/2} = 0. \tag{6.110}$$

For each n and $s \in S_n$, set $T_{n,s} = B(s, D2^{-n})$. Consider $\{X(t), t \in T_{n,s}\}$. Obviously, this process is bounded and so, by Theorem 6.3.4, there exists a probability measure $\mu_{n,s}$ on $T_{n,s}$ such that

$$\sup_{t \in T_{n,s}} \int_0^{D2^{-n}} \left(\log \frac{1}{\mu_{n,s}(B(t,u))} \right)^{1/2} du \leq CE \sup_{t \in T_{n,s}} X(t), \tag{6.111}$$

where $C = 835$.

We now construct a probability measure μ on T by setting

$$\mu = \sum_{n=0}^\infty 2^{-(n+1)} \sum_{s \in S_n} \frac{\mu_{n,s}}{\#S_n}. \tag{6.112}$$

For each $t \in T$ and $k \geq 0$ one of the sets $T_{k,s}$, where $s \in S_k$, contains t. Therefore, by (6.111) with n replaced by k and the obvious inequality

$$\mu(B(t,u)) > 2^{-(k+1)} \frac{\mu_{k,s}(B(t,u))}{\#S_k}, \tag{6.113}$$

which holds for all $s \in S_k$,

$$\int_0^{D2^{-k}} \left(\log \frac{1}{\mu(B(t,u))} \right)^{1/2} du \tag{6.114}$$

$$\leq D2^{-k} \left(\log(\#S_k) \right)^{1/2} + D2^{-k} \left(\log 2^{k+1} \right)^{1/2} + CE \sup_{t \in T_{k,s}} X(t).$$

Consider (6.114) for $k = 0$. Since $\#S_0 = 1$ and, by (5.232), $D <$

$\sqrt{2\pi} E \sup_{t \in T} X(t)$, we get

$$\sup_{t \in T} \int_0^D \left(\log \frac{1}{\mu(B(t,u))} \right)^{1/2} du \le (2\sqrt{\pi} + C) \, E \sup_{t \in T} X(t), \qquad (6.115)$$

which gives us (6.108).

Returning again to (6.114) and using (6.34), we see that

$$\sup_{t \in T} \int_0^{D2^{-k}} \left(\log \frac{1}{\mu(B(t,u))} \right)^{1/2} du \le D2^{-k} \left(\log(\#S_k) \right)^{1/2} \qquad (6.116)$$
$$+ D2^{-k} \left(\log 2^{k+1} \right)^{1/2} + CE \sup_{\substack{d(s,t) \le D2^{-k} \\ s,t \in T}} |X(s) - X(t)|.$$

Using (6.110) and the hypothesis that the process is uniformly continuous, we see that all the terms on the right-hand side of (6.116) go to zero as k goes to infinity. Thus we get (6.109). $\qquad \square$

Considering the significance of having necessary and sufficient conditions for the boundedness and continuity of Gaussian processes, we can afford to be repetitious and restate them in the following theorem.

Theorem 6.3.6 *Let $X = \{X(t), t \in T\}$ be a mean zero Gaussian process, where (T, d_X) is a separable metric or pseudometric space with finite diameter D. There exists a version of X with bounded sample paths if and only if there exists a probability measure μ on T such that*

$$\sup_{t \in T} \int_0^D \left(\log \frac{1}{\mu(B(t,u))} \right)^{1/2} du < \infty. \qquad (6.117)$$

Furthermore, there exists a version of X with bounded uniformly continuous sample paths if and only if there exists a probability measure μ on T such that (6.117) holds and in addition

$$\lim_{\epsilon \to 0} \sup_{t \in T} \int_0^\epsilon \left(\log \frac{1}{\mu(B(t,u))} \right)^{1/2} du = 0. \qquad (6.118)$$

The conditions for continuity and boundedness in Theorem 6.3.6 are rather abstract, since there is no guidance as to what measure to use. Nevertheless, we see in the next example that in some cases it is possible to find these measures.

Example 6.3.7 We return to the process studied in Example 6.1.3 and show that we can prove that it is continuous using Theorem 6.3.1. We also modify the example slightly to show how the majorizing measure condition can show that a process is bounded when it is not also continuous. As an extension of Example 6.1.3 let $X = \{X(t), t \in T\}$, where $T = \{\{t_k\}_{k=4}^{\infty} \cup 0\}$ with $t_4 = 1$, $t_k \downarrow 0$, and $X(t_k) := \xi_k/b_k$, where $\{\xi_k\}_{k=4}^{\infty}$ is a standard normal sequence, and $X(0) = 0$. Thus $d_X(t_j, t_k) = \left(1/b_j^2 + 1/b_k^2\right)^{1/2}$. We define a probability measure μ on T as follows: $\mu(0) = 1/2$ and $\mu(t_k) = (Ck^2)^{-1}$ with the appropriate constant C.

Note that $\mu(B(t_k, u)) = (Ck^2)^{-1}$ when $u < 1/b_k$, since t_k is the only point in $B(t_k, u)$ for these values of u. When $u \geq 1/b_k$, $\mu(B(t_k, u)) > 1/2$ since 0 is now also in $B(t_k, u)$. Consequently,

$$\int_0^1 \left(\log \frac{1}{\mu(B(t_k, u))}\right)^{1/2} du \leq \frac{(2\log k + \log C)^{1/2}}{b_k} + (\log 2)^{1/2},$$

(6.119)

and for $k \geq j$

$$\frac{(2\log k + \log C)^{1/2}}{b_k} \leq \int_0^{1/b_j} \left(\log \frac{1}{\mu(B(t_k, u))}\right)^{1/2} du \quad (6.120)$$

$$\leq \frac{(2\log k + \log C)^{1/2}}{b_k} + \frac{(\log 2)^{1/2}}{b_j}.$$

Let $b_k = (2\log k \log\log k)^{1/2}$ as in Example 6.1.3. It follows from the second inequality in (6.120) and Theorem 6.3.1 that X is continuous at zero. In Example 6.1.3 we showed that this obvious fact does not follow from Theorem 6.1.2.

Now let $b_k = (2\log k)^{1/2}$. One knows from the Borel–Cantelli Lemmas that $\limsup_{k \to \infty} \xi_k/(2\log k)^{1/2} = 1$ almost surely. The inequalities in (6.59) and (6.119) show that in this case X is bounded almost surely. We also see from the first inequality in (6.120) that (6.60) does not hold, which, of course, must be the case.

It is interesting to note that we can actually use Theorem 6.3.4 to show that $\limsup_{k \to \infty} \xi_k/(2\log k)^{1/2} > 0$. Let $Y = \{Y(t), t \in T\}$, where $T = \{\{t_k\}_{k=4}^{\infty} \cup 0\}$ with $t_4 = 1$, $t_k \downarrow 0$, and $X(t_k) := \xi_k/(2\log k)^{1/2}$. Every probability measure on T is of the form $\mu(t_k) = a_k$, $\mu(0) = a_0$, and the measure for which (6.89) holds must assign a positive measure to all points of T. By (6.120), for $k \geq j$,

$$\frac{(\log 1/a_k)^{1/2}}{(2\log k)^{1/2}} \leq \int_0^{1/(2\log j)^{1/2}} \left(\log \frac{1}{\mu(B(t_k, u))}\right)^{1/2} du. \quad (6.121)$$

It is easy to see, since $\sum_{k=4}^{\infty} a_k < 1$, that the limit superior of the left-hand side of (6.121) does not go to zero. Consequently, by Theorem

6.3.4,

$$\lim_{j \to \infty} E \left(\sup_{k \geq j} \frac{\xi_k}{(2 \log k)^{1/2}} \right) > 0. \tag{6.122}$$

However, since $\sup_k \xi_k / (2 \log k)^{1/2} < \infty$ almost surely, we can use Corollary 5.4.7 and the Dominated Convergence Theorem to see that $\limsup_{k \to \infty} \xi_k / (2 \log k)^{1/2} > 0$ with probability greater than zero. Since this is a tail event, it must occur with probability one.

6.4 Simple criteria for continuity

We obtain both necessary and sufficient conditions for the continuity of Gaussian processes with stationary increments and nicely behaved covariance functions or spectral distributions. These results apply, for example, to Gaussian processes associated with symmetric stable processes and Brownian motion and, in fact, to symmetric Lévy processes in general. We achieve this by finding an integral expression that is equivalent to the metric entropy integral in (6.3) but which is easier to work with.

Let $\sigma(x, y)$ be a translation invariant metric or pseudometric on R^n. Since $\sigma(x, y) = \sigma(0, x - y) = \sigma(0, y - x)$, we write $\sigma(x, y) = \sigma(x - y)$. Let $K \subset R^n$ be a compact symmetric neighborhood of zero. Define $K \oplus K = \{x + y | x \in K, y \in K\}$ and in a similar fashion define $\oplus_n K = \{x_1 + \cdots + x_n | x_1 \in K, \ldots, x_n \in K\}$. Let μ denote Lebesgue measure on R^n and let $\mu_n = \mu(\oplus_n K)$. Define

$$m_\sigma(\epsilon) = \mu(\{x \in K \oplus K | \sigma(x) < \epsilon\}) \tag{6.123}$$

and note that m_σ is left continuous and nondecreasing. Set

$$\overline{\sigma}(u) = \sup\{y | m_\sigma(y) < u\}. \tag{6.124}$$

We see that $\overline{\sigma}$ is also left continuous and nondecreasing. Such a function is often referred to as the left continuous inverse of m_σ. Also note that, by the left continuity of m_σ, $m_\sigma(\overline{\sigma}(s)) = s$ for all s for which $\overline{\sigma}(s) > \overline{\sigma}(s - \epsilon)$ for all $0 < \epsilon < s$.

Since $0 \leq m_\sigma(\epsilon) \leq \mu_2$ and $m_\sigma(\infty) = \mu_2$, we can restrict the domain of $\overline{\sigma}$ to $[0, \mu_2]$. Note that $\overline{\sigma}(\cdot)$ considered as a random variable on $[0, \mu_2]$ has the same probability distribution with respect to normalized Lebesgue measure on $[0, \mu_2]$ that $\sigma(\cdot)$ has with respect to normalized Lebesgue measure on $K \oplus K$ (this statement is equivalent to the elementary observation that, given the probability distribution function of a random variable, one can find an increasing function on the unit interval with the same probability distribution function). In keeping with classical terminology, we call $\overline{\sigma}$ the nondecreasing rearrangement of σ (with respect to $K \oplus K$).

In terms of μ, σ, and K, we define

$$I(K, \sigma) = \int_0^{\mu_2} \frac{\overline{\sigma}(s)}{s \left(\log \frac{\mu_4}{s} \right)^{1/2}} \, ds. \qquad (6.125)$$

We also recall the metric entropy integral in (6.3) applied to (K, σ), which we write as

$$J(K, \sigma) = \int_0^D (\log N(K, \sigma, u))^{1/2} \, du, \qquad (6.126)$$

where $D = \sup_{x,y \in K} \sigma(x - y) := \widehat{\sigma}$.

Lemma 6.4.1

$$-\widehat{\sigma} \left(\log \frac{\mu_2}{\mu_1} \right)^{1/2} + \frac{1}{2} I(K, \sigma) \le J(K, \sigma) \le 4\widehat{\sigma} \left(\log \frac{\mu_4}{\mu_2} \right)^{1/2} + 2I(K, \sigma). \qquad (6.127)$$

To prove this lemma we first explore the relationship between the metric entropy of K and the measure m_σ on $K \oplus K$.

Lemma 6.4.2 *Using the notation given above, we have*

$$N(K \oplus K, \sigma, \epsilon) \le \frac{\mu_4}{m_\sigma(\epsilon/2)} \qquad (6.128)$$

$$N(K, \sigma, \epsilon) \ge \frac{\mu_1}{m_\sigma(\epsilon)} \vee 1 \qquad (6.129)$$

and

$$N(K, \sigma, 2\epsilon) \le N(K \oplus K, \sigma, \epsilon). \qquad (6.130)$$

Proof For all $t \in K \oplus K$, we have

$$\mu(\{B_\sigma(t, \epsilon) \cap \oplus_4 K\}) \ge \mu(\{B_\sigma(0, \epsilon) \cap K \oplus K\}). \qquad (6.131)$$

This inequality is elementary since for $t \in K \oplus K$, $t + \{B_\sigma(0, \epsilon) \cap K \oplus K\} \subset \{B_\sigma(t, \epsilon) \cap \oplus_4 K\}$. We also note that

$$\widetilde{D}(K \oplus K, \sigma, \epsilon/2) \ge N(K \oplus K, \sigma, \epsilon), \qquad (6.132)$$

where $\widetilde{D}(K \oplus K, \sigma, \epsilon/2)$ is the maximum number of closed balls of radius $\epsilon/2$ with centers in $K \oplus K$ that are disjoint in R^n. This inequality has the same proof as the first inequality in Lemma 6.1.1. For later use let t_j, $1 \leq j \leq \widetilde{D}(K \oplus K, \sigma, \epsilon/2)$, denote the centers of these balls.

Inequality (6.132) has the same proof as the first inequality in Lemma 6.1.1.

Using (6.131) and (6.132), we see that

$$
\begin{aligned}
\mu_4 & \geq \mu \left(\cup_{j=1}^{\widetilde{D}(K \oplus K, \sigma, \epsilon/2)} \{ B_\sigma(t_j, \epsilon/2) \cap \oplus_4 K \} \right) && (6.133) \\
& \geq \widetilde{D}(K \oplus K, \sigma, \epsilon/2) \mu \left(\{ B_\sigma(0, \epsilon/2) \cap K \oplus K \} \right) \\
& \geq N(K \oplus K, \sigma, \epsilon) m_\sigma(\epsilon/2),
\end{aligned}
$$

which is (6.128).

To prove (6.129) we first note that, analogous to (6.131), for all $t \in K$,

$$
\mu(\{ B_\sigma(0, \epsilon) \cap K \oplus K \}) \geq \mu(\{ B_\sigma(t, \epsilon) \cap K \}). \qquad (6.134)
$$

Also, since $K \subset \cup_{j=1}^{N(K, \sigma, \epsilon)} B_\sigma(t_j, \epsilon)$ for some set of points $\{ t_j \}$ in K (the sets $\{ t_j \}$ are not necessarily the same in each paragraph),

$$
\begin{aligned}
\mu_1 & = \mu \left(\{ \cup_{j=1}^{N(K, \sigma, \epsilon)} B_\sigma(t_j, \epsilon) \cap K \} \right) && (6.135) \\
& \leq N(K, \sigma, \epsilon) \mu(\{ B_\sigma(0, \epsilon) \cap K \oplus K \}),
\end{aligned}
$$

where we use (6.134) in the last step. This gives us (6.129).

Since $0 \in K$, $K \subset K \oplus K$. Thus it may appear that (6.130) is trivial. However, the centers of the balls forming a cover of $K \oplus K$ may not be in K. Let $\{ t_j \}$, $j = 1, \ldots, N(K \oplus K, \sigma, \epsilon)$ be such that $K \oplus K \subset \cup_{j=1}^{N(K \oplus K, \sigma, \epsilon)} B_\sigma(t_j, \epsilon)$. For each nonempty set $B_\sigma(t_j, \epsilon) \cap K$ choose a point $s_j \in B_\sigma(t_j, \epsilon) \cap K$. The balls $\{ B_\sigma(s_j, 2\epsilon) \}$ cover K since, if not, there exists a $u \in K$ such that $\sigma(s_j - u) > 2\epsilon$. This implies that $\sigma(t_j - u) > \epsilon$, which contradicts the fact that the balls $\{ B_\sigma(t_j, \epsilon) \}$ cover K. Thus we get (6.130). $\qquad \square$

Proof of Lemma 6.4.1 It follows from (6.129) and the fact that $D = \hat{\sigma}$ that

$$
\begin{aligned}
J(K, \sigma) & \geq \int_0^{\hat{\sigma}} \left(\log \frac{\mu_4}{m_\sigma(u)} - \log \frac{\mu_4}{\mu_1} \right)^{1/2} du && (6.136) \\
& \geq \int_0^{\hat{\sigma}} \left(\log \frac{\mu_4}{m_\sigma(u)} \right)^{1/2} du - \hat{\sigma} \left(\log \frac{\mu_4}{\mu_1} \right)^{1/2}.
\end{aligned}
$$

By the change of variables $u = \overline{\sigma}(s)$ and the fact that $\overline{\sigma}(\mu_2) = \hat{\sigma}$,

$$
\int_0^{\hat{\sigma}} \left(\log \frac{\mu_4}{m_\sigma(u)} \right)^{1/2} du = \int_0^{\mu_2} \left(\log \frac{\mu_4}{s} \right)^{1/2} d\overline{\sigma}(s), \qquad (6.137)
$$

where we also use the fact that $m_\sigma(\overline{\sigma}(s)) = s$ for all $s \in [0, \mu_2]$ for which $\overline{\sigma}(s) > \overline{\sigma}(s - \epsilon)$ for all $0 < \epsilon < s$, as noted above. Since the right-hand integral in (6.137) only increases at these points, we get the equality as stated.

Assume that $J(K, \sigma) < \infty$. Then, by (6.136) and (6.137), the integral on the right-hand side of (6.137) is finite. We now show that this implies that

$$\lim_{s \to 0} \overline{\sigma}(s) \left(\log \frac{\mu_4}{s} \right)^{1/2} = 0. \tag{6.138}$$

If $\overline{\sigma}(s) \equiv 0$ is $[0, \delta]$, (6.138) is trivial. Assume that this is not the case and the limit in (6.138) is not zero. Then there exists an $\epsilon > 0$ and a sequence $\{t_k\}_{k=1}^\infty$ deceasing to zero such that

$$\lim_{k \to 0} \overline{\sigma}(t_k) \left(\log \frac{\mu_4}{t_k} \right)^{1/2} \geq \epsilon. \tag{6.139}$$

Consequently, for any subsequence $\{s_k\}_{k=1}^\infty$ of $\{t_k\}_{k=1}^\infty$ with $s_1 < \mu_2$,

$$\int_0^{\mu_2} \left(\log \frac{\mu_4}{s} \right)^{1/2} d\overline{\sigma}(s) \geq \sum_{k=1}^\infty \int_{s_{k+1}}^{s_k} \left(\log \frac{\mu_4}{s} \right)^{1/2} d\overline{\sigma}(s) \tag{6.140}$$

$$\geq \sum_{k=1}^\infty \left(\log \frac{\mu_4}{s_k} \right)^{1/2} (\overline{\sigma}(s_k) - \overline{\sigma}(s_{k+1})).$$

We choose $\{s_k\}_{k=1}^\infty$ so that $\overline{\sigma}(s_k) - \overline{\sigma}(s_{k+1}) \geq \overline{\sigma}(s_k)/2$. Here we use the fact that $\overline{\sigma}(s_k) > 0$ and $\lim_{k \to \infty} \overline{\sigma}(s_k) = 0$. Using this subsequence, it follows from (6.139) and (6.140) that the integral on the left-hand side of (6.140) is infinite. This contradiction establishes (6.138). Therefore, by integration by parts,

$$\int_0^{\mu_2} \left(\log \frac{\mu_4}{s} \right)^{1/2} d\overline{\sigma}(s) = \hat{\sigma} \left(\log \frac{\mu_4}{\mu_2} \right)^{1/2} + \frac{1}{2} I(K, \sigma). \tag{6.141}$$

Using this in (6.136) we get the left-hand side of (6.127).

Also, by (6.130) and (6.128),

$$J(K, \sigma) \leq 2J(K \oplus K, \sigma) \leq 4 \int_0^{\hat{\sigma}} \left(\log \frac{\mu_4}{m_\sigma(u)} \right)^{1/2} du. \tag{6.142}$$

Using (6.137) and (6.141) in (6.142), we get the right-hand side of (6.127). $\square$

Remark 6.4.3 If instead of R^n we took $[-1/2, 1/2]^n$ considered as an

Abelian group with addition modulo one, and took $K = [-1/2, 1/2]^n$, then $\oplus_n K = K$ and $\mu_n = \mu_1$. In this setting (6.127) becomes

$$\tfrac{1}{2} I(K, \sigma) \leq J(K, \sigma) \leq 2I(K, \sigma). \qquad (6.143)$$

The next corollary follows immediately from Lemma 6.4.1.

Corollary 6.4.4 *Let* $X = \{X(t), t \in [-1/2, 1/2]^n\}$ *be a Gaussian process with stationary increments and let*

$$\sigma(x - y) = \left(E(X(x) - X(y))^2 \right)^{1/2}. \qquad (6.144)$$

X has a version with continuous sample paths on $([-1/2, 1/2]^n, \sigma)$ *if and only if there exists a* $\delta > 0$ *such that*

$$\int_0^\delta \frac{\overline{\sigma}(s)}{s \left(\log \dfrac{1}{s} \right)^{1/2}} \, ds < \infty, \qquad (6.145)$$

where $\overline{\sigma}(s)$ *is the nondecreasing rearrangement of* σ*, as given in (6.124).*

The condition in (6.145) is appealing because it is a simple functional of the metric that defines X. However, it is illusionary since, in general, finding the nondecreasing rearrangement of an irregular function is neither easier nor harder than figuring out the metric entropy with respect to that function. Nevertheless, (6.145) implies conditions for continuity that can be very easy to verify.

Example 6.4.5 Let X and σ be as in Corollary 6.4.4, and take $n = 1$. Let σ^* and σ_* be a monotone majorant and minorant for σ on $[0, \delta/2]$ for some $\delta > 0$, that is,

$$\sigma_*(u) \leq \sigma(u) \leq \sigma^*(u) \qquad u \in [0, \delta/2]. \qquad (6.146)$$

Since $\sigma(u) = \sigma(-u)$, we have

$$2\sigma_*(u/2) \leq \overline{\sigma}(u) \leq 2\sigma^*(u/2) \qquad u \in [0, \delta]. \qquad (6.147)$$

Therefore, it is obvious, by (6.145), that

$$\int_0^\delta \frac{\sigma^*(s)}{s \left(\log \dfrac{1}{s} \right)^{1/2}} \, ds < \infty \qquad (6.148)$$

is a sufficient condition for the continuity of X on $([-1/2, 1/2], \sigma)$ and

$$\int_0^\delta \frac{\sigma_*(s)}{s \left(\log \frac{1}{s} \right)^{1/2}} \, ds < \infty \qquad (6.149)$$

is a necessary condition for the continuity of X on $([-1/2, 1/2], \sigma)$.

Actually, (6.148) is a continuity condition for any Gaussian process on $[-1/2, 1/2]^n$ in the following sense.

Lemma 6.4.6 *Let $X = \{X(t), t \in [-1/2, 1/2]^n\}$ be a Gaussian process and let*

$$\sigma^+(u) := \sup_{\substack{|x-y| \leq u \\ x,y \in [-1/2, 1/2]^n}} \left(E(X(x) - X(y))^2 \right)^{1/2}. \qquad (6.150)$$

If there exists a $\delta > 0$ such that

$$\int_0^\delta \frac{\sigma^+(s)}{s \left(\log \frac{1}{s} \right)^{1/2}} \, ds < \infty, \qquad (6.151)$$

X has a version with continuous sample paths on $([-1/2, 1/2]^n, d_X)$.

Proof Let d_X be as given in (6.1) and note that $\sigma^+(x, y) := \sigma^+(|x-y|)$ is itself a metric or pseudometric on $K := [-1/2, 1/2]^n$. Therefore, by (6.128),

$$N(K \oplus K, d_X, \epsilon) \leq N(K \oplus K, \sigma^+, \epsilon) \qquad (6.152)$$
$$\leq \frac{\mu_4}{m_{\sigma^+}(\epsilon/2)}.$$

Using this we get the inequalities in (6.142), and, as in the line following (6.142), this implies

$$J(K, d_X) \leq 4\sigma^+(\sqrt{n})(n \log 2)^{1/2} + 2I(K, \sigma^+), \qquad (6.153)$$

which, by (6.151) and Theorem 6.1.2, implies that X has a version with continuous sample paths on $([-1/2, 1/2]^n, d_X)$. $\qquad \square$

Example 6.4.7 Consider (5.256) with $h(\lambda) = \lambda^{-\beta}$, $\lambda > 0$, and $1 < \beta < 3$. By a change of variables, it is easy to see that $\psi(u)$ in (5.256) is equal to

$$\left(4 \int_0^\infty \frac{\sin^2 \pi s}{s^\beta} \, ds \right) |u|^{\beta-1}. \qquad (6.154)$$

This shows that for all $0 < \alpha < 1$, we can obtain a Gaussian process $X = \{X(t), t \in R^1\}$ with stationary increments satisfying

$$\left(E(X(x) - X(y))^2\right)^{1/2} = |x - y|^\alpha. \tag{6.155}$$

These processes are often referred to as fractional Brownian motion and for $\alpha = 1/2$ give Brownian motion itself. It is easy to see that (6.148) holds for these processes, that is, they have continuous sample paths. (Strictly speaking, it shows that the paths are continuous on $(R^1, |x - y|^\alpha)$, but obviously this implies that they are also continuous with respect to the Euclidean metric on R^1. In what follows we may omit this comment when it is obvious.)

The same argument applied to (5.247) with $\mu(d\lambda) = (\lambda^{-\beta} \wedge 1) \, d\lambda$, $\lambda > 0$, and $1 < \beta < 3$ shows that for all $0 < \alpha < 1$ we can obtain stationary Gaussian processes $\widetilde{X} = \{\widetilde{X}(t), t \in R^1\}$ for which

$$C_1 |x - y|^\alpha \le \left(E(\widetilde{X}(x) - \widetilde{X}(y))^2\right)^{1/2} \le C_2 |x - y|^\alpha \tag{6.156}$$

for $x, y \in [-\delta, \delta]$ for some $\delta > 0$ and constants $C_1, C_2 > 0$, which may depend on α and, furthermore, that

$$\left(E(\widetilde{X}(x) - \widetilde{X}(y))^2\right)^{1/2} \sim C|x - y|^\alpha \tag{6.157}$$

as $|x - y| \to 0$ for some constant C (which depends on α).

To complete this discussion we note that there are stationary Gaussian processes for which (6.156) and (6.157) also hold for $\alpha = 1$.

We now give continuity conditions in terms of the spectral distribution of Gaussian processes with stationary increments. We first consider stationary Gaussian processes.

Lemma 6.4.8 *Let $X = \{X(t), t \in [-1/2, 1/2]\}$ be a stationary Gaussian process with*

$$\begin{aligned} \sigma(x - y) &= \left(E(X(x) - X(y))^2\right)^{1/2} \tag{6.158} \\ &= \left(\int_0^\infty (\sin^2 \pi\lambda(x - y)) \, dF(\lambda)\right)^{1/2}. \end{aligned}$$

We have

$$I([-1/2, 1/2], \sigma) \le 21 \left(F^{1/2}(\infty) + \int_{1/(\mu_4\pi)}^\infty \frac{(F(\infty) - F(x))^{1/2}}{x(\log(2\pi x))^{1/2}} \, dx\right). \tag{6.159}$$

Proof Let $\sigma^*(s)$ be the smallest monotone majorant of $\sigma(s)$ for $s \in [0,1]$. Then, as in (6.147), $\overline{\sigma}(s) \le 2\sigma^*(s/2)$. Therefore, by (6.125),

$$I([-1/2, 1/2], \sigma) \le 2 \int_0^{\mu_2/\mu_4} \frac{\sigma^*(\mu_4 s/2)}{s \left(\log \dfrac{1}{s} \right)^{1/2}} \, ds \qquad (6.160)$$

$$= 2 \int_0^{1/2} \frac{\sigma^*(2s)}{s \left(\log \dfrac{1}{s} \right)^{1/2}} \, ds$$

since $\mu_2 = 2$ and $\mu_4 = 4$. Note that

$$\int_0^{1/2} \frac{\sigma^*(2s)}{s \left(\log \dfrac{1}{s} \right)^{1/2}} \, ds \le \frac{1}{(\log 2)^{1/2}} \sum_{n=1}^{\infty} \frac{\sigma^*(2^{-n+1})}{n^{1/2}}. \qquad (6.161)$$

By (6.158),

$$\sigma(2s) = \left(\int_0^{\infty} (\sin^2 vs) \, dF \left(\frac{v}{2\pi} \right) \right)^{1/2}$$

$$:= \left(\int_0^{\infty} (\sin^2 vs) \, d\widetilde{F}(v) \right)^{1/2}. \qquad (6.162)$$

Therefore,

$$\sigma^*(2^{-n+1}) = \sup_{s \le 2^{-n+1}} \left(\int_0^{\infty} (\sin^2 vs) \, d\widetilde{F}(v) \right)^{1/2} \qquad (6.163)$$

$$\le 4 \cdot 2^{-n} \left(\int_0^{2^n} v^2 \, d\widetilde{F}(v) \right)^{1/2} + \left(\widetilde{F}(\infty) - \widetilde{F}(2^n) \right)^{1/2}$$

$$\le 4 \cdot 2^{-n} \left(4\widetilde{F}(\infty) + \sum_{j=1}^{n-1} 2^{2(j+1)} t_j^2 \right)^{1/2}$$

$$+ \left(\widetilde{F}(\infty) - \widetilde{F}(2^n) \right)^{1/2},$$

where we set $t_j^2 = \widetilde{F}(2^{j+1}) - \widetilde{F}(2^j)$ and use the fact that $\int_0^2 v^2 \, d\widetilde{F}(v) \le 4\widetilde{F}(\infty)$.

To obtain a bound for the right-hand side of (6.161) we note that

$$\sum_{n=1}^{\infty} \frac{\sigma^*(2^{-n+1})}{n^{1/2}} \qquad (6.164)$$

$$\le 8F^{1/2}(\infty) + 8 \sum_{n=1}^{\infty} 2^{-n} \sum_{j=1}^{n-1} 2^j t_j + \sum_{n=1}^{\infty} \left(\frac{\widetilde{F}(\infty) - \widetilde{F}(2^n)}{n} \right)^{1/2}$$

and, by (14.76),

$$\sum_{n=1}^{\infty} 2^{-n} \sum_{j=1}^{n-1} 2^j t_j \;\le\; \sum_{j=1}^{\infty} t_j \le 2 \sum_{n=1}^{\infty} \left(\frac{1}{n} \sum_{j=n}^{\infty} t_j^2 \right)^{1/2} \tag{6.165}$$

$$= 2 \sum_{n=1}^{\infty} \left(\frac{\widetilde{F}(\infty) - \widetilde{F}(2^n)}{n} \right)^{1/2}.$$

Since

$$\sum_{n=1}^{\infty} \left(\frac{\widetilde{F}(\infty) - \widetilde{F}(2^n)}{n} \right)^{1/2} \le \int_1^{\infty} \frac{\left(\widetilde{F}(\infty) - \widetilde{F}(x) \right) 1/2}{x(\log x)^{1/2}} \, dx, \tag{6.166}$$

we can complete the proof of (6.159). $\qquad\square$

Remark 6.4.9 When the spectral distribution F in (6.158) is supported on the integers it determines a Gaussian process that is periodic on $[-1/2, 1/2]$ (it is of the form of (5.260)). In this case we consider $[-1/2, 1/2]$ as a group and take $\mu_4 = 1$ in (6.159).

Let $Z = \{Z(t), t \in [-1/2, 1/2]\}$ be a Gaussian process with stationary increments. In this case we can write

$$\psi(x - y) \;=\; \left(E(Z(x) - Z(y))^2 \right)^{1/2} \tag{6.167}$$

$$= \left(\int_0^{\infty} (\sin^2 \pi\lambda(x - y)) \, \nu(d\lambda) \right)^{1/2},$$

where $\int (1 \wedge |\lambda|^2)\nu(d\lambda) < \infty$. Therefore,

$$\psi(\mu_4 s) \le \pi\mu_4 \left(\int_0^1 \lambda^2 \nu(d\lambda) \right)^{1/2} s + \left(\int_1^{\infty} (\sin^2 \pi\lambda\mu_4 s) \, dH(\lambda) \right)^{1/2}, \tag{6.168}$$

where $H(\lambda) = \nu([1, \infty)) - \nu([\lambda, \infty))$. We already dealt with the last term in (6.168) in Lemma 6.4.8. Incorporating the first term into an estimate for $I([-1/2, 1/2], \psi)$ is elementary. Since $H(\infty) - H(x) = \nu([x, \infty))$, we get

$$I([-1/2, 1/2], \psi) \le 5 \left(\mu_4 \left(\int_0^1 \lambda^2 \nu(d\lambda) \right)^{1/2} \right. \tag{6.169}$$

$$\left. + \nu^{1/2}([1, \infty)) + \int_{1/(\mu_4\pi)}^{\infty} \frac{\nu^{1/2}([x, \infty))}{x(\log(\mu_4\pi x))^{1/2}} \, dx \right),$$

where $\mu_4 = 4$. .

The sufficient conditions for continuity in (6.159) and (6.170) are also necessary when $F(\infty) - F(x)$ or $\nu([x, \infty))$ are convex on $[x_0, \infty)$ for some $x_0 < \infty$.

Theorem 6.4.10 *Let* $X = \{X(t), t \in R^1\}$ *be a stationary Gaussian process with*

$$\sigma^2(x - y) = E(X(x) - X(y))^2 = \int_0^\infty (\sin^2 \pi \lambda(x - y)) \, dF(\lambda), \quad (6.170)$$

where $F(\infty) - F(x)$ *is convex on* $[x_0, \infty)$ *for some* $x_0 < \infty$. *Then*

$$\int_2^\infty \frac{(F(\infty) - F(x))^{1/2}}{x(\log(x))^{1/2}} \, dx < \infty \qquad (6.171)$$

is necessary and sufficient for X *to have a continuous version. Let* $Z = \{Z(t), t \in R^1\}$ *be a Gaussian process with stationary increments satisfying*

$$E(Z(x) - Z(y))^2 = \int_0^\infty (\sin^2 \pi \lambda(x - y)) \, \nu(d\lambda), \qquad (6.172)$$

where $\nu([x, \infty))$ *is convex on* $[x_0, \infty)$ *for some* $x_0 < \infty$. *Then (6.171), with* $F(\infty) - F(x)$ *replaced by* $\nu([x, \infty))$, *is necessary and sufficient for* Z *to have a continuous version.*

Proof Sufficiency is given in Lemma 6.4.8. For necessity we note that, by (6.170),

$$\sigma^2(s) \geq \int_{1/4s}^\infty (\sin^2 \pi \lambda s) \, dF(\lambda). \qquad (6.173)$$

Let $k \geq 0$ be an integer. For λ satisfying $k + 1/4 \leq s\lambda \leq k + 3/4$, $\sin^2 \pi \lambda s \geq \sin^2(\pi/4) \geq 1/2$. Therefore,

$$\sigma^2(s) \geq \frac{1}{2} \sum_{k=0}^\infty \left(F((k + 3/4)/s) - F((k + 1/4)/s) \right). \qquad (6.174)$$

For $s \leq 1/(4x_0)$ it follows from the convexity of $F(\infty) - F(x)$ that, for all $k \geq 0$,

$$F((k + 3/4)/s) - F((k + 1/4)/s) \geq F((k + 5/4)/s) - F((k + 3/4)/s). \qquad (6.175)$$

Therefore, we can fill in the missing terms in (6.174) to see that

$$\sigma^2(s) \geq \tfrac{1}{4} \left(F(\infty) - F(1/(4s)) \right). \qquad (6.176)$$

It now follows from (6.149) that (6.171) is a necessary condition for the continuity of X. Exactly the same proof shows that (6.171) is a necessary condition for the continuity of Z.   □

Remark 6.4.11 The spectral distribution of a periodic Gaussian process is supported on the integers, say $F(k) - F(k-) = a_k^2$. Obviously, $F(\infty) - F(x)$ is not convex on $[x_0, \infty)$ for some $x_0 < \infty$. Nevertheless, if a_k^2 is nonincreasing, (6.176) remains true. Thus we have

$$\sum_{n=2}^{\infty} \frac{(\sum_{k=n}^{\infty} a_k^2)^{1/2}}{n(\log n)^{1/2}} < \infty \quad \text{or, equivalently,} \quad \sum_{n=1}^{\infty} \left(\frac{1}{n} \sum_{k=2^n}^{\infty} a_k^2 \right)^{1/2} < \infty$$

$$(6.177)$$

is sufficient for the process to have a continuous version, and when a_k^2 is nonincreasing it is also necessary. Some other simple conditions for the continuity of Gaussian random Fourier series are given in Section 1, Chapter VII in Marcus and Pisier (1981).

6.5 Notes and references

This chapter is the crux of our survey of basic results on Gaussian processes. The necessary and sufficient conditions obtained for the continuity and boundedness of the sample paths of Gaussian processes, in many cases, are also necessary and sufficient conditions for the joint continuity and boundedness of the local times of symmetric Markov processes. We give the conditions in terms of metric entropy and majorizing measures. Even though the latter conditions imply those for metric entropy, it is useful to understand both criteria. We also hope that working through the metric entropy proofs makes the majorizing measure proofs more understandable. Theorem 6.1.2 is due to Dudley (1967); see also Dudley (1973). Our approach is based on 4.3.1 Théorème, 3.4.4 Lemme, and 3.4.6 Lemme in Fernique (1997). Theorem 6.2.2 is due to Fernique (1974). Proofs are given in Fernique (1975) and based on this in Jain and Marcus (1978). Our approach is based on 4.3.6 Théorème, 4.4.2 Théorème, and 4.4.4 Théorème in Fernique (1997).

The ideas behind Theorem 6.3.1 originated in an important early paper by Garsia, Rodemich and Rumsey, Jr. (1970) and were developed further in Preston (1971, 1972), and Fernique (1975). Our proof is based on the treatment in Fernique (1997), his 5.2.6 Théorème, 5.2.7 Lemme, and related material. The shining stars, Talagrand's Theorems 6.3.4 and 6.3.5, first appeared in Talagrand (1987). The proof in Ledoux and Talagrand (1991) is rather difficult. A simpler proof is given in Talagrand (1992). Again we follow Fernique's treatment in Fernique (1997), which is spread out in his Chapter 5, but see in particular 5.3.5 Lemme. Dudley (1973) also follows Fernique's treatment but adds some simplifications. We benefited from them.

We have not yet looked at Talagrand (2005), which was published as we were preparing this book for press. We are sure it has new insights into sample path properties of Gaussian processes.

Lemmas 6.4.1 and 6.4.2 are taken from Chapter I in Marcus and Pisier (1981). Material on continuity conditions in terms of the spectral distribution is from Chapter IV, Section 3 in Jain and Marcus (1978) and Chapter VII, Section 1 in Marcus and Pisier (1981).

7

Moduli of continuity for Gaussian processes

7.1 General results

Let (S, τ) be a metric or pseudometric space and let $K \subset S$ be compact. Under very general conditions, whenever a Gaussian process $\{G(u), u \in K\}$ has continuous sample paths on (S, τ), it also has both an exact uniform and an exact local modulus of continuity. To be more specific, we call $\omega : R_+ \to R_+$, $\omega(0) = 0$, an exact uniform modulus of continuity for $\{G(u), u \in K\}$ on (S, τ) if

$$\lim_{\delta \to 0} \sup_{\substack{\tau(u,v) \leq \delta \\ u,v \in K}} \frac{|G(u) - G(v)|}{\omega(\tau(u,v))} = C \qquad \text{a.s.} \tag{7.1}$$

for some constant $0 < C < \infty$. We call $\rho : R_+ \to R_+$, $\rho(0) = 0$, an exact local modulus of continuity for $\{G(u), u \in S\}$ at some fixed $u_0 \in (S, \tau)$ if

$$\lim_{\delta \to 0} \sup_{\substack{\tau(u,u_0) \leq \delta \\ u \in S}} \frac{|G(u) - G(u_0)|}{\rho(\tau(u,u_0))} = C \qquad \text{a.s.} \tag{7.2}$$

for some constant $0 < C < \infty$. We use the expressions uniform and local moduli of continuity for functions ω and ρ for which the equal signs in (7.1) and (7.2) are replaced by "less than or equal to" signs. We use the terms lower uniform and local moduli of continuity for functions ω and ρ for which the equal signs in (7.1) and (7.2) are replaced by "greater than or equal to" signs.

Actually we are only concerned with two pseudometric spaces: (S, d), where S is a general space, and $d(u, v) = (E(G(u) - G(v))^2)^{1/2}$ (recall $d := d_G$ is defined in (6.1)), and $(R^n, |\cdot|)$, that is, the ordinary Euclidean metric on R^n.

In our discussions of moduli of continuity, we always assume that $\{G(u), u \in K\}$ is continuous on (S, d) and only consider functions ω and

ρ, which are continuous on R_+ and are strictly positive on $(0, \delta']$ for some $\delta' > 0$.

To avoid trivialities in what follows, whenever we consider the local modulus of continuity of $G = \{G(u), u \in S\}$ at a point $u_0 \in S$, we assume that G and (S, d) are such that

$$\sup_{d(u,u_0)\leq\delta} |G(u) - G(u_0)| > 0 \qquad \forall \delta > 0 \quad \text{a.s.,} \tag{7.3}$$

and whenever we consider the uniform modulus of continuity of G, we assume that

$$\sup_{\substack{d(u,v)\leq\delta \\ u,v\in K}} G(u) - G(v) > 0 \qquad \forall \delta > 0 \quad \text{a.s.} \tag{7.4}$$

Note that since (7.4) is symmetric in u, v, it is equivalent to

$$\sup_{\substack{d(u,v)\leq\delta \\ u,v\in K}} |G(u) - G(v)| > 0 \qquad \forall \delta > 0 \quad \text{a.s.} \tag{7.5}$$

We also set $(G(u) - G(v))/d(u,v) = 0$ when $d(u,v) = 0$.

The next simple lemma establishes conditions for a Gaussian process to have a uniform modulus of continuity.

Lemma 7.1.1 *Let $\{G(u), u \in (S, \tau)\}$ be a mean zero Gaussian process. Assume that d is continuous on (S, τ). Let $\omega : R_+ \to R_+$ and $K \subset S$ be a compact set. For the following three statements:*

$$\lim_{\delta\to 0} \sup_{\substack{\tau(u,v)\leq\delta \\ u,v\in K}} \frac{G(u) - G(v)}{\omega(\tau(u,v))} \leq C \qquad a.s. \text{ for some } \quad 0 \leq C < \infty \tag{7.6}$$

$$\lim_{\delta\to 0} \sup_{\substack{\tau(u,v)\leq\delta \\ u,v\in K}} \frac{d(u,v)}{\omega(\tau(u,v))} = 0 \tag{7.7}$$

$$\lim_{\delta\to 0} \sup_{\substack{\tau(u,v)\leq\delta \\ u,v\in K}} \frac{G(u) - G(v)}{\omega(\tau(u,v))} = C' \qquad a.s. \text{ for some } \quad 0 \leq C' \leq \infty \tag{7.8}$$

we have that (7.6) implies (7.7) implies (7.8). (Obviously, if (7.6) holds then $C' < \infty$ in (7.8) but (7.7) implies (7.8) even if the limit superior in (7.6) is not finite almost surely.)

These results are also valid for the local modulus of continuity, that is, (7.6)–(7.8) hold with v replaced by u_0 and with the supremum taken over $u \in K$.

Proof   To show that (7.6) implies (7.7), we show that if (7.7) does not hold, then (7.6) does not hold. Suppose that (7.7) does not hold but that (7.6) does. Then, since we assume that d is continuous on (S, τ), there exists a sequence of pairs $\{(u_k, v_k)\}_{k=1}^{\infty}$ in $K \times K$ for which $d(u_k, v_k) \geq \epsilon \omega(\tau(u_k, v_k))$, $\lim_{k \to \infty} \tau(u_k, v_k) = 0$, and $\lim_{k \to \infty} d(u_k, v_k) = 0$. It follows from (7.6) that almost surely

$$C \geq \lim_{\delta \to 0} \sup_{\substack{\tau(u,v) \leq \delta \\ u,v \in K}} \frac{G(u) - G(v)}{\omega(\tau(u,v))} \geq \limsup_{k \to \infty} \frac{\epsilon(G(u_k) - G(v_k))}{d(u_k, v_k)}, \qquad (7.9)$$

which gives

$$\limsup_{k \to \infty} \frac{G(u_k) - G(v_k)}{d(u_k, v_k)} \leq \frac{C}{\epsilon} \qquad \text{a.s.} \qquad (7.10)$$

But for each $k \geq 1$, $\xi_k := (G(u_k) - G(v_k))/d(u_k, v_k)$ is a normal random variable with mean 0 and variance 1, and (7.10) cannot be finite almost surely for such a sequence, because $\cup_{k \geq n} \{\xi_k > C/\epsilon\} \supset \{\xi_n > C/\epsilon\}$ and the probability of this last set is greater than zero and independent of n since the ξ_n are all $N(0, 1)$. Thus we see that (7.6) implies (7.7).

To show that (7.7) implies (7.8), we express G in terms of its Karhunen–Loève expansion, given in Corollary 5.3.2,

$$G(u) = \sum_{j=1}^{\infty} \xi_j \phi_j(u) \qquad u \in S, \qquad (7.11)$$

where now $\{\xi_j\}_{j=1}^{\infty}$ are independent $N(0, 1)$. Clearly, in this case,

$$d(u, v) = \left(\sum_{j=1}^{\infty} (\phi_j(u) - \phi_j(v))^2 \right)^{1/2}. \qquad (7.12)$$

Set

$$G_N(u) = \sum_{j=1}^{N} \xi_j \phi_j(u) \qquad u \in S \qquad (7.13)$$

and note that

$$|G_N(u) - G_N(v)| \leq \left(\sum_{j=1}^{N} |\xi_j| \right) \sup_{1 \leq j \leq N} |\phi_j(u) - \phi_j(v)| \qquad (7.14)$$

$$\leq \left(\sum_{j=1}^{N} |\xi_j| \right) d(u, v).$$

It follows from (7.7) and (7.14) that

$$\lim_{\substack{\delta \to 0 \\ \substack{\tau(u,v) \le \delta \\ u,v \in K}}} \sup \frac{G_N(u) - G_N(v)}{\omega(\tau(u,v))} = 0 \qquad \text{a.s.} \qquad (7.15)$$

Therefore, the random variable

$$\lim_{\substack{\delta \to 0 \\ \substack{\tau(u,v) \le \delta \\ u,v \in K}}} \sup \frac{G(u) - G(v)}{\omega(\tau(u,v))} \qquad (7.16)$$

is measurable with respect to the tail field of $\{\xi_j\}_{j=1}^\infty$ and hence is constant almost surely. Since, by symmetry,

$$\lim_{\substack{\delta \to 0 \\ \substack{\tau(u,v) \le \delta \\ u,v \in K}}} \sup \frac{G(u) - G(v)}{\omega(\tau(u,v))} \ge 0, \qquad (7.17)$$

this implies (7.8). $\qquad\qquad\square$

The proofs are exactly the same for the local modulus. Just take v and v_k equal to u_0.

Using Theorems 6.3.3 and 6.3.6 and the above lemma, we can find a uniform modulus of continuity for continuous Gaussian processes.

Theorem 7.1.2 *Let $G = \{G(u), u \in T\}$ be a mean zero Gaussian process, where (T, d) is a separable metric or pseudometric space with finite diameter D. Suppose that X has bounded uniformly continuous sample paths. Then there exists a probability measure μ on T such that*

$$\lim_{\epsilon \to 0} \sup_{t \in T} \int_0^\epsilon \left(\log \frac{1}{\mu(B(t,s))} \right)^{1/2} ds = 0. \qquad (7.18)$$

Suppose in addition that

$$\lim_{\epsilon \to 0} \frac{1}{\epsilon} \sup_{t \in T} \int_0^\epsilon \left(\log \frac{1}{\mu(B(t,s))} \right)^{1/2} ds = \infty. \qquad (7.19)$$

Then

$$\lim_{\substack{\delta \to 0 \\ \substack{d(u,v) \le \delta \\ u,v \in T}}} \sup \frac{G(u) - G(v)}{\left(\sup_{t \in T} \int_0^{d(u,v)} \left(\log \frac{1}{\mu(B(t,s))} \right)^{1/2} ds \right)} = C \qquad \text{a.s.} \qquad (7.20)$$

for some constant $0 \le C < \infty$.

Proof The assertion in (7.18) is part of Theorem 6.3.6. Furthermore, by Theorem 6.3.3, the left-hand side of (7.20) is finite almost surely. Finally, (7.19) implies that the denominator of the left-hand side of

(7.20) satisfies (7.7) with $\tau = d$. Therefore, (7.20) follows from Lemma 7.1.1. □

Remark 7.1.3 We note the following points with regard to Theorem 7.1.2:

(1) It follows from the proof that if (7.18) and (7.19) hold for any probability measure ν on T, then (7.20) holds with μ replaced by ν. The distinction here is that the continuity of G implies the existence of some μ for which (7.18) holds, whereas, if (7.18) holds for any probability measure ν on T, then G is continuous.

(2) If the constant C in (7.20) is not zero, $\sup_{t \in T} \int_0^\delta (\log \frac{1}{\mu(B(t,u))})^{1/2} du$ is an exact uniform modulus of continuity for G. Unfortunately, it is not easy to see whether this is the case.

(3) Suppose that (7.18) holds for some probability measure μ on T. Then a sufficient condition that (7.19) holds is that $\lim_{\epsilon \to 0} M(T, d, \epsilon) = \infty$ (or, equivalently, that $\lim_{\epsilon \to 0} N(T, d, \epsilon) = \infty$). To see this, let $M(T, d, \epsilon) = K(\epsilon)$ and note that $K(\epsilon) \geq 1$. Then, since μ is a probability measure, using the definition of $M(T, d, \epsilon)$, we see that there exists a point, say $t_1 \in T$, for which $\mu(B(t_1, \epsilon)) \leq 1/K(\epsilon)$. Consequently,

$$\int_0^\epsilon \left(\log \frac{1}{\mu(B(t_1, u))} \right)^{1/2} du \geq \epsilon \left(\log K(\epsilon) \right)^{1/2} . \qquad (7.21)$$

Therefore, if $\lim_{\epsilon \to 0} M(T, d, \epsilon) = \infty$, (7.19) holds.

If $\lim_{\epsilon \to 0} M(T, d, \epsilon) < \infty$, we can find a probability measure μ on T for which (7.18) holds but (7.19) does not. To see this note that under this condition there exists a K such that $M(T, d, \epsilon) = K$ for all $\epsilon < \epsilon'$, for some $\epsilon' > 0$. Reconsider (7.19) and note that it suffices to take the supremum over those points $t, t' \in T$ for which $d(t, t') > 0$. Since $M(T, d, \epsilon) = K$ for all $\epsilon < \epsilon'$, there are only K points in T with this property. If we choose the probability measure μ to be uniform on these points, we see that for all $\epsilon \leq \epsilon'$,

$$\sup_{t \in T} \int_0^\epsilon \left(\log \frac{1}{\mu(B(t, u))} \right)^{1/2} du \leq \epsilon \left(\log K \right)^{1/2} . \qquad (7.22)$$

Thus, for this measure (7.19) does not hold, but, obviously (7.18) does.

Note that when $M(T, d, \epsilon) = K$ for all $\epsilon < \epsilon'$, for some $\epsilon' > 0$, all the points in T are separated, so (7.20) has no meaning.

We now present an abstract result that gives necessary and sufficient conditions for the existence of exact local and uniform moduli of continuity. We say abstract because it is difficult to express these moduli in a recognizable way.

Theorem 7.1.4 *Let $G = \{G(y), y \in K\}, (K, d)$ a compact metric space, be a mean zero Gaussian process with continuous sample paths. For $\delta > 0$ let*

$$E(\delta) = E\left(\sup_{\substack{d(y,y_0) \geq \delta \\ y \in K}} \frac{|G(y) - G(y_0)|}{d(y, y_0)}\right) \tag{7.23}$$

for some fixed $y_0 \in K$ and

$$\widehat{E}(\delta) = E\left(\sup_{\substack{d(x,y) \geq \delta \\ x,y \in K}} \frac{G(x) - G(y)}{d(x, y)}\right). \tag{7.24}$$

Then

(1) *If $\lim_{\delta \to 0} E(\delta) = \infty$, $\rho(\delta) = \delta E(\delta)$ is an exact local modulus of continuity for G at y_0, that is, (7.2) holds with u_0 replaced by y_0 and $\rho(d(u, u_0))$ replaced by $d(u, y_0)E(d(u, y_0))$.*

(2) *If $\lim_{\delta \to 0} \widehat{E}(\delta) = \infty$, $\omega(\delta) = \delta \widehat{E}(\delta)$ is an exact uniform modulus of continuity for G, that is, (7.1) holds with $\omega(d(u, v))$ replaced by $d(u, v)\widehat{E}(d(u, v))$.*

(3) *If $E(\delta)$ is bounded, G does not have an exact local modulus of continuity at y_0.*

(4) *If $\widehat{E}(\delta)$ is bounded, G does not have an exact uniform modulus of continuity.*

Because of (5.183), this theorem also holds with the expectation replaced by the median, that is, with $E(\delta)$ replaced by $m(\delta)$ and $\widehat{E}(\delta)$ replaced by $\widehat{m}(\delta)$, where

$$m(\delta) = \text{median of } \sup_{\substack{d(y,y_0) \geq \delta \\ y \in K}} \frac{|G(y) - G(y_0)|}{d(y, y_0)} \tag{7.25}$$

for some fixed $y_0 \in K$ and

$$\widehat{m}(\delta) = \text{median of } \sup_{\substack{d(x,y) \geq \delta \\ x,y \in K}} \frac{G(x) - G(y)}{d(x, y)}. \tag{7.26}$$

We use the following general lemma, which is interesting in its own right, in the proof of Theorem 7.1.4.

Lemma 7.1.5 *Let (S, ϕ) be a separable metric or pseudometric space. Let $S' \subset S$ be such that there exists an $s_0 \in S - S'$ with $\phi(s_0, S') = 0$. Assume that G is a mean zero Gaussian process continuous on S', with $\sup_{s \in S'} EG^2(s) \leq 1$. Set*

$$M(u) = \text{median of } \sup_{\substack{\phi(v, s_o) \geq u \\ v \in S'}} G(v). \tag{7.27}$$

(1) *Then $\lim_{u \to 0} M(u) < \infty$ if and only if $\sup_{s \in S'} G(s) < \infty$ almost surely.*

(2) *If $\lim_{u \to 0} M(u) = \infty$,*

$$\lim_{\delta \to 0} \sup_{\substack{\phi(s, s_0) \leq \delta \\ s \in S'}} \frac{G(s)}{M(\phi(s, s_0))} = 1 \qquad a.s. \tag{7.28}$$

Proof Suppose that $\sup_{s \in S'} G(s) < \infty$ almost surely. Then, by Corollary 5.4.7, $E(\sup_{s \in S'} G(s)) < \infty$ and consequently, by (5.183), $\sup_u M(u) < \infty$. Conversely, if $\sup_u M(u) < \infty$, then, by (5.183), $E \sup_{\phi(s, s_0) \geq u, s \in S'} G(s) < \sup_u M(u) + 1$. Therefore, by the Monotone Convergence Theorem, $E(\sup_{s \in S'} G(s)) < \infty$, which implies that $\sup_{s \in S'} G(s) < \infty$ almost surely. This verifies statement (1).

Using the fact that $M(u)$ is decreasing, we now show that to obtain (7.28) it suffices to show that

$$\lim_{\delta \to 0} \sup_{\substack{\phi(s, s_0) \geq \delta \\ s \in S'}} \frac{G(s)}{M(\delta)} = 1 \qquad a.s. \tag{7.29}$$

To see this note that (7.29) implies that, for all $\epsilon > 0$, there exists a sequence $\{\delta_k\}$ decreasing to zero such that

$$\frac{G(s_k)}{M(\delta_k)} \geq 1 - \epsilon \quad \text{for some } s_k \in S, \text{ with } \phi(s_k, s_0) \geq \delta_k. \tag{7.30}$$

Suppose that $\phi(s_k, s_0) = \delta'_k$. Obviously, $\delta'_k \geq \delta_k$ and since M is decreasing, $G(s_k)/M(\delta'_k) \geq 1 - \epsilon$. This implies that

$$\sup_{\substack{\phi(s, s_0) \leq \delta'_k \\ s \in S'}} \frac{G(s)}{M(\phi(s, s_0))} \geq 1 - \epsilon. \tag{7.31}$$

Note that $\delta'_k \to 0$. To see this, assume that for some subsequence k_j and some $\epsilon > 0$ we have $\delta'_{k_j} \geq \epsilon$. Then, as in the first paragraph of this proof, $\sup_{\phi(s, s_0) \geq \epsilon} G(s)$ is bounded almost surely. However, since $\lim_{u \to 0} M(u) = \infty$, this implies that $G(s_{k_j})/M(\delta_{k_j}) = 0$, which contradicts (7.30). Thus (7.29) implies the lower bound in (7.28). It is even easier to see that it also implies the upper bound.

We now show that (7.29) holds. It is easy to see that by (5.152) and the Borel–Cantelli Lemma, we can find a sequence δ_k decreasing to zero such that

$$\lim_{k\to\infty} \sup_{\substack{\phi(s,s_0)\geq\delta_k \\ s\in S'}} \frac{G(s)}{M(\delta_k)} = 1 \qquad \text{a.s.} \tag{7.32}$$

This shows that 1 is a lower bound in (7.29). Showing that it is an upper bound is more delicate.

By Lemma 5.4.3, $M(u)$ is unique. Furthermore, $M(u)$ is decreasing and is left continuous, with right limits (this can be proved using the ideas at the end of the proof Lemma 5.4.3, starting with the paragraph containing (5.169)). Consider also

$$\overline{M}(u) := \text{median of} \sup_{\substack{\phi(s,s_0)>u \\ s\in S'}} G(s), \tag{7.33}$$

which is decreasing and is right continuous, with left limits. We have, for all $u > 0$,

$$\lim_{v\downarrow u} M(v) = \overline{M}(u) \leq M(u) = \lim_{v\uparrow u} \overline{M}(v). \tag{7.34}$$

Also, for all integers n, there exist numbers $\{d_n\}$ such that

$$\{u > 0 : M(u) \geq n\} = (0, d_n] \quad \text{and} \quad \{u > 0 : \overline{M}(u) \leq n\} = [d_n, \infty). \tag{7.35}$$

Set $Z_n = \sup_{\phi(s,s_0)\geq d_n, s\in S'} G(s)$ and $\overline{Z}_n = \sup_{\phi(s,s_0)>d_n, s\in S'} G(s)$. It follows from (5.150) and (5.151), respectively, and the Borel–Cantelli Lemma that there exists an $n_0(\omega)$ such that, for all $n \geq n_0(\omega)$,

$$Z_n \geq M(d_n) - 1 - \sqrt{2\log n} \geq n - 1 - \sqrt{2\log n} \tag{7.36}$$

and using the fact that $\overline{M}(u) \leq M(u)$

$$\overline{Z}_n \leq M(d_n) + 1 + \sqrt{2\log n} \leq n + 1 + \sqrt{2\log n}. \tag{7.37}$$

For any $u \in (0, d_{n_0}]$ there exists some integer n such that $u\in (d_{n+1}, d_n]$. For this u, $M(u) \in [n, n+1)$. Consequently, for this u,

$$\sup_{\substack{\phi(s,s_0)\geq u \\ s\in S'}} G(s) \geq \sup_{\substack{\phi(s,s_0)\geq d_n \\ s\in S'}} G(s) \tag{7.38}$$

$$\geq n - 1 - \sqrt{2\log n}$$

$$\geq M(u) - 2 - \sqrt{2\log M(u)}$$

and

$$\sup_{\substack{\phi(s,s_0)>u \\ s\in S'}} G(s) \;\leq\; \sup_{\substack{\phi(s,s_0)>d_{n+1} \\ s\in S'}} G(s) \tag{7.39}$$

$$\leq\; n+2+\sqrt{2\log(n+1)}$$

$$\leq\; \overline{M}(u)+2+\sqrt{2\log(\overline{M}(u)+1)}.$$

Combining (7.38) and (7.39) we see that

$$M(u)-2-\sqrt{2\log M(u)} \;\leq\; \sup_{\substack{\phi(s,s_0)\geq u \\ s\in S'}} G(s) \tag{7.40}$$

$$\leq\; \lim_{v\uparrow u}\ \sup_{\substack{\phi(s,s_0)>v \\ s\in S'}} G(s)$$

$$\leq\; \lim_{v\uparrow u} \overline{M}(v)+2+\sqrt{2\log(\overline{M}(v)+1)}$$

$$=\; M(u)+2+\sqrt{2\log(M(u)+1)},$$

which gives the upper bound in (7.29). $\qquad\square$

Proof of Theorem 7.1.4 By (5.183) it is enough to prove this theorem with $E(\delta)$ replaced by $m(\delta)$ and $\widehat{E}(\delta)$ replaced by $\widehat{m}(\delta)$. We begin with (1). Consider Lemma 7.1.5 with $(S,\phi)=(K,d)$ and with G replaced by G', which we define as

$$G'(s)=\frac{G(s)-G(s_0)}{d(s,s_0)}. \tag{7.41}$$

We take $K'=\{d(s,s_0)>0\}$. The statement in (1) follows immediately from Lemma 7.1.5 applied to G'.

The statement in (2) also follows from Lemma 7.1.5, but applying it is more complicated. Let $\overline{K}:=K\times K$ and define the pseudometric

$$\overline{\phi}((s,t),(s',t'))=\big(E[[(G(s)-G(t))-(G(s')+G(t'))]^2]\big)^{1/2} \tag{7.42}$$

on $\overline{K}$. Let $\Delta=\{(s,s); s\in K\}$ and define

$$\overline{S}'=\{(s,t):\overline{\phi}((s,t),\Delta)>0\}. \tag{7.43}$$

For some $s_0\in K$ consider the Gaussian process

$$\overline{G}(s,t):=\frac{G(s)-G(t)}{\phi((s,t),(s_0,s_0))} \tag{7.44}$$

on $(\overline{S}', \overline{\phi})$. By Lemma 7.1.5 (2) applied to $\overline{G}$,

$$\lim_{\delta \to 0} \sup_{\substack{\overline{\phi}((s,t),(s_0,s_0)) \leq \delta \\ (s,t) \in \overline{S}'}} \frac{\overline{G}((s,t))}{\widehat{m}(\overline{\phi}((s,t),(s_0,s_0)))} = 1 \qquad \text{a.s.} \qquad (7.45)$$

Since $\overline{\phi}((s,t),(s_0,s_0)) = d(s,t)$, this is precisely statement (2).

To obtain statement (4) let us assume that $\widehat{m}(\delta)$ is bounded but that G does have an exact uniform modulus of continuity ω. By Lemma 7.1.5 (1), the condition that $\widehat{m}(\delta)$ is bounded implies that

$$\sup_{(s,t) \in \overline{K}'} \frac{\overline{G}((s,t))}{\overline{\phi}((s,t),(s_0,s_0))} < \infty \qquad \text{a.s.} \qquad (7.46)$$

or, equivalently, that

$$\sup_{s,t \in K} \frac{G(s) - G(t)}{d(s,t)} < \infty \qquad \text{a.s.} \qquad (7.47)$$

and hence that

$$\lim_{\delta \to 0} \sup_{\substack{d(s,t) \leq \delta \\ s,t \in K}} \frac{G(s) - G(t)}{d(s,t)} < \infty \qquad \text{a.s.} \qquad (7.48)$$

But now, writing

$$\lim_{\delta \to 0} \sup_{\substack{d(s,t) \leq \delta \\ s,t \in K}} \frac{G(s) - G(t)}{\omega(d(s,t))} = \lim_{\delta \to 0} \sup_{\substack{d(s,t) \leq \delta \\ x,y \in K}} \frac{G(s) - G(t)}{d(s,t)} \frac{d(s,t)}{\omega(d(s,t))} \qquad (7.49)$$

and using (7.48) and (7.7), which follows by our assumption that ω is an exact uniform modulus of continuity, we get

$$\lim_{\delta \to 0} \sup_{\substack{d(s,t) \leq \delta \\ s,t \in K}} \frac{G(s) - G(t)}{\omega(d(s,t))} = 0 \qquad \text{a.s.,} \qquad (7.50)$$

which contradicts the assumption that ω is an exact uniform modulus of continuity. Thus we have established statement (4).

The proof of statement (3) is exactly the same as the proof of statement (4), except that we use Lemma 7.1.1 as it applies to the the local modulus of continuity. $\qquad \square$

7.1.1 m-modulus of continuity

We now give another way to describe moduli of continuity that is also commonly used. In analogy with (7.1) we call $\omega : R_+ \to R_+$, $\omega(0) = 0$,

an exact uniform m-modulus of continuity for $\{G(u), u \in K\}$ on (S, τ)
if

$$\limsup_{\delta \to 0} \sup_{\substack{\tau(u,v) \leq \delta \\ u,v \in K}} \frac{|G(u) - G(v)|}{\omega(\delta)} = C \qquad \text{a.s.} \qquad (7.51)$$

for some constant $0 < C < \infty$. We call $\rho : R_+ \to R_+$, $\rho(0) = 0$, an
exact m-local modulus of continuity for $\{G(u), u \in S\}$ at some fixed
$u_0 \in (S, \tau)$ if

$$\limsup_{\delta \to 0} \sup_{\substack{\tau(u,u_0) \leq \delta \\ u \in S}} \frac{|G(u) - G(u_0)|}{\rho(\delta)} = C \qquad \text{a.s.} \qquad (7.52)$$

for some constant $0 < C < \infty$. We define a uniform and local m-modulus
and a lower uniform and local m-modulus similarly to the definitions for
uniform and local moduli on page 282. As with the moduli introduced
on page 282, we assume that the m-moduli are continuous and strictly
positive on $(0, \delta']$ for some $\delta' > 0$.

Lemma 7.1.6

$$\limsup_{\delta \to 0} \sup_{\substack{\tau(u,v) \leq \delta \\ u,v \in K}} \frac{|G(u) - G(v)|}{\omega(\delta)} \leq C \qquad a.s. \qquad (7.53)$$

implies

$$\lim_{\delta \to 0} \sup_{\substack{\tau(u,v) \leq \delta \\ u,v \in K}} \frac{|G(u) - G(v)|}{\omega(\tau(u,v))} \leq C \qquad a.s. \qquad (7.54)$$

Moreover, when ω is nondecreasing, (7.54) implies (7.53). Conversely,

$$\lim_{\delta \to 0} \sup_{\substack{\tau(u,v) \leq \delta \\ u,v \in K}} \frac{|G(u) - G(v)|}{\omega(\tau(u,v))} \geq C' \qquad a.s. \qquad (7.55)$$

implies

$$\limsup_{\delta \to 0} \sup_{\substack{\tau(u,v) \leq \delta \\ u,v \in K}} \frac{|G(u) - G(v)|}{\omega(\delta)} \geq C' \qquad a.s., \qquad (7.56)$$

and when ω is nondecreasing, (7.56) implies (7.55).

Proof Suppose (7.54) fails. Then there exist a sequences $\{\delta_k\}$, $\delta_k \downarrow 0$,
and $\{(u_k, v_k)\}$, with $\tau(u_k, v_k) = \delta_k$ such that

$$\sup_k \frac{|G(u_k) - G(v_k)|}{\omega(\delta_k)} > C. \qquad (7.57)$$

Consequently (7.53) fails. This shows that (7.53) implies (7.54).

Now suppose that (7.54) holds and ω is nondecreasing. Then, since $\tau(u,v) \leq \delta$ implies that $\omega(\tau(u,v)) \leq \omega(\delta)$, we get (7.53). The converse follows immediately (we display it for later reference). $\qquad\square$

The following simple consequences of Lemma 7.1.6 are used often.

Lemma 7.1.7

(1) *If (7.53) and (7.55) hold, with the same constant C, then ω is both an exact uniform modulus and an exact uniform m-modulus of continuity for G, with the same constant C.*

(2) *If ω is nondecreasing, then it is an exact uniform modulus of continuity for G if and only if it is an exact uniform m-modulus of continuity for G.*

(3) *The statements in (7.53)–(7.56) and (1) and (2) of this lemma remain valid when formulated for local moduli and local m-moduli.*

We use the expression m-modulus because whenever $\omega(\delta)$ or $\rho(\delta)$ are exact m-moduli of continuity they can essentially be taken to be monotone. Set $\widetilde{\omega}(\delta) := \inf_{\delta \leq x \leq \epsilon} \omega(x)$. Obviously $\widetilde{\omega}(\delta)$ is an increasing function on $[0, \epsilon]$. We now show that if

$$\limsup_{\substack{\delta \to 0}} \sup_{\substack{\tau(u,v) \leq \delta \\ u,v \in K}} \frac{|G(u) - G(v)|}{\omega(\delta)} \leq C \qquad \text{a.s.,} \qquad (7.58)$$

then

$$\limsup_{\substack{\delta \to 0}} \sup_{\substack{\tau(u,v) \leq \delta \\ u,v \in \widetilde{K}}} \frac{|G(u) - G(v)|}{\widetilde{\omega}(\delta)} \leq C \qquad \text{a.s.} \qquad (7.59)$$

To see this suppose that $\widetilde{\omega}(\delta) < \omega(\delta)$ for some $\delta \in (0, \epsilon]$. Then there exists a $\delta' > \delta$ such that $\widetilde{\omega}(\delta) = \widetilde{\omega}(\delta') = \omega(\delta')$. Note that

$$\sup_{\substack{\tau(u,v) \leq \delta \\ u,v \in \widetilde{K}}} \frac{|G(u) - G(v)|}{\widetilde{\omega}(\delta)} \leq \sup_{\substack{\tau(u,v) \leq \delta' \\ u,v \in K}} \frac{|G(u) - G(v)|}{\widetilde{\omega}(\delta')} \qquad (7.60)$$

$$= \sup_{\substack{\tau(u,v) \leq \delta' \\ u,v \in K}} \frac{|G(u) - G(v)|}{\omega(\delta')}.$$

Thus, if the left-hand side of (7.59) is greater than or equal to C, so is the left-hand side of (7.58). This shows that (7.58) implies (7.59). Using the facts that $\widetilde{\omega}(\,\cdot\,) \leq \omega(\,\cdot\,)$ and is nondecreasing and Lemma 7.1.7 (2), we get the following lemma.

Lemma 7.1.8 *If ω is an exact uniform m-modulus of continuity for G, then $\widetilde{\omega}$ is both an exact uniform m-modulus of continuity for G and an exact uniform modulus of continuity for G and in all three cases the constant C is the same. The same result holds for local moduli of continuity.*

Lemma 7.1.1 is used often to prove zero–one laws for moduli of continuity. Mimicking the proof of Lemma 7.1.1, it is easy to see that the following version of it holds for m-moduli of continuity.

Lemma 7.1.9 *Let $\{G(u), u \in (S, \tau)\}$ be a mean zero Gaussian process. Assume that d is continuous on (S, τ). Let $\omega : R_+ \to R_+$ and $K \subset S$ be a compact set. For the following three statements:*

$$\limsup_{\delta \to 0} \sup_{\substack{\tau(u,v) \le \delta \\ u,v \in K}} \frac{G(u) - G(v)}{\omega(\delta)} \le C \qquad \text{a.s. for some} \quad 0 \le C < \infty$$

$$(7.61)$$

$$\lim_{\delta \to 0} \sup_{\substack{\tau(u,v) \le \delta \\ u,v \in K}} \frac{d(u, v)}{\omega(\delta)} = 0 \qquad\qquad (7.62)$$

$$\limsup_{\delta \to 0} \sup_{\substack{\tau(u,v) \le \delta \\ u,v \in K}} \frac{G(u) - G(v)}{\omega(\delta)} = C' \qquad \text{a.s. for some} \quad 0 \le C' \le \infty$$

$$(7.63)$$

we have that (7.61) implies (7.62) implies (7.63). (Obviously, if (7.61) holds, then $C' < \infty$ in (7.63), but (7.62) implies (7.63) even if the limit superior in (7.61) is not finite almost surely.)

These results are also valid for the local m-modulus of continuity, that is, (7.61)–(7.63) hold with v replaced by u_0 and with the supremum taken over $u \in K$.

The study of the local moduli of continuity of Gaussian processes is simplified by the following corollary of Slepian's Lemma:

Lemma 7.1.10 *Let X and Y be Gaussian processes on (S, τ) and assume that there exists a $u_0 \in S$ such that, for all $s, t \in S$ in some neighborhood of u_0,*

$$E(Y(s) - Y(t))^2 \le E(X(s) - X(t))^2. \qquad (7.64)$$

Suppose that ψ is a local m-modulus of continuity for X at u_0, that is, that

$$\limsup_{\substack{\delta \to 0}} \sup_{\substack{\tau(u,u_0) \leq \delta \\ u \in S}} \frac{|X(u) - X(u_0)|}{\psi(\delta)} \leq C \qquad a.s. \qquad (7.65)$$

Then (7.65) holds with X replaced by Y.

Proof We can assume that $X(u_0) = Y(u_0) = 0$ since $X(t) - X(u_0)$ and $Y(t) - Y(u_0)$ also satisfy (7.64) in the same neighborhood of u_0, which we denote by $N_\delta = \{t | \tau(t, u_0) \leq \delta\}$. Let $t_1, \cdots, t_n \in N_\delta$ and be different from u_0. Set

$$\rho^2(\delta) = \sup_{1 \leq j \leq n} EX^2(t_j) \qquad (7.66)$$

and then set

$$\begin{align}
f^2(t_j) &= \rho^2(\delta) - EX^2(t_j) + EY^2(t_j), \qquad (7.67) \\
\overline{Y}(t_j) &= Y(t_j) + \eta\rho(\delta), \\
\overline{X}(t_j) &= X(t_j) + \eta' f(t_j),
\end{align}$$

where η and η' are independent standard normal random variables independent of $\{X(t_j)\}$ and $\{Y(t_j)\}$. Note that

$$\begin{align}
E(\overline{Y}(t_j) - \overline{Y}(t_k))^2 &= E(Y(t_j) - Y(t_k))^2 \qquad (7.68) \\
&\leq E(X(t_j) - X(t_k))^2 \leq E(\overline{X}(t_j) - \overline{X}(t_k))^2
\end{align}$$

and

$$E\overline{Y}(t_j)^2 = E\overline{X}(t_j)^2. \qquad (7.69)$$

Consequently,

$$E\overline{Y}(t_j)\overline{Y}(t_k) \geq E\overline{X}(t_j)\overline{X}(t_k) \qquad \forall 1 \leq j, k \leq n. \qquad (7.70)$$

By convention, $\rho(\delta) > 0$ for all $\delta > 0$. Therefore, by Slepian's Lemma, Lemma 5.5.1,

$$P\left(\max_j \overline{Y}(t_j) > C\psi(\delta)\right) \leq P\left(\max_j \overline{X}(t_j) > C\psi(\delta)\right). \qquad (7.71)$$

Equivalently,

$$P\left(\max_j \frac{Y(t_j)}{\psi(\delta)} + \frac{\eta\rho(\delta)}{\psi(\delta)} > C\right) \leq P\left(\max_j \frac{X(t_j)}{\psi(\delta)} + \frac{\eta f(t_j)}{\psi(\delta)} > C\right). \qquad (7.72)$$

We take limits to extend (7.72) to a countable set. Then, by separability, we can extend it further so that it holds for all $t \in N_\delta$. We extend the definition of $\rho(\delta)$ so that $\rho^2(\delta) := \sup_{t \in N_\delta} EX^2(t_j)$ and likewise extend the definition of $f(t_j)$ in (7.67) so that it holds for all $t \in N_\delta$.

It follows from (7.65) and the fact that (7.61) implies (7.62), which also holds for the local m-modulus of continuity, that

$$\lim_{\delta \to 0} \frac{\rho(\delta)}{\psi(\delta)} = 0. \tag{7.73}$$

Therefore, since $\sup_{t \in N_\delta} |f(t)| \le \rho(\delta)$, for any $\epsilon > 0$ we can find a δ such that

$$\sup_{t \in N_\delta} \left(\frac{X(t)}{\psi(\delta)} + \frac{\eta f(t)}{\psi(\delta)} \right) \le C + \epsilon \tag{7.74}$$

with probability greater than $1 - \epsilon$. Consequently, by (7.72),

$$\sup_{t \in N_\delta} \left(\frac{Y(t)}{\psi(\delta)} + \frac{\eta \rho(\delta)}{\psi(\delta)} \right) \le C + \epsilon \tag{7.75}$$

with probability greater than $1 - \epsilon$. Using this and (7.73) we see that

$$\limsup_{\delta \to 0} \sup_{\substack{\tau(u, u_0) \le \delta \\ u \in S}} \frac{Y(u)}{\psi(\delta)} \le C \qquad \text{a.s.} \tag{7.76}$$

By Lemma 7.1.9 for the local m-modulus of continuity, and the symmetry of Y, we can replace Y by its absolute value to get that (7.65) holds with X replaced by Y. $\qquad \square$

Remark 7.1.11 We generally use Lemma 7.1.10 in the contrapositive form, that is, if

$$\limsup_{\delta \to 0} \sup_{\substack{\tau(u, u_0) \le \delta \\ u \in S}} \frac{|Y(u) - Y(u_0)|}{\psi(\delta)} \ge C \qquad \text{a.s.} \tag{7.77}$$

then

$$\limsup_{\delta \to 0} \sup_{\substack{\tau(u, u_0) \le \delta \\ u \in S}} \frac{|X(u) - X(u_0)|}{\psi(\delta)} \ge C \qquad \text{a.s.} \tag{7.78}$$

Another implication of Lemma 7.1.10 is that if X and Y have stationary increments and

$$\sigma_X^2(h) \sim \sigma_Y^2(h) \tag{7.79}$$

(see (7.136)) and

$$\limsup_{\delta \to 0} \sup_{\substack{\tau(u, u_0) \le \delta \\ u \in S}} \frac{|X(u) - X(u_0)|}{\psi(\delta)} = C \qquad \text{a.s.,} \tag{7.80}$$

then

$$\limsup_{\substack{\delta \to 0 \\ \substack{\tau(u,u_0) \le \delta \\ u \in S}}} \sup \frac{|Y(u) - Y(u_0)|}{\psi(\delta)} = C \qquad \text{a.s.} \qquad (7.81)$$

This is easy to see since we can use the lemma, first on Y and $(1+\epsilon)X$ and then on X and $(1+\epsilon)Y$, for any $\epsilon > 0$.

The results in Lemma 7.1.10 and this remark are for the local m-modulus of continuity. It follows from Lemma 7.1.7 that when $\psi(\cdot)$ is increasing on $[0, \delta']$, they also apply to the local modulus of continuity. When ψ is not increasing one can replace ψ by its monotone minorant, as in Lemma 7.1.8, and then these results also hold for the local modulus of continuity.

7.2 Processes on R^n

It is difficult to say anything more about general Gaussian processes that sheds light on many of the specific processes we are interested in. To obtain useful results we restrict ourselves to Gaussian processes with stationary increments on R^n and, more particularly, on R^1. Since one of our primary concerns in this book is to study local times of Lévy processes, and since their associated processes are Gaussian processes with stationary increments in R^1, this is not a significant restriction for us.

We consider Gaussian processes $G = \{G(s), s \in [-T,T]^n\}$. As usual, we take $d(s,t) = \left(E(G(s) - G(t))^2\right)^{1/2}$. We also consider monotone majorants for d, that is, strictly increasing functions ϕ, with $\phi(0) = 0$, such that

$$d(s,t) \le \phi(|s-t|). \qquad (7.82)$$

Define

$$\widetilde{\omega}(\delta) = \widetilde{\omega}(\phi,\delta) = \phi(\delta) \left(\log 1/\delta\right)^{1/2} + \int_0^\delta \frac{\phi(u)}{u(\log 1/u)^{1/2}} \, du \qquad (7.83)$$

and

$$\widetilde{\rho}(\delta) = \widetilde{\rho}(\phi,\delta) = \phi(\delta) \left(\log\log 1/\delta\right)^{1/2} + \int_0^{1/2} \frac{\phi(\delta u)}{u(\log 1/u)^{1/2}} \, du. \qquad (7.84)$$

We show that the functions $\widetilde{\omega}(\delta)$ and $\widetilde{\rho}(\delta)$ are uniform or local moduli of continuity for Gaussian processes for which (7.82) holds. Whenever we use $\widetilde{\omega}$, we include the unstated hypothesis that ϕ is such that $\lim_{\delta \to 0} \widetilde{\omega}(\delta) = 0$, and similarly whenever we use $\widetilde{\rho}$. It follows from Lemma 6.4.6 that the finiteness of the integral in (7.83) is a sufficient condition for the continuity of G, and we show in Lemma 7.2.5 that, for a given ϕ such that $\phi(2u) \leq 2\phi(u)$, the integrals in (7.83) and (7.84) are both finite or both infinite.

In the next theorem we see that $\widetilde{\omega}$ is a uniform m-modulus of continuity for G and $\widetilde{\rho}$ is a local m-modulus of continuity for G as defined in Subsection 7.1.1.

Theorem 7.2.1 *Let G be a Gaussian process satisfying (7.82). Set $S = [-T,T]^n$; then*

$$\limsup_{\delta \to 0} \sup_{\substack{|u-v| \leq \delta \\ u,v \in S}} \frac{|G(u) - G(v)|}{\widetilde{\omega}(\delta)} \leq C \qquad a.s., \qquad (7.85)$$

and for each $u_0 \in S$

$$\limsup_{\delta \to 0} \sup_{\substack{|u-u_0| \leq \delta \\ u \in S}} \frac{|G(u) - G(u_0)|}{\widetilde{\rho}(\delta)} \leq C' \qquad a.s., \qquad (7.86)$$

where $0 \leq C, C' < \infty$.

Before proving this theorem we note the following estimates, which are also interesting on their own.

Lemma 7.2.2 *Let G be a Gaussian process satisfying (7.82). Set $S = [-T,T]^n$. Then for all $\delta > 0$ sufficiently small,*

$$E \sup_{\substack{|s-t| \leq \delta \\ s,t \in S}} G(s) - G(t) \leq C_{n,T}\widetilde{\omega}(\delta), \qquad (7.87)$$

and for each $t_0 \in S$

$$E \sup_{\substack{|s-t_0| \leq \delta \\ s \in S}} |G(s) - G(t_0)| \leq C_n \left(\phi(\delta) + \int_0^{1/2} \frac{\phi(\delta u)}{u(\log 1/u)^{1/2}} \, du \right). \qquad (7.88)$$

Proof The inequality in (7.87) follows from (6.61) by finding an upper bound for its right-hand side. We take for the majorizing measure, $\mu = \lambda/(2T)^n$, where λ is Lebesgue measure on R^n. By (7.82), the ball $B_d(s,\epsilon)$ of radius ϵ in the metric d contains a Euclidean ball of radius $\rho \geq \phi^{-1}(\epsilon)$. Taking into account the fact that a Euclidean ball of radius less than T with its center in S has at least 2^{-n} of its volume in S, we see that for all $s \in S$, $\mu(B_d(s,\epsilon)) \geq (D_{n,T}\phi^{-1}(\epsilon))^n$, where $D_{n,T}$ is a constant that depends only on n and T, so that for all $s \in S$,

$$\log \frac{1}{\mu(B_d(s,\epsilon))} \leq n \log \frac{C_{n,T}}{\phi^{-1}(\epsilon)} \qquad (7.89)$$

for some constant $C_{n,T} \geq 1$, which depends only on n and T. Therefore, for all δ' sufficiently small,

$$\int_0^{\delta'} \left(\log \frac{1}{\mu(B_d(s,u))} \right)^{1/2} du \qquad (7.90)$$

$$\leq \sqrt{n} \int_0^{\delta'} \left(\log \frac{C_{n,T}}{\phi^{-1}(u)} \right)^{1/2} du$$

$$= \sqrt{n} \int_0^{\phi^{-1}(\delta')} \left(\log \frac{C_{n,T}}{u} \right)^{1/2} d\phi(u)$$

$$\leq \sqrt{n} \left(\delta' \left(\log \frac{C_{n,T}}{\phi^{-1}(\delta')} \right)^{1/2} + \int_0^{\phi^{-1}(\delta')} \frac{\phi(u)}{u(\log(C_{n,T}/u))^{1/2}} du \right)$$

$$\leq C_{n,T} \left(\delta' \left(\log \frac{1}{\phi^{-1}(\delta')} \right)^{1/2} + \int_0^{\phi^{-1}(\delta')} \frac{\phi(u)}{u(\log(1/u))^{1/2}} du \right).$$

The constants are not necessarily the same at each step. (Note also that it is easy to show that when the integral in (7.83) is finite, $\lim_{u \to 0} \phi(u) \left(\log \frac{C_{n,T}}{u} \right)^{1/2} = 0$. However we do not need to consider this when we do the integration by parts in (7.90) because the contribution of this term is negative.)

Consequently, by (6.61),

$$E \sup_{\substack{d(s,t) \leq \delta' \\ s,t \in S}} G(s) - G(t) \qquad (7.91)$$

$$\leq C_{n,T} \left(\delta' \left(\log \frac{1}{\phi^{-1}(\delta')} \right)^{1/2} + \int_0^{\phi^{-1}(\delta')} \frac{\phi(u)}{u(\log(1/u))^{1/2}} du \right).$$

Setting $\delta' = \phi(\delta)$ and using (7.82), we get (7.87) (to be more precise, it follows from (7.82) that the subset of $S \times S$ for which $d(s,t) \leq \phi(\delta)$ contains the subset of $S \times S$ for which $|s - t| \leq \delta$).

To obtain (7.88), we use (6.59) with $X'(s) = G(s) - G(t_0)$. Choose $\delta' > 0$ such that $\phi^{-1}(\delta') \leq T$ and let $S_{\delta'} = \{s : |s - t_0| \leq \phi^{-1}(\delta')\} \cap S$. Consider the process $\{X'(s), s \in S_{\delta'}\}$. Since $\left(E(X'(s) - X'(t))^2 \right)^{1/2} = d(s,t)$, it follows from (7.82) that the d-diameter of $S_{\delta'}$ is less than or equal to $2\delta'$. By (6.59),

$$E \sup_{s \in S_{\delta'}} |G(s) - G(t_0)| \qquad (7.92)$$

$$\leq \frac{C}{2} \sup_{s \in S_{\delta'}} \int_0^{2\delta'} \left(\log \frac{1}{\mu(B_d(s,u) \cap S_{\delta'})} \right)^{1/2} du$$

$$\leq C \sup_{s \in S_{\delta'}} \int_0^{\delta'} \left(\log \frac{1}{\mu(B_d(s,u) \cap S_{\delta'})} \right)^{1/2} du,$$

where μ is any subprobability measure on $S_{\delta'}$. The second inequality in (7.92) is because the integrand is decreasing. (Note that because X' is symmetric in (6.59), it also holds for $|X'|$ if the constant on the right-hand side is doubled.)

We take $\mu = \lambda/\lambda(B(0, \phi^{-1}(\delta')))$ in (7.92), where λ is Lebesgue measure on R^n and $B(0, \phi^{-1}(\delta'))$ is a Euclidean ball of radius $\phi^{-1}(\delta')$ in R^n. Taking into account again the fact that a Euclidean ball of radius less than T with center in S has at least 2^{-n} of its volume in S, we see that, for all $s \in S$, $\mu(B_d(s, \epsilon) \cap S_{\delta'}) \geq (\phi^{-1}(\epsilon)/(2\phi^{-1}(\delta')))^n$, so that for all $s \in S$ and $\epsilon \leq \delta'$,

$$\log \frac{1}{\mu(B_d(s,\epsilon) \cap S_{\delta'})} \leq n \log \frac{2\phi^{-1}(\delta')}{\phi^{-1}(\epsilon)}. \tag{7.93}$$

As in (7.90), but with $C_{N,T}$ replaced by $2\phi^{-1}(\delta')$ followed by a change of variables,

$$\int_0^{\delta'} \left(\log \frac{1}{\mu(B_d(s,u)) \cap S_{\delta'}} \right)^{1/2} du \tag{7.94}$$

$$\leq \sqrt{n} \int_0^{\delta'} \left(\log \frac{2\phi^{-1}(\delta')}{\phi^{-1}(u)} \right)^{1/2} du$$

$$\leq \sqrt{n} \left(\delta' (\log 2)^{1/2} + \frac{1}{2} \int_0^1 \frac{\phi(\phi^{-1}(\delta')u)}{u(\log 2/u)^{1/2}} du \right).$$

Consequently, by (6.59),

$$E \sup_{\substack{|s-t_0| \leq \phi^{-1}(\delta') \\ s \in S}} |G(s) - G(t_0)| \tag{7.95}$$

$$\leq C_n \left(\delta' (\log 2)^{1/2} + \frac{1}{2} \int_0^1 \frac{\phi(\phi^{-1}(\delta')u)}{u(\log 2/u)^{1/2}} du \right).$$

Setting $\phi^{-1}(\delta') = \delta$ and making a change of variables, we get

$$E \sup_{\substack{|s-t_0| \leq \delta \\ s \in S}} G(s) - G(t) \tag{7.96}$$

$$\leq C_n \left(\phi(\delta) (\log 2)^{1/2} + \frac{1}{2} \int_0^{1/2} \frac{\phi(2\delta u)}{u(\log 1/u)^{1/2}} du \right).$$

Let $\widetilde{\phi}$ be any increasing function for which (7.82) holds ($\widetilde{\phi}$ need not be strictly increasing). By the triangle inequality,

$$d(s,t) \leq d(s,(s+t)/2) + d((s+t)/2,t) \leq 2\widetilde{\phi}(|s-t|/2). \tag{7.97}$$

Now take ϕ^* to be the smallest increasing function satisfying (7.82), which again need not be strictly increasing. By (7.97), $2\phi^*(|s-t|/2)$ is also a monotone majorant for $d(s,t)$. However, since ϕ^* is the smallest such majorant, we have $\phi^*(|s-t|) \le 2\phi^*(|s-t|/2)$ or, equivalently, that

$$\phi^*(2|s-t|) \le 2\phi^*(|s-t|). \tag{7.98}$$

Taking limits, we see that (7.96) holds with ϕ replaced by ϕ^*. Using (7.98) we get (7.88) with ϕ replaced by ϕ^* (we absorb the factor 2 into the constant). Thus it also holds as stated since $\phi^* \le \phi$. □

Proof of Theorem 7.2.1 By Theorem 6.3.3 and (7.90) with δ' replaced by $\phi(\delta)$,

$$\limsup_{\substack{\delta \to 0 \\ \substack{d(u,v) \le \phi(\delta) \\ u,v \in S}}} \sup \frac{|G(u) - G(v)|}{\widetilde{\omega}(\delta)} < \infty \qquad \text{a.s.} \tag{7.99}$$

Since $\phi(\delta) = o(\widetilde{\omega}(\delta))$, (7.99) is a tail random variable (as in the proof of Lemma 7.1.1). Consequently, the limit in (7.99) is equal to some finite constant. This implies (7.85) as we pointed out in the parenthetical remark following (7.91). We now obtain (7.86). Let $\delta_k := \delta_k(\theta) = \theta^{-k}$, $1 < \theta \le 2$ and $a_k = \text{median sup}_{|u-u_0| \le \delta_k, u \in S} G(u) - G(u_0)$. We use Theorem 5.4.3, (5.151), and the Borel–Cantelli Lemma to see that, for all $\epsilon > 0$,

$$\sup_{|u-u_0| \le \delta_k, u \in S} G(u) - G(u_0) \le a_k + \phi(\delta_k)\left((2+\epsilon)\log\log 1/\delta_k\right)^{1/2} \tag{7.100}$$

for all $k \ge k_0(\omega)$ almost surely. By Corollary 5.4.5 and (7.88),

$$a_k \le C_n \left(\phi(\delta_k) + \int_0^{1/2} \frac{\phi(\delta_k u)}{u(\log 1/u)^{1/2}}\, du \right). \tag{7.101}$$

Therefore, the right-hand side of (7.100) can be replaced by

$$\widehat{\rho}(\phi, \delta_k) := \phi(\delta_k)\left(\left((2+\epsilon)\log\log 1/\delta_k\right)^{1/2} + C_n\right) \tag{7.102}$$
$$+ C_n \left(\int_0^{1/2} \frac{\phi(\delta_k u)}{u(\log 1/u)^{1/2}}\, du \right).$$

(We write it this way for use again in Corollary 7.2.3.)

The inequality in (7.100) also holds with ϕ replaced by ϕ^*, and by (7.98) we get

$$\widehat{\rho}(\phi^*, \delta_k) \le 2\widehat{\rho}(\phi^*, \delta_{k+1}). \tag{7.103}$$

This enables us to interpolate for $\delta_{k+1} < \delta < \delta_k$ and obtain (7.86) with ϕ replaced by ϕ^*. But, since $\phi \ge \phi^*$, it also holds as stated. □

In the next corollary we add some conditions that enable us to sharpen Theorem 7.2.1 so that we get upper bounds that are best possible for certain classes of Gaussian processes (see Theorems 7.2.14 and 7.2.15).

Corollary 7.2.3 *Let $G = \{G(s), s \in [-T,T]^n\}$ be a Gaussian process for which (7.82) holds. Suppose furthermore that ϕ is such that, for all $\epsilon > 0$, there exists a $\theta > 1$ for which*

$$\phi(\theta u) \leq (1 + \epsilon)\phi(u) \tag{7.104}$$

uniformly in $[0, u_0]$, for some $u_0 > 0$. Then, if

$$\int_0^{1/2} \frac{\phi(\delta u)}{u(\log 1/u)^{1/2}} \, du = o\left(\phi(\delta) \left(\log \log 1/\delta\right)^{1/2}\right), \tag{7.105}$$

as $\delta \to 0$

$$\limsup_{\delta \to 0} \sup_{\substack{|u-u_0| \leq \delta \\ u \in [-T,T]^n}} \frac{|G(u) - G(u_0)|}{\left(2\phi^2(\delta) \log \log 1/\delta\right)^{1/2}} \leq 1 \qquad a.s., \tag{7.106}$$

and if

$$\int_0^{1/2} \frac{\phi(\delta u)}{u(\log 1/u)^{1/2}} \, du = o\left(\phi(\delta) \left(\log 1/\delta\right)^{1/2}\right), \tag{7.107}$$

as $\delta \to 0$

$$\limsup_{\delta \to 0} \sup_{\substack{|u-v| \leq \delta \\ u,v \in [-T,T]^n}} \frac{|G(u) - G(v)|}{\left(2n\phi^2(\delta) \log 1/\delta\right)^{1/2}} \leq 1 \qquad a.s. \tag{7.108}$$

Proof   The statement in (7.106) follows immediately from (7.102) since we can use (7.104) in the interpolation of (7.102).

We now obtain (7.108). Let $\{t_k\} = \{t_k(\delta)\}$, $t_1 = 0$, be the centers of a covering of $[-T,T]^n$ by closed Euclidean balls of radius δ. The number of balls in this covering is bounded by $C_{T,n}/\delta^n$, where $C_{T,n}$ is a constant depending on T and n. Let $S_k(\delta) = B(t_k, 2\delta)$ in the Euclidean metric. Let $s,t \in [-T,T]^n$ be such that $|s - t| \leq \delta$. It is easy to see that both s and t must lie in one of the balls $S_k(\delta)$. Therefore,

$$\sup_{\substack{|u-v| \leq \delta \\ u,v \in [-T,T]^n}} |G(u) - G(v)| = \sup_k \sup_{\substack{|u-v| \leq \delta \\ u,v \in S_k(\delta)}} , |G(u) - G(v)|, \tag{7.109}$$

and consequently

$$P\left(\sup_{\substack{|u-v| \leq \delta \\ u,v \in [-T,T]^n}} |G(u) - G(v)| \geq a(\delta) \right) \tag{7.110}$$

$$\leq \sum_{k=1}^{C_{T,n}/\delta^n} P\left(\sup_{\substack{|u-v|\leq\delta \\ u,v\in S_k(\delta)}} |G(u) - G(v)| \geq a(\delta)\right).$$

Take $\delta_j = \theta^{-j}$ and let

$$a_k(\delta_j) = \text{median} \sup_{\substack{|u-v|\leq\delta_j \\ u,v\in S_k(\delta_j)}} |G(u) - G(v)| + (2(1+\epsilon)n\phi^2(\delta_j)\log 1/\delta_j)^{1/2}.$$

$$(7.111)$$

It follows from (5.152) that

$$\sum_{k=1}^{C_{T,n}/\delta_j^n} P\left(\sup_{\substack{|u-v|\leq\delta_j \\ u,v\in S_k(\delta_j)}} |G(u) - G(v)| \geq a_k(\delta_j)\right) \qquad (7.112)$$

is a term in a convergent sequence.

We show below that

$$\text{median} \sup_{\substack{|u-v|\leq\delta \\ u,v\in S_k(\delta)}} |G(u) - G(v)| = o(2n\phi^2(\delta)\log 1/\delta)^{1/2}, \qquad (7.113)$$

so that

$$a(\delta_j) := (2(1+2\epsilon)n\phi^2(\delta_j)\log 1/\delta_j)^{1/2} \geq a_k(\delta_j) \qquad (7.114)$$

for all j sufficiently large. It now follows from (7.110) and (7.112) that, for all $j \geq j(\omega)$,

$$\sup_{\substack{|u-v|\leq\delta_j \\ u,v\in[-T,T]^n}} |G(u) - G(v)| \leq a(\delta_j). \qquad (7.115)$$

We can now interpolate as in the first part of this proof to get (7.108).

To complete the proof we verify (7.113). It suffices to show it for the mean. We have

$$E \sup_{\substack{|u-v|\leq\delta \\ u,v\in S_k(\delta)}} |G(u) - G(v)| \leq 2E \sup_{|u-t_k|\leq 2\delta} |G(u) - G(t_k)| \qquad (7.116)$$

$$\leq C_n\left(\phi(2\delta) + \int_0^{1/2} \frac{\phi(2\delta u)}{u(\log 1/u)^{1/2}} du\right)$$

by (7.88). Therefore, by the hypothesis (7.107), we get (7.113). $\qquad\square$

Let

$$I_{loc,\phi}(\delta) := \int_0^{1/2} \frac{\phi(\delta u)}{u(\log 1/u)^{1/2}} du \qquad (7.117)$$

and

$$I_{unif,\phi}(\delta) := \int_0^\delta \frac{\phi(u)}{u(\log 1/u)^{1/2}} \, du. \tag{7.118}$$

As we have already seen, these integrals play an important role in describing local and uniform moduli of continuity for Gaussian processes on R^1. We examine some of their properties in the next two lemmas, after we introduce the concept of regularly varying functions.

A function f is said to be regularly varying at zero with index α if

$$\lim_{x \to 0} \frac{f(ux)}{f(x)} = u^\alpha \tag{7.119}$$

for all $u \geq 0$. If $\alpha = 0$, f is also said to be slowly varying at 0. A regularly varying function f at zero with index α can be written in the form

$$f(x) = x^\alpha \beta(x) \exp\left(\int_1^x \frac{\epsilon(u)}{u} \, du\right), \tag{7.120}$$

where $\lim_{u \to 0} \epsilon(u) = 0$ and $\lim_{x \to 0} \beta(x) = C$ for some constant $C \neq 0$.

A function f is said to be regularly varying at infinity with index α if (7.119) holds at infinity. If $\alpha = 0$, f is also said to be slowly varying at infinity. A regularly varying function f at infinity with index α can be written in the form

$$f(x) = x^\alpha \beta(x) \exp\left(\int_x^\infty \frac{\epsilon(u)}{u} \, du\right), \tag{7.121}$$

where $\lim_{u \to \infty} \epsilon(u) = 0$ and $\lim_{x \to \infty} \beta(x) = C$ for some constant $C \neq 0$.

In general, there are no smoothness properties imposed on β and hence on f. A subclass of regularly varying functions that are easier to work with are those in which $\beta \equiv C$. A function f is called a normalized regularly varying function at zero with index α if it can be written in the form

$$f(x) = Cx^\alpha \exp\left(\int_1^x \frac{\epsilon(u)}{u} \, du\right) \tag{7.122}$$

for some constant $C \neq 0$. When $\alpha = 0$, f is also called a normalized slowly varying function at zero (see Bingham, Goldie and Teugels (1987)). A similar definition applies at infinity.

Let $f(x)$ be a regularly varying function as represented in (7.120) and write $\beta(x) = \beta(0)h(x)$. Then $f(x)$ is a normalized regularly varying function if the function $h(x)$ can be absorbed into the exponential term.

Writing $h(x) = \exp(\int_1^x h'(u)/h(u)\, du)$, we see that this can be done as long as $xh'(x)/h(x) \to 0$ as $x \to 0$ or, equivalently, as long as

$$x\beta'(x) \to 0 \qquad \text{or, equivalently,} \qquad \frac{xL'(x)}{L(x)} \to 0 \qquad (7.123)$$

as $x \to 0$, where

$$L(x) = \beta(x)\exp\left(\int_1^x \frac{\epsilon(u)}{u}\, du\right). \qquad (7.124)$$

Note that if $f(x)$ is a slowly varying function at zero, $f(1/x)$ is slowly varying at infinity.

Lemma 7.2.4 *A regularly varying function at zero, or infinity, with index not equal to zero is asymptotic to a monotonic function at zero, or infinity.*

Proof Suppose f is regularly varying at infinity with index $p \neq 0$. Then we can write $f(x) = x^p\beta(x)\exp\left(\int_1^x (\epsilon(u)/u)\, du\right)$. Clearly $f(x) \sim \widehat{f}(x) := \beta(\infty)x^p \exp\left(\int_1^x (\epsilon(u)/u)\, du\right)\, du$ at infinity. We have

$$(\widehat{f}(x))' = x^{p-1}(p + \epsilon(x))\exp\left(\int_1^x (\epsilon(u)/u)\, du\right), \qquad (7.125)$$

which is strictly positive or negative according to the sign of p, for all x sufficiently large. The behavior at zero is handled in the same way. $\square$

Other properties of regularly varying functions that we use in this book are summarized in Section 14.7.

Lemma 7.2.5 *Let ϕ, $\phi(0) = 0$, be continuous and nondecreasing on $[0, 1/2]$ and satisfy $\phi(2u) \leq 2\phi(u)$. Then, for all $\delta > 0$ sufficiently small,*

$$I_{unif,\phi}(\delta) \leq 2I_{loc,\phi}(\delta) \qquad (7.126)$$

$$I_{loc,\phi}(\delta) \leq 2\sqrt{2}\left(I_{unif,\phi}(\delta) + \phi(\delta)(\log 1/\delta)^{1/2}\right). \qquad (7.127)$$

If ϕ is regularly varying at zero with index $0 < \alpha \leq 1$,

$$I_{loc,\phi}(\delta) \leq O(\phi(\delta)). \qquad (7.128)$$

Proof For (7.126) we note that

$$I_{loc,\phi}(\delta) = \int_0^\delta \frac{\phi(u/2)}{u(\log 1/u - \log 1/2\delta)^{1/2}}\, du \qquad (7.129)$$

$$\geq \frac{1}{2}\int_0^\delta \frac{\phi(u)}{u(\log 1/u - \log 1/2\delta)^{1/2}}\, du > \frac{1}{2}I_{unif,\phi}(\delta).$$

For (7.127), we have

$$
\begin{aligned}
I_{loc,\phi}(\delta) &= \int_0^\delta \frac{\phi(u/2)}{u(\log 1/u - \log 1/2\delta)^{1/2}}\, du \qquad\qquad (7.130)\\
&\leq \int_0^{\delta^2} \frac{\phi(u/2)}{u(\log 1/u - \log 1/2\delta)^{1/2}}\, du \\
&\quad + \phi(\delta) \int_{\delta^2}^\delta \frac{du}{u(\log 1/u - \log 1/2\delta)^{1/2}} \\
&\leq \sqrt{2}\left(\int_0^{\delta^2} \frac{\phi(u/2)}{u(\log 1/u)^{1/2}}\, du + 2\phi(\delta)(\log 1/\delta)^{1/2} \right).
\end{aligned}
$$

Here the estimate of the second integral in (7.130) follows because $\log 1/u \geq 2\log 1/\delta$ on $[0, \delta^2]$. The third integral is easy to evaluate and bound (its integrand is the derivative of $-2(\log 1/u - \log 1/2\delta)^{1/2}$).

Using the fact established at the end of the proof of Lemma 7.2.2, that we can replace $\phi(u/2)$ by $2\phi(u)$, we get (7.127).

By (7.120), if ϕ is regularly varying at zero with index $0 < \alpha \leq 1$,

$$
\begin{aligned}
\frac{\phi(u\delta)}{\phi(\delta)} &= u^\alpha \frac{\beta(u\delta)}{\beta(\delta)} \exp\left(\int_{u\delta}^\delta \frac{\epsilon(u)}{u}\, du \right) \qquad\qquad (7.131)\\
&\leq u^\alpha \frac{\beta(u\delta)}{\beta(\delta)} \exp\left(\sup_{0<s\leq\delta} \epsilon(s) \log 1/u \right). \qquad (7.132)
\end{aligned}
$$

Therefore, given $\eta > 0$ for all $\delta > 0$ sufficiently small,

$$
\frac{\phi(u\delta)}{\phi(\delta)} \leq Cu^{\alpha-\eta} \qquad \forall u \in [0, 1/2]. \qquad\qquad (7.133)
$$

Thus

$$
I_{loc,\phi}(\delta) \leq C\phi(\delta) \int_0^{1/2} \frac{u^{\alpha-\eta}}{u(\log 1/u)^{1/2}}\, du. \qquad\qquad (7.134)
$$

For any $\alpha > 0$ we can choose η so that $\alpha - \eta > 0$. Consequently, (7.128) follows from (7.134). $\square$

Remark 7.2.6 It is not so simple to evaluate $I_{loc,\phi}$. In Lemma 7.6.5 we give examples in which $I_{loc,\phi}(\delta)$ grows faster than $\phi(\delta)(\log\log 1/\delta)^{1/2}$ and more slowly than $\phi(\delta)(\log 1/\delta)^{1/2}$, the familiar growth rates for the local and uniform moduli of continuity of many Gaussian processes. When

$$
\phi(\delta)(\log 1/\delta)^{1/2} = O(I_{unif,\phi}(\delta)) \qquad\qquad (7.135)
$$

it follows from Lemma 7.2.5 that $I_{loc,\phi}(\delta)$ and $I_{unif,\phi}(\delta)$ are comparable.

In this case it is easier to see what $I_{loc,\phi}(\delta)$ looks like since it is generally simpler to estimate $I_{unif,\phi}(\delta)$.

A function f on R_+ with $f(0) = 0$ is said to be of type A if f is regularly varying with index 1 and $f(x)/x$ is nonincreasing on $(0, \delta]$ for some $\delta > 0$.

Let $G = \{G(x), x \in [-1,1]\}$ be a Gaussian process with stationary increments. Set

$$\sigma^2(|x - y|) := \sigma_G^2(|x - y|) = E(G(x) - G(y))^2. \tag{7.136}$$

We obtain information on the lower bounds for uniform moduli of continuity of G under the following conditions:

$$\sigma^2(u) \text{ is concave for } 0 \le u \le h \text{ for some } h > 0. \tag{7.137}$$

$\sigma^2(u)$ is a regularly varying function at 0 with index $0 \le \alpha < 1$
or is of type A . $\tag{7.138}$

$\sigma^2(u)$ is a normalized regularly varying function at 0 with index $0 \le \alpha < 1$ or is a normalized regularly varying function of type A.
$\tag{7.139}$

This is because, under these conditions, the covariance of nonoverlapping increments of G can be made close to zero, and hence, by Slepian's Lemma, the increments can be treated as though they are independent. We develop some technical lemmas that enable us to exploit this idea.

Lemma 7.2.7 *Let $G = \{G(x), x \in [0,1]\}$ be a Gaussian process with stationary increments for which (7.137) holds and let $0 < a < b < c < d \le h$. Then*

$$E(G(d) - G(c))(G(b) - G(a)) \le 0. \tag{7.140}$$

Proof   We have

$$E(G(d) - G(c))(G(b) - G(a)) \tag{7.141}$$
$$= \tfrac{1}{2} \left(\sigma^2(d - a) + \sigma^2(c - b) - \sigma^2(d - b) - \sigma^2(c - a) \right).$$

The lemma follows from the concavity of σ^2 since $d-a > d-b$, $c-b < c-a$ and the midpoint of the line segment between $\sigma^2(d - a)$ and $\sigma^2(c - b)$ has the same x-coordinate as the midpoint of the line segment between $\sigma^2(d - b)$ and $\sigma^2(c - a)$. $\square$

Alternately, we can write

$$E(G(d) - G(c))(G(b) - G(a)) \tag{7.142}$$
$$= \tfrac{1}{2} \left(\{\sigma^2(d - a) - \sigma^2(d - b)\} - \{\sigma^2(c - a) - \sigma^2(c - b)\} \right)$$

and use the fact that for a concave function the increments over intervals of the same length (here $b - a$) is decreasing.

Lemma 7.2.8 *Let* $G = \{G(x), x \in [0, 1]\}$ *be a Gaussian process with stationary increments for which (7.138) holds. Let*

$$\xi_k = \frac{G(\frac{k+1}{N}) - G(\frac{k}{N})}{\sigma(\frac{1}{N})} \qquad k = 0, \dots, qN \tag{7.143}$$

for some $q = q(N) > 0$ *such that* qN *is an integer. Then, for* $j \neq k$,

$$E\xi_j\xi_k \leq h(q) + \sup_{j,k} \left[|j - k + 1|^\alpha \beta^{-1}\left(\frac{1}{N}\right) \tag{7.144} \right.$$
$$\left. \left| \beta\left(\frac{|j - k + 1|}{N}\right) - \beta\left(\frac{|j - k|}{N}\right) \right| \exp\left(\int_{1/N}^{(j-k)/N} \frac{\epsilon(u)}{u} \, du \right) \right]$$

for some function $h(q) = o(q)$ *as* $q \downarrow 0$ *for all* $N = N(q)$ *sufficiently large.*

Proof For $j, k = 0, \dots, qN$; $j \neq k$ we have

$$E\xi_j\xi_k = \frac{\sigma^2(\frac{|(j-k)+1|}{N}) + \sigma^2(\frac{|(j-k)-1|}{N}) - 2\sigma^2(\frac{|j-k|}{N})}{2\sigma^2(\frac{1}{N})}. \tag{7.145}$$

We write $\sigma^2(|u|) = |u|^\alpha L(|u|)$ and make use of the fact that $|u|^\alpha$ is concave to see, as in the previous lemma, that

$$\left| \frac{j - k + 1}{N} \right|^\alpha + \left| \frac{j - k - 1}{N} \right|^\alpha - 2\left| \frac{j - k}{N} \right|^\alpha \leq 0. \tag{7.146}$$

For $a = \pm 1$ we write

$$\sigma^2\left(\frac{|(j - k) + a|}{N} \right) = \left| \frac{|(j - k) + a|}{N} \right|^\alpha \left[L\left(\frac{|(j - k)|}{N} \right) \tag{7.147} \right.$$
$$\left. + \left(L\left(\frac{|(j - k) + a|}{N} \right) - L\left(\frac{|(j - k)|}{N} \right) \right) \right]$$

and use (7.146) to see that, for $j, k = 0, \dots, qN$; $j \neq k$,

$$E\xi_j\xi_k \leq \sum_{a=\pm 1} \frac{|j - k + a|^\alpha \left(L\left(\frac{|j-k+a|}{N} \right) - L\left(\frac{|j-k|}{N} \right) \right)}{2L(1/N)}. \tag{7.148}$$

We write $L(x) = \beta(x) \exp\left(\int_1^x \frac{\epsilon(u)}{u}\, du\right) := \beta(x)\widetilde{L}(x)$ (see (7.120)) and use the rearrangement

$$\frac{\beta(x)\widetilde{L}(x) - \beta(y)\widetilde{L}(y)}{\beta(c)\widetilde{L}(c)} = \left(\beta(x)\Big(\frac{\widetilde{L}(x)}{\widetilde{L}(y)} - 1\Big) + (\beta(x) - \beta(y))\right) \frac{\widetilde{L}(y)}{\beta(c)\widetilde{L}(c)}. \tag{7.149}$$

to see that, for $j > k$ and $a = 1$,

$$\frac{L\left(\frac{|j-k+1|}{N}\right) - L\left(\frac{|j-k|}{N}\right)}{L(1/N)} \tag{7.150}$$

$$= \beta^{-1}\Big(\frac{1}{N}\Big)\left[\beta\Big(\frac{|j-k+1|}{N}\Big)\left(\exp\Big(\int_{|j-k|/N}^{|j-k+1|/N} \frac{\epsilon(u)}{u}\, du\Big) - 1\right)\right.$$

$$\left. + \beta\Big(\frac{|j-k+1|}{N}\Big) - \beta\Big(\frac{|j-k|}{N}\Big)\right] \exp\Big(\int_{1/N}^{|j-k|/N} \frac{\epsilon(u)}{u}\, du\Big)$$

$$:= \mathcal{F} + \mathcal{G}.$$

We note that

$$\exp\Big(\int_{|j-k|/N}^{|j-k+1|/N} \frac{\epsilon(u)}{u}\, du\Big) - 1 \leq \frac{\epsilon^*(q)}{|j-k|}, \tag{7.151}$$

where $\epsilon^*(q) := \sup_{0 < u \leq q} |\epsilon(u)|$ and

$$\exp\Big(\int_{1/N}^{|j-k|/N} \frac{\epsilon(u)}{u}\, du\Big) \leq |j-k|^{\epsilon^*(q)}. \tag{7.152}$$

Therefore

$$|j - k + 1|^\alpha \,\mathcal{F} \leq C\frac{\epsilon^*(q)}{|j-k|^{1-\alpha-\epsilon^*(q)}}. \tag{7.153}$$

For $\alpha < 1$ this goes to zero as q goes to zero. When $\alpha = 1$, using the additional hypothesis, we see that the left-hand side of (7.152), which is $L(|j-k|/N)/L(1/N)$, is bounded by one for all N sufficiently large. Thus, in this case, $|j - k + 1|\,\mathcal{F}$ also goes to zero as q goes to zero.

When $a = -1$ we consider $k > j$ and get exactly the same terms as above. Thus we obtain (7.144). □

We can use Lemmas 7.2.7 and 7.2.8 to give conditions under which a Gaussian process with stationary increments has an exact uniform modulus of continuity.

Theorem 7.2.9 *Let $G = \{G(x), x \in [0, 1]\}$ be a Gaussian process with stationary increments for which (7.137) or (7.138) holds. Then G has* *an exact uniform modulus of continuity on* $[0,1]$ *with respect to the Euclidean metric, that is, (7.1) holds for some function* ω *with* $K = [0,1]$ *and* $\tau(u,v) = |u-v|$.

Proof We show that $\lim_{\delta \to 0} \widehat{E}(\delta) = \infty$ (see Theorem 7.1.4 (2)). By Lemma 5.5.6, we need only show that there is an unbounded number of 1-distinguishable points in $\{(G(u) - G(v))/d(u,v), u,v \in [0,1]\}$. Let ξ_k be as given in Lemma 7.2.8. Let $q(N) = M/N$ for some fixed integer M. Since

$$E(\xi_j - \xi_k)^2 = 2(1 - E\xi_j\xi_k), \qquad (7.154)$$

we see that $\{\xi_k\}_{k=1}^M$ are certainly 1-distinguishable when the last term in (7.144) goes to zero as $N \to \infty$. To see this, note that the left-hand side of (7.152) is bounded by $M^{\epsilon^*(q)}$ and hence the last term in (7.144) is bounded by

$$C\left|\beta\left(\frac{|j-k+1|}{N}\right) - \beta\left(\frac{|j-k|}{N}\right)\right| M^{\alpha + \epsilon^*(q)}.$$

For M fixed, this goes to 0 as $N \to \infty$ because $\beta(x)$ is continuous at 0. Since M can be taken as large as we like, the proof is completed in the case when (7.138) holds. However, under (7.137), by Lemma 7.2.7, $E\xi_j\xi_k \le 0$, so, of course, the proof also holds in this case. $\qquad \square$

We now take up the question of finding conditions under which we have equality in (7.106) and (7.108).

Theorem 7.2.10 *Let* $G = \{G(x), x \in [0,1]\}$ *be a Gaussian process with stationary increments for which (7.137) holds or (7.139) holds. Then*

$$\lim_{\delta \to 0} \sup_{\substack{|u-v| \le \delta \\ u,v \in [0,1]}} \frac{|G(u) - G(v)|}{(2\sigma^2(|u-v|)\log 1/|u-v|)^{1/2}} \ge 1 \qquad a.s. \qquad (7.155)$$

Proof We give the proof under the assumption that (7.139) holds. The proof under assumption (7.137) is essentially the same, but slightly easier. We consider the setup in Lemma 7.2.8. For some $\epsilon > 0$ we take q to be a positive constant that is small enough so that the right-hand side of (7.144) is less than ϵ. This is easy because, by hypothesis, the complicated term in (7.144) is identically zero.

Let $\{Z_n\}$ be a standard normal sequence. By Theorem 5.1.4,

$$\lim_{n \to \infty} P\left(\sup_{1 \le k \le n} \frac{Z_k}{(2\log n)^{1/2}} \ge 1 - \epsilon'\right) = 1 \qquad (7.156)$$

for all $\epsilon' > 0$. Then, by Slepian's Lemma, as in the proof of Corollary 5.5.2,

$$\lim_{N\to\infty} P\left(\sup_{1\le k\le qN} \frac{\xi_k}{(2\log qN)^{1/2}} \ge (1-\epsilon)^{1/2}(1-\epsilon')\right) = 1 \qquad (7.157)$$

for all $\epsilon, \epsilon' > 0$. Therefore, by first taking q small enough and then taking N large enough, we see that for all $\epsilon'' > 0$,

$$\lim_{N\to\infty} P\left(\sup_{1\le k\le qN} \frac{\xi_k}{(2\log N)^{1/2}} \ge (1-\epsilon'')\right) = 1. \qquad (7.158)$$

It follows from (7.158) that

$$\limsup_{|u-v|\to 0} \frac{|G(u)-G(v)|}{(2\sigma^2(|u-v|)\log 1/|u-v|)^{1/2}} \ge 1 \qquad \text{a.s.,} \qquad (7.159)$$

which is a different way of writing (7.155). $\qquad\square$

Remark 7.2.11 Apropos of requiring that $\sigma^2(h)$ is of type A when it is a regularly varying function of index one, when σ^2 is concave on $[0,\delta]$, it also has the property that $\sigma^2(x)/x$ is nonincreasing on $[0,\delta]$. Actually, we need something less than this for the proofs. Writing $\sigma^2(x) = xL(x)$, where L is slowly varying, we require that $\limsup_{h\to 0}\sup_{h\le x\le\delta}(L(x)/L(h)) < \infty$ for some $\delta > 0$.

Adding still more conditions on the slowly varying part of σ^2, we can go beyond regularly varying functions of index one. For example, suppose that $\{G(x), x \in [0,1]\}$ is a Gaussian process for which $\sigma^2(h) = |h|^\alpha$ for some $1 < \alpha < 2$ (the existence of such a process is shown in Example 6.4.7). Define

$$\xi_k = \frac{G(\frac{k\ell+1}{N}) - G(\frac{k\ell}{N})}{\sigma(\frac{1}{N})} \qquad k = 0,\ldots,qN \qquad (7.160)$$

for some integer ℓ. Then, for $j \ne k$ and ℓ sufficiently large,

$$E\xi_j\xi_k \le \frac{\alpha(\alpha-1)}{(|j-k|\ell)^{2-\alpha}}. \qquad (7.161)$$

Obviously this can be made arbitrarily close to zero by taking ℓ large enough. Thus we get (7.157) with q replaced by q/ℓ. This does not affect the proof of Theorem 7.2.10, so (7.155) holds in this case also.

We obtain information on lower bounds for local moduli of continuity of G under conditions that are slightly different from those given in (7.137)–(7.139). We use the following conditions:

$\sigma^2(t+h) - \sigma^2(t) \leq \sigma^2(h)$ for $t, h > 0,$ and for all $\epsilon > 0$ there exists
$\theta > 0$ such that $\sigma^2(\theta u) \leq \epsilon \sigma^2(u)$ for all $|u| \leq u_0;$ $\qquad$ (7.162)
$\sigma^2(u)$ is a normalized regularly varying function at 0 with index
$0 < \alpha < 2.$ $\qquad$ (7.163)

Theorem 7.2.12 *Let $G = \{G(x), x \in [0,1]\}$ be a Gaussian process with stationary increments for which (7.162) or (7.163) holds. Then*

$$\limsup_{\substack{\delta \to 0 \\ u \leq \delta}} \frac{|G(u) - G(0)|}{(2\sigma^2(u) \log\log 1/u)^{1/2}} \geq 1 \qquad a.s. \qquad (7.164)$$

Proof Suppose that (7.163) holds. Let $0 < \theta < 1$ and $t_k = \theta^k/N$, $k = 0, \ldots, \infty$ and set

$$\widetilde{\xi}_k = \frac{G(t_k) - G(0)}{\sigma(t_k)}. \qquad (7.165)$$

We show that

$$E\widetilde{\xi}_j \widetilde{\xi}_k = \mathrm{o}(\theta, N), \qquad (7.166)$$

where $\mathrm{o}(\theta, N)$ is a function such that, for all $\delta > 0$, $\mathrm{o}(\theta, N) \leq \delta$ for all $\theta > 0$ sufficiently small and for $N \geq N_0(\theta, \delta)$ for some $N_0(\theta, \delta)$ sufficiently large.

We write

$$E\widetilde{\xi}_j \widetilde{\xi}_k = \frac{\sigma^2(t_k) - \sigma^2(t_k - t_j) - \sigma^2(t_j)}{2\sigma(t_j)\sigma(t_k)} + \frac{\sigma(t_j)}{\sigma(t_k)} \qquad (7.167)$$

and represent $\sigma^2(h) = h^\alpha \widetilde{L}(h)$, where $\widetilde{L}(h)$ is given as in (7.124) with $\beta \equiv C$. Note that for $j > k$,

$$\frac{\widetilde{L}(t_j)}{\widetilde{L}(t_k)} \leq \exp\left(\int_{t_j}^{t_k} \frac{|\epsilon(u)|}{u} \, du\right) \qquad (7.168)$$
$$\leq \theta^{-(j-k)\epsilon^*(t_k)},$$

where $\epsilon^*(t_k) = \sup_{u \leq t_k} |\epsilon(u)|$. Therefore, for $j > k$ we have

$$\frac{\sigma(t_j)}{\sigma(t_k)} \leq (\theta^{j-k})^{\alpha/2 - \epsilon^*(t_k)/2}. \qquad (7.169)$$

Thus, for N sufficiently large,

$$\frac{\sigma(t_j)}{\sigma(t_k)} \leq \theta^{\alpha'} \qquad (7.170)$$

for some $\alpha' > 0$.

Considering (7.170), to obtain (7.166) we need only consider

$$\frac{\sigma^2(t_k) - \sigma^2(t_k - t_j)}{\sigma(t_j)\sigma(t_k)} \tag{7.171}$$

$$\leq \frac{t_k^\alpha - (t_k - t_j)^\alpha}{\sigma(t_j)\sigma(t_k)}\widetilde{L}(t_k) + \frac{(t_k - t_j)^\alpha|\widetilde{L}(t_k) - \widetilde{L}(t_k - t_j)|}{\sigma(t_j)\sigma(t_k)}$$

for $j > k$. Note that for θ sufficiently small

$$t_k^\alpha - (t_k - t_j)^\alpha \leq 2\alpha t_j t_k^{\alpha-1}. \tag{7.172}$$

Using this and (7.168), we see that the first term on the right-hand side in (7.171) is less than or equal to

$$2\alpha \frac{t_j^{1-\alpha/2}\widetilde{L}^{1/2}(t_k)}{t_k^{1-\alpha/2}\widetilde{L}^{1/2}(t_j)} \leq 2\alpha\theta^{(j-k)(1-\alpha/2-\epsilon^*(t_k))}. \tag{7.173}$$

The second term on the right-hand side in (7.171) is less than or equal to

$$\frac{\sigma(t_k)}{\sigma(t_j)}\left(1 - \frac{\widetilde{L}(t_k - t_j)}{\widetilde{L}(t_k)}\right) \leq 2\frac{\sigma(t_k)}{\sigma(t_j)}\frac{t_j\epsilon^*(t_k)}{t_k}$$

$$= \alpha\frac{t_j^{1-\alpha/2}\widetilde{L}^{1/2}(t_k)}{t_k^{1-\alpha/2}\widetilde{L}^{1/2}(t_j)}\epsilon^*(t_k)$$

since

$$1 - \frac{\widetilde{L}(t_k - t_j)}{\widetilde{L}(t_k)} \leq 2\int_{t_k-t_j}^{t_k}\frac{|\epsilon(u)|}{u}\,du \tag{7.174}$$

$$\leq 2\frac{t_j\epsilon^*(t_k)}{t_k}.$$

Thus this term is even smaller than the right-hand side of (7.173). Since θ can be taken arbitrarily close to zero, we get (7.166).

The hypotheses in (7.162) immediately give

$$E\widetilde{\xi}_j\widetilde{\xi}_k \leq o(\theta) \tag{7.175}$$

as is easily seen, since they imply that the first term on the right-hand side of (7.167) is less than or equal to zero, and, for all $\epsilon > 0$, θ can be chosen such that $\sigma(t_j)/\sigma(t_k) \leq \epsilon$.

The proof of this theorem now follows the proof of Theorem 7.2.10 since for $\{Z_k\}$ a standard normal sequence

$$\limsup_{k\to\infty}\frac{Z_k}{(2\log\log\theta^k/N)^{1/2}} = 1 \quad \text{a.s.} \tag{7.176}$$

for all $N > 1$ and $0 < \theta < 1$. $\square$

We now give a lower bound for the local modulus of continuity that may seem artificial but which is very useful in the study of moduli of continuity of associated Gaussian processes in Section 7.4.

Lemma 7.2.13 *Let $G = \{G(u), u \in [0,1]\}$ be a Gaussian process and let $f(u)$, $u \in [0,1]$ be a continuous real-valued function. Assume that, for all $s, t > 0$ sufficiently small,*

$$E(G(t) - G(s))^2 \geq c^2 |f^2(t) - f^2(s)| \tag{7.177}$$

for some $c > 0$. Let $\widetilde{f}^2(u) := \sup_{0 \leq s \leq u} |f^2(s) - f^2(0)|$. Then

$$\limsup_{\delta \downarrow 0} \sup_{u \leq \delta} \frac{|G(u) - G(0)|}{(2\widetilde{f}^2(u) \log \log 1/\widetilde{f}^2(u))^{1/2}} \geq c \qquad a.s. \tag{7.178}$$

Proof   Note that (7.177) can be written as

$$E(G(t) - G(s))^2 \geq E\left(cB(f^2(t)) - cB(f^2(s))\right)^2, \tag{7.179}$$

where B is standard Brownian motion. By Theorem 7.2.12,

$$\limsup_{\delta \downarrow 0} \sup_{u \leq \delta} \frac{|B(\widetilde{f}^2(u))|}{(2\widetilde{f}^2(u) \log \log 1/\widetilde{f}(u))^{1/2}} \geq 1 \qquad a.s. \tag{7.180}$$

Let U_δ denote the flat spots of $\widetilde{f}^2$, that is, $U_\delta = \{u \in (0, \delta] : \exists u' < u, \widetilde{f}^2(u') = \widetilde{f}^2(u)\}$. Let $T_\delta = (0, \delta] - U_\delta$. The supremum in (7.180) is effectively taken over T_δ. To see this, suppose that $v' \in U_\delta$. Let $v'' = \inf\{v : \widetilde{f}^2(v) = \widetilde{f}^2(v')\}$. Then $v'' \in T_\delta$ and $\widetilde{f}^2(v'') = \widetilde{f}^2(v')$. Consequently, (7.180) also holds if we replace $\sup_{u \leq \delta}$ by $\sup_{\{u \leq \delta\} \cap T_\delta}$.

Consider $\widetilde{f}^2(u)$. For each $u \in T_\delta$, either $f^2(u) > f^2(0)$ or $f^2(u) < f^2(0)$. Let $T_{\delta,1}$ denote those $u \in T_\delta$ for which $f^2(u) > f^2(0)$, and let $T_{\delta,2} = T_\delta - T_{\delta,1}$. It follows from the Blumenthal zero–one law for Brownian motion, Lemma 2.2.9, that

$$\limsup_{\substack{\delta \downarrow 0 \\ u \leq \delta \\ u \in T_{\delta,i}}} \frac{|B(\widetilde{f}^2(u))|}{(2\widetilde{f}^2(u) \log \log 1/\widetilde{f}(u))^{1/2}} = C_i \qquad a.s. \qquad i = 1, 2 \tag{7.181}$$

for some constants C_1 and C_2. Furthermore, by (7.180), at least one of the constants C_i is greater than or equal to one. Suppose $C_1 \geq 1$. In this case, $\widetilde{f}^2(u) = f^2(u) - f^2(0)$. Since

$$\{B(f^2(u) - f^2(0)); \ u \in T_{\delta,1}\} \overset{law}{=} \{B(f^2(u)) - B(f^2(0)); \ u \in T_{\delta,1}\},$$

we have

$$\limsup_{\substack{\delta \downarrow 0 \\ u \in \widetilde{T}_{\delta,1}}} \sup_{u \le \delta} \frac{|B(f^2(u)) - B(f^2(0))|}{(2\widetilde{f}^2(u) \log \log 1/\widetilde{f}(u))^{1/2}} \ge 1 \qquad \text{a.s.,} \qquad (7.182)$$

so that, obviously,

$$\limsup_{\delta \downarrow 0} \sup_{u \le \delta} \frac{|B(f^2(u)) - B(f^2(0))|}{(2\widetilde{f}^2(u) \log \log 1/\widetilde{f}(u))^{1/2}} \ge 1 \qquad \text{a.s.} \qquad (7.183)$$

The same argument gives (7.183) when $C_2 \ge 1$.

Finally we note that $x \log \log 1/x$ is increasing on $[0, \delta']$ for some $\delta' > 0$. Therefore by Remark 7.1.11 (see, in particular, the last paragraph), we get (7.178). □

We now list conditions under which we obtain exact uniform moduli and uniform m-moduli of continuity with the precise value of the constant.

Theorem 7.2.14 *Let* $G = \{G(x), x \in [-1, 1]\}$ *be a Gaussian process with stationary increments and let* σ^2 *be as defined in (7.136). If any of the following three sets of conditions holds:*

(1) $\sigma^2(h)$ *is a normalized regularly varying function at zero with index* $0 < \alpha < 1$ *or is a normalized regularly varying function of type A;*

(2) $\sigma^2(h)$ *is a normalized slowly varying function at zero that is asymptotic to an increasing function near zero and*

$$\int_0^{1/2} \frac{\sigma(\delta u)}{u(\log 1/u)^{1/2}} \, du = o\left(\sigma(\delta) (\log 1/\delta)^{1/2}\right); \qquad (7.184)$$

(3) $\sigma^2(u)$ *is concave for* $u \le h$ *for some* $h > 0$, *and (7.184) holds;*

then

$$\limsup_{\delta \to 0} \sup_{\substack{|u-v| \le \delta \\ u,v \in [-1,1]}} \frac{|G(u) - G(v)|}{(2\sigma^2(\delta) \log 1/\delta)^{1/2}} = 1 \qquad \text{a.s.} \qquad (7.185)$$

and

$$\lim_{\delta \to 0} \sup_{\substack{|u-v| \le \delta \\ u,v \in [-1,1]}} \frac{|G(u) - G(v)|}{(2\sigma^2(|u-v|) \log 1/|u-v|)^{1/2}} = 1 \qquad \text{a.s.} \qquad (7.186)$$

Proof The lower bound in (7.186) for each of the three sets of conditions is given in Theorem 7.2.10.

To obtain the upper bound in (7.185) we first note that we can replace

ϕ in (7.108) by $(1+\epsilon)\sigma$ for any $\epsilon > 0$. That we can do this in condition (1) follows from the fact that, by Lemma 7.2.4, a regularly varying function of index $\alpha > 0$ is asymptotic to an increasing function near zero. In condition (2) it is an hypothesis and it is obvious in condition (3). Thus, to complete the proof, we need only show that σ satisfies (7.104) and (7.107). That (7.104) holds under conditions (1) and (2) follows from (7.119). It is easy to see that it holds for concave functions. That (7.107) holds under condition (1) follows from (7.128). Finally, note that (7.107) is (7.184), so we simply add this as a hypothesis for conditions (2) and (3).

Using Lemma 7.1.7 (1), we get equality in both (7.185) and (7.186).

$\square$

Theorem 7.2.15 *Let $G = \{G(x), x \in [-1, 1]\}$ be a Gaussian process with stationary increments and let σ^2 be as defined in (7.136). If either of the following two sets of conditions holds:*

(1) *$\sigma^2(h)$ is a normalized regularly varying function at zero with index $0 < \alpha < 2$;*

(2) *$\sigma^2(h)$ is asymptotic to an increasing function near zero; $\sigma^2(t + h) - \sigma^2(t) \le \sigma^2(h)$, for all $t, h > 0$ sufficiently small; for all $\epsilon > 0$ there exists $\theta > 0$ such that $\sigma^2(\theta u) \le \epsilon \sigma^2(u)$, $|u| \le u_0$; and*

$$\int_0^{1/2} \frac{\sigma(\delta u)}{u(\log 1/u)^{1/2}} \, du = o\left(\sigma(\delta)(\log \log 1/\delta)^{1/2}\right); \quad (7.187)$$

then

$$\limsup_{\delta \to 0} \sup_{|u| \le \delta} \frac{|G(u) - G(0)|}{(2\sigma^2(\delta) \log \log 1/\delta)^{1/2}} = 1 \qquad a.s. \qquad (7.188)$$

and

$$\limsup_{\delta \to 0} \sup_{|u| \le \delta} \frac{|G(u) - G(0)|}{(2\sigma^2(|u|) \log \log 1/|u|)^{1/2}} = 1 \qquad a.s. \qquad (7.189)$$

Proof The lower bound in (7.189) for each of the sets of conditions is given in Theorem 7.2.12.

The upper bound in (7.188) follows from (7.106) once we show that (7.104) and (7.105) hold. That (7.104) holds under condition (1) follows from (7.119) and the fact that a regularly varying function of index $\alpha > 0$ is asymptotic to an increasing function near zero. That it holds under condition (2) can be seen by taking $h = \epsilon' t$ and noting that $\sigma^2((1 + \epsilon')t) \le \sigma^2(t) + \sigma^2(\epsilon' t)$ and $\sigma^2(\epsilon' t) \le \epsilon \sigma^2(t)$. That (7.105) holds under condition (1) follows from (7.128). Finally, note that (7.187) is (7.105), so we simply add this as a hypothesis for condition (2).

The proof is completed using Lemma 7.1.7. □

Example 7.2.16 It is easy to see that Theorems 7.2.15 and 7.2.14 give (2.14) and (2.15) for standard Brownian motion.

7.3 Processes with spectral densities

We often require that the increments variance of a Gaussian process with stationary increments, that is, σ^2 in (7.136), is a normalized regularly varying function at zero. In the next theorem we show that when the process has a spectral density, which is a normalized regularly varying function at infinity, σ^2 has this property. To be more explicit, we take

$$\sigma^2(h) = \frac{4}{\pi} \int_0^\infty \frac{\sin^2 \lambda h/2}{\theta(\lambda)} \, d\lambda, \qquad (7.190)$$

where

$$\int_0^\infty \frac{1 \wedge \lambda^2}{\theta(\lambda)} \, d\lambda < \infty. \qquad (7.191)$$

Theorem 7.3.1 *When θ is a regularly varying function at infinity with index $1 < p < 3$,*

$$\sigma^2(h) \sim C_p \frac{1}{h\,\theta(1/h)} \qquad \text{as } h \to 0, \qquad (7.192)$$

where

$$C_p = \frac{4}{\pi} \int_0^\infty \frac{\sin^2 s/2}{s^p} \, ds. \qquad (7.193)$$

When θ is a regularly varying at infinity with index 1,

$$\sigma^2(h) \sim \frac{2}{\pi} \int_{1/h}^\infty \frac{1}{\theta(\lambda)} \, d\lambda \qquad \text{as } h \to 0, \qquad (7.194)$$

and it is a slowly varying function at zero.

Furthermore, if θ is a normalized regularly varying function at infinity with index $1 \leq p < 3$, then σ^2 is a normalized regularly varying function.

It follows from (7.192) that when θ is a regularly varying function at infinity with index $1 < p < 3$, $\sigma^2(h)$ is a regularly varying function at zero with index $p - 1$.

Proof To obtain the asymptotic result (7.192), it suffices to show that

$$\lim_{h \to 0} h\,\theta(1/h) \frac{4}{\pi} \int_K^\infty \frac{\sin^2 \lambda h/2}{\theta(\lambda)} \, d\lambda = C_p \qquad (7.195)$$

because the contribution of the integral from 0 to K is $O(h^2)$. We write $\theta(\lambda) = \lambda^p L(\lambda)$, where L is a slowly varying function at infinity. By a change of variables,

$$h\,\theta(1/h)\,\frac{4}{\pi}\int_K^\infty \frac{\sin^2 \lambda h/2}{\theta(\lambda)}\,d\lambda \;=\; \frac{4}{\pi}\int_{Kh}^\infty \frac{\sin^2 s/2}{s^p}\frac{L(1/h)}{L(s/h)}\,ds \qquad (7.196)$$

$$=\; \frac{4}{\pi}\int_0^\infty \frac{\sin^2 s/2}{s^p}\frac{L(1/h)}{L(s/h)}I_{(Kh,\infty)}(s)\,ds.$$

By (7.124), for $Kh < 1$,

$$\left|\frac{L(1/h)}{L(s/h)}\right|I_{(Kh,\infty)}(s) \;=\; \frac{\beta(1/h)}{\beta(s/h)}\exp\left(\int_{s/h}^{1/h}\frac{\epsilon(u)}{u}\,du\right)I_{(Kh,\infty)}(s)$$

$$\leq\; 2\left(s^{\epsilon^*} + s^{-\epsilon^*}\right), \qquad (7.197)$$

where $\epsilon^* = \sup_{u>K}|\epsilon(u)|$. We take K large enough so that $p - \epsilon^* > 1$, $p + \epsilon^* < 3$, and the ratio of the two β terms is bounded by two. For this choice of K we can use the Dominated Convergence Theorem in (7.196) to get (7.195).

We now assume that θ is a normalized regularly varying function with index $1 < p < 3$ and show that $\sigma^2(h)$ is a normalized regularly varying function. It follows from (7.123) that this is the case if and only if

$$\lim_{h\downarrow 0}\frac{h(\sigma^2(h))'}{\sigma^2(h)} = p - 1. \qquad (7.198)$$

We write

$$\sigma^2(h) \;=\; \frac{4}{\pi}\int_0^M \frac{\sin^2 \lambda h/2}{\theta(\lambda)}\,d\lambda + \frac{4}{\pi}\int_M^\infty \frac{\sin^2 \lambda h/2}{\theta(\lambda)}\,d\lambda \quad (7.199)$$

$$:=\; \sigma_1^2(h) + \sigma_2^2(h)$$

and note that

$$(\sigma_1^2(h))' = \frac{2}{\pi}\int_0^M \frac{\lambda \sin \lambda h}{\theta(\lambda)}\,d\lambda \sim h\frac{2}{\pi}\int_0^M \frac{\lambda^2}{\theta(\lambda)}\,d\lambda \qquad (7.200)$$

as $h \downarrow 0$. Using integration by parts, we write

$$\sigma_2^2(h) \;=\; -\frac{4}{\pi}\int_M^\infty \int_0^\lambda \sin^2\frac{uh}{2}\,du\,\frac{d}{d\lambda}\frac{1}{\theta(\lambda)}\,d\lambda \qquad (7.201)$$

$$-\frac{1}{\theta(M)}\frac{4}{\pi}\int_0^M \sin^2\frac{uh}{2}\,du$$

$$:=\; \sigma_3^2(h) + \sigma_4^2(h).$$

Integrating we see that

$$\sigma_3^2(h) = -\frac{2}{\pi} \int_M^\infty \left(\lambda - \frac{\sin \lambda h}{h}\right) \frac{d}{d\lambda} \frac{1}{\theta(\lambda)} d\lambda. \tag{7.202}$$

Differentiating under the integral, we get

$$(\sigma_3^2(h))' = \frac{2}{\pi h^2} \int_M^\infty (\lambda h \cos \lambda h - \sin \lambda h) \frac{d}{d\lambda} \frac{1}{\theta(\lambda)} d\lambda \tag{7.203}$$

(we justify this step and show that this integral exists at the end of the proof). We express $\theta(\lambda) = \lambda^p L(\lambda)$, where L is a slowly varying function at infinity. We have

$$\frac{d}{d\lambda} \frac{1}{\theta(\lambda)} = -\frac{p + \delta(\lambda)}{\lambda^{p+1} L(\lambda)}, \tag{7.204}$$

where $\delta(\lambda) = \lambda L'(\lambda)/L(\lambda)$. Note that the hypothesis that θ is a normalized regularly varying function at infinity implies that $\lim_{\lambda \to \infty} \delta(\lambda) = 0$ (see (7.123)). Using (7.204) and a change of variables in (7.203), we see that

$$(\sigma_3^2(h))' = \frac{2h^{p-2}}{\pi} \int_{Mh}^\infty \frac{(s \cos s - \sin s)}{s^{p+1}} \frac{(p + \delta(s/h))}{L(s/h)} ds. \tag{7.205}$$

Therefore, by (7.192),

$$\begin{aligned} \lim_{h \downarrow 0} \frac{h(\sigma_3^2(h))'}{\sigma^2(h)} &= \lim_{h \to 0} \frac{2}{C_p \pi} \int_{Mh}^\infty \frac{(s \cos s - \sin s)}{s^{p+1}} \frac{(p + \delta(s/h)) L(1/h)}{L(s/h)} ds \\ &= \frac{2p}{C_p \pi} \int_0^\infty \frac{(s \cos s - \sin s)}{s^{p+1}} ds. \end{aligned} \tag{7.206}$$

That we can take the limit under the integral sign follows from the same argument used in (7.197) and the fact that, for $s \geq 0$, $|(s \cos s - \sin s)| = O(s^3 \wedge s)$.

We leave it to the reader to show that $(\sigma_4^2(h))' = O(h)$, and we show in (7.200) that the same is true for $(\sigma_1^2(h))'$. Consequently,

$$\lim_{h \downarrow 0} \frac{h(\sigma^2(h))'}{\sigma^2(h)} = \frac{2p}{C_p \pi} \int_0^\infty \frac{(s \cos s - \sin s)}{s^{p+1}} ds. \tag{7.207}$$

It is easy to see that the constant on the right-hand side of (7.207) is equal to $p - 1$ since, when $\theta(\lambda) = \lambda^p$, $\sigma^2(h) = C_p h^{p-1}$, and in this case $h(\sigma^2(h))'/\sigma^2(h) \equiv p - 1$.

As for the differentiation under the integral sign in (7.202), to obtain (7.203), let $F(\lambda, h) = \lambda - ((\sin \lambda h)/h)$. Using the facts that $|(1 - \cos x)/x| \leq 1$ and $|(\sin x)/x| \leq 1$, we see that $|(F(\lambda, h + \Delta) - F(\lambda, h))/\Delta| \leq 2\lambda/h$. Using this and (7.204), we see that we can use the Dominated Convergence Theorem to justify differentiating under the integral.

To obtain (7.194) we write

$$
\sigma^2(h) = \frac{4}{\pi} \int_0^{M/h} \frac{\sin^2 \lambda h/2}{\theta(\lambda)} d\lambda + \frac{4}{\pi} \int_{M/h}^\infty \frac{\sin^2 \lambda h/2}{\theta(\lambda)} d\lambda
$$

$$
:= \sigma_5^2(h) + \sigma_6^2(h). \tag{7.208}
$$

As above, we express $\theta(\lambda) = \lambda L(\lambda)$, where L is a slowly varying function at infinity. By Theorem 14.7.1,

$$
\sigma_5^2(h) \le Ch^2 \int_0^{M/h} \frac{\lambda^2}{\theta(\lambda)} d\lambda \sim C \frac{M^2}{L(M/h)} \tag{7.209}
$$

and

$$
\sigma_6^2(h) = \frac{2}{\pi} \int_{M/h}^\infty \frac{1}{\theta(\lambda)} d\lambda - \frac{2}{\pi} \int_{M/h}^\infty \frac{\cos \lambda h}{\theta(\lambda)} d\lambda \tag{7.210}
$$

$$
:= \sigma_7^2(h) + \sigma_8^2(h).
$$

We show below that, for any $\epsilon > 0$, we can choose M sufficiently large so that $|\sigma_8^2(h)| \le \epsilon \sigma_7^2(h)$, for $h = h(\epsilon)$ sufficiently small. Also, it follows from Theorem 14.7.2 that $\sigma_5^2(h) = o(\sigma_7^2(h))$ as $h \to 0$. Thus we get (7.194) but with the lower limit of the integral replaced by M/h. It also follows from Theorem 14.7.2 that this integral is a slowly varying function of h at zero. Consequently,

$$
\left(\frac{2}{\pi} \int_{M/h}^\infty \frac{1}{\theta(\lambda)} d\lambda \right) \left(\frac{2}{\pi} \int_{1/h}^\infty \frac{1}{\theta(\lambda)} d\lambda \right)^{-1} \to 1 \tag{7.211}
$$

as $h \downarrow 0$. Thus we get (7.194), as stated.

Let $M = (k_0 + 1/2)\pi$ in σ_8^2, where k_0 is odd. Then, by a change of variables, we have

$$
|\sigma_8^2| = \frac{2}{\pi h} \int_M^\infty \frac{\cos s}{\theta(s/h)} ds \tag{7.212}
$$

$$
= \sum_{k=k_0} \frac{2}{\pi h} \int_{(k+1/2)\pi}^{(k+3/2)\pi} \frac{\cos s}{\theta(s/h)} ds
$$

$$
= \sum_{k=k_0,\, k \text{ odd}}^\infty \frac{2}{\pi h} \int_{(k+1/2)\pi}^{(k+3/2)\pi} \frac{|\cos s|}{\theta(s/h)} \left(1 - \frac{\theta(s/h)}{\theta((s+\pi)/h)} \right) ds.
$$

Writing $\theta(s/h)/\theta((s+\pi)/h) = ((\theta(s/h)/\theta(1/h))\, (\theta(1/h)/\theta((s+\pi)/h)))$, we see that $\lim_{h\downarrow 0} \theta(s/h)/\theta((s+\pi)/h) = s/(s+\pi)$. Thus, for all $h_0, \epsilon > 0$, we can take $k_0 = k_0(h_0, \epsilon)$ sufficiently large such that

$$
|\sigma_8^2| \le \sum_{k=k_0,\, k \text{ odd}}^\infty \frac{2\epsilon}{\pi h} \int_{(k+1/2)\pi}^{(k+3/2)\pi} \frac{1}{\theta(s/h)} ds \tag{7.213}
$$

$$< \frac{2\epsilon}{\pi h} \int_{(k_0+1/2)\pi}^{\infty} \frac{1}{\theta(s/h)} \, ds = \epsilon \sigma_7^2(h) \qquad (7.214)$$

for all $h \in [0, h_0]$.

Finally, we show that (7.198) holds when θ is a normalized regularly varying function with index $p = 1$. We return to the paragraph containing (7.199)–(7.206) and replace M by M/h and p by one. The critical term in determining the asymptotic behavior of the derivative of σ^2 is (7.205). We have

$$\begin{aligned} h(\sigma_3^2(h))' &= \frac{4}{\pi h} \int_M^{\infty} \frac{\cos s}{\theta(s/h)} (1 + \delta(s/h)) \, ds \qquad (7.215) \\ &\quad - \frac{4}{\pi} \int_M^{\infty} \frac{\sin s}{s^2} \frac{(1 + \delta(s/h))}{L(s/h)} \, ds \\ &:= \sigma_9^2(h) + \sigma_{10}^2(h). \end{aligned}$$

We showed in the previous paragraph that $|\sigma_9^2(h)| = o(\sigma^2(h))$. By the same argument used in the first paragraph of this proof we see that $|\sigma_{10}^2(h)| = O(1/L(1/h))$, which is $o(\sigma^2(h))$ by Theorem 14.7.2. The other terms are not difficult to deal with using ideas already contained in this proof. □

Theorem 7.3.1 allows us to extend the lower bounds for the moduli of continuity of Gaussian processes with stationary increments given in Theorems 7.2.10 and 7.2.12 to Gaussian processes that have regularly varying spectral densities.

Theorem 7.3.2 *Let $G = \{G(x), x \in [0, 1]\}$ be a Gaussian process with stationary increments that has a spectral density that is regularly varying at infinity with index $-p$ for $1 < p < 3$. Then*

$$\limsup_{\delta \to 0} \sup_{u \leq \delta} \frac{|G(u) - G(0)|}{(2\sigma^2(u) \log \log 1/u)^{1/2}} \geq 1 \qquad a.s. \qquad (7.216)$$

Proof Let $1/\theta(\lambda)$ denote the spectral density of G, that is, $1/\theta(\lambda) = h(\lambda/\pi)$ in (5.256), and assume that (7.191) holds. The increments variance σ^2 is given by (7.190) and, as in (7.192),

$$\sigma^2(h) \sim C_p \frac{1}{h \, \theta(1/h)} \qquad \text{as } h \to 0. \qquad (7.217)$$

Since θ is regularly varying, it has the form

$$\frac{1}{\theta(\lambda)} = \frac{\beta(\lambda)}{\lambda^p} \tilde{L}(\lambda), \qquad (7.218)$$

where $\widetilde{L}$ is a normalized slowly varying function at infinity. Define

$$\frac{1}{\widetilde{\theta}(\lambda)} = \frac{\beta(\lambda)}{\lambda^p}\widetilde{L}(\lambda)I_{\{\lambda<\lambda_0\}} + \frac{\beta(\lambda_0)}{\lambda^p}\widetilde{L}(\lambda)I_{\{\lambda\geq\lambda_0\}}. \tag{7.219}$$

Clearly, for all $\epsilon > 0$, we can take $\lambda_0 = \lambda_0(\epsilon)$ sufficiently large so that

$$(1-\epsilon)\frac{1}{\theta(\lambda)} \leq \frac{1}{\widetilde{\theta}(\lambda)} \leq (1+\epsilon)\frac{1}{\theta(\lambda)} \tag{7.220}$$

for all $\lambda > 0$.

Note that $\widetilde{\theta}$ is a normalized regularly varying function at infinity of index $1 < p < 3$. Furthermore, since $1/\theta(\lambda)$ is the spectral density of a Gaussian process with stationary increments, $1/\widetilde{\theta}$ can also be taken to be the spectral density of a Gaussian process with stationary increments, say $\widetilde{G}$ (this is because $1/\widetilde{\theta}$ is equal to $1/\theta$ on $[0,\lambda_0)$ and is integrable on $[\lambda_0,\infty)$). Let $\widetilde{\sigma}^2$ be the increments variance of this process. By (7.192),

$$(1-2\epsilon)\sigma^2(h) \leq \widetilde{\sigma}^2(h) \leq (1+2\epsilon)\sigma^2(h) \qquad \text{for all } h \in [0,h_0] \tag{7.221}$$

for some $h_0 > 0$ sufficiently small.

By Theorem 7.3.1, $\widetilde{\sigma}^2$ is a normalized regularly varying function at zero with index $0 < \alpha < 2$. Therefore, by Theorem 7.2.12,

$$\lim_{\delta\to 0}\sup_{u\leq\delta} \frac{|\widetilde{G}(u) - \widetilde{G}(0)|}{(2\widetilde{\sigma}^2(u)\log\log 1/u)^{1/2}} \geq 1 \qquad \text{a.s.} \tag{7.222}$$

The inequality in (7.216) now follows from (7.221), Lemma 7.1.10, and Remark 7.1.11. $\qquad\square$

Theorem 7.3.3 *Let $G = \{G(x), x \in [0,1]\}$ be a Gaussian process with stationary increments that has a spectral density that is regularly varying at infinity with index $-p$ for $1 \leq p < 2$. Then*

$$\lim_{\delta\to 0}\sup_{|u-v|\leq\delta} \frac{|G(u) - G(v)|}{(2\sigma^2(|u-v|)\log 1/|u-v|)^{1/2}} \geq 1 \qquad a.s. \tag{7.223}$$

Let $1/\theta(\lambda)$ denote the spectral density of G. If $\theta(\lambda)$ is regularly varying at infinity with index 2 and $\theta(\lambda)/\lambda^2$ is nonincreasing as $\lambda \to \infty$, (7.223) continues to hold.

Proof We use the notation in the first paragraph of the proof of Theorem 7.3.2. If $\beta(\lambda) \geq \beta(\infty)$ for all $\lambda \geq \lambda_0$, for some λ_0, we write

$$\frac{1}{\theta(\lambda)} = \left(\frac{\beta(\lambda)}{\lambda^p}\widetilde{L}(\lambda)I_{\{\lambda<\lambda_0\}} + \frac{\beta(\infty)}{\lambda^p}\widetilde{L}(\lambda)I_{\{\lambda\geq\lambda_0\}}\right) \tag{7.224}$$

$$+ \frac{\beta(\lambda) - \beta(\infty)}{\lambda^p}\widetilde{L}(\lambda)I_{\{\lambda\geq\lambda_0\}}.$$

If there is no λ_0 with this property, there exists a sequence of points $\{\lambda_k\}$, with $\lim_{k\to\infty}\lambda_k = \infty$, such that $\beta(\lambda) - \beta(\lambda_k) \geq 0$ for all $\lambda \geq \lambda_k$. Choose one of these points, relabel it λ_0, and write

$$\frac{1}{\theta(\lambda)} = \frac{1}{\widetilde{\theta}(\lambda)} + \frac{\beta(\lambda) - \beta(\lambda_0)}{\lambda^p} \widetilde{L}(\lambda) I_{\{\lambda \geq \lambda_0\}}. \qquad (7.225)$$

We use this formulation because we want to write $1/\theta(\lambda)$ as a sum of positive terms. We now see that whichever way we write $1/\theta$ we can decompose it as $1/\theta(\lambda) = 1/\theta_1(\lambda) + 1/\theta_2(\lambda)$, where both $1/\theta_1(\lambda)$ and $1/\theta_2(\lambda)$ can be spectral densities of independent Gaussian processes with stationary increments. Denote these process G_1 and G_2. By checking their covariances, we see that

$$\{G(x), x \in [0,1]\} \stackrel{law}{=} \{G_1(x) + G_2(x), x \in [0,1]\}. \qquad (7.226)$$

Let σ_1^2 and σ_2^2 denote the increments variance of G_1 and G_2. It is easy to see, using (7.192) and (7.194), that for all $\epsilon > 0$ we can take λ_0 such that

$$(1-\epsilon)\sigma^2(h) \leq \sigma_1^2(h) \leq \sigma^2(h) \quad \text{and} \quad \sigma_2^2(h) \leq \epsilon\sigma^2(h) \quad \text{for all } h \in [0, h_0]. \qquad (7.227)$$

By Theorem 7.3.1, σ_1^2 is a normalized regularly varying function at zero with index $0 \leq \alpha < 1$. Therefore, by Theorem 7.2.10,

$$\lim_{\delta \to 0} \sup_{\substack{|u-v| \leq \delta \\ u,v \in [0,1]}} \frac{|G_1(u) - G_1(v)|}{(2\sigma_1^2(|u-v|)\log 1/|u-v|)^{1/2}} \geq (1-\epsilon)^{1/2} \quad \text{a.s.} \quad (7.228)$$

It follows from (7.228) that there exists an infinite sequence of points (u_k, v_k) with $\lim_{k\to\infty}|u_k - v_k| = 0$ for which the inequality is realized. Clearly

$$|G(u_k) - G(v_k)| = |(G_1(u_k) - G_1(v_k)) + (G_2(u_k) - G_2(v_k))|. \qquad (7.229)$$

Since G_1 and G_2 are independent and symmetric, we see that, at least with probability $1/2$, $|G(u_k) - G(v_k)| \geq |G_1(u_k) - G_1(v_k)|$ infinitely often. Consequently, (7.228) holds with probability at least $1/2$ when G_1 is replaced by G. Since ϵ can be taken arbitrarily small, we see that (7.223) holds with probability at least $1/2$. It follows from Lemma 7.1.1 that (7.223) holds almost surely.

We can handle the case when the spectral density is regularly varying of index 2 as long as $\sigma_1^2(h)$ is also of type A. We leave it to the reader to show that this is the case when $\theta(\lambda)/\lambda^2$ is nonincreasing as $\lambda \to \infty$. $\square$

Remark 7.3.4 In Theorem 7.3.3, if one restricts the hypothesis to a Gaussian process G with stationary increments that has a spectral density that is regularly varying at infinity with index $-p$ for $1 < p \leq 2$, there is equality in (7.223). This follows because, by (7.192) and Lemma 7.2.4, the increments variance of G is asymptotic to an increasing regularly varying function with index $0 < \alpha \leq 1$. Since it is regularly varying, (7.104) is satisfied and, by (7.128), so is (7.107). Therefore, by Corollary 7.2.3, we get that 1 is an upper bound in (7.223).

7.4 Local moduli of associated processes

In this section we obtain a lower bound for the local modulus of continuity of associated, and related, Gaussian processes that hold under weaker conditions than similar results for more general Gaussian processes. This is important for us because this is used in Section 9.5 to obtain local moduli of continuity of local times. An associated Gaussian process $G = \{G(t), t \in S\}$ is a mean zero Gaussian process with a covariance that is the 0-potential density of a strongly symmetric Borel right process X with continuous 0-potential densities (S is the state space of X). By Lemma 3.4.3, the covariance of G has the remarkable property that

$$EG(s)G(t) \leq (EG^2(s)) \wedge (EG^2(t)). \tag{7.230}$$

It follows from (7.230) that

$$\begin{aligned} E(G(t) - G(s))^2 &= EG^2(t) + EG^2(s) - 2EG(s)G(t) \quad (7.231) \\ &\geq |EG^2(t) - EG^2(s)|. \end{aligned}$$

This inequality is not interesting when G is stationary since $EG^2(t) = EG^2(s)$, but it is very interesting for processes with $G(0) = 0$ for some $0 \in S$, in which case it can be written as

$$E(G(t) - G(s))^2 \geq |\sigma_G^2(t) - \sigma_G^2(s)|, \tag{7.232}$$

where

$$\sigma_G^2(t) := E(G(t) - G(0))^2. \tag{7.233}$$

Remark 7.4.1 Let X be a strongly symmetric Borel right process with state space S and assume that X has local times. Let T be a terminal time for X. Let $u_T(x, y)$ be as give in (2.138) and assume that it is continuous on $S \times S$. Suppose G is a Gaussian process with covariance

$u_T(x, y)$. Then, by (3.201) and symmetry,

$$u_T(x, y) \leq u_T(x, x) \wedge u_T(y, y). \tag{7.234}$$

In other words, the covariance of G satisfies (7.230) and consequently (7.232).

We see by (7.232) that the hypotheses of Lemma 7.2.13 are satisfied and we have a lower bound for the local modulus of continuity of G. Before pursuing this we show how to obtain a result similar to (7.232) for stationary associated Gaussian processes.

Lemma 7.4.2 *Let $\widehat{G} = \{\widehat{G}(t), t \in S\}$, where S is an Abelian group, be a mean zero associated stationary Gaussian process with covariance $u(x, y) = u(x - y)$. Then*

$$E(\widehat{G}(t) - \widehat{G}(s))^2 \geq \frac{u(t) + u(s)}{2u(0)} |\sigma_{\widehat{G}}^2(t) - \sigma_{\widehat{G}}^2(s)|. \tag{7.235}$$

Proof We begin by noting that

$$\sigma_{\widehat{G}}^2(x) = 2(u(0) - u(x)) \tag{7.236}$$

so that

$$u(x) - u(y) = \frac{\sigma_{\widehat{G}}^2(y) - \sigma_{\widehat{G}}^2(x)}{2}. \tag{7.237}$$

Consider

$$\overline{G}(x) := \widehat{G}(x) - \frac{u(x)}{u(0)} \widehat{G}(0). \tag{7.238}$$

Note that

$$E\left(\overline{G}(x)\overline{G}(y)\right) = u(x - y) - \frac{u(x)u(y)}{u(0)}. \tag{7.239}$$

By Remark 3.8.3, $\overline{G}$ is the mean zero Gaussian process with covariance $u_{T_0}(x, y)$. Therefore, by Remark 7.4.1 and (7.237),

$$E(\overline{G}(x) - \overline{G}(y))^2 \geq |E(\overline{G}^2(x)) - E(\overline{G}^2(y))| \tag{7.240}$$

$$= \frac{1}{u(0)} |u^2(x) - u^2(y)| = \frac{(u(x) + u(y))}{2u(0)} |\sigma_{\widehat{G}}^2(x) - \sigma_{\widehat{G}}^2(y)|.$$

Also,

$$E(\overline{G}(x) - \overline{G}(y))^2 = E(\widehat{G}(x) - \widehat{G}(y))^2 - \frac{(u(x) - u(y))^2}{u(0)} \leq E(\widehat{G}(x) - \widehat{G}(y))^2. \tag{7.241}$$

Using (7.240) and (7.241), we get (7.235). $\square$

Remark 7.4.3 Let $\widehat{G}$ be as in Lemma 7.4.2 and $\overline{G}$ be as defined in (7.238). It follows from (7.237) and (7.235) that

$$\frac{(u(x) - u(y))^2}{u(0)} = \frac{\left(\sigma_{\widehat{G}}^2(y) - \sigma_{\widehat{G}}^2(x)\right)^2}{4u(0)} \tag{7.242}$$

$$\leq \frac{u(0)\left(E(\widehat{G}(x) - \widehat{G}(y))^2\right)^2}{(u(x) + u(y))^2}.$$

Furthermore, by stationarity,

$$E(\widehat{G}(x) - \widehat{G}(y))^2 = \sigma_{\widehat{G}}^2(x - y). \tag{7.243}$$

It follows from the equality in (7.241), (7.242), and (7.243) that

$$E(\overline{G}(x) - \overline{G}(y))^2 \geq \sigma_{\widehat{G}}^2(x - y)\left(1 - \frac{u(0)\sigma_{\widehat{G}}^2(x - y)}{(u(x) + u(y))^2}\right) \tag{7.244}$$

and from (7.244) that

$$E(\overline{G}(t + x) - \overline{G}(t))^2 \sim \sigma_{\widehat{G}}^2(x) \qquad \text{as} \quad |x| \to 0. \tag{7.245}$$

We define $\mathcal{G}$ to be the class of Gaussian processes with a continuous covariance that has at least one of the following properties:

(1) It is associated and stationary.
(2) It has covariance of the form $u_T(x, y)$ for some terminal time T and is equal to zero at 0.

When a Gaussian process has property (1), we sometimes say that it is Category (1) and similarly when it has property (2).

For Gaussian processes in class $\mathcal{G}$, we obtain the lower bound on the local modulus of continuity given in Theorem 7.2.12 more simply and with less stringent conditions on the covariance of the processes. We first note the following lemma.

Lemma 7.4.4 Let $G = \{G(x), x \in R^1\}$ be a Gaussian process in class $\mathcal{G}$ and let $\sigma^2(x) := E(G(x) - G(0))^2$. Let $\widetilde{\sigma}^2(u) := \sup_{|s| \leq u} \sigma^2(|s|)$. Then

$$\limsup_{\delta \downarrow 0} \frac{|G(u) - G(0)|}{(2\widetilde{\sigma}^2(u) \log \log 1/\widetilde{\sigma}^2(u))^{1/2}} \geq 1 \qquad a.s. \tag{7.246}$$

Proof Suppose that G is in Category (2). Then (7.177) holds with $c = 1$ by (7.231) and the fact that $G(0) = 0$. Thus, by Lemma 7.2.13, we get (7.246).

When G is in Category (1), it follows from (7.235) and the continuity of $u(x)$ that, for any $\epsilon > 0$, we have (7.177) for some $c = c(\epsilon)$ such that $\lim_{\epsilon \to 0} c(\epsilon) \uparrow 1$. As above, this leads to (7.246) but with the right-hand side replaced with $c(\epsilon)$. Since this is true for all $\epsilon > 0$, we obtain (7.246). $\square$

Theorem 7.4.5 *Let $G = \{G(x), x \in R^1\}$ be a Gaussian process in class $\mathcal{G}$ and let $\sigma^2(x) := E(G(x) - G(0))^2$. Suppose that $\sigma(u) = O(|u|^\alpha)$ for some $\alpha > 0$. Then*

$$\limsup_{\delta \downarrow 0} \frac{|G(u) - G(0)|}{u \leq \delta \, (2\sigma^2(u) \log \log 1/u)^{1/2}} \geq 1 \qquad a.s. \qquad (7.247)$$

Proof This follows from Lemma 7.4.4 and the fact that $u \log \log 1/u$ is increasing for $u \in [0, \delta]$ for some $\delta > 0$. Therefore, since $\widetilde{\sigma}(u) \geq \sigma(u)$, we can replace $\widetilde{\sigma}$ by σ in the denominator of (7.246) and then use the bound on σ in the hypothesis of this theorem to obtain (7.247). $\square$

The advantage of this result over Theorem 7.2.12 is that here we have a simple bound on σ at zero rather than a smoothness condition. (One can weaken the condition on σ a bit more since we only require that $\log \log 1/u < (1 + \epsilon) \log \log 1/\sigma(u)$ for all $\epsilon > 0$.)

Remark 7.4.6 All we used about processes G in Category (2) is (7.230) and the fact that $G(0) = 0$. It is not necessary for the covariance of G to be related to the potential density of a terminal time. We can also weaken the condition that $G(0) = 0$. It follows from Lemma 7.2.13 and (7.231) that when (7.230) holds

$$\limsup_{\delta \downarrow 0} \frac{|G(u) - G(0)|}{u \leq \delta \, (2\widetilde{\sigma}^2(u, 0) \log \log 1/\widetilde{\sigma}(u, 0))^{1/2}} \geq 1 \qquad a.s., \qquad (7.248)$$

where

$$\widetilde{\sigma}^2(u, 0) = \sup_{0 \leq s \leq u} |EG^2(s) - EG^2(0)| \qquad (7.249)$$

as long as $\sigma^2(u, 0) > 0$ for $u > 0$. When this does not hold we can get interesting bounds by taking the limsup in (7.248) over the set where it does hold.

It is important to understand how special are the Gaussian processes in class $\mathcal{G}$. If G satisfies (7.230) we show in (7.231) that

$$|EG^2(t) - EG^2(s)| \leq E(G(t) - G(s))^2. \qquad (7.250)$$

Suppose that we don't know whether G is an associated process. In this case all we can do is write $G^2(t) - G^2(s) = (G(t) - G(s))(G(t) + G(s))$ and use the Schwarz inequality to get

$$|EG^2(t) - EG^2(s)| \leq (E(G(t) - G(s))^2)^{1/2}(E(G(t) + G(s))^2)^{1/2}. \quad (7.251)$$

When $\widehat{G}$ is an associated stationary Gaussian process with covariance $u(t, s) = u(t - s)$, we have, by (7.235),

$$|\sigma_{\widehat{G}}^2(t) - \sigma_{\widehat{G}}^2(s)| \leq \frac{2u(0)}{u(t) + u(s)} \sigma_{\widehat{G}}^2(t - s). \quad (7.252)$$

Treating $\widehat{G}$ simply as a Gaussian process, we can apply the Schwarz inequality to $(\widehat{G}(x) - \widehat{G}(y))\widehat{G}(0)$ to obtain

$$|u(t) - u(s)| \leq u^{1/2}(0)\,\widehat{\sigma}_G(t - s). \quad (7.253)$$

By (7.237), this is equivalent to

$$|\widehat{\sigma}_G^2(t) - \widehat{\sigma}_G^2(s)| \leq 2u^{1/2}(0)\,\widehat{\sigma}_G(t - s). \quad (7.254)$$

Here are some more properties that distinguish the class $\mathcal{G}$ from general Gaussian processes.

Lemma 7.4.7 *Let* $G = \{G(x); x \in R^1\}$ *be a Gaussian process with stationary increments in class* $\mathcal{G}$ *for which* $\sigma_G^2(x) \not\equiv 0$. *Then*

$$|x| = O(\sigma_G^2(x)) \quad as \quad x \to 0, \quad (7.255)$$

and for all $\epsilon > 0$

$$\sigma_G(2x) \leq \sqrt{2}\,(1 + \epsilon)\,\sigma_G(x) \quad \forall\, |x| \quad sufficiently\ small. \quad (7.256)$$

Proof Suppose that $\liminf_{x \to 0} \sigma_G^2(x)/|x| = 0$, that is, (7.255) does not hold. Fix $y \in R^1$. Since $\sigma_G^2(y) = \sigma_G^2(-y)$, we may assume that $y > 0$. Fix $\epsilon > 0$ and choose $x > 0$ such that $\sigma_G^2(x)/x \leq \epsilon$ and $\sup_{0 \leq z \leq x} \sigma_G^2(z) \leq \epsilon y$. Write $y = kx + z$ for some positive integer k and $0 \leq z \leq x$.

Now suppose that G is in Category (2) in the definition of $\mathcal{G}$. By (7.232), which implies that σ_G^2 is subadditive, $\sigma_G^2(y) \leq k\sigma_G^2(x) + \sigma_G^2(z)$. Hence

$$\frac{\sigma_G^2(y)}{y} \leq \frac{k\sigma_G^2(x)}{kx} + \frac{\sigma_G^2(z)}{y} \leq 2\epsilon. \quad (7.257)$$

Since this is true for any $\epsilon > 0$, we see that $\sigma_G^2 \equiv 0$, which is a contradiction.

Suppose now that G is in Category (1). We use the same construction as in the first paragraph of this proof, except that now we choose $y \in R^1$ so that $\sup_{|z| \leq |y|} \sigma^2(z) \leq u(0)$. It follows from (7.235) and (7.236) that

$$\sigma_G^2(y) \leq \sigma_G^2((k-1)x + z) + \left(\frac{2u(0)}{u((k-1)x + z) + u(x)} \right) \sigma_G^2(x)$$

$$= \sigma_G^2((k-1)x + z)$$

$$+ \left(\frac{2u(0)}{2u(0) - (1/2)(\sigma_G^2((k-1)x + z) + \sigma_G^2(x))} \right) \sigma_G^2(x)$$

$$= \sigma_G^2((k-1)x + z)$$

$$+ \left(\frac{1}{1 - (1/4u(0))(\sigma_G^2((k-1)x + z) + \sigma_G^2(x))} \right) \sigma_G^2(x)$$

$$\leq \sigma_G^2((k-1)x + z) + 2\sigma_G^2(x). \tag{7.258}$$

Hence, by induction,

$$\sigma_G^2(y) \leq 2k\sigma_G^2(x) + \sigma_G^2(z). \tag{7.259}$$

As in (7.257), we see that $\sigma_G^2(x) = 0$ for $x \in [0, \delta]$ for some $\delta > 0$ (this is because of the restriction on y). However, since G has stationary increments, we see that $\sigma_G^2 \equiv 0$, again a contradiction.

The inequality in (7.256) follows easily from (7.232) when G is in Category (2) and holds with $\epsilon = 0$. To obtain (7.256) when G is in Category (1), let $z = x$ and $k = 1$ in (7.258) and use the first equality to see that, for any $\epsilon > 0$,

$$\sigma_G^2(2x) \leq \sigma_G^2(x) + \left(\frac{1}{1 - \epsilon} \right) \sigma_G^2(x) \tag{7.260}$$

for all $|x|$ sufficiently small. This gives (7.256). $\qquad \square$

For general Gaussian processes we can only say that $|x|^2 = O(\sigma_G^2(x))$ as $x \to 0$ and $\sigma_G(2x) \leq 2\sigma_G(x)$ for all x.

7.4.1 Gaussian processes associated with Lévy processes

Gaussian processes in class $\mathcal{G}$, defined on page 326, can be obtained from Lévy processes with various stopping times. Let $X = \{X(t), t \in R_+\}$ be a real-valued symmetric Lévy process, that is,

$$Ee^{i\lambda X(t)} = e^{-t\psi(\lambda)}, \tag{7.261}$$

where

$$\psi(\lambda) = 2 \int_0^\infty (1 - \cos \lambda u)\nu(du) \tag{7.262}$$

for ν a symmetric Lévy measure, that is, ν is symmetric, and

$$\int_0^\infty (1 \wedge x^2)\, \nu(dx) < \infty. \tag{7.263}$$

We assume that

$$\int_1^\infty \frac{1}{\psi(\lambda)}\, d\lambda < \infty. \tag{7.264}$$

Let $u^\alpha(x, y)$ denote the α-potential density of X,

$$u^\alpha(x, y) = \frac{1}{\pi} \int_0^\infty \frac{\cos \lambda(x - y)}{\alpha + \psi(\lambda)}\, d\lambda \tag{7.265}$$

(as given in (4.84)). For $\alpha \geq 0$, we define $G_\alpha = \{G_\alpha(x), x \in R^1\}$ to be a mean zero stationary Gaussian process with covariance $u^\alpha(x, y)$, with the understanding that we only consider G_0 when $\int_0^1 \psi^{-1}(\lambda) < \infty$. When $\alpha > 0$, $u^\alpha(x, y)$ is the 0-potential density of X killed at the end of an independent exponential time with mean $1/\alpha$, as in Section 3.5. When $u^0(x, y)$ is finite, X is a transient process. Set

$$\begin{aligned}
\sigma_\alpha^2(x - y) &:= E(G_\alpha(x) - G_\alpha(y))^2 \tag{7.266} \\
&= u^\alpha(x, x) + u^\alpha(y, y) - 2u^\alpha(x, y) \\
&= \frac{4}{\pi} \int_0^\infty \sin^2 \frac{\lambda(x - y)}{2}\, \frac{1}{\alpha + \psi(\lambda)}\, d\lambda
\end{aligned}$$

and note that

$$EG_\alpha^2(x) = \frac{1}{\pi} \int_0^\infty \frac{1}{\alpha + \psi(\lambda)}\, d\lambda. \tag{7.267}$$

It follows from (4.77) that $\sigma_0^2(\cdot)$ exists for all Lévy processes satisfying (7.264).

We define $\widetilde{G}_0 = \{\widetilde{G}_0(x), x \in R^1\}$ to be a mean zero Gaussian process with covariance

$$E\widetilde{G}_0(x)\widetilde{G}_0(y) = \tfrac{1}{2}\left(\sigma_0^2(x) + \sigma_0^2(y) - \sigma_0^2(x - y)\right), \tag{7.268}$$

(it follows from (5.252) that the right-hand side of (7.268) is positive definite). Using (7.268), we see that $\widetilde{G}(0) = 0$ and that $E(\widetilde{G}_0(x) - \widetilde{G}_0(y))^2 = \sigma_0^2(x - y)$, so that $\widetilde{G}_0$ has stationary increments. We only consider $\widetilde{G}_0$ when $\int_0^1 \psi^{-1}(\lambda) = \infty$. In this case, X is recurrent and $\widetilde{G}_0$ has covariance $u_{T_0}(x, y)$ (see (4.86) and note that $\sigma_0^2(x) = 2\phi(x)$ in (4.87)).

We also consider

$$\overline{G}_\alpha(x) = G_\alpha(x) - \frac{u^\alpha(x)}{u^\alpha(0)} G_\alpha(0) \tag{7.269}$$

for $\alpha \geq 0$. When $\alpha = 0$, X is transient, and, by Remark 3.8.3, $\overline{G}_0$ has covariance $u_{T_0}(x, y)$. When $\alpha > 0$, by Section 3.5, u^α is the 0-potential of X killed at the end of an independent exponential time with mean $1/\alpha$, so $\overline{G}_\alpha$ has covariance $u^\alpha_{T_0}(x, y)$.

It is clear that the processes G_α, $\alpha \geq 0$ are in Category (1) of $\mathcal{G}$, and the processes $\overline{G}_\alpha$, $\alpha \geq 0$, and $\widetilde{G}_0$ are in Category (2) of $\mathcal{G}$ (note that $\overline{G}_\alpha$ does not have stationary increments).

It follows from (7.235), that for $\alpha > 0$,

$$|\sigma_\alpha^2(x) - \sigma_\alpha^2(y)| \leq \frac{2u^\alpha(0)}{u^\alpha(x) + u^\alpha(y)} \sigma_\alpha^2(x - y) \tag{7.270}$$

and also for $\alpha = 0$ when $\int_0^1 \psi^{-1}(\lambda) < \infty$. When $\int_0^1 \psi^{-1}(\lambda) = \infty$, by (7.231) applied to $\widetilde{G}_0$,

$$|\sigma_0^2(x) - \sigma_0^2(y)| = |E\widetilde{G}_0^2(x) - E\widetilde{G}_0^2(y)| \leq E(\widetilde{G}_0(x) - \widetilde{G}_0(y))^2 = \sigma_0^2(x - y) \tag{7.271}$$

and similarly

$$|E\overline{G}_\alpha^2(x) - E\overline{G}_\alpha^2(y)| \leq E(\overline{G}_\alpha(x) - \overline{G}_\alpha(y))^2. \tag{7.272}$$

Lemma 7.4.8 *For all $\alpha \geq 0$,*

$$|x| = o(\sigma_\alpha^2(x)) \quad as \quad x \to 0. \tag{7.273}$$

For all $\epsilon > 0$ and $\alpha' > 0$ there exists an $h(\alpha') > 0$ such that

$$\sigma_\alpha^2(x) \leq \sigma_0^2(x) \leq (1 + \epsilon) \sigma_\alpha^2(x) \quad \forall \alpha \in [0, \alpha'] \text{ and } |x| \leq h(\alpha') \tag{7.274}$$

and

$$\sigma_\alpha(2x) \leq \sqrt{2}(1 + \epsilon) \sigma_\alpha(x) \quad \forall \alpha \in [0, \alpha'] \text{ and } |x| \leq h(\alpha'). \tag{7.275}$$

To appreciate the significance of (7.275) compare it with (7.97). Also compare (7.273) with (7.255).

Proof It suffices to prove (7.273) for $\alpha > 0$. By (4.76) for all $\epsilon > 0$ there exists a λ_0 such that $\psi(\lambda) \leq \epsilon\lambda^2$ for all $\lambda \geq \lambda_0$. Let $N \geq \lambda_0$ be such that $\alpha/N^2 \leq \epsilon$. Using the fact that, for $0 \leq x \leq \pi/2$, $\sin x = \int_0^x \cos u \, du \geq x \cos x$, we see that, for $x < 1/N$,

$$\sigma_\alpha^2(x) = \frac{4}{\pi} \int_0^\infty \sin^2 \frac{\lambda x}{2} \frac{1}{\alpha + \psi(\lambda)} d\lambda \tag{7.276}$$

$$\geq \frac{x^2 \cos^2(1/2)}{\pi} \int_N^{1/x} \frac{\lambda^2}{\alpha + \epsilon \lambda^2} \, d\lambda$$

$$\geq \frac{x^2 \cos^2(1/2)}{\pi} \int_N^{1/x} \frac{1}{\alpha/\lambda^2 + \epsilon} \, d\lambda$$

$$\geq \frac{x^2 \cos^2(1/2)}{2\epsilon \, \pi} \left(\frac{1}{x} - N \right),$$

which implies (7.273).

The first inequality in (7.274) is trivial. For the second we note that

$$\sigma_0^2(x) = \sigma_\alpha^2(x) + \frac{4}{\pi} \int_0^\infty \sin^2 \frac{\lambda x}{2} \frac{\alpha}{\psi(\lambda)(\alpha + \psi(\lambda))} \, d\lambda. \tag{7.277}$$

Let λ_0 be such that $\psi(\lambda) \geq N_0$ for $\lambda \geq \lambda_0$. This exists by (4.78) and (4.74). Then

$$\int_0^\infty \sin^2 \frac{\lambda x}{2} \frac{\alpha}{\psi(\lambda)(\alpha + \psi(\lambda))} \, d\lambda \leq x^2 \int_0^{\lambda_0} \frac{\lambda^2}{\psi(\lambda)} \, d\lambda + \frac{\alpha}{N_0} \sigma_\alpha^2(x). \tag{7.278}$$

The second inequality in (7.274) now follows from (7.277), (7.278), (4.77), and (7.273).

The inequality in (7.275) follows from (7.270) and (7.271), each used in the different cases that we must consider. We also use the fact that $\sigma_\alpha(x)$ increases as $\alpha \downarrow 0$.   □

In (7.274) we showed that the σ_α are asymptotically equivalent at zero. Using this and Theorem 7.3.1, we get the following useful estimates.

Lemma 7.4.9 *When $\psi(\lambda)$ in (7.261) is regularly varying at infinity with index $1 < p \leq 2$,*

$$\sigma_\alpha^2(h) \sim C_p \frac{1}{h \, \psi(1/h)} \qquad \text{as } h \to 0 \tag{7.279}$$

for all $0 \leq \alpha < \infty$, where

$$C_p = \frac{4}{\pi} \int_0^\infty \frac{\sin^2 s/2}{s^p} \, ds. \tag{7.280}$$

When $\psi(\lambda)$ in (7.261) is regularly varying at infinity with index 1,

$$\sigma_\alpha^2(h) \sim \frac{2}{\pi} \int_{1/h}^\infty \frac{1}{\psi(\lambda)} \, d\lambda \qquad \text{as } h \to 0 \tag{7.281}$$

for all $0 \leq \alpha < \infty$.

The next lemma, which is essentially a version of (7.192) and (7.194) of Theorem 7.3.1 at infinity, shows that we can find regularly varying Lévy exponents.

Lemma 7.4.10 *Let $\psi(\lambda)$ be as given in (7.261) and (7.262), and let the Lévy measure ν be of the form*

$$\nu([u, \infty)) = \int_u^\infty \frac{1}{\theta(x)} \, dx, \tag{7.282}$$

where θ is a regularly varying function at zero with index $1 < p < 3$. Then

$$\psi(\lambda) \sim \frac{\pi C_p}{\lambda \, \theta(1/\lambda)} \qquad \text{as } \lambda \to \infty, \tag{7.283}$$

where C_p is given in (7.280). When θ is a regularly varying at zero with index 3,

$$\psi(\lambda) \sim 2 \int_0^{1/\lambda} \frac{u^2}{\theta(u)} \, du \qquad \text{as } \lambda \to \infty \tag{7.284}$$

and is a slowly varying function at infinity.

Furthermore, if θ is a normalized regularly varying function at infinity with index $1 < p \leq 3$, then ψ is a normalized regularly varying function.

Proof Using (7.262) and (7.283), we see that

$$\psi(\lambda) = 4 \int_0^\infty \frac{\sin^2 \lambda u/2}{\theta(u)} \, du. \tag{7.285}$$

Compare this with (7.190). The proofs of (7.283) and (7.284) are essentially the same as the proofs of (7.192) and (7.194). We simply consider the behavior of θ at zero rather than at infinity. In Theorem 7.3.1, the critical case is when $\theta(\lambda)$ is regularly varying at infinity with index 1. Here, because of (7.284), it is when $\theta(\lambda)$ is regularly varying at zero with index 3. $\qquad \square$

In this section we define G_α to be a stationary Gaussian process and $\widetilde{G}_0$ to be a Gaussian process with stationary increments, with $\widetilde{G}_0(0) = 0$. We see in the next lemma, which we use in Section 9.5, that these processes are closely related.

Lemma 7.4.11 *For $\alpha > 0$, let G_α be as defined on page 330. Set $\widetilde{G}_\alpha(x) = \{G_\alpha(x) - G_\alpha(0), x \in R^1\}$ and let $H_\alpha = \{H_\alpha(x), x \in R^1\}$ be a* *mean zero Gaussian process with stationary increments, independent of* $\widetilde{G}_\alpha$, *with* $H_\alpha(0) = 0$, *satisfying*

$$E(H_\alpha(x) - H_\alpha(y))^2 = \frac{4}{\pi} \int_0^\infty \sin^2 \frac{\lambda(x-y)}{2} \frac{\alpha}{\psi(\lambda)(\alpha + \psi(\lambda))} \, d\lambda. \quad (7.286)$$

Then, for $\widetilde{G}_0$, *as defined on page 330,*

$$\{\widetilde{G}_0(x), x \in R^1\} \overset{law}{=} \{\widetilde{G}_\alpha(x), x \in R^1\} + \{H_\alpha(x), x \in R^1\} \quad (7.287)$$

and

$$E(H_\alpha(x+h) - H_\alpha(x))^2 = o(\sigma_0^2(|h|)) \qquad as \ h \to 0. \quad (7.288)$$

Proof The processes $\widetilde{G}_0$, $\widetilde{G}_\alpha$, and H_α can be represented as in (5.257) with $h(\lambda)$ equal, respectively, to $1/\psi(\lambda)$, $1/(\alpha + \psi(\lambda))$, $\alpha/(\psi(\lambda)(\alpha + \psi(\lambda)))$, and other obvious minor modifications. The equality in law is then easily verified by computing the covariances of the three processes.

By (7.287), the left-hand side of (7.288) is equal to $\sigma_0^2(|h|) - \sigma_\alpha^2(|h|)$, so (7.288) follows from (7.274). $\qquad \square$

For the Gaussian processes associated with Lévy processes that we consider in this subsection, we get the following simplification of Theorem 7.4.5.

Theorem 7.4.12 *Let* X *be a real-valued symmetric Lévy process as defined in (7.261) satisfying (7.264). Let* G *denote any of the Gaussian processes associated with* X *considered in this subsection, that is,* $\widetilde{G}_0$, G_α *for* $\alpha \geq 0$, *and* $\overline{G}_\alpha$ *for* $\alpha \geq 0$. *Let*

$$\sigma^2(x) = \frac{4}{\pi} \int_0^\infty \sin^2 \frac{\lambda(x-y)}{2} \frac{1}{\psi(\lambda)} \, d\lambda \quad (7.289)$$

and assume that $\sigma(u) = O|u|^\alpha$ *for some* $\alpha > 0$. *Then*

$$\limsup_{\delta \downarrow 0} \sup_{u \leq \delta} \frac{|G(u) - G(0)|}{(2\sigma^2(u) \log \log 1/u)^{1/2}} \geq 1 \qquad a.s. \quad (7.290)$$

Proof This theorem is a simple consequence of Theorem 7.4.5 since all these processes are in class $\mathcal{G}$. The only difference between (7.247) and (7.290) is that in (7.247) the increments variance of the different Gaussian processes enters into the denominator. We get (7.290) because they are all asymptotically equivalent at zero. Note that $\sigma^2(x)$ is the increments variance of $\widetilde{G}_0$ or G_0, according to whether $\int_0^1 \psi^{-1}(\lambda) = \infty$ or 0. Its equivalence with the increments variance of G_α for $\alpha > 0$ is given in (7.274). Its equivalence with the increments variance of $\overline{G}_\alpha$, $\alpha \geq 0$, is given in (7.245). □

Combining Theorem 7.2.14 and Lemma 7.4.11, we can find uniform and local moduli of continuity for different Gaussian processes related to symmetric p-stable Lévy processes.

Example 7.4.13 Consider the Gaussian processes of the type G_α, $\alpha > 0$ and $\widetilde{G}_0$ with increments variance σ_α^2, $\alpha \geq 0$ and with $\psi(\lambda) = \lambda^p$, $1 < p < 2$. For $\alpha > 0$, processes of the type G_α are associated with symmetric p-stable processes killed at the end of an independent exponential time with mean $1/\alpha$. Processes of the type $\widetilde{G}_0$ have covariance $u_{T_0}(x, y)$.

We show here that

$$\lim_{\delta \to 0} \sup_{\substack{|x| \leq \delta \\ x \in I}} \frac{|G_\alpha(x) - G_\alpha(0)|}{(|x - y|^{p-1} \log \log 1/|x - y|)^{1/2}} = \sqrt{2C_p} \qquad \text{a.s.} \qquad (7.291)$$

and

$$\lim_{\delta \to 0} \sup_{\substack{|x-y| \leq \delta \\ x,y \in I}} \frac{|G_\alpha(x) - G_\alpha(y)|}{(|x - y|^{p-1} \log 1/|x - y|)^{1/2}} = \sqrt{2C_p} \qquad \text{a.s.}, \qquad (7.292)$$

where C_p is given in (7.280) and similarly with G_α replaced by $\widetilde{G}_0$.

We point out in Example 7.2.16 that (7.291) and (7.292) also hold for Brownian motion, in which case $p = 2$ (we show in Remark 4.2.6 that $C_2 = 1$). This is because the Gaussian process associated with the 2-stable process, killed the first time it hits zero, is Brownian motion. It follows from Lemma 2.5.1 that the Gaussian process associated with Brownian motion, killed the first time it hits zero, is the 2-stable process.

To verify these statements we first note that, by (7.268), $\sigma_{\widetilde{G}_0}^2(h) = \sigma_0^2(h)$, and by Example 4.2.5,

$$\sigma_0^2(h) = C_p |h|^{p-1}. \qquad (7.293)$$

It follows from Theorem 7.2.14 that (7.292) holds for $\widetilde{G}_0$ because σ_0^2 is concave.

Since (7.292) holds for $\widetilde{G}_0$, it follows Lemma 7.4.11 that it holds for $G_\alpha(x) - G_\alpha(y) + H_\alpha(x) - H_\alpha(y)$. It follows from Theorem 7.2.1 and (7.288) that the uniform modulus of continuity of H_α is "little o" of the uniform modulus of continuity of G_α. Thus, by the triangle inequality, (7.292) holds for G_α, for all $\alpha > 0$.

The proof of (7.291) is similar, except that we use Theorem 7.2.15. (Concave functions satisfy condition (2).)

We also note that (7.291) and (7.292) hold for $\widetilde{G}_0$ when $2 < p < 3$. For (7.292), the lower bound follows from Remark 7.2.11 and the upper bound, from Corollary 7.2.3. For (7.292), the upper bound also is given by Corollary 7.2.3, and one can easily check that the proof of Theorem 7.2.7 goes through in this special case.

7.5 Gaussian lacunary series

The limit laws in (7.186) and (7.189) are direct generalizations of the moduli of continuity results for Brownian motion stated in (2.14) and (2.15). However, this is far from the whole story. Other classes of Gaussian processes exhibit different types of moduli of continuity. In particular, there are Gaussian processes with the same exact local and uniform moduli of continuity. We study Gaussian lacunary series to obtain examples that show how varied are the many types of moduli of continuity that can arise. But, more significantly, they allow us to find much larger lower bounds for the local moduli of continuity of certain Gaussian processes than those given in Theorem 7.2.12.

A lacunary series is a trigonometric series,

$$f(t) = \sum_{n=0}^{\infty} (a_n \cos 2\pi\lambda_n t + b_n \sin 2\pi\lambda_n t) \qquad t \in [0, 1], \qquad (7.294)$$

where $\lambda_n \in \mathcal{N}$ and $\lambda_{n+1}/\lambda_n \geq q > 1$, for all $n \in \mathcal{N}$. The important property of lacunary series that we use (see, e.g., Katzenelson (1968, page 108)) is that

$$A_q \sum_{n=0}^{\infty} \left(|a_n|^2 + |b_n|^2\right)^{1/2} \leq \sup_{t \in [0,1]} |f(t)| \leq \sum_{n=0}^{\infty} \left(|a_n|^2 + |b_n|^2\right)^{1/2},$$

$$(7.295)$$

where $A_q > 0$ is a constant depending only on q.

We restrict our attention to Gaussian lacunary series of the form

$$X(t) = \sum_{n=0}^{\infty} a_n \left(\eta_n \cos 2\pi 2^n t + \eta_n' \sin 2\pi 2^n t\right) \qquad t \in [0, 1], \qquad (7.296)$$

where $\{\eta_n\}$ and $\{\eta_n'\}$ are independent standard normal sequences. Since the Gaussian random variables are symmetric, without loss of generality we assume that the $a_n \geq 0$. The reason we take $\lambda_n = 2^n$ is explained just before (7.334). Note that $X := \{X(t), t \in [0, 1]\}$ is a stationary process.

A necessary and sufficient condition for the continuity of X is that $\{a_n\} \in \ell_1$. We assume that this is the case. Under this condition, the series in (7.296) converges uniformly almost surely and thus gives a nice representation for X. By (7.295),

$$A_q \sum_{n=0}^{\infty} a_n(|\eta_n|^2 + |\eta_n'|^2)^{1/2} \le \sup_{t \in [0,1]} |X(t)| \le \sum_{n=0}^{\infty} a_n(|\eta_n|^2 + |\eta_n'|^2)^{1/2},$$

(7.297)

where A_q is a constant depending only on q. We assume that, for all $m \in N$,

$$\sum_{n=0}^{m} a_n^i \left(\frac{2^n}{2^m}\right)^i \le C \sum_{n=m+1}^{\infty} a_n^i \qquad i = 1, 2.$$

(7.298)

In particular, this requires that $a_n > 0$ infinitely often.

For $h > 0$, let $N(h) = \max\{n : 2^n h \le 1\}$. Consider

$$\Phi(h) := \max \left[\left(\log N(h) \sum_{n=N(h)+1}^{\infty} a_n^2 \right)^{1/2}, \sum_{n=N(h)+1}^{\infty} a_n \right].$$

(7.299)

Theorem 7.5.1 *Let X be a Gaussian lacunary series as described in (7.296) and assume that (7.298) holds. Then, if either*

(1) $\limsup_{j \to \infty} \dfrac{\sum_{n \ge j} a_n}{\left(\log j \sum_{n \ge j} a_n^2 \right)^{1/2}} > \delta$ *for some $\delta' > 0$*

or

(2) a_n *is nonincreasing as $n \to \infty$*

there exist constants $0 < C_0 \le C_1 < \infty$ so that both

$$\lim_{\delta \to 0} \sup_{|u| \le \delta} \frac{|X(t+u) - X(t)|}{\Phi(|u|)} = C_0 \qquad a.s.$$

(7.300)

for all $t \in [0,1]$, and

$$\lim_{\delta \to 0} \sup_{\substack{|t-s| \le \delta \\ s,t \in [0,1]}} \frac{|X(t) - X(s)|}{\Phi(|t-s|)} = C_1 \qquad a.s.$$

(7.301)

Furthermore, (7.300) and (7.301) hold with $\Phi(|\cdot|)$ replaced by $\Phi(\delta)$ and lim replaced by $\limsup$ (and possibly different nonzero constants whem Φ is not nondecreasing).

The fact that X has the same local and uniform moduli of continuity shows that processes such as these behave quite differently from those studied in the Section 7.2.

To prove Theorem 7.5.1, we need several inequalities, which we incorporate into the following lemma:

Lemma 7.5.2

(1)

$$\limsup_{j \to \infty} \frac{\sum_{n \geq j} a_n |\eta_n|}{\max\{\sum_{n \geq j} a_n, (\log j \sum_{n \geq j} a_n^2)^{1/2}\}} \leq 3 \qquad a.s. \tag{7.302}$$

(2)

$$\limsup_{j \to \infty} \frac{\sum_{n=0}^{j} a_n |\eta_n|}{\max\{\sum_{n=0}^{j} a_n, (\log j \sum_{n=0}^{j} a_n^2)^{1/2}\}} \leq 3 \qquad a.s. \tag{7.303}$$

(3)

$$\limsup_{j \to \infty} \frac{\sum_{n \geq j} a_n |\eta_n|}{\sum_{n \geq j} a_n} \geq C \qquad a.s. \tag{7.304}$$

for some constant $C > 0$.

(4) *When $a_n \downarrow$ and $(\log j \sum_{n \geq j} a_n^2)^{1/2} \geq 2 \sum_{n \geq j} a_n$ for all j suffi-ciently large,*

$$\limsup_{j \to \infty} \frac{\sum_{n \geq j} a_n \eta_n}{(\log j \sum_{n \geq j} a_n^2)^{1/2}} \geq C \qquad a.s. \tag{7.305}$$

for some constant $C > 0$.

Proof Let $N_1 = [0, \ldots, j]$ and $N_2 = [j, \ldots \infty]$ and set

$$f_i(j) = \max\{\sum_{n \in N_i} a_n, (\log j \sum_{n \in N_i} a_n^2)^{1/2}\} \qquad i = 1, 2. \tag{7.306}$$

By Chebyshev's inequality,

$$P\left(\alpha_j \sum_{n \in N_i} a_n |\eta_n| \geq c f_i(j) \alpha_j\right) \tag{7.307}$$

$$\leq \exp(-c f_i(j) \alpha_j) E\left(\exp\left(\alpha_j \sum_{n \in N_i} a_n |\eta_n|\right)\right).$$

Also,

$$E\left(\exp\left(\alpha_j \sum_{n \in N_i} a_n |\eta_n|\right)\right) \leq \prod_{n \in N_i} e^{\alpha_j^2 a_n^2/2} \sqrt{2/\pi} \int_0^\infty e^{-(y - \alpha_j a_n)^2/2} \, dy \tag{7.308}$$

and

$$\sqrt{2/\pi} \int_0^\infty e^{-(y - \alpha_j a_n)^2/2} \, dy \leq 1 + \sqrt{2/\pi} \alpha_j a_n. \tag{7.309}$$

Consequently,

$$P\left(\alpha_j \sum_{n \in N_i} a_n |\eta_n| \geq c f_i(j) \alpha_j\right) \tag{7.310}$$

$$\leq \exp\left(-c f_i(j)\alpha_j + \frac{\alpha_j^2}{2} \sum_{n \in N_i} a_n^2 + \sqrt{2/\pi}\alpha_j \sum_{n \in N_i} a_n\right).$$

Let $\alpha_j = \log j / f_i(j)$. Then, for $c \geq 3$, the right-hand side of (7.310) is a term of a convergent series so that both (7.302) and (7.303) follow by the Borel–Cantelli Lemma.

To obtain (7.304), we note that since $E|\eta_n| = \sqrt{2/\pi}$,

$$\left(E \sum_{n \geq j} a_n |\eta_n|\right)^2 = \frac{2}{\pi}\left(\sum_{n \geq j} a_n\right)^2 \tag{7.311}$$

$$\geq \frac{2}{\pi} E \left(\sum_{n \geq j} a_n |\eta_n|\right)^2.$$

Therefore, by the Paley–Zygmund Lemma (Lemma 14.8.2), for all $0 < \lambda < 1$,

$$P\left(\sum_{n \geq j} a_n |\eta_n| \geq \lambda \sqrt{2/\pi} \sum_{n \geq j} a_n\right) \geq (1 - \lambda)^2 \frac{2}{\pi}, \tag{7.312}$$

which implies that

$$P\left(\limsup_{j \to \infty} \frac{\sum_{n \geq j} a_n |\eta_n|}{\sum_{n \geq j} a_n} \geq \lambda \sqrt{2/\pi}\right) \geq (1 - \lambda)^2 \frac{2}{\pi}. \tag{7.313}$$

Since the event in (7.313) is a tail event, we get (7.304).

We now obtain (7.305). We define a subsequence of the integers as follows: Let $g(x) = x + \log x$. For some integer $m > 3$, define the sequence $n(0) = m, n(1) = [g(m)], \ldots, n(i + 1) = [g(n(i))], \ldots$. We first show that $\sum 1/n(i) = \infty$. Let $N_k := 2^k n(0)$. Observe that there are at least $\dfrac{N_k - \log N_k}{\log 2N_k}$ of the terms $n(i)$ between N_k and $N_{k+1} = 2N_k$. (To see this, note that the first term, $n(i) \geq N_k$ is less than or equal to $N_k + \log N_k$. Each subsequent $n(j)$ between N_k and N_{k+1} is such that $n(j) < n(j-1) + \log n(j-1) \leq n(j-1) + \log 2N_k$.) Consequently,

$$\sum_{i=0}^{\infty} \frac{1}{n(i)} \geq \sum_{k=0}^{\infty} \frac{2^k n(0) - \log 2^k n(0)}{\log 2^{k+1} n(0)} 2^{-k-1} n^{-1}(0) \tag{7.314}$$

$$\geq \frac{1}{4} \sum_{k=0}^{\infty} \frac{1}{\log 2^{k+1} n(0)} = \infty.$$

Let

$$Y_{n(j)} := \frac{\sum_{n \geq n(j)} a_n \eta_n}{\left(\sum_{n \geq n(j)} a_n^2\right)^{1/2}}. \tag{7.315}$$

We have

$$EY_{n(j)}^2 = 1 \quad \text{and} \quad EY_{n(j)}Y_{n(k)} \leq 1/2 \quad j \neq k. \tag{7.316}$$

The equality is obvious. To prove the inequality, it is enough to show that

$$\frac{\sum_{n \geq n(i+1)} a_n^2}{\sum_{n \geq n(i)} a_n^2} \leq \frac{1}{4}. \tag{7.317}$$

By the hypothesis and Lemma 14.8.3,

$$\left(\log n(i) \sum_{n \geq n(i)} a_n^2\right)^{1/2} \tag{7.318}$$

$$\geq 2 \sum_{n \geq n(i)} a_n > \sum_{k=n(i)}^{\infty} \frac{\left(\sum_{n \geq k} a_n^2\right)^{1/2}}{(k+1-n(i))^{1/2}}$$

$$> \left(\sum_{n \geq n(i)+[\log n(i)]} a_n^2\right)^{1/2} \sum_{k=n(i)}^{n(i)+[\log n(i)]} (k+1-n(i))^{-1/2}$$

$$> 2 \left(\log n(i) \sum_{n \geq n(i)+[\log n(i)]} a_n^2\right)^{1/2},$$

which gives the inequality in (7.317). Let $\{\xi_i\}$ be a standard normal sequence. By (7.316), as in the proof of Corollary 5.5.2,

$$P\left(\cup_{j=k}^{\infty} \{Y_{n(j)}/(2\log n(j))^{1/2}\} \geq 1/\sqrt{2}\right) = 1. \tag{7.319}$$

Using the Monotone Convergence Theorem, we get (7.305). □

We also use the following simple observation.

Lemma 7.5.3 *Suppose that* $\{a_n\} \in \ell_1$, $a_n \downarrow$, *and*

$$\left(\log j \sum_{n=j}^{\infty} a_n^2\right)^{1/2} \geq \sum_{n=j}^{\infty} a_n \tag{7.320}$$

for all j sufficiently large. Then

$$\log j \sum_{n=j}^{\infty} a_n^2 \qquad (7.321)$$

is decreasing in j for all j sufficiently large.

Proof The monotonicity of a_n implies that $a_n \geq \sum_{n=m}^{\infty} a_n^2 / \sum_{n=m}^{\infty} a_n$. Using this and (7.320), we see that

$$a_j^2 \log j \geq \frac{\log j (\sum_{n=j}^{\infty} a_n^2)^2}{(\sum_{n=j}^{\infty} a_n)^2} \geq \sum_{n=j}^{\infty} a_n^2. \qquad (7.322)$$

Also,

$$\log j \sum_{n=j}^{\infty} a_n^2 \geq \log(j+1) \sum_{n=j+1}^{\infty} a_n^2 + \log j \, a_j^2 - \log \frac{j+1}{j} \sum_{n=j+1}^{\infty} a_n^2. \quad (7.323)$$

Using (7.322) in (7.323), we get (7.321). $\qquad\square$

Proof of Theorem 7.5.1 In order to use Lemma 7.1.7, we prove the lower bound in (7.300) as given and the upper bound in (7.301) with lim replaced by lim sup and $\Phi(|t - s|)$ replaced by $\Phi(\delta)$.

We start with the upper bound. We have

$$X(t+u) - X(t) \qquad (7.324)$$

$$= \sum_{n=0}^{\infty} a_n \Big([\, \eta_n (\cos 2\pi 2^n u - 1) + \eta_n' \sin 2\pi 2^n u] \cos 2\pi 2^n t$$

$$+ [\, \eta_n' (\cos 2\pi 2^n u - 1) - \eta_n \sin 2\pi 2^n u] \sin 2\pi 2^n t \Big).$$

By (7.295),

$$\sup_{\substack{t \in [0,1] \\ |u| \leq |h|}} |X(t+u) - X(t)| \qquad (7.325)$$

$$\leq \sup_{|u| \leq |h|} 2 \sum_{n=0}^{\infty} a_n \Big| \sin \frac{2\pi 2^n u}{2} \Big| \left(\eta_n^2 + (\eta_n')^2 \right)^{1/2}$$

$$\leq 2\pi \sum_{n=0}^{N(h)} a_n 2^n h \left(|\eta_n| + |\eta_n'| \right) + 2 \sum_{N(h)+1}^{\infty} a_n \left(|\eta_n| + |\eta_n'| \right).$$

$$:= I_1(h) + I_2(h).$$

Let

$$\widetilde{\Phi}(h) := \max\left[\left(\log N(h) \sum_{n=0}^{N(h)} a_n^2 \left(\frac{2^n}{2^{N(h)}} \right)^2 \right)^{1/2}, \sum_{n=0}^{N(h)} a_n \frac{2^n}{2^{N(h)}} \right].$$

$$(7.326)$$

By Lemma 7.5.2 (2) and the fact that $h \leq 2^{-N(h)}$,

$$\limsup_{h \to 0} \frac{I_1(h)}{\widetilde{\Phi}(h)} \leq 12\pi \qquad \text{a.s.} \qquad (7.327)$$

Also, by Lemma 7.5.2 (1),

$$\limsup_{h \to 0} \frac{I_2(h)}{\widetilde{\Phi}(h)} \leq 12 \qquad \text{a.s.} \qquad (7.328)$$

By (7.298), $\widetilde{\Phi}(h) \leq C\Phi(h)$. Thus we have

$$\limsup_{\delta \to 0} \sup_{\substack{|t-s| \leq \delta \\ s,t \in [0,1]}} \frac{|X(t) - X(s)|}{\Phi(\delta)} \leq C \qquad \text{a.s.} \qquad (7.329)$$

for some constant $C < \infty$.

Using the relationship (7.53) implies (7.54), we see that we also have

$$\lim_{\delta \to 0} \sup_{\substack{|t-s| \leq \delta \\ s,t \in [0,1]}} \frac{|X(t) - X(s)|}{\Phi(|t-s|)} \leq C \qquad \text{a.s.} \qquad (7.330)$$

By Lemmas 7.1.1 and 7.1.9, we actually have equality in (7.329) and (7.330) for constants $0 \leq C', C'' \leq C$.

We next obtain a lower bound for the left-hand side of (7.300), which is greater than zero. Using this and the relationship (7.55) implies (7.56), which also holds for the local modulus of continuity, we see that C' and C'' are both greater than zero, which completes the proof.

By (7.324),

$$X(u) - X(0) = \sum_{n=0}^{\infty} a_n \sin 2\pi 2^n u \, \eta_n + \sum_{n=0}^{\infty} a_n (\cos 2\pi 2^n u - 1) \, \eta_n'. \quad (7.331)$$

Since the two sums in (7.331) are independent and symmetric, on a set with probability greater than or equal to $1/2$,

$$\sup_{0 \leq u \leq 2^{-m}} |X(u) - X(0)| \qquad (7.332)$$

$$\geq \sup_{0 \leq u \leq 2^{-m}} \left| \sum_{n=0}^{\infty} a_n \sin 2\pi 2^n u \, \eta_n \right|$$

$$= \sup_{0 \leq u \leq 2^{-m}} \left| \sum_{n=0}^{m-1} a_n \sin 2\pi 2^n u \, \eta_n + \sum_{n=m}^{\infty} a_n \sin 2\pi 2^n u \, \eta_n \right|.$$

By the same argument, we see that on a set with probability greater than or equal to 1/4,

$$\sup_{0 \le u \le 2^{-m}} |X(u) - X(0)| \ge \sup_{0 \le u \le 2^{-m}} \left| \sum_{n=m}^{\infty} a_n \sin 2\pi 2^n u \, \eta_n \right|. \quad (7.333)$$

Note that the random Fourier series in (7.333) has period 2^{-m}. Thus the suprema taken over $[0, 2^{-m}]$ and $[0, 2\pi]$ in the next equation are the same (this is why we take $\lambda_n = 2^n$ in (7.296)). Consequently, by (7.297),

$$\sup_{0 \le u \le 2^{-m}} \left| \sum_{n=m}^{\infty} a_n \sin 2\pi 2^n u \, \eta_n \right| = \sup_{0 \le u \le 1} \left| \sum_{n=m}^{\infty} a_n \sin 2\pi 2^n u \, \eta_n \right|$$

$$\ge A_q \sum_{n=m}^{\infty} a_n |\eta_n|. \quad (7.334)$$

Suppose that condition (1) holds. Let $\{n_j\}$ be a sequence along which

$$\sum_{n \ge n_j} a_{n_j} \ge \delta' \left(\log n_j \sum_{n \ge n_j} a_n^2 \right)^{1/2}. \quad (7.335)$$

Then, by (7.334), we have

$$P \left(\sup_{0 \le u \le 2^{-m}} \left| \sum_{n=n_j}^{\infty} a_n \sin 2\pi 2^n u \, \eta_n \right| \ge C \sum_{n=n_j}^{\infty} a_n |\eta_n| \right) \ge \frac{1}{4} \quad (7.336)$$

for some constant $C > 0$, which implies, by Lemma 7.5.2 (3) that

$$P \left(\limsup_{j \to \infty} \frac{\sup_{0 \le u \le 2^{-n_j}} |\sum_{n=n_j}^{\infty} a_n \sin 2\pi 2^n u \, \eta_n|}{\sum_{n=n_j}^{\infty} a_n} \ge C \right) = 1 \quad (7.337)$$

for some constant $C > 0$, where we also use the fact that this is a tail event.

Note that, for $2^{-n_j} \le u \le 2^{-n_j+1}$, $\Phi(u) = \sum_{n=n_j}^{\infty} a_n$. Also, $\sum_{n=n_j}^{\infty} a_n \sin 2\pi 2^n u \, \eta_n$ is periodic with period 2^{-n_j}. Thus, for each path in the stochastic process $\sum_{n=n_j}^{\infty} a_n \sin 2\pi 2^n u \, \eta_n$ there is a $u \in (2^{-n_j}, 2^{-n_j+1})$, depending on the path, for which

$$u = \sup_{0 \le u \le 2^{-n_j}} \left| \sum_{n=n_j}^{\infty} a_n \sin 2\pi 2^n u \, \eta_n \right|. \quad (7.338)$$

This shows that

$$P \left(\limsup_{j \to \infty} \frac{\sup_{0 \le u \le 2^{-n_j+1}} |\sum_{n=n_j}^{\infty} a_n \sin 2\pi 2^n u \, \eta_n|}{\Phi(u)} \ge C \right) = 1 \quad (7.339)$$

for some constant $C > 0$. Using (7.333), we see that, under condition (1)

$$\lim_{\delta \to 0} \sup_{|u| \leq \delta} \frac{|X(u) - X(0)|}{\Phi(|u|)} \geq C \qquad (7.340)$$

for some constant $C > 0$, with probability at least $1/4$. We show in the next paragraph that the left-hand side of (7.340) is a tail event.

Let $X_M(u) - X_M(0)$ be given by the sum in (7.331) taken from $n = 0$ to $n = M$. It is easy to see that $\sup_{0 \leq u \leq 2\pi/2^m} (E(|X_M(u) - X_M(0)|)^2)^{1/2} \leq C_M 2^{-m}$. To show that the left-hand side of (7.340) is a tail event, it suffices to show that

$$\lim_{j \to \infty} 2^m \Phi(2\pi/2^m) = \infty, \qquad (7.341)$$

which implies that the event in (7.337) is a tail event. By (7.298),

$$
\begin{aligned}
2^m \Phi(2^{-m}) &\geq C 2^m \left(\log m \sum_{n=m}^{\infty} a_n^2 \right)^{1/2} \qquad (7.342) \\
&\geq C 2^m \left(\log m \sum_{n=0}^{m-1} a_n^2 2^{-2(m-n-1)} \right)^{1/2} \\
&\geq \left(\log m \sum_{n=0}^{m-1} a_n^2 2^{2(n-1)} \right)^{1/2}.
\end{aligned}
$$

Note that $\limsup_{n \to \infty} a_n 2^n \leq 1$ violates condition (1). When $\limsup_{n \to \infty} a_n 2^n > 1$, the last term in (7.342) goes to infinity as $j \to \infty$. This verifies (7.341). Therefore, (7.340) holds almost surely when condition (1) is satisfied.

When condition (1) is not satisfied

$$\left(\log j \sum_{n=j}^{\infty} a_n^2 \right)^{1/2} > C \sum_{n=j}^{\infty} a_n \qquad \text{for all } j \geq j' \text{ for some } j' \qquad (7.343)$$

and some $C > 0$. Since multiplication by a constant only changes the constant in (7.300) and (7.301), we take $C = 2$. In this case we use Lemma 7.5.2 (4) in (7.336), which requires that $\{a_n\}$ is monotonic, and proceed as above to obtain

$$\limsup_{\delta \to 0} \sup_{0 \leq u \leq \delta} \frac{|X(t+u) - X(t)|}{\Phi(\delta)} \geq C \qquad \text{a.s.} \qquad (7.344)$$

with the same constant $C > 0$. By Lemma 7.5.3, under (7.343), $\Phi(|u|)$

is increasing, and this implies that

$$\lim_{\delta \to 0} \sup_{0 \le u \le \delta} \frac{|X(t+u) - X(t)|}{\Phi(|u|)} \ge C \qquad \text{a.s.} \qquad (7.345)$$

We have already shown that (7.345) holds under condition (1). □

Corollary 7.5.4 *Let X be a Gaussian lacunary series as described in (7.296) and assume that (7.298) holds. Then*

$$\psi(|u|) = \sum_{n=N(u)+1}^{\infty} a_n \qquad (7.346)$$

is a lower bound for the local modulus of continuity of X, that is, (7.300) holds with a greater than or equal to sign when $\Phi(|u|)$ *is replaced by* $\psi(|u|)$ *and also when* $\lim$ *is replaced by* $\lim \sup$ *and* $\Phi(|u|)$ *is replaced by* $\psi(\delta)$.

The significance of this result in relation to Theorem 7.5.1 is that we do not require that either condition (1) or condition (2) of Theorem 7.5.1 holds.

Proof By (7.337) and the monotonicity of $\psi(|u|)$, we see that it is a lower bound for the local modulus of continuity of X with probability 1/4. To replace 1/4 by 1, we need show that the event in (7.337) is a tail event. As in (7.341), this comes down to showing that

$$\lim_{N \to \infty} 2^N \psi(2^{-N}) = \infty. \qquad (7.347)$$

By (7.298),

$$2^N \psi(2^{-N}) \ge C \sum_{n=0}^{N} a_n 2^n. \qquad (7.348)$$

If $\liminf_{N \to \infty} 2^N \psi(2^{-N}) < \infty$ for all $\epsilon > 0$, $a_n < \epsilon/2^n$ for all $n \ge N(\epsilon)$. Therefore,

$$\sum_{n=N(\epsilon)+1}^{\infty} a_n \le \frac{\epsilon}{2^{N(\epsilon)}}, \qquad (7.349)$$

and, consequently, by (7.298)

$$\sum_{n=0}^{N(\epsilon)} a_n 2^n \le C\epsilon. \qquad (7.350)$$

That this cannot hold for all $\epsilon > 0$ verifies (7.348). □

Example 7.5.5 We give a specific example of a Gaussian lacunary series that has the same local and uniform moduli of continuity and shows that there is no analog of Lemma 7.1.10 for the uniform modulus of continuity. Let

$$X(t) = \sum_{n=0}^{\infty} 2^{-(n+1)/2} \left(\eta_n \cos 2\pi 2^n t + \eta'_n \sin 2\pi 2^n t \right) \qquad t \in [0, 1],$$

$$(7.351)$$

where $\{\eta_n\}$ and $\{\eta'_n\}$ are independent standard normal sequences. Clearly,

$$E(X(t)X(s)) = \sum_{n=0}^{\infty} 2^{-(n+1)} \cos 2\pi 2^n (t - s). \qquad (7.352)$$

It follows from Theorem 7.5.1 that $(h \log \log 1/h)^{1/2}$ is both an exact local and exact uniform modulus of continuity for X (i.e., $\omega(|u - v|) = (|u-v| \log \log 1/|u-v|)^{1/2}$ in (7.1) and $\rho(|u-u_o|) = (|u-u_o| \log \log 1/|u-u_o|)^{1/2}$ in (7.2)), since

$$\log N(h) \sim \log \log 1/h \qquad (7.353)$$

and

$$h/2 \leq \sum_{n=N(h)+1}^{\infty} 2^{-(n+1)} \leq h. \qquad (7.354)$$

We now show that

$$\frac{\pi|t - s|}{2} \leq E(X(t) - X(s))^2 \leq 16\pi|t - s|. \qquad (7.355)$$

By (7.352),

$$E(X(t) - X(s))^2 = 4 \sum_{n=0}^{\infty} 2^{-(n+1)} \sin^2 \frac{2^{n+1}\pi|t - s|}{2}. \qquad (7.356)$$

For $2^{-(k+1)} < 2\pi|t - s| \leq 2^{-k}$,

$$4 \sum_{n=0}^{\infty} 2^{-(n+1)} \sin^2 \frac{2^{n+1}\pi|t - s|}{2} \qquad (7.357)$$

$$\leq 4 \sum_{n=0}^{k+1} 2^{-(n+1)} \frac{2^{2n}2^{-2k}}{4} + 4 \sum_{n=k+2}^{\infty} 2^{-(n+1)}$$

$$\leq \frac{8}{2^{k+1}} \leq 16\pi|t - s|,$$

and, using the fact that $\sin^2 \theta \geq \theta^2/2$ for $|\theta| \leq 1$,

$$4 \sum_{n=0}^{\infty} 2^{-(n+1)} \sin^2 \frac{2^{n+1}\pi|t-s|}{2} \geq 2^{-(k-1)} \sin^2 \frac{2^{k+1}\pi|t-s|}{2} \geq \pi|t-s|/2.$$

$$(7.358)$$

Let B be standard Brownian motion and set $B_1 = \pi B/2$ and $B_2 = 16\pi B$. We see that, for $t \in [0, 2\pi]$,

$$E(B_1(t) - B_1(s))^2 \leq E(X(t) - X(s))^2 \leq E(B_2(t) - B_2(s))^2. \quad (7.359)$$

X and B have the same exact local modulus of continuity, consistent with Lemma 7.1.10. But X and B do not have the same exact uniform modulus of continuity. This shows that Slepian's Lemma does not generalize to cover uniform moduli of continuity.

7.6 Exact moduli of continuity

The results on Gaussian lacunary series in the previous section may seem very specialized. However, coupled with the comparison lemma, Lemma 7.1.10, they enable us to obtain exact local moduli of continuity for a much wider class of Gaussian processes, including those that are associated with Lévy processes.

Let $G = \{G(t), t \in R^1\}$ be a mean zero Gaussian process. Some results in this section are expressed in terms of a monotone minorant for the increments variance of G. For $t_0 \in R^1$ and $\delta > 0$, we consider functions ω such that

$$\omega^2(h) = \omega^2(h; t_0, \delta) \leq \min_{\substack{|t-s| \geq h \\ s,t \in [t_0 - \delta, t_0 + \delta]}} E(G(t) - G(s))^2. \quad (7.360)$$

Restricting the definition of ω to some neighborhood of t_0 enables us to avoid possible zeros of $G(t) - G(s)$ when $s \neq t$.

Theorem 7.6.1 *Let* $G = \{G(t), t \in R^1\}$ *be a mean zero Gaussian process and* ω *be as defined in (7.360) and satisfying*

$$\sum_{n=1}^{\infty} \left(\omega^2(2^{-n}) - \omega^2(2^{-n-1})\right)^{1/2} < \infty. \quad (7.361)$$

Let $N(h) = \max\{n : 2^n h \leq 1\}$ *and*

$$\widetilde{\psi}(h) = \sum_{n=N(h)+1}^{\infty} \left(\omega^2(2^{-n}) - \omega^2(2^{-n-1})\right)^{1/2} \quad (7.362)$$

and assume that $\sup_{|u|\leq h} E(G(t_0 + u) - G(t_0))^2 = o(\widetilde{\psi}^2(h))$ *for some* $t_0 \in R^1$ *and*

$$\sum_{n=k}^{M} \left(\omega^2(2^{-n}) - \omega^2(2^{-n-1})\right)^{p/2} 2^{-p(M-n)} \leq C\omega^p(2^{-p(M+1)}) \qquad p = 1, 2$$

(7.363)

for some k *sufficiently large and all* $M \geq k$. *Then*

$$\limsup_{\substack{\delta \to 0 \\ |u| \leq \delta}} \frac{|G(t_0 + u) - G(t_0)|}{\widetilde{\psi}(u)} \geq C \qquad a.s. \qquad (7.364)$$

for some constant $C > 0$.

Proof To simplify the notation we take $t_0 = 0$. Consider

$$Y(t) := \sum_{n=k}^{\infty} a_n \left(\eta_n \cos 2\pi 2^n t + \eta_n' \sin 2\pi 2^n t\right) \qquad t \in [0, 1], \qquad (7.365)$$

where $a_n^2 = \omega^2(2^{-n}) - \omega^2(2^{-n-1})$, $\{\eta_n\}$ and $\{\eta_n'\}$ are independent standard normal sequences and k is such that (7.363) holds. It follows from Corollary 7.5.4 that (7.364) holds with G replaced by Y (it is easy to see that (7.363) implies (7.298)). Arguing as in (7.356) and (7.357) with $2^{-(n+1)/2}$ replaced by a_n and using the telescoping nature of a_n^2 and (7.363), we see that

$$E(Y(t) - Y(s))^2 \qquad (7.366)$$

$$\leq C \left(\sum_{n=0}^{N(|t-s|)} a_n^2 2^{-2(N(|t-s|)-n)} + \omega^2(2^{-N(|t-s|)-1}) \right)$$

$$\leq C\omega^2(2^{-N(|t-s|)-1})$$

$$\leq C\omega^2(|t-s|)$$

for $|t - s|$ sufficiently small. Therefore, by (7.360), Lemma 7.1.10, and Remark 7.1.11, and the fact that $\widetilde{\psi}$ is increasing, we get (7.364). $\quad\square$

The next lemma gives a simple sufficient condition for (7.363).

Lemma 7.6.2 *If*

$$\omega^2(2^{-n}) \leq 2^{\alpha}\omega^2(2^{-n-1}) \qquad (7.367)$$

for some $\alpha < 2$, *for all* n *sufficiently large,* ω *satisfies* (7.363).

Proof Let k to be large enough so that (7.367) holds for all $n \geq k$.

Then

$$\sum_{n=k}^{M} \left(\omega^2(2^{-n}) - \omega^2(2^{-n-1})\right)^{p/2} 2^{-p(M-n)} \qquad (7.368)$$

$$\leq (2^\alpha - 1)^{p/2} \sum_{n=k}^{M} \omega^p(2^{-n-1}) 2^{-p(M-n)}$$

$$= (2^\alpha - 1)^{p/2} \sum_{j=0}^{M-k} \omega^p(2^{-(M+1-j)}) 2^{-pj}$$

$$\leq (2^\alpha - 1)^{p/2} \omega^p(2^{-(M+1)}) \sum_{j=0}^{M-k} 2^{-pj(1-\alpha/2)}$$

$$\leq C\omega^p(2^{-(M+1)}),$$

where we use (7.367) for the second and fourth inequalities. □

The next theorem relates the bounds in Theorem 7.6.1 to those in Theorem 7.2.1.

Theorem 7.6.3 *In addition to the hypotheses of Theorem 7.6.1, assume that $\omega^2(2^{-n}) - \omega^2(2^{-n-1})$ is decreasing for all n sufficiently large. Then (7.364) holds with $\widetilde{\psi}(h)$ replaced by*

$$\widehat{\psi}(h) = \max\left\{\omega(h)(\log\log 1/h)^{1/2}, \int_0^{1/2} \frac{\omega(hu)}{u(\log 1/u)^{1/2}}\, du\right\}. \qquad (7.369)$$

Proof Consider $\{Y(t), t \in [0, 2\pi]\}$ given in (7.365). Since $\{a_n\}$ is nonincreasing for n sufficiently large, it follows from Theorem 7.5.1 that (7.364) holds with $\widetilde{\psi}(h)$ replaced by

$$\overline{\psi}(h) = \max\left[\left(\omega^2(2^{-N(h)-1})\log N(h)\right)^{1/2},\right. \qquad (7.370)$$

$$\left. \sum_{n=N(h)+1}^{\infty} \left(\omega^2(2^{-n}) - \omega^2(2^{-n-1})\right)^{1/2}\right].$$

By Boas's Lemma (Lemma 14.8.3),

$$\sum_{j=1}^{\infty} \frac{\omega(2^{-N(h)-j})}{\sqrt{j}} \leq 2 \sum_{n=N(h)+1}^{\infty} \left(\omega(2^{-n}) - \omega(2^{-n-1})\right)^{1/2}. \qquad (7.371)$$

Also, since $\omega(2^{-N(h)-j}) \geq \omega(h2^{-j})$,

$$\sum_{j=1}^{\infty} \frac{\omega(2^{-N(h)-j})}{\sqrt{j}} \geq \sum_{j=1}^{\infty} \frac{\omega(h2^{-j})}{\sqrt{j}} \qquad (7.372)$$

$$\geq \int_{j=1}^{\infty} \frac{\omega(h2^{-j})}{\sqrt{j}} \, dj$$

$$= C \int_0^{1/2} \frac{\omega(hu)}{u(\log 1/u)^{1/2}} \, du.$$

Using this in (7.370) justifies the presence of the integral in (7.369). The iterated logarithm term in (7.369) comes from the similar term in (7.370). □

Combining various results, we now get a complete description of the exact local moduli of continuity of a wide class of Gaussian processes with fairly regular increments variances.

Theorem 7.6.4 *Let G be a mean zero real-valued Gaussian process that is continuous in some neighborhood of t_0. Let $\phi(h)$ be an increasing function such that*

(1)

$$C_0\phi(|h|) \leq \left(E(G(t+h) - G(t))^2\right)^{1/2} \leq C_1\phi(|h|) \qquad (7.373)$$

for all t in some neighborhood of t_0 and all $|h|$ sufficiently small;
(2) $\phi^2(2^{-n}) - \phi^2(2^{-n-1})$ is nonincreasing in n for all n sufficiently large;
(3) $\phi(2^{-n}) \leq 2^{\alpha}\phi(2^{-n-1})$ for all n sufficiently large for some $\alpha < 1$.

Set

$$\Phi_1(h) = \phi(h) \left(\log\log 1/h\right)^{1/2} + \int_0^{1/2} \frac{\phi(hu)}{u(\log 1/u)^{1/2}} \, du. \qquad (7.374)$$

Then

$$\lim_{\delta \to 0} \sup_{|u| \leq \delta} \frac{|G(t_0 + u) - G(t_0)|}{\Phi_1(u)} = C \qquad a.s. \qquad (7.375)$$

and

$$\limsup_{\delta \to 0} \sup_{|u| \leq \delta} \frac{|G(t_0 + u) - G(t_0)|}{\Phi_1(\delta)} = C' \qquad a.s. \qquad (7.376)$$

for constants $0 < C \leq C' < \infty$.

Proof Using Theorem 7.2.1, we obtain

$$\limsup_{\delta \to 0} \sup_{|u| \leq \delta} \frac{|G(t_0 + u) - G(t_0)|}{\Phi_1(\delta)} \leq C'' \qquad a.s. \qquad (7.377)$$

for some constant $C'' < \infty$. Let C' be the smallest constant for which (7.377) holds. By Lemma 7.1.6,

$$\lim_{\substack{\delta \to 0 \\ |u| \le \delta}} \sup \frac{|G(t_0 + u) - G(t_0)|}{\Phi_1(u)} \le C' \qquad \text{a.s.} \qquad (7.378)$$

It then follows from Lemma 7.1.1 that

$$\lim_{\substack{\delta \to 0 \\ |u| \le \delta}} \sup \frac{|G(t_0 + u) - G(t_0)|}{\Phi_1(u)} = C \qquad \text{a.s.} \qquad (7.379)$$

for some constant $C \le C'$ and from Theorem 7.6.3 that in fact $C > 0$. This gives us (7.375), and, since $C' \ge C$, we also get (7.376). (The event in (7.376) is a tail event.) $\qquad \square$

We take up the question of when the constants C and C' in Theorem 7.6.4, and in similar results, are equal, in Remark 7.6.7.

Note that by Lemma 6.4.6 the condition that G is continuous implies that the integral in (7.374) is finite. Also, condition (2) is satisfied if $\phi^2(2^{-x})$ is convex for all x sufficiently large.

In order to give some concrete examples of the types of local moduli of continuity we get from Theorem 7.6.3, we need some estimates of $I_{loc,\phi}$. The next lemma gives us an idea of what $I_{loc,\phi}$ looks like when ϕ is slowly varying but larger than powers of a logarithm.

Lemma 7.6.5 *Let* $\phi(u) = \exp(-g(\log 1/u))$, *where* $g'(s)$ *is a normalized regularly varying function at infinity of index* $-\gamma$, $0 < \gamma < 1$. *Then there exist constants* $0 < C_0 \le C_1 < \infty$ *such that*

$$C_0 \phi(\delta)(1/g'(\log 1/\delta))^{1/2} \le I_{loc,\phi}(\delta) \le C_1 \phi(\delta)(1/g'(\log 1/\delta))^{1/2}, \qquad (7.380)$$

that is,

$$I_{loc,\phi}(\delta) \approx \phi(\delta)(\log 1/\delta)^{\gamma/2} \widetilde{L}(\log 1/\delta), \qquad (7.381)$$

for all $\delta > 0$ *sufficiently small and*

$$\phi(\delta) \approx \exp\left(-\frac{(\log 1/\delta)^{1-\gamma}}{1-\gamma} \widetilde{L}(\log 1/\delta)\right) \qquad (7.382)$$

for all $\delta > 0$ *sufficiently small, where* $\widetilde{L}(u)$ *is a normalized slowly varying function at infinity.*

Proof By Theorem 14.7.1,

$$g(s) \sim \frac{s^{1-\gamma}}{1-\gamma} \exp\left(\int_1^s \frac{\epsilon(t)}{t}\, dt\right) \qquad (7.383)$$

as $s \to \infty$, where $\lim_{t \to \infty} \epsilon(t) = 0$. This gives (7.382).

To obtain (7.380), let $\log 1/\delta = v$ so that $\phi(\delta u) = \exp(-g((\log 1/u) + v))$. Then, making the change of variables $s = (\log 1/u) + v$, we see that

$$
\begin{aligned}
I_{loc,\phi}(\delta) &= \int_{\log 2 + v}^{\infty} e^{-g(s)} \frac{ds}{(s-v)^{1/2}} \qquad (7.384) \\
&\leq \int_{\log 2 + v}^{\log 2 + 1/g'(v) + v} e^{-g(s)} \frac{ds}{(s-v)^{1/2}} \\
&\quad + (g'(v))^{1/2} \int_{\log 2 + 1/g'(v) + v}^{\infty} e^{-g(s)} \, ds.
\end{aligned}
$$

Since $g'(s)$ is a normalized regularly varying function at infinity of negative index, by Lemma 7.2.4, it is decreasing for all s sufficiently large. Writing g as the integral of its derivative, we see that $e^{-g(s)}$ evaluated at the lower and upper limits of the second integral in (7.384) are both comparable to $e^{-g(v)}$. Therefore, we can remove this term from the integral and integrate the remainder to see that the first integral after the second equal sign in (7.384) is comparable to $e^{-g(v)} 1/(g'(v))^{1/2}$ at infinity. This gives the lower bound in (7.380). To get the upper bound in (7.380), we need only show that the last term in (7.384) is bounded above by a constant times $e^{-g(v)} 1/(g'(v))^{1/2}$ at infinity. To see this we write

$$
\int_{\alpha(v)}^{\infty} e^{-g(s)} \, ds = \frac{e^{-g(\alpha(v))}}{g'(\alpha(v))} + \int_{\alpha(v)}^{\infty} e^{-g(s)} \frac{d}{ds} \left(\frac{1}{g'(s)} \right) ds, \qquad (7.385)
$$

where $\alpha(v) = \log 2 + 1/g'(v) + v$. Using the fact that $g'(s)$ is a normalized regularly varying function at infinity of index $-\gamma$, we see that $\frac{d}{ds} \left(\frac{1}{g'(s)} \right)$ is dominated by a regularly varying function of index $\gamma - 1$. Thus, for v sufficiently large, the last integral in (7.385) is little "o" of the first one. We already remarked that $e^{-g(\alpha(v))}$ is comparable to $e^{-g(v)}$. Since $v < \alpha(v) < 2v$ for v sufficiently large, $1/g'(\alpha(v))$ is comparable to $1/g'(v)$. Thus we get the desired upper bound for the last term in (7.384). $\qquad \square$

Example 7.6.6 Here are some examples of functions ϕ that satisfy conditions (2) and (3) of Theorem 7.6.4 and the corresponding functions Φ_1 for which (7.375) and (7.376) hold. The constants are not necessarily the same in each example.

(1) ϕ is regularly varying with index $0 < a < 1$
$$\Phi_1(u) = \phi(u)(\log \log 1/|u|)^{1/2};$$

(2) $\phi(u) = \exp(-(\log 1/|u|)^\alpha (\log \log 1/|u|)^\beta)$, $0 < \alpha < 1$, $-\infty < \beta < \infty$
  $\Phi_1(u) = \phi(|u|)((\log 1/|u|)^{1-\alpha}(\log \log 1/|u|)^{-\beta})^{1/2}$;
(3) $\phi(u) = (\log 1/|u|)^{-\alpha}$, $\alpha > 1/2$
  $\Phi_1(u) = \phi(|u|)(\log 1/|u|)^{1/2}$;
(4) $\phi(u) = ((\log 1/|u|)(\log \log 1/|u|)^\beta)^{-1/2}$, $\beta > 2$
  $\Phi_1(u) = \phi(|u|)(\log 1/|u|)^{1/2} \log \log 1/|u|$.

Proof The functions ϕ in (2)–(4) are explicit. One can check that they satisfy conditions (2) and (3) of Theorem 7.6.4. To check this when ϕ is regularly varying with index $0 < a < 1$, use (7.119). To verify that Φ_1 is as stated, we use Lemma 7.2.5, (7.128), for (1); Lemma 7.6.5 for (2); Lemma 7.2.5, (7.126), and (7.127) for (3); and Remark 7.2.6 for (4). □

Comparing Example 7.6.6 (1) with Theorem 7.2.15, we see that by giving up the precise value of the constant in (7.375) and (7.376) we can extend the result from normalized regularly varying σ^2 to regularly varying σ^2.

Remark 7.6.7 We point out in Lemma 7.1.7 that a nondecreasing function is an exact modulus if and only if it is an exact m-modulus. In this context, it is interesting to note that $\widetilde{\omega}(\delta)$ in (7.83) is increasing for δ in some interval $[0, \delta_0]$, where $\delta_0 > 0$ (its derivative is positive). It is not clear whether $\widetilde{\rho}(\delta)$ in (7.84) always has this property; it certainly does in the examples above (note that $\widetilde{\rho}(\delta) = \Phi_1(\delta)$).

More generally, if ϕ in (7.84) satisfies $\phi(x) \leq \exp(-(\log 1/x)^{1-\epsilon})$ for x in some interval $[0, x_0]$, for $x_0 > 0$, for all $\epsilon > 0$, then

$$\phi(\delta) (\log \log 1/\delta)^{1/2} \sim \phi(\delta) (\log \log 1/\phi(x))^{1/2} \qquad (7.386)$$

at zero. Since the second term in (7.386) is increasing at zero, $\widetilde{\rho}(\delta)$ is asymptotic to an increasing function at zero in this case (here we also use the obvious inequality $\phi(x) \geq x^3$ in some interval $[0, x_0]$, $x_0 > 0$). On the other hand, if $\phi(x) \geq \exp(-(\log 1/x)^{1-\epsilon})$, for some $0 < \epsilon < 1$, then, loosely speaking, the integral term in $\widetilde{\rho}(\delta)$ tends to dominate the iterated logarithm term (see Example 7.6.6 (2)), and, of course, the integral term is increasing.

Because of Lemma 7.1.10, we can study the local modulus of continuity in terms of certain monotone minorants for its increments variance. At the end of Section 7.5 we showed that this technique cannot be used for the uniform modulus of continuity. That is why Theorem 7.2.10 is expressed in terms of the increments variance itself. We begin our discussion of exact uniform moduli of continuity by combining Theorems 7.2.1 and 7.2.10 and Lemma 7.1.1.

Theorem 7.6.8 *Let* $G = \{G(x), x \in [0,1]\}$ *be a Gaussian process with stationary increments for which (7.137) holds or (7.139) holds with* $\alpha > 0$. *Assume furthermore that*

$$\int_0^\delta \frac{\sigma(u)}{u(\log 1/u)^{1/2}} \, du \le C\sigma(\delta)(\log 1/\delta)^{1/2}. \qquad (7.387)$$

Then

$$\lim_{\delta \to 0} \sup_{\substack{|u-v| \le \delta \\ u,v \in [0,1]}} \frac{|G(u) - G(v)|}{(2\sigma^2(|u-v|) \log 1/|u-v|)^{1/2}} = C' \qquad a.s. \qquad (7.388)$$

for some constant $0 < C' < \infty$. *When* $\alpha = 0$ *in (7.139), this result holds as long as* σ *is asymptotic to a monotonic function at zero.*

These statements also hold when $|u-v|$ *is replaced by* δ *in the denominator of the fraction in (7.388),* lim *is replaced by* lim sup, *and* C' *by some constant* C'', *where* $C' \le C'' < \infty$.

Note that in Theorem 7.2.1 the modulus is given in terms of a monotonic majorant ϕ of σ (see (7.82)). When σ^2 is regularly varying at zero with index $\alpha > 0$, it is asymptotic to an increasing function at zero, so we can replace ϕ by σ. When σ^2 is slowly varying at zero, we need to include the hypothesis that it is asymptotic to a monotonic function at zero. Note also that if the left-hand side of (7.387) is "little o" of the right-hand side, then $C' = C''$. We obtain this by showing that (7.388) holds with $\widetilde{\omega}(|u-v|)$ in the denominator and then using Remark 7.6.7.

The results in this section about the local modulus of continuity give us some free information about the uniform modulus of continuity, simply because a lower bound for the local modulus of continuity is also a lower bound for the uniform modulus of continuity. Combining Theorems 7.2.1 and 7.6.4, Remark 7.2.6, and Lemma 7.1.1, we get larger exact uniform moduli of continuity for certain Gaussian processes than is suggested by the analog of the result for Brownian motion.

Theorem 7.6.9 *Let* $G = \{G(t), t \in [0,1]\}$ *be a mean zero real-valued continuous Gaussian process. Let* $\phi(h)$ *be an increasing function that satisfies conditions (2) and (3) of Theorem 7.6.4 (with* $t_0 \in [0,1]$) *and is such that*

$$C_0\phi(|h|) \le (E(G(t+h) - G(t)))^{1/2} \le C_1\phi(|h|) \qquad (7.389)$$

for all $t \in [0,1]$ *and all* $|h|$ *sufficiently small. Let*

$$\Phi_2(h) = \int_0^h \frac{\phi(u)}{u(\log 1/u)^{1/2}} \, du \qquad (7.390)$$

and assume that $\phi(h)(\log 1/h)^{1/2} = O(\Phi_2(h))$. *Then*

$$\lim_{\substack{\delta \to 0}} \sup_{\substack{|u-v| \leq \delta \\ u,v \in [0,1]}} \frac{|G(u) - G(v)|}{\Phi_2(|u - v|)} = C \qquad a.s. \tag{7.391}$$

for some constant $0 < C < \infty$.

Furthermore, (7.391) holds with $\Phi_2(|u - v|)$ replaced by $\Phi_2(|\delta|)$ and $\lim$ replaced by $\limsup$ (with the same constant C).

Example 7.6.10 We consider what the exact uniform modulus of continuity is in the four cases of Example 7.6.6. If (7.137) or (7.139) holds, $\sigma(\delta)$ is asymptotic to a monotone function, and $\sigma(u) \leq C(\log 1/|u|)^{-\alpha}$, $\alpha > 1/2$, it follows from Theorem 7.6.8 that (7.388) holds. This is the case if $\sigma = \phi$ in Example 7.6.6 (2) and (3) since ϕ^2 is concave in these cases. This is also the case in Example 7.6.6 (1) if ϕ also satisfies (7.137) or (7.139).

If $\sigma(u) = ((\log 1/|u|)(\log \log 1/|u|)^\beta)^{-1/2}$, $\beta > 2$, it follows from Theorem 7.6.9 that (7.391) holds with $\Phi_2(u) = \phi(|u|)(\log 1/|u|)^{1/2} \log \log 1/|u|$, as in Example 7.6.6 (4).

Note that in cases (3) and (4) the Gaussian process has the same exact local and uniform moduli of continuity, although we do not know if the constants are the same.

Remark 7.6.11 To put condition (7.367) of Lemma 7.6.2 in perspective, suppose that ω is the maximal monotone minorant for the increments variance, that is, there is equality in (7.360). Then, when G is a continuous Gaussian process with stationary increments, there exists an $\epsilon > 0$ such that

$$\omega^2(2h) \leq 4\omega^2(h) \qquad \forall \, h \leq \epsilon/2. \tag{7.392}$$

Thus (7.367) only requires a slight strengthening of the most general case. (In the paragraph containing (7.97) we show that the smallest monotone majorant of $E(G(t) - G(s))^2$ also satisfies (7.392).)

To obtain (7.392) let $\rho^2(u) = E(G(u) - G(0))^2$. Assume that ρ has a smallest strictly positive zero (if not, $\omega(h; 0, \delta) \equiv 0$ and there is nothing to prove). We first note that there is an $\epsilon > 0$ for which

$$0 < \rho(\epsilon/2) = \min\{\rho(u) : \epsilon/2 \leq u \leq \epsilon\}, \tag{7.393}$$

that is, the minimum of ρ on $[\epsilon/2, \epsilon]$ is taken at $\epsilon/2$. To see that such an ϵ exists, let z be the smallest strictly positive zero of ρ, or any positive number if no such zero exists, and let $m(u) = \min\{\rho(x) : u \leq x \leq z\}$. Choose ϵ as the largest number less than or equal to z such that

$\rho(\epsilon/2) = m(z/2)$. Such a number exists because $0 < m(z/2) \leq \rho(z/2)$ and ρ goes to zero at zero. We now observe that $0 < \rho(\epsilon/2) \leq \rho(x)$ for $\epsilon/2 < x \leq z/2$, since this is how $\epsilon/2$ was chosen. Also, $\rho(\epsilon/2) \leq \rho(x)$ for $z/2 \leq x \leq \epsilon$ because $\rho(\epsilon/2) = m(z/2)$ and $\epsilon \leq z$. Thus (7.393) holds.

By (7.360) (with equality),

$$
\begin{aligned}
\omega^2(2h) \;=\; & \omega^2(2h; 0, \epsilon) = \min_{2h \leq |u| \leq \epsilon} \rho^2(u) & (7.394) \\
\leq\; & 4 \min_{2h \leq |u| \leq \epsilon} \rho^2(u/2) \leq 4 \min_{h \leq |s| \leq \epsilon/2} \rho^2(s) \\
\leq\; & 4 \min_{h \leq |s| \leq \epsilon} \rho^2(s) = 4\omega^2(h).
\end{aligned}
$$

Here we use (7.97) for the first inequality and (7.393) for the last.

7.7 Moduli of continuity for squares of Gaussian processes

Suppose that a Gaussian process G has a local or uniform modulus of continuity. We raise the question, what is the local or uniform modulus of continuity of G^2? This question may seem artificial, but answering it allows us, in Section 9.5, to obtain local and uniform moduli of continuity for the local times of the Markov process associated with G (when G is an associated process).

In order to extend the results of Section 7.1 so that they apply to squares of Gaussian processes, we slightly generalize the class of Gaussian processes that we consider. We drop the condition that the Gaussian process $G = \{G(y), y \in S\}$ has mean zero and instead require that

$$
EG(y) = C \qquad \forall\, y \in S. \tag{7.395}
$$

where C is a constant. Obviously, $d_G(s, t)$, defined in (6.1), is invariant under the translation $\{G(y), y \in S\} \to \{G(y) + a, y \in S\}$ for some constant a, and Lemma 7.1.1 continues to hold for Gaussian processes satisfying (7.395).

The result for the local modulus of continuity of squares of Gaussian processes is easy to obtain.

Theorem 7.7.1 *Let $G = \{G(y), y \in (S, \tau)\}$ be a continuous Gaussian process satisfying (7.395). Let $y_0 \in S$. Assume that G satisfies (7.2) with $C = 1$. Then*

$$
\lim_{\delta \to 0} \sup_{\substack{\tau(y, y_0) \leq \delta \\ y \in S}} \frac{|G^2(y) - G^2(y_0)|}{2\rho(\tau(y, y_0))} = |G(y_0)| \qquad a.s. \tag{7.396}
$$

*Furthermore, if (7.2) holds with a less than or equal to sign, then (7.396)
holds with a less than or equal to sign, and if (7.2) holds with a greater
than or equal to sign, then (7.396) holds with a greater than or equal to
sign.*

Proof This is immediate since

$$\sup_{\tau(y,y_0)\leq\delta} \frac{|G(y)-G(y_0)|}{\rho(\tau(y,y_0))} \left(|G(y_0)| - \frac{1}{2} \sup_{\tau(y,y_0)\leq\delta} |G(y)-G(y_0)| \right) \quad (7.397)$$

$$\leq \sup_{\tau(y,y_0)\leq\delta} \frac{|G^2(y)-G^2(y_0)|}{2\rho(\tau(y,y_0))}$$

$$\leq \sup_{\tau(y,y_0)\leq\delta} \frac{|G(y)-G(y_0)|}{\rho(\tau(y,y_0))} \left(|G(y_0)| + \frac{1}{2} \sup_{\tau(y,y_0)\leq\delta} |G(y)-G(y_0)| \right).$$

We see that (7.396) follows by continuity. It is also clear from (7.397)
that the one-sided results also hold.

Remark 7.7.2 When $G(y_0) \equiv 0$, (7.396) is not very interesting since,
obviously,

$$\lim_{\delta\to0} \sup_{\substack{\tau(y,y_0)\leq\delta \\ y\in S}} \frac{G^2(y)}{\rho^2(\tau(y,y_0))} = 1 \qquad \text{a.s.} \qquad (7.398)$$

It is also easy to find an upper bound for the uniform modulus of
continuity of G^2.

Theorem 7.7.3 *Let $G = \{G(y), y \in S\}$ be a continuous Gaussian
process satisfying (7.395) and (7.1) with "$= C$" replaced by "≤ 1."
Then*

$$\lim_{\delta\to0} \sup_{\substack{\tau(u,v)\leq\delta \\ u,v\in K}} \frac{G^2(u)-G^2(v)}{2\omega(\tau(u,v))} \leq \sup_{u\in K} |G(u)| \qquad \text{a.s.} \qquad (7.399)$$

Proof This is immediate since

$$\lim_{\delta\to0} \sup_{\substack{\tau(u,v)\leq\delta \\ u,v\in K}} \frac{G^2(u)-G^2(v)}{2\omega(\tau(u,v))} \qquad (7.400)$$

$$\leq \lim_{\delta\to0} \sup_{\substack{\tau(u,v)\leq\delta \\ u,v\in K}} \frac{|G(u)-G(v)|}{\omega(\tau(u,v))} \lim_{\delta\to0} \sup_{\substack{\tau(u,v)\leq\delta \\ u,v\in K}} \frac{|G(u)+G(v)|}{2}.$$

The first term on the right side of the inequality in (7.400) is less than
or equal to 1 by hypothesis. The rest is obvious. $\qquad \square$

Finding an exact uniform modulus of continuity for G^2 is more complicated. We begin with a technical lemma.

Lemma 7.7.4 *Let $G = \{G(y), y \in S\}$ be a continuous Gaussian process satisfying (7.395) and (7.1) with "$= C$" replaced by "≥ 1." Assume that ω (in (7.1)) satisfies (7.7). Then there exists a $u_0 \in K$ such that, for all $\epsilon > 0$,*

$$\lim_{\delta \to 0} \sup_{\substack{\tau(u,v) \leq \delta \\ u,v \in B(u_0, \epsilon) \cap K}} \frac{G(u) - G(v)}{\omega(\tau(u,v))} \geq 1 \qquad a.s. \qquad (7.401)$$

and

$$\lim_{\delta \to 0} \sup_{\substack{\tau(u,v) \leq \delta \\ u,v \in K}} \frac{G^2(u) - G^2(v)}{2\omega(\tau(u,v))} \geq |G(u_0)| \qquad a.s. \qquad (7.402)$$

Proof We first show that for all $\epsilon_n > 0$ there exists a $u_0(\epsilon_n) \in K$ such that

$$\lim_{\delta \to 0} \sup_{\substack{\tau(u,v) \leq \delta \\ u,v \in B(u_0(\epsilon_n), \epsilon_n) \cap K}} \frac{G(u) - G(v)}{\omega(\tau(u,v))} \geq 1 \qquad a.s. \qquad (7.403)$$

Obviously, $K \subset \cup_{x \in K} B_\tau(x, \epsilon_n/2)$. Let $B_\tau(x_1, \epsilon_n/2), \ldots, B_\tau(x_m, \epsilon_n/2)$ be a finite cover of K. Note that if $u, v \in K$ such that $\tau(u,v) < \epsilon_n/2$, then both u and v are contained in $B_\tau(x_j, \epsilon_n)$ for some $1 \leq j \leq m$. Thus, for $0 < \delta < \epsilon_n/2$,

$$\sup_{1 \leq j \leq m} \sup_{\substack{\tau(u,v) \leq \delta \\ u,v \in B(x_j, \epsilon_n) \cap K}} \frac{G(u) - G(v)}{\omega(\tau(u,v))} = \sup_{\substack{\tau(u,v) \leq \delta \\ u,v \in K}} \frac{G(u) - G(v)}{\omega(\tau(u,v))}. \qquad (7.404)$$

Suppose there exists an $\epsilon' > 0$ and a $1 \leq j \leq m$ such that

$$\lim_{\delta \to 0} \sup_{\substack{\tau(u,v) \leq \delta \\ u,v \in B_\tau(x_j, \epsilon_n) \cap K}} \frac{G(u) - G(v)}{\omega(\tau(u,v))} \leq 1 - \epsilon' \qquad (7.405)$$

on a set of positive measure. Then, since (7.7) holds on $B_\tau(x_j, \epsilon_n) \cap K$, it follows from Lemma 7.1.1 that the event in (7.405) holds almost surely. If the event in (7.405) holds almost surely for all $1 \leq j \leq m$, then, for any $\gamma > 0$, we can find a $\delta'' > 0$ such that, for each j and $\delta' \leq \delta''$,

$$\sup_{\substack{\tau(u,v) \leq \delta' \\ u,v \in B_\tau(x_j, \epsilon_n) \cap K}} \frac{G(u) - G(v)}{\omega(\tau(u,v))} \leq 1 - \epsilon'/2 \qquad (7.406)$$

with probability greater that $1 - \gamma$. Therefore, by (7.404),

$$\sup_{\substack{\tau(u,v)\leq\delta \\ u,v\in K}} \frac{G(u) - G(v)}{\omega(\tau(u,v))} \tag{7.407}$$

$$= \sup_{1\leq j\leq m} \sup_{\substack{\tau(u,v)\leq\delta' \\ u,v\in B_\tau(x_j,\epsilon_n)\cap K}} \frac{G(u) - G(v)}{\omega(\tau(u,v))} \leq 1 - \epsilon'/2$$

with probability greater that $1 - m\gamma$. Consequently, by Lemma 7.1.1,

$$\lim_{\delta\to 0} \sup_{\substack{\tau(u,v)\leq\delta \\ u,v\in K}} \frac{G(u) - G(v)}{\omega(\tau(u,v))} \leq 1 - \epsilon' \qquad \text{a.s.,} \tag{7.408}$$

which contradicts the fact that ω is a lower uniform modulus of continuity for G with constant greater than or equal to one.

Thus the limit superior in (7.405) is greater than or equal to one almost surely for some $1 \leq j \leq m$. We set $u_0(\epsilon_n) = x_j$ for some j for which this occurs and obtain (7.403).

Now choose a sequence $\epsilon_n \to 0$ and consider the balls $B_\tau(u(\epsilon_n), \epsilon_n)$ for which (7.403) holds. Since K is compact there exists a sequence $\{u_{n_k}\}_{k=1}^\infty$ such that $\lim_{k\to\infty} u_{n_k} = u_0$ for some $u_0 \in K$. It is easy to see that for all $\epsilon > 0$ there exists a $k_0(\epsilon)$ such that, for $k \geq k_0(\epsilon)$, $B_\tau(u(\epsilon_{n_k}), \epsilon_{n_k}) \subset B_\tau(u_0, \epsilon)$. This observation and (7.403) give (7.401).

Note that

$$\sup_{\substack{\tau(u,v)\leq\delta \\ u,v\in B_\tau(u_0,\epsilon)\cap K}} \frac{G^2(u) - G^2(v)}{2\omega(\tau(u,v))} \tag{7.409}$$

$$\geq \sup_{\substack{\tau(u,v)\leq\delta \\ u,v\in B_\tau(u_0,\epsilon)\cap K}} \frac{|G(u) - G(v)|}{2\omega(\tau(u,v))} \inf_{\substack{\tau(u,v)\leq\delta \\ u,v\in B_\tau(u_0,\epsilon)\cap K}} |G(u) + G(v)|.$$

Taking the limit as $\delta \to 0$ and using (7.401), we get (7.402). $\qquad\square$

We now introduce homogeneity conditions that allow us to obtain (7.402) for all $u_0 \in K$. We say that a metric or pseudometric space $(\mathcal{S}, \tau)$ is locally homogeneous if any two points in $\mathcal{S}$ have isometric neighborhoods in the metric τ.

Theorem 7.7.5 *Let (S, d_G) be locally homogeneous and let $K \subset S$ be a compact set that is the closure of its interior. Let $G = \{G(u), u \in S\}$ be a continuous Gaussian process satisfying (7.395) and (7.1) with "$= C$"*

replaced by "≥ 1." Assume that ω (in (7.1)) satisfies (7.7). Then

$$\lim_{\substack{\delta \to 0 \\ d_G(u,v) \leq \delta \\ u,v \in K}} \sup \frac{G^2(u) - G^2(v)}{2\omega(d_G(u,v))} \geq \sup_{u \in K} |G(u)| \qquad a.s. \qquad (7.410)$$

Proof Since K is the closure of its interior, for every $u_0 \in K$ there exists an $\epsilon > 0$ such that $B_{d_G}(u_0, \epsilon) \subset K$. It then follows from the homogeneity of (S, d_G) that (7.401) is satisfied for this u_0 and this ϵ. In fact, for every $u_0 \in \text{int } K$, (7.401) is satisfied for some $\epsilon > 0$ and, consequently, so is (7.402). In particular, if $\{x_j\}_{j=1}^n$ are points in the interior of K, then

$$\lim_{\substack{\delta \to 0 \\ d_G(u,v) \leq \delta \\ u,v \in K}} \sup \frac{G^2(u) - G^2(v)}{2\omega(d_G(u,v))} \geq \sup_{1 \leq j \leq n} |G(x_j)| \qquad a.s. \qquad (7.411)$$

Since this inequality is independent of n, it also holds for $\{x_j\}_{j=1}^\infty$ contained in a countable dense subset of K. Therefore, since G is uniformly continuous on K, we get (7.410).

Combining Theorems 7.7.3 and 7.7.5, we get

Theorem 7.7.6 *Let (S, d_G) be locally homogeneous and let $K \subset S$ be a compact set that is the closure of its interior. Let $G = \{G(y), y \in S\}$ be a continuous Gaussian process satisfying (7.395) and (7.1) with $C = 1$. Then*

$$\lim_{\substack{\delta \to 0 \\ d_G(u,v) \leq \delta \\ u,v \in K}} \sup \frac{G^2(u) - G^2(v)}{2\omega(d_G(u,v))} = \sup_{u \in K} |G(u)| \qquad a.s. \qquad (7.412)$$

Note that by Lemma 7.1.1, the hypothesis of Theorem 7.7.5, that ω satisfies (7.7), follows from the hypotheses of this theorem.

Remark 7.7.7 All the results in this section hold for m-moduli, in the sense that results for the m-moduli of G extend to the squares of G in analogy with the results in Theorems 7.7.1–7.7.6. The lower bounds follow because, by Lemma 7.1.6, the m-moduli are larger than the corresponding moduli considered in this section. It is easy to see that the proofs for the upper bounds in Theorems 7.7.1 and 7.7.3 also work for m-moduli. There is only one minor point to consider. The hypothesis of the analog of Theorem 7.7.5 requires that the uniform m-modulus satisfies (7.7). This is shown in Lemma 7.1.9.

7.8 Notes and references

Much of the material in Sections 7.1–7.6 is summarized in Marcus and Rosen (1992a), which deals with local and uniform moduli of continuity of local times. We consider local and uniform moduli of continuity of local times in Section 9.5. Theorem 7.1.4 and Lemma 7.1.5 are constructed from material in Sections 6.2, 6.4, and 6.5 of Fernique (1997). Lemma 7.1.10 is by Marcus and Shepp (1972).

The first half of Section 7.2 contains more modern proofs of some classical results that were obtained before the introduction of metric entropy and majorizing measures into the study of Gaussian processes. Many of the old papers containing these results are still interesting. In chronological order we mention some of them: Chung, Erdös and Sirao (1959); Nisio (1967); Marcus (1968, 1970); Garsia, Rodemich and Rumsey, Jr. (1970); Kono (1970); and Sirao and Watanabe (1970). The material in Lemma 7.2.5 is taken from Marcus (1972). The continuity conditions involving regular variation and normalized regular variation develop an idea introduced in Marcus and Rosen (1992a). The material in Section 7.3 may appear to deal with too restricted a class of Gaussian processes but it is what we need when applying these results to study the moduli of continuity of local times in Section 9.5.

Theorem 7.4.5 and its corollary Theorem 7.4.12, which are new, exploit special properties of the covariances of associated Gaussian processes. Theorem 7.4.5 may seem peculiar because it requires an upper bound on σ whereas we know that for larger σ the iterated log term in (7.247) is replaced by a larger function of u. The problem here is that we are using the local modulus of time changed Brownian motion as a lower bound. This technique is less effective for processes with increments variance much larger than the increments variance of Brownian motion.

Sections 7.5 and 7.6 are taken from Marcus (1972). Section 7.7 is taken from Marcus and Rosen (1992a).

8

Isomorphism Theorems

The relationship between strongly symmetric Borel right processes X and their associated mean zero Gaussian processes G (G is the Gaussian process with covariance equal to the 0-potential of X) is given by several isomorphism theorems that relate squares of G to the local times of X.

8.1 The isomorphism theorems of Eisenbaum and Dynkin

We begin with an isomorphism theorem due to N. Eisenbaum. This theorem plays an important role in this book.

Theorem 8.1.1 (Eisenbaum Isomorphism Theorem) *Let* $X = (\Omega, \mathcal{G}, \mathcal{G}_t, X_t, \theta_t, P^x)$ *be a strongly symmetric Borel right process with continuous 0-potential density* $u(x, y)$ *and state space* S. *Let* $L = \{L_t^y \, ; \, (y, t) \in S \times R_+\}$ *denote the local time for* X *normalized so that* $E^x(L_\infty^y) = u(x, y)$. *Let* $G = \{G_y \, ; \, y \in S\}$ *denote the mean zero Gaussian process with covariance* $u(x, y)$. *Then, for any countable subset* $D \subseteq S$, *any* $x \in S$, *and any* $s \neq 0$,

$$\left\{ L_\infty^y + \frac{1}{2}(G_y + s)^2 \, ; \, y \in D, \, P^x \times P_G \right\} \tag{8.1}$$

$$\overset{law}{=} \left\{ \frac{1}{2}(G_y + s)^2 \, ; \, y \in D, \, (1 + \frac{G_x}{s})P_G \right\}.$$

Equivalently, for all $x, x_1, \ldots, x_n$ *in* S *and bounded measurable functions* F *on* R_+^n, *for all* n, *and all* $s \neq 0$,

$$E^x E_G \left(F \left(L_\infty^{x_i} + \frac{(G_{x_i} + s)^2}{2} \right) \right) = E_G \left(\left(1 + \frac{G_x}{s} \right) F \left(\frac{(G_{x_i} + s)^2}{2} \right) \right). \tag{8.2}$$

Here we use the notation $F(f(x_i)) := F(f(x_1), \ldots, f(x_n))$.

362

Proof   We first show that

$$E^x E_G \exp \left(\sum_{i=1}^n \lambda_i \left(L_\infty^{x_i} + \frac{(G_{x_i} + s)^2}{2} \right) \right) \tag{8.3}$$

$$= E_G \left(\left(1 + \frac{G_x}{s} \right) \exp \left(\sum_{i=1}^n \frac{\lambda_i (G_{x_i} + s)^2}{2} \right) \right)$$

for all $\lambda_1, \ldots, \lambda_n$ sufficiently small, where $x = x_1$. We write (8.3) as

$$E^x \exp \left(\sum_{i=1}^n \lambda_i L_\infty^{x_i} \right) = 1 + \frac{E \left(G_x \exp \left(\sum_{i=1}^n \lambda_i (G_{x_i} + s)^2 / 2 \right) \right)}{s E \exp \left(\sum_{i=1}^n \lambda_i (G_{x_i} + s)^2 / 2 \right)}. \tag{8.4}$$

Let Σ denote the covariance matrix $\{u(x_i, x_j)\}_{i,j=1}^n$. It follows from (5.59) that

$$\frac{E \left(G_x \exp \left(\sum_{i=1}^n \lambda_i (G_{x_i} + s)^2 / 2 \right) \right)}{s E \exp \left(\sum_{i=1}^n \lambda_i (G_{x_i} + s)^2 / 2 \right)} = \{\widetilde{\Sigma} \Lambda \mathbf{1}^t\}_1, \tag{8.5}$$

where $\widetilde{\Sigma} = (I - \Sigma \Lambda)^{-1} \Sigma$ and $\mathbf{1}^t$ is the transpose of the row vector $\mathbf{1} = (1, \ldots, 1)$.

Therefore, to obtain (8.3) it suffices to show that

$$E^x \exp \left(\sum_{i=1}^n \lambda_i L_\infty^{x_i} \right) = 1 + \{\widetilde{\Sigma} \Lambda \mathbf{1}^t\}_1. \tag{8.6}$$

This is indeed the case since, by (3.246) with $x = x_1$,

$$\begin{aligned}
E^{x_1} \exp \left(\sum_{i=1}^n \lambda_i L_\infty^{x_i} \right) &= \{(I - \Sigma \Lambda)^{-1} \mathbf{1}^t\}_1 \tag{8.7} \\
&= \{I \mathbf{1}^t + (I - \Sigma \Lambda)^{-1} \Sigma \Lambda \mathbf{1}^t\}_1 \\
&= 1 + \{(I - \Sigma \Lambda)^{-1} \Sigma \Lambda \mathbf{1}^t\}_1. \\
&= 1 + \{\widetilde{\Sigma} \Lambda \mathbf{1}^t\}_1.
\end{aligned}$$

Let μ_1 and μ_2 be the measures on R_+^n defined by

$$\int F(\cdot) \, d\mu_1 = E^x E_G \left(F \left(L_\infty^{x_i} + \frac{(G_{x_i} + s)^2}{2} \right) \right) \tag{8.8}$$

and

$$\int F(\cdot) \, d\mu_2 = E_G \left(\left(1 + \frac{G_x}{s} \right) F \left(\frac{(G_{x_i} + s)^2}{2} \right) \right) \tag{8.9}$$

for all nonnegative measurable functions F on R_+^n. The measure μ_1 is determined by its moment generating function, the left-hand side of (8.3). Furthermore, (8.3) shows that the measures μ_1 and μ_2 are equal.

Therefore, although it is not clear to begin with that μ_2 is a positive measure, this argument shows that it is.

The two sides of (8.1) determine measures on R_+^∞ equipped with the σ-algebra of cylinder sets. We have just seen that these measures have the same finite-dimensional distributions; hence they are equal. This gives (8.1). $\square$

The isomorphism in (8.1) is given for the total accumulated local time of a strongly symmetric Borel right process with continuous 0-potential density. Exactly the same proof (see in particular the use of (3.246)) shows that it also holds for terminal times and inverse local times. Thus we have the following corollary of the proof of Theorem 8.1.1.

Corollary 8.1.2 *Let* $X = (\Omega, \mathcal{G}, \mathcal{G}_t, X_t, \theta_t, P^x)$ *be a strongly symmetric Borel right process with continuous α-potential densities $u^\alpha(x,y)$ and state space S. Let $L = \{L_t^y \, ; \, (y,t) \in S \times R_+\}$ denote the local time for X normalized so that*

$$E^x \left(\int_0^\infty e^{-\alpha t} \, dL_t^y \right) = u^\alpha(x,y), \qquad (8.10)$$

and let T be a terminal time for X. Assume that there exists a mean zero Gaussian process $G = \{G_y \, ; \, y \in S\}$ with covariance $u_T(x,y)$. Then, for any countable subset $D \subseteq S$, any $x \in S$, and any $s \neq 0$,

$$\left\{ L_T^y + \frac{1}{2}(G_y + s)^2 \, ; \, y \in D \, , \, P^x \times P_G \right\} \qquad (8.11)$$

$$\overset{law}{=} \left\{ \frac{1}{2}(G_y + s)^2 \, ; \, y \in D \, , \, (1 + \frac{G_x}{s})P_G \right\}.$$

This also holds when T is replaced by the inverse local time $\tau_A(\lambda)$, for any CAF, $A = \{A_t, t \in R_+\}$, and P^x is replaced by P_λ^x.

The first isomorphism theorem relating Markov local times and Gaussian processes is due to E. B. Dynkin. It relates local times for the h-transform of a strongly symmetric Markov process X to the Gaussian process associated with X.

Theorem 8.1.3 (Dynkin Isomorphism Theorem) *Let* $X = (\Omega, \mathcal{G}, \mathcal{G}_t, X_t, \theta_t, P^x)$ *be a strongly symmetric Borel right process with continuous 0-potential density $u(x,y)$. Let 0 denote a fixed element of S. Assume that*

$$h_x = P^x(T_0 < \infty) = \frac{u(x,0)}{u(0,0)} > 0 \qquad (8.12)$$

for all $x \in S$. Let $\widetilde{X} = \left(\Omega_h, \mathcal{F}^h, \mathcal{F}_t^h, X_t, \theta_t, P^{x,0}\right)$ denote the h-transform of X as described in Section 3.9 and let $\overline{L} = \{\overline{L}_t^y ; (y,t) \in S \times R_+\}$ denote the local time for $\widetilde{X}$, normalized so that

$$E^{x,0}\left(\overline{L}_\infty^y\right) = \frac{u(x,y)h(y)}{h(x)} \tag{8.13}$$

(see (3.247)). Let $G = \{G_y ; y \in S\}$ denote the mean zero Gaussian process with covariance $u(x,y)$. Then, for any countable subset $D \subseteq S$,

$$\left\{\overline{L}_\infty^y + \frac{1}{2}G_y^2 ; y \in D, P^{x,0} \times P_G\right\} \overset{law}{=} \left\{\frac{1}{2}G_y^2 ; y \in D, \frac{G_x G_0}{u(0,x)}P_G\right\}. \tag{8.14}$$

Equivalently, for all $x, x_1, \ldots, x_n$ in S and bounded measurable functions F on R_+^n, for all n,

$$E^{x,0}E_G\left(F\left(\overline{L}_\infty^{x_i} + \frac{G_{x_i}^2}{2}\right)\right) = E_G\left(\frac{G_x G_0}{u(0,x)}F\left(\frac{G_{x_i}^2}{2}\right)\right). \tag{8.15}$$

Here we use the notation $F(f(x_i)) := F(f(x_1), \ldots, f(x_n))$.

Proof To prove this theorem it suffices to show that

$$u(0,x)E^{x,0}\exp\left(\sum_{i=1}^n \lambda_i \overline{L}_\infty^{x_i}\right) = \frac{E\left(G_x G_0 \exp\left(\sum_{i=1}^n \lambda_i G_{x_i}^2/2\right)\right)}{E\exp\left(\sum_{i=1}^n \lambda_i G_{x_i}^2/2\right)} \tag{8.16}$$

for all $\lambda_1, \ldots, \lambda_n$ sufficiently small where $x = x_1$ and $0 = x_n$. (As in the proof of Theorem 8.1.1, this implies that $\dfrac{G_x G_0}{u(0,x)}P_G$ is a positive measure on the paths of $\{G_y^2 ; y \in D\}$.) Let Σ denote the covariance matrix $\{u(x_i, x_j)\}_{i,j=1}^n$. It follows from (5.58) that

$$\frac{E\left(G_{x_1}G_{x_n}\exp\left(\sum_{i=1}^n \lambda_i G_{x_i}^2/2\right)\right)}{E\exp\left(\sum_{i=1}^n \lambda_i G_{x_i}^2/2\right)} = \{\widetilde{\Sigma}\}_{1,n} \tag{8.17}$$

$$= \{(I - \Sigma\Lambda)^{-1}\Sigma\}_{1,n}$$

$$= \sum_{j=1}^n \{(I - \Sigma\Lambda)^{-1}\}_{1,j}u(x_j, 0),$$

where $\widetilde{\Sigma} = (I - \Sigma\Lambda)^{-1}\Sigma$ and $\mathbf{1}^t$ is the transpose of the row vector $\mathbf{1} = (1, \ldots, 1)$. On the other hand, by (3.254), we see that

$$u(0,x)E^{x,0}\exp\left(\sum_{i=1}^n \lambda_i \overline{L}_\infty^{x_i}\right) = \sum_{j=1}^n \{(I - \Sigma\Lambda)^{-1}\}_{1,j}u(x_j, 0). \tag{8.18}$$

Comparing this with (8.17) gives (8.16). $\qquad\square$

There is some similarity between this proof and the one in Rogers and Williams (2000b), which they refer to as a "caricature" of Dynkin's Isomorphism Theorem.

Remark 8.1.4 We now present a relationship that combines Theorems 8.1.1 and 8.1.3. To simplify the expressions we make the following change of notation. Let ν be a finite discrete measure on S of the form $\sum_{i=1}^{n} \lambda_i \delta_{x_i}(\,\cdot\,)$, where the λ_i are such that the relationships given exist. In this notation we write (8.16) as

$$u(x,y)E^{x,y} \exp\left(\int \overline{L}_\infty^r \, d\nu(r)\right) = \frac{E\left(G_x G_y \exp\left(\int G_r^2 \, d\nu(r)/2\right)\right)}{E \exp\left(\int G_r^2 \, d\nu(r)/2\right)} \tag{8.19}$$

and (8.6) as

$$E^x \exp\left(\int L_\infty^r \, d\nu(r)\right) = 1 + \int \widetilde{\Sigma}_{1,y}\, \nu(dy), \tag{8.20}$$

where $x=x_1$. Therefore, by (8.19) and (8.17),

$$E^x \exp\left(\int L_\infty^r \, d\nu(r)\right) \tag{8.21}$$

$$= 1 + \int \left(E^{x,y} \exp\left(\int L_\infty^r \, d\nu(r)\right)\right) u(x,y) \, d\nu(y).$$

Here we use $\{L_\infty^s, s \in S\}$ to indicate the total accumulated local time of the Markov process under consideration. Which process that is is indicated by the probability measure. Thus, on the left-hand side of (8.21) we are considering the total accumulated local times of a strongly symmetric Borel right process X with 0-potential density $u(x,y)$, whereas on the right-hand side of (8.21) we are considering the total accumulated local times of a family of h-transforms of X with $h(y) = u(x,y)/u(x,x)$.

8.1.1 The Generalized First Ray–Knight Theorem

Theorem 8.1.1 applied to the local times of processes killed at T_0, where 0 is some fixed point in S, gives a generalization of the First Ray–Knight Theorem. Recall that, by Lemma 3.8.1, if $P^x(T_0 < \infty) > 0$ for all $x \in S$, then $u_{T_0}(x,y)$ is finite and positive definite. The next theorem is simply a special case of Corollary 8.1.2.

Theorem 8.1.5 (Generalized First Ray–Knight Theorem) *Let $X = (\Omega, \mathcal{G}, \mathcal{G}_t, X_t, \theta_t, P^x)$ be a strongly symmetric Borel right process with continuous α-potential density $u^\alpha(x,y)$ and state space S. Assume that*

$P^x(T_0 < \infty) > 0$ *for all* $x \in S$. *Let* $L = \{L_t^y\,;\,(y,t) \in S \times R_+\}$ *denote the local times for* X *normalized so that* $E^x(\int_0^\infty e^{-\alpha t}\,dL_t^y) = u^\alpha(x,y)$. *Let* $G = \{G_y\,;\,y \in S\}$ *denote the mean zero Gaussian process with covariance* $u_{T_0}(x,y)$. *Then, for any countable subset* $D \subseteq S$, *any* $x \in S$, *and any* $s \neq 0$,

$$\left\{ L_{T_0}^y + \frac{1}{2}(G_y + s)^2 \,;\, y \in D\,,\, P^x \times P_G \right\} \tag{8.22}$$

$$\overset{law}{=} \left\{ \frac{1}{2}(G_y + s)^2 \,;\, y \in D\,,\, (1 + \frac{G_x}{s})P_G \right\}.$$

Remark 8.1.6 We justify calling Theorem 8.1.5 the Generalized First Ray–Knight Theorem by showing that for Brownian motion it is equivalent to the classic First Ray–Knight Theorem (Theorem 2.6.3). Since both theorems describe the law of $\{L_{T_0}^y\,;\,y \in R^1\}$, they must be equivalent, but we now show how the much simpler statement of Theorem 2.6.3 follows directly from (8.22).

Recall from (2.139) that $u_{T_0}(x,y) = 2(|x| \wedge |y|)$ for $xy > 0$ and 0 otherwise. Thus $G = \{G_y\,;\,y \in R_+\}$ is $\{\sqrt{2}\,B_y\,;\,y \in R_+\}$, where $\{B_y\,;\,y \in R_+\}$ is standard Brownian motion. Then, using continuity we can write (8.22) for $x > 0$ and $s > 0$ as

$$\{ L_{T_0}^y + (B_y + s)^2 \,;\, y \in R_+\,,\, P^x \times P_B \} \tag{8.23}$$

$$\overset{law}{=} \{ (B_y + s)^2 \,;\, y \in R_+\,,\, (1 + \frac{B_x}{s})P_B \}.$$

We simplify this expression by changing the measure. Let P_B^s denote the probability measure of Brownian motion starting at $s > 0$. Then we can rewrite (8.23) as

$$\{ L_{T_0}^y + B_y^2 \,;\, y \in R_+\,,\, P^x \times P_B^s \} \overset{law}{=} \{ B_y^2 \,;\, y \in R_+\,,\, (\frac{B_x}{s})P_B^s \}. \tag{8.24}$$

As in Theorem 2.6.1, we first consider the restricted case where all $y \in [0,x]$, so that (8.24) becomes

$$\{ L_{T_0}^y + B_y^2 \,;\, y \in [0,x]\,,\, P^x \times P_B^s \} \overset{law}{=} \{ B_y^2 \,;\, y \in [0,x]\,,\, (\frac{B_x}{s})P_B^s \} \tag{8.25}$$

$s > 0$. We claim that in fact

$$\{ L_{T_0}^y + B_y^2 \,;\, y \in [0,x]\,,\, P^x \times P_B^s \} \overset{law}{=} \{ B_y^2 \,;\, y \in [0,x]\,,\, 1_{\{T_0 > x\}}(\frac{B_x}{s})P_B^s \}. \tag{8.26}$$

To see this it suffices to show that for any $0 \leq y_1 < \ldots < y_n \leq x$ and any $\lambda_1, \ldots, \lambda_n$ sufficiently small

$$E^s\left(1_{\{T_0 < x\}} B_x \exp\left(\sum_{i=1}^n \lambda_i B_{y_i}^2\right)\right) = 0. \qquad (8.27)$$

Heuristically, this is an obvious consequence of the strong Markov property at T_0, since after hitting 0, B_x is symmetric about 0, and all the other terms in B_{y_i} are squared. We give the details at the conclusion of this remark.

Let $\widetilde{X}_t$ denote Brownian motion killed at T_0 and $\widetilde{P}^s$ denote the probabilities with respect to this process. We can write (8.26) as

$$\{L_{T_0}^y + B_y^2 \; ; \; y \in [0, x] \,, \, P^x \times P_B^s\} \quad \overset{law}{=} \quad \{\widetilde{X}_y^2 \; ; \; y \in [0, x] \,, \, (\frac{\widetilde{X}_x}{s})\widetilde{P}^s\}$$

$$\overset{law}{=} \quad \{Y_y^2 \; ; \; y \in [0, x] \,, \, P_Y^s\}, \quad (8.28)$$

where Y_t is a three-dimensional Bessel process and for the last equality we use (4.185). We now note that under P_B^s, B_y^2 is a $BESQ^1(s^2)$ process and under P_Y^s, Y_y^2 is a $BESQ^3(s^2)$ process. Therefore, it follows from (8.28) and Theorem 14.2.2 that under P^x, $\{L_{T_0}^y, y \in [0, x]\}$ has the law of a $BESQ^2(0)$ process. This gives us Theorem 2.6.1, the restricted form of the First Ray–Knight Theorem.

To describe the evolution of $L_{T_0}^y$ for $y > x$ we note that, by (8.24),

$$\{L_{T_0}^{x+y} + B_{x+y}^2 \; ; \; y \in R_+ \,, \, P^x \times P_B^s\} \overset{law}{=} \{B_{x+y}^2 \; ; \; y \in R_+ \,, \, (\frac{B_x}{s})P_B^s\}. \qquad (8.29)$$

Writing $B_{x+y} = B_x + \bar{B}_y$, where $\bar{B}_y$ is an independent Brownian motion starting at 0, we can rewrite (8.29) as

$$\{L_{T_0}^{x+y} + (B_x + \bar{B}_y)^2 \; ; \; y \in R_+ \,, \, P^x \times P_B^s \times P_{\bar{B}}^0\} \qquad (8.30)$$

$$\overset{law}{=} \{(B_x + \bar{B}_y)^2 \; ; \; y \in R_+ \,, \, (\frac{B_x}{s})P_B^s \times P_{\bar{B}}^0\}$$

$$\overset{law}{=} \{(\sqrt{L_{T_0}^x + B_x^2} + \bar{B}_y)^2 \; ; \; y \in R_+ \,, \, P^x \times P_B^s \times P_{\bar{B}}^0\},$$

where in the last step we use (8.29) again, but with $y = 0$.

Note that conditional on B_x, $(B_x + \bar{B}_y)^2$ is a $BESQ^1(B_x^2)$ process under $P_{\bar{B}}^0$. Also conditional on B_x and $L_{T_0}^x$, $\left(\sqrt{L_{T_0}^x + B_x^2} + \bar{B}_y\right)^2$ is a $BESQ^1(L_{T_0}^x + B_x^2)$ process under $P_{\bar{B}}^0$. Using these two observations in the first and third lines of (8.30) along with Theorem 14.2.2, we see that conditional on $L_{T_0}^x$, $\{L_{T_0}^{x+y}, y \in R_+\}$ has the law of a $BESQ^0(L_{T_0}^x)$ process under P^x. This together with the restricted form of the First Ray–Knight Theorem just proved gives us the equivalent formulation of the First Ray–Knight Theorem given in the second paragraph of Theorem 2.6.3, except for the final statement about independence. This is obvious, as we point out in Remark 14.2.3.

To prove (8.27) let E_ω^z denote the expectation operator for canonical Brownian motion on the space of continuous paths $\omega = \{\omega(t), t \geq 0\}$ starting at z. We first note that for any $F \in \mathcal{F}_{T_0} \times \mathcal{F}$

$$E_\omega^s \left(F(\omega, \theta_{T_0}\omega) 1_{\{T_0 < \infty\}} \right) = E_\omega^s \left(E_{\omega'}^0 \left(F(\omega, \omega') \right) 1_{\{T_0 < \infty\}} \right). \qquad (8.31)$$

When $F = F_1 F_2$ with $F_1 \in \mathcal{F}_{T_0}$, $F_2 \in \mathcal{F}$, this is just the strong Markov property. Its extension to general functions $F \in \mathcal{F}_{T_0} \times \mathcal{F}$ follows by a monotone class argument. Let $f, g \in C(R_+)$ and define $T_0(f) = \inf\{t \mid f(t) = 0\}$. Let

$$F(f, g) = \exp\left(\sum_i 1_{\{y_i \leq T_0(f)\}} \lambda_i f^2(y_i) \right) \qquad (8.32)$$

$$g(x - T_0(f)) \exp\left(\sum_i 1_{\{y_i > T_0(f)\}} \lambda_i g^2(y_i - T_0(f)) \right).$$

Clearly $F \in \mathcal{F}_{T_0} \times \mathcal{F}$ and

$$F(B, \theta_{T_0} B) = B(x) \exp\left(\sum_{i=1}^n \lambda_i B^2(y_i) \right). \qquad (8.33)$$

Using this and (8.31) we see that the left-hand side of (8.27) is equal to

$$E_B^s \left(\exp\left(\sum_i 1_{\{y_i \leq T_0\}} \lambda_i B^2(y_i) \right) H(x) \right), \qquad (8.34)$$

where

$$H(x) = E_{B'}^0 \left(B'(x - T_0(B)) \right) \qquad (8.35)$$

$$\exp\left(\sum_i 1_{\{y_i > T_0(B)\}} \lambda_i B'^2(y_i - T_0(B)) \right) 1_{\{T_0(B) < x\}} \right).$$

Using the fact that $T_0(B)$, inside the expectation $E_{B'}^0$, is a fixed constant, and the symmetry of Brownian motion starting at 0, we see that $H(x) = 0$, which gives (8.27).

Remark 8.1.7

(1) Dynkin's Isomorphism Theorem (Theorem 8.1.3) cannot be used to study L_{T_0}. This is because $G(0)$, which appears on the right-hand side of (8.14), is equal to 0.

(2) The main result in Subsection 4.5.1, (4.185), is itself an isomorphism theorem. As in Subsection 4.5.1, let $\{\widetilde{X}_t; t \in (0, \infty)\}$ denote Brownian motion, starting at $x > 0$, killed the first time it hits 0, and let $\{|W_t|; t \in (0, \infty)|\}$ denote a three-dimensional Bessel process. Then it follows from (4.185) that

$$\left\{ |W_u|; u \in (0, t], P_W^x \right\} \overset{law}{=} \left\{ \widetilde{X}_u; u \in (0, t], \frac{\widetilde{X}_t}{x} \widetilde{P}^x \right\}, \qquad (8.36)$$

where P_W^x is the law of three-dimensional Brownian motion starting at x and $\widetilde{P}^x$ is the law of $\widetilde{X}$ starting at x.

8.2 The Generalized Second Ray–Knight Theorem

Theorems 8.1.1, 8.1.3, and 8.1.5 are isomorphisms between processes but with changes of measure. It is much more convenient to have isomorphisms between processes in which the measures are the natural measures of the processes themselves. We can do this for processes that start and terminate at the same point in the state space. The results can be considered to be extensions of the classical Second Ray–Knight Theorem (Theorem 2.7.1) to general strongly symmetric Borel right processes.

Recall that a strongly symmetric Borel right process X with continuous α-potential density for some (and hence all) $\alpha > 0$ has local times $\{L_t^y, (t, y) \in (R_+ \times S)\}$ (see Theorem 3.6.3). Let T be a terminal time for X (see page 42) and τ the inverse local time of X at 0, (see (3.132)). Consider u_T and $u_{\tau(\lambda)}$ as defined in (2.138) and (2.146), where λ is an exponential random variable independent of X. In Lemma 3.10.2 we obtain moment generating functions for $L_T := \{L_T^x, x \in S\}$ and $L_{\tau(\lambda)} := \{L_{\tau(\lambda)}^x, x \in S\}$. In the next lemma we show that when u_T and $u_{\tau(\lambda)}$ have a certain form we can obtain simple isomorphism theorems for L_T and $L_{\tau(\lambda)}$.

Lemma 8.2.1 *Let $X = (\Omega, \mathcal{G}, \mathcal{G}_t, X_t, \theta_t, P^x)$ be a strongly symmetric Borel right process with continuous α-potential densities $u^\alpha(x, y)$, $\alpha > 0$. Let $\{L_t^y, (t, y) \in (R_+ \times S)\}$ denote the local times of X and 0 denote some fixed element in S. Let T be a terminal time for X such that*

$$u_T(x, y) = v(x, y) + \gamma, \qquad (8.37)$$

where $v(x, y)$ is a symmetric positive definite function with $v(0, y) = 0$ for all $y \in S$ and $\gamma > 0$ is a constant. Let $\eta = \{\eta_x \, ; \, x \in S\}$ denote the mean zero Gaussian process with covariance $v(x, y)$ and Y be an exponential random variable with mean γ. Then, for any countable subset

$D \subseteq S$,

$$\left\{ L_T^x + \tfrac{1}{2}\eta_x^2; \, x \in D, P^0 \times P_\eta \right\} \overset{law}{=} \left\{ \tfrac{1}{2}\left(\eta_x + \sqrt{2Y}\right)^2; \, x \in D, P_\eta \times P_Y \right\}. \tag{8.38}$$

Let τ be the inverse local time of X at 0 and let λ be an exponential random variable independent of $(\Omega, \mathcal{G}, \mathcal{G}_t, P^x)$. Then, if (8.37) holds with u_T replaced by $u_{\tau(\lambda)}$

$$\left\{ L_{\tau(\lambda)}^x + \frac{1}{2}\eta_x^2; \, x \in D, P^0 \times P_\eta \times P_\lambda \right\} \tag{8.39}$$

$$\overset{law}{=} \left\{ \frac{1}{2}\left(\eta_x + \sqrt{2Y}\right)^2; \, x \in D, P_\eta \times P_Y \right\}.$$

Proof Let $\{\overline{\eta}_x; \, x \geq 0\}$ be an independent copy of $\{\eta_x; \, x \geq 0\}$ and let ξ and $\overline{\xi}$ be independent normal random variables with mean zero and variance γ that are independent of everything else. By (5.72),

$$\{(\eta_x + \xi)^2 + (\overline{\eta}_x + \overline{\xi})^2; \, x \in D\} \overset{law}{=} \{\eta_x^2 + (\overline{\eta}_x + \sqrt{2Y})^2; \, x \in D\}. \tag{8.40}$$

Therefore, to obtain (8.38) it suffices to show that under the measure $P^0 \times P_{\eta, \overline{\eta}, \xi, \overline{\xi}}$,

$$\left\{ L_T^x + \frac{\eta_x^2}{2} + \frac{\overline{\eta}_x^2}{2}; \, x \in D \right\} \overset{law}{=} \left\{ \frac{(\eta_x + \xi)^2}{2} + \frac{(\overline{\eta}_x + \overline{\xi})^2}{2}; \, x \in D \right\}. \tag{8.41}$$

To prove (8.41) it suffices to show that for $x_1 = 0$, and arbitrary $x_2, \ldots, x_n$, and sufficiently small $\lambda_1, \lambda_2, \ldots, \lambda_n$, for all n,

$$E^{x_1} \exp\left(\sum_{i=1}^n \lambda_i L_T^{x_i} \right) = \left(\frac{E\left(\exp \sum_{i=1}^n \lambda_i (\eta_{x_i} + \xi)^2 / 2 \right)}{E\left(\exp \sum_{i=1}^n \lambda_i \eta_{x_i}^2 / 2 \right)} \right)^2. \tag{8.42}$$

By Lemma 5.2.1, the right-hand side of (8.42) is

$$= \frac{\det(I - \Sigma\Lambda)}{\det(I - \overline{\Sigma}\Lambda)}, \tag{8.43}$$

where $\Sigma_{i,j} = v(x_i, x_j)$, $\overline{\Sigma}_{i,j} = v(x_i, x_j) + \gamma$, and $\Lambda_{i,j} = \lambda_i \delta_{i,j}$.

By (8.37) and Lemma 3.10.2

$$E^{x_1} \exp\left(\sum_{i=1}^n \lambda_i L_T^{x_i} \right) = \frac{\det(I - \widehat{\Sigma}\Lambda)}{\det(I - \overline{\Sigma}\Lambda)}, \tag{8.44}$$

where $\widehat{\Sigma}_{i,j} = \overline{\Sigma}_{i,j} - \overline{\Sigma}_{1,j} = \Sigma_{i,j}$ since $v(0, y) = 0$ for all $y \in S$. Thus (8.43) and (8.44) are equal and (8.42) is established.

The key step in the proof of (8.42) is Lemma 3.10.2, which also holds for $u_{\tau(\lambda)}$. Therefore the same proof as above gives (8.39).   □

We prove our generalization of the Second Ray–Knight Theorem in two versions, first when 0 is recurrent for X and then when it is transient for X.

Theorem 8.2.2 (Second Ray–Knight Theorem, recurrent case)
Let $X = (\Omega, \mathcal{G}, \mathcal{G}_t, X_t, \theta_t, P^x)$ be a strongly symmetric Borel right process with continuous α-potential densities $u^\alpha(x, y)$, $\alpha > 0$. Let $\{L_t^y, (t, y) \in (R_+ \times S)\}$ denote the local times of X and 0 denote some fixed element in S. Let $\tau(t) = \inf\{s; L_s^0 > t\}$. Assume that $u(0, 0) = \infty$ and that $P^x(T_0 < \infty) > 0$ for all $x \in S$. As usual, let

$$u_{T_0}(x, y) = E^x(L_{T_0}^y). \tag{8.45}$$

Let $\eta = \{\eta_x ; x \in S\}$ denote the mean zero Gaussian process with covariance $u_{T_0}(x, y)$. Then, under the measure $P^0 \times P_\eta$, for any $t > 0$ and countable subset $D \subseteq S$,

$$\left\{L_{\tau(t)}^x + \tfrac{1}{2}\eta_x^2; x \in D\right\} \stackrel{law}{=} \left\{\tfrac{1}{2}\left(\eta_x + \sqrt{2t}\right)^2; x \in D\right\}. \tag{8.46}$$

Proof Let Y be an exponential random variable with mean γ that is independent of everything else. Since $u(0, 0) = \infty$, it follows from Lemma 3.8.4 that

$$u_{\tau(Y)}(x, y) = u_{T_0}(x, y) + \gamma. \tag{8.47}$$

Consequently, we can apply (8.39) of Lemma 8.2.1 with $v(x, y) = u_{T_0}(x, y)$ to obtain that, for any countable subset $D \subseteq S$,

$$\{L_{\tau(Y)}^x + \tfrac{1}{2}\eta_x^2; x \in D, P_Y^0 \times P_\eta\} \stackrel{law}{=} \{\tfrac{1}{2}\left(\eta_x + \sqrt{2Y}\right)^2; x \in D, P_\eta \times P_Y\}. \tag{8.48}$$

This implies that

$$E_Y^0 E_\eta \exp\left(\sum_{i=1}^n \lambda_i \left(L_{\tau(Y)}^{x_i} + \frac{\eta_{x_i}^2}{2}\right)\right) \tag{8.49}$$

$$= E_Y E_\eta \exp\left(\sum_{i=1}^n \frac{\lambda_i}{2}\left(\eta_{x_i} + \sqrt{2Y}\right)^2\right)$$

for all $(x_1, \ldots, x_n)$ in S, for all n. Writing out the expectation with respect to Y, we get

$$\int_0^\infty E_{P^0} E_\eta \left(\exp \sum_{i=1}^n \lambda_i \left(L_{\tau(t)}^{x_i} + \frac{\eta_{x_i}^2}{2}\right)\right) e^{-t/\gamma}\, dt \tag{8.50}$$

$$= \int_0^\infty E_\eta \left(\exp \sum_{i=1}^n \frac{\lambda_i}{2} \left(\eta_{x_i} + \sqrt{2t} \right)^2 \right) e^{-t/\gamma} \, dt.$$

Since this is true for all $\gamma > 0$, we see that

$$\mathcal{L}\left(E_{P^0} E_\eta \exp \sum_{i=1}^n \lambda_i \left(L^{x_i}_{\tau(t)} + \frac{\eta_{x_i}^2}{2} \right) \right) = \mathcal{L}\left(E_\eta \exp \sum_{i=1}^n \frac{\lambda_i}{2} \left(\eta_{x_i} + \sqrt{2t} \right)^2 \right),$$
(8.51)

where $\mathcal{L}$ indicates Laplace transform. Using the fact that $\tau(t)$ and hence $L^{x_i}_{\tau(t)}$ are right continuous as a function of t, this shows that the moment generating functions of

$$\sum_{i=1}^n \lambda_i \left(L^{x_i}_{\tau(t)} + \frac{\eta_{x_i}^2}{2} \right) \qquad \text{and} \qquad \sum_{i=1}^n \frac{\lambda_i}{2} \left(\eta_{x_i} + \sqrt{2t} \right)^2$$

are equal for all $(x_1, \ldots, x_n)$ in S, for all n. This gives us (8.46).   □

Note that when 0 is transient for X, the 0-potential density of X is finite and

$$u_{T_0}(x, y) = E^x(L^y_{T_0}) = u(x, y) - \frac{u(x, 0) u(y, 0)}{u(0, 0)} \tag{8.52}$$

by Lemma 3.8.1. The transient case is more subtle than the recurrent case because in the transient case L^0_∞ the total accumulated local time of X at 0 is finite. Consequently, $L^0_{\tau(Y)} = L^0_\infty$ when $Y > L^0_\infty$. We deal with this by first considering the h-transform of X.

Theorem 8.2.3 (Second Ray–Knight Theorem, transient case)
Let $X = (\Omega, \mathcal{G}, \mathcal{G}_t, X_t, \theta_t, P^x)$ be a strongly symmetric Borel right process with continuous 0-potential densities $u(x, y)$. Let $\{L^y_t, (t, y) \in (R_+ \times S)\}$ denote the local times of X and 0 denote some fixed element in S. Assume that $h_x = P^x(T_0 < \infty) > 0$ for all $x \in S$. Let $\{\eta_x \, ; \, x \in S\}$ denote the mean zero Gaussian process with covariance $u_{T_0}(x, y)$. Then, for any countable subset $D \subseteq S$,

$$\left\{ L^x_{\tau^-(L^0_\infty)} + \tfrac{1}{2}\eta_x^2; \, x \in D, P^0 \times P_\eta \right\} \tag{8.53}$$
$$\stackrel{law}{=} \left\{ \tfrac{1}{2} \left(\eta_x + h_x \sqrt{2\rho} \right)^2 ; \, x \in D, P_\eta \times P_\rho \right\},$$

where ρ is an exponential random variable with mean $u(0, 0)$ that is independent of $\{\eta_x \, ; \, x \in S\}$. Furthermore, for any $t > 0$,

$$\left\{ L^x_{\tau^-(t \wedge L^0_\infty)} + \tfrac{1}{2}\eta_x^2; x \in D, P^0 \times P_\eta \right\} \tag{8.54}$$
$$\stackrel{law}{=} \left\{ \tfrac{1}{2} \left(\eta_x + h_x \sqrt{2(t \wedge \rho)} \right)^2 ; x \in D, P_\eta \times P_\rho \right\}.$$

Proof Let $\widetilde{X}$ denote the h-transform of X as described in Section 3.9. We use a tilde ($\widetilde{\ }$) to distinguish objects associated with $\widetilde{X}$. We are assuming that $u(x,y)$ is the 0-potential density of X with respect to the reference measure $m(y)$. Therefore, by (3.217),

$$\widetilde{u}(x,y) = \frac{u(x,y)}{h(x)h(y)} \qquad (8.55)$$

is the 0-potential density of $\widetilde{X}$ with respect to the reference measure $\widetilde{m}(y)$, where $d\widetilde{m}(y) = h^2(y)\,dm(y)$.

Recall that $h(x) = u(x,0)/u(0,0)$; see (3.204). Consequently, $\widetilde{u}(x,0) = u(0,0)$ for all $x \in S$ and

$$\widetilde{h}(x) := \widetilde{P}^x(T_0 < \infty) = \frac{\widetilde{u}(x,0)}{\widetilde{u}(0,0)} = 1. \qquad (8.56)$$

Therefore, by Remark 3.8.3,

$$
\begin{aligned}
\widetilde{u}_{T_0}(x,y) &= \widetilde{u}(x,y) - \frac{\widetilde{u}(x,0)\,\widetilde{u}(0,y)}{\widetilde{u}(0,0)} \qquad (8.57)\\
&= \frac{u(x,y)}{h(x)h(y)} - u(0,0)\\
&= \frac{u_{T_0}(x,y)}{h(x)h(y)}.
\end{aligned}
$$

In particular, using (8.55) and the second line of (8.57), we obtain

$$\widetilde{u}(x,y) = \widetilde{u}_{T_0}(x,y) + u(0,0). \qquad (8.58)$$

Note that by (8.57), $\{\eta_x/h(x)\,;\, x \in S\}$ is a mean zero Gaussian process with covariance $\widetilde{u}_{T_0}(x,y)$. Hence, by (8.38) with $T \equiv \infty$,

$$\{\widetilde{L}^x_\infty + \tfrac{1}{2}(\eta_x/h(x))^2\,;\, x \in D, P^{0,0} \times P_\eta\} \qquad (8.59)$$
$$\overset{law}{=} \{\tfrac{1}{2}\left(\eta_x/h(x) + \sqrt{2\rho}\right)^2\,;\, x \in D, P_\eta \times P_\rho\},$$

where $\{\widetilde{L}^x_t\,;\, (x,t) \in S \times R_+\}$ are the local times for $\widetilde{X}$ normalized so that

$$E^{y,0}\left(\int_0^\infty e^{-\alpha t}d\widetilde{L}^x_t\right) = \widetilde{u}^\alpha(y,x) \qquad \forall \alpha \geq 0. \qquad (8.60)$$

Note that by Lemma 3.9.5, since $\mathcal{L} < \infty$, P^0 almost surely,

$$\{h^2(x)\widetilde{L}^x_\infty\,;\, x \in D, P^{0,0}\} \overset{law}{=} \{L^x_\infty \circ k_{\mathcal{L}}\,;\, x \in D, P^0\}. \qquad (8.61)$$

Therefore, multiplying by $h^2(x)$ in (8.59) and using Remark 3.9.7, we get (8.53).

We now prove (8.54). Let Y be an exponential random variable with mean γ that is independent of X. By (3.192) we see that

$$\widetilde{u}_{\widetilde{\tau}(Y)}(x,y) = \widetilde{u}_{T_0}(x,y) + \gamma', \tag{8.62}$$

where $1/\gamma' = 1/\gamma + 1/u(0,0)$. By Remark 5.2.8, $Y \wedge \rho$ is an exponential random variable with mean γ'. Note again that by (8.57), $\{\eta_x/h(x)\,;\,x \in S\}$ is a mean zero Gaussian process with covariance $\widetilde{u}_{T_0}(x,y)$. Therefore, by (8.39) we see that

$$\{\widetilde{L}^x_{\widetilde{\tau}(Y)} + \tfrac{1}{2}(\eta_x/h(x))^2;\ x \in D,\ P^{0,0} \times P_\eta \times P_Y\} \tag{8.63}$$

$$\overset{law}{=} \{\tfrac{1}{2}\left(\eta_x/h(x) + \sqrt{2(Y \wedge \rho)}\right)^2;\ x \in D,\ P_\eta \times P_Y \times P_\rho\}.$$

We can rewrite (8.63) as

$$\{h^2(x)\widetilde{L}^x_{\widetilde{\tau}(Y)} + \tfrac{1}{2}\eta_x^2;\ x \in D\} \overset{law}{=} \{\tfrac{1}{2}\left(\eta_x + h(x)\sqrt{2(Y \wedge \rho)}\right)^2;\ x \in D\}. \tag{8.64}$$

Therefore, as in the proof of Theorem 8.2.2, we have that for each $t > 0$,

$$\{h^2(x)\widetilde{L}^x_{\widetilde{\tau}(t)} + \tfrac{1}{2}\eta_x^2;\ x \in D,\ P^{0,0} \times P_\eta\} \tag{8.65}$$

$$\overset{law}{=} \{\tfrac{1}{2}\left(\eta_x + h(x)\sqrt{2(t \wedge \rho)}\right)^2;\ x \in D,\ P_\eta \times P_\rho\}.$$

Using Lemma 3.9.6, we get (8.54). $\qquad\square$

Remark 8.2.4 We have the following interesting variation of Theorem 8.2.2. As we pointed out in (8.41), an equivalent form of (8.48) is

$$\left\{L^x_{\tau(Y)} + \frac{\eta_x^2}{2} + \frac{\overline{\eta}_x^2}{2};\ x \in D\right\} \overset{law}{=} \left\{\frac{(\eta_x + \xi)^2}{2} + \frac{(\overline{\eta}_x + \overline{\xi})^2}{2};\ x \in D\right\}, \tag{8.66}$$

where $\{\overline{\eta}_x;\ x \geq 0\}$ is an independent copy of $\{\eta_x;\ x \geq 0\}$ and ξ and $\overline{\xi}$ are independent $N(0,\gamma)$ random variables that are independent of everything else. By (5.72) and the strong Markov property,

$$\{L^x_{\tau(Y)};\ x \in D,\ P^0_Y\} \overset{law}{=} \{L^x_{\tau(\xi^2)} + \overline{L}^x_{\tau(\overline{\xi}^2)};\ x \in D,\ P^0 \times P_{\xi,\xi'}\}, \tag{8.67}$$

where $\overline{L}$ and $\overline{\xi}$ are independent copies of L and ξ. Using this in (8.66), we get

$$\{L^x_{\tau(\xi^2)} + \frac{\eta_x^2}{2};\ x \in D,\ P^0 \times P_\eta \times P_\xi\} \overset{law}{=} \{\frac{(\eta_x + \xi)^2}{2};\ x \in D,\ P_\eta \times P_\xi\}. \tag{8.68}$$

Note that $(\eta_x + \xi)^2 \overset{law}{=} \left(\eta_x + \sqrt{\xi^2}\right)^2$. Taking the expectation in (8.68) with respect to ξ^2 and considering it as a function of γ, we can derive (8.46) from (8.68).

Example 8.2.5 Let X be a symmetric real-valued recurrent Lévy process with characteristic exponent ψ as in (4.63). We show in Theorem 4.2.4 that when X does not have a Gaussian component,

$$u_{T_0}(x, y) = \phi(x) + \phi(y) - \phi(x - y), \tag{8.69}$$

where

$$\phi(x) = \frac{1}{\pi} \int_0^\infty \frac{1 - \cos \lambda x}{\psi(\lambda)} \, d\lambda \qquad x \in R^1. \tag{8.70}$$

We see in (5.256) that the mean zero Gaussian process $\eta = \{\eta(x), x \in R^1\}$ with covariance $u_{T_0}(x, y)$ has stationary increments and $\eta(0) = 0$. When X is Brownian motion, u_{T_0} is as in (8.69) with $\phi(x) = 2|x|$; see (2.139). In this case, $\eta(x) = \sqrt{2}\left(B(x) + \overline{B}(-x)\right)$, where B and $\overline{B}$ are independent standard Brownian motions on R_+.

When $\psi(\lambda) = |\lambda|^p$, $1 < p \leq 2$ in (4.63), X is a symmetric stable process and η is fractional Brownian motion (or Brownian motion when $p = 2$); see Examples 4.2.5 and 6.4.7.

One can construct many examples of transient processes to apply Theorem 8.2.3. We might consider the processes just discussed but killed at the end of an independent exponential time with mean α. In this case, the 0-potential of the killed process is the α-potential of the original process

$$u^\alpha(x) = \frac{1}{\pi} \int_0^\infty \frac{\cos \lambda x}{\alpha + \psi(\lambda)} \, d\lambda \qquad x \in R^1. \tag{8.71}$$

By Remark 3.8.3,

$$\widetilde{u}_{T_0}(x, y) = u^\alpha(x - y) - \frac{u^\alpha(x)u^\alpha(y)}{u^\alpha(0)} \tag{8.72}$$

(we use $\sim$ to distinguish this from (8.69)). Let $G = \{G(x), x \in R^1\}$ be a mean zero Gaussian process with covariance $u^\alpha(x, y)$. Then

$$\eta(x) = G(x) - \frac{u^\alpha(x)}{u^\alpha(0)}G(0), \tag{8.73}$$

the orthogonal compliment of the projection of $G(x)$ on $G(0)$.

Here is an another example of a transient process that is a little more esoteric. Once again we consider the processes discussed in the first paragraph of this example but, this time it is killed the first time it hits some fixed element a in R^1, where $a \neq 0$. Following the proof of Theorem 4.2.4, we see that (4.85) holds with $0 = a$ and

$$u_{T_a}(x, y) = \phi(x - a) + \phi(y - a) - \phi(x - y) \qquad (8.74)$$

for ϕ as given in (8.70). In this case, by Remark 3.8.3, as in (8.72),

$$\overline{u}_{T_0}(x, y) = u_{T_a}(x, y) - \frac{u_{T_a}(x, 0)\, u_{T_a}(y, 0)}{2\phi(a)} \qquad (8.75)$$

since $u_{T_a}(0, 0) = 2\phi(a)$. Let $\{\eta(x), x \in R^1\}$ be a mean zero Gaussian process with covariance $u_{T_0}(x, y)$ as in the first paragraph. Then, clearly $\{\eta(a - x), x \in R^1\}$ is a mean zero Gaussian process with covariance $u_{T_a}(x, y)$. The mean zero Gaussian process with covariance $\overline{u}_{T_0}(x, y)$ is

$$\overline{\eta}(x) = \eta(a - x) - \frac{u_{T_a}(x, 0)}{2\phi(a)}\eta(a). \qquad (8.76)$$

Theorem 8.2.3 holds with η replaced by $\overline{\eta}$ and with $h(x) = u_{T_a}(x, 0)/(2\phi(a))$. Perhaps it is more useful in this case to give an equivalent form of (8.53), which is

$$\{L^x_{\tau^-(L^0_\infty)} + \frac{\overline{\eta}^2_x}{2} + \frac{(\overline{\eta}'_x)^2}{2}; x \in D, P^0 \times P_{\eta,\eta'}\} \qquad (8.77)$$

$$\overset{law}{=} \{\frac{\eta^2_{a-x}}{2} + \frac{(\eta'_{a-x})^2}{2}; x \in D, P_{\eta,\eta'}\},$$

where $\overline{\eta}'$ and η' are independent copies of $\overline{\eta}$ and η (see (8.41)).

We remind the reader that we are considering Lévy processes without a Gaussian component, so these results do not apply to Brownian motion. It is easy to give corresponding results for the local times of Brownian motion. We include them in the next subsection, where we consider the more general case of recurrent diffusions. We consider transient diffusions in Chapter 12.

One interesting difference between (8.77) and its counterpart for Brownian motion is that for Brownian motion, if $a > 0$, there no local time for $x > a$. That is not the case in (8.77) since we are considering pure jump process, which consequently can take values greater than a without hitting a (in (8.77), $L^a_{\tau^-(L^0_\infty)} = 0$).

8.2.1 Complete Second Ray–Knight Theorem for recurrent diffusions

Diffusions, which are Borel right processes with continuous paths, are introduced in Section 4.3. For diffusions we can give a version of Theorem 8.2.2 that does not require that the process starts and stops at the same point in its state space. What is needed to extend Theorem 8.2.2 is to describe the local time starting at some point other than 0, and killed the first time it hits 0.

Let X be a recurrent diffusion in R^1 with continuous α-potential densities. Assume that $P^x(T_y < \infty) = 1$ for all $x, y \in R^1$. As in (2.138), let

$$u_{T_0}(y, x) = E^y(L_{T_0}^x). \tag{8.78}$$

We show in Lemma 4.3.2 that

$$u_{T_0}(x, y) = \begin{cases} u_{T_0}(x, x) \wedge u_{T_0}(y, y) & xy \geq 0 \\ 0 & xy < 0, \end{cases} \tag{8.79}$$

where $u_{T_0}(x, x)$ is strictly increasing in $|x|$.

Theorem 8.2.6 (First Ray–Knight Theorem for Diffusions) *Let* $\{L_t^y, (t, y) \in (R_+ \times S)\}$ *denote the local times of X and let $y > 0$. Let* $\rho(x) = u_{T_0}(x, x)/2$. *Let* $\{B_r, r \in R_+\}$ *and* $\{\overline{B}_r, r \in R_+\}$ *be independent standard Brownian motions starting at 0. Then, under* $P^y \times P_{B, \overline{B}}$,

$$\left\{ L_{T_0}^r + (B_{\rho(r)-\rho(y)}^2 + \overline{B}_{\rho(r)-\rho(y)}^2) 1_{\{r \geq y\}} : r \in R_+ \right\} \tag{8.80}$$

$$\overset{law}{=} \left\{ B_{\rho(r)}^2 + \overline{B}_{\rho(r)}^2 : r \in R_+ \right\} \quad \text{on } C(R_+).$$

Proof The proof is the same as the proof of Theorem 2.6.3 for Brownian motion. By Lemma 3.10.2 (which is an immediate generalization of Lemma 2.6.2),

$$E^{x_l} \exp\left(\sum_{i=1}^n \lambda_i L_{T_0}^{x_i} \right) = \frac{\det((I - \widehat{\Sigma}\Lambda))}{\det(I - \Sigma\Lambda)}, \tag{8.81}$$

where, by (8.79), $\Sigma_{i,j} = 2(\rho(x_i) \wedge \rho(x_j))$, $i, j = 1, \ldots n$ and

$$\widehat{\Sigma}_{i,j} = 2\left((\rho(x_i) \wedge \rho(x_j)) - (\rho(x_l) \wedge \rho(x_j)) \right) \quad i, j = 1, \ldots, n. \tag{8.82}$$

The proof proceeds exactly as the proof of Theorem 2.6.3 with $\rho(x_i)$ replacing x_i, $i = 1, \ldots, n$. $\qquad \square$

For diffusions we can generalize the Second Ray–Knight Theorem so that the starting point is not necessarily 0. We use the word "complete" to describe this.

Theorem 8.2.7 (Complete Second Ray–Knight Theorem for Diffusions) *Let* $\{L_t^y, (t, y) \in (R_+ \times S)\}$ *denote the local times of X and* let $y > 0$. Let $\{B_r, r \in R_+\}$ be independent standard Brownian motions starting at 0. Then, under $P^y \times P_{B,\overline{B}}$,

$$\{L^r_{\tau(t)} + (B^2_{\rho(r)-\rho(y)})1_{\{r \geq y\}} + (\overline{B}^2_{\rho(r)-\rho(y)})1_{\{r \geq y\}} : r \in R_+\} \quad (8.83)$$

$$\overset{law}{=} \{\overline{B}^2_{\rho(r)} + (B_{\rho(r)} + \sqrt{t})^2 : r \in R_+\}$$

on $C(R_+)$ (recall that $\tau(t) := \inf\{s \mid L^0_s > t\}$).

Note that $\{\sqrt{2}B_{\rho(r)}; r \in R_+\}$ is a mean zero Gaussian process with covariance $u_{T_0}(x, y)$. Therefore, since $\rho(0) = 0$, this is Theorem 8.2.2 when $y = 0$ (cancel $\overline{B}^2_{\rho(r)}$ from each side). To be more precise, it is Theorem 8.2.2 restricted to R_+. However, (8.80) also holds on the negative half-line with independent copies of everything. Combined, the two parts give Theorem 8.2.2.

Proof By (2.139), $\{\sqrt{2}B_{\rho(r)}, r \in R_+\}$ is a mean zero Gaussian process with covariance $u_{T_0}(r, s)$. Therefore, (8.48) holds with $\{\eta_x, x \in D\}$ replaced by $\{\sqrt{2}B_{\rho(r)}, r \in R_+\}$. Adding an independent copy of $B^2_{\rho(r)}$ to each side of the equality in law, we see that under $P^0 \times P_{B,\overline{B},Y}$,

$$\{L^r_{\tau(Y)} + B^2_{\rho(r)} + \overline{B}^2_{\rho(r)} : r \in R_+\} \overset{law}{=} \{\overline{B}^2_{\rho(r)} + (B_{\rho(r)} + \sqrt{Y})^2 : r \in R_+\} \quad (8.84)$$

on $C(R_+)$, where Y is an exponential random variable with mean γ, which is independent of everything else. (We can pass from the countable set D to R_+ because everything is continuous.) By the strong Markov property, this implies that under $P^y \times P_{B,\overline{B},Y}$,

$$\{L^r_{\tau(Y)} \circ \theta_{T_0} + B^2_{\rho(r)} + \overline{B}^2_{\rho(r)} : r \in R_+\} \overset{law}{=} \{\overline{B}^2_{\rho(r)} + (B_{\rho(r)} + \sqrt{Y})^2 : r \in R_+\}. \quad (8.85)$$

By the strong Markov property, $L^r_{T_0}$ and $L^r_{\tau(Y)} \circ \theta_{T_0}$ are independent. We now substitute an independent copy of the left-hand side of (8.80) for the term $B^2_{\rho(r)} + \overline{B}^2_{\rho(r)}$ on the left-hand side of (8.84) to get that, under $P^y \times P_{B,\overline{B},Y}$,

$$\{L^r_{T_0} + L^r_{\tau(Y)} \circ \theta_{T_0} + (B^2_{\rho(r)-\rho(y)})1_{\{r \geq y\}} + (\overline{B}^2_{\rho(r)-\rho(y)})1_{\{r \geq y\}} : r \in R_+\}$$

$$\overset{law}{=} \{\overline{B}^2_{\rho(r)} + (B_{\rho(r)} + \sqrt{Y})^2 : r \in R_+\}. \quad (8.86)$$

Using the additivity of local time, $L^r_{\tau(Y)} = L^r_{T_0} + L^r_{\tau(Y)} \circ \theta_{T_0}$, we then have that under $P^y \times P_{B,\overline{B},Y}$,

$$\{L^r_{\tau(Y)} + (B^2_{\rho(r)-\rho(y)})1_{\{r \geq y\}} + (\overline{B}^2_{\rho(r)-\rho(y)})1_{\{r \geq y\}} : r \in R_+\}$$

$$\overset{law}{=} \{\overline{B}^2_{\rho(r)} + (B_{\rho(r)} + \sqrt{Y})^2 : r \in R_+\}. \quad (8.87)$$

Following the proof of Theorem 8.2.2, starting from (8.48), we get (8.83).

$\square$

We can get a more symmetric form of (8.83) by writing its right-hand side as $\{(B_{\rho(r)} + \sqrt{t/2})^2 + (\overline{B}_{\rho(r)} + \sqrt{t/2})^2 : r \in R_+\}$.

8.3 Combinatorial proofs

Dynkin's proof of Theorem 8.1.3 uses a combinatorial argument and a calculation of moments modeled on "Feynman diagrams." Eisenbaum's proof of Theorem 8.1.1 also follows this approach. Furthermore, the initial approach to the Generalized Second Ray–Knight Theorem was also combinatoric. In this section we present these proofs, both for their intrinsic interest and because understanding them may be fruitful in future research.

8.3.1 Dynkin Isomorphism Theorem

We first show that

$$E^{x,0} E_G \left(\prod_{i=1}^n \left(\overline{L}_\infty^{x_i} + \frac{G_{x_i}^2}{2} \right) \right) = E_G \left(\frac{G_x G_0}{u(x,0)} \left(\prod_{i=1}^n \frac{G_{x_i}^2}{2} \right) \right) \qquad (8.88)$$

for any $x_1, \ldots, x_n \in S$, not necessarily distinct.

Using Lemma 5.2.6 we see that

$$E_G \left(G_x G_0 \prod_{i=1}^n \left(\frac{G_{v_i} G_{v_i'}}{2} \right) \right) = \frac{1}{2^n} \sum_{\mathcal{D} = (D_1, \ldots, D_{n+1})} u(D_1) \cdots u(D_{n+1}),$$

$$(8.89)$$

where the sum is over all pairings $\mathcal{D} = (D_1, \ldots, D_{n+1})$ of the $2n + 2$ indices $\{v_i\}_{i=1}^n$, $\{v_i'\}_{i=1}^n$, x, and 0 and, for example, $u(\{v_i, v_j'\}) = u(v_i, v_j')$ and $u(\{v_i, 0\}) = u(v_i, 0)$. We now rewrite the right-hand side of (8.89). The reader should keep in mind that, eventually, we will set $v_i = v_i' = x_i$ and obtain the left-hand side of (8.88).

Given a specific pairing $\mathcal{D}$, we create an ordering of a certain subset of the elements of $\mathcal{D}$, starting with the element of $\mathcal{D}$ that contains x and ending with the element of $\mathcal{D}$ that contains 0, in the following way: Let $D^{(1)}$ be the unique element of $\mathcal{D}$ that contains x. If x is paired with either v_i or v_i', set $\pi(1) = i$ and define $(y_{\pi(1)}, z_{\pi(1)})$ to be (v_i, v_i') if x is paired with v_i, but (v_i', v_i) if x is paired with v_i'. Next let $D^{(2)}$ be the unique element of $\mathcal{D}$ that contains $z_{\pi(1)}$. If $z_{\pi(1)}$ is paired with either v_j or v_j', set $\pi(2) = j$ and define $(y_{\pi(2)}, z_{\pi(2)})$ to be (v_j, v_j') if $z_{\pi(1)}$ is paired with v_j or (v'_j, v_j) if $z_{\pi(1)}$ is paired with v'_j. We proceed in this manner, getting $D^{(1)}, \ldots, D^{(l)}$ until we get to $D^{(l+1)}$, the unique element of $\mathcal{D}$ that contains $z_{\pi(l)}$ and 0. Clearly $l \leq n$, but it is important to note that $l < n$ is often the case.

Let $C(\mathcal{D}) = \{\pi(1), \ldots, \pi(l)\}$ and $\mathcal{D}' = (D^{(1)}, \ldots, D^{(l+1)})$. Clearly $\mathcal{D}'$ is a pairing of the $2l + 2$ elements $\{v_i\}_{i \in C(\mathcal{D})}$, $\{v'_i\}_{i \in C(\mathcal{D})}$, x, and 0. Let $B(\mathcal{D}) = \{1, \ldots, n\}/C(\mathcal{D})$ and $\mathcal{D}'' = \mathcal{D}/\mathcal{D}'$. It is also clear that $\mathcal{D}''$ is a pairing of the set of $2(n - l)$ indices consisting of $\{v_i\}_{i \in B(\mathcal{D})}$ and $\{v'_i\}_{i \in B(\mathcal{D})}$. Taking this into account, we see that we can rewrite (8.89) as

$$E_G \left(G_x G_0 \prod_{i=1}^{n} \left(\frac{G_{u_i} G_{v_i}}{2} \right) \right) \tag{8.90}$$

$$= \frac{1}{2^n} \sum_{B \cup C = \{1,\ldots,n\}} \left(\sum_{\substack{\text{pairings of} \\ \{v_i\}_{i \in B} \cup \{v'_i\}_{i \in B}}} u(B_1) \cdots u(B_{|B|}) \right)$$

$$\times \sum u(x, y_{\pi(1)}) u(z_{\pi(1)}, y_{\pi(2)}) \cdots u(z_{\pi(i)}, y_{\pi(i+1)}) \cdots u(z_{\pi(|C|)}, 0),$$

where the last sum is over all permutations $(\pi(1), \ldots, \pi(|C|))$ of C, and over all ways of assigning $(v_{\pi(i)}, v'_{\pi(i)})$ to $(y_{\pi(i)}, z_{\pi(i)})$. Of course there are $2^{|C|}$ ways to make these assignments. Thus, if we set $v_i = v'_i = x_i$, the last sum in (8.90) is

$$2^{|C|} \sum_{\pi(C)} u(x, x_{\pi(1)}) u(x_{\pi(1)}, x_{\pi(2)}) \cdots u(x_{\pi(i)}, x_{\pi(i+1)}) \cdots u(x_{\pi(|C|)}, 0),$$

$$\tag{8.91}$$

where now the sum is over all permutations π of C.

Using (5.62), we see that

$$\sum_{\substack{\text{pairings of} \\ \{v_i\}_{i \in B} \cup \{v'_i\}_{i \in B}}} u(B_1) \cdots u(B_{|B|}) = E_G \left(\prod_{i \in B} G_{v_i} G_{v'_i} \right). \tag{8.92}$$

Therefore, setting $v_i = v'_i = x_i$ in (8.90) and using (8.91) and (8.92), we have

$$E_G \left(G_x G_0 \prod_{i=1}^{n} \frac{G_{x_i}^2}{2} \right) = \sum_{B \cup C = \{1,\ldots,n\}} E_G \left(\prod_{i \in B} \frac{G_{x_i}^2}{2} \right) \tag{8.93}$$

$$\sum_{\pi(C)} u(x, x_{\pi(1)}) u(x_{\pi(1)}, x_{\pi(2)}) \cdots u(x_{\pi(|C|)}, 0).$$

The left-hand side of (8.88) is

$$E^{x,0} E_G \left(\prod_{i=1}^{n} \left(\bar{L}_\infty^{x_i} + \frac{G_{x_i}^2}{2} \right) \right) \tag{8.94}$$

$$= \sum_{B \cup C = \{1,\ldots,n\}} E_G \left(\prod_{i \in B} \frac{G_{x_i}^2}{2} \right) E^{x,0} \left(\prod_{i \in C} \bar{L}_\infty^{x_i} \right),$$

and by (3.248) we see that (8.93) and $u(0,x)$ times (8.94) are identical. This establishes (8.88).

Let $z_1, \ldots, z_n$ be fixed and let μ_1 and μ_2 be the measures on R_+^n defined by

$$\int F(\cdot)\, d\mu_1 = E^{x,0} E_G \left(F \left(\bar{L}_\infty^{z_1} + \frac{G_{z_1}^2}{2}, \ldots, \bar{L}_\infty^{z_n} + \frac{G_{z_n}^2}{2} \right) \right) \tag{8.95}$$

and

$$\int F(\cdot)\, d\mu_2 = E_G \left(\frac{G_x G_0}{u(x,0)} F \left(\frac{G_{z_1}^2}{2}, \ldots, \frac{G_{z_n}^2}{2} \right) \right) \tag{8.96}$$

for all bounded measurable functions F on R_+^n. The measure μ_1 is determined by its characteristic function

$$\varphi_1(\lambda_1, \ldots, \lambda_n) = E^{x,0} E_G \left(\exp \left(i \sum_{i=1}^{n} \lambda_i (\bar{L}_\infty^{z_i} + G_{z_i}^2 / 2) \right) \right). \tag{8.97}$$

For $\lambda_1, \ldots, \lambda_n$ fixed, $\varphi_1(\lambda_1, \ldots, \lambda_n)$ is determined by the distribution function of the real-valued random variable $\xi = \sum_{i=1}^{n} \lambda_i (\bar{L}_\infty^{z_i} + G_{z_i}^2 / 2)$. Let μ_{2k} denote the $2k$-th moment of ξ. If $\sum_{k=1}^{\infty} \mu_{2k} t^k / (2k)!$ converges for $t \in [0, \delta]$ for some $\delta > 0$, then the distribution function of ξ is uniquely determined by its moments (see, e.g., Feller (1971, page 224)). Considering (8.88), we see that this sum converges if $E_G(\exp(\sum_{i=1}^{n} s_i G_{z_i}^2) |G_x| |G_0|) < \infty$ for sufficiently small $s_i > 0$, $i = 1, \ldots, n$. By repeated use of the Schwarz inequality, it is easy to see that this is the case. Hence the measure μ_1 is uniquely determined by the moments of ξ or, equivalently, by the terms in the left-hand side of (8.88).

Set

$$\varphi_2(\lambda_1, \ldots, \lambda_n) = E_G \left(\frac{G_x G_0}{u(x,0)} \exp \left(i \sum_{i=1}^{n} \lambda_i G_{z_i}^2 \right) \right). \tag{8.98}$$

We see by (8.88) and the above argument that $\varphi_1(\lambda_1, \ldots, \lambda_n) = \varphi_2(\lambda_1, \ldots, \lambda_n)$. Hence $\mu_1 = \mu_2$, so (8.95) and (8.96) give (8.15). Note that, although it is not clear to begin with that μ_2 is a positive measure, this argument shows that it is. $\qquad \square$

8.3.2 Eisenbaum Isomorphism Theorem

As in the combinatorial proof of the Dynkin Isomorphism Theorem, it suffices to show that

$$
E^x E_G \left(\prod_{i=1}^n \left(L_\infty^{x_i} + \frac{(G_{x_i} + s)^2}{2} \right) \right) \tag{8.99}
$$

$$
= E_G \left(\left(1 + \frac{G_x}{s} \right) \left(\prod_{i=1}^n \frac{(G_{x_i} + s)^2}{2} \right) \right)
$$

for any $x_1, \ldots, x_n \in S$, not necessarily distinct, and all n. We show below that

$$
E_G \left(\frac{G_x}{s} \left(\prod_{i=1}^n \frac{(G_{x_i} + s)^2}{2} \right) \right) \tag{8.100}
$$

$$
= \sum_{\substack{B \cup C = \{1, \ldots, n\} \\ C \neq \emptyset}} E_G \left(\prod_{i \in B} \frac{(G_{x_i} + s)^2}{2} \right)
$$

$$
\sum_{\pi(C)} u(x, x_{\pi(1)}) u(x_{\pi(1)}, x_{\pi(2)}) \cdots u(x_{\pi(|C|-1)}, x_{\pi(|C|)}),
$$

where the last sum is over all permutations π of C (note the similarity of this equation and (8.93)). It follows immediately from (8.100) that

$$
E_G \left(\left(1 + \frac{G_x}{s} \right) \left(\prod_{i=1}^n \frac{(G_{x_i} + s)^2}{2} \right) \right) \tag{8.101}
$$

$$
= \sum_{B \cup C = \{1, \ldots, n\}} E_G \left(\prod_{i \in B} \frac{(G_{x_i} + s)^2}{2} \right)
$$

$$
\sum_{\pi(C)} u(x, x_{\pi(1)}) u(x_{\pi(1)}, x_{\pi(2)}) \cdots u(x_{\pi(|C|-1)}, x_{\pi(|C|)})
$$

if we define the last sum to be identically 1 when $C = \emptyset$.

The left-hand side of (8.99) is

$$
E^x E_G \left(\prod_{i=1}^n \left(L_\infty^{x_i} + \frac{(G_{x_i} + s)^2}{2} \right) \right) \tag{8.102}
$$

$$
= \sum_{B \cup C = \{1, \ldots, n\}} E_G \left(\prod_{i \in B} \frac{(G_{x_i} + s)^2}{2} \right) E^x \left(\prod_{i \in C} L_\infty^{x_i} \right),
$$

and by (3.242) we see that the sum on the right-hand side of (8.102) is identical to (8.101). This establishes (8.99).

To prove (8.100), we first expand

$$E_G \left(\frac{G_x}{s} \left(\prod_{i=1}^{n} (G_{v_i} + s)(G_{v'_i} + s) \right) \right) = \sum_{\mathcal{G}} E_G \left(G_x \prod_{r \in \mathcal{G}} G_r \right) s^{2n - (|\mathcal{G}| + 1)},$$

(8.103)

where the sum runs over all possible subsets $\mathcal{G}$ of $\{v_i, v'_i, i = 1, \dots, n\}$. The reader should keep in mind that, eventually, we will set $v_i = v'_i = x_i$. Note that $|\mathcal{G}|$ must be odd for a summand on the right-hand side of (8.103) to be nonzero. Let $m_{\mathcal{G}} = (|\mathcal{G}| + 1)/2$.

Using Lemma 5.2.6, we see that

$$E_G \left(G_x \prod_{r \in \mathcal{G}} G_r \right) = \sum_{\mathcal{D} = (D_1, \dots, D_{m_{\mathcal{G}}})} u(D_1) \cdots u(D_{m_{\mathcal{G}}}),$$

(8.104)

where the sum is over all pairings $\mathcal{D} = (D_1, \dots, D_{m_{\mathcal{G}}})$ of the $2m_{\mathcal{G}}$ indices $\mathcal{G} \cup \{x\}$, and for example, $u(\{v_i, v'_j\}) = u(v_i, v'_j)$ and $u(\{v_i, x\}) = u(v_i, x)$. We now rewrite the right-hand side of (8.104).

As in the combinatorial proof of the Dynkin Isomorphism Theorem, we divide each pairing into two sets. In the first set, we start with the pairing containing x and continue with the pairings as in the scheme of the previous proof, except now we stop according to a different rule, that is, let $D^{(1)}$ be the unique element of $\mathcal{D}$ that contains x. If x is paired with either v_i or v'_i, set $\pi(1) = i$. If both v_i and v'_i are in $\mathcal{G}$, define $(y_{\pi(1)}, z_{\pi(1)})$ to be (v_i, v'_i) if x is paired with v_i, but (v'_i, v_i) if x is paired with v'_i. If only one of the indices v_i and v'_i are in $\mathcal{G}$, let $y_{\pi(1)}$ be that index and proceed to the next paragraph. Otherwise, let $D^{(2)}$ be the unique element of $\mathcal{D}$ that contains $z_{\pi(1)}$. If $z_{\pi(1)}$ is paired with either v_j or v'_j, set $\pi(2) = j$. If both v_j and v'_j are in $\mathcal{G}$, define $(y_{\pi(2)}, z_{\pi(2)})$ to be (v_j, v'_j) if $z_{\pi(1)}$ is paired with v_j, or (v'_j, v_j) if $z_{\pi(1)}$ is paired with v'_j. If only one of v_j and v'_j is in $\mathcal{G}$, let $y_{\pi(2)}$ be that index and proceed to the next paragraph. We proceed in this manner, getting $D^{(1)}, \dots, D^{(l)}$ until our procedure stops. Clearly $l \leq m_{\mathcal{G}}$, but it is important to note that $l < m_{\mathcal{G}}$ is often the case.

Let $C = C(\mathcal{D}) = \{\pi(1), \dots, \pi(l)\}$ and $\mathcal{D}' = (D^{(1)}, \dots, D^{(l)})$. Clearly $\mathcal{D}'$ is a pairing of the $2l$ elements x, $y_{\pi(i)}, z_{\pi(i)}, i = 1, \dots, l - 1$, and $y_{\pi(l)}$. Let $\mathcal{G}' = \{y_{\pi(i)}, z_{\pi(i)}, i = 1, \dots, l - 1\} \cup \{y_{\pi(l)}\}$. Set $\mathcal{G}'' = \mathcal{G}/\mathcal{G}'$ and $\mathcal{D}'' = \mathcal{D}/\mathcal{D}'$. Clearly $\mathcal{D}''$ is a pairing of the set of indices in $\mathcal{G}''$. We see that we can rewrite (8.103) as

$$E_G \left(\frac{G_x}{s} \prod_{i=1}^{n} (G_{v_i} + s)(G_{v'_i} + s) \right)$$

(8.105)

$$= \sum_{\mathcal{G}} \sum_{\substack{\mathcal{G}' \cup \mathcal{G}''=\mathcal{G} \\ \mathcal{G}' \neq \emptyset}} \left(\sum_{\substack{\text{pairings of} \\ \mathcal{G}''}} u(B_1) \cdots u(B_{|\mathcal{G}''|/2}) \right) s^{2n-(|\mathcal{G}|+1)}$$

$$\times \left(\sum u(x, y_{\pi(1)}) u(z_{\pi(1)}, y_{\pi(2)}) \cdots u(z_{\pi(l-1)}, y_{\pi(l)}) \right)$$

where the last sum is over all permutations $(\pi(1), \ldots, \pi(|C|))$ of C and over all ways of assigning $\{v_{\pi(i)}, v'_{\pi(i)}\}$ to $\{y_{\pi(i)}, z_{\pi(i)}\}$.

Let $B = \{1, \ldots, n\}/C$. The sum in the large parentheses in (8.105) runs over all possible subsets $\mathcal{G}''$ of $\{v_i, v'_i, i \in B\}$. As in (8.103), we have

$$E_G \left(\left(\prod_{i \in B} (G_{v_i} + s)(G_{v'_i} + s) \right) \right) \tag{8.106}$$

$$= \sum_{\mathcal{G}''} E_G \left(\prod_{r \in \mathcal{G}''} G_r \right) s^{2|B|-|\mathcal{G}''|}$$

$$= \sum_{\mathcal{G}''} E_G \left(\prod_{r \in \mathcal{G}''} G_r \right) s^{2n-(|\mathcal{G}|+1)},$$

where, for the last line, we use the facts that $|B| + |C| = n$, $|\mathcal{G}'| + |\mathcal{G}''| = |\mathcal{G}|$ and $|\mathcal{G}'| = 2|C| - 1$. Therefore, using (5.62) again, we see that

$$E_G \left(\left(\prod_{i \in B} (G_{v_i} + s)(G_{v'_i} + s) \right) \right) \tag{8.107}$$

$$= \sum_{\mathcal{G}''} \left(\sum_{\substack{\text{pairings of} \\ \mathcal{G}''}} u(B_1) \cdots u(B_{|\mathcal{G}''|/2}) \right) s^{2n-(|\mathcal{G}|+1)}.$$

Thus (8.105) can be written as

$$E_G \left(\frac{G_x}{s} \prod_{i=1}^n (G_{v_i} + s)(G_{v'_i} + s) \right) \tag{8.108}$$

$$= \sum_{\substack{B \cup C = \{1, \ldots, n\} \\ C \neq \emptyset}} E_G \left(\prod_{i \in B} (G_{v_i} + s)(G_{v'_i} + s) \right)$$

$$\times \left(\sum u(x, y_{\pi(1)}) u(z_{\pi(1)}, y_{\pi(2)}) \cdots u(z_{\pi(l-1)}, y_{\pi(l)}) \right),$$

where the last sum is over all permutations $(\pi(1), \ldots, \pi(|C|))$ of C and over all ways of assigning $\{v_{\pi(i)}, v'_{\pi(i)}\}$ to $\{y_{\pi(i)}, z_{\pi(i)}\}$. Of course there are $2^{|C|}$ ways to make these assignments. Thus, if we set $v_i = v_i' = x_i$, the last sum in (8.108) is

$$2^{|C|} \sum_{\pi(C)} u(x, x_{\pi(1)})u(x_{\pi(1)}, x_{\pi(2)}) \cdots u(x_{\pi(l-1)}, x_{\pi(l)}), \qquad (8.109)$$

where now the sum is over all permutations π of C. Finally, setting $v_i = v_i' = x_i$ in (8.108) establishes (8.100). □

8.3.3 Generalized Second Ray–Knight Theorem (recurrent case)

Under the hypotheses and notation of Theorem 8.2.2, it suffices to show that, for all $t \geq 0$,

$$E^0 E_\eta \left(\prod_{i=1}^n (L_{\tau(t)}^{x_i} + \frac{\eta_{x_i}^2}{2}) \right) = E_\eta \left(\prod_{i=1}^n \frac{(\eta_{x_i} + \sqrt{2t})^2}{2} \right) \qquad (8.110)$$

for an arbitrary sequence $x_1, x_2, \ldots, x_n$ of (not necessarily distinct) elements in S and any n, as in the previous two proofs. To do this we need to develop some material on combinatorics.

Fix n and a sequence $x_1, x_2, \ldots, x_n$ of (not necessarily distinct) elements in S. By a $2+k$ chain on $[n] := \{1, 2, \ldots, n\}$ we mean an unordered pair $\{i_1, i_2\}$ of integers in $[n]$ referred to as the endpoints together with an unordered set $\{j_1, j_2, \ldots, j_k\}$ of k integers in $[n]$ referred to as the intermediate points. It is assumed that the $2+k$ elements $(i_1, i_2, j_1, j_2, \ldots, j_k)$ are distinct. We write such a chain as $(i_1, i_2; j_1, j_2, \ldots, j_k)$. Here we allow $k = 0, 1, \ldots, n-2$. (Note that two chains are the same if they have the same endpoints and the same intermediate points.)

In addition, we also refer to a 1–tuple (i) as a trivial chain.

Let $(i_1, i_2; j_1, j_2, \ldots, j_k)$ be a $2+k$ chain. We define

$$ch(i_1, i_2; j_1, j_2, \ldots, j_k) \qquad (8.111)$$

$$= 2 \sum_\sigma u_{T_0}(x_{i_1}, x_{j_{\sigma(1)}})u_{T_0}(x_{j_{\sigma(1)}}, x_{j_{\sigma(2)}}) \cdots u_{T_0}(x_{j_{\sigma(k)}}, x_{i_2})$$

$$= \sum_\sigma u_{T_0}(x_{i_1}, x_{j_{\sigma(1)}})u_{T_0}(x_{j_{\sigma(1)}}, x_{j_{\sigma(2)}}) \cdots u_{T_0}(x_{j_{\sigma(k)}}, x_{i_2})$$

$$+ \sum_\sigma u_{T_0}(x_{i_2}, x_{j_{\sigma(1)}})u_{T_0}(x_{j_{\sigma(1)}}, x_{j_{\sigma(2)}}) \cdots u_{T_0}(x_{j_{\sigma(k)}}, x_{i_1}),$$

where the sum runs over all permutations σ of $\{1, 2, \ldots, k\}$. For a trivial chain (i) we simply set $ch(i) = 1$. The use of trivial chains helps us with the "bookkeeping."

Lemma 8.3.1 (The Chain Decomposition) *For all $t \geq 0$,*

$$E^0 \left(\prod_{i=1}^{n} L_{\tau(t)}^{x_i} \right) = \sum_{C \in \mathcal{C}(1,\ldots,n)} t^{|C|} \prod_{j=1}^{|C|} ch(C_j), \qquad (8.112)$$

where the sum runs over the set $\mathcal{C}(1,\ldots,n)$ of all partitions $C = \{C_1, \ldots, C_{|C|}\}$ of $[n]$ into an unordered collection $C_1,\ldots,C_{|C|}$ of chains.

Proof Let λ be an exponential random variable with mean γ that is independent of X. By Lemma 3.8.4,

$$u_{\tau(\lambda)}(x,y) = u_{T_0}(x,y) + \gamma. \qquad (8.113)$$

By (3.242) we have

$$\int_0^\infty \frac{1}{\gamma} e^{-t/\gamma} E^0 \left(\prod_{i=1}^n L_{\tau(t)}^{x_i} \right) dt = E^0 \left(\prod_{i=1}^n L_{\tau(\lambda)}^{x_i} \right) \qquad (8.114)$$

$$= \sum_\pi u_{\tau(\lambda)}(0, x_{\pi(1)}) u_{\tau(\lambda)}(x_{\pi(1)}, x_{\pi(2)}) \cdots u_{\tau(\lambda)}(x_{\pi(n-1)}, x_{\pi(n)})$$

$$= \sum_\pi \gamma \prod_{j=2}^n \left(u_{T_0}(x_{\pi(j-1)}, x_{\pi(j)}) + \gamma \right)$$

$$= \sum_\pi \sum_{\substack{A,B \\ 1 \in B}} \left(\prod_{j \in A} u_{T_0}(x_{\pi(j-1)}, x_{\pi(j)}) \right) \gamma^{|B|},$$

where we use the fact that $u_{T_0}(0, x) = 0$ for all x. The last sum runs over all partitions A, B of $[n]$ with $1 \in B$.

Fix π and a partition A, B of $[n]$ with $1 \in B$. With each $l \in B$ we associate a chain $C(l)$ as follows. If l is the largest element in B, set $l' = n$. Otherwise, let l' be the next largest element of B after l. If $l' = l + 1$, we take $C(l)$ to be the trivial chain $C(l) = (\pi(l))$. Otherwise, we set $C(l) = (\pi(l), \pi(l'-1); \pi(l+1), \ldots, \pi(l'-2))$. Thus we can write

$$\prod_{j \in A} u_{T_0}(x_{\pi(j-1)}, x_{\pi(j)}) \qquad (8.115)$$

$$= \prod_{\substack{l \in B \\ C(l) \text{ nontrivial}}} u_{T_0}(x_{\pi(l)}, x_{\pi(l+1)}) \cdots u_{T_0}(x_{\pi(l'-2)}, x_{\pi(l'-1)}).$$

In this way, to each permutation π and $B \subseteq [n]$ with $1 \in B$ we associate a partition $C = \{C(l)\}_{l \in B} \in \mathcal{C}(1,\ldots,n)$. How many permutations π will give rise to the same partition $C = \{C_1,\ldots,C_{|C|}\}$? Clearly we can permute the endpoints and intermediate points in each chain C_j, but in addition we can also permute the $|C|$ chains $C_1,\ldots,C_{|C|}$. Note that by (8.111), the sum of the right-hand side of (8.115) over all permutations of the endpoints and intermediate points in each chain $C(l)$ is $\prod_{l=1}^{|B|} ch(C(l))$. Here we also use the fact that $ch(C(j)) = 1$ for a trivial chain. Finally, since there are $|C|!$ ways of permuting the chains $C_1, \ldots, C_{|C|}$, we get

$$\int_0^\infty \frac{1}{\gamma} e^{-t/\gamma} E^0 \left(\prod_{i=1}^n L_{\tau(t)}^{x_i} \right) dt \tag{8.116}$$

$$= \sum_{C \in \mathcal{C}(1,\ldots,n)} |C|! \gamma^{|C|} \prod_{j=1}^{|C|} ch(C_j)$$

$$= \int_0^\infty \frac{1}{\gamma} e^{-t/\gamma} \left(\sum_{C \in \mathcal{C}(1,\ldots,n)} t^{|C|} \prod_{j=1}^{|C|} ch(C_j) \right) dt,$$

from which the lemma follows. $\square$

Proof of Theorem 8.2.2 We use Lemma 8.3.1 to obtain (8.110). We begin by writing

$$E_\eta \left(\prod_{i=1}^n \frac{\left(\eta_{x_i} + \sqrt{2t} \right)^2}{2} \right) \tag{8.117}$$

$$= E_\eta \left(\prod_{i=1}^n \left(\frac{\eta_{x_i}^2}{2} + \eta_{x_i} \sqrt{2t} + t \right) \right)$$

$$= \sum_{A,B,D} E_\eta \left(\prod_{i \in A} \frac{\eta_{x_i}^2}{2} \prod_{j \in B} \left(\eta_{x_j} \sqrt{2t} \right) \right) t^{|D|},$$

where the sum runs over all partitions $A \cup B \cup D = \{1, 2, \ldots, n\}$. Because the odd moments of a mean zero normal random variable are all zero, a summand on the right-hand side of (8.117) is zero unless $|B|$ is even. Thus we can write the right-hand side of (8.117) as

$$\sum_{\substack{A,B,D \\ |B| \text{ even}}} E_\eta \left(\prod_{i \in A} \frac{\eta_{x_i}^2}{2} \prod_{j \in B} \eta_{x_j} \right) 2^{|B|/2} t^{|D|+|B|/2}. \tag{8.118}$$

We claim that when $|B|$ is even,

$$2^{|B|/2} E_\eta \left(\prod_{i \in A} \frac{\eta_{x_i}^2}{2} \prod_{j \in B} \eta_{x_j} \right) \tag{8.119}$$

$$= \sum_{A',A''} \sum_{C \in \mathcal{C}'(B;\,A')} \left(\prod_{j=1}^{|B|/2} ch(C_j) \right) E_\eta \left(\prod_{i \in A''} \frac{\eta_{x_i}^2}{2} \right),$$

where the first sum runs over all partitions $A' \cup A'' = A$ and the second sum runs over the collection $\mathcal{C}'(B; A')$ of all partitions $C = \{C_1, \ldots, C_{|B|/2}\}$ of $A' \cup B$ into $|B|/2$ nontrivial chains $C_1, \ldots, C_{|B|/2}$ with endpoints from B and intermediate points from A'. Indeed, when $|B| = 2$, this is precisely (8.93). The proof for general $|B|$ is a straightforward extension of the proof of (8.93).

Combining (8.117)–(8.119) we have

$$E_\eta \left(\prod_{i=1}^n \frac{(\eta_{x_i} + \sqrt{2t})^2}{2} \right) \tag{8.120}$$

$$= \sum_{\substack{A',A'',B,D \\ |B|\,\text{even}}} \sum_{C \in \mathcal{C}'(B;\,A')} \left(\prod_{j=1}^{|B|/2} ch(C_j) \right) E_\eta \left(\prod_{i \in A''} \frac{\eta_{x_i}^2}{2} \right) t^{|D|+|B|/2}$$

$$= \sum_{F \subseteq \{1,\ldots,n\}} \sum_{C \in \mathcal{C}(F)} t^{|C|} \prod_{j=1}^{|C|} ch(C_j) E_\eta \left(\prod_{i \in F^c} \frac{\eta_{x_i}^2}{2} \right),$$

where now the first sum runs over all subsets $F \subseteq \{1, \ldots, n\}$ and the second sum runs over the collection $\mathcal{C}(F)$ of all partitions $C = \{C_1, \ldots, C_{|C|}\}$ of F into chains $C_1, \ldots, C_{|C|}$ (we include trivial chains that correspond to the elements in D). Using Lemma 8.3.1, we now see that

$$E_\eta \left(\prod_{i=1}^n \frac{(\eta_{x_i} + \sqrt{2t})^2}{2} \right) \tag{8.121}$$

$$= \sum_{F \subseteq \{1,\ldots,n\}} E^0 \left(\prod_{i \in F} L_{\tau(t)}^{x_i} \right) E_\eta \left(\prod_{i \in F^c} \frac{\eta_{x_i}^2}{2} \right)$$

$$= E^0 E_\eta \left(\prod_{i=1}^n \left(L_{\tau(t)}^{x_i} + \frac{\eta_{x_i}^2}{2} \right) \right),$$

which gives (8.110). $\qquad\qquad\qquad\qquad\qquad\qquad\qquad\qquad\square$

Remark 8.3.2 Assume that instead of (8.113) we have

$$u_{\tau(\lambda)}(x, y) = v(x, y) + \gamma, \tag{8.122}$$

where $v(x, y)$ is a symmetric positive definite function with $v(0, y) = 0$ for all $y \in S$ and λ is an exponential random variable with mean α

that is independent of X. In this case the proof of Lemma 8.3.1 yields (8.116) but with γ on the left-hand side replaced by α. If we let Y be an exponential random variable with mean γ, this can be written as

$$E_\lambda^0 \left(\prod_{i=1}^n L_{\tau(\lambda)}^{x_i} \right) = E_Y \left(\sum_{C \in \mathcal{C}(1,\ldots,n)} Y^{|C|} \prod_{j=1}^{|C|} ch(C_j) \right). \qquad (8.123)$$

Now let $\eta = \{\eta_x \, ; \, x \in S\}$ be a mean zero Gaussian process with covariance $v(x,y)$ that is independent of Y and λ. The proof of Theorem 8.2.2, just above, shows that

$$E_\lambda^0 E_\eta \left(\prod_{i=1}^n (L_{\tau(\lambda)}^{x_i} + \frac{\eta_{x_i}^2}{2}) \right) = E_{\eta,\,Y} \left(\prod_{i=1}^n \frac{\left(\eta_{x_i} + \sqrt{2Y} \right)^2}{2} \right), \qquad (8.124)$$

which gives (8.39). The proof of the Generalized Second Ray–Knight Theorem in the transient case (8.54) uses only (8.39) and some facts about h-transforms.

8.4 Additional proofs

We show how the Generalized Second Ray Knight Theorems can be obtained from Dynkin's and Eisenbaum's Isomorphism Theorems. To begin, we note the following lemma, which is simply a reformulation of (8.53).

Lemma 8.4.1 *Let X be as given in Theorem 8.2.3 and let $\widetilde{X}$ denote its h-transform for $h(x) = u(x,0)/u(0,0)$. Let $\overline{L} = \{\overline{L}_t^x, (x,t) \in S \times R_+\}$ denote the local times of $\widetilde{X}$ normalized so that*

$$E^x \left(\overline{L}_\infty^y \right) = \frac{u(x,y)h(y)}{h(x)} \qquad (8.125)$$

(see Remark 3.9.3). Let $G = \{G_x, x \in S\}$ be a mean zero Gaussian process with covariance $u(x,y)$ and $\eta = \{\eta_x, x \in S\}$ be a mean zero Gaussian process with covariance

$$u_{T_0}(x,y) = u(x,y) - \frac{u(x,0)u(y,0)}{u(0,0)}. \qquad (8.126)$$

Let $\overline{G}$ and $\overline{\eta}$ be independent copies of G and η. Then, for any countable subset $D \subseteq S$,

$$\left\{\overline{L}_\infty^x + \frac{\eta_x^2}{2} + \frac{\overline{\eta}_x^2}{2}; x \in D, P^{0,0} \times P_{\eta,\overline{\eta}}\right\} \overset{law}{=} \left\{\frac{G_x^2}{2} + \frac{\overline{G}_x^2}{2}; x \in D, P_{G,\overline{G}}\right\}.$$
(8.127)

Equivalently, let ρ be an exponential random variable with mean $u(0,0)$. Then

$$\left\{\overline{L}_\infty^x + \frac{\eta_x^2}{2}; x \in D, P^{0,0} \times P_\eta\right\} \overset{law}{=} \left\{\frac{(\eta_x + h(x)\sqrt{2\rho})^2}{2}; x \in D, P_\eta \times P_\rho\right\}.$$
(8.128)

Proof We obtain (8.128) by multiplying each side of (8.59) by $h^2(x)$ and using (3.219). We have often pointed out that (8.127) and (8.128) are equivalent. □

Lemma 8.4.1 is simply a restatement of (8.53). It the context of Section 8.2, (8.53) seems like one of several different formulations of the Generalized Second Ray–Knight Theorems. Actually, all the results in Section 8.2 can be obtained from (8.53) or, equivalently, Lemma 8.4.1. We do not pursue this point here. What we do is show how to derive (8.127) or, equivalently, (8.128) from the Dynkin and Eisenbaum Isomorphism Theorems. It is remarkable how easily (8.128) follows from the Dynkin Isomorphism Theorem. To see this, we note the following lemma.

Lemma 8.4.2 *Let $G = \{G_x, x \in S\}$ be a mean zero Gaussian process with covariance $u(x,y)$. Let $\eta = \{\eta_x, x \in S\}$ be a mean zero Gaussian process with covariance given by (8.126). Let $\chi = \sum_{i=1}^3 G_{0,i}^2$, where $G_{0,i}^2$ are independent identically distributed copies of G_0^2. Then*

$$E \exp\left(\sum_{i=1}^n \lambda_i (\eta_{x_i} + h(x_i)\chi^{1/2})^2/2\right) = E\left(\frac{G_0^2}{u(0,0)} \exp\left(\sum_{i=1}^n \lambda_i G_{x_i}^2/2\right)\right).$$
(8.129)

Proof The η_{x_i}, $i = 1, \ldots, n$ are independent of χ. Take the expectation of the argument in the left-hand side of (8.129) with respect to χ. In general,

$$E_\chi f(\chi) = E f\left(u^{1/2}(0,0)(\xi_1^2 + \xi_2^2 + \xi_3^2)^{1/2}\right) \qquad (8.130)$$

$$= E\left(|\xi|^2 f\left(u^{1/2}(0,0)|\xi|\right)\right)$$

for any function f for which the first term exists. Here, ξ is $N(0,1)$ and ξ_i, $i = 1, \ldots, 3$, are independent identically distributed copies of ξ.

To obtain the second equality in (8.130), convert the integral to polar coordinates, integrate out the angle terms, and reinterpret the integral with respect to the radial term as the expectation of an $N(0,1)$ random variable. Using (8.130), we see that the left-hand side of (8.129) is

$$E\left(\frac{G_0^2}{u(0,0)}\exp\left(\sum_{i=1}^n \lambda_i(\eta_{x_i}+h_{x_i}|G_0|)^2/2\right)\right). \qquad (8.131)$$

The second equality in (8.129) follows from this since $G(x_i)=\eta_{x_i}+h(x_i)\,G(0)$, $i=1,\dots,n$, and $\{(\eta_{x_i}+h(x_i)\,G(0))^2\}_{i=1}^n \overset{law}{=} \{(\eta_{x_i}+h(x_i)|G(0)|)^2\}_{i=1}^n$. □

Using Lemma 8.4.2 and Dynkin's Isomorphism Theorem (Theorem 8.1.3), we see that

$$\left\{\overline{L}_\infty^x+\frac{G_x^2}{2};\, x\in D,\, P^{0,0}\times P_G\right\} \overset{law}{=} \left\{\frac{(\eta_x+h(x)\sqrt{\chi}\,)^2}{2};\, x\in D,\, P_\eta\times P_\chi\right\}. \qquad (8.132)$$

Thus we get a direct isomorphism theorem involving the natural measures of the processes.

It is easy to see that (8.132) is equivalent to (8.128). Let η' and G' be independent copies of η and G. By Corollary 5.2.2 and (5.71),

$$\left\{\frac{(\eta_x+h(x)\sqrt{\chi}\,)^2}{2}+\frac{\eta_x'^2}{2};\, x\in D\right\} \overset{law}{=} \left\{\frac{(\eta_x+h(x)\sqrt{2\rho}\,)^2}{2}+\frac{G_x'^2}{2};\, x\in D\right\}, \qquad (8.133)$$

where ρ is an exponential random variable with mean $u(0,0)$. We now add $\eta_x'^2/2$ to each side of (8.132) and use (8.133) on the resulting right-hand side. Then we cancel $G_x^2/2$ and $G_x'^2/2$ and get (8.128).

We can also derive (8.127) from the Eisenbaum Isomorphism Theorem (Theorem 8.1.1), although it is not as simple as the above. To begin with, as stated, Theorem 8.1.1 does not apply to the h-transform process because this process does not have symmetric potential densities with respect to the reference measure. We can easily deal with this, as we did in the proof of Lemma 3.10.4, by first applying Theorem 8.1.1 to $\widetilde{X}$ in Theorem 3.9.2, for which the associated process is $G./h(\cdot)$, and then multiplying by $h(\cdot)$ and using (3.219). We then get that, under the hypotheses of the Dynkin Isomorphism Theorem (Theorem 8.1.3),

$$E^{x,0}E_G\left(F\left(\overline{L}_\infty^{x_i}+\frac{(G_{x_i}+h(x_i)s)^2}{2}\right)\right) \qquad (8.134)$$

$$=E_G\left(\left(1+\frac{G_x}{h(x)s}\right)F\left(\frac{(G_{x_i}+h(x_i)s)^2}{2}\right)\right).$$

That is, (8.134) and (8.15) deal with the same processes (we are actually interested in this when $x = 0$ and use the fact that $h(0) = 1$).

We now derive (8.127), which is equivalent to (8.128), from (8.134). We take the limit as s goes to zero in (8.134) and obtain

$$E^{0,0} E_G \left(F \left(\overline{L}_\infty^{x_i} + \frac{G_{x_i}^2}{2} \right) \right) \tag{8.135}$$

$$= E \left(F \left(\frac{G_{x_i}^2}{2} \right) \right) + \frac{d}{ds} E \left(G_0 F \left(\frac{(G_{x_i} + h(x_i)s)^2}{2} \right) \right) \Bigg|_{s=0},$$

where we use the fact that $E \left(G_0 F \left(\frac{G_{x_i}^2}{2} \right) \right) = 0$. As usual, we write $G_{x_i} = \eta_{x_i} + h(x_i) G_0$. We see from Lemma 5.2.1 that

$$E_\eta \exp \left(\sum_{i=1}^n \lambda_i (G_{x_i} + h(x_i)s)^2 / 2 \right) \tag{8.136}$$

$$= E_\eta \exp \left(\sum_{i=1}^n \lambda_i (\eta_{x_i} + h(x_i)(G_0 + s))^2 / 2 \right)$$

$$= E_\eta \exp \left(\sum_{i=1}^n \lambda_i \eta_{x_i}^2 / 2 \right) \exp(K(G_0 + s)^2)/2),$$

where $K = K(h, \widetilde{\Sigma}, \Lambda) = h \Lambda h^t + h \Lambda \widetilde{\Sigma} \Lambda h^t$ for Λ and $\widetilde{\Sigma}$ as defined in Lemma 5.2.1 and $h = (h_{x_1} \ldots, h_{x_n})$.

One consequence of (8.136), which is obtained by setting $s = 0$ and taking the expectation with respect to G_0, is that

$$E \exp \left(\sum_{i=1}^n \lambda_i G_{x_i}^2 / 2 \right) = E \exp \left(\sum_{i=1}^n \lambda_i \eta_{x_i}^2 / 2 \right) \frac{1}{(1 - u(0,0)K)^{1/2}}. \tag{8.137}$$

We also use (8.136) to see that

$$\frac{d}{ds} E G_0 \exp \left(\sum_{i=1}^n \lambda_i (G_{x_i} + h(x_i)s)^2 / 2 \right) \tag{8.138}$$

$$= \frac{d}{ds} E G_0 \exp \left(\sum_{i=1}^n \lambda_i (\eta_{x_i} + h(x_i)(G_0 + s))^2 / 2 \right)$$

$$= E_\eta \exp \left(\sum_{i=1}^n \lambda_i \eta_{x_i}^2 / 2 \right) \frac{d}{ds} E G_0 \exp(K(G_0 + s)^2)/2).$$

Furthermore,

$$\frac{d}{ds} E G_0 \exp(K(G_0 + s)^2)/2) \Bigg|_{s=0} = K E G_0^2 \exp(K G_0^2 / 2)$$

$$= \frac{Ku(0,0)}{(1 - Ku(0,0))^{3/2}}. \quad (8.139)$$

Therefore, by (8.137)–(8.139), we have

$$E \exp \left(\sum_{i=1}^{n} \lambda_i G_{x_i}^2 / 2 \right) + \frac{d}{ds} E G_0 \exp \left(\sum_{i=1}^{n} \lambda_i (G_{x_i} + h(x_i)s)^2 \right) \Bigg|_{s=0}$$

$$= E \exp \left(\sum_{i=1}^{n} \lambda_i \eta_{x_i}^2 / 2 \right) \frac{1}{(1 - u(0,0)K)^{3/2}}. \quad (8.140)$$

We now use (8.135) to compute the moment generating functions. It follows from (8.137) and (8.140) that

$$E^{0,0} \exp \left(\sum_{i=1}^{n} \lambda_i \overline{L}_{\infty}^{x_i} \right) E \exp \left(\sum_{i=1}^{n} \lambda_i G_{x_i}^2 / 2 \right) \quad (8.141)$$

$$= E \exp \left(\sum_{i=1}^{n} \lambda_i \eta_{x_i}^2 / 2 \right) \frac{1}{(1 - u(0,0)K)^{3/2}},$$

which by (8.137) again gives

$$E^{0,0} \exp \left(\sum_{i=1}^{n} \lambda_i \overline{L}_{\infty}^{x_i} \right) = \frac{1}{(1 - u(0,0)K)}. \quad (8.142)$$

Therefore, by (8.137) again, we have

$$E^{0,0} \exp \left(\sum_{i=1}^{n} \lambda_i \overline{L}_{\infty}^{x_i} \right) \left(E \exp \left(\sum_{i=1}^{n} \lambda_i \eta_{x_i}^2 / 2 \right) \right)^2 \quad (8.143)$$

$$= \left(E \exp \left(\sum_{i=1}^{n} \lambda_i G_{x_i}^2 / 2 \right) \right)^2,$$

which gives (8.127).

8.5 Notes and references

Dynkin's Isomorphism Theorem (Theorem 8.1.3) is given in Dynkin (1984); see also Dynkin (1983). It was the starting point of our research on the connection between Gaussian processes and the local times of strongly symmetric Borel right processes and was used extensively in our initial paper on this subject, Marcus and Rosen (1992d). Eisenbaum's Isomorphism Theorem (Theorem 8.1.1) is given in Eisenbaum (1995). After it appeared, we started using it instead of Dynkin's Theorem. It seems to be easier to apply because one does not have to worry about h-transform processes. Our proofs of Theorems 8.1.1 and 8.1.3 in Section 8.1 first appeared in Marcus and Rosen (2001). Their original proofs were combinatorial, similar to our proofs in Section 8.3. (Our proof of Dynkin's Isomorphism Theorem and some generalizations in Marcus and Rosen (1992d) are also combinatoric.)

The results in Section 8.2 first appeared in Eisenbaum, Kaspi, Marcus, Rosen and Shi (2000).

9

Sample path properties of local times

The concept of associated Gaussian process, defined on page 324, is fundamental in this book. Recall that, given a strongly symmetric Borel right process X with 0-potential density $u(x, y)$, the associated Gaussian process G_0 is the mean zero Gaussian process with covariance $u(x, y)$. Even when X does not have a 0-potential density, it does have α-potential densities $u^\alpha(x, y)$ for all $\alpha > 0$. As we explain in Section 3.5, $u^\alpha(x, y)$ is the 0-potential density of $\widehat{X}$, which is X killed at the end of an independent exponential time with mean $1/\alpha$. Therefore, the mean zero Gaussian process G_α with covariance $u^\alpha(x, y)$ is the associated Gaussian process for $\widehat{X}$.

In a slightly different way of looking at things, we may say that there is an infinite family of Gaussian processes associated with X, the Gaussian processes G_α, for all $\alpha > 0$ and also G_0 if X has a 0-potential density. We call these the family of Gaussian processes associated with X.

Using the isomorphism theorems of Chapter 8, we show that the local times of X and the family of Gaussian processes associated with X have similar sample paths properties and zero–one laws. Our proofs are soft, in the sense that we do not give concrete conditions for determining whether the local times have certain properties. We merely say that the local times have certain properties if and only if the family of associated Gaussian processes does. But this is the strength of our approach, because in Chapters 5–7 we have a complete catalog of conditions for the Gaussian processes to have these properties.

9.1 Bounded discontinuities

We give conditions for local times to be bounded that do not require that they are also continuous. We begin with a simple example of this

396

phenomenon which is an immediate consequence of the Second Ray–Knight Theorem, Theorem 8.2.2.

Lemma 9.1.1 *Let $X = (\Omega, \mathcal{G}, \mathcal{G}_t, X_t, \theta_t, P^x)$ be a strongly symmetric Borel right process with continuous α-potential densities $u^\alpha(x, y)$, $\alpha > 0$, and state space (S, τ), where S is a locally compact separable metric space. Let $L = \{L_t^y \,; (y, t) \in S \times R_+\}$ denote the local times of X normalized by (3.91). Let 0 denote some fixed element in S and let $\tau_0(t) = \inf\{s \,| L_s^0 > t\}$. Assume that $u(0, 0) = \infty$ and that $P^x(T_0 < \infty) > 0$ for all $x \in S$. Let $\eta = \{\eta_x \,; x \in S\}$ denote the mean zero Gaussian process with covariance $u_{T_0}(x, y)$ and assume that $u_{T_0}(x, y)$ is continuous. Assume further that η has oscillation function $0 \leq \beta(0) < \infty$ at 0. Then, for any $t > 0$ and any countable dense set $C \subset S$,*

$$\frac{\beta(0)\sqrt{t}}{\sqrt{2}} \leq \lim_{\delta \to 0} \sup_{x \in C \cap B(0, \delta)} (L_{\tau_0(t)}^x - t) \leq \frac{\beta(0)\sqrt{t}}{\sqrt{2}} + \frac{\beta^2(0)}{8} \quad P^0 \quad a.s.,$$
(9.1)

where $B(0, \delta)$ denotes a closed ball of radius δ at 0 in the metric τ.

Note that the existence of u_{T_0} follows from Lemma 3.8.1.

Proof Since $u_{T_0}(x, y)$ is continuous, we can take C as a separability set for η. We note that $\eta(0) \equiv 0$. Therefore, the hypothesis that η has oscillation function $\beta(0)$ at 0 implies, by Theorem 5.3.7, that

$$\lim_{\delta \to 0} \sup_{x \in C \cap B(0, \delta)} \eta(x) = \frac{\beta(0)}{2} \quad a.s. \tag{9.2}$$

Consequently,

$$\lim_{\delta \to 0} \sup_{x \in C \cap B(0, \delta)} \frac{1}{2}\left(\eta(x) + \sqrt{2t}\right)^2 = \frac{\beta(0)\sqrt{t}}{\sqrt{2}} + \frac{\beta^2(0)}{8} + t \quad a.s. \tag{9.3}$$

Therefore, by Theorem 8.2.2,

$$\lim_{\delta \to 0} \sup_{x \in C \cap B(0, \delta)} \left(L_{\tau_0(t)}^x + \frac{\eta^2(x)}{2}\right) = \frac{\beta(0)\sqrt{t}}{\sqrt{2}} + \frac{\beta^2(0)}{8} + t \quad P^0 \times P_\eta \quad a.s. \tag{9.4}$$

The inequalities in (9.1) follows immediately from this. The upper bound is obvious. For the lower bound, we use the triangle inequality and (9.3) with $t = 0$. □

We now obtain results in the form of (9.1) that hold for any strongly symmetric Borel right process that has a local time and with $L_{\tau(t)}^\cdot$ replaced by $L_t^\cdot$. To do this we first develop an important corollary of Theorem 8.1.1 that enables us to prove that almost sure events for members of the family of Gaussian processes associated with a strongly symmetric Markov process X are also almost sure events for the local times of X.

For any set C, let $F(C)$ denote the set of real-valued functions f on C. Define the evaluations $i_x : F(C) \mapsto R^1$ by $i_x(f) = f(x)$. We use $\mathcal{M}(F(C))$ to denote the smallest σ-algebra for which the evaluations i_x are Borel measurable for all $x \in C$. $\mathcal{M}(F(C))$ is generally referred to as the σ-algebra of cylinder sets in $F(C)$.

Lemma 9.1.2 *Let* $X = (\Omega, \mathcal{G}, \mathcal{G}_t, X_t, \theta_t, P^x)$ *be a strongly symmetric Borel right process with continuous α-potential densities $u^\alpha(x, y)$, $\alpha > 0$, and state space (S, τ), where S is a locally compact separable metric space. Let $L = \{L_t^y ; (y, t) \in S \times R_+\}$ denote the local time of X normalized by (3.91). Let $G_\alpha = \{G_\alpha(y) ; y \in S\}$ denote a mean zero Gaussian process with covariance $u^\alpha(x, y)$. Let C be a countable dense subset of S. Let $B \in \mathcal{M}(F(C))$ and assume that, for some $s \neq 0$,*

$$P\left((G_\alpha(\,\cdot\,) + s)^2 /2 \in B\right) = 1. \tag{9.5}$$

Let Leb denote Lebesgue measure on R_+. Then, for almost all $(\omega', t) \in \Omega_{G_\alpha} \times R_+$ with respect to $P_{G_\alpha} \times$ Leb and all $x \in S$,

$$P^x\left(L_t^{\cdot} + \frac{(G_\alpha(\,\cdot\,, \omega') + s)^2}{2} \in B\right) = 1, \tag{9.6}$$

and for almost all $\omega' \in \Omega_{G_\alpha}$ with respect to P_{G_α} and all $x \in S$,

$$P^x\left(L_t^{\cdot} + \frac{(G_\alpha(\,\cdot\,, \omega') + s)^2}{2} \in B \quad \text{for almost all } t \in R_+\right) = 1. \tag{9.7}$$

Also, we can choose a countable dense set $Q \subset R_+$ such that, for almost all $\omega' \in \Omega_{G_\alpha}$ with respect to P_{G_α} and all $x \in S$,

$$P^x\left(L_t^{\cdot} + \frac{(G_\alpha(\,\cdot\,, \omega') + s)^2}{2} \in B \quad \text{for all } t \in Q\right) = 1. \tag{9.8}$$

Proof Following Section 3.5, we construct the process $\widehat{X} = (\widehat{\Omega}, \widehat{\mathcal{G}}, \widehat{\mathcal{G}}_t, \widehat{X}_t, \widehat{\theta}_t, \widehat{P}^x)$, which is X killed at the end of an independent exponential time with mean $1/\alpha$. The 0-potential density of $\widehat{X}$ is u^α and $\widehat{P}^x = P^x \times \alpha e^{-\alpha u}\, du$. We apply Theorem 8.1.1 to $\widehat{X}$. We first note that $I_B((G_\alpha(\,\cdot\,) + s)^2/2) = 1$ almost surely, by (9.5). Therefore, since G_α has mean zero, we have

$$E\left(G_\alpha(x)\, I_B((G_\alpha(\,\cdot\,) + s)^2/2)\right) = 0. \tag{9.9}$$

(To make this observation obvious, write $I_B = 1 - I_{B^c}$ and use the Schwarz inequality on $E(G_\alpha(x)I_{B^c})$.) Consequently, it follows from Theorem 8.1.1 and (9.5) that

$$\widehat{P}^x \times P_{G_\alpha} \left(L_\lambda + \tfrac{1}{2}(G_\alpha(\,\cdot\,) + s)^2 \in B \right) = 1, \qquad (9.10)$$

where λ is an exponential random variable with mean $1/\alpha$.

Writing out the expectation with respect to the independent exponential time, (9.10) becomes

$$\alpha \int_0^\infty P^x \times P_{G_\alpha} \left(L_t + (G_\alpha(\,\cdot\,) + s)^2/2 \in B \right) e^{-\alpha t}\, dt = 1, \qquad (9.11)$$

which shows that $P^x \times P_{G_\alpha} \left(L_t + (G_\alpha(\,\cdot\,) + s)^2/2 \in B \right) = 1$ for almost all $t \in R_+$. This gives both (9.6) and (9.7). Equation (9.8) follows immediately from (9.6). $\qquad \square$

As an immediate application of Lemma 9.1.2, we relate the bounded discontinuities of the local times of X to the bounded discontinuities of the family of Gaussian processes associated with X.

Theorem 9.1.3 Let $X = (\Omega, \mathcal{G}, \mathcal{G}_t, X_t, \theta_t, P^x)$ be a strongly symmetric Borel right process with continuous α-potential densities $u^\alpha(x, y)$, $\alpha > 0$, and state space (S, τ), where S is a locally compact separable metric space. Let $L = \{L_t^y\,;\, (y,t) \in S \times R_+\}$ denote the local time of X normalized by (3.91). Let $G_\alpha = \{G_\alpha(y)\,;\, y \in S\}$ be a real-valued Gaussian process with mean zero and covariance $u^\alpha(x, y)$. Assume that G_α has oscillation function $0 \le \beta(x_0) < \infty$ at x_0. Then, for any countable dense set $C \subset S$,

$$\frac{\beta(x_0)\sqrt{L_t^{x_0}}}{\sqrt{2}} \le \lim_{\delta \to 0} \sup_{x \in C \cap B(x_0, \delta)} L_t^x - L_t^{x_0} \le \frac{\beta^2(x_0)}{8} + \frac{\beta(x_0)\sqrt{L_t^{x_0}}}{\sqrt{2}} \qquad (9.12)$$

for all $t \in R_+$, P^y almost surely for all $y \in S$, and

$$\lim_{\delta \to 0} \sup_{x \in C \cap B(x_0, \delta)} |L_t^x - L_t^{x_0}| \le \left(\frac{\beta^2(x_0)}{4} + \frac{\beta(x_0)\sqrt{L_t^{x_0}}}{\sqrt{2}} \right) \qquad (9.13)$$

for all $t \in R_+$, P^y almost surely for all $y \in S$. Here $B(0, \delta)$ denotes a closed ball of radius δ at 0 in the metric τ.

Since $0 < L_t^{x_0}$, P^{x_0} almost surely for all $t > 0$, we see from (9.12) and (9.13) that the local times of a Markov process associated with a Gaussian process that has a bounded discontinuity at x_0 (a probability 0 or 1 event by Corollary 5.3.6) itself has a bounded discontinuity at x_0 almost surely with respect to P^{x_0}.

If the process is started at $y \neq x_0$, there could be paths for which $L_t^{x_0} = 0$ for sufficiently small t. In this case, both sides of (9.12) should be equal to zero. Consequently, we suspect that the term containing $\beta^2(x_0)$ can be eliminated from (9.12) and (9.13), and also from (9.1), but we do not know how to do it.

Proof Let C be a countable separating set for $\{G_\alpha(x), x \in S\}$ and note that

$$(G_\alpha(x) + s)^2 - (G_\alpha(x_0) + s)^2 \tag{9.14}$$
$$= (G_\alpha(x) - G_\alpha(x_0))^2 + 2(G_\alpha(x_0) + s)(G_\alpha(x) - G_\alpha(x_0)).$$

It follows from (9.14) and Theorem 5.3.7 that

$$\lim_{\delta \to 0} \sup_{x \in C \cap B(x_0, \delta)} \frac{(G_\alpha(x) + s)^2 - (G_\alpha(x_0) + s)^2}{2} \tag{9.15}$$
$$= \frac{\beta^2(x_0)}{8} + \frac{\beta(x_0)|G_\alpha(x_0) + s|}{2} \quad \text{a.s} \quad P_{G_\alpha}.$$

Let

$$B = \left\{ f \in F(C) \middle| \lim_{\delta \to 0} \sup_{x \in C \cap B(x_0, \delta)} f(x) - f(x_0) \right. \tag{9.16}$$
$$\left. = \frac{\beta^2(x_0)}{8} + \frac{\beta(x_0)|f(x_0)|^{1/2}}{\sqrt{2}} \right\}.$$

Then, by (9.15),

$$P\left((G_\alpha(\cdot) + s)^2 / 2 \in B\right) = 1. \tag{9.17}$$

It follows from Lemma 9.1.2, that for almost all $\omega' \in \Omega_{G_\alpha}$ with respect to P_{G_α},

$$\lim_{\delta \to 0} \sup_{x \in C \cap B(x_0, \delta)} L_t^x - L_t^{x_0} + \frac{(G_\alpha(x, \omega') + s)^2 - (G_\alpha(x_0, \omega') + s)^2}{2} \tag{9.18}$$
$$= \frac{\beta^2(x_0)}{8} + \frac{\beta(x_0)}{\sqrt{2}} \sqrt{L_t^{x_0} + \frac{(G_\alpha(x_0, \omega') + s)^2}{2}} \quad \text{for all } t \in Q, \ P^y \text{ a.s.}$$

for all $y \in S$, where Q is a countable dense subset of R_+. Therefore, for almost all $\omega' \in \Omega_{G_\alpha}$ with respect to P_{G_α},

$$\lim_{\delta \to 0} \sup_{x \in C \cap B(x_0, \delta)} L_t^x - L_t^{x_0} \tag{9.19}$$
$$+ \lim_{\delta \to 0} \sup_{x \in C \cap B(x_0, \delta)} \frac{(G_\alpha(x, \omega') + s)^2 - (G_\alpha(x_0, \omega') + s)^2}{2}$$
$$\geq \frac{\beta^2(x_0)}{8} + \frac{\beta(x_0)}{\sqrt{2}} \sqrt{L_t^{x_0} + \frac{(G_\alpha(x_0, \omega') + s)^2}{2}} \quad \text{for all } t \in Q, \ P^y \text{ a.s.}$$

for all $y \in S$. It follows from (9.17) that, for almost all $\omega' \in \Omega_{G_\alpha}$ with respect to P_{G_α},

$$\lim_{\delta \to 0} \sup_{x \in C \cap B(x_0, \delta)} L_t^x - L_t^{x_0} \geq \frac{\beta(x_0)}{\sqrt{2}} \sqrt{L_t^{x_0}}$$
$$- \frac{\beta(x_0)|G_\alpha(x_0, \omega') + s|}{2} \qquad \text{for all } t \in Q, \qquad P^y \text{ a.s. (9.20)}$$

By Lemma 5.3.5, for all $\epsilon > 0$, we can find an $\omega' \in \Omega_{G_\alpha}$ such that (9.20) holds and $|G_\alpha(x_0, \omega')| < \epsilon$. Let us also take $s = \epsilon$. For this ω' we have

$$\lim_{\delta \to 0} \sup_{x \in C \cap B(x_0, \delta)} L_t^x - L_t^{x_0} \geq \frac{\beta(x_0)}{\sqrt{2}} \left(\sqrt{L_t^{x_0}} - \sqrt{2}\epsilon \right) \qquad \text{for all } t \in Q$$
(9.21)

P^y almost surely, and since this holds for all $\epsilon > 0$, we get

$$\lim_{\delta \to 0} \sup_{x \in C \cap B(x_0, \delta)} L_t^x - L_t^{x_0} \geq \frac{\beta(x_0)}{\sqrt{2}} \sqrt{L_t^{x_0}} \qquad \text{for all } t \in Q \qquad P^y \text{ a.s.}$$
(9.22)

Let $\widetilde{\Omega}$, $P^y(\widetilde{\Omega}) = 1$, be the set on which (9.22) holds. For any $t > 0$, choose a sequence $t_i \in Q$ such that $t_i \uparrow t$. For $\omega \in \widetilde{\Omega}$, we have, by the monotonicity of the local time,

$$\lim_{\delta \to 0} \sup_{x \in C \cap B(x_0, \delta)} L_t^x(\omega) \geq L_{t_i}^{x_0}(\omega) + \frac{\beta(x_0)\sqrt{L_{t_i}^{x_0}(\omega)}}{\sqrt{2}}. \qquad (9.23)$$

Since $L_t^{x_0}$ is continuous in t, we see that

$$\lim_{\delta \to 0} \sup_{x \in C \cap B(x_0, \delta)} L_t^x(\omega) \geq L_t^{x_0}(\omega) + \frac{\beta(x_0)\sqrt{L_t^{x_0}(\omega)}}{\sqrt{2}}, \qquad (9.24)$$

and since this is valid for all t and all $\omega \in \widetilde{\Omega}$, we get the lower bound in (9.12).

To obtain the upper bound in (9.12), we use (9.18) to immediately obtain

$$\lim_{\delta \to 0} \sup_{x \in C \cap B(x_0, \delta)} L_t^x - L_t^{x_0} \qquad (9.25)$$
$$\leq \frac{\beta^2(x_0)}{8} + \frac{\beta(x_0)}{\sqrt{2}} \sqrt{L_t^{x_0} + \frac{(G_\alpha(x_0, \omega') + s)^2}{2}} + \frac{(G_\alpha(x_0, \omega') + s)^2}{2}$$

for all $t \in Q$, P^y almost surely, for almost all $\omega' \in \Omega_{G_\alpha}$ with respect to P_{G_α}. As above, $|G_\alpha(x_0, \omega')|$ and s can be made as small as we like and we get

$$\lim_{\delta \to 0} \sup_{x \in C \cap B(x_0, \delta)} L_t^x - L_t^{x_0} \leq \frac{\beta^2(x_0)}{8} + \frac{\beta(x_0)\sqrt{L_t^{x_0}}}{\sqrt{2}} \qquad (9.26)$$

$$\text{for all } t \in Q, \qquad P^y \text{ a.s.}$$

For all t choose $t_i \in Q$ such that $t_i \downarrow t$. Following the argument given in (9.23) and (9.24), we get the upper bound in (9.12).

To obtain (9.13) we repeat the above argument with a minor variation. Analogous to (9.16), we consider the set

$$B' = \left\{ f \in F(C) \,\middle|\, \lim_{\delta \to 0} \sup_{x \in C \cap B(x_0, \delta)} |f(x) - f(x_0)| \qquad (9.27) \right.$$

$$\left. = \frac{\beta^2(x_0)}{8} + \frac{\beta(x_0)|f(x_0)|^{1/2}}{\sqrt{2}} \right\}.$$

Then, essentially the same argument used in (9.14) and (9.15) shows that

$$P\left((G_\alpha(\cdot) + s)^2 / 2 \in B' \right) = 1. \qquad (9.28)$$

Therefore, analogous to (9.25), we have

$$\lim_{\delta \to 0} \sup_{x \in C \cap B(x_0, \delta)} |L_t^x - L_t^{x_0}|$$

$$\leq \frac{\beta^2(x_0)}{8} + \frac{\beta(x_0)}{\sqrt{2}} \sqrt{L_t^{x_0} + \frac{(G_\alpha(x_0, \omega') + s)^2}{2}}$$

$$+ \lim_{\delta \to 0} \sup_{x \in C \cap B(x_0, \delta)} \frac{|(G_\alpha(x, \omega') + s)^2 - (G_\alpha(x_0, \omega') + s)^2|}{2}$$

$$= \frac{\beta^2(x_0)}{4} + \frac{\beta(x_0)}{\sqrt{2}} \sqrt{L_t^{x_0} + \frac{(G_\alpha(x_0, \omega') + s)^2}{2}} + \frac{\beta(x_0)|G(x_0) + s|^{1/2}}{\sqrt{2}}$$

for all $t \in Q$, P^y almost surely, for almost all $\omega' \in \Omega_{G_\alpha}$ with respect to P_{G_α}. As above, $|G_\alpha(x_0, \omega')|$ and s can be made as small as we like and we get

$$\lim_{\delta \to 0} \sup_{x \in C \cap B(x_0, \delta)} |L_t^x - L_t^{x_0}| \leq \frac{\beta^2(x_0)}{4} + \frac{\beta(x_0)\sqrt{L_t^{x_0}}}{\sqrt{2}} \qquad (9.29)$$

$$\text{for all } t \in Q, \qquad P^y \text{ a.s.}$$

We now show that this holds for all $t \in R_+$. It is trivial for $t = 0$. For $t > 0$, pick $u, v \in Q$, $u < t < v$ so that, for a fixed ω, $L_v^{x_0} - L_t^{x_0} < \epsilon$ and $L_t^{x_0} - L_u^{x_0} < \epsilon$. Note that $|L_t^x - L_t^{x_0}| \leq |L_v^x - L_u^{x_0}| \vee |L_u^x - L_v^{x_0}|$. Then

$$|L_v^x - L_u^{x_0}| \leq |L_v^x - L_v^{x_0}| + |L_v^{x_0} - L_t^{x_0}| + |L_t^{x_0} - L_u^{x_0}| \qquad (9.30)$$

with a similar bound for $|L_u^x - L_v^{x_0}|$. Using the fact that $u, v \in Q$ and (9.29), we see that

$$\lim_{\delta \to 0} \sup_{x \in C \cap B(x_0, \delta)} |L_t^x - L_t^{x_0}| \le 4\epsilon + \frac{\beta^2(x_0)}{4} + \frac{\beta(x_0)\sqrt{L_v^{x_0}}}{\sqrt{2}}. \qquad (9.31)$$

Since $L_v^{x_0} \le L_t^{x_0} + \epsilon$ and ϵ can be taken as small as we like, we get (9.13). $\square$

9.2 A necessary condition for unboundedness

Let X be a strongly symmetric Borel right process with continuous α-potential density $u^\alpha(x, y)$. To obtain a necessary condition for the local times of X to be unbounded almost surely, we first consider the process $\widehat{X} = (\widehat{\Omega}, \widehat{\mathcal{G}}, \widehat{\mathcal{G}}_t, \widehat{X}_t, \widehat{\theta}_t, \widehat{P}^x)$, introduced in Section 3.5, which is X killed at the end of an independent exponential time with mean $1/\alpha$, $\alpha > 0$. The 0-potential density of $\widehat{X}$ is u^α and $\widehat{P}^x = P^x \times \alpha e^{-\alpha u} du$. Let $\widehat{L} = \{\widehat{L}_t^y \, ; \, (y, t) \in S \times R_+\}$ denote the local time of $\widehat{X}$.

Consider $\widehat{X}$ killed the first time it hits zero and denote its potential density by $\widehat{u}_{T_0}^\alpha(x, y)$. Since the 0-potential density of $\widehat{X}$ is u^α, it follows from Remark 3.8.3 that

$$\widehat{u}_{T_0}^\alpha(x, y) = u^\alpha(x, y) - h_\alpha(x)h_\alpha(y)u^\alpha(0, 0), \qquad (9.32)$$

where

$$h_\alpha(x) = \frac{u^\alpha(x, 0)}{u^\alpha(0, 0)}. \qquad (9.33)$$

Let $\eta_\alpha(x)$ be a mean zero Gaussian process with covariance $\widehat{u}_{T_0}^\alpha(x, y)$. By Theorem 8.2.3, the Second Ray–Knight Theorem, under the measure $P^0 \times P_{\eta_\alpha}$, for any $t > 0$ and countable subset $D \subseteq S$,

$$\left\{ \widehat{L}_{\tau^-(t \wedge \widehat{L}_\infty^0)}^x + \tfrac{1}{2}\eta_\alpha^2(x) \, ; \, x \in D \right\} \qquad (9.34)$$

$$\overset{law}{=} \left\{ \tfrac{1}{2}\left(\eta_\alpha(x) + h_\alpha(x)\sqrt{2(t \wedge \rho)} \right)^2 \, ; \, x \in D \right\}$$

where ρ is an exponential random variable with mean $u^\alpha(0, 0)$ that is independent of $\eta_\alpha = \{\eta_\alpha(x) \, ; \, x \in S\}$.

Theorem 9.2.1 *Let* $X = (\Omega, \mathcal{G}, \mathcal{G}_t, X_t, \theta_t, P^x)$ *be a strongly symmetric Borel right process with continuous α-potential densities* $u^\alpha(x, y)$, $\alpha > 0$, *and state space* (S, τ), *where* S *is a locally compact separable metric space. Let* $L = \{L_t^y \, ; \, (y, t) \in S \times R_+\}$ *denote the local time of* X

normalized by (3.91). Let $G_\alpha = \{G_\alpha(y)\,;\,y \in S\}$ be a real-valued Gaussian process with mean zero and covariance $u^\alpha(x,y)$, and let 0 denote a fixed point in S. Suppose there exists a countable dense set $C \subset S$ for which

$$\lim_{\delta \to 0} \sup_{x \in C \cap B(0,\delta)} G_\alpha(x) = \infty \qquad a.s. \qquad (9.35)$$

Then

$$\lim_{\delta \to 0} \sup_{x \in C \cap B(0,\delta)} L_t^x = \infty \qquad \forall t > 0 \quad a.s. \quad P^0. \qquad (9.36)$$

Proof We use (9.34) applied to $\widehat{X}$. Let $\eta_\alpha(x) = G_\alpha(x) - h_\alpha(x)G_\alpha(0)$. $\eta_\alpha(x)$ is a mean zero Gaussian process with covariance $\widehat{u}_{T_0}^\alpha(x,y)$ given in (9.32). Since $G_\alpha(0)$ is finite almost surely and $u^\alpha(x,y)$ is continuous, we see that

$$\lim_{\delta \to 0} \sup_{x \in C \cap B(0,\delta)} \eta_\alpha(x) = \infty \qquad (9.37)$$

and

$$\lim_{\delta \to 0} \sup_{x \in C \cap B(0,\delta)} E\eta_\alpha^2(x) = \lim_{\delta \to 0} \sup_{x \in C \cap B(0,\delta)} \left(u^\alpha(x,x) - \frac{(u^\alpha(x,0))^2}{u^\alpha(0,0)} \right) = 0. \qquad (9.38)$$

Choose some $t > 0$. For $\delta > 0$, which we choose below, let $T \in C \cap B(0,\delta)$ be a finite set. Note that $\lim_{x \to 0} h_\alpha(x) = 1$. Let δ' be such that $\inf_{x \in [0,\delta')} h_\alpha(x) > 0.9$. We require that $\delta \leq \delta'$. Let ρ be an exponential random variable with mean $u^\alpha(0,0)$ that is independent of η_α. We first note that it follows from (5.150) that

$$P_{\eta_\alpha} \left(\sup_{x \in T} \left(\eta_\alpha(x) + h_\alpha(x)\sqrt{2(t \wedge \rho)} \right) \geq a - s\sigma_T + 0.9\sqrt{2(t \wedge \rho)} \right)$$
$$\geq 1 - \Psi(s), \quad (9.39)$$

where a is the median of $\sup_{x \in T} \eta_\alpha(x)$, $\sigma_T = \sup_{x \in T}(E\eta_\alpha^2(x))^{1/2}$ and $\Psi(s) = 1 - \Phi(s)$. We assume that $a - s\sigma_T > 0$. It is obvious, by (9.37) and (9.38), that whatever the value of δ and s, we can find finite sets T for which this is the case. Since $\eta_\alpha(0) = 0$, both terms in the argument of P_{η_α} in (9.39) are positive. Consequently,

$$P_{\eta_\alpha} \left(\sup_{x \in T} \frac{(\eta_\alpha(x) + h_\alpha(x)\sqrt{2(t \wedge \rho)}\,)^2}{2} \geq \frac{(a - s\sigma_T + 0.9\sqrt{2(t \wedge \rho)})^2}{2} \right)$$
$$\geq 1 - \Psi(s). \qquad (9.40)$$

For any $\epsilon > 0$, we choose k so that $P(.9\sqrt{2(t \wedge \rho)} \geq k) \geq 1 - \epsilon$. Then

$$P\left(\sup_{x \in T} \frac{(\eta_\alpha(x) + h_\alpha(x)\sqrt{2(t \wedge \rho)}\,)^2}{2} \geq \frac{(a - s\sigma_T + k)^2}{2}\right) \geq 1 - \Psi(s) - \epsilon.$$
(9.41)

It follows from (9.34) and the triangle inequality that

$$P\left(\sup_{x \in T} \widehat{L}^x_{\tau^-(t \wedge \widehat{L}^0_\infty)} \geq \frac{(a - s\sigma_T + k)^2}{2} - \sup_{x \in T} \frac{\eta_\alpha(x)^2}{2}\right) \geq 1 - \Psi(s) - \epsilon.$$
(9.42)

By (5.151), since $\eta_\alpha(0) = 0$,

$$P\left(\sup_{x \in T} \frac{\eta_\alpha^2(x)}{2} \leq \frac{(a + s\sigma_T)^2}{2}\right) \geq 1 - \Psi(s). \qquad (9.43)$$

Using this in (9.42) we see that

$$P^0\left(\sup_{x \in T} \widehat{L}^x_{\tau^-(t \wedge \widehat{L}^0_\infty)} \geq a(k - 2s\sigma) + \frac{k}{2}(k - 2s\sigma)\right) \geq 1 - 2\Psi(s) - \epsilon,$$
(9.44)

where $\sigma = \sup_{x \in C \cap B(0,\delta)}(E\eta_\alpha^2(x))^{1/2} \geq \sigma_T$.

We now choose s sufficiently large so that $\Psi(s) \leq \epsilon$. Next, we note that by (9.38), we can choose $0 < \delta < \delta'$ so that $s\sigma \leq k/8$. With these choices we have

$$P^0\left(\sup_{x \in T} \widehat{L}^x_{\tau^-(t \wedge \widehat{L}^0_\infty)} \geq \frac{ak}{2}\right) \geq 1 - 3\epsilon. \qquad (9.45)$$

Because of (9.37) we can take T to be large enough set so that $ak/2 > M$ for any number M. Therefore, for any $\epsilon > 0$,

$$P^0\left(\sup_{x \in C \cap B(0,\delta)} \widehat{L}^x_{\tau^-(t \wedge \widehat{L}^0_\infty)} = \infty\right) \geq 1 - 3\epsilon, \qquad (9.46)$$

and this holds for any $t > 0$. It also holds for all $0 < \delta'' \leq \delta$, because we can always restrict T to be contained in $[0, \delta'']$. Therefore,

$$P^0\left(\lim_{\delta \to 0} \sup_{x \in C \cap B(0,\delta)} \widehat{L}^x_{\tau^-(t \wedge \widehat{L}^0_\infty)} = \infty\right) = 1 \qquad (9.47)$$

for all $t > 0$.

Now choose any $t' > 0$. By the definition of local time, we can find a $t > 0$ such that $P^0(\tau^-(t \wedge \widehat{L}^0_\infty) < t') > 1 - \epsilon$. Using this observation and (9.47), we see that

$$P^0\left(\lim_{\delta \to 0} \sup_{x \in C \cap B(0,\delta)} \widehat{L}^x_t = \infty\right) = 1. \qquad (9.48)$$

Since $\widehat{L}_t^x \leq L_t^x$, we get (9.36). $\qquad\qquad\qquad\qquad\qquad\square$

9.3 Sufficient conditions for continuity

In (3.180) and (3.182) of Corollary 3.7.4 we give moment conditions that imply that the local times of a strongly symmetric Borel right process X with continuous α-potential density has a version that is jointly continuous. In the next two lemmas we show that these conditions are satisfied when some member of the family of Gaussian processes associated with X is continuous.

Lemma 9.3.1 *Let $X = (\Omega, \mathcal{G}, \mathcal{G}_t, X_t, \theta_t, P^x)$ be a strongly symmetric Borel right process with continuous α-potential densities $u^\alpha(x, y)$, $\alpha > 0$, and state space (S, τ), where S is a locally compact separable metric space. Let $L = \{L_t^y, (t, y) \in R_+ \times S\}$ be local times for X and let λ be an exponential random variable with mean $1/\alpha$ that is independent of X. Let K be a compact subset of S and $G_\alpha = \{G_\alpha(y); y \in K\}$ be a mean zero Gaussian process with covariance $u^\alpha(x, y)$ such that*

$$\sup_{y \in K} |G_\alpha(y)| < \infty \qquad a.s. \qquad (9.49)$$

Let

$$u_\alpha^* := \sup_{y \in K} u^\alpha(y, y), \qquad (9.50)$$

and let D be a countable subset of K. Then, for all $x \in S$,

$$E_\lambda^x \exp\left(\frac{\sup_{y \in D} L_\lambda^y}{c\, u_\alpha^*}\right) < \infty \qquad (9.51)$$

for all $c > 1$.

Proof We use the Eisenbaum Isomorphism Theorem (Theorem 8.1.1). Let $\sup_{y \in D} |\cdot| := \|\cdot\|$. It follows from (8.2) that

$$E_\lambda^x \exp\left(\frac{\|L_\lambda\|}{c\, u_\alpha^*}\right) \qquad (9.52)$$

$$\leq E\left(\left(1 + \frac{G_\alpha(x)}{s}\right) \exp\left(\frac{\|(G_\alpha(\cdot) + s)^2\|}{2c\, u_\alpha^*}\right)\right) \qquad (9.53)$$

$$\leq \left(E\left(1 + \frac{|G_\alpha(x)|}{s}\right)^p\right)^{1/p} \left(E \exp\frac{q\,\|(G_\alpha(\cdot) + s)^2\|}{2c\, u_\alpha^*}\right)^{1/q},$$

where $1/p + 1/q = 1$.

For real numbers, $(a+b)^2 \leq (1+a^2)(1+b^2)$. Using this we see that

$$
\begin{aligned}
\|(G_\alpha(\cdot)+s)^2\| &\leq \|(1+s^2)+(1+s^2)G_\alpha^2(\cdot)\| &\quad (9.54)\\
&\leq (1+s^2)+(1+s^2)\|G_\alpha^2(\cdot)\|.
\end{aligned}
$$

Consequently,

$$
E \exp \frac{q\,\|(G_\alpha(\cdot)+s)^2\|}{2c\,u_\alpha^*} \leq \exp\left(\frac{q(1+s^2)}{2c\,u_\alpha^*}\right) E \exp \frac{q(1+s^2)\,\|G_\alpha(\cdot)\|^2}{2c\,u_\alpha^*}. \tag{9.55}
$$

By (5.190), the expectation on the right-hand side of (9.55) is finite for $c/q(1+s^2) > 1$. Since we can take q arbitrarily close to one and s arbitrarily close to zero, we get (9.51). $\qquad\square$

For a stochastic process Z set

$$
\|Z\|_\infty = \sup_{y\in D} |Z(y)| \quad\text{and}\quad \|Z\|_\delta = \sup_{\substack{\tau(y,z)\leq\delta \\ y,z\in D}} |Z(y)-Z(z)|. \tag{9.56}
$$

Lemma 9.3.2 *Fix $x \in S$. Under the same hypotheses as in Lemma 9.3.1, but with the additional assumption that $x \in D$, we have that, for all $p \geq 1$,*

$$
\left(E_\lambda^x \sup_{\substack{\tau(y,z)\leq\delta \\ y,z\in D}} |L_\lambda^y - L_\lambda^z|^p\right)^{1/p} \leq C\,p\,E(\|G_\alpha\|_\delta)E(\|G_\alpha\|_\infty), \tag{9.57}
$$

where C is an absolute constant. Furthermore,

$$
E_\lambda^x \exp\left(C'\,\frac{\sup_{\tau(y,z)\leq\delta;\,y,z\in D} |L_\lambda^y - L_\lambda^z|}{E(\|G_\alpha\|_\delta)E(\|G_\alpha\|_\infty)}\right) < \infty \tag{9.58}
$$

for some $C' > 0$.

Proof Since

$$
\left(E_\lambda^x \left(\|L_\lambda^\cdot\|_\delta^p\right)\right)^{1/p} \leq \sup_{v\in D} \left(E_\lambda^v\left(\|L_\lambda^\cdot\|_\delta^p\right)\right)^{1/p}, \tag{9.59}
$$

it follows from (8.2) that

$$
\begin{aligned}
\left(E_\lambda^x\left(\|L_\lambda^\cdot\|_\delta^p\right)\right)^{1/p} \leq\; & \sup_{v\in D}\left(E\left\{\left(1+\frac{G_\alpha(v)}{s}\right)\left(\frac{\|(G_\alpha+s)^2\|_\delta}{2}\right)^p\right\}\right)^{1/p}\\
& + \left(E\left\{\left(\frac{\|(G_\alpha+s)^2\|_\delta}{2}\right)^p\right\}\right)^{1/p}. \tag{9.60}
\end{aligned}
$$

By the Cauchy-Schwarz inequality,

$$\left(E\left\{ \left(1 + \frac{G_\alpha(v)}{s}\right) \left(\frac{\|(G_\alpha + s)^2\|_\delta}{2}\right)^p \right\} \right)^{1/p} \tag{9.61}$$

$$\le \frac{1}{2} \left(E\left\{ \left(1 + \frac{|G_\alpha(v)|}{s}\right)^2 \right\} E\left\{ \left(\|(G_\alpha + s)^2\|_\delta\right)^{2p} \right\} \right)^{1/(2p)}.$$

Using this and applying the Cauchy-Schwarz inequality to the last term in (9.60), we see that

$$\left(E_\lambda^x\left(\|L_\lambda^\cdot\|_\delta^p\right)\right)^{1/p} \le \sup_{v \in D} \frac{1}{2} \left(\left(E\left\{\left(1 + \frac{|G_\alpha(v)|}{s}\right)^2\right\}\right)^{1/2} + 1 \right)$$

$$\left(E\left\{ \left(\|(G_\alpha + s)^2\|_\delta\right)^{2p} \right\} \right)^{1/(2p)}. \tag{9.62}$$

Since

$$(G_\alpha(y) + s)^2 - (G_\alpha(z) + s)^2 = (G_\alpha(y) + G_\alpha(z) + 2s)(G_\alpha(y) - G_\alpha(z)), \tag{9.63}$$

we get

$$\|(G_\alpha + s)^2\|_\delta \le 2\|G_\alpha\|_\delta \left(\|G_\alpha\|_\infty + s\right). \tag{9.64}$$

Let $s = E\|G_\alpha\|_\infty$. It follows from Corollary 5.4.7 that

$$\left(E\left(\|(G_\alpha + s)^2\|_\delta^{2p}\right) \right)^{1/(2p)} \tag{9.65}$$

$$\le 2 \left(E\left(\|G_\alpha\|_\delta^{4p}\right) \right)^{1/(4p)} \left(\left(E\left(\|G_\alpha\|_\infty^{4p}\right)\right)^{1/(4p)} + s \right)$$

$$\le C'' p E(\|G_\alpha\|_\delta) E(\|G_\alpha\|_\infty).$$

Also, since $E(|G_\alpha(v)|^2) \le (\pi/2)(E|G_\alpha(v)|)^2$,

$$E\left(1 + \frac{|G_\alpha(v)|}{s}\right)^2 \le 3 + \frac{E|G_\alpha(v)|^2}{s^2} \le 3 + \frac{\pi}{2}. \tag{9.66}$$

Therefore,

$$\sup_{v \in D} \left(E\left\{ \left(1 + \frac{|G_\alpha(v)|}{s}\right)^2 \right\} \right)^{1/2} \le \left(3 + \frac{\pi}{2}\right)^{1/2}. \tag{9.67}$$

Using (9.65) and (9.67) in (9.62), we get (9.57).

To get (9.58) we write out the series expansion for the exponential and use (9.57). $\qquad \square$

We can now show that (3.180) and (3.182) of Corollary 3.7.4 are satisfied and thus get a sufficient condition for the joint continuity of the local times.

Theorem 9.3.3 *Let* $X = (\Omega, \mathcal{G}, \mathcal{G}_t, X_t, \theta_t, P^x)$ *be a strongly symmetric Borel right process with continuous α-potential density $u^\alpha(x, y)$ and state space (S, ρ), where S is a locally compact separable metric space. Let $L = \{L_t^y, (t, y) \in R_+ \times S\}$ be local times for X. Let $G_\alpha = \{G_\alpha(y); y \in S\}$ be a mean zero Gaussian process with covariance $u^\alpha(x, y)$. Suppose that G_α has continuous sample paths; then we can find a version of L that is continuous on $R_+ \times S$.*

Proof Let $K \subseteq S$ be compact. To prove this theorem we need only verify that (3.180) and (3.182) of Corollary 3.7.4 are satisfied. That (3.180) holds follows immediately from (9.51). Considering (9.57), to show that (3.182) holds we need only show that

$$E \sup_{y \in K} |G_\alpha(y)| < \infty \qquad (9.68)$$

and

$$\lim_{\delta \to 0} E \sup_{\substack{\rho(y,z) \leq \delta \\ y,z \in K}} |G_\alpha(y) - G_\alpha(z)| = 0. \qquad (9.69)$$

Since G_α has a continuous version on K, it is bounded almost surely on K. Thus, (9.68) follows from Corollary 5.4.7.

It follows from (9.68) that

$$E \sup_{\substack{\rho(y,z) \leq \delta \\ y,z \in K}} |G_\alpha(y) - G_\alpha(z)| < \infty. \qquad (9.70)$$

We take the limit of the left-hand side of (9.70) as $\delta \to 0$. Considering (9.68) we can use the Dominated Convergence Theorem to take the limit inside the expectation. This gives us (9.69) since continuous functions on compact sets are uniformly continuous. $\qquad \square$

Remark 9.3.4 Note that (9.51), as stated, does not mention Gaussian processes. Similarly, using results from Chapter 6, we can write (9.57) is terms of majorizing measure or metric entropy conditions determined by the α-potential of X.

9.4 Continuity and boundedness of local times

The results in the first three sections of this chapter show that local times inherit continuity and boundedness properties from their associated Gaussian processes. We summarize these results in this section. There are some subtle points that require careful explanation. They center around the fact that a local time at a point does not begin to increase until the Markov process hits the point.

Throughout this section, $X = (\Omega, \mathcal{G}, \mathcal{G}_t, X_t, \theta_t, P^x)$ is a strongly symmetric Borel right process with continuous α-potential density $u^\alpha(x, y)$, $\alpha > 0$, and state space (S, τ), where S is a locally compact separable metric space. $L = \{L_t^y, (t, y) \in R_+ \times S\}$ are local times for X and $G_\alpha = \{G_\alpha(y), y \in S\}$ is a mean zero Gaussian process with covariance $u^\alpha(x, y)$.

We first state several results and then give their proofs. Most of the work has already been done, so the proofs are all short. To simplify statements, we use the expression "a process is continuous almost surely" to mean that we can find a continuous version of the process, and similarly for other properties of the process.

Theorem 9.4.1 (Continuity of local times)

(1) $\{L_t^y, (t, y) \in R_+ \times S\}$ *is continuous almost surely if and only if* $\{G_\alpha(y); y \in S\}$ *is continuous almost surely;*

(2) *Let* K *be a compact subset of* S. $\{L_t^y, (t, y) \in R_+ \times K\}$ *is continuous almost surely if and only if* $\{G_\alpha(y); y \in K\}$ *is continuous almost surely.*

(3) *Let* $D \subset S$ *be countable. Then, for any compact subset* K *of* S, $\{L_t^y, (t, y) \in [0, T] \times D \cap K\}$ *is bounded for all* $T < \infty$ *almost surely if and only if* $\{G_\alpha(y), y \in D \cap K\}$ *is bounded almost surely.*

In the proof of statement (1), we show that when $\{G_\alpha(y); y \in S\}$ is continuous we can find a continuous version of $\{L_t^y, (t, y) \in R_+ \times S\}$. As for the converse, we assume that $\{L_t^y, (t, y) \in R_+ \times S\}$ has a continuous version and show that its associated Gaussian process must be continuous.

Of course, statement (1) implies that if the Gaussian process is continuous on S, then the local time of the associated Markov process is continuous almost surely for all compact subsets K of S. However, the relationship between G and L is a local one. Thus, for example, all we need to know is that G is continuous on K to determine that $\{L_t^y, (t, y) \in R_+ \times K\}$ is continuous, and conversely. Most of the results in this section emphasize the local nature of the relationship between the local time of a Markov process and the Gaussian process associated with the Markov process.

As far as we know, there is no general theory that allows one to assume that the local time of a Markov process has a separable version. In the proof of statement (1) we construct such a version. However, we do not know how to do this, for example, for local times that are bounded but not necessarily continuous. For this reason, we consider the local time process $\{L_t^y, (t, y) \in R_+ \times D\}$, where D is some countable subset of S in statement (3). Since D is arbitrary, the results obtained are still quite strong.

Remark 9.4.2 To keep the statement of Theorem 9.4.1 from becoming even more cumbersome, it is given in terms of some fixed G_α. A more precise way to state (1) would be

(1a) If $\{L_t^y, (t, y) \in R_+ \times S\}$ is continuous almost surely, then $\{G_\alpha(y); y \in S\}$ is continuous almost surely for each $\alpha > 0$.

(1b) If $\{G_\alpha(y); y \in S\}$ is continuous almost surely for some $\alpha > 0$, $\{L_t^y, (t, y) \in R_+ \times S\}$ is continuous almost surely.

Clearly the same remark applies to statement (2) and to many of the statements made in the rest of this section.

Theorem 9.4.1 (1) provides an interesting fact about the family of Gaussian processes associated with a strongly symmetric Markov process X, namely, if G_α is continuous for some $\alpha > 0$, then the G_α are continuous for all $\alpha > 0$. If X is a Lévy process, this follows easily from Gaussian considerations, using Lemma 5.5.3, since, by (7.274), $\sigma_\alpha^2(h) \sim \sigma_\beta^2(h)$ as $h \to 0$ for all $\alpha, \beta \geq 0$. However, we do not know how to show this by Gaussian process theory in the general case. This remark also applies to many of the other path properties considered in this section.

In the next theorem we consider the behavior of the local time at a fixed point of S.

Theorem 9.4.3 *Let $D \subset S$ be countable and $y_0 \in D$. Consider the processes $\{L_t^y, (t, y) \in R_+ \times D\}$ and $\{G_\alpha(y); y \in D\}$. We have*

(1) L_t^y is continuous at y_0 for each $t > 0$, P^{y_0} almost surely, if and only if $G_\alpha(y)$ is continuous at y_0 almost surely.

(2) L_t^y has a bounded discontinuity at y_0 for each $t > 0$, P^{y_0} almost

surely, if and only if $G_\alpha(y)$ has a bounded discontinuity at y_0 almost surely.

(3) L_t^y is unbounded at y_0 for each $t > 0$, P^{y_0} almost surely, if and only if $G_\alpha(y)$ is unbounded at y_0 almost surely;

and for each $y_0 \in D$ precisely one of these three cases holds. Furthermore, this theorem remains valid with the term "each t" replaced by "some t" in (1)–(3).

Remark 9.4.4 By Corollary 5.3.6, continuity, boundedness, and unboundedness, both globally and locally, are probability 0 or 1 properties for Gaussian processes. Thus, by the above results, they are probability 0 or 1 properties for the local times of the associated Markov processes. However, a certain degree of care is necessary in expressing this phenomenon. For example, if a Gaussian process is unbounded almost surely on some compact set $K \subset S$, then there exists a point $y_0 \in K$ such that the process is unbounded almost surely at y_0. Roughly speaking, this implies that the local time of the associated Markov process will also be unbounded at y_0, but only if the Markov process hits y_0. Thus we can say that each of the events

$$\{L_t^y \text{ is continuous at } y_0 \text{ for each } t > 0\} \qquad (9.71)$$

$$\{L_t^y \text{ has a bounded discontinuity at } y_0 \text{ for each } t > 0\} \qquad (9.72)$$

$$\{L_t^y \text{ is unbounded at } y_0 \text{ for each } t > 0\} \qquad (9.73)$$

has P^{y_0} probability 0 or 1. Furthermore, these statements are also true with the term "each t" replaced by "some t."

To clarify some of the implications of Theorems 9.4.1 and 9.4.3 we give some of their immediate consequences in the next theorem.

Theorem 9.4.5 *Let $D \subset S$ be countable and let $K \subset S$ be a compact set. Then:*

(1) *Either $\{L_t^y, (t, y) \in R_+ \times K\}$ is continuous almost surely or else there exists an $x_0 \in K$ such that, for any countable dense set $D \subset K$, with $x_0 \in D$, the event "$\{L_t^y, (t, y) \in R_+ \times D\}$ is continuous" has P^{x_0} measure zero.*

(2) *Either $\{L_t^y, (t, y) \in [0, T] \times D \cap K\}$ is bounded for each $T < \infty$ almost surely or else there exists an $x_0 \in D \cap K$ such that the event "$\{L_t^y, (t, y) \in [0, T] \times D \cap K\}$ is bounded for some $T < \infty$" has P^{x_0} measure zero.*

(3) If $\{L_t^y, y \in K\}$ is continuous for some $t > 0$ almost surely, then $\{L_t^y, (t, y) \in R_+ \times K\}$ is continuous almost surely.

(4) If $\{L_t^y, y \in D \cap K\}$ is bounded for some $t > 0$ almost surely, then $\{L_t^y, (t, y) \in [0, T] \times D \cap K\}$ is bounded for each $T < \infty$, almost surely.

The next theorem shows that the continuity of the local time at each point in the state space almost surely implies that the local time is jointly continuous. (This is not the case for many processes. Let $\{Z_t; t \in [0, 1]\}$ be a Lévy process without continuous component. Z_t is continuous almost surely for each $t \in [0, 1]$, but except for some very simple examples, a Lévy process is discontinuous almost surely on [0,1].)

Theorem 9.4.6 L_t^y *is continuous at* y_0 *for each* $t > 0$ *almost surely for all* $y_0 \in K$ *if and only if* $\{L_t^y, (t, y) \in R_+ \times K\}$ *is continuous almost surely. Furthermore, this theorem remains valid with the term "each* t*" replaced by "some* t*."*

The information we need to prove these results is given in the next lemma. We note again that, by Corollary 5.3.6, the Gaussian properties referred to in this lemma have probability 0 or 1.

Lemma 9.4.7 *Let* $D \subset S$ *be countable and let* $K \subset S$ *be a compact set.*

(1) If $\{G_\alpha(y); y \in K\}$ has a continuous version, we can find a version of L that is continuous on $R_+ \times K$.

(2) If $\{G_\alpha(y); y \in K\}$ has a bounded version, then $\sup_{y \in D \cap K} L_t^y < \infty$ for all $t < \infty$ almost surely.

(3) If $\{G_\alpha(y); y \in K\}$ is unbounded, $\sup_{y \in D \cap K} L_t^y = \infty$, for all $t > 0$, P^{x_0} almost surely for some $x_0 \in K$.

(4) If G_α is continuous at $x_0 \in S$, then $\{L_t^y, y \in D \cup \{x_0\}\}$ is continuous at x_0 for all $t \geq 0$, P^{x_0} almost surely.

(5) If G_α has a bounded discontinuity at $x_0 \in S$, then $\{L_t^y, y \in D\}$ has a bounded discontinuity at x_0 for all $t > 0$, P^{x_0} almost surely.

(6) If G_α is unbounded at $x_0 \in S$, then $\{L_t^y, y \in D\}$ is unbounded at x_0 for all $t > 0$, P^{x_0} almost surely.

Proof Statement (1) follows from Theorem 9.3.3.

By (9.51), if $G_\alpha = \{G_\alpha(y); y \in K\}$ has a bounded version, then

$$\int_0^\infty E^x \left(\sup_{y \in D \cap K} L_t^y \right) e^{-\alpha t} \, dt < \infty. \qquad (9.74)$$

Therefore, there exists a sequence $t_i \uparrow \infty$ for which $E^x(\sup_{y \in D \cap K} L_{t_i}^y) < \infty$. Since L_t is increasing, this gives (2).

If G_α is unbounded on K, it follows from Corollary 5.3.8 (3) that there is a point $x_0 \in K$ for which

$$\lim_{\delta \to 0} \sup_{x \in C \cap B(x_0, \delta)} G_\alpha(x) = \infty \qquad \text{a.s.} \qquad (9.75)$$

Statement (3) now follows from Theorem 9.2.1. (To be more precise, in Theorem 9.2.1 we choose a generic point 0, which we take here to be x_0.)

Statements (4) and (5) follow from (9.12) since $\beta(x_0) = 0$ when G_α is continuous at x_0 and $0 < \beta(x_0) < \infty$ when G_α has a bounded discontinuity at x_0. Also, $L_t^{x_0} > 0$, P^{x_0} almost surely for all $t > 0$.

Statement (6) is actually what is proved in the proof of (3). $\qquad \square$

We now prove all the results in this section.

Proof of Theorem 9.4.1 Suppose that $\{G_\alpha(y); y \in K\}$ is continuous almost surely; then, by Lemma 9.4.7, $\{L_t^y, (t, y) \in R_+ \times K\}$ is continuous almost surely. Now suppose that $\{G_\alpha(y); y \in K\}$ is not continuous almost surely; then, by Theorem 5.3.7, there is an $x_0 \in K$ for which the oscillation function of $\{G_\alpha(y); y \in K\}$ is greater than zero. This implies, by (9.12), that L_t^y is not continuous at x_0, P^{x_0} almost surely for all $t > 0$. (Recall that the statement "$\{L_t^y, (t, y) \in R_+ \times K\}$ is continuous almost surely" means that it is P^x continuous almost surely, for all $x \in K$.) Therefore, $\{L_t^y, (t, y) \in R_+ \times K\}$ continuous almost surely implies that $\{G_\alpha(y); y \in K\}$ is continuous almost surely. This proves statement (2).

To obtain statement (1) suppose that $\{G_\alpha(y); y \in S\}$ is continuous almost surely; then, for all compact sets $K \subset S$, $\{G_\alpha(y); y \in K\}$ is continuous almost surely. Therefore, by statement (2) of this theorem, $\{L_t^y, (t, y) \in R_+ \times K\}$ is continuous almost surely. Since this holds for all $K \subset S$, $\{L_t^y, (t, y) \in R_+ \times S\}$ is continuous almost surely. The converse has exactly the same proof.

Statement (3) follows from statements (2) and (3) of Lemma 9.4.7. $\square$

Proof of Theorem 9.4.3 All these statements follow from statements (4), (5), and (6) of Lemma 9.4.7. It is clear that if these results hold for some t, they hold for each $t > 0$, because everything used in the proof is valid for all $t > 0$. $\qquad \square$

Proof of Theorem 9.4.5

(1) If $\{L_t^y, (t, y) \in R_+ \times K\}$ is not continuous almost surely, then

by Lemma 9.4.7, G_α is not continuous almost surely. Since this is a probability zero–one property of G_α, we are in the situation covered by statement (5) or (6) of Lemma 9.4.7.

(2) Same proof as (1) with obvious modifications.

(3) We use Theorem 9.4.3 (1) with each t replaced by some t, to see that $\{G_\alpha(y); y \in K\}$ is continuous almost surely. The statement now follows from Theorem 9.4.1 (1).

(4) We use both (1) and (2) of Theorem 9.4.3, with each t replaced by some t, to see that $\{G_\alpha(y); y \in K\}$ is bounded almost surely. The statement now follows from Theorem 9.4.1 (3). $\qquad\square$

Proof of Theorem 9.4.6 By (9.12), the oscillation function of the associated Gaussian process is zero. Hence the associated Gaussian process is continuous and so is the local time. $\qquad\square$

As corollaries of the results in this section we give some concrete continuity conditions for the local times of different strongly symmetric Borel right processes.

Corollary 9.4.8 *Let* $X = \{X(t), t \in R_+\}$ *be a real-valued symmetric Lévy process with characteristic exponent* $\psi(\lambda)$ *as given in (7.261) that satisfies (7.264). Let* $\{L_t^x(x, t); (x, t) \in R^1 \times R_+\}$ *denote the local times of* X *and let*

$$\sigma^2(x) = \int_0^\infty \frac{\sin^2(x/2)}{\psi(\lambda)} \, d\lambda. \tag{9.76}$$

Then $\{L_t^x(x, t); (x, t) \in R^1 \times R_+\}$ *is continuous almost surely, if and only if, for some* $\delta > 0$,

$$\int_0^\delta \frac{\overline{\sigma}(u)}{u(\log 1/u)^{1/2}} \, du < \infty, \tag{9.77}$$

where $\overline{\sigma}(u)$ *is the nondecreasing rearrangement of* $\sigma(u)$ *on* $[0, \delta]$.

Proof The α-potential of X is given by (7.265). Therefore, the increments variance of the associated Gaussian process, σ_α, is given by (7.266). It follows from Corollary 6.4.4 that

$$\int_0^\delta \frac{\overline{\sigma}_\alpha(u)}{u(\log 1/u)^{1/2}} \, du < \infty \tag{9.78}$$

for some $\delta > 0$ is a necessary and sufficient condition for the continuity of the associated Gaussian process on (R^1, σ_α). Furthermore, since σ_α is continuous in the Euclidean metric and is equal to zero only at zero,

(9.78) is a necessary and sufficient condition for the continuity of the associated Gaussian process on R^1 in the Euclidean metric. It now follows from Theorem 9.4.1 (1) that (9.78) is a necessary and sufficient condition for the continuity of $\{L_t^x(x,t); (x,t) \in R^1 \times R_+\}$. The proof is completed by noting that by (7.274), (9.77), and (9.78) are equivalent.

$\square$

The next corollary is not as precise as Corollary 9.4.8 but is easier to verify when its hypotheses are satisfied.

Corollary 9.4.9 *Let* $X = \{X(t), t \in R_+\}$ *be a real-valued symmetric Lévy process with characteristic exponent* $\psi(\lambda)$ *as given in (7.261) that satisfies (7.264). Let* $\{L_t^x(x,t); (x,t) \in R^1 \times R_+\}$ *denote the local times of* X. *Then* $\{L_t^x(x,t); (x,t) \in R^1 \times R_+\}$ *is continuous almost surely, if there exists an* x_0 *such that*

$$\int_{x_0}^{\infty} \frac{\left(\int_x^{\infty} \psi^{-1}(\lambda) \, d\lambda\right)^{1/2}}{x(\log x)^{1/2}} \, dx < \infty. \tag{9.79}$$

Furthermore, if there exists an x_0 *such that* ψ *is increasing on* $[x_0, \infty)$, *(9.79) is a necessary condition for the continuity of* $\{L_t^x(x,t); (x,t) \in R^1 \times R_+\}$.

Proof This follows from Theorem 6.4.10. $\square$

The next result is a simple sufficient condition for the continuity of local times of any Borel right process on R^n.

Corollary 9.4.10 *Let* $X = \{X(t), t \in R_+\}$ *be a strongly symmetric Borel right process with state space* R^n *and* α-potential density $u^\alpha(x,y)$ *for some* $\alpha > 0$. *Let* $\{L_t^x(x,t); (x,t) \in [-T,T]^n \times R_+\}$ *denote the local times of* X *on the interval* $[-T,T]^n$. *Let*

$$\rho(\delta) := \sup_{\substack{|x-y| \leq \delta \\ x,y \in [-T,T]^n}} \left(u^\alpha(x,x) + u^\alpha(y,y) - 2u^\alpha(x,y)\right)^{1/2}. \tag{9.80}$$

Then $\{L_t^x(x,t); (x,t) \in [-T,T]^n \times R_+\}$ *is continuous almost surely, if, for some* $\delta' > 0$,

$$\int_0^{\delta'} \frac{\rho(u)}{u(\log 1/u)^{1/2}} \, du < \infty. \tag{9.81}$$

Proof This corollary follows immediately from Lemma 6.4.6 and Theorem 9.4.1 (2). $\square$

The next theorem gives our most general result on necessary and sufficient conditions for the continuity and boundedness of the local times of strongly symmetric Borel right processes.

Theorem 9.4.11 *Let $X = \{X(t), t \in R_+\}$ be a strongly symmetric Borel right process with continuous α-potential density $u^\alpha(x, y)$ for some $\alpha > 0$ and a locally compact separable state space (S, d_X), where*

$$d_X(x, y) = (u^\alpha(x, x) + u^\alpha(y, y) - 2u^\alpha(x, y))^{1/2}. \qquad (9.82)$$

Let K be a compact subset of S and let D denote the diameter of K. Let $\{L_t^x; (x, t) \in K \times R_+\}$ denote the local times of X on $K \times R_+$.

Let $K' \subset K$ be countable. Then $\{L_t^x; (x, t) \in K' \cap K \times [0, T]\}$ is bounded almost surely for all $T < \infty$, if and only if there exists a probability measure μ on K such that

$$\sup_{y \in K} \int_0^D \left(\log \frac{1}{\mu(B_{d_X}(y, u))} \right)^{1/2} du < \infty. \qquad (9.83)$$

Furthermore, $\{L_t^x; (x, t) \in K \times R_+\}$ has bounded uniformly continuous sample paths if and only if there exists a probability measure μ on K such that (9.83) holds and, in addition,

$$\lim_{\epsilon \to 0} \sup_{y \in K} \int_0^\epsilon \left(\log \frac{1}{\mu(B_{d_X}(y, u))} \right)^{1/2} du = 0. \qquad (9.84)$$

Proof Let $G_\alpha = \{G_\alpha(x), x \in K\}$ be a mean zero Gaussian process with $(E(G_\alpha(x) - G_\alpha(y))^2)^{1/2} = d_X(x, y)$. By Theorem 9.4.1, these statements hold if and only if G_α has bounded sample paths or G_α is continuous. It follows from Theorems 6.3.1 and 6.3.4 that (9.83) is a necessary and sufficient condition for G_α to be bounded and from Theorems 6.3.1 and 6.3.5 that (9.83) and (9.84) give necessary and sufficient conditions for G_α to be continuous (we also use Corollary 5.4.7). □

9.5 Moduli of continuity

In this section $X = (\Omega, \mathcal{G}, \mathcal{G}_t, X_t, \theta_t, P^x)$ is a strongly symmetric Borel right process on a locally compact metric space (S, τ) with continuous α-potential density $u^\alpha(x, y)$. $L = \{L_t^y; (y, t) \in S \times R_+\}$ denotes the local times of X normalized by (3.91). $G_\alpha = \{G_\alpha(y); y \in S\}$ is a real-valued Gaussian process with mean zero and covariance $u^\alpha(x, y)$. We obtain local and uniform moduli of continuity for L in terms of the corresponding quantities for G_α.

Theorem 9.5.1 *For $\alpha > 0$, let ρ_α be an exact local modulus of continuity for $G_\alpha = \{G_\alpha(y), y \in S\}$ at $y_0 \in S$ (i.e., (7.2) holds). Let $\mathcal{C}$ be a countable separating set for G_α. Then*

$$\lim_{\delta \to 0} \sup_{\substack{\tau(y,y_0) \le \delta \\ y \in \mathcal{C}}} \frac{|L_t^y - L_t^{y_0}|}{\rho_\alpha(\tau(y,y_0))} = \sqrt{2}\, C_\alpha \left(L_t^{y_0}\right)^{1/2} \quad \text{for almost all } t \text{ a.s.,}$$

$$(9.85)$$

where C_α is the same constant as in (7.2). If ρ_α is simply a local modulus of continuity for G_α, (9.85) holds with the equality sign replaced by a "less than or equal to sign" and if ρ_α is a lower local modulus of continuity for G_α, (9.85) holds with the equality sign replaced by a "greater than or equal to sign."

Proof We use Lemma 9.1.2. Let

$$B = \left\{ f \in F(\mathcal{C}) \,\middle|\, \lim_{\delta \to 0} \sup_{\substack{\tau(y,y_0) \le \delta \\ y \in \mathcal{C}}} \frac{|f(y) - f(y_0)|}{\rho_\alpha(\tau(y,y_0))} = \sqrt{2}\, C_\alpha |f(y_0)|^{1/2} \right\}.$$

$$(9.86)$$

Then, using Theorem 7.7.1 on $G_\alpha(\,\cdot\,) + s$, we see that

$$P\left((G_\alpha(\,\cdot\,) + s)^2 / 2 \in B \right) = 1. \tag{9.87}$$

It follows from the Lemma 9.1.2 that for almost all $\omega' \in \Omega_{G_\alpha}$, with respect to P_{G_α},

$$\lim_{\delta \to 0} \sup_{\substack{\tau(y,y_0) \le \delta \\ y \in \mathcal{C}}} \left| \frac{L_t^y - L_t^{y_0}}{\rho_\alpha(\tau(y,y_0))} + \frac{(G_\alpha(y,\omega') + s)^2 - (G_\alpha(y_0,\omega') + s)^2}{2\rho_\alpha(\tau(y,y_0))} \right| \quad (9.88)$$

$$= \sqrt{2}\, C_\alpha \left(L_t^{y_0} + \frac{(G_\alpha(y_0,\omega') + s)^2}{2} \right)^{1/2} \quad \text{for almost all } t \in R_+$$

P^x almost surely, for all $x \in S$. Writing

$$(G_\alpha(y,\omega') + s)^2 - (G_\alpha(y_0,\omega') + s)^2 \tag{9.89}$$

$$= G_\alpha^2(y,\omega') - G_\alpha^2(y_0,\omega') + 2s(G_\alpha(y,\omega') - G_\alpha(y_0,\omega'))$$

and using Theorem 7.7.1 again, we see that for almost all $\omega' \in \Omega_{G_\alpha}$, with respect to P_{G_α},

$$\lim_{\delta \to 0} \sup_{\substack{\tau(y,y_0) \le \delta \\ y \in \mathcal{C}}} \left| \frac{(G_\alpha(y,\omega') + s)^2 - (G_\alpha(y_0,\omega') + s)^2}{2\rho_\alpha(\tau(y,y_0))} \right| \tag{9.90}$$

$$\le C_\alpha \left(|G_\alpha(y_0,\omega')| + 2s \right).$$

As in the proof of Theorem 9.1.3, we can take $|G_\alpha(y_0,\omega')|$ arbitrarily close to zero on a set of positive probability, and, of course, we can also take s arbitrarily close to zero. Using this in (9.88) along with the triangle inequality, we get (9.85).

The proof of the one-sided results when ρ_α is either a local or lower local modulus of continuity is simply a one-sided version of the proof of (9.85). $\square$

Remark 9.5.2 In Theorem 9.5.1, if G_α is continuous on a compact set $K \subset S$ that contains a neighborhood of y_0, it follows from Theorem 9.4.1 (2) that there is a version of L that is continuous on K. In this case, using the continuity of L, we obtain (9.85) with the lim sup taken on $y \in K$, provided that $\rho(y, y_0)$ is also continuous in y. Consequently, we can extend (9.85) from $\mathcal{C}$ to S. In particular, for real-valued Lévy processes, the associated Gaussian processes have stationary increments; therefore, if they are continuous at a point, they are continuous on all of R^1 (see Theorem 5.3.10). Thus we can always make the above extension for the local times of Lévy processes.

The hypotheses of the next three theorems imply that the Gaussian process is continuous on a compact set $K \subset S$. As we just remarked, it follows from Theorem 9.4.1 (2) that there is a version of L that is continuous on K. The theorems refer to this version. The proofs of the next three theorems are very similar to the proof of Theorem 9.5.1.

Theorem 9.5.3 *Let $K \subset S$ be compact. For $\alpha > 0$, let ω_α be a uniform modulus of continuity for $\{G_\alpha(y), y \in K\}$ (i.e., (7.1) holds with a "less than or equal to" sign). Then*

$$\lim_{\delta \to 0} \sup_{\substack{\tau(x,y) \leq \delta \\ x,y \in K}} \frac{|L_t^x - L_t^y|}{\omega_\alpha(\tau(x,y))} \leq \sqrt{2}\, C_\alpha \sup_{y \in K}(L_t^y)^{1/2} \quad \text{for almost all } t \text{ a.s.,}$$

(9.91)

where C_α is the same constant as in (7.1).

Proof As in the proof of Theorem 9.5.1, we use Lemma 9.1.2. Let $\mathcal{C}$ be a countable separating set for G_α. Let

$$B' = \left\{ f \in F(\mathcal{C}) \Big| \lim_{\delta \to 0} \sup_{\substack{\tau(x,y) \leq \delta \\ x,y \in \mathcal{C}}} \frac{|f(x) - f(y)|}{\omega_\alpha(\tau(x,y))} \leq \sqrt{2}\, C_\alpha \sup_{x \in K} |f(x)|^{1/2} \right\}.$$

(9.92)

Then, by Theorem 7.7.3,

$$P\left((G_\alpha(\cdot) + s)^2 / 2 \in B' \right) = 1.$$

(9.93)

It follows from the Lemma 9.1.2 that for almost all $\omega' \in \Omega_{G_\alpha}$, with respect to P_{G_α},

$$\lim_{\delta \to 0} \sup_{\substack{\tau(x,y) \leq \delta \\ x,y \in \mathcal{C}}} \left| \frac{L_t^x - L_t^y}{w_\alpha(\tau(x,y))} + \frac{(G_\alpha(x,\omega')+s)^2 - (G_\alpha(y,\omega')+s)^2}{2w_\alpha(\tau(x,y))} \right| \quad (9.94)$$

$$\leq \sqrt{2}\, C_\alpha \sup_{z \in K} \left(L_t^z + \frac{(G_\alpha(z,\omega')+s)^2}{2} \right)^{1/2} \quad \text{for almost all } t \in R_+$$

P^x almost surely, for all $x \in S$. Using Theorem 7.7.3 again, we see that for almost all $\omega' \in \Omega_{G_\alpha}$, with respect to P_{G_α},

$$\lim_{\delta \to 0} \sup_{\substack{d(x,y) \leq \delta \\ x,y \in \mathcal{C}}} \left| \frac{(G_\alpha(x,\omega')+s)^2 - (G_\alpha(y,\omega')+s)^2}{2w_\alpha(\tau(x,y))} \right| \quad (9.95)$$

$$\leq C_\alpha \left(\sup_{z \in K} |G_\alpha(z,\omega')| + 2s \right).$$

By Lemma 5.3.5 we can take $\sup_{z \in K} |G_\alpha(z,\omega')|$ arbitrarily close to zero on a set of positive measure. This enables us to complete the proof. $\square$

The next theorem shows that if w_α is an exact uniform modulus of continuity for G_α, then it is the "best possible" in (9.91).

Theorem 9.5.4 *Let $K \subset S$ be compact. For $\alpha > 0$ let w_α be an exact uniform modulus of continuity for $\{G_\alpha(y), y \in K\}$ (i.e., (7.1) holds). Then there exists a $y_0 \in K$ such that*

$$\lim_{\delta \to 0} \sup_{\substack{\tau(x,y) \leq \delta \\ x,y \in K}} \frac{|L_t^x - L_t^y|}{w_\alpha(\tau(x,y))} \geq \sqrt{2}\, C_\alpha \, (L_t^{y_0})^{1/2} \quad \text{for almost all } t \text{ a.s.,} \quad (9.96)$$

where C_α is the same constant as in (7.1).

Proof We use the fact that (7.1) holds with a "greater than or equal to" sign and the ideas in the beginning of the proof of Theorem 9.5.3 along with Lemma 7.7.4 to get (analogously to (9.94)) that

$$\lim_{\delta \to 0} \sup_{\substack{\tau(x,y) \leq \delta \\ x,y \in \mathcal{C}}} \left| \frac{L_t^x - L_t^y}{w_\alpha(\tau(x,y))} + \frac{(G_\alpha(x,\omega')+s)^2 - (G_\alpha(y,\omega')+s)^2}{2w_\alpha(\tau(x,y))} \right| \quad (9.97)$$

$$\geq \sqrt{2}\, C_\alpha \left(L_t^{y_0} + \frac{(G_\alpha(y_0,\omega')+s)^2}{2} \right)^{1/2} \quad \text{for almost all } t \in R_+$$

P^x almost surely, for some $y_0 \in K$. (We add the point y_0, which is designated by Lemma 7.7.4, to $\mathcal{C}$.) We use the fact that (7.1) holds with a "less than or equal to" sign and Theorem 7.7.3 to control the term in G_α, in the left-hand side of (9.97). This enables us to complete the proof as in the previous two theorems. □

In order to get the result we want for the exact uniform modulus of continuity, we need (S, d_{G_α}) to be locally homogeneous.

Theorem 9.5.5 *Assume that (S, d_{G_α}) is a locally homogeneous metric space, and let $K \subset S$ be a compact set that is the closure of its interior. For $\alpha > 0$ let ω_α be an exact uniform modulus of continuity for $\{G_\alpha(y), y \in K\}$ (i.e., (7.1) holds with $\tau(u,v) = d_{G_\alpha}(u,v)$). Then*

$$\lim_{\delta \to 0} \sup_{\substack{d_{G_\alpha}(x,y) \le \delta \\ x,y \in K}} \frac{|L_t^x - L_t^y|}{\omega_\alpha(d_{G_\alpha}(x,y))} = \sqrt{2}\, C_\alpha \sup_{y \in K}(L_t^y)^{1/2} \quad \text{for almost all } t \text{ a.s.,}$$

(9.98)

where C_α is the same constant as in (7.1).

Proof Let B'' be the set defined in (9.92) but with an equal sign. By Theorem 7.7.6, the equality in (9.93) holds for B'' and we get (9.94) with an equal sign. The rest of the proof is the same as the proof of Theorem 9.5.3. □

In dealing with local times it is natural to express the moduli with respect to Euclidean distance on R^1. We can do this when the G_α are Gaussian processes with stationary increments.

Theorem 9.5.6 *Assume that G_α is a real-valued Gaussian process with stationary increments. Let $K \subset R^1$ be an interval. For $\alpha > 0$ let ω_α be an exact uniform modulus of continuity for $\{G_\alpha(y), y \in K\}$ (i.e., (7.1) holds with $\tau(u,v) = |u - v|$). Then*

$$\lim_{\delta \to 0} \sup_{\substack{|x-y| \le \delta \\ x,y \in K}} \frac{|L_t^x - L_t^y|}{\omega_\alpha(|x-y|)} = \sqrt{2}\, C_\alpha \sup_{y \in K}(L_t^y)^{1/2} \quad \text{for almost all } t \text{ a.s.,}$$

(9.99)

where C_α is the same constant as in (7.1).

Proof By hypothesis, Theorem 7.7.4 holds with $\tau(u,v) = |u - v|$. Because G_α has stationary increments, $d_{G_\alpha}(x,y)$ is a function of $|x - y|$. Therefore, the neighborhoods $B_{d_{G_\alpha}}(u_0, \epsilon)$ are isometric for all $u_0 \in K$. Consequently, the proof of Theorem 7.7.5 and hence of Theorem 7.7.6 goes through exactly as written. This is all we need, as in the proof of Theorem 9.5.5. □

Remark 9.5.7 It follows from Remark 7.7.7 that if ω_α or ρ_α is an m-modulus, Theorems 9.5.1–9.5.6 hold with $\omega_\alpha(\tau(\,\cdot\,,\,\cdot\,))$ replaced by $\omega_\alpha(\delta)$ and with $\lim_{\delta\to 0}$ replaced by $\limsup_{\delta\to 0}$, and similarly for ρ_α.

Remark 9.5.8 The fact that we can only obtain Theorems 9.5.1–9.5.6 for almost all t, rather than for all t, is a weakness of our method. What we actually obtain in the critical Lemma 9.1.2 is (9.10), in which λ is an exponential random variable. In using Fubini's Theorem we can only get (9.7) for almost all $t \in R_+$.

Using Theorems 9.5.1–9.5.6 and material in Chapter 7, we now give some specific results about the modulus of continuity of local times. We begin with a general result for Borel right processes on R^1. For X and G_α as in the first paragraph of this section, we define the metric

$$
\begin{aligned}
d_\alpha(x,y) &:= \left(E(G_\alpha(x) - G_\alpha(y))^2\right)^{1/2} \\
&= \left(u^\alpha(x,x) + u^\alpha(y,y) - 2u^\alpha(x,y)\right)^{1/2}.
\end{aligned}
\tag{9.100}
$$

Theorem 9.5.9 *Let X be a strongly symmetric Borel right process on R^1 with α-potential density $u^\alpha(x,y)$, $\alpha > 0$ and local time $L = \{L_t^y \,;\, (y,t) \in R^1 \times R_+\}$. Let $y_0 \in R^1$ and $\phi_\alpha(\,\cdot\,)$ be an increasing function such that:*

(1) There exist constants $0 < C_0 \le C_1 < \infty$ for which

$$
C_0 \phi_\alpha(|h|) \le d_\alpha(y, y+h) \le C_1 \phi_\alpha(|h|)
\tag{9.101}
$$

for all y in some neighborhood of y_0 and all $|h|$ sufficiently small.

(2) $\phi_\alpha^2(2^{-n}) - \phi_\alpha^2(2^{-n-1})$ is nonincreasing in n for all n sufficiently large.

(3) $\phi_\alpha(2^{-n}) \le 2^\beta \phi_\alpha(2^{-n-1})$ for all n sufficiently large for some $\beta < 1$.

Set

$$
\rho_\alpha(h) = \phi_\alpha(h)\,(\log\log 1/h)^{1/2} + \int_0^{1/2} \frac{\phi_\alpha(hu)}{u(\log 1/u)^{1/2}}\,du
\tag{9.102}
$$

and assume that the integral is finite for $h \in [0,\delta]$ for some $\delta > 0$. Then

$$
\lim_{\delta\to 0}\ \sup_{\substack{|y-y_0|\le\delta \\ y\in\mathcal{C}}} \frac{|L_t^y - L_t^{y_0}|}{\rho_\alpha(|y-y_0|)} = C(y_0,\alpha)\,(L_t^{y_0})^{1/2} \quad \textit{for almost all } t \textit{ a.s.,}
$$

$$
\tag{9.103}
$$

where $C(y_0,\alpha) > 0$ is a constant depending on y_0 and α, and $\mathcal{C}$ is any countable separating set for G_α.

Furthermore, (9.103) also holds with $\rho_\alpha(|y-y_0|)$ replaced by $\rho_\alpha(\delta)$, $\lim_{\delta\to 0}$ replaced by $\limsup_{\delta\to 0}$, and $C(y_0,\alpha)$ replaced by a finite constant $C'(y_0,\alpha) \ge C(y_0,\alpha)$.

Proof It follows from Lemma 6.4.6 and (7.126) that G is continuous in a neighborhood of y_0. Therefore, by Theorem 7.6.4 and Remark 9.5.7, ρ_α is both an exact local modulus and an exact local m-modulus of continuity of the associated Gaussian process G_α. The results now follow from Theorem 9.5.1. □

Remark 9.5.10 By Remark 9.5.2, if L, or equivalently G_α, is continuous on a compact neighborhood of y_0, we can replace $\mathcal{C}$ by R^1 in (9.103).

Using (7.274), we can simplify Theorem 9.5.9 when X is a symmetric Lévy process. Recall the function $\sigma_0^2(h)$, defined on page 330:

$$\sigma_0^2(h) = \frac{4}{\pi} \int_0^\infty \sin^2 \frac{\lambda h}{2} \frac{1}{\psi(\lambda)} \, d\lambda. \tag{9.104}$$

Theorem 9.5.11 *Let X be a symmetric Lévy process as defined in (7.261) satisfying (7.264) with local time $L = \{L_t^y \,;\, (y, t) \in R^1 \times R_+\}$. Let $y_0 \in R^1$ and $\phi(\cdot)$ be an increasing function such that:*

(1) There exist constants $0 < C_0 \leq C_1 < \infty$ for which

$$C_0 \phi(|h|) \leq \sigma_0(h) \leq C_1 \phi(|h|) \tag{9.105}$$

for all y in some neighborhood of y_0 and all $|h|$ sufficiently small.
(2) $\phi^2(2^{-n}) - \phi^2(2^{-n-1})$ is nonincreasing in n for all n sufficiently large.
(3) $\phi(2^{-n}) \leq 2^\beta \phi(2^{-n-1})$ for all n sufficiently large for some $\beta < 1$.
Set

$$\rho(h) = \phi(h) \left(\log \log 1/h\right)^{1/2} + \int_0^{1/2} \frac{\phi(hu)}{u(\log 1/u)^{1/2}} \, du \tag{9.106}$$

and assume that the integral is finite for $h \in [0, \delta]$ for some $\delta > 0$. Then

$$\lim_{\delta \to 0} \sup_{\substack{|y - y_0| \leq \delta \\ y \in R^1}} \frac{|L_t^y - L_t^{y_0}|}{\rho(|y - y_0|)} = C \, (L_t^{y_0})^{1/2} \quad \text{for almost all } t \text{ a.s.} \tag{9.107}$$

for some constant $0 < C < \infty$.
Furthermore, (9.107) also holds with $\rho(|y - y_0|)$ replaced by $\rho(\delta)$, $\lim_{\delta \to 0}$ replaced by $\lim\sup_{\delta \to 0}$, and C replaced by a constant $C \leq C' < \infty$.

Note that, in general, it is easier to estimate the integral that expresses σ_0 than the corresponding integrals for σ_α, $\alpha > 0$.

Proof Consider X killed at the end of an independent exponential time with mean 1. If one could replace σ_0 by σ_1 in (9.105), this theorem would follow immediately from Theorem 9.5.9. By (7.274), we can do precisely this. □

Example 9.5.12 The conditions on ϕ in Theorem 9.5.11 are the same as those in Theorem 7.6.4, and Φ_1 in Example 7.6.6 is the same as ρ in (9.106). Thus, everything in Example 7.6.6 carries over to ϕ and ρ in Theorem 9.5.11. However, they are of no interest unless we can find functions σ_0 for which (9.105) holds for these functions ϕ. In Example 7.6.6 (1), ϕ is a regularly varying function of index $0 < \alpha < 1$. By (7.279) and (7.283), we can find σ_0 asymptotic to any regularly varying function of index $0 < \alpha < 1$.

In Example 7.6.6 (2)–(4), ϕ is slowly varying at zero. Thus, by (7.281) it is asymptotic to $C \int_{1/h}^{\infty} \psi^{-1}(\lambda) \, d\lambda$ at zero. By (7.283), we can take ψ asymptotic to any regularly varying function of index one at infinity. Consequently, we can find functions σ_0 asymptotic at zero, to the ϕ in (2)–(4).

We illustrate this in the case of Example 7.6.6 (2). Suppose that $\phi(h) = \exp(-g(\log 1/h))$ for $h \in [0, h_0]$ for some $h_0 > 0$, where g is a differentiable normalized regularly varying functions at infinity of index $0 < \delta < 1$. Set

$$
\begin{aligned}
\phi(h) &= \int_{1/h}^{\infty} d \exp(-g(\log \lambda)) \qquad (9.108) \\
&= \int_{1/h}^{\infty} \frac{g'(\log \lambda)}{\lambda} \exp(-g(\log \lambda)) \, d\lambda.
\end{aligned}
$$

By (7.283), we can take ψ asymptotic to $\frac{\lambda}{g'(\log \lambda)} \exp(g(\log \lambda))$ at infinity.

The next theorem gives an iterated logarithm law for the local times of Lévy processes with the precise value of the constant under minimal conditions.

Theorem 9.5.13 *Let X be a symmetric Lévy process as defined in (7.261) satisfying (7.264), and let $\sigma_0(u)$ be as given in (9.104). Let $L = \{L_t^y ; (y, t) \in R^1 \times R_+\}$ denote the local times of X. Let $\sigma_0^*(u) = \sup_{s \le u} \sigma_0(s)$. Assume that:*

(1) For all $\epsilon > 0$ there exists $\theta > 1$ such that $\sigma_0^*(\theta u) \le (1 + \epsilon)\sigma_0^*(u)$, $|u| \le u_0$ for some $u_0 > 0$ and

$$
\int_0^{1/2} \frac{\sigma_0^*(\delta u)}{u(\log 1/u)^{1/2}} \, du = o\left(\sigma_0^*(\delta) (\log \log 1/\delta)^{1/2}\right). \qquad (9.109)
$$

(2) $\sigma_0^*(|u|) = O(|u|^{\alpha})$ for some $\alpha > 0$.

Then, for all y_0 in R^1,

$$\lim_{\delta \to 0} \sup_{\substack{|y-y_0| \le \delta \\ y \in R^1}} \frac{|L_t^y - L_t^{y_0}|}{\sigma_0^*(|y-y_0|) \, (\log \log 1/|y-y_0|)^{1/2}} = 2 \, (L_t^{y_0})^{1/2} \qquad (9.110)$$

for almost all t almost surely. Indeed, without condition (1), the lower bound in (9.110) still holds, and without requiring condition (2) and the first half of condition (1), replacing 2 by 4 gives an upper bound.

Furthermore, this theorem remains valid with $\sigma_0^*(|y-y_0|)(\log \log 1/ |y-y_0|)^{1/2}$ in (9.110) replaced by $\sigma_0^*(\delta) \, (\log \log 1/\delta)^{1/2}$ and with $\lim_{\delta \to 0}$ replaced by $\lim \sup_{\delta \to 0}$.

Proof As in the proof of Theorem 9.5.11, we consider X killed at the end of an independent exponential time with mean 1 and the associated Gaussian process, which has increments variance σ_1^2. By (7.274), Lemma 7.4.4, and Theorem 9.5.1, we get the lower bound in (9.110) with σ_0^* replaced by σ_1^*. Using (7.274) again, we can replace σ_1^* by σ_0^*.

We use the same procedure for the upper bound along with Lemma 7.1.6, Corollary 7.2.3, Theorem 9.5.1, and Remark 9.5.2.

The statement about replacing 2 by 4 follows from the proof of Corollary 7.2.3 except that (7.103) is used in the interpolation of (7.102).

The statement about the m-modulus also uses Lemma 7.1.6. $\qquad\square$

Note that condition (1) is satisfied when σ_0^* is a regularly varying function of index $0 < \beta \le 1$.

Remark 9.5.14 A portion of Theorem 9.5.13 holds in great generality. Let X be a strongly symmetric Borel right process with continuous zero potential density $u(x, y)$ and set

$$\sigma^*(y, y_0) = \sup_{y_0 \le s \le y} (u(s, s) - u(y_0, y_0)). \qquad (9.111)$$

If $\sigma^*(y, y_0) > 0$ for $y > y_0$, it follows from Remark 7.4.6 that

$$\lim_{\delta \downarrow 0} \sup_{\substack{y_0 < y \le y_0 + \delta \\ y \in \mathcal{C}}} \frac{|L_t^y - L_t^{y_0}|}{\sigma^*(y, y_0) \, (\log \log 1/\sigma^*(y, y_0))^{1/2}} \ge 2 \, (L_t^{y_0})^{1/2} \qquad (9.112)$$

for almost all $t \in R_+$ almost surely, for all countable sets $\mathcal{C} \subset R^1$.

We now consider uniform moduli of continuity.

Theorem 9.5.15 *Let X be a symmetric Lévy process as defined in (7.261) satisfying (7.264) and let $\sigma_0(u)$ be as given in (9.104). Let*

$L = \{L_t^y \; ; \; (y,t) \in R^1 \times R_+\}$ *denote the local times of* X. *Consider the following conditions:*

(1) $\sigma_0^2(h)$ *is a normalized regularly varying function at zero with index* $0 < \alpha < 1$ *or a normalized regularly varying function of type A.*

(2) $\sigma_0^2(h)$ *is a normalized slowly varying function at zero that is asymptotic to an increasing function near zero and*

$$\int_0^{1/2} \frac{\sigma_0(\delta u)}{u(\log 1/u)^{1/2}} \, du = o\left(\sigma_0(\delta)(\log 1/\delta)^{1/2}\right). \qquad (9.113)$$

(3) $\sigma_0^2(u)$ *is concave for* $0 \leq u \leq h$ *for some* $h > 0$, *and* (9.113) *holds.*

Let I *be a closed interval in* R^1. *Then, if any of the conditions* (1)–(3) *hold*

$$\lim_{\delta \to 0} \sup_{\substack{|x-y| \leq \delta \\ x,y \in I}} \frac{|L_t^x - L_t^y|}{\sigma_0(|x-y|)(\log 1/|x-y|)^{1/2}} = 2 \sup_{y \in I}(L_t^y)^{1/2} \qquad (9.114)$$

for almost all t *almost surely .*

Furthermore, this theorem remains valid with $\sigma_0(|x-y|)(\log 1/|x-y|)^{1/2}$ *in* (9.114) *replaced by* $\sigma_0(\delta)(\log 1/\delta)^{1/2}$ *and with* $\lim_{\delta \to 0}$ *replaced by* $\limsup_{\delta \to 0}$.

Proof It follows from Theorem 7.2.14 that $(2\sigma_0(|x-y|) \log 1/|x-y|)^{1/2}$ is an exact uniform modulus of continuity for the Gaussian process $\widetilde{G}_0$ defined on page 330. By (7.287),

$$\widetilde{G}_0(x) - \widetilde{G}_0(y) \stackrel{law}{=} \widetilde{G}_1(x) - \widetilde{G}_1(y) + H_1(x) - H_1(y) \qquad (9.115)$$

as stochastic processes on $I \times I$ ($\widetilde{G}_1$ and H_1 are defined in Lemma 7.4.11). By (7.288) and Theorem 7.2.1, the uniform modulus of continuity of H_1 is "little o" of the uniform modulus of continuity of $\widetilde{G}_0$. Therefore, by the triangle inequality, $(2\sigma_0(|x-y|) \log 1/|x-y|)^{1/2}$ is also an exact uniform modulus of continuity for $\widetilde{G}_1$. Using this in Theorem 9.5.6, we get (9.114). The last statement in the theorem follows from Remark 9.5.7. □

Condition (3) is interesting. Every function that is concave near zero can be the increments variance of a Gaussian process with stationary increments, but it is not so easy to see whether associated Gaussian processes have this property. One class of associated processes, for which it is easy to see that σ_0 is concave, is the class of symmetric stable processes of index $1 < p < 2$. We point this out in Example 7.4.13. Because of the significance of stable processes, we give this special case of Theorems 9.5.13 and 9.5.15 in the next example. In the next section we introduce what we call stable mixtures, which also have associated Gaussian processes with concave σ_0.

Example 9.5.16 *Let X be a symmetric stable process of index $1 < p < 2$ with local time $L = \{L_t^y \, ; \, (y, t) \in R^1 \times R_+\}$. Then, for all $y_0 \in R^1$,*

$$\lim_{\delta \to 0} \sup_{\substack{(|y - y_0| \le \delta) \\ y \in R^1}} \frac{|L_t^y - L_t^{y_0}|}{(|y - y_0|^{p-1} \log \log 1/|y - y_0|)^{1/2}} = (2C_p L_t^{y_0})^{1/2} \quad (9.116)$$

for almost all $t \in R_+$ almost surely, and

$$\lim_{\delta \to 0} \sup_{\substack{|x - y| \le \delta \\ x, y \in I}} \frac{|L_t^x - L_t^y|}{(|x - y|^{p-1} \log 1/|x - y|)^{1/2}} = \left(2C_p \sup_{y \in I} L_t^y\right)^{1/2} \quad (9.117)$$

for almost all $t \in R^+$ almost surely. If X is standard Brownian motion, both (9.116) and (9.117) hold with $p = 2$ (C_p is given in (7.280); see also Example 7.4.13).

Furthermore, (9.117) remains valid with $|x - y|^{p-1}(\log 1/|x - y|)^{1/2}$ replaced by $\delta^{p-1}(\log 1/\delta)^{1/2}$ and similarly for (9.116) and with $\lim_{\delta \to 0}$ replaced by $\lim \sup_{\delta \to 0}$.

Condition (1) of Theorem 9.5.15 is also difficult to work with. The hypothesis of the next corollary is easier to verify.

Corollary 9.5.17 *Let X be a symmetric Lévy process as defined in (7.261) with Lévy exponent ψ. Let $L = \{L_t^y \, ; \, (y, t) \in R^1 \times R_+\}$ denote the local times of X. Assume that ψ is regularly varying at infinity with index $1 < p < 2$ or is regularly varying at infinity with index 2 and $\psi(\lambda)/\lambda^2$ is nonincreasing as $\lambda \to \infty$. Then (9.114) holds, with $\sigma_0(u)$ as given in (9.104).*

Lemma 7.4.10, shows that $\psi(\lambda)$ can be taken to be asymptotic to any regularly varying function of index $1 < p < 2$ or of index 2 as long as $\psi(\lambda)/\lambda^2$ is nonincreasing as $\lambda \to \infty$.

Proof Consider the stationary Gaussian process G_1 defined on page 330, with spectral density $(1 + \psi(\lambda))^{-1}$. By Remark 7.3.4, (7.223), with equality, holds for G_1. Therefore the proof follows from Theorem 9.5.6 and (7.274). □

For a rather narrow class of processes with continuous local times a uniform modulus of the type given in (9.114) in not large enough. We consider this in the next theorem.

Theorem 9.5.18 *Let X be a strongly symmetric Borel right process on R^1 with α-potential density $u^\alpha(x,y)$, $\alpha > 0$ and local time $L = \{L_t^y ; (y,t) \in R^1 \times R_+\}$. Let $K \subset R^1$ be an interval. Let $\phi_\alpha(\cdot)$ be an increasing function such that:*

(1) There exist constants $0 < C_0 \le C_1 < \infty$ for which

$$C_0 \phi_\alpha(|x - y|) \le d_\alpha(x,y) \le C_1 \phi_\alpha(|x - y|) \tag{9.118}$$

for all $x, y \in K$ and all $|h|$ sufficiently small.

(2) $\phi_\alpha^2(2^{-n}) - \phi_\alpha^2(2^{-n-1})$ is nonincreasing in n for all n sufficiently large.

(3) $\phi_\alpha(2^{-n}) \le 2^\beta \phi_\alpha(2^{-n-1})$ for all n sufficiently large for some $\beta < 1$.

Let

$$\Phi_{2,\alpha}(h) = \int_0^h \frac{\phi_\alpha(u)}{u(\log 1/u)^{1/2}} \, du \tag{9.119}$$

and assume that $\phi_\alpha(h)(\log 1/h)^{1/2} = O(\Phi_{2,\alpha}(h))$; then there exist constants $0 < C_\alpha < \infty$ such that

$$\lim_{\delta \to 0} \sup_{\substack{|x-y| \le \delta \\ x,y \in K}} \frac{|L_t^x - L_t^y|}{\Phi_{2,\alpha}(|x - y|)} = C_\alpha \sup_{y \in K} (L_t^y)^{1/2} \tag{9.120}$$

for almost all t almost surely.

Furthermore, (9.103) also holds with $\Phi_{2,\alpha}(|x - y|)$ replaced by $\Phi_{2,\alpha}(\delta)$ and with $\lim_{\delta \to 0}$ replaced by $\limsup_{\delta \to 0}$.

Note that it follows from Lemma 7.2.5 that the hypotheses of this theorem are not satisfied if ϕ is regularly varying at zero with index $\alpha > 0$.

Proof This follows immediately from Theorems 7.6.9 and 9.5.6. □

Remark 9.5.19 Just as Theorem 9.5.9 gives a simplified version of Theorem 9.5.11 for Lévy processes, we can use (7.274) to simplify Theorem 9.5.18 when X is a symmetric Lévy process by replacing $d_\alpha(x,y)$ by $\sigma_0(|x - y|)$, ϕ_α by ϕ and C_α by C. (One can then denote the function $\Phi_{2,\alpha}$ in (9.119) by Φ_2.)

Example 9.5.20 We consider examples of exact uniform moduli of continuity for local times that correspond to the four cases considered in Example 7.6.6.

(1) Corresponding to this case we have Corollary 9.5.17. When ψ is regularly varying at infinity with index $1 < p < 2$, σ_0^2 is regularly varying at zero with index $0 < \alpha < 1$, by (7.279).

(2) Using the argument in the paragraph containing (9.108) along with Lemma 7.4.10, we can actually take ψ to be a normalized regularly varying function of index 1 and asymptotic to F at infinity. Thus σ_0 can be taken to be a normalized slowly varying function asymptotic to $\exp(-g(\log 1/h))$ at zero. It follows from Theorem 9.5.15 under condition (2) that (9.114) continues to hold.

(3) In this case, by Lemma 7.2.5, Φ_1 in (7.374) and $\Phi_{2,0}$ in (9.119) are comparable and we get (9.120) by Theorem 9.5.18 and Remark 9.5.19.

(4) Same as (3).

Remark 9.5.21 The key result in establishing the correspondences between limit laws for local times and their associated Gaussian processes given in Theorems 9.5.1–9.5.6 is Lemma 9.1.2. In the proof of this lemma we initially consider $\widehat{X}$, which is X killed at the end of an independent exponential time with mean $1/\alpha$ and its 0-potential density u^α. It is clear that we could have considered any Borel right process X with 0-potential density $u(x,y)$ and associated Gaussian process G and in place of (9.10) obtained that when (9.5) holds, (with G_α replaced by G)

$$P^x \times P_G \left(L_\infty^{\cdot} + \frac{1}{2}(G(\cdot) + s)^2 \in B \right) = 1, \qquad (9.121)$$

where L_∞ is the total accumulated local time of X. This is all we need to obtain Theorems 9.5.1–9.5.6 with $L_t^{\cdot}$ replaced by $L_\infty^{\cdot}$, with the various moduli of continuity being those of G.

As an example suppose that X is a Borel right process with continuous α-potential densities $\alpha > 0$ that is killed the first time it hits 0. Then, using Corollary 8.1.2, we see that Theorems 9.5.1–9.5.6 hold with $L_t^{\cdot}$ replaced by $L_{T_0}^{\cdot}$, with the various moduli of continuity being those of a mean zero Gaussian process with covariance u_{T_0}. We give some of these results, more explicitly, for stable processes in the next example.

Example 9.5.22 Let X be a symmetric stable process of index $1 < p < 2$ with local time $L = \{L_t^y \, ; \, (y,t) \in R^1 \times R_+\}$. Then (9.116) and (9.117) hold with $L_t^{\cdot}$ replaced by $L_{T_0}^{\cdot}$. Let us focus on the local result

$$\lim_{\delta \to 0} \sup_{\substack{|y - y_0| \le \delta \\ y \in R^1}} \frac{|L_{T_0}^y - L_{T_0}^{y_0}|}{(|y - y_0|^{p-1} \log \log 1/|y - y_0|)^{1/2}} = (2C_p L_{T_0}^{y_0})^{1/2} \quad \text{a.s.}$$

$$(9.122)$$

Note that when $y_0 = 0$, the right-hand side is zero. This is correct but not very interesting since for Brownian motion we have the much stronger result given in (2.217). We prove a generalization of (2.217) for diffusions in Chapter 12. We do not know what the left-hand side of (9.122) is when $y_0 = 0$, for processes that do not have continuous paths. However, we can obtain an upper bound for it, which holds for all strongly symmetric Borel right processes and is probably the best possible. We do this in Subsection 9.5.1.

Example 9.5.23 Here is a final example of the application of these ideas. Suppose that X is a Borel right process with continuous α-potential densities $\alpha > 0$ that is killed at $\tau(\lambda)$. Here τ is the inverse local time at 0 and λ is an independent exponential random variable. Suppose also that 0 is recurrent for X. Then, using Corollary 8.1.2 and the Laplace transform technique, as in (9.11), we see that Theorems 9.5.1–9.5.6 hold with $L_t^\cdot$ replaced by $L_{\tau(t)}^\cdot$ with the various moduli of continuity being those of a mean zero Gaussian process with covariance u_{T_0}. (Actually, the covariance of the associated process, before passing from λ to t, is $u_{\tau(\lambda)}$, as given in (3.193), but the constant does not contribute anything to the moduli of G.)

In this case, the local modulus of continuity has a neat form. For the same processes as in the previous example, we get

$$\lim_{\delta \to 0} \sup_{\substack{|y| \le \delta \\ y \in R^1}} \frac{|L_{\tau(t)}^y - t|}{(t\,|y|^{p-1} \log \log 1/|y|)^{1/2}} = (2C_p)^{1/2} \text{ for almost all } t \text{ a.s.}$$

(9.123)

(actually, this result is also easily obtained from Theorem 8.2.2 and Example 7.4.13, which actually show that it holds for all $t \ge 0$).

9.5.1 Local behavior of L_{T_0}

In Section 9.5, all the results on moduli of continuity of local times are lifted from corresponding results for the associated Gaussian processes. We cannot do this in general for the behavior of $L_{T_0}^\cdot$ near zero because of the presence of the constant s in Theorem 8.1.1. Nevertheless, we can still use Theorem 8.1.1 by applying it to $\sup_{|y| \le \delta} |L_{T_0}^y|$ directly. Before we do this we consider the much simpler case of the local times of recurrent diffusions, since in this case we can use the First Ray–Knight Theorem for diffusions (Theorem 8.2.6) (we do not have a simple version of the first Ray–Knight Theorem for Markov processes without continuous sample paths).

We begin with the following simple lemma.

Lemma 9.5.24 *Let* $\{B(t), t \in R_+\}$ *and* $\{\overline{B}(t), t \in R_+\}$ *be independent standard Brownian motions. Then*

$$\limsup_{t \downarrow 0} \frac{B^2(t) + \overline{B}^2(t)}{2t \log \log(1/t)} = 1 \qquad a.s. \qquad (9.124)$$

It follows from Khintchine's law of the iterated logarithm, (2.15), that the left-hand side of (9.124) is between one and two. It is remarkable that the limit is actually one.

Proof By (2.15) we need only obtain the upper bound. Let $H(t) = B^2(t) + \overline{B}^2(t)$, $a(t) = 2t \log \log(1/t)$, and $t_k = \theta^k$. We take $\theta < 1$ and note that for any $\epsilon > 0$ we can choose θ sufficiently close to one so that $1 \leq a(t_{k-1})/a(t_k) \leq 1 + \epsilon$. The left-hand side of (9.124) is bounded above by

$$\limsup_{k \to \infty} \sup_{t_k \leq t \leq t_{k-1}} \frac{H(t) - H(t_k)}{a(t_k)} + \limsup_{k \to \infty} \frac{H(t_k)}{a(t_k)}. \qquad (9.125)$$

By (5.71),

$$P(H(t) \geq x) = \exp\left(-\frac{x}{2t}\right). \qquad (9.126)$$

Therefore, by the Borel–Cantelli Lemma, the second term in (9.125) is less than or equal to one.

We now show that

$$\limsup_{k \to \infty} \sup_{t_k \leq t \leq t_{k-1}} \frac{B^2(t) - B^2(t_k)}{a(t_k)} = 0 \qquad a.s. \qquad (9.127)$$

which completes the proof of the lemma. The left-hand side of (9.127) is bounded by

$$\limsup_{k \to \infty} \sup_{t_k \leq t \leq t_{k-1}} \frac{|B(t) + B(t_k)|}{\sqrt{a(t_k)}} \limsup_{k \to \infty} \sup_{t_k \leq t \leq t_{k-1}} \frac{|B(t) - B(t_k)|}{\sqrt{a(t_k)}}. \qquad (9.128)$$

It is clear from (2.14) that the first $\limsup$ in (9.128) is bounded almost surely by some constant. It is easy to see, again by the Borel–Cantelli Lemma, that the second $\limsup$ in (9.128) is zero. This is because, by Lemma 2.2.11 for all k sufficiently large,

$$P\left(\sup_{t_k \leq t \leq t_{k-1}} \frac{|B(t) - B(t_k)|}{\sqrt{a(t_k)}} \geq \delta\right) \qquad (9.129)$$

$$\leq 2P\left(\sup_{t_k \leq t \leq t_{k-1}} \frac{B(t) - B(t_k)}{\sqrt{a(t_k)}} \geq \delta\right)$$

$$= 4P\left(\frac{B(t_{k-1}) - B(t_k)}{\sqrt{a(t_k)}} \geq \delta\right)$$

$$\leq C(\theta)(\log k)^{1/2} \exp\left(-\frac{\delta^2 \log k}{(1/\theta) - 1}\right),$$

where $C(\theta)$ depends only on θ (so that, for any $\delta > 0$, we can choose θ such that the last term in (9.129) is a term of a convergent sequence). Thus we get (9.127). □

Theorem 9.5.25 *Let X be a recurrent diffusion in R^1 with continuous α-potential densities as in Subsection 8.2.1. Let $L^r_{T_0}$ denote the local time of X starting at $y > 0$ and killed the first time it hits 0. Then*

$$\limsup_{x \downarrow 0} \frac{L^x_{T_0}}{u_{T_0}(x, x) \log\log(1/u_{T_0}(x, x))} = 1 \qquad P^y \ a.s. \qquad (9.130)$$

Proof This follows immediately from Lemma 9.5.24 and Theorem 8.2.6, since, in applying (8.80), the term $(B^2_{\rho(r) - \rho(y)})1_{\{r \geq y\}} + (\overline{B}^2_{\rho(r) - \rho(y)})1_{\{r \geq y\}}$ contributes nothing. □

In the general case we can only obtain upper bounds for the local behavior $L^{\cdot}_{T_0}$ in the neighborhood of 0. We use the notation $\|f.\|_\delta := \sup_{|y| \leq \delta} |f_y|$ and $\|g(\cdot, \cdot)\|_\delta := \sup_{|y| \leq \delta} |g(y, y)|$, where f and g are continuous functions.

Lemma 9.5.26 *Let $X = (\Omega, \mathcal{G}, \mathcal{G}_t, X_t, \theta_t, P^x)$ be a strongly symmetric Borel right process with continuous α-potential density $u^\alpha(x, y)$ and state space R^1. Assume that $P^x(T_0 < \infty) > 0$ for all $x \in S$. Let $L = \{L^y_t ; (y, t) \in R^1 \times R_+\}$ denote the local time for X normalized so that*

$$E^x\left(\int_0^\infty e^{-\alpha t} dL^y_t\right) = u^\alpha(x, y). \qquad (9.131)$$

Let $G = \{G_y ; y \in S\}$ denote the mean zero Gaussian process with covariance $u_{T_0}(x, y)$. Then, for all $\theta < 1$ and $\gamma > 0$,

$$\limsup_{n \to \infty} \sup_{|y| \leq \theta^n} \frac{L^y_{T_0}}{(E\|G.\|_{\theta^n})^2 + \gamma\|u_{T_0}(\cdot, \cdot)\|_{\theta^n} \log\log 1/\theta^n} \leq \frac{1 + \gamma}{\gamma} \quad a.s. \qquad (9.132)$$

The proof depends on an inequality for the moment generating function of the square of the supremum of a Gaussian process.

Lemma 9.5.27 *Let $\{G(y), y \in [0, \delta]\}$ be a continuous Gaussian process. Set $\sigma_\delta^2 = \|EG^2\|_\delta$ and let a denote the median of $\|G.\|_\delta$. Then, for $\lambda < 1/(2\sigma_\delta^2)$ and all $\epsilon > 0$ sufficiently small,*

$$E \exp\left(\lambda \|G. + \epsilon a\|_\delta^2\right) \leq e^{\lambda(1+\epsilon)^2 a^2} \tag{9.133}$$

$$+ 6\lambda \left(\frac{\sigma_\delta^2}{(1 - 2\sigma_\delta^2\lambda)} + \frac{a\sigma_\delta}{(1 - 2\sigma_\delta^2\lambda)^{3/2}}\right) \exp\left(\frac{\lambda(1+\epsilon)^2 a^2}{1 - 2\sigma_\delta^2\lambda}\right).$$

Proof Let $\lambda' = \lambda/(1 + \epsilon)^2$. For any positive random variable X, by integration by parts, we have

$$
\begin{aligned}
E e^{\lambda' X} &\leq 1 + \int_0^\infty \lambda' e^{\lambda' y} P(X > y)\, dy \tag{9.134} \\
&= e^{\lambda a^2} + \int_{((1+\epsilon)a)^2}^\infty \lambda' e^{\lambda' y} P(X > y)\, dy \\
&= e^{\lambda a^2} + 2 \int_a^\infty \lambda u e^{\lambda u^2} P(X > ((1 + \epsilon)u)^2)\, du.
\end{aligned}
$$

Take $X = \|G. + \epsilon a\|_\delta^2$. By (5.151) and (5.18), for $u \geq a$,

$$
\begin{aligned}
P(X > ((1 + \epsilon)u)^2) &\leq P(\|G.\|_\delta - a \geq (1 + \epsilon)(u - a)) \tag{9.135} \\
&\leq \exp\left(-\frac{(1 + \epsilon)^2 (u - a)^2}{2\sigma_\delta^2}\right).
\end{aligned}
$$

Let $\widetilde{\sigma}_\delta^2 := \sigma_\delta^2/(1 + \epsilon)^2$ and $\rho^2 := \widetilde{\sigma}_\delta^2/(1 - 2\widetilde{\sigma}_\delta^2\lambda)$. Using (9.135), we see that the last integral in (9.134) is less than or equal to

$$
\begin{aligned}
& 2\lambda \int_a^\infty u \exp\left(-\frac{u^2(1 - 2\widetilde{\sigma}_\delta^2\lambda) - 2ua}{2\widetilde{\sigma}_\delta^2}\right) du \exp\left(-\frac{a^2}{2\widetilde{\sigma}_\delta^2}\right) \\
&= 2\lambda \int_a^\infty u \exp\left(-\frac{u^2 - 2ua\rho^2/\widetilde{\sigma}_\delta^2}{2\rho^2}\right) du \exp\left(-\frac{a^2}{2\widetilde{\sigma}_\delta^2}\right) \tag{9.136} \\
&= 2\lambda \int_a^\infty u \exp\left(-\frac{(u - a\rho^2/\widetilde{\sigma}_\delta^2)^2}{2\rho^2}\right) du \, \exp\left(\frac{\lambda a^2 \rho^2}{\widetilde{\sigma}_\delta^2}\right).
\end{aligned}
$$

Note that

$$2 \int_\alpha^\infty u \exp\left(-\frac{(u - \alpha)^2}{2\rho^2}\right) du = 2\rho^2 + \sqrt{2\pi}\alpha\rho. \tag{9.137}$$

Using this in (9.136) along with (9.134), we get (9.133). $\qquad \square$

Proof of Theorem 9.5.26 We use Theorem 8.1.1, as in the proof of Lemma 9.3.1, applied to $\{L_{T_0}^x, x \in [-\delta, \delta]\}$. We consider X under P^x. Let T_δ denote the first hitting time of $[-\delta, \delta]$ by X. Then, by the strong Markov property,

$$E^x \exp\left(w\|L_{T_0}.\|_\delta\right) = E^x \left(E^{X_{T_\delta}} \exp\left(w\|L_{T_0}.\|_\delta\right)\right). \tag{9.138}$$

It follows from (8.2) that, for any $s \neq 0$,

$$E^{X_{T_\delta}} \exp\left(w \|L_{T_0}\|_\delta\right) \tag{9.139}$$

$$\leq E\left(\left(1 + \frac{G'_{X_{T_\delta}}}{s}\right) \exp\left(w \|(G'. + s)^2\|_\delta\right)\right)$$

$$\leq \left(E\left(1 + \frac{G'_{X_{T_\delta}}}{s}\right)^p\right)^{1/p} \left(E \exp(q\, w \|(G'. + s)^2\|_\delta)\right)^{1/q},$$

where $G' = G/\sqrt{2}$ and $1/p + 1/q = 1$. We now take $s = \epsilon a$ to see that the right-hand side of (9.139) is less than or equal to

$$\left(E\left(1 + \frac{G'_{X_{T_\delta}}}{\epsilon a}\right)^p\right)^{1/p} \left(E \exp(q\, w \|G'. + \epsilon a\|_\delta^2)\right)^{1/q}. \tag{9.140}$$

Using this and (5.184), we see that for any $\epsilon' > 0$, we can choose q sufficiently close to one such that

$$E^{X_{T_\delta}} \exp\left(w \|L_{T_0}\|_\delta\right) \leq C_{p,\epsilon} E \exp(w(1 + \epsilon') \|G'. + \epsilon a\|_\delta^2). \tag{9.141}$$

Denote the right-hand side of (9.133) by $H_\epsilon(a, \lambda, \sigma_\delta)$. Then, by Chebyshev's inequality,

$$P^x\left(\|L_{T_0}\|_\delta > y\right) \leq C_{p,\epsilon} H_\epsilon(a, \lambda, \sigma_\delta) \exp\left(-\frac{\lambda y}{1 + \epsilon'}\right), \tag{9.142}$$

where

$$\sigma_\delta^2 = \|E\left(G'.\right)^2\|_\delta = \frac{\|u_{T_0}(\cdot, \cdot)\|_\delta}{2}. \tag{9.143}$$

Let $\lambda = (a^2 b(1 + \epsilon)^2 + 2\sigma_\delta^2)^{-1}$, where $b > 0$. Then

$$H_\epsilon(a, \lambda, \sigma_\delta) \leq e^{1/b}\left(1 + 3\left(\frac{\sigma_\delta^2}{a^2 b(1 + \epsilon)^2} + \frac{\sigma_\delta(a^2 + 2\sigma_\delta^2)^{1/2}}{a^2(b(1 + \epsilon))^{3/2}}\right)\right). \tag{9.144}$$

Now let $b = \beta/((1 + \epsilon)^2 \log n)$ and $y = \dfrac{c \log n}{\lambda}$, where

$$c = (1 + \epsilon')^2 + \frac{(1 + \epsilon')(1 + \epsilon)^2}{\beta}. \tag{9.145}$$

With these choices and (5.184) we see from (9.142) that

$$P^x\left(\|L_{T_0}\|_\delta > y\right) \leq C_{p,\epsilon}\left(\frac{\log n}{\beta}\right)^{3/2} \frac{1}{n^{1+\epsilon'}}. \tag{9.146}$$

Using this, the Borel–Cantelli Lemma, and (5.183) we get (9.132), in which $\gamma = 1/\beta$. $\square$

Under some mild additional conditions we can write (9.132) in a more understandable form. Consider the following conditions imposed on the quantities in Lemma 9.5.26:

(1) $(E\|G.\|_\delta)^2 = o(\|u_{T_0}(\,\cdot\,,\,\cdot\,)\|_\delta \log\log 1/\delta)$ for all δ sufficiently small.
(2) $(\|u_{T_0}(\,\cdot\,,\,\cdot\,)\|_\delta \log\log 1/\delta) = o(E\|G.\|_\delta)^2$ for all δ sufficiently small.
(3) For all $\epsilon > 0$ there exists a $\theta < 1$ such that $\|u_{T_0}(\,\cdot\,,\,\cdot\,)\|_{\theta^n} \leq (1+\epsilon)\|u_{T_0}(\,\cdot\,,\,\cdot\,)\|_{\theta^{n+1}}$ for all n sufficiently large.
(4) For all $\epsilon > 0$ there exists a $\theta < 1$ such that $E\|G.\|_{\theta^n} \leq (1+\epsilon)E\|G.\|_{\theta^{n+1}}$ for all n sufficiently large.

Theorem 9.5.28 *In addition to the hypotheses of Lemma 9.5.28:*
If conditions (1) and (3) hold, then

$$\limsup_{\delta\to 0}\ \sup_{|y|\leq\delta}\ \frac{L^y_{T_0}}{\|u_{T_0}(\,\cdot\,,\,\cdot\,)\|_\delta \log\log 1/\delta} \leq 1 \qquad a.s. \tag{9.147}$$

If conditions (2) and (4) hold, then

$$\limsup_{\delta\to 0}\ \sup_{|y|\leq\delta}\ \frac{L^y_{T_0}}{(E\|G.\|_\delta)^2} \leq 1 \qquad a.s. \tag{9.148}$$

If conditions (3) and (4) hold then, for all $\gamma > 0$,

$$\limsup_{\delta\to 0}\ \sup_{|y|\leq\delta}\ \frac{L^y_{T_0}}{(E\|G.\|_\delta)^2 + \gamma\|u_{T_0}(\,\cdot\,,\,\cdot\,)\|_\delta \log\log 1/\delta} \leq \frac{1+\gamma}{\gamma} \qquad a.s. \tag{9.149}$$

If G has stationary increments, then without conditions (3) and (4), (9.147)–(9.149) still hold if their right-hand sides are multiplied by four.

Furthermore, all these results hold with δ replaced by y in the denominators of the fractions.

The next to last statement of the theorem implies that (9.149), with the right-hand side multiplied by four, holds for all Lévy processes with continuous local times.

Proof All these results are easy consequences of Lemma 9.5.26. Conditions (3) and (4) enable the interpolation between θ^n and θ^{n+1}. Condition (1) enables us to ignore the expectation in the denominator of (9.132). We can then cancel the γ term in each denominator and take the remaining γ to be arbitrarily close to zero. Condition (2) enables us the ignore the u_{T_0} in the denominator of (9.132). We then take γ arbitrarily large. Without condition (3), if G has stationary increments, we still have

$$\sup_{|x|\le\theta^n} u_{T_0}(x,x) \;\le\; 8 \sup_{|x|\le\theta^n} EG^2(x/2) \tag{9.150}$$

$$= \; 8 \sup_{|x|\le\theta^n/2} EG^2(x)$$

$$\le \; 4 \sup_{|x|\le\theta^{n+1}} u_{T_0}(x,x)$$

for θ sufficiently close to one. Without condition (4) the same argument works for $E\|G.\|_{\theta^n}$. $\square$

Corollary 9.5.29 *Let X be a symmetric Lévy process, without Gaussian component, with Lévy exponent ψ, and let σ_0 be as defined in (7.266). Assume that σ_0 is regularly varying at 0 with positive index. Let $\{L_t^x,(x,t)\in R^1\times R_+\}$ denote the local times of X. Then*

$$\limsup_{\delta\to 0}\;\sup_{|y|\le\delta}\;\frac{L_{T_0}^y}{\sigma_0^2(\delta)\log\log 1/\delta}\le 1 \qquad a.s. \tag{9.151}$$

and

$$\lim_{\delta\to 0}\;\sup_{|y|\le\delta}\;\frac{L_{T_0}^y}{\sigma_0^2(y)\log\log 1/y}\le 1 \qquad a.s. \tag{9.152}$$

Proof   By (4.86) and (4.87), $\sigma_0^2(x)=u_{T_0}(x,x)$. Let G be the Gaussian process with covariance $u_{T_0}(x,y)$. It follows from (7.88) and (7.128) that condition (1), prior to Theorem 9.5.28, holds for G. The regular variation of σ_0 implies that condition (3) is satisfied. Thus (9.151) follows from (9.147) and (9.152) follows from Lemma 7.1.6. $\square$

Example 9.5.30   Using (2.139) and (9.147), we get the same upper bound for L_{T_0} of standard Brownian motion as in (2.217). (It is clear that for Markov processes with continuous paths, if $x>0$, $L_{T_0}^x$ is non-zero only if the process starts above zero. Therefore, we must modify (9.147) by writing P^x almost surely, $x>0$ and similarly when $x<0$.)

For symmetric stable processes with index $1<p<2$, by Corollary 9.5.29, we get

$$\lim_{\delta\to 0}\;\sup_{\substack{|y|\le\delta\\ y\in R^1}}\;\frac{L_{T_0}^y}{C_p|y|^{p-1}\log\log 1/|y|}\le 1 \qquad a.s. \tag{9.153}$$

(We show in Remark 4.2.6 that $C_2=1$, so there appears to be a discrepancy between (9.153)) and the result for standard Brownian motion. But remember that the Lévy exponent of standard Brownian motion is $|\lambda|^2/2$, whereas the Lévy exponent of a symmetric p-stable process, $0 \le p \le 2$, is $|\lambda|^p$.)

Finally, we note that there are many interesting cases in which condition (2) holds. It follows from (7.88) that for some constant C, $C\widetilde{\rho}(\delta)$ (see (7.84)) is an upper bound for $E\|G.\|_\delta$. Using the arguments in the proof of Lemma 6.4.6, we can show that, up to a constant multiple greater than zero, it is also a lower bound when the integral term in (7.84) exceeds the other term and $EG^2(x)$ is asymptotic to a monotonic function. Therefore, the examples in Example 7.6.6 are applicable and give many cases in which condition (2) is satisfied and an idea of what the denominator in (9.147) looks like.

9.6 Stable mixtures

Let $X = \{X(t), t \in R_+\}$ be a canonical stable process, that is,

$$\psi(\lambda) = |\lambda|^p \tag{9.154}$$

for some $0 < p \le 2$ for ψ as defined in (4.63) (see also page 141). X has local times only for $1 < p \le 2$. We restrict ourselves to this range for p.

We call a Lévy process a stable mixture if its Lévy exponent $\psi(\lambda)$ can be represented as

$$\psi(\lambda) = \int_1^2 |\lambda|^s \, d\mu(s), \tag{9.155}$$

where μ is a finite positive measure on $(1, 2]$ such that

$$\int_1^2 \frac{d\mu(s)}{2 - s} < \infty. \tag{9.156}$$

By the observations in the paragraph containing (4.94) we see that $\psi(\lambda) = 2 \int_0^\infty (1 - \cos(\lambda u)) \, d\nu(u)$ with

$$\frac{d\nu(u)}{du} = \int_1^2 \frac{1}{2\pi C_{s+1} u^{s+1}} \, d\mu(s). \tag{9.157}$$

To verify that ψ is the Lévy exponent of a Lévy process it suffices, by (7.263), to show that ν is a Lévy measure, that is,

$$\int_0^\infty (u^2 \wedge 1) \int_1^2 \frac{1}{u^{s+1}} \, d\mu(s) \, du < \infty. \tag{9.158}$$

This is simple to do by integrating first with respect to u.

In the next two lemmas we show that the Lévy exponent of a stable mixture and the increments variance of its associated Gaussian process have many of the smoothness properties that are hypotheses in our results on moduli of continuity of Gaussian processes and local times.

Lemma 9.6.1 *Suppose that μ in (9.155) is supported on $[1, \beta]$ for some $1 < \beta \leq 2$. Then ψ is a normalized regularly varying function at infinity of index β.*

Proof Let $1 < \beta' < \beta$. Since the support of μ is $[1, \beta]$, it puts positive mass on $[b, \beta]$, where $b = (\beta' + \beta)/2$. Thus

$$\psi(\lambda) \sim \int_b^\beta \lambda^s \, d\mu(s) \qquad \text{as} \quad \lambda \to \infty. \tag{9.159}$$

Taking β' arbitrarily close to β shows that $\psi(2\lambda) \sim 2^\beta \psi(\lambda)$ as $\lambda \to \infty$, which, by definition, means that ψ is regularly varying at infinity with index β. By similar reasoning we see that

$$\lambda \psi'(\lambda) \sim \int_b^\beta s\lambda^s \, d\mu(s) \sim \beta \psi(\lambda) \qquad \text{as} \quad \lambda \to \infty. \tag{9.160}$$

Since $\lambda \psi'(\lambda)/\psi(\lambda) \to \beta$ as $\lambda \to \infty$, we see that ψ is a normalized regularly varying function at infinity of index β. $\square$

Remark 9.6.2 Continuing the argument that gives (9.160), we see that for $n = 1, 2, \ldots$,

$$\lambda^n \psi^{(n)}(\lambda)/\psi(\lambda) \to \beta(\beta - 1) \ldots (\beta - n + 1) \qquad \text{as} \quad \lambda \to \infty. \tag{9.161}$$

This, by definition, shows that ψ is actually a "smoothly" varying function at infinity (see Bingham, Goldie and Teugels (1987, page 44)).

Lemma 9.6.3 *Let $\psi(\lambda)$ be given as in (9.155). Then*

$$\sigma^2(x) = \int_0^\infty \frac{1 - \cos \lambda x}{\psi(\lambda)} \, d\lambda \tag{9.162}$$

is concave on $[0, \infty]$.

Proof Note that

$$2\sigma^2(x) - \sigma^2(x - h) - \sigma^2(x + h) = \int_0^\infty (1 - \cos v) \tag{9.163}$$

$$\left(\frac{2}{x\psi(v/x)} - \frac{1}{(x + h)\psi(v/x + h)} - \frac{1}{(x - h)\psi(v/x - h)} \right) dv$$

for all $|h| \leq |x|$. Therefore, to show that σ^2 is concave on $[0, \infty]$, it suffices to show that the term in the bracket is positive, that is, $1/(x\psi(v/x))$

is concave in x for all $x > 0$ and $v > 0$. Clearly this is equivalent to showing that $g(x) = 1/(x\psi(1/x))$ is concave for $x > 0$.

By definition,

$$g(x) = \frac{1}{\int_1^2 x^{1-s}\, d\mu(s)}, \qquad (9.164)$$

so that

$$g'(x) = \frac{\int_1^2 (s-1)x^{-s}\, d\mu(s)}{\left(\int_1^2 x^{1-s}\, d\mu(s)\right)^2} \qquad (9.165)$$

and

$$g''(x) = \frac{2\left(\int_1^2 (s-1)x^{-s}\, d\mu(s)\right)^2}{\left(\int_1^2 x^{1-s}\, d\mu(s)\right)^3} - \frac{\int_1^2 s(s-1)x^{-s-1}\, d\mu(s)}{\left(\int_1^2 x^{1-s}\, d\mu(s)\right)^2}. \qquad (9.166)$$

Thus, $g'' \le 0$ if

$$2\left(\int_1^2 (s-1)x^{-s}\, d\mu(s)\right)^2 \le \int_1^2 s(s-1)x^{-s-1}\, d\mu(s) \int_1^2 x^{1-s}\, d\mu(s) \qquad (9.167)$$

or, equivalently, if

$$2\left(\int_1^2 (s-1)x^{-s}\, d\mu(s)\right)^2 \le \int_1^2 s(s-1)x^{-s}\, d\mu(s) \int_1^2 x^{-s}\, d\mu(s). \qquad (9.168)$$

This follows from the Schwarz inequality applied to

$$2\left(\int_1^2 (s-1)x^{-s}\, d\mu(s)\right)^2 \qquad (9.169)$$

$$= 2\left(\int_1^2 [x^{-s}(s-1)/s]^{1/2}\, [x^{-s}(s-1)s]^{1/2}\, d\mu(s)\right)^2$$

since $2(s-1)/s \le 1$. $\qquad \square$

Stable mixtures give rise to a large class of regularly varying characteristic exponents of Lévy processes.

Lemma 9.6.4 *Let $\rho(s)$ be a bounded continuous increasing function on $[0, \beta - 1]$, $1 < \beta < 2$. Then we can find a stable mixture with characteristic exponent*

$$\psi(\lambda) = \lambda^\beta \hat\rho(\log \lambda) \qquad (9.170)$$

for all $\lambda > 0$, where

$$\hat{\rho}(v) = \int_0^\infty e^{-vs}\, d\rho(s). \tag{9.171}$$

If, in addition,

$$\int_0^1 \frac{d\rho(s)}{s} < \infty, \tag{9.172}$$

the above statements are also valid when $\beta = 2$.

Proof Note that $R(s) := \rho(\beta - 1) - \rho(\beta - s)$ is an increasing function on $[1, \beta]$. Let $\mu(s)$ in (9.155) be a measure with distribution function R. Then, for $1 < \beta < 2$,

$$
\begin{aligned}
\psi(\lambda) &= -\int_1^\beta \lambda^s\, d\rho(\beta - s) & (9.173)\\
&= \lambda^\beta \int_0^{\beta-1} \lambda^{-s}\, d\rho(s)\\
&= \lambda^\beta \int_0^{\beta-1} e^{-(\log \lambda)s}\, d\rho(s)\\
&= \lambda^\beta \hat{\rho}(\log \lambda).
\end{aligned}
$$

When $\beta = 2$, (9.172) shows that μ satisfies (9.156). $\square$

Corollary 9.6.5 *Let h be any function that is regularly varying at infinity with positive index or is slowly varying at infinity and increasing. Then, for any $1 < \beta < 2$, there exists a Lévy process for which $\sigma^2(x)$ in (9.162) is concave and satisfies*

$$\sigma^2(x) \sim |x|^{\beta-1} h(\log 1/|x|) \qquad as \qquad x \to 0. \tag{9.174}$$

If, in addition,

$$\int_0^1 \frac{dx}{h(x)} < \infty, \tag{9.175}$$

the above statement is also valid when $\beta = 2$.

Proof Let $h(x) = x^p L(x)$, where $p > 0$ and L is slowly varying at infinity. Take $\rho(s) \sim \dfrac{s^p}{L(1/s)\Gamma(1 + 1/p)}$, as $s \to 0+$, in Lemma 9.6.4. (Since $p > 0$, we can take $\rho(s)$ to be increasing.) By Theorem 14.7.6, $\hat{\rho}(x) \sim 1/h(x)$ at infinity. Consider the stable mixture with characteristic exponent given by (9.170). By Lemma 7.4.9,

$$\sigma^2(x) \sim C_\beta |x|^{\beta-1} \frac{1}{\hat{\rho}(\log 1/|x|)} \qquad as \qquad x \to 0. \tag{9.176}$$

This gives (9.174).

If h is slowly varying at infinity and increasing and is equal to $L(x)$, we take $\rho(s) = 1/L(1/s)$ at 0, which is increasing, and the same proof suffices.

When $\beta = 2$ we must make sure that (9.172) holds or, equivalently, that

$$\int_0^1 \frac{\rho(s)}{s^2} \, ds < \infty. \tag{9.177}$$

This in turn is equivalent to

$$\int_1^\infty \hat{\rho}(s) \, ds < \infty \tag{9.178}$$

from which we get (9.175). $\qquad\square$

9.7 Local times for certain Markov chains

In Section 9.1 we show that when the Gaussian process associated with a strongly symmetric Borel right process X has a bounded discontinuity at some point, the local time of X has a bounded discontinuity at that point. Naturally we would like to give concrete examples of such processes. To do so we must leave the realm of Lévy processes, which have been our source of examples up to this point. All the various killed Lévy processes that we can think of have associated Gaussian process that have stationary increments or are minor variations of such Gaussian processes. It follows from Theorem 5.3.10 that such processes either have continuous paths almost surely on all open subsets of their domain or else are unbounded almost surely on all open subsets of their domain. To find examples of local times with bounded discontinuities we turn to a certain class of Markov chains with a single instantaneous state, which are variations of an example of Kolmogorov (1951). (An element $x \in S$ is an instantaneous state for a Markov chain X_t with state space S, if $\inf\{t \mid X_t \neq x\} = 0$, P^x almost surely.)

Focusing on Markov chains gives us an opportunity to discuss the local time of a Borel right process with a countable state space. Theorem 3.6.3, which establishes the existence of local times for Borel right processes X with continuous α-potential densities, is formulated with respect to approximate δ-functions with respect to the reference measure m of X at a point, say y, in the state space of X. When y is an isolated point in the state space of X, we take the δ-function at y to be a unit point mass at y divided by $m(\{y\})$. Then the local time at y

(see (3.92)) is simply the amount of time that X spends at y divided by $m(\{y\})$.

We use Theorem 4.1.3 to define the Markov chains that we consider in this section by their α-potentials. The state space of the chains is the sequence $S = \{\frac{1}{2}, \frac{1}{3}, \ldots, \frac{1}{n}, \ldots, 0\}$ with the topology inherited from the real line. Clearly S is a compact metric space with one limit point. Let $\{q_n\}_{n=2}^{\infty}$ and $\{r_n\}_{n=2}^{\infty}$ be strictly positive real numbers such that

$$\sum_{n=2}^{\infty} \frac{q_n}{r_n} < \infty \quad \text{and} \quad \lim_{n \to \infty} q_n = \infty. \tag{9.179}$$

We define a finite measure m on S by $m(1/n) = q_n/r_n$ and $m(0) = 1$. We also define an α-potential $\{U^{\alpha}, \alpha > 0\}$ on $C(S)$ in terms of its density $u^{\alpha}(x, y), x, y \in S$ with respect to m. That is, for all $f \in C_b(S)$,

$$U^{\alpha} f(x) = \int_S u^{\alpha}(x, y) f(y) m(dy), \tag{9.180}$$

where

$$u^{\alpha}(0,0) \;=\; \frac{1}{\alpha + \displaystyle\sum_{j=2}^{\infty} \frac{\alpha q_j}{\alpha + r_j}}, \tag{9.181}$$

$$u^{\alpha}(0, 1/i) \;=\; u^{\alpha}(1/i, 0) = u^{\alpha}(0,0) \frac{r_i}{\alpha + r_i},$$

$$u^{\alpha}(1/i, 1/j) \;=\; \delta_{ij} \frac{r_j}{q_j(\alpha + r_j)} + u^{\alpha}(0,0) \frac{r_i}{\alpha + r_i} \frac{r_j}{\alpha + r_j}.$$

It is clear that $u^{\alpha}(x, y)$ is symmetric and continuous on S. Furthermore, one can check that $U^{\alpha} : C(S) \to C(S)$ and that

$$\alpha U^{\alpha} 1 \;=\; 1, \tag{9.182}$$

$$\|\alpha U^{\alpha}\| \;\leq\; 1,$$

$$U^{\alpha} - U^{\beta} + (\alpha - \beta) U^{\alpha} U^{\beta} \;=\; 0,$$

$$\lim_{\alpha \to \infty} \alpha U^{\alpha} f(x) \;=\; f(x) \qquad \forall x \in S.$$

It is easy to verify the first equation in (9.182), for $u^{\alpha}(0, \cdot)$. Then, using the fact that

$$u^{\alpha}(1/j, 1/k) = \delta_{j,k} \frac{u^{\alpha}(0, 1/k)}{q_k u^{\alpha}(0,0)} + \frac{u^{\alpha}(0, 1/j) u^{\alpha}(0, 1/k)}{u^{\alpha}(0,0)}, \tag{9.183}$$

it is easy to verify it for $u^{\alpha}(1/j, \cdot)$, $j = 2, \ldots$. The second equation in (9.182) follows immediately from the proof of the first one. For the third equation it suffices to show that

$$u^\alpha(x,y) - u^\beta(x,y) = (\beta - \alpha) \int_S u^\alpha(x,z) u^\beta(z,y) m(dz) \qquad (9.184)$$

for all $x, y \in S$. This is easy to do for $x = y = 0$. Then, using (9.183) one can show it for all $x, y \in S$. To obtain the limit in the last equation in (9.182) we use the Dominated Convergence Theorem.

It follows from Theorem 4.1.3 that $\{U^\alpha, \alpha > 0\}$ is the resolvent of a Feller process X with state space S. By Theorem 3.6.3, X has local times, which we denote by $L = \{L_t^{1/n}, (t, 1/n) \in R_+ \times S\}$.

We now examine the behavior of the local time in the neighborhood of L_t^0. Note that by Example 3.10.5, under P^0, L_∞^0, the total amount of time that X spends at 0 is an exponential random variable with mean $u^\alpha(0,0)$.

Theorem 9.7.1 *Let X be a Markov chain as defined above, with state space S and $\alpha = 1$. Let $L = \{L_t^{1/n}, (t, 1/n) \in R_+ \times S\}$ denote the local times of X. Let*

$$\beta = \beta(\{q_n\}) = \limsup_{n \to \infty} \left(\frac{2 \log n}{q_n^*} \right)^{1/2}, \qquad (9.185)$$

where $\{q_n^*\}_{n=2}^\infty$ is a nondecreasing rearrangement of $\{q_n\}_{n=2}^\infty$. Then

$$(2L_t^0)^{1/2}\beta \le \limsup_{n \to \infty} L_t^{1/n} - L_t^0 \le \frac{\beta^2}{2} + (2L_t^0)^{1/2}\beta \qquad (9.186)$$

for all $t \ge 0$ almost surely.
When $\beta = 0$, $\{L_t^{1/n}\}_{n=2}^\infty$ is continuous at 0 and

$$\limsup_{n \to \infty} \frac{|L_t^{1/n} - L_t^0|}{\left(\dfrac{\log n}{q_n} \right)^{1/2}} = 2(L_t^0)^{1/2} \qquad \text{for almost all} \quad t \ge 0 \quad \text{a.s.}$$

$$(9.187)$$

When $\beta = \infty$ and $\{q_n\}_{n=2}^\infty$ is nondecreasing,

$$\limsup_{n \to \infty} \sup_{1 \le k \le n} \frac{L_t^{1/k}}{\left(\dfrac{\log n}{q_n} \right)^{1/2}} \ge 2(L_t^0)^{1/2} \qquad (9.188)$$

for all $t \ge 0$ almost surely.

To prove the theorem we first examine the behavior at 0 of the Gaussian process associated with X. The following lemma is used to study this process.

Lemma 9.7.2 *Let $\{\xi_n\}_{n=1}^\infty$ be a sequence of mean zero normal random variables. Let $\{a_n\}_{n=1}^\infty$ and $\{v_n\}_{n=1}^\infty$ be strictly positive real numbers such that*

$$\sum_{n=1}^\infty \frac{v_n}{a_n} < \infty \qquad \text{and} \qquad \lim_{n\to\infty} v_n = \infty. \qquad (9.189)$$

Suppose that

$$E\xi_n^2 = \frac{1}{v_n} \qquad \text{and} \qquad E\xi_n\xi_m = \frac{1}{a_n a_m} \qquad n \neq m. \qquad (9.190)$$

Let $\{v_n^*\}_{n=1}^\infty$ be a nondecreasing rearrangement of $\{v_n\}_{n=1}^\infty$. Then

$$\limsup_{n\to\infty} \xi_n = \limsup_{n\to\infty} \left(\frac{2\log n}{v_n^*}\right)^{1/2} \qquad a.s. \qquad (9.191)$$

and

$$\limsup_{n\to\infty} \frac{|\xi_n|}{\left(\dfrac{2\log n}{v_n}\right)^{1/2}} = 1 \qquad a.s. \qquad (9.192)$$

Furthermore, if v_n is nondecreasing and

$$\limsup_{n\to\infty} \left(\frac{2\log n}{v_n}\right)^{1/2} = \infty, \qquad (9.193)$$

then, for any integer m and $0 < \epsilon < 10^{-2}$, there exists an $n_0 = n_0(m, \epsilon)$ such that, for all $n \geq n_0$,

$$median\left(\sup_{m \leq k \leq n} \xi_k\right) \geq (1-\epsilon)\left(\frac{2\log n}{v_n}\right)^{1/2}. \qquad (9.194)$$

Proof Let n^* be a rearrangement of n such that $\{v_{n^*}\}_{n^*=1}^\infty$ is nondecreasing. Clearly the sum in (9.189) remains unchanged if we sum on n^* rather than on n. To simplify the notation, we take $\{v_n\}_{n=1}^\infty$ to be nondecreasing. Let $\{\eta_n\}_{n=0}^\infty$ be independent normal random variables with mean 0 and variance 1. By (9.189), we can choose N sufficiently large such that $(1/v_n - 1/a_n^2) > 0$ for all $n \geq N$. For $n \geq N$ define

$$\tilde\eta_n = \left(\frac{1}{v_n} - \frac{1}{a_n^2}\right)^{1/2}\eta_n + \frac{1}{a_n}\eta_0. \qquad (9.195)$$

We see by (10.12) that $\{\tilde\eta_n\}_{n=N}^\infty$ and $\{\xi_n\}_{n=N}^\infty$ are equivalent Gaussian sequences. Therefore, with $b_n := (1/v_n - 1/a_n^2)^{1/2}$, we have

$$\limsup_{n\to\infty} \xi_n \overset{law}{=} \limsup_{n\to\infty} \tilde\eta_n = \limsup_{n\to\infty} b_n\eta_n. \qquad (9.196)$$

The last equality follows since (9.189) implies that $\lim_{n \to \infty} 1/a_n = 0$. Using

$$P(\sup_{N \leq n \leq N_1} b_n \eta_n > \lambda) = 1 - \prod_{n=N}^{N_1} (1 - P(b_n \eta_n > \lambda)) \qquad (9.197)$$

and (5.20) without the absolute value, along with the fact that, by (9.189), $\lim_{n \to \infty} b_n = 0$, we see that

$$\limsup_{n \to \infty} \xi_n = \lambda^*, \qquad (9.198)$$

where

$$\lambda^* = \inf \left\{ \lambda : \sum_{n=1}^{\infty} e^{-\lambda^2/(2b_n^2)} < \infty \right\}. \qquad (9.199)$$

Furthermore, we see by (9.189) that λ^* does not change if we replace $1/b_n^2$ in (10.21) by v_n for $n \geq 1$. To complete the proof we need only show that

$$\limsup_{n \to \infty} \left(\frac{2 \log n}{v_n} \right)^{1/2} = \inf \left\{ \lambda : \sum_{n=1}^{\infty} e^{-\lambda^2 v_n/2} < \infty \right\}. \qquad (9.200)$$

Let

$$\delta = \limsup_{n \to \infty} \frac{2 \log n}{v_n} \qquad (9.201)$$

and assume that $\delta < \infty$. Then, given $\varepsilon > 0$, there exists an $N(\varepsilon)$ such that, for all $n \geq N(\varepsilon)$,

$$\frac{v_n}{(2 \log n)} \delta \geq (1 - \varepsilon/2) \qquad (9.202)$$

and there exists an infinite subsequence $\{n_j\}$ of $\{n\}$ such that

$$\frac{v_{n_j}}{2 \log n_j} \delta \leq (1 + \varepsilon/2). \qquad (9.203)$$

It follows from (9.202) that $\lambda^* \leq ((1+\varepsilon)\delta)^{1/2}$. To obtain a lower bound for λ^* we recall the well-known fact that if $\{c_n\}_{n=1}^{\infty}$ is a nonincreasing sequence of real numbers such that $\sum_{n=1}^{\infty} c_n < \infty$, then $c_n = o(1/n)$. If $\lambda = ((1 - \varepsilon)\delta)^{1/2}$, it follows from (9.203) that

$$e^{-\lambda^2 v_{n_j}/2} \geq \frac{1}{n_j^{1-\varepsilon/2}}, \qquad (9.204)$$

which implies that $\lambda^* \geq ((1 - \varepsilon)\delta)^{1/2}$. Since ε can be made arbitrarily small, we get (9.200) and hence (9.191) when it is finite.

If $\delta = \infty$, for all $\lambda > 0$ there exists an infinite subsequence n_j such that

$$\frac{v_{n_j}}{2 \log n_j} \leq \frac{1}{\lambda}. \tag{9.205}$$

Thus

$$e^{-\lambda^2 v_{n_j}/2} \geq \frac{1}{n_j}, \tag{9.206}$$

which implies that $\lambda^* \geq \lambda$, and since this is true for all λ, we get $\lambda^* = \infty$. Thus we have established (9.191) in this case also.

It follows from Theorem 5.1.4 that

$$\limsup_{n \to \infty} \frac{b_n \eta_n}{(2 b_n^2 \log n)^{1/2}} = 1 \qquad \text{a.s.} \tag{9.207}$$

Thus (9.192) follows from (9.196). (It is easy to see that (9.192) holds for both $\{\xi_n\}$ and $\{|\xi_n|\}$.)

To show that (9.193) implies (9.194), we note that by (9.189) we can choose an $m_1 \geq m$ such that

$$\sup_{k \geq m_1} \frac{v_k}{a_k^2} < \frac{\epsilon}{2}. \tag{9.208}$$

Then, by (9.195) and the comment immediately following it, we see that

$$P\left(\sup_{m_1 \leq k \leq n} \xi_k \geq (1 - \epsilon) \left(\frac{2 \log n}{v_n} \right)^{1/2} \right) \tag{9.209}$$

$$\geq P\left(\sup_{m_1 \leq k \leq n} \eta_k \geq (1 - \epsilon')(2 \log n)^{1/2} + \epsilon' |\eta_0| \right)$$

for some $\epsilon' > 0$. Since $\{\eta_k\}$ are independent normal random variables with mean 0 and variance 1, it is easy to check (see (9.197)) that the last term in (9.209) goes to 1 as n goes to infinity. This gives us (9.194). $\square$

Proof of Theorem 9.7.1 By Theorem 9.1.3 we obtain (9.186) once we show that 2β is the oscillation function of a mean zero Gaussian process $\{G(x), x \in S\}$ with covariance $u^1(x, y)$ given in (9.181). Consider

$$\xi_n = G(1/n) - G(0) \qquad n = 2, 3, \dots \tag{9.210}$$

Then for $n \geq 2$ we have

$$\begin{aligned} E\xi_n^2 &= u^1(1/n, 1/n) - 2u^1(1/n, 0) + u^1(0, 0) \tag{9.211} \\ &= \frac{r_n}{q_n(1 + r_n)} + u^1(0, 0) \left(1 - \frac{r_n}{1 + r_n} \right)^2 \end{aligned}$$

$$= \frac{1 + \alpha(n)}{q_n},$$

where $\lim_{n \to \infty} \alpha(n) = 0$. Also, for $n \neq m$, $n, m \geq 2$,

$$E\xi_n\xi_m = u^1(0,0) \left(1 - \frac{r_n}{1 + r_n}\right) \left(1 - \frac{r_m}{1 + r_m}\right) \qquad (9.212)$$

$$= \frac{u^1(0,0)}{(1 + r_n)(1 + r_m)}.$$

For $n \geq 2$, set

$$v_n = \frac{q_n}{1 + \alpha(n)} \qquad (9.213)$$

$$a_n = \frac{1 + r_n}{(u^1(0,0))^{1/2}}.$$

We see that (9.189) is satisfied. It now follows from Lemma 9.7.2 that the oscillation function of $G(x)$ at $x = 0$ is twice the right-hand side of (9.191), that is, it is equal to 2β. This gives us (9.186).

To obtain (9.187) we note that it follows from (9.192) that

$$\limsup_{n \to \infty} \frac{|G(1/n) - G(0)|}{\left(\frac{2(1 + \alpha(n)) \log n}{q_n}\right)^{1/2}} = 1 \qquad \text{a.s.} \qquad (9.214)$$

This is equivalent to

$$\lim_{\delta \to 0} \sup_{\substack{d(1/n,0) \leq \delta \\ n \geq 2}} \frac{|G(1/n) - G(0)|}{(2d(1/n,0) \log n)^{1/2}} = 1 \qquad \text{a.s} \qquad (9.215)$$

$(d := d_G$ is defined in (6.1)). It now follows from Theorem 9.5.1 that

$$\lim_{\delta \to 0} \sup_{\substack{d(1/n,0) \leq \delta \\ n \geq 2}} \frac{|L_t^{1/n} - L_t^0|}{(2d(1/n,0) \log n)^{1/2}} = \sqrt{2}(L_t^0)^{1/2} \quad \text{for almost all } t \geq 0 \text{ a.s.,}$$

$$(9.216)$$

which is equivalent to (9.187).

Finally, (9.188) is an immediate application of Theorem 9.8.1 in the next section, (9.194), and Corollary 5.4.5. $\qquad \square$

9.8 Rate of growth of unbounded local times

Suppose that the local time of X in Theorem 9.2.1 has an unbounded discontinuity at a point, say y_0, in its state space. We now examine the rate of growth of the local time in the neighborhood of y_0. In the notation of Theorem 9.2.1, let $G := G_1$ and set

$$d(x,y) = (E(G(x) - G(y))^2)^{1/2} = (u^1(x,x) + u^1(y,y) - 2u^1(x,y))^{1/2}. \tag{9.217}$$

We make the following assumptions about $u^1(x,y)$ or, equivalently, about G. Let $Y \subset S$ be countable and let $y_0 \in Y$. Assume that y_0 is a limit point of Y with respect to d and

$$\sup_{\substack{y \in Y \\ d(y,y_0) \geq \delta}} G(y) < \infty \qquad \text{a.s.} \quad \forall \delta > 0 \tag{9.218}$$

and

$$\lim_{\delta \to 0} \sup_{\substack{y \in Y \\ d(y,y_0) \geq \delta}} G(y) = \infty \qquad \text{a.s.} \tag{9.219}$$

Let

$$a(\delta) = E \left(\sup_{\substack{y \in Y \\ d(y,y_0) \geq \delta}} G(y) \right) \tag{9.220}$$

and note that by (9.219), $\lim_{\delta \to 0} a(\delta) = \infty$.

Theorem 9.8.1 *Let X and G be associated processes as described above, so that, in particular, (9.218) and (9.219) are satisfied on a countable subset Y of S. Let $L = \{L_t^y, (t,y) \in R_+ \times S\}$ be the local time of X. Then*

$$\limsup_{\delta \to 0} \sup_{\substack{y \in Y \\ d(y,y_0) \geq \delta}} \frac{L_t^y}{a(\delta)} \geq (2L_t^{y_0})^{1/2} \qquad \forall t \in R_+ \quad \text{a.s.} \tag{9.221}$$

and

$$\limsup_{\delta \to 0} \sup_{\substack{y \in Y \\ d(y,y_0) \geq \delta}} \frac{L_t^y}{a^2(\delta)} \leq 1 \qquad \forall t \in R_+ \quad \text{a.s.,} \tag{9.222}$$

where $a(\delta)$ is given in (9.220).

Obviously there is a big gap between (9.221) and (9.222). This is caused by the same technical difficulty that gives the β^2 term in (9.12). We conjecture that the lower bound is the correct limit.

To prove this theorem we use Theorem 8.1.1 on $\widehat{X}$, which is X killed at the end of an independent exponential time with mean one. The 0-potential density of $\widehat{X}$ is $u^1(x,y)$. For convenience we denote y_0 by 0. Let $h(y) = u^1(y,0)/u^1(0,0)$.

Set

$$\eta(y) = G(y) - h(y)G(0). \tag{9.223}$$

Let $Y' \subset Y$ be finite. Let

$$\sigma = \sup_{y \in Y'} \left(\eta^2(y) \right)^{1/2} \tag{9.224}$$

and

$$a = \operatorname{med} \left(\sup_{y \in Y'} \eta(y) \right). \tag{9.225}$$

Set $h_1 = \min_{y \in Y'} h(y)$. Assume that $h_1 > 0$ and note that $\max_{y \in Y'} h(y) \leq 1$.

In preparation for the use of Theorem 8.1.1, we consider the probability distribution of

$$A(y, G(0), s) := \tfrac{1}{2} \left((G(y) + s)^2 - h^2(y)(G(0) + s)^2 \right). \tag{9.226}$$

Let

$$B(y, G(0) + s, s) = (1 - h(y))s \left(h(y)(G(0) + s) + \frac{(1 - h(y))s}{2} \right). \tag{9.227}$$

Writing

$$G(y) + s = \eta(y) + h(y)(G(0) + s) + (1 - h(y)), \tag{9.228}$$

we see that

$$A(y, G(0), s) \tag{9.229}$$
$$= \frac{\eta^2(y)}{2} + \eta(y) \left(h(y)(G(0) + s) + (1 - h(y))s \right) + B(y, G(0) + s, s)$$
$$= \frac{\eta^2(y)}{2} + \eta(y) \left(h(y)G(0) + s \right) + B(y, G(0) + s, s).$$

Note that $\eta = \{\eta(y), y \in Y\}$ and $G(0)$ are independent and that, for any $y \in Y$,

$$B(y, G(0) + s, s) \leq (1 - h_1)s \left(|G(0) + s| + \frac{(1 - h_1)s}{2} \right) \tag{9.230}$$
$$:= \widetilde{B}(h_1, |G(0) + s|, s).$$

Lemma 9.8.2 *Let $t > 0$ and assume that $a - \sigma t > 0$. Then*

$$P_\eta \left(\sup_{y \in Y'} A(y, G(0), s) > \frac{(a - \sigma t)^2}{2} \right) \tag{9.231}$$

$$+ (a - \sigma t)\left(h_1|G(0) + s| - (1 - h_1)s\right) - \widetilde{B}(h_1, |G(0) + s|, s)\Big)$$

$$\geq 2\Phi(t) - 1$$

and

$$P_\eta\left(\sup_{y \in Y'} A(y, G(0), s) < \frac{(a + \sigma t)^2}{2}\right) \tag{9.232}$$

$$+ (a + \sigma t)\left(|G(0) + s| + (1 - h_1)s\right) + \widetilde{B}(h_1, |G(0) + s|, s)\Big)$$

$$\geq 2\Phi(t) - 1,$$

where $\Phi(t)$ is given in (5.16).

Proof We first obtain (9.231). One sees from (5.150) that, on a set of measure greater than or equal to $(2\Phi(t) - 1)$, both $\sup_{y \in Y'} \eta(y) > (a - \sigma t)$ and $\sup_{y \in Y'} -\eta(y) > (a - \sigma t)$. Therefore, whatever the sign of $(h(y)G(0) + s)$,

$$\sup_{y \in Y'} \eta(y)(h(y)G(0) + s) \geq \inf_{y \in Y'}(a - \sigma t)|h(y)G(0) + s| \tag{9.233}$$

$$= \inf_{y \in Y'}(a - \sigma t)|h(y)(G(0) + s) + (1 - h(y))s|$$

$$\geq (a - \sigma t)\left(h_1|G(0) + s| - |(1 - h_1)s|\right)$$

on a set of measure greater than or equal to $(2\Phi(t) - 1)$. Using this and (9.229) we get (9.231).

The inequality in (9.232) is less subtle than (9.231) because we can replace $\eta(z)(h(y)G(0) + s)$ by $|\eta(z)(h(y)G(0) + s)|$ in taking the upper bound. $\qquad \square$

Lemma 9.8.3 *For the event*

$$C = \left\{ \sup_{y \in Y'} A(y, G(0), s) \leq \frac{(a - \sigma t)^2}{2} \right. \tag{9.234}$$

$$+ (a - \sigma t)(h_1|G(0) + s| - (1 - h_1)s) - \widetilde{B}(h_1, |G(0) + s|, s) \Big\}$$

we have

$$E\left(\left(1 + \frac{G(x)}{s}\right)1_C\right) \leq \left(\left(1 + \frac{u^1(x, x)}{s^2}\right)(2(1 - \Phi(t)))\right)^{1/2} \tag{9.235}$$

$$:= H(x, s, t).$$

Proof By the Schwarz inequality,

$$E\left(\left(1 + \frac{G(x)}{s}\right)1_C\right) \leq \left(\left(1 + \frac{u^1(x, x)}{s^2}\right)E(1_C)\right)^{1/2}. \tag{9.236}$$

Since η and $G(0)$ are independent, $E(1_C) = E_{G(0)}E_\eta(1_C)$. We see from (9.231) that

$$E_\eta(1_{C^c}) \geq 2\Phi(t) - 1. \tag{9.237}$$

Thus $E(1_{C^c}) \geq 2\Phi(t) - 1$ and hence $E(1_C) \leq 2(1 - \Phi(t))$. Using this in (9.236) we get (9.235). □

Proof of Theorem 9.8.1 We use (8.2) with $F\left((G_{x_i} + s)^2/2\right) = 1_C$. By (9.235), the right-hand side of (8.2) is bounded by $H(x, s, t)$. The left-hand side of (8.2) replaces $(G_{x_i} + s)^2/2$ by $(G_{x_i} + s)^2/2 + L_\infty^{x_i}$ and $|G(0) + s|$ by $(2((G(0) + s)^2/2 + L_\infty^0))^{1/2}$, which gives us

$$\widehat{P}^x P_{G(0)} P_\eta \left(\sup_{y \in Y'} \left\{ L_\infty^y - h^2(y)L_\infty^0 + A(y, G(0), s) \right\} \leq W \right) \leq H(x, s, t),$$

where

$$W = \frac{(a - \sigma t)^2}{2} + \sqrt{2}h_1(a - \sigma t)\left(\sqrt{L_\infty^0 + \frac{|G(0) + s|^2}{2}} - \frac{(1 - h_1)}{\sqrt{2}h_1} \right)$$
$$- \widetilde{B}(h_1, (2((G(0) + s)^2/2 + L_\infty^0))^{1/2}, s). \tag{9.238}$$

Obviously,

$$\widehat{P}^x P_{G(0)} P_\eta \left(\sup_{y \in Y'} \left\{ L_\infty^y - h^2(y)L_\infty^0 \right\} + \sup_{y \in Y'} A(y, G(0), s) \leq W \right)$$
$$\leq H(x, s, t).$$

Now note that for any random variables U, V, W, Z,

$$P(U + V \leq W) \tag{9.239}$$
$$= P(U + V \leq W, Z \leq V) + P(U + V \leq W, Z > V)$$
$$\leq P(U + Z \leq W) + P(Z > V).$$

We apply this with

$$U = \sup_{y \in Y'} \left\{ L_\infty^y - h_1^2 L_\infty^0 \right\},$$
$$V = \frac{(a + \sigma t)^2}{2} + (a + \sigma t)\left(|G(0) + s| + (1 - h_1)s \right) + \widetilde{B}(h_1, |G(0) + s|, s),$$
$$W \text{ as in (9.238)},$$
$$Z = \sup_{y \in Y'} A(y, G(0), s).$$

Using (9.232) to handle the last term in (9.239), we obtain

$$\widehat{P}^x P_{G(0)} P_\eta \left(\sup_{y \in Y'} \left\{ L_\infty^y - h^2(y)L_\infty^0 \right\} \right. \tag{9.240}$$

$$\leq \frac{(a - \sigma t)^2}{2} + \sqrt{2}h_1(a - \sigma t)\left(\sqrt{L_\infty^0 + \frac{|G(0) + s|^2}{2}} - \frac{(1 - h_1)s}{\sqrt{2}h_1}\right)$$

$$- \widetilde{B}(h_1, (2((G(0) + s)^2/2 + L_\infty^0))^{1/2}, s)$$

$$- \frac{(a + \sigma t)^2}{2} - (a + \sigma t)(|G(0) + s| + (1 - h_1)s)$$

$$- \widetilde{B}(h_1, |G(0) + s|, s)\Bigg) \leq H(x, s, t) + 2(1 - \Phi(t)).$$

Note that

$$\widetilde{B}(h_1, (2((G(0) + s)^2/2 + L_\infty^0))^{1/2}, s)$$

$$\leq \frac{(1 - h_1)^2 s^2}{2} + 2(1 - h_1)s\left(\sqrt{L_\infty^0} + |G(0) + s|\right)$$

$$:= D(h_1, s, \sqrt{L_\infty^0}, |G(0) + s|, s). \tag{9.241}$$

Using this in (9.240) we see that

$$\widehat{P}^x P_{G(0)}\Bigg(\sup_{y \in Y'} L_\infty^y - h_1^2 L_\infty^0 \leq (a - \sigma t)\sqrt{2}h_1\sqrt{L_\infty^0} - (a + \sigma t)$$

$$|G(0) + s| - 2a(\sigma t + (1 - h_1)s) - 2D(h_1, s, \sqrt{L_\infty^0}, |G(0) + s|, s)\Bigg)$$

$$\leq H(x, s, t) + 2(1 - \Phi(t)). \tag{9.242}$$

Let $0 < \epsilon < 1/2$ and take $s = \epsilon$. Then choose t so that the right-hand side of (9.242) is less than ϵ. We then note the following key point: We can choose a $\delta = \delta(\epsilon) > 0$ and $Y' \subset \{d(y, 0) \geq \delta\}$ such that, for all $a(\delta)$ sufficiently large,

$$a - \sigma t \geq (1 - \epsilon)a(\delta), \tag{9.243}$$

$$a + \sigma t \leq (1 + \epsilon)a(\delta),$$

and

$$h_1 \geq 1 - \epsilon. \tag{9.244}$$

To accomplish this we first choose δ_1 so that $\sigma_1 := (E \sup_{y \in S, d(y,0) \leq \delta_1} \eta^2(y))^{1/2}$ satisfies $\sigma_1 t < \epsilon$ and (9.244) holds. We then, at the same time, choose $\delta > 0$ sufficiently small and $Y' \subset \{y | \delta \leq d(y, 0) \leq \delta_1\}$ with enough elements so that a is sufficiently close to $a(\delta)$ and large enough so that the inequalities in (9.243) hold as stated. (We also use (5.183).)

Using these inequalities in (9.242) we see that

$$P_{G(0)}\widehat{P}^x\Bigg(\sup_{y \in Y'} L_\infty^y - h_1^2 L_\infty^0 \leq a(\delta)(1 - \epsilon)^2\sqrt{2L_\infty^0} - a(\delta)(1 + \epsilon) \tag{9.245}$$

$$|G(0) + \epsilon| - 4\epsilon a(\delta) - \left(\epsilon^2 + 4\epsilon\left(\sqrt{L_\infty^0} + |G(0) + \epsilon|\right)\right)\right) \le \epsilon.$$

$P_{G(0)}\widehat{P}^x$ is a product measure on $(R^1 \times \Omega)$. Let us denote the elements of this space by (v, ω). We see from (9.245) that the set

$$D(v, \omega) := \sup_{y \in Y'} L_\infty^y - h_1 L_\infty^0 > a(\delta)(1 - \epsilon)^2 \sqrt{2L_\infty^0} \qquad (9.246)$$

$$- a(\delta)(1 + \epsilon)|G(0) + \epsilon| - 4\epsilon a(\delta) - \left(\epsilon^2 + 4\epsilon\left(\sqrt{L_\infty^0} + |G(0) + \epsilon|\right)\right)$$

has $P_{G(0)}\widehat{P}^x$ measure greater than or equal to $\widehat{P}^x(\widehat{\Omega}) - \epsilon$.

Let $p(x)$ denote the probability density function of $|G(0)|$. Let $b = b(\epsilon)$ be such that

$$\int_0^{b(\epsilon)} p(x)\, dx = \epsilon^{1/2}. \qquad (9.247)$$

We claim that there exists a $v \le b$ such that

$$\int_{\widehat{\Omega}} I_D(v, \omega)\widehat{P}^x(d\omega) \ge \widehat{P}^x(\widehat{\Omega}) - \epsilon^{1/2}, \qquad (9.248)$$

since, if not, by (9.247) we have

$$\int_0^\infty \int_{\widehat{\Omega}} I_D(v, \omega)\widehat{P}^x(d\omega)p(v)\, dv \qquad (9.249)$$

$$= \int_0^b \int_{\widehat{\Omega}} I_D(v, \omega)\widehat{P}^x(d\omega)p(v)\, dv + \int_b^\infty \int_{\widehat{\Omega}} I_D(v, \omega)\widehat{P}^x(d\omega)p(v)\, dv$$
$$< \epsilon^{1/2}(\widehat{P}^x(\widehat{\Omega}) - \epsilon^{1/2}) + (1 - \epsilon^{1/2})\widehat{P}^x(\widehat{\Omega}) = \widehat{P}^x(\widehat{\Omega}) - \epsilon,$$

which contradicts the fact that the set D has $P_{G(z_0)}\widehat{P}^x$ measure greater than or equal to $\widehat{P}^x(\widehat{\Omega}) - \epsilon$. Therefore,

$$\widehat{P}^x\left(\sup_{y \in Y'} L_\infty^y - h_1^2 L_\infty^0 \le a(\delta)\left((1 - \epsilon)^2\sqrt{2L_\infty^0} - (1 + \epsilon)|b(\epsilon) + \epsilon|\right.\right.$$

$$\left.\left. -4\epsilon\right) - \left(\epsilon^2 + 4\epsilon\left(\sqrt{L_\infty^0} + |b(\epsilon) + \epsilon|\right)\right)\right) \le \epsilon^{1/2}. \quad (9.250)$$

Clearly, this holds for $Y' = \{d(y, 0) \ge \delta = \delta(\epsilon)\}$, so that

$$\widehat{P}^x\left(\sup_{\substack{y \in Y \\ d(y,0) \ge \delta}} L_\infty^y - h_1^2 L_\infty^0 \le a(\delta)\left((1 - \epsilon)^2\sqrt{2L_\infty^0} - (1 + \epsilon)|b(\epsilon) + \epsilon|\right.\right.$$

$$\left.\left. -4\epsilon\right) - \left(\epsilon^2 + 4\epsilon\left(\sqrt{L_\infty^0} + |b(\epsilon) + \epsilon|\right)\right)\right) \le \epsilon^{1/2}. \quad (9.251)$$

Taking $\epsilon_n = n^{-4}$, $\delta_n = \delta(\epsilon_n)$, we see from the Borel–Cantelli Lemma that

$$\sup_{\substack{y \in Y \\ d(y,0) \geq \delta_n}} L_\infty^y - h_1^2 L_\infty^0 \leq a(\delta_n)\Big(\big(1 - \epsilon_n\big)^2 \sqrt{2L_\infty^0} - (1 + \epsilon_n)|b(\epsilon_n) + \epsilon_n|$$

$$- 4\epsilon_n\Big) - \Big(\epsilon_n^2 + 4\epsilon_n\Big(\sqrt{L_\infty^0} + |b(\epsilon_n) + \epsilon_n|\Big)\Big) \quad \forall\, n \geq n(\omega) \quad \text{a.s.}$$

Furthermore, since $\lim_{\delta \to 0} a(\delta) = \infty$ and $\lim_{n \to \infty} b(\epsilon_n) = 0$, we see that

$$\limsup_{\delta \to 0} \sup_{\substack{y \in Y \\ d(y,0) \geq \delta}} \frac{L_\infty^y}{a(\delta)} \geq (2L_\infty^0)^{1/2} \qquad \text{a.s.} \tag{9.252}$$

The almost sure in (9.252) is with respect to $\widehat{P}^x = P^x P_\lambda$, where P^x is the probability for X and P_λ refers to an independent exponential random variable with mean one. Thus we can write (9.252) as

$$\int_0^\infty P^x \left(\limsup_{\delta \to 0} \sup_{\substack{y \in Y \\ d(y,0) \geq \delta}} \frac{L_t^y}{a(\delta)} \geq (2L_t^0)^{1/2} \right) e^{-t}\, dt = 1. \tag{9.253}$$

This implies that

$$\limsup_{\delta \to 0} \sup_{\substack{y \in Y \\ d(y,0) \geq \delta}} \frac{L_t^y}{a(\delta)} \geq (2L_t^0)^{1/2} \qquad \text{a.s.} \tag{9.254}$$

for almost all $t \in R_+$. Using the monotonicity of $L_t^{\cdot}$ and the continuity of L_t^0 we get (9.221).

The proof of (9.222) is much simpler; we leave it to the reader. □

Remark 9.8.4 It is clear that Theorem 9.8.1 implies Theorem 9.2.1, although the proof of Theorem 9.2.1 is much simpler. We could have used Theorem 9.2.1 to get a lower bound for the rate of growth of $L_{\tau(t)}^{\cdot}$, but it seems more natural to give it for $L_t^{\cdot}$. Also note that the proof of Theorem 9.8.1 also goes through when $\lim_{\delta \to 0} a(\delta) < \infty$. In this case we get the lower bound in (9.12), although, obviously, the proof of (9.12) is much simpler.

9.9 Notes and references

This chapter presents all the results of Marcus and Rosen (1992d) and (1992a). The proofs are simplified in Sections 9.1 and 9.3 by using Eisenbaum's Isomorphism Theorem (Theorem 8.1.1), instead of Theorem 8.1.3, which was used in the original proofs. The simplification is considerable in Section 9.2, in which the proof of Theorem 9.2.1, using the Generalized Second Ray–Knight Theorem, is much easier than the proof of this result in Marcus and Rosen (1992d). On the other hand, the more precise Theorem 9.8.1 is similar to the proof (of our Theorem 9.2.1) given in Marcus and Rosen (1992d), except that, once again, Theorem 8.1.1 is used in place of Theorem 8.1.3. Theorem 9.8.1 is not explicitly stated in Marcus and Rosen (1992d), although some remarks in Marcus and Rosen (1992d) about how it is obtained are given in the discussion of the Markov chains considered in our Section 9.7. Theorem 9.8.1 is given in Marcus (1991), which contains examples of how it can be applied to Lévy processes with unbounded local times.

The necessary and sufficient condition for the continuity of local times of Lévy processes by Barlow and Hawkes, mentioned in the Introduction, that does not require that the Lévy processes are symmetric, is precisely our Corollary 9.4.8 with $\psi(\lambda)$ replaced by $Re\,(\psi(\lambda))$.

In Remark 9.5.8 we point out that we can only obtain Theorems 9.5.1–9.5.6 for almost all t. In Barlow (1988), he obtains many of these limit laws for Lévy processes that hold for all $t \in R_+$.

Stable mixtures are introduced in Marcus and Rosen (1993) and are considered again in Marcus and Rosen (1999). Several of the results on moduli of continuity of local times in Section 9.5 have not appeared elsewhere.

The Markov chains considered in Section 9.7.1 are symmetrized versions of an extension by Reuter (1969) of Kolmogorov's example. Kolmogorov's example is treated extensively in Chung (1967), pages 278–283, where one can find a lucid description of the sample paths of the Markov chains that we are considering. These processes have Q matrix

$$
\begin{array}{c}
0 \\
1/2 \\
1/3 \\
1/4 \\
\vdots
\end{array}
\left(
\begin{array}{ccccc}
-\infty & q_2 & q_3 & q_4 & \cdots \\
r_2 & -r_2 & 0 & 0 & \\
r_3 & 0 & -r_3 & 0 & \\
r_4 & 0 & 0 & -r_4 & \\
\vdots & \vdots & & & \ddots
\end{array}
\right).
$$

Walsh (1978) gives an example of a diffusion with a discontinuous local time in which the 1-potential density is discontinuous. Also, the process he considers is not symmetric. If one changes the speed measure to obtain a symmetric process, then both the 1-potential and the local time become continuous.

10

p-variation of Gaussian processes and local times

10.1 Quadratic variation of Brownian motion

Interest in the p-variation of stochastic processes was initiated, no doubt, by the elegant result of Lévy (1940) on the quadratic, or 2-variation, of standard Brownian motion $\{B(t), t \in R_+\}$,

$$\lim_{n \to \infty} \sum_{i=0}^{2^n - 1} \left(B\left(\frac{i}{2^n}\right) - B\left(\frac{i+1}{2^n}\right) \right)^2 = 1 \qquad \text{a.s.} \qquad (10.1)$$

Generalizations of this result lead to complications. Let $\pi = \{0 = x_0 < x_1 \cdots < x_{k_\pi} = a\}$ denote a partition of $[0, a]$, and let $m(\pi) = \sup_{1 \le i \le k_\pi} (x_i - x_{i-1})$ denote the length of the largest interval in π. ($m(\pi)$ is called the mesh of π.) Dudley (1973) showed that, for any sequence $\{\pi(n)\}$ of partitions of $[0, a]$ such that $m(\pi(n)) = o\left(\frac{1}{\log n}\right)$,

$$\lim_{n \to \infty} \sum_{x_i \in \pi(n)} (B(x_i) - B(x_{i-1}))^2 = a \qquad \text{a.s.} \qquad (10.2)$$

(To clarify the notation in (10.2) and in all that follows, note that in the expression $\sum_{x_i \in \pi} f(x_{i-1}, x_i)$, for some function f and partition π, we mean that the sum is taken over all the terms in which both x_{i-1} and x_i are contained in π.)

de la Vega (1973) showed that (10.2) no longer holds if the condition on $m(\pi(n))$ is relaxed to $m(\pi(n)) = O\left(\frac{1}{\log n}\right)$. (In fact, he showed more, as we point out below.)

It is natural to ask what happens if one considers taking the limit over all partitions. Let $Q_a(\delta) = \{\text{partitions } \pi \text{ of } [0, a] \mid m(\pi) \le \delta\}$. Taylor (1972) showed that

$$\lim_{\delta \to 0} \sup_{\pi \in Q_a(\delta)} \sum_{x_i \in \pi} \overline{\psi}(|B(x_i) - B(x_{i-1})|) = a \qquad \text{a.s.,} \qquad (10.3)$$

where $\overline{\psi}(x) = |x/\sqrt{2\log^+ \log 1/x}\,|^2$ and $(\log^+ u \equiv 1 \vee \log u)$.

It may be helpful to give a heuristic explanation of where the function $\overline{\psi}$ comes from. By (2.14) we may say that, for $|x_i - x_{i-1}|$ small,

$$|B(x_i) - B(x_{i-1})| \sim (2|x_i - x_{i-1}|\log\log(1/|x_i - x_{i-1}|))^{1/2}. \quad (10.4)$$

Note that $\overline{\psi}(x)$ is, effectively, the inverse of $(2|x|\log\log(1/|x|))^{1/2}$. Thus, approximating $|B(x_i) - B(x_{i-1})|$ as in (10.4), we see that the sum on the left-hand side of (10.3) is just $\sum_{x_i \in \pi}(x_i - x_{i-1}) = a$.

All these results are generally referred to as results about the quadratic or 2-variation of Brownian motion.

It is a short step from results about the quadratic variation of Brownian motion to results about the quadratic variation of squares of Brownian motion. Then, employing techniques we have used repeatedly, we can obtain results about the quadratic variation of the local times of Brownian motion in the spatial variable. Furthermore, as in the preceding chapters, we can approach this question in much greater generality. This is done in Marcus and Rosen (1993). In this chapter, for technical reasons we discuss in Section 10.6, we restrict ourselves to studying the variation of local times of symmetric stable processes of index $1 < \beta \leq 2$ in the spatial variable. It turns out that the appropriate power to use to get nice limit theorems like (10.2) is $2/(\beta - 1)$. We refer to results like (10.2) about the variation of stochastic processes, with 2 replaced by p, as results about the p-variation of these processes.

Our results on the p-variation of local times of symmetric stable processes follow from Theorem 8.1.1. In order to use it we need to know about the p-variation of squares of the associated Gaussian processes. As in the case of Brownian motion, these follow easily from results about the p-variation of the Gaussian processes themselves. We take this up in the next section.

10.2 p-variation of Gaussian processes

We begin with some technical lemmas. Let $B = (B_{ij})_{i,j=1}^n$ be an $n \times n$ positive definite symmetric matrix and let $\|B\|$ denote the operator norm of B as an operator from $\ell_2^n \to \ell_2^n$, that is,

$$\|B\| := \sup_{\substack{\|x\|_2 \leq 1 \\ x \in \ell_2^n}} \|Bx\|_2, \quad (10.5)$$

where $\|x\|_2$ denotes the norm of x in ℓ_2^n.

Lemma 10.2.1 *Let B be as above and suppose that*

$$\sum_{j=1}^{n} |B_{ij}| \le C \qquad \forall \, 1 \le i \le n. \tag{10.6}$$

Then $\|B\| \le C$.

Proof Let $x = (x_1, \ldots, x_n)$. Then, using the fact that B is a symmetric matrix, we have

$$
\begin{aligned}
\|Bx\|_2^2 &= \sum_{i=1}^{n} \left(\sum_{j=1}^{n} B_{ij} x_j \right)^2 \tag{10.7} \\
&\le \sum_{i=1}^{n} \left(\sum_{j=1}^{n} |B_{ij}|^{1/2} \left(|B_{ij}|^{1/2} |x_j| \right) \right)^2 \\
&\le \sum_{i=1}^{n} \left(\sum_{j=1}^{n} |B_{ij}| \right) \left(\sum_{j=1}^{n} |B_{ij}| \, |x_j|^2 \right) \\
&\le C \sum_{i=1}^{n} \sum_{j=1}^{n} |B_{ij}| \, |x_j|^2 \\
&= C \sum_{j=1}^{n} \left(\sum_{i=1}^{n} |B_{ij}| \right) |x_j|^2 \le C^2 \|x\|_2^2.
\end{aligned}
$$

$\square$

Let $\{a_k\} \in \ell_p$ and $\{b_k\} \in \ell_q$, where $1/p + 1/q = 1$. It is well known that

$$\|\{a_k\}\|_p = \sup_{\|\{b_k\}\|_q \le 1} \sum_{k=1}^{\infty} b_k a_k. \tag{10.8}$$

Using (10.8), we see that the operator norm of B defined in (10.5) can be written as

$$\|B\| = \sup_{\|\{a_j\}\|_2 \le 1, \|\{b_k\}\|_2 \le 1} \sum_{j,k=1}^{n} a_j b_k B_{j,k}. \tag{10.9}$$

We now consider a slightly broader definition of partition than the one given in Section 10.1. We let $\pi = \{b_0 = x_0 < x_1 \cdots < x_{k_\pi} = b_1\}$ denote a partition of $[b_0, b_1]$, with the understanding that b_0 and b_1 can be different for the different partitions considered. For $G = \{G(x), x \in$

$R^1\}$ a real-valued Gaussian process, we associate with a partition π the covariance matrix

$$\rho_{ij}(\pi) = E(G(x_i) - G(x_{i-1}))(G(x_j) - G(x_{j-1})) \qquad i,j = 1, \ldots, k_\pi.$$
$$(10.10)$$

We denote the median of a real-valued random variable Z by $\mathrm{med}(Z)$.

Lemma 10.2.2 *Let* $G = \{G(x), x \in R^1\}$ *be a real-valued Gaussian process and let* $\{\pi(m)\}_{m=1}^\infty$ *be partitions of* $\{[b_0(m), b_1(m)]\}_{m=1}^\infty$. *For* $p > 1$ *define*

$$\|G\|_{\pi(m),p} = \left(\sum_{x_i \in \pi(m)} |G(x_i) - G(x_{i-1})|^p \right)^{1/p}. \qquad (10.11)$$

Then

$$P\left(\left| \sup_m \|G\|_{\pi(m),p} - \mathrm{med}\left(\sup_m \|G\|_{\pi(m),p} \right) \right| > t \right) \leq 2e^{-t^2/(2\hat\sigma^2)}, \qquad (10.12)$$

where

$$\hat\sigma^2 = \sup_m \sup_{\{\{a_k\}: \sum |a_k|^q \leq 1\}} \sum_{i,j=1}^{k_{\pi(m)}} a_i a_j \rho_{i,j}(\pi(m)) \qquad (10.13)$$

and $1/p + 1/q = 1$. *If* $p \geq 2$, *then*

$$\hat\sigma^2 \leq \sup_m \|\rho(\pi(m))\|, \qquad (10.14)$$

where $\|\rho(\pi)\|$ *denotes the operator norm of* ρ *as an operator from* $\ell_2^{k_\pi} \to \ell_2^{k_\pi}$. *Also,*

$$\left| E(\sup_m \|G\|_{\pi(m),p}) - \mathrm{med}\left(\sup_m \|G\|_{\pi(m),p} \right) \right| \leq \frac{\hat\sigma}{\sqrt{2\pi}}. \qquad (10.15)$$

Proof Let $U = \Pi \times B_q$, where B_q is a countable dense subset of the unit ball of ℓ_q and $\Pi = \{\pi(m)\}_{m=1}^\infty$. For $\pi(m) \in \Pi$ and $\{a_i\} \in B_q$, set

$$H(\pi(m), \{a_i\}) = \sum_{x_i \in \pi(m)} a_i(G(x_i) - G(x_{i-1})). \qquad (10.16)$$

We see that

$$\sup_{(\pi(m), \{a_i\}) \in U} H(\pi(m), \{a_i\}) = \sup_{\pi(m) \in \Pi} \left(\sum_{i=1}^{k_{\pi(m)}} |G(x_i) - G(x_{i-1})|^p \right)^{1/p}$$
$$= \sup_{\pi(m) \in \Pi} \|G\|_{\pi(m),p}.$$

Furthermore,

$$\sup_{(\pi(m),\{a_i\})\in U} E(H^2(\pi(m),\{a_i\})) \tag{10.17}$$

$$= \sup_{(\pi(m),\{a_i\})\in U} \sum_{i,j=1}^{k_{\pi(m)}} a_i a_j \rho_{i,j}(\pi(m))$$

$$= \sup_m \sup_{\{\{a_i\}:\sum |a_i|^q \le 1\}} \sum_{i,j=1}^{k_{\pi(m)}} a_i a_j \rho_{i,j}(\pi(m)).$$

The statements in (10.12) and (10.13) now follow from (5.152).

When $p \ge 2$ we have $q \le 2$ and since, in this case, the unit ball of ℓ_q is contained in the unit ball of ℓ_2, we see that the last line of (10.17) is less than or equal to

$$\sup_m \sup_{\{\{a_i\}:\sum |a_i|^2 \le 1\}} \sum_{i,j=1}^{k_{\pi(m)}} a_i a_j \rho_{i,j}(\pi(m)) \tag{10.18}$$

$$\le \sup_m \|\rho(\pi(m))\|$$

(see (10.9)).

The statement in (10.15) follows from (5.183).   □

Let $\{G(x), x \in R^1\}$ be a mean zero Gaussian process with stationary increments. Recall the definition

$$\sigma^2(h) = E(G(x+h) - G(x))^2. \tag{10.19}$$

Theorem 10.2.3 *Let $\{G(x), x \in R^1\}$ be a mean zero Gaussian process with stationary increments and assume that $\sigma^2(h)$ is concave for $h \in [0,\delta]$, for some $\delta > 0$, and satisfies $\lim_{h\to 0} \sigma(h)/h^{1/p} = \alpha$ for some $p \ge 2$ and $0 \le \alpha < \infty$. Let $\{\pi(n)\}_{n=1}^\infty$ be partitions of $\{[b_0(n), b_1(n)]\}_{n=1}^\infty$ with $[b_0(n), b_1(n)] \subseteq [0,\delta]$ for all n such that $m(\pi(n)) = o\left(\frac{1}{\log n}\right)^{p/2}$, $\lim_{n\to\infty} b_0(n) = b_0$, and $\lim_{n\to\infty} b_1(n) = b_1$, where $b_1 - b_0 > 0$. Then*

$$\lim_{n\to\infty} \sum_{x_i \in \pi(n)} |G(x_i) - G(x_{i-1})|^p = E|\eta|^p \alpha^p (b_1 - b_0) \quad a.s., \tag{10.20}$$

where η is a normal random variable with mean 0 and variance 1.

Proof In order to use the concavity of $\sigma^2(h)$ on $[0,\delta]$, we initially take $b_1 - b_0 < \delta/2$. We now consider Lemma 10.2.2 with the sequence of partitions $\{\pi(m)\}_{m=1}^\infty$ replaced by the single partition $\pi = \pi(n)$. In this case (10.12) and (10.14) tell us that

$$P\left(\left|\|G\|_{\pi(n),p} - \mathrm{med}(\|G\|_{\pi(n),p})\right| > t\right) \leq 2e^{-t^2/(2\hat{\sigma}_n^2)} \qquad (10.21)$$

where

$$\hat{\sigma}_n^2 \leq \|\rho(\pi(n))\|. \qquad (10.22)$$

We show below that

$$\|\rho(\pi(n))\| = o\left(\frac{1}{\log n}\right) \qquad (10.23)$$

as $n \to \infty$. Assuming this, we see from (10.21), (10.22), and the Borel–Cantelli Lemma that

$$\lim_{n \to \infty} \left(\|G\|_{\pi(n),p} - \mathrm{med}\|G\|_{\pi(n),p}\right) = 0 \qquad \text{a.s.} \qquad (10.24)$$

Let $\mathrm{med}(\|G\|_{\pi(n),p}) = M_n$ and note that (see page 224)

$$M_n \quad \leq \quad 2E(\|G\|_{\pi(n),p}) \leq 2(E\|G\|_{\pi(n),p}^p)^{1/p} \qquad (10.25)$$

$$= \quad 2(E|\eta|^p)^{1/p}\left(\sum_{x_i \in \pi(n)} \sigma^p(x_i - x_{i-1})\right)^{1/p}, \qquad (10.26)$$

where we use the fact that $G(x_i) - G(x_{i-1}) \stackrel{law}{=} \sigma(x_i - x_{i-1})\eta$.

It follows from the hypotheses on σ^2 that

$$M_n \leq C(E|\eta|^p)^{1/p}(b_1 - b_0)^{1/p} \qquad \forall n, \qquad (10.27)$$

where C is an absolute constant. Choose some convergent subsequence $\{M_{n_i}\}_{i=1}^\infty$ of $\{M_n\}_{n=1}^\infty$ and suppose that

$$\lim_{i \to \infty} M_{n_i} = \overline{M}. \qquad (10.28)$$

It then follows from (10.24) and (10.28), that

$$\lim_{i \to \infty} \|G\|_{\pi(n_i),p} = \overline{M} \qquad \text{a.s.} \qquad (10.29)$$

Let us also note that it follows from (10.12), (10.23) and (10.27) that for all $r > 0$ there exist finite constants $C(r)$ such that

$$E\|G\|_{\pi(n),p}^r \leq C(r) \qquad \forall n \geq 1. \qquad (10.30)$$

Thus, in particular, $\{\|G\|_{\pi(n),p}^p; \ n = 1, \ldots\}$ is uniformly integrable. This, together with (10.29), shows that

$$\lim_{i \to \infty} E\|G\|_{\pi(n_i),p}^p = \overline{M}^p. \qquad (10.31)$$

Since it is obvious because of our assumption on σ^2 that

$$\lim_{n\to\infty} E\|G\|^p_{\pi(n),p} = (b_1 - b_0)\alpha^p E|\eta|^p, \tag{10.32}$$

we have

$$\overline{M}^p = (b_1 - b_0)\alpha^p E|\eta|^p. \tag{10.33}$$

Thus the bounded set $\{M_n\}_{n=1}^\infty$ has a unique limit point $\overline{M}$. It now follows from (10.24) that

$$\lim_{n\to\infty} \|G\|^p_{\pi(n),p} = (b_1 - b_0)\alpha^p E|\eta|^p. \tag{10.34}$$

We now obtain (10.23). This follows from Lemma 10.2.1 and our hypothesis on $m(\pi(n))$, once we show that

$$\sum_j |\rho_{ij}(\pi(n))| \le 2\sigma^2(x_i - x_{i-1}) \le 2 \max_{x_i \in \pi(n)} \sigma^2(x_i - x_{i-1}). \tag{10.35}$$

To obtain (10.35) we first note from (10.10) that

$$\begin{aligned} \rho_{ij} &= -\tfrac{1}{2}[\sigma^2(x_{j-1} - x_{i-1}) - \sigma^2(x_{j-1} - x_i)] \\ &\quad + \tfrac{1}{2}[\sigma^2(x_j - x_{i-1}) - \sigma^2(x_j - x_i)] \\ &= -A_{j-1,i} + A_{j,i}, \end{aligned} \tag{10.36}$$

where

$$A_{j,i} := \tfrac{1}{2}[\sigma^2(x_j - x_{i-1}) - \sigma^2(x_j - x_i)]. \tag{10.37}$$

Assume that $j > i$. Using the assumption that σ^2 is concave and monotonically increasing on $[0,\delta]$, we see that that $A_{j,i} \ge 0$ and also that $A_{j-1,i} \ge A_{j,i}$ for all $j \ge i$. Therefore,

$$\sum_{j=i+1}^k |\rho_{ij}| = \sum_{j=i+1}^k (A_{j-1,i} - A_{j,i}) = A_{i,i} - A_{k,i} \le A_{i,i} = \tfrac{1}{2}\sigma^2(x_i - x_{i-1}), \tag{10.38}$$

where $k = k_{\pi(n)}$ is the number of partition points in $\pi(n)$.

When $j < i$, we rewrite (10.36) as

$$\rho_{ij} = D_{j-1,i} - D_{j,i}, \tag{10.39}$$

where

$$D_{j,i} = -A_{j,i} = \tfrac{1}{2}[\sigma^2(x_i - x_j) - \sigma^2(x_{i-1} - x_j)]. \tag{10.40}$$

(For the last equality we use the fact that $\sigma^2(h) = \sigma^2(-h)$ since the Gaussian process has stationary increments.) Using the monotonicity and concavity of σ^2 once more, we see that $D_{j,i} \geq 0$ and also that $D_{j,i} \geq D_{j-1,i}$ for all $j < i$. Therefore,

$$\sum_{j=1}^{i-1} |\rho_{ij}| = \sum_{j=1}^{i-1} (D_{j,i} - D_{j-1,i}) = D_{i-1,i} - D_{0,i} \leq D_{i-1,i} = \tfrac{1}{2}\sigma^2(x_i - x_{i-1}),$$

(10.41)

and, of course,

$$\rho_{i,i} = \sigma^2(x_i - x_{i-1}).$$

(10.42)

Using (10.38), (10.41), and (10.42) we obtain (10.35). Thus we get (10.20) under the assumption that $b_1 - b_0 < \delta/2$.

We now extend the result so that it holds for $b_1 - b_0 = a$, for any $a < \infty$. For clarity, for a given partition π, we write

$$\pi = [0 = x_0(\pi) < \cdots < x_{k_\pi}(\pi) = a].$$

(10.43)

We divide $[0, a]$ into m equal subintervals $I_{j,m}(a) := [((j-1)/m)a, (j/m)a]$, $j = 1, \ldots, m$, where m is chosen so that $1/m < \delta/4$. Using the partition points of π we define

$$x_{k(j)}(\pi) = \sup_k \left\{ x_k(\pi) \colon x_k(\pi) \leq \frac{j}{m}a \right\} \qquad j = 0, \ldots, m.$$

(10.44)

Consider the subset of π given by the increasing sequence of points

$$\pi(I_{j,m}(a)) = \{x_{k(j-1)}(\pi) < x_{k(j-1)+1}(\pi) < \cdots < x_{k(j)}(\pi)\},$$

(10.45)

$j = 1, \ldots, m$. We write

$$\sum_{x_i \in \pi} |G(x_i) - G(x_{i-1})|^p = \sum_{j=1}^{m} \sum_{x_i \in \pi(I_{j,m}(a))} |G(x_i) - G(x_{i-1})|^p.$$

(10.46)

Taking the limit as n goes to infinity and using our prior result, which holds on intervals of length less than $\delta/2$, we get (10.20). □

We now show that Theorem 10.2.3 is best possible when $p = 2$ and also pretty strong when $p > 2$.

Theorem 10.2.4 *Let $\{G(x), x \in R^1\}$ be as in Theorem 10.2.3. For any $b > 0$ we can find a sequence of partitions $\{\pi(n)\}_{n=1}^\infty$ with $m(\pi(n)) \leq \dfrac{b}{\log n}$, for all n sufficiently large, such that (10.20) is false.*

Proof To simplify matters we take $\alpha = 1$ and $b_1 - b_0 = 1$. That the proof is valid for all $0 < \alpha < \infty$ and b_1 and b_0 should be obvious from the proof in this case.

For each integer $q \geq 1$, let Π_q be the set of those partitions of $[0, 1]$, each of which contains for each integer k, $0 \leq k \leq 2^{q-1} - 1$, either the interval $J_q^k = [2k/2^q, (2k+2)/2^q]$ or both intervals $I_q^{2k} = [2k/2^q, (2k+1)/2^q]$ and $I_q^{2k+1} = [(2k+1)/2^q, 2k+2/2^q]$. There are $2^{2^{q-1}}$ partitions in Π_q. One of them has mesh 2^{-q}; all the others have mesh 2^{1-q}. (Note that $\Pi_q \cap \Pi_{q+1}$ consists of a single partition, the unique partition in Π_q with mesh 2^{-q}.)

Consider the sequence of partitions $\{\pi(n)\}$ constructed as follows: $\pi(0)$ is the partition consisting of the interval $[0, 1]$. Then, for each $q \geq 0$, set

$$\mathcal{A}_{q+1} = \left\{ n \, \middle| \, 1 + \sum_{0 \leq r \leq q-1} 2^{2^r} < n \leq 1 + \sum_{0 \leq r \leq q} 2^{2^r} \right\}. \tag{10.47}$$

Since $\mathcal{A}_{q+1}$ and Π_{q+1} have the same number of elements, we can choose a bijection π from $\mathcal{A}_{q+1}$ onto Π_{q+1}. This defines $\pi(n)$ for $n \in \mathcal{A}_{q+1}$. For these values of n, $m(\pi(n)) \leq 2^{-q}$. Since $n \leq 2^{1+2^q}$ for $n \in \mathcal{A}_{q+1}$, it follows that $m(\pi(n)) \leq 1/\log n$ as long as $n > 10$.

For the Gaussian processes considered in Theorem 10.2.3 we define, for $0 \leq k \leq 2^{q-1} - 1$,

$$L(I_q^{2k}) = G\left(\frac{2k+1}{2^q}\right) - G\left(\frac{2k}{2^q}\right), \tag{10.48}$$

$$L(I_q^{2k+1}) = G\left(\frac{2k+2}{2^q}\right) - G\left(\frac{2k+1}{2^q}\right),$$

$$L(J_q^k) = G\left(\frac{2k+2}{2^q}\right) - G\left(\frac{2k}{2^q}\right),$$

and

$$M_q^k = \max\{|L(I_q^{2k})|^p + |L(I_q^{2k+1})|^p, |L(J_q^k)|^p\}. \tag{10.49}$$

We have

$$EM_q^k = \sigma^p \left(2^{-q}\right) E \max\{|\xi_q|^p + |\eta_q|^p, |\xi_q + \eta_q|^p\}, \tag{10.50}$$

where ξ_q and η_q are normal random variables with mean zero, variance 1, and

$$E\xi_q \eta_q = -\left(1 - \frac{\sigma^2(1/2^{q-1})}{2\sigma^2(1/2^q)}\right); \tag{10.51}$$

see (10.36) with $j = i + 1$.

We now show that

$$\lim_{q \to \infty} E \sup_{\pi \in \Pi_q} \|G\|_{\pi,p}^p = (1 + c_p) E|\eta|^p \tag{10.52}$$

for some $c_p > 0$, where η is a normal random variable with mean zero and variance one. To see this note that

$$E \sup_{\pi \in \Pi_q} \|G\|_{\pi,p}^p = \sum_{k=1}^{2^{q-1}-1} E(M_q^k), \tag{10.53}$$

and so, by (10.50) and the hypothesis on σ,

$$\lim_{q \to \infty} E \sup_{\pi \in \Pi_q} \|G\|_{\pi,p}^p = \tfrac{1}{2} \lim_{q \to \infty} E \max\{|\xi_q|^p + |\eta_q|^p, |\xi_q + \eta_q|^p\}. \tag{10.54}$$

To evaluate the right-hand side of (10.54), set $h_q = E\xi_q\eta_q$ and note that

$$h := \lim_{q \to \infty} E\xi_q\eta_q = -(1 - 2^{(2/p)-1}). \tag{10.55}$$

(ξ_q, η_q) has covariance matrix

$$A_q = \begin{pmatrix} 1 & h_q \\ h_q & 1 \end{pmatrix}. \tag{10.56}$$

Since $|h| < 1$, we see that for sufficiently large q, A_q is invertible and

$$A_q^{-1} = \frac{1}{1 - h_q^2} \begin{pmatrix} 1 & -h_q \\ -h_q & 1 \end{pmatrix}. \tag{10.57}$$

Let $f_q(\,\cdot\,, \,\cdot\,)$ be the joint density of ξ_q and η_q. Then, for sufficiently large q,

$$\begin{aligned} f_q(x, y) &= \left(2\pi\sqrt{1 - h_q^2}\right)^{-1} \exp\left(-(x^2 - h_q 2xy + y^2)/2(1 - h_q^2)\right) \\ &\leq C \exp\left(-D(x^2 + y^2)\right) \end{aligned} \tag{10.58}$$

for some constants C and D independent of q. (Here we again use the fact that $|h| < 1$.)

Let $g(x, y) = \max\{|x|^p + |y|^p, |x + y|^p\}$. Then

$$E \max\{|\xi_q|^p + |\eta_q|^p, |\xi_q + \eta_q|^p\} \tag{10.59}$$

$$= \int_0^\infty \iint_{g(x,y) \geq \lambda} f_q(x, y)\, dx\, dy\, d\lambda$$

$$= \int_0^\infty \lambda^{2/p} \iint_{g(u,v) \geq 1} f_q(\lambda^{1/p}u, \lambda^{1/p}v)\, du\, dv\, d\lambda.$$

Note that $f(\,\cdot\,, \,\cdot\,) := \lim_{q \to \infty} f_q(\,\cdot\,, \,\cdot\,)$ is the joint density of normal random variables with mean zero and variance 1 and with covariance h. Clearly, $f(\,\cdot\,, \,\cdot\,)$ is strictly positive. We use the Dominated Convergence Theorem to take the limit as q goes to infinity in (10.59) and get

$$\lim_{q \to \infty} E \max\{|\xi_q|^p + |\eta_q|^p, |\xi_q + \eta_q|^p\} \tag{10.60}$$

$$= \iint_{g(u,v) \geq 1} \int_0^\infty \lambda^{2/p} f(\lambda^{1/p} u, \lambda^{1/p} v) \, d\lambda \, du \, dv.$$

By the same reasoning we have

$$2E|\eta|^p = \lim_{q \to \infty} E(|\xi_q|^p + |\eta_q|^p) \tag{10.61}$$

$$= \iint_{|u|^p + |v|^p \geq 1} \int_0^\infty \lambda^{2/p} f(\lambda^{1/p} u, \lambda^{1/p} v) \, d\lambda \, du \, dv.$$

Because of the different areas of integration in the (u, v) plane we see that the right-hand side of (10.60) is equal to $(1 + c_p)$ times the right-hand side of (10.61) for some $c_p > 0$. Using this in (10.54) we get (10.52).

We now show that

$$\lim_{q \to \infty} \sup_{\pi \in \Pi_q} \|G\|_{\pi,p}^p = (1 + c_p)E|\eta|^p \tag{10.62}$$

for c_p given in (10.52). To do this we use Lemma 10.2.2 exactly as it was used in the proof of Theorem 10.2.3, but with $\|G\|_{\pi(n),p}$ replaced by $\sup_{\pi \in \Pi_q} \|G\|_{\pi,p}$. Analogous to (10.24) we have

$$\lim_{q \to \infty} \left(\sup_{\pi \in \Pi_q} \|G\|_{\pi,p} - \text{med}(\sup_{\pi \in \Pi_q} \|G\|_{\pi,p}) \right) = 0 \quad \text{a.s.} \tag{10.63}$$

because, in this case, for q fixed, $\hat{\sigma}^2 \leq 4^{1-q/p}$ for all q sufficiently large (see (10.35)).

Let $\widetilde{M}_q = \text{med} \left(\sup_{\pi \in \Pi_q} \|G\|_{\pi,p} \right)$. By (10.50),

$$EM_q^k \leq C_p \sigma^p(2^{-q}) \leq \frac{\alpha^p C_p}{2^{q-1}} \tag{10.64}$$

for all q sufficiently large, where C_p is a constant depending only on p. Therefore, it follows from (10.53) that

$$\widetilde{M}_q \leq 2E \left(\sup_{\pi \in \Pi_q} \|G\|_{\pi,p} \right) \leq C_p' \tag{10.65}$$

for some constant C_p', which is independent of q. Using this in (10.12), we see that there exist finite constants $C(r, p)$ such that

$$E \sup_{\pi \in \Pi_q} \|G\|_{\pi,p}^r \leq C(r, p) \quad \forall q \geq 1. \tag{10.66}$$

Following the proof of Theorem 10.2.3, we can show that

$$\lim_{q \to \infty} E \sup_{\pi \in \Pi_q} \|G\|_{\pi,p}^p = \lim_{q \to \infty} \widetilde{M}_q^p. \tag{10.67}$$

Using (10.52), (10.67), and (10.63), we get (10.62).

We see from (10.62) that

$$\limsup_{n \to \infty} \|G\|_{\pi(n),p}^p = (1 + c_p)E|\eta|^p. \tag{10.68}$$

Considering the condition on $(\pi(n))$, we now have that (10.20) does not hold for a sequence of partitions $\{\pi_n\}$ for which $m(\pi_n) \leq 1/\log n$. It is easy to improve this to get that (10.20) does not hold for certain sequences of partitions $\{\pi_n\}$ for which $m(\pi_n) \leq b/\log n$ for all $b > 0$. For q fixed we create $2^{2^{q-1}}$ different partitions on $[0, 2^{-j}]$ just as we did above on $[0, 1]$. To each of these we add $2^q(2^j - 1)$ intervals of size $2^{-(j+q)}$ to complete the partition on $[2^{-j}, 1]$. We still have $2^{2^{q-1}}$ partitions, but now one of them has mesh $2^{-(j+q)}$ and all the others have mesh $2^{1-(j+q)}$. Since the partitions on $[2^{-j}, 1]$ are simply the dyadic partitions, the sum of the p-th powers of the increments of the Gaussian process converges (as q goes to infinity) to $(1 - 2^{-j})E|\eta|^p$. However, on $[0, 2^{-j}]$, as above, it converges to $2^{-j}(1 + c_p)E|\eta|^p$. Thus we still have a counterexample to (10.20), but now $m(\pi_n) \leq 2^{-j}/\log n$. Since this holds for all $j \geq 1$, we obtain Theorem 10.2.4, as stated. $\qquad\square$

10.3 Additional variational results for Gaussian processes

We first consider the p-variation of squares of Gaussian processes. This may seem artificial at first thought, but it is precisely what we need to use the Eisenbaum Isomorphism Theorem to obtain results about the p-variation of local times of the Markov processes associated with these Gaussian processes.

Theorem 10.3.1 *Let* $\{G(x), x \in R^1\}$ *be a mean zero Gaussian process with stationary increments and increments variance* σ^2 *as defined in (10.19). If* $\sigma^2(h)$ *is concave for* $h \in [0, \delta]$*, for some* $\delta > 0$*, and satisfies* $\lim_{h \to 0} \sigma(h)/h^{1/p} = \alpha$ *for some* $p \geq 2$ *and* $0 \leq \alpha < \infty$*, then, for any sequence of partitions* $\{\pi(n)\}$ *of* $[0, a]$ *such that* $m(\pi(n)) = o\left(\dfrac{1}{\log n}\right)^{p/2}$,

$$\lim_{n \to \infty} \sum_{x_i \in \pi(n)} |G^2(x_i) - G^2(x_{i-1})|^p = E|\eta|^p (2\alpha)^p \int_0^a |G(x)|^p \, dx \quad a.s.,$$
$$\tag{10.69}$$

where η *is a normal random variable with mean 0 and variance 1.*

Furthermore, for any $b > 0$*, there exist partitions of* $[0, a]$ *with* $m(\pi(n)) < \dfrac{b}{\log n}$ *for which (10.69) does not hold.*

Proof Using the notation introduced in the last paragraph of the proof of Theorem 10.2.3, in analogy with (10.46), we have

$$\sum_{x_i \in \pi} |G^2(x_i) - G^2(x_{i-1})|^p \tag{10.70}$$

$$= \sum_{j=1}^{m} \sum_{x_i \in \pi(I_{j,m}(a))} |G^2(x_i) - G^2(x_{i-1})|^p$$

$$\leq 2^p \sum_{j=1}^{m} \sum_{x_i \in \pi(I_{j,m}(a))} |G(x_i) - G(x_{i-1})|^p \sup_{x_{k(j-1)}(\pi) \leq x \leq x_{k(j)}(\pi)} |G(x)|^p.$$

(The clarification of notation given following (10.2) is particularly relevant in the last two lines of (10.70) as well as to some similar statements involving subpartions that are given below.) It follows from (6.148) that the Gaussian process G has continuous sample paths almost surely. Using this fact and Theorem 10.2.3, we can take the limit, as n goes to infinity, of the terms to the right of the inequality in (10.70) to obtain

$$\limsup_{n \to \infty} \sum_{x_i \in \pi(n)} |G^2(x_i) - G^2(x_{i-1})|^p \tag{10.71}$$

$$\leq E|\eta|^p (2\alpha)^p \sum_{j=1}^{m} \frac{a}{m} \sup_{x \in I_{j,m}(a)} |G(x)|^p \qquad \text{a.s.}$$

Taking the limit of the right-hand side of (10.71) as m goes to infinity, and using the definition of Riemann integration, we get the upper bound in (10.69).

The argument that gives the lower bound is slightly more subtle. Let $B_m(a) := \{j | G(x) \text{ does not change sign on } I_{j,m}(a)\}$. Similarly to the way we obtain (10.71) we get

$$\liminf_{n \to \infty} \sum_{x_i \in \pi(n)} |G^2(x_i) - G^2(x_{i-1})|^p \tag{10.72}$$

$$\geq E|\eta|^p (2\alpha)^p \sum_{j \in B_m(a)} \frac{a}{m} \inf_{x \in I_{j,m}(a)} |G(x)|^p \qquad \text{a.s.}$$

Taking the limit of the right-hand side of (10.72) as m goes to infinity, we get the lower bound in (10.69). (We know that the set of zeros of each path of G on $[0, a]$ has measure zero. But we need not worry about this since this set, whatever its size, contributes nothing to the integral. This is because, by the uniform continuity of G, $|G|$ is arbitrarily small on sufficiently small intervals containing its zeros.)

As we remarked at the beginning of the proof of Theorem 10.2.4, the example we give also works for all $0 < \alpha, a < \infty$. Thus, if the partitions are imposed on $[0, a]$ and $\lim_{h \to \infty} \sigma(h)/h^{1/p} = \alpha$ in place of (10.52), we have

$$\lim_{q \to \infty} \sup_{\pi \in \Pi_q} \|G\|_{\pi,p}^p = (1 + c_p)\alpha^p a E |\eta|^p. \tag{10.73}$$

By the comments made at the end of Theorem 10.2.4, there are partitions with mesh size less than $b/\log n$ for which

$$\lim_{q \to \infty} \sup_{\pi \in \Pi_q} \|G\|_{\pi,p}^p = (1 + c_{p,b})\alpha^p a E |\eta|^p \qquad \text{a.s.} \tag{10.74}$$

for some $c_{p,b} > 0$, where $c_{p,b}$ is a constant depending on p and b. We now show that this implies that

$$\lim_{q \to \infty} \sup_{\pi \in \Pi_q} \|G^2\|_{\pi,p}^p = (1 + c_{p,b})(2\alpha)^p E |\eta|^p \int_0^a |G(x)|^p \, dx \qquad \text{a.s.,} \tag{10.75}$$

which implies the statement in the second paragraph of this theorem.

Let q' and r be positive integers. For each $1 \leq j \leq 2^r$, let $\Pi_{q',j}$ be the set of partitions of the form of $\Pi_{q'}$ on the interval $[(j-1)/2^r, j/2^r] \equiv I_{j,r}$ (rather than on $[0, 1]$ as we did in the proof of Theorem 10.2.4). Note that for $q = q' + r$, Π_q is the set of partitions of $[0, 1]$ formed by putting together one partition from each $\{\Pi_{q',j}\}_{j=1}^{2^r}$.

It follows from (10.74) that

$$\lim_{q \to \infty} \sup_{\pi \in \Pi_q} \|G^2\|_{\pi,p}^p = \sum_{j=1}^{2^r} \lim_{q \to \infty} \sup_{\pi \in \Pi_{q',j}} \|G^2\|_{\pi,p}^p \tag{10.76}$$

$$\leq \sum_{j=1}^{2^r} \lim_{q' \to \infty} \sup_{\pi \in \Pi_{q',j}} \|G\|_{\pi,p}^p \sup_{x \in I_{j,r}} |2G(x)|^p$$

$$= (1 + c_{p,b})(2\alpha)^p E |\eta|^p \sum_{j=1}^{2^r} \sup_{x \in I_{j,r}} |G(x)|^p \frac{1}{2^r}.$$

Passing to the limit as $r \to \infty$, we get the upper bound in (10.75). Similarly,

$$\lim_{q \to \infty} \sup_{\pi \in \Pi_q} \|G^2\|_{\pi,p}^p \tag{10.77}$$

$$\geq \sum_{j=1}^{2^r} \lim_{q' \to \infty} \sup_{\pi \in \Pi_{q',j}} \|G\|_{\pi,p}^p \inf_{x \in I_{j,r}} |2G(x)|^p$$

$$= (1 + c_{p,b})(2\alpha)^p E |\eta|^p \sum_{j=1}^{2^r} \inf_{x \in I_{j,r}} |G(x)|^p \frac{1}{2^r}.$$

Passing to the limit as $r \to \infty$, we get the lower bound in (10.75). $\square$

We now generalize (10.3) so that it holds for the Gaussian processes we are considering.

Theorem 10.3.2 *Let* $G = \{G(x), x \in R^1\}$ *be a mean zero Gaussian process with stationary increments. If* $\sigma^2(h)$ *is concave for* $h \in [0, \delta]$ *for some* $\delta > 0$, *and satisfies* $\lim_{h \to 0} \sigma(h)/h^{1/p} = \alpha$ *for some* $p \geq 2$ *and* $0 \leq \alpha < \infty$, *then, for* $\varphi(x) = \left| x / \sqrt{2 \log^+ \log 1/x} \right|^p$,

$$\lim_{\delta \to 0} \sup_{\pi \in Q_a(\delta)} \sum_{x_i \in \pi} \varphi \left(|G(x_i) - G(x_{i-1})| \right) = \alpha^p a \qquad a.s. \qquad (10.78)$$

Also,

$$\lim_{\delta \to 0} \sup_{\pi \in Q_a(\delta)} \sum_{x_i \in \pi} \varphi \left(|G^2(x_i) - G^2(x_{i-1})| \right) = (2\alpha)^p \int_0^a |G(x)|^p \, dx \qquad a.s.$$
$$(10.79)$$

Note that by Theorem 7.2.15 we can give the same heuristic explanation for the choice of φ as we give for the choice of $\overline{\psi}$ on page 456.

The proof depends on the following slight generalization of the local iterated logarithm law for certain Gaussian processes.

Lemma 10.3.3 *Let* $G = \{G(x), x \in R^1\}$ *be a mean zero Gaussian process with stationary increments. If* $\lim_{h \to 0} \sigma(h)/h^{1/p} = \alpha$ *for some* $p \geq 2$ *and* $0 \leq \alpha < \infty$, *then, for each* $t \in R^1$,

$$\lim_{\delta \to 0} \sup_{u,v \in S_\delta} \frac{|G(t-u) - G(t+v)|}{|u+v|^{1/p} \left(2 \log \log 1/(u+v) \right)^{1/2}} \leq \alpha \qquad a.s., \qquad (10.80)$$

where $S_\delta = \{(u, v) | 0 < u + v \leq \delta\}$.

Proof For fixed $t \in R^1$, we consider the Gaussian process $H(u, v) = G(t - u) - G(t + v)$. Let

$$a_\delta = med \sup_{u,v \in S_\delta} H(u, v) \qquad (10.81)$$

and

$$\sigma_\delta^* = \left(\sup_{u,v \in S_\delta} E(G(t-u) - G(t+v))^2 \right)^{1/2} \qquad (10.82)$$
$$\leq (1 + \epsilon_\delta) \alpha \delta^{1/p},$$

where $\lim_{\delta \to 0} \epsilon_\delta = 0$. By Theorem 5.4.3 with $\delta = \theta^n$ for some $\theta < 1$ and all n sufficiently large, and the Borel–Cantelli Lemma,

$$\lim_{n \to \infty} \sup_{u,v \in S_{\theta^n}} \frac{|H(u,v) - a_{\theta^n}|}{\theta^{n/p} \left(2 \log \log 1/\theta^n\right)^{1/2}} \le \alpha \qquad \text{a.s.} \tag{10.83}$$

By stationarity and Lemma 7.2.2,

$$\alpha_\delta \;\le\; E \sup_{u,v \in S_\delta} |H(u,v)| \tag{10.84}$$

$$\le\; 4E \sup_{|u-t| \le \delta} |G(u) - G(t)| \le C\delta^{1/p}.$$

Thus we see that the terms a_{θ^n} are negligible in (10.83) and can be removed. Doing this and interpolating, we get

$$\lim_{\delta \to 0} \sup_{u,v \in S_\delta} \frac{|G(t-u) - G(t+v)|}{\delta^{1/p} \left(2 \log \log 1/\delta\right)^{1/2}} \le \alpha \qquad \text{a.s.} \tag{10.85}$$

The inequality (10.80) follows from the fact that (7.53) implies (7.54) together with Lemma 7.1.7 (3). $\qquad\square$

Proof of Theorem 10.3.2 Let $\psi(x)$ denote the function that is the inverse of $\alpha|x|^{1/p}(2 \log \log 1/x)^{1/2}$ on $[0,h]$. Note that $\psi(x) \sim \varphi(x)/\alpha^p$ at zero. It is easy to see that

$$\psi\left(\sigma(x)(\log 1/x)^{1/2}\right) \le x(\log 1/x)^p \tag{10.86}$$

for all $x > 0$ sufficiently small. For $\epsilon > 0$ and $t \in [0,a]$, set

$$A_\delta = \left\{(t,\omega) : \sup_{u,v \in S_\delta} \frac{|G(t-u) - G(t+v)|}{\alpha|u+v|^{1/p} \left(2 \log \log 1/(u+v)\right)^{1/2}} \le (1+\epsilon)\right\}. \tag{10.87}$$

Note that A_δ is measurable. Let $1_\delta(t,\omega)$ be the indicator function of A_δ. It follows from Lemma 10.3.3 that, for each $t \in R^1$, $\lim_{\delta \to 0} 1_\delta(t,\omega) = 1$ almost surely. Hence, by Fubini's Theorem, for almost every ω, we have $\lim_{\delta \to 0} 1_\delta(t,\omega) = 1$ for almost every $t \in [0,a]$. Therefore, by the Dominated Convergence Theorem,

$$\lim_{\delta \to 0} \int_0^a 1_\delta(t,\omega)\, dt = a \qquad \text{a.s.} \tag{10.88}$$

This means that there exists a set Ω' of measure 1 in Ω such that, for any $\epsilon > 0$, there exists $\delta_0 = \delta_0(\omega,\epsilon)$ such that for all $\delta \le \delta_0$,

$$\int_0^a 1_\delta(t,\omega)\, dt \ge a(1-\epsilon) \qquad \forall \omega \in \Omega'. \tag{10.89}$$

Let $\pi = \{0 = x_0 < x_1 \cdots < x_{k_\pi} = a\}$ denote a partition in $Q_\delta(a)$. For a given path of $G(\,\cdot\,, \omega)$, if the interval $[x_{i-1}, x_i]$ contains a t such that $(t, \omega) \in A_\delta$, we have

$$\psi(|G(x_i, \omega) - G(x_{i-1}, \omega)|) \tag{10.90}$$
$$\leq \psi\left((1 + \epsilon)\alpha|x_i - x_{i-1}|^{1/p} \left(2 \log \log 1/(x_i - x_{i-1})\right)^{1/2}\right)$$
$$\leq ((1 + \epsilon)^p + \epsilon_\delta)\psi\left(\alpha|x_i - x_{i-1}|^{1/p} \left(2 \log \log 1/(x_i - x_{i-1})\right)^{1/2}\right)$$
$$= ((1 + \epsilon)^p + \epsilon_\delta)(x_i - x_{i-1}),$$

where $\lim_{\delta \to 0} \epsilon_\delta = 0$. (Here we use the property that ψ is regularly varying at zero with index p.) This is clearly what we want since summing over all intervals of the partition π containing a $t \in A_\delta$ for this ω, and taking the limit as δ goes to zero, we get the upper bound in (10.78). Thus, for a given ω, we must show that the sum over the intervals of π that do not contain any values of t that are also in A_δ, does not contribute anything when we take the limit as δ goes to zero.

Let

$$\Lambda(\omega) = \left\{i : \text{there is no value of } t \in (x_{i-1}, x_i) \text{ satisfying } (t, \omega) \in A_\delta\right\}, \tag{10.91}$$

and for some constant $D > p + 4$ let

$$\Lambda'(\omega) = \left\{i : |G(x_i, \omega) - G(x_{i-1}, \omega)| \tag{10.92}\right.$$
$$\left. > \alpha|x_i - x_{i-1}|^{1/p} \left(2D \log \log 1/(x_i - x_{i-1})\right)^{1/2} \right\}.$$

Let $\omega \in \Omega'$ and assume that δ is small enough so (10.89) holds for this ω. Then

$$\sum_{i \in \Lambda(\omega)} (x_i - x_{i-1}) < a\epsilon. \tag{10.93}$$

Therefore, for this $\omega \in \Omega'$,

$$\sum_{i \in \Lambda \cap (\Lambda')^c} \psi\left(|G(x_i, \omega) - G(x_{i-1}, \omega)|\right) \tag{10.94}$$
$$\leq \sum_{i \in \Lambda} \psi\left(\alpha|x_i - x_{i-1}|^{1/p} \left(2D \log \log 1/(x_i - x_{i-1})\right)^{1/2}\right)$$
$$\leq \sum_{i \in \Lambda} (D^p + \epsilon_\delta)\psi\left(\alpha|x_i - x_{i-1}|^{1/p} \left(2 \log \log 1/(x_i - x_{i-1})\right)^{1/2}\right)$$
$$= (D^p + \epsilon_\delta) \sum_{i \in \Lambda} |x_i - x_{i-1}| < (D^p + \epsilon_\delta)a\epsilon.$$

(We simplify the notation by writing Λ for $\Lambda(\omega)$ and similarly for Λ'.) Thus the sum in (10.78) can be made arbitrarily small on $\Lambda \cap (\Lambda')^c$.

To estimate $\sum_{i \in \Lambda \cap \Lambda'} \psi \left(|G(x_i, \omega) - G(x_{i-1}, \omega)| \right)$, we consider the random variable

$$Z_n(\omega) := \mathrm{card}\Big\{ j : \sup_{t,s \in J_{n,j}} |G(t, \omega) - G(s, \omega)| \qquad (10.95)$$
$$> \alpha h_n^{1/p} \left(2D \log \log 1/h_n \right)^{1/2} \Big\},$$

where $h_n = e^{-n}$ and $J_{n,j} = [jh_n/2, (j/2+1)h_n]$, $0 \le j \le 2e^n - 1$. Let

$$A_{n,j} := \Big\{ \omega : \sup_{t,s \in J_{n,j}} |G(t, \omega) - G(s, \omega)| > \alpha h_n^{1/p} \left(2D \log \log 1/h_n \right)^{1/2} \Big\}.$$
$$(10.96)$$

By the same argument used in the proof of Lemma 10.3.3, we see that for all $\epsilon > 0$ sufficiently small,

$$P(A_{n,j}) \le Cn^{-(D-\epsilon)} \le Cn^{-(p+4)} \qquad (10.97)$$

if $n = n(\epsilon)$ is sufficiently large. Since $Z_n = \sum_{j=0}^{2e^n} 1_{A_{n,j}}$, we see from Chebyshev's inequality that

$$P\left(Z_n > e^n \, n^{-(p+3/2)} \right) \le Cn^{-2} \qquad (10.98)$$

for all $n = n(p)$ sufficiently large. Therefore, by the Borel–Cantelli Lemma, for all $\omega \in \Omega''$, with $P(\Omega'') = 1$ there exists an $n_0(\omega)$ such that, for all $n \ge n_0(\omega)$,

$$Z_n(\omega) \le e^n \, n^{-(p+3/2)}. \qquad (10.99)$$

We now order the partitions in π according to their size. Let $m_0 = [\log 1/(2\delta)]$. For all $m \ge m_0$ set

$$P_m = \Big\{ i : h_{m+1}/2 \le x_i - x_{i-1} < h_m/2 \Big\}. \qquad (10.100)$$

Note that for each $i \in P_m$ there is a j for which $(x_{i-1}, \Lambda(\omega)) \subset J_{m,j}$. We write

$$\sum_{i \in \Lambda \cap \Lambda'} \psi \left(|G(x_i, \omega) - G(x_{i-1}, \omega)| \right) \qquad (10.101)$$
$$\le \sum_{m=m_0}^{\infty} \sum_{i \in P_m \cap \Lambda'} \psi \left(|G(x_i, \omega) - G(x_{i-1,\omega})| \right).$$

For our purposes we can assume that δ is small enough so that for this ω, $m_0 \ge n_0(\omega)$ and also

$$\sup_{\substack{|s-t| \le h_m/2 \\ s,t \in [0,a]}} |G(s, \omega) - G(t, \omega)| \le \sigma(h_m/2) \left(2(1 + \epsilon) \log 2/h_m \right)^{1/2} \quad (10.102)$$

for all $m \geq m_0$ (see (7.108)). Then, using (10.101), (10.102), (10.99), and (10.86), we have

$$\sum_{i \in \Lambda \cap \Lambda'} \psi\left(|G(x_i, \omega) - G(x_{i-1}, \omega)|\right) \tag{10.103}$$

$$\leq \sum_{m=m_0}^{\infty} Z_m(\omega) \psi\left(\sigma(h_m/2)\left(2(1+\epsilon)\log 2/h_m\right)^{1/2}\right)$$

$$\leq C_p \sum_{m=m_0}^{\infty} e^m\, m^{-(p+3/2)} h_m (\log 1/h_m)^p$$

$$\leq C_p \sum_{m=m_0}^{\infty} m^{-3/2}.$$

It follows from (10.94) and (10.103) that

$$\lim_{\delta \to 0} \sup_{\pi \in Q_a(\delta)} \sum_{i \in \Lambda} \psi\left(|G(x_i) - G(x_{i-1})|\right) = 0 \qquad \text{a.s.,} \tag{10.104}$$

which, together with the comments following (10.90), gives us the upper bound in (10.78).

The proof of lower bound in (10.78) is easier. For fixed $\epsilon > 0$ let

$$E_\delta = E_\delta(\omega) := \Big\{ s \in (0, a) : \psi(|G(s + h, \omega) - G(s, \omega)|)$$

$$> (1 - \epsilon)h \quad \text{for some } h \in (0, \delta) \Big\}.$$

It follows from Theorem 7.2.15 that for each fixed t and δ, $P(t \in E_\delta) = 1$. Therefore, by Fubini's Theorem we see that

$$P(|E_\delta| = a) = 1, \tag{10.105}$$

where $|\cdot|$ indicates Lebesgue measure.

Let $E := \cap_{0 < \delta \leq 1} E_\delta = \cap_{n=1}^{\infty} E_{1/n}$. It follows from (10.105) that

$$P(|E| = a) = 1. \tag{10.106}$$

Let Ω' be the subset of measure 1 of the probability space of G on which $|E| = a$, and let $\omega' \in \Omega'$. Note that for each $t \in E(\omega')$ there are arbtrarily small intervals $[t, t + h]$ for which

$$\psi\left(|G(t + h, \omega') - G(t, \omega')|\right) > (1 - \epsilon)h. \tag{10.107}$$

Consequently, we can find a finite set of disjoint intervals of length less than δ on which (10.107) holds, such that the sum of their lengths is greater than $(1-\epsilon)a$. Let $\pi = \pi(\omega')$ be a partition in $Q(\delta)$ that includes all these intervals, which we label $[t_j, t + h_j]$. We have

$$\sum_{x_i \in \pi} \psi\left(|G(x_i) - G(x_{i-1})|\right) \geq \sum_j \psi\left(|G(t_j + h_j) - G(t_j)|\right)$$

$$\geq (1-\epsilon)\sum_j h_j = (1-\epsilon)^2 a.$$

Therefore, for each $\epsilon, \delta > 0$ we have

$$P\left(\sup_{\pi \in Q(\delta)} \sum_{x_i \in \pi} \psi\left(|G(x_i) - G(x_{i-1})|\right) > (1-\epsilon)^2 a\right) = 1. \qquad (10.108)$$

Letting first ϵ and then δ decrease to zero through a countable set gives the lower bound in (10.78).

We now prove (10.79). Continuing the notation introduced in the last paragraph of the proof of Theorem 10.2.3, in addition to the subpartitions of π given by $\pi(I_{j,m}(a))$, $j = 1, \ldots, m-1$, we define the larger partition

$$\sigma(\pi)(I_{j,m}(a)) = \left\{\frac{j-1}{m}a < x_{k(j-1)+1}(\pi) < \cdots < x_{k(j)}(\pi) \leq \frac{j}{m}a\right\}, \qquad (10.109)$$

$j = 1, \ldots, m$ (note that $\dfrac{j-1}{m}a$ and $\dfrac{j}{m}a$ are points in the partition given in (10.109)). We then have

$$\sum_{x_i \in \pi} \varphi(|G^2(x_i) - G^2(x_{i-1})|) \qquad (10.110)$$

$$= \sum_{j=1}^m \sum_{x_i \in \pi(I_{j,m}(a))} \varphi(|G^2(x_i) - G^2(x_{i-1})|)$$

$$\leq \sum_{j=1}^m \sum_{x_i \in \sigma(\pi)(I_{j,m}(a))} \varphi(|G^2(x_i) - G^2(x_{i-1})|)$$

$$+ \sum_{j=1}^{m-1} \varphi(|G^2(x_{k(j)}(\pi)) - G^2(x_{k(j)+1}(\pi))|).$$

To get the inequality in (10.110) we added partition points at $\{\frac{j-1}{m}a\}_{j=2}^m$. These points are included in the first term after the inequality sign in (10.110). In the second term after the inequality sign we have written the partitions that were present that bracketed the added points.

Fix $v > 0$. It is easy to see that, for any $\epsilon > 0$, we can find a $c(\epsilon) > 0$ such that for all $c \in [0, c(\epsilon)]$,

$$\varphi(cb) \leq (1+\epsilon)\varphi(c)|b|^p \qquad \forall\, b \in [0, 2v] \qquad (10.111)$$

(this comes down to noting that for any $\epsilon' > 0$, $(\log(1/cb))^{(1+\epsilon')} \geq \log(1/c)$ for all c sufficiently small). Let $I(A)$ denote the indicator function of the set A. Since $G(x)$ is uniformly continuous almost surely on $[0, a]$, we can find a δ sufficiently small, depending on ϵ and ω, such that for all ω in a set of measure 1,

$$\sup_{\pi \in Q_a(\delta)} \sum_{x_i \in \pi} \varphi\left(|G^2(x_i) - G^2(x_{i-1})|\right) I\left(\sup_{x \in [0,a]} |G(x)| \leq v\right) \quad (10.112)$$

$$\leq (1+\epsilon)2^p \sum_{j=1}^{m} \sup_{\pi \in Q_a(\delta)} \sum_{x_i \in \sigma(\pi)(I_{j,m}(a))} \varphi(|G(x_i) - G(x_{i-1})|)$$

$$\sup_{x \in I_{j,m}(a)} |G(x)|^p + m \sup_{\substack{|x-y| \leq \delta \\ x,y \in [0,a]}} \varphi\left(|G(x) - G(y)| \, 2v\right).$$

It follows from Theorem 7.2.15 that

$$\limsup_{\delta \to 0} \sup_{|x-y| \leq \delta} \frac{|G(x) - G(y)|}{(\delta^{2/p}(\log 1/\delta))^{1/2}} = \alpha \qquad \text{a.s.} \quad (10.113)$$

Thus the last term in (10.112) is almost surely $o(\delta^r)$ as δ goes to zero, for all $r < 1$. Using this fact and taking the limit as δ goes to 0 in (10.112), we get by (10.78) that

$$\lim_{\delta \to 0} \sup_{\pi \in Q_a(\delta)} \sum_{x_i \in \pi} \varphi(|G^2(x_i) - G^2(x_{i-1})|) I\left(\sup_{x \in [0,a]} |G(x)| \leq v\right)$$

$$\leq (1+\epsilon)(2\alpha)^p \sum_{j=1}^{m} \frac{a}{m} \sup_{x \in I_{j,m}(a)} |G(x)|^p \qquad \text{a.s.} \quad (10.114)$$

Finally, taking the limit as m goes to infinity, we get

$$\lim_{\delta \to 0} \sup_{\pi \in Q_a(\delta)} \sum_{x_i \in \pi} \varphi(|G^2(x_i) - G^2(x_{i-1})|) I\left(\sup_{x \in [0,a]} |G(x)| \leq v\right)$$

$$\leq (1+\epsilon)(2\alpha)^p \int_0^a |G(x)|^p \, dx \qquad \text{a.s.,} \quad (10.115)$$

and since this holds for all $\epsilon > 0$ and all v, we get (10.79) but with a less than or equal to sign.

To get the opposite inequality we note that

$$\sup_{\pi \in Q_a(\delta)} \sum_{x_i \in \pi} \varphi(|G^2(x_i) - G^2(x_{i-1})|) \geq \sum_{j \in B_m(a)} \sup_{\pi \in Q_a(\delta)} \quad (10.116)$$

$$\sum_{x_i \in \sigma(\pi)(I_{j,m}(a))} \varphi\left(|G(x_i) - G(x_{i-1})| \inf_{x \in I_{j,m}(a)} |2G(x)|\right)$$

for $B_m(a)$ as in (10.72). Without loss of generality we assume that $\sup_i |G^2(x_i) - G^2(x_{i-1})| < e^{-1}$, so that the iterated log term is well defined. Similarly to (10.111), it is easy to see that when $bc < e^{-1}$, for any $u > 0$, and $\epsilon > 0$ sufficiently small, for c sufficiently small, we have $\varphi(cb) \geq (1 - \epsilon)\varphi(c)|b|^p$ for all $b \geq u$ (this comes down to noting that for any $\epsilon' > 0$, $(\log(1/cb))^{(1-\epsilon')} \leq \log(1/c)$ for all c sufficiently small). Therefore, for any $\epsilon > 0$ we can find a $\delta = \delta(\epsilon)$ sufficiently small, such that the term to the right of the inequality in (10.116) is greater than or equal to

$$(1 - \epsilon)2^p \sum_{j \in B_m(a)} \sup_{\pi \in Q_a(\delta)} \sum_{x_i \in \sigma(\pi)(I_{j,m}(a))} \varphi(|G(x_i) - G(x_{i-1})|)$$

$$\inf_{x \in I_{j,m}(a)} |G(x)|^p I \left(\inf_{x \in I_{j,m}(a)} |G(x)| \geq u \right). \qquad (10.117)$$

Taking the limit in (10.116) first as δ goes to zero and then as m goes to infinity, we get that the left-hand side of (10.79) is greater than or equal to

$$(1 - \epsilon)(2\alpha)^p \int_0^a |G(x)|^p I(G(x) \geq u) \, dx. \qquad (10.118)$$

Since this is true for all $\epsilon > 0$ and all $u > 0$, we obtain (10.79) but with a greater than or equal to sign. $\square$

Example 10.3.4 Let $\widetilde{G}_{0,\beta} = \{\widetilde{G}_{0,\beta}(x), x \in R^1\}$ be a Gaussian process with stationary increments and increments variance

$$\sigma_{0,\beta}^2(h) = \frac{4}{\pi} \int_0^\infty \frac{\sin^2 \lambda h/2}{\lambda^\beta} \, d\lambda \qquad (10.119)$$

$$= h^{\beta-1} \frac{4}{\pi} \int_0^\infty \frac{\sin^2 s/2}{s^\beta} \, ds = C_\beta h^{\beta-1},$$

as defined on page 330, where it is labeled simply $\widetilde{G}_0$ (see also (7.280)). Clearly, (10.20), (10.69), (10.78), and (10.79) hold for $\widetilde{G}_{0,\beta}$.

In the next section we apply the variational results about Gaussian processes to obtain variational results about local times of symmetric β stable processes. In order to use Lemma 9.1.2, we need to extend the results of the last two sections to Gaussian processes associated with symmetric stable processes killed at the end of an indendent exponential time with mean 1. Let $G_{1,\beta} = \{G_{1,\beta}(x), x \in R^1\}$ be a stationary

Gaussian process with increments variance

$$\sigma_{1,\beta}^2(h) = \frac{4}{\pi} \int\limits_0^\infty \frac{\sin^2 \lambda h/2}{1 + \lambda^\beta} \, d\lambda. \qquad (10.120)$$

These processes are first considered on page 330, where they are labeled G_1. The reason we cannot apply the results of the preceding two sections directly is because we have not established that $\sigma_{1,\beta}^2(h)$ is concave in $[0, \delta]$ for some $\delta > 0$.

Theorem 10.3.5 *Let $G_{1,\beta} = \{G_{1,\beta}(x), x \in R^1\}$ be a mean zero stationary Gaussian processes with increments variance given by (10.120). These processes satisfy (10.20), (10.69), (10.78), and (10.79), where $p = 2/(\beta - 1)$ and $\alpha = \sqrt{C_\beta}$.*

Proof This follows from Lemma 7.4.11. Let H_β be defined as in Lemma 7.4.11 with $\psi(\lambda) = \lambda^\beta$, and let $\widetilde{G}_{1,\beta} = \{G_{1,\beta}(x) - G_{1,\beta}(0), x \in R^1\}$. Then, in the notation that we are using now, we have

$$\{\widetilde{G}_{0,\beta}(x), x \in R^1\} \stackrel{law}{=} \{\widetilde{G}_{1,\beta}(x) + H_\beta(x), x \in R^1\}, \qquad (10.121)$$

where $\widetilde{G}_{1,\beta}(x)$ and H_β are independent. It follows from Example 10.3.4 that $\widetilde{G}_{1,\beta}(x) + H_\beta$ satisfy (10.20), (10.69), (10.78), and (10.79). Furthermore, it is easy to see that

$$E(H_\beta(x + h) - H_\beta(x))^2 = O(h^\gamma) \qquad (10.122)$$

as h goes to zero, for all $\gamma < (2\beta - 1) \wedge 2$. It follows from (10.122) and Theorem 7.2.1 that

$$\limsup_{\delta \to 0} \sup_{|x-y| \le \delta} \frac{|H_\beta(x) - H_\beta(y)|}{(\delta^\gamma (\log 1/\delta))^{1/2}} \le C \qquad \text{a.s.} \qquad (10.123)$$

It is easy to see that this implies that $\lim_{n \to \infty} \|H_\beta\|_{\pi(n),p} = 0$ for the partitions $\pi(n)$ defined in Theorem 10.2.3. Therefore it follows from the triangle inequality applied to $\|\widetilde{G}_{1,\beta}(x) + H_\beta\|_{\pi(n),p}$ that $\widetilde{G}_{1,\beta}$ satisfies (10.20). Therefore $G_{1,\beta}$ also satisfies (10.20).

$G_{1,\beta}$ also satisfies (10.69), since the proof of (10.69) only requires that the Gaussian process is continuous and satisfies (10.20).

We now show that $G_{1,\beta}$ satisfies (10.78). It is clear that $\widetilde{G}_{0,\beta}$ satisfies (10.78) and therefore so does $\widetilde{G}_{1,\beta}(x) + H_\beta$. For $x \in R^1$ define $\bar{\varphi}(x) = \varphi(|x|)$. Clearly $\bar{\varphi}(x)$ is convex for $x \in [-\delta, \delta]$ for some $\delta > 0$. Therefore, for any $\epsilon > 0$, for all $|a|$ and $|b|$ sufficiently small, depending on ϵ, we have

$$\bar{\varphi}\left(\frac{a}{1-\epsilon}\right) \geq \frac{1}{1-\epsilon}\bar{\varphi}\left(a+b\right) - \frac{\epsilon}{1-\epsilon}\bar{\varphi}\left(\frac{b}{\epsilon}\right) \qquad (10.124)$$

and

$$\bar{\varphi}\left(a\right) \leq (1-\epsilon)\bar{\varphi}\left(\frac{a+b}{1-\epsilon}\right) + \epsilon\bar{\varphi}\left(\frac{b}{\epsilon}\right). \qquad (10.125)$$

We use these inequalities in (10.78) with φ replaced by $\bar{\varphi}$ and with $a = \widetilde{G}_{1,\beta}(x_i) - \widetilde{G}_{1,\beta}(x_{i-1})$ and $b = H_\beta(x_i) - H_\beta(x_{i-1})$. Since all these processes are uniformly continuous, there is no problem in taking the terms arbitrarily small. It should be clear now that, in order to show that $\widetilde{G}_{1,\beta}$ satisfies (10.78), we need only show that for all $\epsilon > 0$,

$$\lim_{\delta \to 0} \sup_{\pi \in Q_a(\delta)} \sum_{x_i \in \pi} \varphi\left(\frac{|H_\beta(x_i) - H_\beta(x_{i-1})|}{\epsilon}\right) = 0 \qquad \text{a.s.} \qquad (10.126)$$

This follows immediately from (10.123) since $((2\beta - 1) \wedge 2)(p/2) > 1$. Thus we see that $\widetilde{G}_{1,\beta}$ satisfies (10.78) and hence so does $G_{1,\beta}$. This also implies that $G_{1,\beta}$ satisfies (10.79) since in the proof of Theorem 10.3.2 we show that any uniformly continuous process that satisfies (10.78) satisfies (10.79). $\qquad \square$

Remark 10.3.6 For later use let us note that (10.75) and the final comment in the above proof imply that, for any $b > 0$, we can find a sequence of partitions $\{\pi_n\}$ with $m(\pi_n) < \frac{b}{\log n}$ such that

$$\limsup_{n \to \infty} \|G_{1,\beta}^2\|_{\pi_n,p}^p = (1 + c_{p,b})(4C_\beta)^{p/2}E|\eta|^p \int_0^a |G_{1,\beta}(x)|^p \, dx \qquad \text{a.s.} \qquad (10.127)$$

for some $c_{p,b} > 0$, where $p = 2/(\beta - 1)$.

10.4 *p*-variation of local times

Theorem 10.4.1 *Let* $X = \{X(t), t \in R_+\}$ *be a real-valued symmetric stable process of index* $1 < \beta \leq 2$ *and let* $\{L_t^x, (t,x) \in R_+ \times R^1\}$ *be the local time of* X. *If* $\{\pi(n)\}$ *is any sequence of partitions of* $[0,a]$ *such that* $m(\pi(n)) = o\left(\frac{1}{\log n}\right)^{1/(\beta-1)}$, *then, for all* $t \in R_+$,

$$\lim_{n \to \infty} \sum_{x_i \in \pi(n)} |L_t^{x_i} - L_t^{x_{i-1}}|^{2/(\beta-1)} = c(\beta) \int_0^a |L_t^x|^{1/(\beta-1)} dx \qquad \text{a.s.,} \qquad (10.128)$$

where

$$c(\beta) = (2C_\beta)^{p/2} E|\eta|^p \tag{10.129}$$

$$= \frac{2^{2/(\beta-1)}}{\sqrt{\pi}} \Gamma\left(\frac{1}{\beta-1} + \frac{1}{2}\right) \left(\frac{1}{\Gamma(\beta)\sin\left(\frac{\pi}{2}(\beta-1)\right)}\right)^{1/(\beta-1)}.$$

Proof In this section we obtain (10.128) for almost all $t \in R_+$. We complete the proof at the end of Section 10.5.

Let $G_{1,\beta}$ be the Gaussian process associated with the symmetric stable process of index β killed at the end of an independent exponential time with mean 1. It follows from Theorem 10.3.5 that, under the condition on $m(\pi(n))$ given in Theorem 10.3.1, we have

$$\lim_{n\to\infty} \|G_{1,\beta}^2/2\|_{\pi(n),p} = (2C_\beta)^{p/2} (E|\eta|^p)^{1/p} \left(\int_0^a |G_{1,\beta}^2(x)/2|^{p/2}\,dx\right)^{1/p} \tag{10.130}$$

almost surely, where C_β is given in (10.119) and $p = 2/(\beta-1)$. The same proof but with a very minor modification gives

$$\lim_{n\to\infty} \|(G_{1,\beta}+s)^2/2\|_{\pi(n),p} \tag{10.131}$$

$$= (2C_\beta)^{p/2} (E|\eta|^p)^{1/p} \left(\int_0^a |(G_{1,\beta}(x)+s)^2/2|^{p/2}\,dx\right)^{1/p} \quad \text{a.s.}$$

for all $s \neq 0$. Therefore, by Lemma 9.1.2, for almost all $\omega \in \Omega_{G_{1,\beta}}$,

$$\lim_{n\to\infty} \|L_t + (G_{1,\beta}(\omega)+s)^2/2\|_{\pi(n),p} \tag{10.132}$$

$$= (2C_\beta)^{p/2} (E|\eta|^p)^{1/p} \left(\int_0^a |L_t^x + (G_{1,\beta}(x,\omega)+s)^2/2|^{p/2}\,dx\right)^{1/p}$$

for almost all $t \in R_+$. It follows that for almost all $\omega \in \Omega_{G_{1,\beta}}$,

$$\limsup_{n\to\infty} \|L_t\|_{\pi(n),p} \leq (2C_\beta)^{p/2} (E|\eta|^p)^{1/p} \tag{10.133}$$

$$\left(\left(\int_0^a |L_t^x|^{p/2}\,dx\right)^{1/p} + \left(\int_0^a |(G_{1,\beta}(x,\omega)+s)^2/2|^{p/2}\,dx\right)^{1/p}\right)$$

$$+ \limsup_{n\to\infty} \|(G_{1,\beta}(\omega)+s)^2/2\|_{\pi(n),p}$$

for almost all $t \in R_+$. Using (10.131) on the last term in (10.133) we see that for almost all $\omega \in \Omega_{G_{1,\beta}}$,

$$\limsup_{n\to\infty} \|L_t\|_{\pi(n),p} \leq (2C_\beta)^{p/2} (E|\eta|^p)^{1/p} \tag{10.134}$$

$$\left(\left(\int_0^a |L_t^x|^{p/2}\,dx\right)^{1/p} + 2\left(\int_0^a |(G_{1,\beta}(x,\omega)+s)^2/2|^{p/2}\,dx\right)^{1/p}\right)$$

for almost all t almost surely. And finally, since this holds for all $s \neq 0$, we get

$$\limsup_{n \to \infty} \|L_t\|_{\pi(n),p} \leq (2C_\beta)^{p/2} \, (E|\eta|^p)^{1/p} \tag{10.135}$$

$$\left(\left(\int_0^a |L_t^x|^{p/2} \, dx \right)^{1/p} + 2 \left(\int_0^a |G_{1,\beta}^2(x,\omega)/2|^{p/2} \, dx \right)^{1/p} \right).$$

Since $G_{1,\beta}$ has continuous sample paths, it follows from Lemma 5.3.5 that for all $\epsilon > 0$,

$$P \left(\sup_{x \in [0,a]} |G_\beta(x)| \leq \epsilon \right) > 0. \tag{10.136}$$

Therefore we can choose ω in (10.135) so that the integral involving the Gaussian process can be made arbitrarily small. Thus

$$\limsup_{n \to \infty} \|L_t\|_{\pi(n),p} \leq (2C_\beta)^{p/2} \, (E|\eta|^p)^{1/p} \left(\int_0^a |L_t^x|^{p/2} \, dx \right)^{1/p} \tag{10.137}$$

for almost all t almost surely. By the same methods we can obtain the reverse of (10.137) for the limit inferior.

Since $c(\beta) = (2C_\beta)^{p/2} \, E|\eta|^p$, we obtain (10.128). Using (5.74) and (4.102) we get (10.129). $\qquad \square$

Theorem 10.4.2 *Under the hypotheses of Theorem 10.4.1, for all $b > 0$ we can find a sequence of partitions $\{\pi(n)\}_{n=1}^\infty$ with $m(\pi(n)) \leq \frac{b}{\log n}$ such that*

$$\limsup_{n \to \infty} \sum_{x_i \in \pi(n)} |L_t^{x_i} - L_t^{x_{i-1}}|^{2/(\beta-1)} \tag{10.138}$$

$$= (1 + \delta_{\beta,b}) c(\beta) \int_0^a |L_t^x|^{1/(\beta-1)} \, dx \qquad a.s.$$

for some constant $\delta_{\beta,p} > 0$.

This shows that for Brownian motion, the restriction on $m(\pi(n))$ in Theorem 10.4.1 cannot be strengthened.

Proof We follow the proof of Theorem 10.4.1 precisely, except that, instead of (10.131), we have the same expression as in (10.131), but lim is replaced by lim sup and with an additional factor of $(1 + c_{p,b})$ on the right-hand side. This follows from Remark 10.3.6. With this change the rest of the proof of Theorem 10.4.1 gives (10.137) with an additional factor of $(1 + c_{p,b})$ on the right-hand side. By essentially the same methods we can obtain (10.137) with a greater than or equal to sign. □

10.5 Additional variational results for local times

Theorem 10.5.1 *Let* $X = \{X(t), t \in R_+\}$ *be a real-valued symmetric stable process of index* $1 < \beta \leq 2$ *and let* $\{L_t^x, (t,x) \in R_+ \times R^1\}$ *be the local time of* X. *If* $\varphi(x) = |x/\sqrt{2\log^+\log 1/x}\,|^{2/(\beta-1)}$, *then*

$$\lim_{\delta \to 0} \sup_{\pi \in Q_a(\delta)} \sum_{x_i \in \pi} \varphi(|L_t^{x_i} - L_t^{x_{i-1}}|) = c'(\beta) \int_0^a |L_t^x|^{1/(\beta-1)}\, dx \quad (10.139)$$

almost surely for each $t \in R_+$, *where*

$$c'(\beta) = \left(\frac{2}{\Gamma(\beta)\sin\left(\frac{\pi}{2}(\beta-1)\right)} \right)^{1/(\beta-1)}. \quad (10.140)$$

Proof To simplify the notation we denote, for real-valued functions $\{\tau(x), x \in R_+\}$ and $\{f(x), x \in [0, a]\}$,

$$V_{\tau,a}(f) = \lim_{\delta \to 0} \sup_{\pi \in Q_a(\delta)} \sum_{x_i \in \pi} \tau(|f(x_i) - f(x_{i-1})|). \quad (10.141)$$

The first part of the proof is basically exactly the same as the proof of Theorem 10.5.1. It follows from Theorem 10.3.5 that

$$V_{\varphi,a}\left(G_{1,\beta}^2/2\right) = (2C_\beta)^{p/2} \int_0^a |(G_{1,\beta}^2(x)/2|^{p/2}\, dx \qquad \text{a.s.,} \quad (10.142)$$

where $p = 2/(\beta - 1)$. The same proof with minor modifications gives

$$V_{\varphi,a}\left((G_{1,\beta} + s)^2/2\right) = (2C_\beta)^{p/2} \int_0^a |(G_{1,\beta}(x) + s)^2/2|^{p/2}\, dx \qquad \text{a.s.} \quad (10.143)$$

for all $s \neq 0$. Therefore, by Lemma 9.1.2, for almost all $\omega \in \Omega_{G_{1,\beta}}$,

$$V_{\varphi,a}(L_t + (G_{1,\beta}(\cdot, \omega) + s)^2/2) \quad (10.144)$$

$$= (2C_\beta)^{p/2} \int_0^a |L_t^x + (G_{1,\beta}(x, \omega) + s)^2/2|^{p/2}\, dx$$

for almost all t almost surely.

We note that by (10.111), for all $c > 0$, $\varphi(c|x|) \leq (1 + \delta')c^p\varphi(|x|)$ for any $\delta' > 0$, for all x sufficiently small. Using this and (10.125) we see that for almost all $\omega \in \Omega_{G_{1,\beta}}$ and $0 < \epsilon \leq 1/2$,

$$
\begin{aligned}
V_{\varphi,a}(L_t) \quad \leq \quad & (1+\delta')(1-\epsilon)^{1-p}V_{\varphi,a}((L_t + (G_{1,\beta}(\,\cdot\,,\omega)+s)^2/2) \\
& +(1+\delta')\epsilon^{1-p}V_{\varphi,a}((G_{1,\beta}(\,\cdot\,,\omega)+s)^2/2)
\end{aligned}
$$

for almost all t almost surely. (We assume that the partition size δ is sufficiently small so that the increments of $(L_t + (G_{1,\beta}(\,\cdot\,,\omega)+s))^2$ are also sufficiently small.) Therefore, by (10.143) and (10.144) for almost all $\omega \in \Omega_{G_{1,\beta}}$,

$$V_{\varphi,a}(L_t) \tag{10.145}$$

$$
\begin{aligned}
\leq \; & (1+\delta')(1-\epsilon)^{1-p}\,(2C_\beta)^{p/2}\left(\int_0^a |L_t^x + (G_{1,\beta}(x,\omega)+s)^2/2|^{p/2}\,dx\right) \\
& + (1+\delta')\epsilon^{1-p}\,(2C_\beta)^{p/2}\int_0^a |(G_{1,\beta}(x,\omega)+s)^2/2|^{p/2}\,dx
\end{aligned}
$$

for almost all t almost surely. Using (10.136), we can choose an ω and s such that $\sup_{x\in[0,a]}|G_{1,\beta}(x,\omega)+s|$ can be made arbitrarily small. It follows that

$$V_{\varphi,a}(L_t) \leq (2C_\beta)^{p/2}\int_0^a |L_t^x|^{p/2}\,dx \qquad \text{for almost all } t \text{ a.s.} \tag{10.146}$$

A similar argument gives (10.146) with a greater than or equal to sign. Thus we have

$$V_{\varphi,a}(L_t) = (2C_\beta)^{p/2}\int_0^a |L_t^x|^{p/2}\,dx \qquad \text{for almost all } t \text{ a.s.} \tag{10.147}$$

To show that (10.139) holds for all $t \in R_+$, we need the following lemma on scaling the local times of symmetric stable processes.

Lemma 10.5.2 *Let $X = \{X(t), t \in R_+\}$ be a real-valued symmetric stable process of index $1 < \beta \leq 2$ and let $\{L_t^x, (t,x) \in R_+ \times R^1\}$ be the local time of X. For fixed $s,t \in R_+$,*

$$\{L_t^x; x \in R^1\} \overset{law}{=} \{s^{1/\overline{\beta}}L_{t/s}^{x/s^{1/\beta}}; x \in R^1\}, \tag{10.148}$$

where $1/\beta + 1/\overline{\beta} = 1$.

Proof By (4.63) for fixed $s \in R^1$,

$$\{X(t); t \in R_+\} \overset{law}{=} \{s^{1/\beta}X(t/s); t \in R_+\}. \tag{10.149}$$

Therefore, by (2.92), which also holds for Lévy processes,

$$L_t^a = \lim_{\epsilon\to 0}\frac{1}{\epsilon}\int_0^t I(\{a \leq s^{1/\beta}X_{u/s} \leq a+\epsilon\})\,du \tag{10.150}$$

$$= \lim_{\epsilon \to 0} \frac{s}{\epsilon} \int_0^{t/s} I(\{s^{-1/\beta}a \le X_v \le s^{-1/\beta}(a + \epsilon)\}) \, dv$$

$$= s^{1/\beta} \lim_{\epsilon' \to 0} \frac{1}{\epsilon'} \int_0^{t/s} I(\{s^{-1/\beta}a \le X_v \le s^{-1/\beta}a + \epsilon'\}) \, dv.$$

This gives (10.148). $\qquad\qquad\qquad\qquad\qquad\qquad\qquad\qquad\qquad\qquad$ $\square$

Proof of Theorem 10.5.1 continued By (10.148) with $s = t/t_0$ we have

$$\{L_t^x; x \in R^1\} \overset{law}{=} \{(t/t_0)^{1/\bar{\beta}} L_{t_0}^{x(t_0/t)^{1/\beta}}; x \in R^1\}. \tag{10.151}$$

Let Q be a countable dense subset of R_+. It follows from (10.147) and Fubini's Theorem that there exists a $t_0 \in R_+$ such that

$$V_{\varphi,b}(L_{t_0}) = (2C_\beta)^{p/2} \int_0^b |L_{t_0}^x|^{p/2} \, dx \qquad \forall b \in Q \quad \text{a.s.} \tag{10.152}$$

Therefore, since $V_{\varphi,b}(L_{t_0})$ is monotone in b, (10.152) holds for all $b \in R_+$ almost surely. Furthermore, by (9.117) and Fubini's Theorem, we can choose t_0 so that $\{L_{t_0}^x; x \in [0, b]\}$ is uniformly continuous almost surely for any $b < \infty$.

By (10.151) we have

$$\left\{ (L_t^x, \int_0^c |L_t^y|^{p/2} \, dy); x \in R^1 \right\} \tag{10.153}$$

$$\overset{law}{=} \left\{ ((t/t_0)^{1/\bar{\beta}} L_{t_0}^{x(t_0/t)^{1/\beta}}, \int_0^c |(t/t_0)^{1/\bar{\beta}} L_{t_0}^{y(t_0/t)^{1/\beta}}|^{p/2} \, dy); x \in R^1 \right\}$$

$$= \left\{ ((t/t_0)^{1/\bar{\beta}} L_{t_0}^{x(t_0/t)^{1/\beta}}, (t/t_0)^{p/\bar{\beta}} \int_0^{c(t_0/t)^{1/\beta}} |L_{t_0}^z|^{p/2} \, dz); x \in R^1 \right\}$$

where the last equality uses the change of variables $z = y(t_0/t)^{1/\beta}$ in the integral and the fact that $p = 2/(\beta - 1)$, so that $p/(2\bar{\beta}) + 1/\beta = p/\bar{\beta}$.

Using the uniform continuity almost surely of $L_{t_0}^{\cdot}$ and the fact that φ is regularly varying at zero, we see that for any $t \in R_+$

$$V_{\varphi,c(t_0/t)^{1/\beta}}((t/t_0)^{1/\bar{\beta}} L_{t_0}) = (t/t_0)^{p/\bar{\beta}} V_{\varphi,c(t_0/t)^{1/\beta}}(L_{t_0}) \quad \text{a.s.} \tag{10.154}$$

It then follows from (10.153) that

$$(V_{\varphi,c}(L_t), \int_0^c |L_t^y|^{p/2} \, dy) \tag{10.155}$$

$$\overset{law}{=} ((t/t_0)^{p/\bar{\beta}} V_{\varphi,c(t_0/t)^{1/\beta}}(L_{t_0}), (t/t_0)^{p/\bar{\beta}} \int_0^{c(t_0/t)^{1/\beta}} |L_{t_0}^z|^{p/2} \, dz).$$

It now follows from (10.152), which we have seen holds for all $b \in R^1$, that for any $c > 0$

$$V_{\varphi,c}(L_t) = (2C_\beta)^{p/2} \int_0^c |L_t^x|^{p/2} \, dx \quad \text{a.s.} \tag{10.156}$$

Thus we get (10.139); (10.140) follows from (4.102). $\square$

Remark 10.5.3 We cannot use an argument similar to the one used to prove Theorem 10.5.1 to show that Theorem 10.4.1 holds for each $t \in R_+$ almost surely. This is because in Theorem 10.4.1, as stated, the subsets of R_+ of measure zero for which (10.128) may not hold could depend on the particular sequence of partitions $\{\pi(n)\}$. Thus we cannot use scaling because we do not know if a sequence of partitions for which (10.128) holds for L_{t_0} will allow (10.128) to hold when they are rescaled, as we did above, to consider L_t.

We give one final result on the p-variation of local times that mixes properties of Theorems 10.4.1 and 10.5.1, in that it is the same as (10.128) but it holds over all partitions refining to zero. This is possible because we consider a weaker form of convergence.

Theorem 10.5.4 *Let* $X = \{X(t), t \in R_+\}$ *be a real-valued symmetric stable process of index* $1 < \beta \le 2$ *and let* $\{L_t^x, (t, x) \in R_+ \times R^1\}$ *be the local time of* X. *If* $\{\pi(n)\}$ *is any sequence of partitions of* $[0, a]$ *with* $\lim_{n \to \infty} m(\pi(n)) = 0$, *then*

$$\lim_{n \to \infty} \sum_{x_i \in \pi(n)} |L_t^{x_i} - L_t^{x_{i-1}}|^{2/(\beta-1)} = c(\beta) \int_0^a |L_t^x|^{1/(\beta-1)} dx \tag{10.157}$$

in L^r *uniformly in* t *on any bounded interval of* R_+ *for all* $r > 0$, *where* $c(\beta)$ *is given in (10.129).*

We cannot prove this theorem using any of the isomorphism theorems in this book, particularly the statement about uniformity in t. The proof follows from several lemmas on moments of the L^r norm of various functions of the local times.

Lemma 10.5.5 *Let* $X = \{X(t), t \in R_+\}$ *be a symmetric stable process of index* $1 < \beta \le 2$ *and let* $\{L_t^x, (t, x) \in R_+ \times R^1\}$ *be the local time of* X. *Then, for all* $x, y, z \in R^1$, $t \in R_+$ *and integers* $m \ge 1$,

$$E^z \left((L_t^x)^m \right) = m! \int \cdots \int_{0 < t_1 < \cdots < t_m < t} p_{t_1}(x - z) \prod_{i=2}^m p_{\Delta t_i}(0) \, dt, \tag{10.158}$$

where p_t is the probability density function of $X(t)$.

Furthermore,

$$E^z\left((L_t^x - L_t^y)^{2m}\right) \tag{10.159}$$

$$= (2m)! \int \cdots \int_{0<t_1<\cdots<t_{2m}<t} (p_{t_1}(x-z) + p_{t_1}(y-z))$$

$$\prod_{i=2}^{2m} \left(p_{\Delta t_i}(0) - (-1)^{2m-i} p_{\Delta t_i}(x-y)\right) dt,$$

where $\Delta t_i = t_i - t_{i-1}$.

Proof   We show that for any $y_i \in R^1$,

$$E^z\left(\prod_{i=1}^n L_t^{y_i}\right) \tag{10.160}$$

$$= \sum_\pi \int_{0<t_1<\cdots<t_n<t} \prod_{i=1}^n p_{t_i - t_{i-1}}(y_{\pi_{i-1}}, y_{\pi_i}) \prod_{i=1}^n dt_i$$

where $t_0 = \pi_0 = 0$, $y_0 = z$ and the sum is taken over all permutations π of $\{1, \ldots, n\}$. Equation (10.158) follows immediately from this.

Note that both sides of (10.160) are continuous functions of t (we show this for p_t in the paragraph containing (4.74)). Consequently, to obtain (10.160) it suffices to show that both sides of the equation have the same Laplace transform. That is, it suffices to show that, for any $\alpha > 0$,

$$\alpha \int_0^\infty e^{-\alpha t} E^z\left(\prod_{i=1}^n L_t^{y_i}\right) dt \tag{10.161}$$

$$= \sum_\pi \alpha \int_0^\infty e^{-\alpha t} \int_{0<t_1<\cdots<t_n<t} \prod_{i=1}^n p_{t_i - t_{i-1}}(y_{\pi_{i-1}}, y_{\pi_i}) \prod_{i=1}^n dt_i\, dt.$$

By Fubini's Theorem, the right-hand side of (10.161) equals

$$\sum_\pi \int_{0<t_1<\cdots<t_n<\infty} \left(\int_{t_n}^\infty \alpha e^{-\alpha t}\, dt\right) \prod_{i=1}^n p_{t_i - t_{i-1}}(y_{\pi_{i-1}}, y_{\pi_i}) \prod_{i=1}^n dt_i$$

$$= \sum_\pi \int_{0<t_1<\cdots<t_n<\infty} e^{-\alpha t_n} \prod_{i=1}^n p_{t_i - t_{i-1}}(y_{\pi_{i-1}}, y_{\pi_i}) \prod_{i=1}^n dt_i$$

$$= \sum_\pi \int_{0<t_1<\cdots<t_n<\infty} \prod_{i=1}^n e^{-\alpha(t_i - t_{i-1})} p_{t_i - t_{i-1}}(y_{\pi_{i-1}}, y_{\pi_i}) \prod_{i=1}^n dt_i$$

$$= \sum_\pi \prod_{i=1}^n u^\alpha(y_{\pi_{i-1}}, y_{\pi_i}), \tag{10.162}$$

where u^α is the α potential density of X. Since the left-hand side of (10.161) can be written as $E^z\left(\prod_{i=1}^n L_\lambda^{y_i}\right)$, where λ is an independent exponential random variable with mean $1/\alpha$, we see that (10.161) follows from Kac's moment formula, Theorem 3.10.1. This completes the proof of (10.158).

To prove (10.159) we use the notation $\Delta_\delta h(z) = h(z+\delta) - h(z)$. When h is a function of several variables, say $(z_1, \ldots, z_{2n})$, we write Δ_{δ, z_i} for the operator Δ_δ applied to the variable z_i. Recall that $p_t(x, x') = p_t(x - x')$, so that the terms in p_t are actually functions of a single variable. Using (10.160) we have

$$E^z\left(\prod_{i=1}^{2m}(L_t^{y_i} - L_t^{x_i})\right) \tag{10.163}$$

$$= \left(\prod_{i=1}^{2m}\Delta_{y_i - x_i, x_i}\right) E^z\left(\prod_{i=1}^{2m} L_t^{x_i}\right)$$

$$= \sum_\pi \left(\prod_{i=1}^{2m}\Delta_{y_{\pi_i} - x_{\pi_i}, x_{\pi_i}}\right) \int_{0 < t_1 < \cdots < t_n < t} \prod_{i=1}^{2m} p_{\Delta t_i}(x_{\pi_{i-1}}, x_{\pi_i}) \prod_{i=1}^{2m} dt_i,$$

where $t_0 = \pi_0 = 0$, $x_0 = z$ and the sum is taken over all permutations π of $\{1, \ldots, 2m\}$. Ultimately we set all the $x_i = x$ and all the $y_i = y$, $1 \le i \le 2m$. The subscripts are an aid to understanding the proof.

We first consider the case when π is the identity permutation. We show that

$$\left\{\left(\prod_{i=1}^{2m}\Delta_{y_i - x_i, x_i}\right) \prod_{i=1}^{2m} p_{\Delta t_i}(x_{i-1}, x_i)\right\}\Bigg|_{x_i = x,\ y_i = y} \tag{10.164}$$

$$= (p_{t_1}(x - z) + p_{t_1}(y - z)) \prod_{i=2}^{2m} \left(p_{\Delta t_i}(0) - (-1)^{2m-i} p_{\Delta t_i}(x - y)\right).$$

It is easy to see that we get the same result for all permutations π. Using this in (10.163) we get (10.159).

Set $D_i = \Delta_{y_i - x_i, x_i}$ and consider the effect of $D_{2m-1} D_{2m}$ on the product $\prod_{i=1}^{2m} p_{\Delta t_i}(x_{i-1}, x_i)$. Since it operates only on x_{2m-1} and x_{2m}, we need only consider

$$D_{2m-1} D_{2m}\, p_{\Delta t_{2m-1}}(x_{2m-2}, x_{2m-1}) p_{\Delta t_{2m}}(x_{2m-1}, x_{2m}) \tag{10.165}$$

$$= D_{2m-1}\, p_{\Delta t_{2m-1}}(x_{2m-2}, x_{2m-1})$$

$$\qquad \left[p_{\Delta t_{2m}}(x_{2m-1}, y_{2m}) - p_{\Delta t_{2m}}(x_{2m-1}, x_{2m})\right]$$

$$= p_{\Delta t_{2m-1}}(x_{2m-2}, y_{2m-1})\left[p_{\Delta t_{2m}}(y_{2m-1}, y_{2m}) - p_{\Delta t_{2m}}(y_{2m-1}, x_{2m})\right]$$

$$\qquad - p_{\Delta t_{2m-1}}(x_{2m-2}, x_{2m-1})$$

$$[p_{\Delta t_{2m}}(x_{2m-1}, y_{2m}) - p_{\Delta t_{2m}}(x_{2m-1}, x_{2m})].$$

Set $x_{2m-1} = x_{2m} = x$ and $y_{2m-1} = y_{2m} = y$. The expression to the right of the last equal sign in (10.165) becomes

$$p_{\Delta t_{2m-1}}(x_{2m-2}, y)[p_{\Delta t_{2m}}(0) - p_{\Delta t_{2m}}(y - x)] \tag{10.166}$$
$$- p_{\Delta t_{2m-1}}(x_{2m-2}, x)[p_{\Delta t_{2m}}(x - y) - p_{\Delta t_{2m}}(0)]$$
$$= (p_{\Delta t_{2m-1}}(x_{2m-2}, y) + p_{\Delta t_{2m-1}}(x_{2m-2}, x))$$
$$[p_{\Delta t_{2m}}(0) - p_{\Delta t_{2m}}(y - x)].$$

Since $x_0 = z$, this gives (10.164) when $m = 1$.

Assume now that $m \geq 2$. One more application of the difference operator reveals the pattern that gives (10.164):

$$D_{2m-2}\, p_{\Delta t_{2m-2}}(x_{2m-3}, x_{2m-2})[p_{\Delta t_{2m-1}}(x_{2m-2}, y) + p_{\Delta t_{2m-1}}(x_{2m-2}, x)]$$
$$= p_{\Delta t_{2m-2}}(x_{2m-3}, y_{2m-2})[p_{\Delta t_{2m-1}}(y_{2m-2}, y) + p_{\Delta t_{2m-1}}(y_{2m-2}, x)]$$
$$- p_{\Delta t_{2m-2}}(x_{2m-3}, x_{2m-2})[p_{\Delta t_{2m-1}}(x_{2m-2}, y) + p_{\Delta t_{2m-1}}(x_{2m-2}, x)].$$

Setting $x_{2m-2} = x$ and $y_{2m-2} = y$, the right-hand side of the above equation is

$$(p_{\Delta t_{2m-2}}(x_{2m-3}, y) - p_{\Delta t_{2m-2}}(x_{2m-3}, x)) \tag{10.167}$$
$$[p_{\Delta t_{2m-1}}(0) + p_{\Delta t_{2m-1}}(y - x)]$$
$$= \left\{ D_{2m-2}\, p_{\Delta t_{2m-2}}(x_{2m-3}, x_{2m-2}) \right\} \Big|_{x_{2m-2}=x,\ y_{2m-2}=y}$$
$$[p_{\Delta t_{2m-1}}(0) + p_{\Delta t_{2m-1}}(y - x)].$$

Combining (10.166) and (10.167), we obtain

$$\left\{ \left(\prod_{i=1}^{2m} \Delta_{y_i - x_i, x_i} \right) \prod_{i=1}^{2m} p_{\Delta t_i}(x_{i-1}, x_i) \right\} \Bigg|_{x_i=x,\ y_i=y} \tag{10.168}$$
$$= \left\{ \left(\prod_{i=1}^{2m-2} \Delta_{y_i - x_i, x_i} \right) \prod_{i=1}^{2m-2} p_{\Delta t_i}(x_{i-1}, x_i) \right\} \Bigg|_{x_i=x,\ y_i=y}$$
$$(p_{\Delta t_{2m-1}}(0) + p_{\Delta t_{2m-1}}(y - x))$$
$$[p_{\Delta t_{2m}}(0) - p_{\Delta t_{2m}}(y - x)].$$

It should be clear that by continuing this procedure we get (10.164). $\square$

Let Z be a random variable on the probability space of the stable process X. We denote the L^r norm of Z with respect to P^0 by $\|Z\|_r$.

Lemma 10.5.6 *Let* $X = \{X(t), t \in R_+\}$ *be a real-valued symmetric stable process of index* $1 < \beta \le 2$ *and let* $\{L_t^x, (t, x) \in R_+ \times R^1\}$ *be the local time of* X. *Then, for all* $x, y \in R^1$, $s, t \in R_+$ *and integers* $m \ge 1$,

$$\|L_t^x - L_t^y\|_{2m} \le C(\beta, m) t^{(\beta-1)/(2\beta)} |x - y|^{(\beta-1)/2}, \tag{10.169}$$

$$\|L_t^x - L_s^x\|_m \le C'(\beta, m) |t - s|^{(\beta-1)/\beta}, \tag{10.170}$$

and

$$\|L_t^x\|_m \le C'(\beta, m) t^{(\beta-1)/\beta}, \tag{10.171}$$

where $C(\beta, m)$ *and* $C'(\beta, m)$ *are constants depending only on* β *and* m.

It is clear that the inequality in (10.170) is unchanged if we take the norm with respect to P^z for any $z \in R^1$. The same observation applies to (10.169) since it only depends on $|x - y|$.

Proof Let $p_t(u) := p_t(x, x + u)$ denote the transition probability densities of X. By Lemma 10.5.5 with $z = 0$ we see that

$$\|L_t^x - L_t^y\|_{2m}^{2m} = (2m)! \int \cdots \int_{0 < t_1 < \cdots < t_{2m} < t} (p_{t_1}(x) + p_{t_1}(y)) \tag{10.172}$$

$$\prod_{i=2}^{2m} \left(p_{\Delta t_i}(0) - (-1)^{2m-i} p_{\Delta t_i}(x - y) \right) dt,$$

where $\Delta t_i = t_i - t_{i-1}$. Writing $p_t(x)$ as the Fourier transform of its characteristic function, it is easy to see that

$$p_t(x) \le p_t(0) \quad \text{and} \quad p_s(0) = d_\beta \, s^{-1/\beta} \tag{10.173}$$

for some constant d_β. Thus we have that (10.172) is less than or equal to

$$(2m)! \, 2^m \int \cdots \int_{0 < t_1 < \cdots < t_{2m} < t} p_{t_1}(0) \prod_{j=1}^{m-1} p_{\Delta t_{2j+1}}(0) \tag{10.174}$$

$$\prod_{i=2}^{m} \left(p_{\Delta t_{2i}}(0) - (-1)^{2m-i} p_{\Delta t_{2i}}(x - y) \right) dt$$

$$\le (2m)! \, 2^m \left(\int_0^t p_s(0) \, ds \right)^m \left(\int_0^\infty (p_t(0) - p_t(x - y)) \, dt \right)^m.$$

We obtain (10.169) from (10.174) by using (4.90), (4.95), and (10.173). To obtain (10.170) we note that

$$\|L_t^x - L_s^x\|_m = \|L_{t-s}^x \circ \theta_s\|_m = (E^0 \{E^{X_s} (L_{t-s}^x)^m\})^{1/m}, \tag{10.175}$$

where θ denotes the shift operator on the space of paths of X and without loss of generality we assume that $t \geq s$. Note that by Lemma 10.5.5, for any z,

$$
\begin{aligned}
E^z (L_{t-s}^x)^m &= m! E^z \left(\int \cdots \int_{0 < t_1 < \cdots < t_m < t-s} \prod_{i=1}^m dL_{t_i}^x \right) \\
&= m! \int \cdots \int_{0 < t_1 < \cdots < t_m < t-s} p_{t_1}(x-z) \qquad (10.176) \\
&\qquad p_{t_2-t_1}(0) \cdots p_{t_m - t_{m-1}}(0)\, dt_1, \ldots, dt_m \\
&\leq m! \left(\int_0^{t-s} p_r(0) dr \right)^m .
\end{aligned}
$$

Using (10.175), (10.176), and (10.173), we get (10.170).

The bound (10.171) follows from (10.170) by taking $s = 0$.   □

Lemma 10.5.7 *Let* $X = \{X(t), t \in R_+\}$ *be a real-valued symmetric stable process of index* $1 < \beta \leq 2$ *and let* $\{L_t^x, (t,x) \in R_+ \times R^1\}$ *be the local time of* X. *Let* $p = 2/(\beta - 1)$. *Then, for all partitions* π *of* $[0,a]$, $s, t \in R_+$, *with* $s \leq t$, *and integers* $m \geq 1$,

$$
\big\| \|L_t\|_{\pi,p}^p - \|L_s\|_{\pi,p}^p \big\|_m \leq C(\beta,m) t^{(p-1)(\beta-1)/(2\beta)} |t-s|^{(\beta-1)/(2\beta)} a. \tag{10.177}
$$

In particular,

$$
\big\| \|L_t\|_{\pi,p}^p \big\|_m^{1/p} \leq C'(\beta,m) t^{(\beta-1)/(2\beta)} a^{1/p}, \tag{10.178}
$$

where $C(\beta,m)$ *and* $C'(\beta,m)$ *are constants depending only on* β *and* m. *Similarly, for any* $r \geq 1$,

$$
\left\| \int_0^a |L_t^x|^r \, dx - \int_0^a |L_s^x|^r \, dx \right\|_m \leq D(\beta,r,m) t^{(r-1)(\beta-1)/\beta} |t-s|^{(\beta-1)/\beta} a. \tag{10.179}
$$

In particular,

$$
\left\| \int_0^a |L_t^x|^r \, dx \right\|_m \leq D'(\beta,r,m) t^{r(\beta-1)/\beta} a, \tag{10.180}
$$

where $D(\beta,r,m)$ *and* $D'(\beta,r,m)$ *are constants depending only on* β, r *and* m.

Proof For a partition π we set

$$
\Delta L_t^{x_i} = L_t^{x_i} - L_t^{x_{i-1}} \qquad i = 1, \ldots, k_\pi. \tag{10.181}
$$

Suppose that $u \geq v \geq 0$. Writing $u^p - v^p$ as the integral of its derivative, we see that

$$u^p - v^p \leq p(u - v)u^{p-1}. \tag{10.182}$$

Therefore, it follows from (10.182) and the Schwarz inequality that

$$\left\| \|L_t\|_{\pi,p}^p - \|L_s\|_{\pi,p}^p \right\|_m \tag{10.183}$$

$$\leq \sum_{x_i \in \pi} \left\| |\Delta L_t^{x_i}|^p - |\Delta L_s^{x_i}|^p \right\|_m$$

$$\leq \sum_{x_i \in \pi} p \left(\left\| |\Delta L_t^{x_i}|^{p-1} \right\|_{2m} + \left\| |\Delta L_s^{x_i}|^{p-1} \right\|_{2m} \right) \left\| \Delta L_t^{x_i} - \Delta L_s^{x_i} \right\|_{2m}.$$

Let r be the smallest even integer greater than or equal to $2m(p-1)$. Then, by Hölder's inequality and (10.169), we see that

$$\left\| |\Delta L_t^{x_i}|^{p-1} \right\|_{2m} \leq \left\| \Delta L_t^{x_i} \right\|_r^{p-1} \tag{10.184}$$

$$\leq D(\beta, m) t^{(p-1)(\beta-1)/(2\beta)} (x_i - x_{i-1})^{(p-1)(\beta-1)/2},$$

where $D(\beta, m) = (C(\beta, r))^{p-1}$ and $C(\beta, r)$ is the constant in (10.169) (clearly this inequality also holds with t replaced by s).

We also have

$$\left\| \Delta L_t^{x_i} - \Delta L_s^{x_i} \right\|_{2m} = \left\| \Delta L_{t-s}^{x_i} \circ \theta_s \right\|_{2m} \tag{10.185}$$

$$= \left(E^0 \{ E^{X_s} (\Delta L_{t-s}^{x_i})^{2m} \} \right)^{1/2m}.$$

It follows from (10.169) and the remark immediately following the statement of Lemma 10.5.6 that, for all $z \in R^1$,

$$\left(E^z (\Delta L_{t-s}^{x_i})^{2m} \right)^{1/2m} = \left\| \Delta L_{t-s}^{x_i} \right\|_{2m} \tag{10.186}$$

$$\leq C(\beta, m) |t - s|^{(\beta-1)/(2\beta)} |x_i - x_{i-1}|^{(\beta-1)/2}.$$

Combining (10.183)–(10.186) and using the fact that $s \leq t$, we see that

$$\left\| \|L_t\|_{\pi,p}^p - \|L_s\|_{\pi,p}^p \right\|_m \tag{10.187}$$

$$\leq 2pD(\beta, m, p)C(\beta, m) t^{(p-1)(\beta-1)/(2\beta)} |t - s|^{(\beta-1)/(2\beta)}$$

$$\sum_{x_i \in \pi} (x_i - x_{i-1})^{(\beta-1)/2} (x_i - x_{i-1})^{(p-1)(\beta-1)/2}.$$

This gives (10.177) since the sum in (10.187) is equal to a. The statement in (10.178) follows from (10.177) by setting $s = 0$.

To prove (10.179), we take $s < t$ and note that

$$\left\| \int_0^a |L_t^x|^r \, dx - \int_0^a |L_s^x|^r \, dx \right\|_m \tag{10.188}$$

$$\leq \int_0^a \left\| |L_t^x|^r - |L_s^x|^r \right\|_m dx \leq a \sup_x \left\| |L_t^x|^r - |L_s^x|^r \right\|_m.$$

It follows from (10.182) with p replaced by $r \geq 1$, followed by the Cauchy-Schwarz inequality, that

$$\| |L_t^x|^r - |L_s^x|^r \|_m \leq r \| L_t^x - L_s^x \|_{2m} \| |L_t^x|^{r-1} \|_{2m}, \qquad (10.189)$$

and as in (10.184) we have that

$$\| |L_t^x|^{r-1} \|_{2m} \leq \| L_t^x \|_q^{r-1}, \qquad (10.190)$$

where q is the smallest even integer greater than or equal to $2m(r-1)$. The inequality in (10.179) now follows from (10.170) and (10.171). The statement in (10.180) follows from (10.179) by setting $s = 0$. $\qquad \square$

Proof of Theorem 10.5.4 Although we are dealing with a weaker form of convergence than in Theorem 10.4.1, the only way that we know to prove this theorem is by using Theorem 10.4.1. Fix $a > 0$ and let $\mathcal{P}[0, a]$ denote the partitions of $[0, a]$. For $\pi \in \mathcal{P}[0, a]$, let

$$H_\pi(t) = \sum_{x_i \in \pi} |L_t^{x_i} - L_t^{x_{i-1}}|^{2/(\beta-1)} - c(\beta) \int_0^a |L_t^x|^{1/(\beta-1)} dx, \quad (10.191)$$

where c_β is the constant in Theorem 10.4.1. By Lemma 10.5.7 we have that for any m,

$$\| H_\pi(t) \|_m \leq C(m, a, t) < \infty, \qquad (10.192)$$

where $C(m, a, t)$ is independent of π. In particular, for each t the collection $\{ H_\pi(t) ; \pi \in \mathcal{P}[0, a] \}$ is uniformly integrable. Assume that $\{ \pi'(n) \}$ is a sequence in $\mathcal{P}[0, a]$ with $m(\pi'(n)) = o\left((\log n)^{-1/(\beta-1)} \right)$. It follows from Theorem 10.4.1 and Fubini's Theorem that for some dense subset $D \subseteq R_+$, we have that for each $t \in D$, $H_{\pi'(n)}(t)$ converges to 0 almost surely. Since $\{ H_\pi(t) ; \pi \in \mathcal{P}[0, a] \}$ is uniformly integrable, we have that for any $r > 0$,

$$\lim_{n \to \infty} \| H_{\pi'(n)}(t) \|_r = 0 \qquad \forall t \in D. \qquad (10.193)$$

Fix $t \in D$ and let $\{ \pi(n) \}$ be a sequence in $\mathcal{P}[0, a]$ for which we assume only that $\lim_{n \to \infty} m(\pi(n)) = 0$. If $\| H_{\pi(n)}(t) \|_r$ did not converge to 0, we could find some $\epsilon > 0$ and a subsequence $n_j \to \infty$ with

$$\| H_{\pi(n_j)}(t) \|_r \geq \epsilon \qquad j = 1, 2, \dots \qquad (10.194)$$

By taking a further subsequence, if necessary, which we again denote by n_j, we can assume that $m(\pi(n_j)) = o\left((\log n(j))^{-1/(\beta-1)} \right)$. This gives a contradiction between (10.193) and (10.194). Thus we see that (10.193) holds whenever $\lim_{n \to \infty} m(\pi'(n)) = 0$.

Now fix $T > 0$ and a sequence $\{\pi(n)\}$ in $\mathcal{P}[0, a]$ with $\lim_{n \to \infty} m(\pi(n)) = 0$. By Lemma 10.5.7, we have that for any $r > 0$ and any $\epsilon > 0$ we can find a $\delta > 0$ such that

$$\sup_{\substack{0 \leq s, t \leq T \\ |s-t| \leq \delta}} \|H_{\pi(n)}(s) - H_{\pi(n)}(t)\|_r \leq \epsilon \qquad \forall n \geq 1. \qquad (10.195)$$

Choose a finite set $\{t_1, \ldots, t_k\}$ in $D \cap [0, T]$ such that $\cup_{j=1}^{k} [t_j - \delta, t_j + \delta]$ covers $[0, T]$. By (10.193) we can choose an N_ϵ such that

$$\sup_{j=1,\ldots,k} \|H_{\pi(n)}(t_j)\|_r \leq \epsilon \qquad \forall n \geq N_\epsilon. \qquad (10.196)$$

Combined with (10.195) this shows that

$$\sup_{0 \leq s \leq T} \|H_{\pi(n)}(s)\|_r \leq 2\epsilon \qquad \forall n \geq N_\epsilon. \qquad (10.197)$$

$\square$

Using Theorem 10.5.4, we can complete the proof of Theorem 10.4.1.

Proof of Theorem 10.4.1 continued In Section 10.4 we show that (10.128) holds for almost all $t \in R_+$ almost surely. Therefore, by Fubini's Theorem, we can find a dense subset $T' \in R_+$ such that

$$\lim_{n \to \infty} \sum_{x_i \in \pi(n)} |L_t^{x_i} - L_t^{x_{i-1}}|^{2/(\beta-1)} = c(\beta) \int_0^a |L_t^x|^{1/(\beta-1)} dx \qquad \text{a.s.,}$$
$$(10.198)$$

for all $t \in T'$. Fix $t > 0$, and let s_k, $k = 1, \ldots$ be a sequence in T' with $s_k \uparrow t$ and

$$\sum_{k=1}^{\infty} (t - s_k)^{(\beta-1)/2\beta} < \infty. \qquad (10.199)$$

By the additivity of local times,

$$L_t^x - L_{s_k}^x = L_{t-s_k}^x \circ \theta_{s_k}. \qquad (10.200)$$

Set $p = 2/(\beta - 1)$ and note that

$$\begin{aligned} A_k &:= \limsup_{n \to \infty} \left| \|L_t\|_{\pi(n),p} - \|L_{s_k}\|_{\pi(n),p} \right| \qquad (10.201) \\ &\leq \limsup_{n \to \infty} \|L_t - L_{s_k}\|_{\pi(n),p} \\ &= \limsup_{n \to \infty} \|L_{t-s_k} \circ \theta_{s_k}\|_{\pi(n),p}. \end{aligned}$$

Let $\overline{X}_r = X_{r+s_k} - X_{s_k}$, $r \geq 0$. Then $\overline{X} = \{\overline{X}_r; r \geq 0\}$ is a copy of $\{X_r; r \geq 0\}$ that is independent of X_{s_k}. Let $\{\overline{L}_r^x; (x, r) \in R^1 \times R_+\}$

denote the local time for the process $\{\overline{X}_r; \, r \geq 0\}$. It is easy to check that

$$L_{t-s_k}^x \circ \theta_{s_k} = \overline{L}_{t-s_k}^{x-X_{s_k}}. \tag{10.202}$$

Therefore,

$$
\begin{aligned}
\|L_{t-s_k} \circ \theta_{s_k}\|_{\pi(n),p} &= \|\overline{L}_{t-s_k}^{\cdot - X_{s_k}}\|_{\pi(n),p} \tag{10.203} \\
&= \|\overline{L}_{t-s_k}^x\|_{\pi(n)-X_{s_k},p},
\end{aligned}
$$

where $\pi(n) - X_{s_k}$ is the partition of $[0,a] - X_{s_k} = [-X_{s_k}, a - X_{s_k}]$ obtained by subtracting X_{s_k} from every element of the partition $\pi(n)$.

For fixed X_{s_k}, it follows from Theorem 10.5.4 that

$$\lim_{n \to \infty} \|\overline{L}_{t-s_k}\|_{\pi(n)-X_{s_k},p} = c(\beta)^{1/p}\|\overline{L}_{t-s_k}\|_{p/2,[0,a]-X_{s_k}}^{1/2} \quad \text{in } L_{\overline{X}}^1, \tag{10.204}$$

where $\|L_s\|_{r,[a,b]} = \left(\int_a^b |L_s^x|^r \, dx\right)^{1/r}$ and $L_{\overline{X}}^1$ denotes the L^1 space with respect to $\overline{X}$. Using (10.201)–(10.204) and Hölder's inequality, we have

$$
\begin{aligned}
E(A_k \mid X_{s_k}) \qquad &\tag{10.205} \\
\leq c(\beta)^{1/p} E(\|\overline{L}_{t-s_k}\|_{p/2,[0,a]-X_{s_k}}^{1/2} \mid X_{s_k}) & \\
\leq c(\beta)^{1/p} \left|E(\|\overline{L}_{t-s_k}\|_{p/2,[0,a]-X_{s_k}}^{p/2} \mid X_{s_k})\right|^{1/p} & \\
\leq \left(c(\beta)D'(\beta,p/2,1)a\right)^{1/p} (t-s_k)^{(\beta-1)/2\beta}, &
\end{aligned}
$$

where in the last line we use (10.180). Therefore,

$$E(A_k) \leq C\,(t-s_k)^{(\beta-1)/2\beta} \tag{10.206}$$

for some finite constant C independent of k. Then, by (10.199) and the Borel–Cantelli Lemma,

$$A_k \to 0 \quad \text{a.s.} \tag{10.207}$$

The proof is completed by observing that, for each k,

$$
\begin{aligned}
\limsup_{n \to \infty} \|\overline{L}_t\|_{\pi(n),p} \qquad &\tag{10.208} \\
\leq \limsup_{n \to \infty} \|L_{s_k}\|_{\pi(n),p} + A_k & \\
= c(\beta)^{1/p}\|L_{s_k}\|_{p/2,[0,a]}^{1/2} + A_k, &
\end{aligned}
$$

and

$$
\begin{aligned}
\liminf_{n \to \infty} \|L_t\|_{\pi(n),p} \qquad &\tag{10.209} \\
\geq \liminf_{n \to \infty} \|L_{s_k}\|_{\pi(n),p} - A_k &
\end{aligned}
$$

$$= c(\beta)^{1/p}\|L_{s_k}\|_{p/2,[0,a]}^{1/2} - A_k.$$

Therefore,

$$\lim_{k\to\infty}\lim_{n\to\infty}\|L_{s_k}\|_{\pi(n),p} = \lim_{n\to\infty}\|L_t\|_{\pi(n),p} \quad \text{a.s.} \qquad (10.210)$$

It is easy to see that

$$\lim_{k\to\infty}\int_0^a |L_{s_k}^x|^{p/2}\,dx = \int_0^a |L_t^x|^{p/2}\,dx \quad \text{a.s.,} \qquad (10.211)$$

since $\{L_s^x\,;(s,x)\in[0,t]\times[0,a]\}$ is uniformly continuous almost surely. Since (10.198) holds with $t=s_k$ for all k, taking the limit as k goes to infinity we extend (10.128) to all $t\in R_+$. $\qquad\square$

10.6 Notes and references

The material in this chapter is taken from Marcus and Rosen (1992c).

Several earlier papers deal with the quadratic variation of Brownian motion and its local times and some with generalizations to larger classes of processes. Theorem 10.2.3, for Brownian motion, is due to Dudley (1973). Giné and Kline (1975) consider the quadratic variation of a larger class of Gaussian processes. Theorem 10.2.4, for Brownian motion, is due to de la Vega (1973). He only states the result for $b\geq 3$; however, a minor modification of his proof gives all $b>0$ (the description of the partitions used to obtain the examples that establish Theorem 10.2.4 is taken, almost verbatim, from de la Vega (1973)). Other results about various aspects of the p-variation of Gaussian processes appear in Kono (1969), Jain and Monrad (1983), and Adler and Pyke (1993).

There is nothing special about p-variation. Essentially, the function that operates on the increments of the Gaussian process is the inverse of $\sigma(h)$ in (10.19). Our approach to the study of p-variation is to express the ℓ_p norm as the supremum of a linear functional on its dual space. This is simple because σ^{-1} in Theorem 10.2.3 is effectively a power. When it isn't we can apply the same ideas used in the proof of Theorem 10.2.3, but we must deal with Orlicz spaces. This is done in Marcus and Rosen (1993), but it does not seem important to bother with the technicalities of Orlicz spaces in this book.

Theorem 10.3.2, for Brownian motion, is due to Taylor (1972). It is generalized to a larger class of Gaussian processes, including those we consider, by Kawada and Kono (1973). In contrast to the technical difficulties encountered in extending Theorem 10.2.3, it is easy to see how to extend Theorem 10.3.2. The proof of the lower bound only requires the existence of an iterated logarithm law for local behavior of the Gaussian process at a point. Therefore, the lower bound holds under the hypotheses of Theorem 7.4.12. (It is true that this only applies to associated processes, but since we are ultimately concerned with applying these results to local times, this should not discourage us.) The upper bound should hold under the general conditions of Corollary 7.2.3 and some asymptotic monotonicity condition on σ at zero (see Theorem 9.5.13 for an example of how these conditions are combined). Actually one should be able to extend these results even further. We know from Theorem 7.6.4 that the local modulus of continuity does not always contain an iterated logarithm term. Theorem 10.3.2 ought to hold nevertheless, with the function ϕ being the inverse of the local modulus of continuity. (It may be difficult to get the correct constant in general. Also, we should emphasize that we have not worked out the details. There may be unanticipated difficulties along the way.)

Some results along the lines of Theorem 10.5.4 were obtained prior to Marcus and Rosen (1992c). Note that when $\beta = 1 + \frac{1}{k}$, where k is an integer, then (10.157) is

$$\lim_{n \to \infty} \sum_{x_i \in \pi(n)} (L_t^{x_i} - L_t^{x_{i-1}})^{2k} = c(\beta) \int_0^a |L_t^x|^k dx, \qquad (10.212)$$

where the right-hand side is a k-fold self-intersection local time for intersections of the underlying stable process in $[0, a]$. In particular, for the local time of Brownian motion we have

$$\lim_{n \to \infty} \sum_{x_i \in \pi(n)} (L_t^{x_i} - L_t^{x_{i-1}})^2 = 4 \int_0^a L_t^x dx \qquad (10.213)$$

(keep in mind the different characteristic functions of Brownian motion and 2-stable processes). Formula (10.213), but with convergence in probability, was obtained by Bouleau and Yor (1981) and Perkins (1982), and allows one to develop stochastic integration with respect to the space parameter of Brownian local time; see also Walsh (1983). Formula (10.212), with convergence in L^2, was established in Rosen (1993).

Lemma 10.5.5 is given in Rosen (1991).

11

Most visited sites of symmetric stable processes

Let X be a strongly symmetric Borel right process with continuous α-potential densities $u^\alpha(x, y)$, $\alpha > 0$ and state space R^1. Let $\{L_t^y, (t, y) \in (R_+ \times R^1)\}$ denote the local times of X and let

$$\mathcal{V}_t = \{x \in R^1 \mid L_t^x = \sup_y L_t^y\}. \tag{11.1}$$

Following the discussion in Subsection 2.9, we refer to $\mathcal{V}_t$ as the most visited sites, or the favorite points, of X up to time t. Set

$$V_t = \inf_{x \in \mathcal{V}_t} |x|. \tag{11.2}$$

$V = \{V_t, t \in R_+\}$ is a R_+ valued stochastic process. We investigate the behavior of V_t as $t \to \infty$ and also as $t \to 0$. We restrict ourselves to the case in which X is a symmetric β-stable process, $1 < \beta \leq 2$. We discuss some generalizations to a larger class of processes in Section 11.7.

In Section 11.2 we restrict our attention to Brownian motion because the problem is considerably simpler in that case. Indeed, the major inequality, Lemma 11.5.1, that permits us to get results for symmetric β-stable processes for $1 < \beta < 2$ is a triviality for Brownian motion (see the comment following Lemma 11.5.1).

11.1 Preliminaries

Many of the preliminary results we need to study V for Brownian motion are special cases of results for symmetric stable processes. To be efficient we make some preliminary observations that are valid for all these processes. The Gaussian processes associated with symmetric stable processes with index $1 < \beta \leq 2$ are fractional Brownian motions.

Let $\eta = \{\eta(s), s \in R^1\}$ be a fractional Brownian motion of index $\beta - 1$

as defined on page 276, that is, η is a mean zero Gaussian process with stationary increments, $\eta(0) = 0$, and

$$E\left(|\eta(s) - \eta(t)|^2\right) = C_\beta |s - t|^{\beta - 1}, \qquad (11.3)$$

where $\beta \in (1, 2]$ and C_β is given in (7.280). We use this normalization because these are the Gaussian processes associated with the canonical symmetric stable process of index β for $\beta \in (1, 2]$, that is, $\psi(\lambda) = |\lambda|^\beta$ in (7.262), killed the first time it hits zero (see Theorem 4.2.4, Remark 4.2.6, and Lemma 7.4.9). These processes are used in the isomorphism theorems later in this chapter.

It follows from (11.3) that the covariance of fractional Brownian motion of index $\beta - 1$ is

$$\Gamma(s, t) = \frac{C_\beta}{2}\left(|s|^{\beta - 1} + |t|^{\beta - 1} - |s - t|^{\beta - 1}\right). \qquad (11.4)$$

We point out in Remark 4.2.6 that when $\beta = 2$, $\eta(s) = \sqrt{2}B(s)$, where $B = \{B(s), s \in R^1\}$ is standard two-sided Brownian motion. It is easy to see by (11.4) that when $\beta = 2$ and $st < 0$, $\Gamma(s, t) = 0$, consistent with the fact that for two-sided Brownian motion, $\{B(s), s \in R_+\}$ and $\{B(s), s \in (-\infty, 0)\}$ are independent. However, for $1 < \beta < 2$, $\Gamma(s, t) > 0$ whenever s and t are not both equal to zero.

It also follows from (11.3) that

$$\left\{s^{-1/2}\eta\left(s^{1/(\beta-1)}x\right), \, x \in R^1\right\} \stackrel{law}{=} \left\{\eta(x), \, x \in R^1\right\} \qquad (11.5)$$

for any $s > 0$. We refer to this as the scaling property of fractional Brownian motion.

In Lemma 10.5.2 we give the scaling property of the local times of symmetric β-stable processes. This carries over to the total accumulated local time up to τ, the inverse local time at zero (see (3.132)).

Lemma 11.1.1 *Let $X = \{X_t \, ; \, t \geq 0\}$ be a symmetric stable process of index $1 < \beta \leq 2$ in R^1 and denote its local times by $L = \{L_t^x, (x, t) \in R^1 \times R_+\}$. Let $\tau(\,\cdot\,)$ denote the inverse local time of L^0. Then, for all $h > 0$,*

$$\left\{L_{\tau(hr)}^x \, ; \, (x, r) \in R^1 \times R_+\right\} \stackrel{law}{=} \left\{hL_{\tau(r)}^{x/h^{1/(\beta-1)}} \, ; \, (x, r) \in R^1 \times R_+\right\}. \qquad (11.6)$$

Proof Using Lemma 10.5.2 with $s = h^{\beta/(\beta-1)}$, we see that

$$\left\{L_t^x \, ; \, (x, t) \in R^1 \times R_+\right\} \qquad (11.7)$$

$$\overset{law}{=} \left\{ hL_{t/h^{\beta/(\beta-1)}}^{x/h^{1/(\beta-1)}} ; (x,t) \in R^1 \times R_+ \right\}.$$

Under this equivalence in law it follows that

$$\tau(hr) = \inf\{t > 0 \mid L_t^0 > hr\} \tag{11.8}$$
$$\overset{law}{=} \inf\{t > 0 \mid hL_{t/h^{\beta/(\beta-1)}}^0 > hr\}$$
$$= \inf\{t > 0 \mid L_{t/h^{\beta/(\beta-1)}}^0 > r\} = h^{\beta/(\beta-1)}\tau(r).$$

Therefore,

$$\left\{ (L_t^x, \tau(hr)) ; (x,t,r) \in R^1 \times R_+ \times R_+ \right\} \tag{11.9}$$
$$\overset{law}{=} \left\{ \left(hL_{t/h^{\beta/(\beta-1)}}^{x/h^{1/(\beta-1)}}, h^{\beta/(\beta-1)}\tau(r) \right) ; (x,t,r) \in R^1 \times R_+ \times R_+ \right\}.$$

The equivalence (11.6) follows from this. □

To study the most visited sites of a process we consider the supremum of its local time over all of its range. We also need to consider the range in which the local time is not too big, the less frequently visited sites of the process. We do this in the next lemma. We consider $L_{\tau(t)}^{\cdot}$ rather than $L_t^{\cdot}$ in order to use the the Generalized Second Ray–Knight Theorem, Theorem 8.2.2.

Lemma 11.1.2 *Let* $X = \{X_t ; t \geq 0\}$ *be a symmetric stable process of index* $1 < \beta \leq 2$ *in* R^1 *and denote its local times by* $L = \{L_t^x, (x,t) \in R^1 \times R_+\}$. *Let*

$$h_a(t) = \left(\frac{t}{|\log t|^a} \right)^{1/(\beta-1)}. \tag{11.10}$$

Then, for all $a > 0$,

$$\lim_{t \to \infty} \sup_{0 \leq |x| \leq h_a(t)} \frac{|\log t|^{a/2}(L_{\tau(t)}^x - t)}{t(\log|\log t|)^{1/2}} \leq 2\sqrt{C_\beta (1 + a/2)} \qquad a.s. \tag{11.11}$$

In particular, for all $\gamma < a/2$,

$$\lim_{t \to \infty} \sup_{0 \leq |x| \leq h_a(t)} \frac{|\log t|^\gamma (L_{\tau(t)}^x - t)}{t} = 0 \qquad a.s. \tag{11.12}$$

Furthermore, both (11.11) and (11.12) hold with $\lim_{t \to \infty}$ *replaced by* $\lim_{t \to 0}$.

Proof It follows from Theorem 8.2.2 that, for t fixed and $\epsilon > 0$ sufficiently small,

$$P\left(\sup_{0 \le |x| \le h_a(t)} L^x_{\tau(t)} - t \ge (1+\epsilon)\lambda\right) \tag{11.13}$$

$$\le P\left(\sup_{0 \le |x| \le h_a(t)} L^x_{\tau(t)} - t + \frac{\eta^2(x)}{2} \ge (1+\epsilon)\lambda\right)$$

$$\le P\left(\sup_{0 \le |x| \le h_a(t)} \eta^2(x)/2 + \sqrt{2t}\eta(x) \ge (1+\epsilon)\lambda\right)$$

$$\le P\left(\sup_{0 \le |x| \le h_a(t)} \sqrt{2t}\eta(x) \ge \lambda\right) + P\left(\sup_{0 \le |x| \le h(t)} \eta^2(x) \ge 2\epsilon\lambda\right).$$

Let

$$\sigma^2(t) = \sup_{0 \le |x| \le h_a(t)} E\eta^2(x) = C_\beta |h_a(t)|^{\beta-1} = C_\beta \left(\frac{t}{|\log t|^a}\right) \tag{11.14}$$

and

$$a(t) = \text{median} \sup_{0 \le |x| \le h_a(t)} \eta(x). \tag{11.15}$$

It follows from Corollary 5.4.5 and (7.88) that

$$a(t) \le D_\beta |h_a(t)|^{(\beta-1)/2} = D'_\beta \sigma(t) \tag{11.16}$$

for finite constants D_β and D'_β.

We now use (5.151) to get

$$P\left(\sup_{0 \le |x| \le h_a(t)} \sqrt{2t}\eta(x) \ge \lambda\right) \tag{11.17}$$

$$= P\left(\sup_{0 \le |x| \le h_a(t)} \eta(x) - a(t) \ge \frac{\lambda}{\sqrt{2t}} - a(t)\right)$$

$$\le 1 - \Phi\left(\frac{\lambda}{\sqrt{2t}\sigma(t)} - \frac{a(t)}{\sigma(t)}\right).$$

For $\epsilon > 0$ and $0 < \delta < 1$, let

$$\lambda = \left(\frac{(1+\epsilon)4t\,\sigma^2(t)\log|\log t|}{\delta}\right)^{1/2}. \tag{11.18}$$

Using this (11.14) and (11.16) we see that the last term in (11.17) is less than or equal to

$$|\log t|^{-(1+\epsilon')/\delta} \tag{11.19}$$

for some $\epsilon' > 0$ and all t sufficiently large, and also for all t sufficiently small. Similarly

$$P\left(\sup_{0\le|x|\le h_a(t)} \eta^2(x) \ge 2\epsilon\lambda\right) \tag{11.20}$$

$$\le 2P\left(\sup_{0\le|x|\le h_a(t)} \eta(x) - a(t) \ge \sqrt{2\epsilon\lambda} - a(t)\right)$$

$$\le 1 - \Phi\left(\frac{\sqrt{2\epsilon\lambda}}{\sigma(t)} - \frac{a(t)}{\sigma(t)}\right)$$

$$\le \exp\left(-|\log t|^{a/2}\right)$$

for all t sufficiently large and sufficiently small.

Let $t_k = \exp(k^\delta)$. When $t = t_k$, the right-hand sides of (11.20) and (11.17) (see (11.19)) are terms in a sequence, indexed by k, which converges. Therefore, by (11.13), we see that for $k \ge k_0(\omega)$ for $k_0(\omega)$ sufficiently large,

$$\sup_{k\ge k_0(\omega)}\sup_{0\le|x|\le h_a(t_k)} \frac{|\log t_k|^{a/2}(L^x_{\tau(t_k)} - t_k)}{t_k(\log|\log t_k|)^{1/2}} \le (1+\epsilon)2\sqrt{\frac{C_\beta}{\delta}} \tag{11.21}$$

almost surely, where $\epsilon \to 0$ as $k_0(\omega) = k_0(\omega, \epsilon) \uparrow \infty$.

Let $t_{k-1} < t \le t_k$. We have

$$\sup_{0\le|x|\le h_a(t)} \frac{|\log t_k|^{a/2}(L^x_{\tau(t)} - t)}{t_k(\log|\log t_k|)^{1/2}} \le (1+\epsilon)2\sqrt{\frac{C_\beta}{\delta}} + \frac{(t_k - t_{k-1})|\log t_k|^{a/2}}{t_k(\log|\log t_k|)^{1/2}} \tag{11.22}$$

for $k \ge k_0(\omega)$, for $k_0(\omega)$ sufficiently large, almost surely. The last term in (11.22) is less than

$$\frac{\delta}{k^{1-\delta(1+a/2)}(\log k)^{1/2}}. \tag{11.23}$$

We take $\delta = (1+a/2)^{-1}$ and obtain that, for all $\epsilon > 0$ and $t_{k-1} < t < t_k$,

$$\sup_{0\le|x|\le h_a(t)} \frac{|\log t_k|^{a/2}(L^x_{\tau(t)} - t)}{t_k(\log|\log t_k|)^{1/2}} \le 2(1+\epsilon)\sqrt{C_\beta(1+a/2)} \tag{11.24}$$

for $k \ge k_0(\omega)$, for $k_0(\omega)$ sufficiently large, almost surely.

It is easy to extrapolate between t_k and t_{k-1} on the left-hand side of (11.24) to get (11.11). The limit (11.12) is obvious.

Repeating the argument of the preceding three paragraphs with $t_k = \exp(-k^\delta)$, we see that (11.11) and (11.12) also hold when the limit is taken as t goes to zero. $\qquad\square$

We use L_t^0 as a reference for $\sup_x L_t^x$. The next lemma gives a lower bound for the rate of growth of L_t^0.

Lemma 11.1.3 *Let $X = \{X_t \, ; t \geq 0\}$ be a symmetric stable process of index $1 < \beta \leq 2$ in R^1 and denote its local times by $L = \{L_t^x, (x,t) \in R^1 \times R_+\}$. Then, for all $\epsilon > 0$,*

$$\lim_{t \to \infty} \frac{|\log t|^{1+\epsilon} L_t^0}{t^{1/\bar{\beta}}} = \infty \qquad a.s., \qquad (11.25)$$

where $1/\beta + 1/\bar{\beta} = 1$.

Furthermore, (11.25) also holds with $\lim_{t \to \infty}$ replaced by $\lim_{t \to 0}$.

The proof of this lemma is given in the next subsection.

11.1.1 Inverse local time of symmetric stable processes

We continue studying $X = \{X_t \, ; t \geq 0\}$, a symmetric stable process in R^1 of index $1 < \beta \leq 2$, and denote its local times by $L = \{L_t^x, (x,t) \in R^1 \times R_+\}$ and its inverse local time at zero by $\tau = \{\tau(r), r \in R_+\}$. As in Lemma 2.4.5, which is proved for Brownian motion, we can show that τ, which is a positive increasing stochastic process, has stationary and independent increments. In particular, it is a Lévy process.

We show in Lemma 3.6.10 that for all $\alpha > 0$,

$$E^0 \left(e^{-\alpha \tau(r)} \right) = e^{-r/u^\alpha(0)}, \qquad (11.26)$$

where

$$u^\alpha(0) = \frac{1}{\pi} \int \frac{1}{\alpha + |\lambda|^\beta} \, d\lambda \qquad (11.27)$$

(see (4.84)). Let

$$D_\beta = \frac{1}{\pi} \int \frac{1}{1 + |\lambda|^\beta} \, d\lambda. \qquad (11.28)$$

Thus $u^\alpha(0) = D_\beta \alpha^{-1/\beta}$ and we can write (11.26) as

$$E^0 \left(e^{-\alpha \tau(r)} \right) = e^{-r D_\beta^{-1} \alpha^{1/\beta}}. \qquad (11.29)$$

Lemma 11.1.4 *Let $X = \{X_t \, ; t \geq 0\}$ be a symmetric stable process of index $1 < \beta \leq 2$ in R^1 and denote its local times by $L = \{L_t^x, (x,t) \in R^1 \times R_+\}$. Let τ denote the inverse local time of $\{L_t^0, t \in R_+\}$. Then, for any $\epsilon > 0$,*

$$\limsup_{r \to \infty} \frac{\tau(r)}{r^{\bar{\beta}} (\log r)^{\bar{\beta}+\epsilon}} = 0 \qquad P^0 \quad a.s. \qquad (11.30)$$

Also,

$$\liminf_{r \to \infty} \frac{(\log \log r)^{1/(\beta-1)} \tau(r)}{r^{\overline{\beta}}} \geq D_{\beta}^{-\overline{\beta}} \left(\frac{1}{\overline{\beta}}\right) \left(\frac{1}{\beta}\right)^{\overline{\beta}/\beta} \qquad P^0 \quad a.s.$$

(11.31)

Proof We have that

$$P^0(\tau(r) \geq x) = P^0 \left(e^{-1} \geq e^{-\tau(r)/x}\right)$$

(11.32)

$$\leq P^0 \left(1 - e^{-x^{-1}\tau(r)} \geq 1 - e^{-1}\right)$$

$$= \frac{e}{e-1} E^0 \left(1 - e^{-\tau(r)/x}\right)$$

$$= \frac{e}{e-1} \left(1 - e^{-rD_{\beta}^{-1}/x^{1/\overline{\beta}}}\right),$$

where, for the last equality, we use (11.29). Since $1 - e^{-y} \leq y$ for all $y \geq 0$, we have

$$P^0(\tau(r) \geq x) \leq \left(\frac{e}{e-1}\right) \frac{r}{D_{\beta} x^{1/\overline{\beta}}}.$$

(11.33)

Consequently,

$$P^0 \left(\tau(r) \geq r^{\overline{\beta}} (\log r)^{\overline{\beta}+\epsilon}\right)$$

(11.34)

$$\leq \left(\frac{e}{e-1}\right) \frac{1}{D_{\beta} (\log r)^{(1+\epsilon')}},$$

where $\epsilon' = \epsilon/\overline{\beta}$. Let $r_n = e^n$. It follows from the Borel–Cantelli Lemma that

$$\limsup_{n \to \infty} \frac{\tau(r_n)}{r_n^{\overline{\beta}} (\log r_n)^{\overline{\beta}+\epsilon}} = 0 \qquad P^0 \quad a.s.$$

(11.35)

Using the fact that $\tau(r)$ is increasing and $\epsilon > 0$ is arbitrary, we obtain (11.30).

To obtain (11.31) we use (11.29) again and Chebyshev's inequality to see that for any $\lambda > 0$,

$$P^0(\tau(r) \leq x) = P^0 \left(e^{-\lambda \tau(r)} \geq e^{-\lambda x}\right)$$

(11.36)

$$\leq e^{\lambda x} e^{-rD_{\beta}^{-1} \lambda^{1/\overline{\beta}}}.$$

This is minimized by taking $\lambda = \left(\frac{r}{D_{\beta} x \overline{\beta}}\right)^{\beta}$, which gives

$$P^0(\tau(r) \leq x) \leq e^{-d_{\beta} r^{\beta}/x^{\beta-1}},$$

(11.37)

where $d_\beta^{1/\beta} = D_\beta^{-1} \left(\dfrac{1}{\overline{\beta}}\right)^{1/\overline{\beta}} \left(\dfrac{1}{\beta}\right)^{1/\beta}$. Hence

$$P^0\left(\tau(r) \le \frac{(1+\epsilon)d_\beta^{1/(\beta-1)}r^{\overline{\beta}}}{(\log\log r)^{1/(\beta-1)}}\right) \le \frac{1}{(\log r)^{1+\epsilon}}. \tag{11.38}$$

Using the Borel–Cantelli Lemma as in (11.35) and the fact that τ is increasing, we get (11.31). $\qquad\square$

Proof of Lemma 11.1.3 $\tau(t)$ is a generalized inverse of L_t^0. Therefore, when $\tau(t) < \rho(t)$ for some strictly increasing function ρ, $L_t^0 \ge \rho^{-1}(t)$. We get (11.25) from (11.30) because

$$\left(t^{\overline{\beta}}(\log t)^{\overline{\beta}+\epsilon}\right)^{-1} > \frac{t^{1/\overline{\beta}}}{(\log t)^{1+\epsilon}} \tag{11.39}$$

for all t sufficiently large. We get the last sentence of Lemma 11.1.3 by noting that $\{L_t^0, t \in R_+\} \overset{law}{=} \{t^{2/\overline{\beta}}L_{1/t}^0, t \in R_+\}$ (see Lemma 10.5.2).

Remark 11.1.5 The same argument used just above in the proof of Lemma 11.1.3 gives, for the symmetric β-stable process,

$$\limsup_{t\to\infty} \frac{L_t^0}{t^{1/\overline{\beta}}(\log|\log t|)^{1/\beta}} \le \widetilde{D}_\beta \qquad P^0 \quad \text{a.s.}, \tag{11.40}$$

where

$$\widetilde{D}_\beta = D_\beta \overline{\beta}^{1/\overline{\beta}}\beta^{1/\beta}. \tag{11.41}$$

Furthermore, by scaling, as in the proof of Lemma 11.1.3, (11.40) also holds as $t \to 0$.

In fact, in both cases, (11.40) holds with an equal sign. References are given in Section 11.7.

11.2 Most visited sites of Brownian motion

We begin with a simple upper bound for Brownian motion.

Lemma 11.2.1 *Let* $\{W_t, t \ge 0\}$ *be a standard Brownian motion. Then*

$$P\left(\sup_{0\le t\le 1} W_t < \lambda\right) \le \sqrt{\frac{2}{\pi}}\lambda. \tag{11.42}$$

Proof By the reflection principle, Lemma 2.2.11, we have

$$P\left(\sup_{0\le t\le 1} W_t < \lambda\right) = \sqrt{\frac{2}{\pi}}\int_0^\lambda e^{-s^2/2}\,ds, \tag{11.43}$$

which gives (11.42). □

A key element in the proof of Lemma 11.2.3 is the following probability that Brownian motion, starting from zero, lies below a given line.

Lemma 11.2.2 *Let* $\{W_t \,;\, t \geq 0\}$ *be a standard Brownian motion. Then, for all* $a, b > 0$,

$$P\left(W_s < a + bs; \; \forall s \geq 0\right) = 1 - e^{-cab} \qquad (11.44)$$

for some constant $0 < c < \infty$ *independent of a and b.*

In fact $c = 2$, but we skip this point since in our applications of this lemma we do not need to know the value of c.

Proof Let $T_a = \inf\{s \,|\, W_s \geq a + s\}$ and set

$$g(a) = P(W_s \geq a + s; \text{ for some } s \geq 0) = P(T_a < \infty). \qquad (11.45)$$

Let $a_1, a_2 > 0$. A path that hits the line $s + a_1 + a_2$ must first hit the line $s + a_1$ and then move up to meet a line with slope 1 that is a_2 units above the point where it first hit the line $s + a_1$. This implies that

$$T_{a_1 + a_2} = T_{a_2} \circ \theta_{T_{a_1}}. \qquad (11.46)$$

It follows from the strong Markov property that for all $a_1, a_2 > 0$,

$$\begin{aligned} g(a_1 + a_2) &= P(T_{a_1 + a_2} < \infty) & (11.47) \\ &= P(T_{a_1} < \infty, T_{a_2} \circ \theta_{T_{a_1}} < \infty) = g(a_1)g(a_2). \end{aligned}$$

Note that $g(1) \geq P(W(1) \geq 2) > 0$ and $g(a)$ is a nonincreasing function in a. The only solution of (11.47) with these properties is $g(a) = e^{-ca}$ for some constant $0 < c < \infty$. Therefore,

$$P\left(W_s < a + s; \; \forall s \geq 0\right) = 1 - g(a) = 1 - e^{-ca}. \qquad (11.48)$$

Since $\{W_t \,;\, t \geq 0\} \stackrel{law}{=} \{b^{-1}W_{sb^2} \,;\, t \geq 0\}$, we have

$$\begin{aligned} P\left(W_s < a + bs, \; \forall s \geq 0\right) &= P(b^{-1}W_{sb^2} < a + bs, \; \forall s \geq 0) & (11.49) \\ &= P\left(W_{sb^2} < ab + sb^2, \; \forall s \geq 0\right) \\ &= P\left(W_t < ab + t, \; \forall t \geq 0\right), \end{aligned}$$

and (11.44) now follows from (11.48). □

We can now use the Second Ray–Knight Theorem to study the behavior of the local time of Brownian motion near zero.

Lemma 11.2.3 *Let $\{W_t \, ; \, t \geq 0\}$ be a standard Brownian motion and denote its local times by $L = \{L_t^x, (x,t) \in R^1 \times R_+\}$. Let $\tau(\cdot)$ denote the inverse local time of $L_\cdot^0$. Then*

$$P^0\left(\sup_{|x|\leq 1} L_{\tau(1)}^x \leq 1 + \lambda\right) \leq d^2\lambda^2|\log\lambda|^4 \qquad \forall \lambda > 0, \qquad (11.50)$$

where d^2 is a constant.

Proof Let $\eta = \{\eta(x) \, ; \, x \in R^1\}$ be a two-sided Brownian motion independent of $\{W_t \, ; \, t \geq 0\}$. Recall that $\sqrt{2}\eta$ is the mean zero Gaussian process with covariance $u_{T_0}(x, y)$ given in Lemma 2.5.1. Since $\{\eta(x) \, ; \, x > 0\}$ and $\{\eta(-x) \, ; \, x > 0\}$ are independent and identically distributed, it follows from the Generalized Second Ray–Knight Theorem (Theorem 8.2.2) that $\{L_{\tau(1)}^x + \eta^2(x) \, ; \, x > 0\}$ and $\{L_{\tau(1)}^{-x} + \eta^2(-x) \, ; \, x > 0\}$ are independent and identically distributed. This implies that $\{L_{\tau(1)}^x \, ; \, x > 0\}$ and $\{L_{\tau(1)}^{-x} \, ; \, x > 0\}$ are independent and identically distributed (this can be seen easily using characteristic functions). Since $L_{\tau(1)}^0 = 1$, we see that to prove (11.50) it suffices to show that

$$P^0\left(\sup_{0\leq x\leq 1} L_{\tau(1)}^x \leq 1 + \lambda\right) \leq d\lambda|\log\lambda|^2. \qquad (11.51)$$

Using (11.5) we have

$$h_\lambda \; := \; P\left(\sup_{0\leq x\leq \lambda} \eta^2(x) \leq c\lambda\right) \qquad (11.52)$$

$$= \; P\left(\sup_{0\leq x\leq 1} \eta^2(x) \leq c\right).$$

Therefore, for $c > 1$ sufficiently large,

$$h_\lambda = h_1 > 2/3. \qquad (11.53)$$

Also, by Lévy's uniform modulus of continuity for Brownian motion (see (2.15) and Example 7.2.16) we can choose c sufficiently large so that

$$P\left(\eta^2(x) \leq cx\,|\log\lambda|; \; \lambda < x \leq 1\right) > 2/3. \qquad (11.54)$$

Set

$$f_\lambda(x) = \begin{cases} \lambda & 0 \leq x \leq \lambda \\ x\,|\log\lambda| & \lambda < x \leq 1. \end{cases} \qquad (11.55)$$

Using (11.53) and (11.54) we have that, for all $0 < \lambda < 1$,

$$m_\lambda := P\left(\eta^2(x) \leq cf_\lambda(x); \; 0 \leq x \leq 1\right) > 1/3. \qquad (11.56)$$

Since L and η are independent, it follows from (11.56) that for all $\lambda > 0$ sufficiently small,

$$P^0 \left(\sup_{0 \le x \le 1} L^x_{\tau(1)} \le 1 + \lambda \right) \qquad (11.57)$$

$$= m_\lambda^{-1} P^0 \times P_\eta \left(L^x_{\tau(1)} \le 1 + \lambda \text{ and } \eta^2(x) \le c f_\lambda(x); 0 \le x \le 1 \right)$$

$$\le 3 P^0 \times P_\eta \left(L^x_{\tau(1)} + \eta^2(x) \le 1 + 2c f_\lambda(x); 0 \le x \le 1 \right).$$

(Note that the constant d in (11.50) is arbitrary. Without loss of generality we can take $\lambda < 1/e$ so that $\lambda \le f_\lambda(x)$ on $[0,1]$.) It then follows from the Generalized Second Ray–Knight Theorem (Theorem 8.2.2) that

$$P^0 \left(\sup_{0 \le x \le 1} L^x_{\tau(1)} \le 1 + \lambda \right) \qquad (11.58)$$

$$\le 3 P \left(\left(\eta(x) + \sqrt{1} \right)^2 \le 1 + 2c f_\lambda(x); 0 \le x \le \lambda \right)$$

$$\le 3 P \left(\eta^2(x) + 2\eta(x) \le 2c f_\lambda(x); 0 \le x \le 1 \right)$$

$$\le 3 P \left(\eta(x) < 2c f_\lambda(x); 0 \le x \le 1 \right).$$

(Note that η in (8.46) is $\sqrt{2}$ times the η we are using here.)

If $\eta(\lambda) > -\sqrt{\lambda} |\log \lambda|$, the event $\{\eta(x) < 2c f_\lambda(x); 0 \le x \le 1\}$ is contained in the event

$$\{\eta(x) < 2c\lambda; 0 \le x \le \lambda\} \cap \{\eta(x) - \eta(\lambda) < 2c f_\lambda(x) + \sqrt{\lambda} |\log \lambda|; \lambda \le x \le 1\}. \qquad (11.59)$$

Since by (5.18)

$$P \left(\eta(\lambda) > -\sqrt{\lambda} |\log \lambda| \right) \le e^{-|\log \lambda|^2/2}, \qquad (11.60)$$

we have

$$P \left(\eta(x) < 2c f_\lambda(x); 0 \le x \le 1 \right) \qquad (11.61)$$

$$\le e^{-|\log \lambda|^2/2} + P \left(\sup_{0 \le x \le \lambda} \eta(x) < 2c\lambda \right)$$

$$P \left(\eta(x) - \eta(\lambda) < 2cx |\log \lambda| + \sqrt{\lambda} |\log \lambda|; \lambda \le x \le 1 \right).$$

It follows from Lemma 11.2.1 and scaling that

$$P \left(\sup_{0 \le x \le \lambda} \eta(x) < 2c\lambda \right) = O \left(\sqrt{\lambda} \right) \qquad (11.62)$$

at zero. Finally we show below that

$$P \left(\eta(x) - \eta(\lambda) < 2cx |\log \lambda| + \sqrt{\lambda} |\log \lambda|; \lambda \le x \le 1 \right) \le d\sqrt{\lambda} |\log \lambda|^2 \qquad (11.63)$$

for λ sufficiently small. Using this, (11.58), (11.61), and (11.62), we get (11.51).

To obtain (11.63) we note that when $\lambda < 1/2$, the left-hand side of (11.63) equals

$$P\left(\eta(x-\lambda) < 2cx \,|\log \lambda| + \sqrt{\lambda}|\log \lambda| \,;\, \lambda < x \le 1\right) \qquad (11.64)$$
$$\le P\left(\eta(x) < 2cx \,|\log \lambda| + 3c\sqrt{\lambda}|\log \lambda| \,;\, 0 < x \le 1/2\right).$$

Let A denote the event $\{\eta(x) < 2cx \,|\log \lambda| + 3c\sqrt{\lambda}|\log \lambda| \,;\, 0 < x \le 1/2\}$. Note that

$$P(A)P\left(\eta(x) - \eta(1/2) < (x - 1/2) \,|\log \lambda| + 1 \,;\, 1/2 < x\right) \qquad (11.65)$$
$$= P\left(A \cap \{\eta(x) - \eta(1/2) < (x - 1/2) \,|\log \lambda| + 1 \,;\, 1/2 < x\}\right)$$
$$\le P\left(A \cap \{\eta(x) < (x - 1/2 + c) \,|\log \lambda| + 1 + 3c\sqrt{\lambda}|\log \lambda| \,;\, 1/2 < x\}\right)$$

because $\{\eta(1/2) \in A\}$ implies that $\eta(1/2) < c \,|\log \lambda| + 3c\sqrt{\lambda}|\log \lambda|$.

We can absorb the constant 1 into the term $c \,|\log \lambda|$ so that the last line in (11.65) is less than or equal to

$$P_\eta\left(\eta(x) < 4cx \,|\log \lambda| + 4c\sqrt{\lambda}|\log \lambda| \,;\, \forall x \ge 0\right) \qquad (11.66)$$
$$a = 1 - e^{-d\sqrt{\lambda}|\log \lambda|^2}$$
$$\le d\sqrt{\lambda}|\log \lambda|^2,$$

where we use Lemma 11.2.2 to get the second line. Lemma 11.2.2 also implies that

$$P_\eta\left(\eta(x) - \eta(1/2) < (x - 1/2) \,|\log \lambda| + 1 \,;\, 1/2 < x\right) \quad (11.67)$$
$$= 1 - e^{-d|\log \lambda|}$$
$$\ge 1/2$$

for $d > 2$. The inequality (11.63) follows from (11.64)–(11.67). $\qquad \square$

Lemma 11.2.4 *Let $\{W_t \,;\, t \ge 0\}$ be a standard Brownian motion and denote its local times by $L = \{L_t^x, (x,t) \in R^1 \times R_+\}$. Let $\tau(\cdot)$ denote the inverse local time of $L_\cdot^0$. Then, for any $\epsilon > 0$ and all $v > 0$,*

$$P^0\left(\sup_{|s| \le v\lambda} L_{\tau(v)}^s - v \le \lambda v\right) \le d^2\lambda^2 |\log \lambda|^4 \qquad (11.68)$$

for some constant $d < \infty$. Therefore, in particular,

$$P^0\left(L_{\tau(v)}^* - v \le \lambda v\right) \le d^2\lambda^2 |\log \lambda|^4, \qquad (11.69)$$

where $L^* := \sup_x L^x_\cdot$ is the maximal local time.

Proof By (11.6),

$$L^x_{\tau(1)} \overset{law}{=} \frac{1}{v} L^{xv}_{\tau(v)}. \tag{11.70}$$

Using this in (11.50) we get (11.68). □

Lemma 11.2.5 *Let* $\{W_t \, ; \, t \geq 0\}$ *be a standard Brownian motion and denote its local times by* $L = \{L^x_t, (x,t) \in R^1 \times R_+\}$. *Let* $\tau(\cdot)$ *denote the inverse local time of* $L^0_\cdot$. *Then, for all* $b > 1$,

$$\lim_{r \to \infty} \frac{(\log r)^b \left(\sup_x L^x_{\tau(r)} - r \right)}{r} = \infty \tag{11.71}$$

and

$$\lim_{r \to 0} \frac{|\log r|^b \left(\sup_x L^x_{\tau(r)} - r \right)}{r} = \infty. \tag{11.72}$$

To better appreciate these results, recall that $r = L^0_{\tau(r)}$, so we are actually looking at the deviation of the maximal local time from its value at zero.

Proof Let $v_n = e^{n^a}$ for $0 < a < 1$ and $\lambda = n^{-(1+\epsilon)/2}$ in (11.69). It follows from the Borel–Cantelli Lemma that

$$L^*_{\tau(v_n)} - v_n \geq \frac{v_n}{n^{(1+\epsilon)/2}} \tag{11.73}$$

for all $n \geq n_0(\omega)$ on a set of probability one, where ω denotes the path of Brownian motion. Let $v_n \leq r \leq v_{n+1}$. Then, restricted to this set,

$$L^*_{\tau(r)} - r \geq -(r - v_n) + \frac{v_n}{n^{(1+\epsilon)/2}} \tag{11.74}$$

for all $n \geq n_0(\omega)$ and

$$\frac{L^*_{\tau(r)} - r}{r} \geq -\frac{v_{n+1} - v_n}{v_n} + \frac{v_n}{v_{n+1}} \frac{1}{n^{(1+\epsilon)/2}} \tag{11.75}$$

for all $n \geq n_0(\omega)$. It now follows by a simple calculation that

$$\frac{L^*_{\tau(r)} - r}{r} \geq -\frac{2}{n^{1-a}} + \frac{(1-\delta)}{n^{(1+\epsilon)/2}} \tag{11.76}$$

for all $\delta > 0$ and for all $n \geq n_0(\omega)$. It is easy to see that if $1 - a > 1/2$, then for ϵ small enough, $1 - a > (1 + \epsilon)/2$. Thus

$$\frac{L^*_{\tau(r)} - r}{r} \geq \frac{(1-\delta)}{n^{(1+\epsilon)/2}} \tag{11.77}$$

for all $n \geq n_0'(\omega)$ and $v_n \leq r \leq v_{n+1}$ on a set of probability one. For r in this range, $n \sim (\log r)^{1/a}$. Since we can take a arbitrarily close to $1/2$, we get (11.71).

We get (11.155) from (11.158) by taking $v_n = e^{-n^a}$ for $0 < a < 1$ and repeating the rest of the above argument.   □

We can now prove the "most visited sites" limit theorem for Brownian motion discussed in Section 2.9.

Theorem 11.2.6 *Let* $\{W_t \, ; \, t \geq 0\}$ *be a standard Brownian motion and denote its local times by* $L = \{L_t^x, (x,t) \in R^1 \times R_+\}$. *Let* V_t *be given by (11.2). Then, for any* $\gamma > 3$, *we have*

$$\lim_{t \to \infty} \frac{(\log t)^\gamma}{t^{1/2}} V_t = \infty \qquad P^0 \quad a.s. \qquad (11.78)$$

and

$$\lim_{t \to 0} \frac{|\log t|^\gamma}{t^{1/2}} V_t = \infty \qquad P^0 \quad a.s. \qquad (11.79)$$

In particular, (11.78) tells us that $\lim_{t \to \infty} V_t = \infty$, that is, the process $\{V_t, t \in R_+\}$ is transient. Display (11.79) shows that although $\lim_{t \to 0} V_t = 0$, there is a positive lower bound on the rate at which it decreases to zero. We see that starting from zero, at least for a little while, zero is not the favorite point of a Brownian motion.

Proof We first prove (11.78). Let $t \in [\tau(r^-), \tau(r)]$. By (11.71) we have

$$\sup_x L_t^x > r + r/(\log r)^b \qquad (11.80)$$

for all r sufficiently large, for any $b > 1$.

On the other hand, let h_a be as defined in (11.10) with $a = 2(1+\epsilon)b$, for some $\epsilon > 0$. It follows from (11.12) with $\gamma = b$ and using $L_{\tau(r)}^0 = r$ that

$$\sup_{0 \leq |x| \leq h_a(L_t^0)} L_t^x \quad \leq \quad \sup_{0 \leq |x| \leq h_a(r)} L_{\tau(r)}^x \qquad (11.81)$$

$$< \quad r + r/(\log r)^b$$

for all r sufficiently large.

Comparing (11.80) and (11.81), we see that for all t sufficiently large,

$$V_t > h_a(L_t^0). \qquad (11.82)$$

Using Lemma 11.1.3 and the fact that h_a is increasing, this shows that

$$V_t > h_a \left(\frac{t^{1/2}}{(\log t)^{(1+\epsilon)}} \right) \tag{11.83}$$

$$\geq \frac{t^{1/2}}{(\log t)^{(2b+1)(1+\epsilon)}}.$$

Since we can take b arbitrarily close to 1 and ϵ arbitrarily close to 0, we get (11.78).

The main ingredients in the above proof, (11.71), (11.12), and Lemma 11.1.3 have corresponding versions as $t \to 0$. Using them and following the above proof, we get (11.79). □

11.3 Reproducing kernel Hilbert spaces of fractional Brownian motion

Our goal is to obtain a version of Theorem 11.2.6 for the local times of symmetric stable processes of index $1 < \beta < 2$. The main difficulty in extending the results of the preceding section is to find an analog of the elementary Lemma 11.2.1. This is not simple and requires a detour through reproducing kernel Hilbert spaces of fractional Brownian motion and the Cameron–Martin formula.

In Theorem 5.3.1 we define and prove the existence of reproducing kernel Hilbert spaces. In Section 5.3 they are used to obtain the Karhunen–Loéve expansion of Gaussian processes. In this section we obtain explicit descriptions of the reproducing kernel Hilbert spaces of fractional Brownian motion.

In order to motivate the work in this section it is useful to obtain the well-known explicit description of the reproducing kernel Hilbert space of Brownian motion, $B = \{B(t), t \in R_+\}$. Let $L^2 = L^2(R_+, \lambda)$ of real-valued functions, where λ indicates Lebesgue measure, and denote its norm by $\| \cdot \|_2$, that is, $g \in L^2$ is a measurable function with

$$\|g\|_2 = \left(\int_0^\infty |g(\lambda)|^2 \, d\lambda \right)^{1/2} < \infty. \tag{11.84}$$

For $g \in L^2$ let

$$Ig(x) := \int_0^x g(\lambda) \, d\lambda. \tag{11.85}$$

It is easy to see by the Cauchy–Schwarz inequality that Ig exists and is uniformly continuous.

Consider the linear space $H = \{Ig, g \in L^2\}$ equipped with the norm

$\|Ig\| := \|g\|_2$. This norm is well defined since $Ig \equiv 0$ implies that $\|g\|_2 = 0$. To see this, note that restricted to $[0, x]$, for any $x > 0$, g is integrable so that Ig is absolutely continuous and has g as a derivative almost everywhere. This implies that $g = 0$ almost everywhere.

In fact, $(H, \|Ig\|)$ is a separable Hilbert space of real-valued continuous functions. To see that it is complete we note that if $\{Ig_n\}$ is a Cauchy sequence in H, then $\{g_n\}$ is a Cauchy sequence in L^2. Therefore, there exist some $g \in L^2$ such that $g_n \to g$ in L^2. This means that $Ig_n \to Ig$ in H. Also, H is separable because L^2 is separable.

We now show that

$$\Gamma(t, \cdot) \in H \qquad \forall t \in R^1 \tag{11.86}$$

and

$$(Ig(\cdot), \Gamma(t, \cdot)) = Ig(t) \qquad \forall Ig \in H, \tag{11.87}$$

where $(\cdot, \cdot)$ denotes the inner product on H.

This is elementary since the covariance of B,

$$\Gamma(t, x) = t \wedge x = \int_0^x 1_{[0,t]}(s)\, ds, \tag{11.88}$$

and obviously $1_{[0,t]}(\cdot) \in L^2$. Furthermore,

$$(Ig(\cdot), \Gamma(t, \cdot)) = \big(g(\cdot), 1_{[0,t]}(\cdot)\big)_2 = Ig(t), \tag{11.89}$$

where $(\cdot, \cdot)_2$ denotes the inner product on L^2. Therefore, by Theorem 5.3.1, $H = \{Ig, g \in L^2\}$ equipped with the norm $\|Ig\| := \|g\|_2$ is the reproducing kernel Hilbert space of B. It is customary to refer to H as the space of continuous functions f, with $f(0) = 0$, that are absolutely continuous with respect to Lebesgue measure, equipped with the norm $\left(\int |f'(s)|^2\, ds\right)^{1/2}$, where f' denotes the derivative of f.

The key point in the above description of the reproducing kernel Hilbert space of Brownian motion is that the covariance of Brownian motion has a bounded derivative almost everywhere. This property is not shared by most of the Gaussian processes we are interested in. Nevertheless, we get a hint about what norm might define the reproducing kernel Hilbert space of Gaussian processes in general. We note that if Ig above is also in $L^2(R_+)$, then its Fourier transform $\widehat{Ig}$ exists and, by Parseval's Theorem,

$$\int_0^\infty |g(s)|^2\, ds = \frac{1}{2\pi} \int_{-\infty}^\infty |\lambda|^2 |\widehat{Ig}(s)|^2\, ds. \tag{11.90}$$

This suggests that the reproducing kernel Hilbert space norms of Gaussian processes might be function spaces defined by different weightings of their Fourier transforms. We now develop this insight.

Consider fractional Brownian motion as defined in (11.3). We give a function space description of its reproducing kernel Hilbert space.

Consider the complex Hilbert space $L_\beta^2 = L_C^2(R^1, (|\lambda|^\beta/(2\pi))\,d\lambda)$, $1 < \beta < 2$, and denote its norm by $\|\cdot\|_{2,\beta}$, that is, $g \in L_\beta^2$ is a measurable function with

$$\|g\|_{2,\beta} = \left(\frac{1}{2\pi} \int |g(\lambda)|^2 |\lambda|^\beta \, d\lambda\right)^{1/2} < \infty \qquad (11.91)$$

(we use the subscript C to indicate that the functions in L_C^2 are complex valued). If $g \in L_\beta^2$, let

$$Ig(x) := \frac{1}{2\pi} \int (1 - e^{-ix\lambda}) g(\lambda) \, d\lambda. \qquad (11.92)$$

It follows from the Cauchy–Schwarz inequality that Ig exists and is uniformly continuous, since

$$|Ig(x) - Ig(y)| = \left|\frac{1}{2\pi} \int (e^{-iy\lambda} - e^{-ix\lambda}) g(\lambda) \, d\lambda\right| \qquad (11.93)$$

$$\leq \|g\|_{2,\beta} \left(\frac{1}{2\pi} \int \frac{|e^{-iy\lambda} - e^{-ix\lambda}|^2}{\lambda^\beta} \, d\lambda\right)^{1/2}$$

and

$$\int \frac{|e^{-iy\lambda} - e^{-ix\lambda}|^2}{\lambda^\beta} \, d\lambda \quad \leq \quad \int \frac{(|x-y||\lambda| \wedge 1)^2}{\lambda^\beta} \, d\lambda \qquad (11.94)$$

$$\leq \quad C_\beta |x-y|^{\beta-1}.$$

Let $f \in C_0^\infty$ and let $\widehat{f}$ denote its Fourier transform. Recall that the Fourier transform of a C_0^∞ function is rapidly decreasing, that is, $\lim_{|\lambda|\to\infty} |\lambda|^k |\widehat{f}(\lambda)| = 0$ for all $k > 0$. Therefore, $\widehat{f} \in L_\beta^2$ and

$$I\widehat{f}(x) = \frac{1}{2\pi} \int (1 - e^{-ix\lambda}) \widehat{f}(\lambda) \, d\lambda = f(0) - f(x). \qquad (11.95)$$

Lemma 11.3.1 *The linear space*

$$H = \{Ig : g \in L_\beta^2\} \qquad (11.96)$$

equipped with the norm

$$\|Ig\|_H := \|g\|_{2,\beta} \qquad (11.97)$$

is a separable complex Hilbert space of continuous functions.

Furthermore,

$$\Gamma(t, \cdot) \in H \qquad \forall t \in R^1 \tag{11.98}$$

and

$$(f(\cdot), \Gamma(t, \cdot))_H = f(t) \qquad \forall f \in H, \tag{11.99}$$

where $(\cdot, \cdot)_H$ denotes the inner product on H, and Γ, which was given in (11.4), is the covariance of fractional Brownian motion of index $\beta - 1$.

Let $\mathcal{K}_{\beta-1}$ denote the reproducing kernel Hilbert space of fractional Brownian motion of order $\beta - 1$, $1 < \beta < 2$, and let $\| \cdot \|_{\mathcal{K}_{\beta-1}}$ denote the norm on $\mathcal{K}_{\beta-1}$. Then

$$\mathcal{K}_{\beta-1} = \{f \in H \mid f = \bar{f}\} \tag{11.100}$$

and for $f \in \mathcal{K}_{\beta-1}$

$$\|f\|_{\mathcal{K}_{\beta-1}} = \|f\|_H. \tag{11.101}$$

Proof We first show that $I : L^2_\beta \mapsto C(R^1)$ is one–one. Since I is linear, to show this it suffices to show that if

$$\int (1 - e^{-ix\lambda}) g(\lambda) \, d\lambda = 0, \qquad \forall x \in R^1, \tag{11.102}$$

then $\|g\|_{2,\beta} = 0$. Given (11.102), it follows by (11.95) and Fubini's Theorem that for all $f \in C_0^\infty$,

$$\int (f(x) - f(0)) g(x) \, dx \; = \; \frac{1}{2\pi} \int\int (1 - e^{-ix\lambda}) \widehat{f}(\lambda) g(x) \, d\lambda \, dx$$
$$= \; 0. \tag{11.103}$$

In particular, taking $f(\lambda) = \lambda h(\lambda)$ for $h \in C_0^\infty$, we have

$$\frac{1}{2\pi} \int \lambda h(\lambda) g(\lambda) \, d\lambda = 0 \tag{11.104}$$

for all $h \in C_0^\infty$. This implies that $\lambda g(\lambda) = 0$ almost everywhere, so that $\|g\|_{2,\beta} = 0$.

To see that H is complete we note that if $\{Ig_n\}$ is a Cauchy sequence in H, then $\{g_n\}$ is a Cauchy sequence in L^2_β. Therefore, there exist some $g \in L^2_\beta$ such that $g_n \to g$ in L^2_β. This means that $Ig_n \to Ig$ in H. Also, H is separable because L^2_β is separable.

Since $|\lambda|^\beta$ is symmetric we have

$$\frac{1}{2\pi} \int \frac{(1 - e^{is\lambda})}{|\lambda|^\beta} \, d\lambda = \frac{C_\beta}{2} |s|^{\beta-1}, \tag{11.105}$$

where C_β is given in (7.193). Consequently, by (11.4),

$$\Gamma(s,t) = \frac{C_\beta}{2} \left(|s|^{\beta-1} + |t|^{\beta-1} - |t-s|^{\beta-1} \right) \tag{11.106}$$

$$= \frac{1}{2\pi} \left(\int \frac{(1-e^{is\lambda})}{|\lambda|^\beta} \, d\lambda + \int \frac{(1-e^{it\lambda})}{|\lambda|^\beta} \, d\lambda - \int \frac{(1-e^{i(s-t)\lambda})}{|\lambda|^\beta} \, d\lambda \right)$$

$$= \frac{1}{2\pi} \int \frac{(1-e^{is\lambda})(1-e^{-it\lambda})}{|\lambda|^\beta} \, d\lambda.$$

Let

$$g_s(\lambda) = \frac{(1-e^{is\lambda})}{|\lambda|^\beta}. \tag{11.107}$$

Clearly $g_s \in L_\beta^2$ and, by (11.106), $\Gamma(s,t) = Ig_s(t)$. Thus we get (11.98).
Furthermore,

$$\begin{aligned}
(Ig, \Gamma(s,\cdot))_H &= (Ig, Ig_s)_H & (11.108)\\
&= (g, g_s)_{2,\beta}\\
&= \frac{1}{2\pi} \int \overline{g_s(\lambda)} g(\lambda) \, |\lambda|^\beta \, d\lambda\\
&= Ig(s).
\end{aligned}$$

This gives us (11.99).

Given (11.98), (11.99), and the fact that for any complex Hilbert space of functions H, $\{f \in H \mid f = \bar{f}\}$ is a real Hilbert space, the last part of the lemma follows from Theorem 5.3.1. $\qquad \square$

The following Corollary of Lemma 11.3.1 is used in Section 11.5.

Corollary 11.3.2 *Let $f \in C_0^\infty$ be a real-valued function and let $\widehat{f}$ denote its Fourier transform. Then $f - f(0) \in \mathcal{K}_{\beta-1}$ for all $1 < \beta < 2$ and*

$$\|f - f(0)\|_{\mathcal{K}_{\beta-1}} = \left(\frac{1}{2\pi} \int |\lambda|^\beta \, |\widehat{f}(\lambda)|^2 \, d\lambda \right)^{1/2} < \infty. \tag{11.109}$$

Proof This follows from (11.95) and (11.101). $\qquad \square$

Remark 11.3.3 When $1 < \beta < 2$,

$$|\lambda|^\beta = \frac{1}{\pi \, C_{\beta+1}} \int_{R^1} \frac{1 - \cos(\lambda y)}{|y|^{1+\beta}} \, dy, \tag{11.110}$$

where $C_{\beta+1}$ is given in (7.193). Consequently,

$$\int |\lambda|^\beta \, |\widehat{f}(\lambda)|^2 \, d\lambda = \frac{1}{\pi \, C_{\beta+1}} \int \int \frac{1 - \cos(\lambda y)}{|y|^{1+\beta}} |\widehat{f}(\lambda)|^2 \, d\lambda \, dy \tag{11.111}$$

$$= \frac{1}{2\pi \, C_{\beta+1}} \int \frac{1}{|y|^{1+\beta}} \int |(e^{i\lambda y/2} - e^{-i\lambda y/2})\widehat{f}(\lambda)|^2 \, d\lambda \, dy$$

$$= \frac{1}{C_{\beta+1}} \int\!\!\int \frac{|f(y) - f(x)|^2}{|y - x|^{1+\beta}} \, dy \, dx,$$

where for the last equation we use Parseval's Theorem and a change of variables. This shows that that the norm $\|\widehat{f}\|_{2,\beta}$ is equivalent to a smoothness condition on f.

11.4 The Cameron–Martin Formula

The Cameron–Martin Formula is an isomorphism theorem that expresses a Gaussian process translated by an element of its reproducing kernel Hilbert space in terms of the untranslated process, but with a change of measure, as in the isomorphisms of Chapter 8.

Let $\{G_t \, ; \, t \in T\}$ be a Gaussian process on the probability space $(\Omega, \mathcal{F}, P)$ with continuous covariance

$$\Gamma(s,t) = E(G_s G_t) \tag{11.112}$$

and set $f_s(t) = \Gamma(s,t)$. In the proof of Theorem 5.3.1 we show that the reproducing kernel Hilbert space $H(\Gamma)$ is the closure of S, the span of the functions $\{f_s \, ; \, s \in R^1\}$ with respect to the inner product

$$(f_s, f_t) = \Gamma(s,t) = f_s(t). \tag{11.113}$$

By (5.76) we have that, for any $f \in H(\Gamma)$,

$$f(s) = (f, f_s). \tag{11.114}$$

In the proof of Theorem 5.3.2 we show that the linear map $\Theta_P : S \mapsto L^2(P)$ defined by setting $\Theta_P(f_r) = G_r$ extends to an isometry that maps $H(\Gamma)$ onto the Hilbert space obtained by taking the closure of the span of $\{G_t \, ; \, t \in T\}$ in $L^2(\Omega, \mathcal{F}, P)$ with respect to expectation.

For any $f \in H(\Gamma)$ let $G(f) := \Theta_P(f)$. Thus

$$EG^2(f) = (f, f) \tag{11.115}$$

and, in particular

$$EG(f)G_s = (f, f_s) = f(s). \tag{11.116}$$

Theorem 11.4.1 (Cameron–Martin Formula) *Let $\{G_t \, ; \, t \in T\}$ be a mean zero Gaussian process with continuous covariance Γ and let $H(\Gamma)$*

be the reproducing kernel Hilbert space generated by Γ. Let $f \in H(\Gamma)$. Then, for any measurable functional $F(G.)$,

$$E(F(G. + f(\cdot))) = e^{-\|f\|^2/2} E(F(G.) e^{G(f)}), \qquad (11.117)$$

where $\| \cdot \|$ is the norm on $H(\Gamma)$.

Proof It suffices to prove that for any finite subset $A \subset T$,

$$E\left(\prod_{t \in A}(G_t + f(t))\right) = e^{-\|f\|^2/2} E\left(e^{G(f)} \prod_{t \in A} G_t\right). \qquad (11.118)$$

Clearly

$$E\left(\prod_{t \in A}(G_t + f(t))\right) = \sum_{B \subseteq A}\left(\prod_{t \in B} f(t)\right) E\left(\prod_{t \in A-B} G_t\right). \qquad (11.119)$$

On the other hand, by Lemma 5.2.6 and (11.116),

$$E\left(e^{G(f)} \prod_{t \in A} G_t\right) = \sum_{k=0}^{\infty} \frac{1}{k!} E\left(G^k(f) \prod_{t \in A} G_t\right) \qquad (11.120)$$

$$= \sum_{k=0}^{\infty} \frac{1}{k!} \sum_{j=0}^{k} \binom{k}{j} E(G^j(f)) \sum_{B \subseteq A, |B|=k-j} (k-j)!$$

$$\left(\prod_{t \in B} f(t)\right) E\left(\prod_{t \in A-B} G_t\right).$$

(Note that Lemma 5.2.6 does apply here since $E(G^j(f)) = 0$ when j is odd and $E\left(\prod_{t \in A-B} G_t\right) = 0$ when j is even and k is odd.) Since $E(G^{2j}(f)) = \frac{(2j)!}{j! 2^j} \|f\|^{2j}$, we can write (11.120) as

$$E(e^{G(f)} \prod_{t \in A} G_t) \qquad (11.121)$$

$$= \sum_{k=0}^{\infty} \sum_{j=0}^{k/2} \frac{1}{(2j)!} E(G^{2j}) \sum_{B \subseteq A, |B|=k-2j}\left(\prod_{t \in B} f(t)\right) E\left(\prod_{t \in A-B} G_t\right)$$

$$= \sum_{j=0}^{\infty} \frac{\|f\|^{2j}}{j! 2^j} \sum_{k=2j}^{\infty} \sum_{B \subseteq A, |B|=k-2j}\left(\prod_{t \in B} f(t)\right) E\left(\prod_{t \in A-B} G_t\right)$$

$$= \sum_{j=0}^{\infty} \frac{\|f\|^{2j}}{j! 2^j} \sum_{B \subseteq A}\left(\prod_{t \in B} f(t)\right) E\left(\prod_{t \in A-B} G_t\right).$$

Comparing (11.119) and (11.121), we see that we have obtained (11.118).

$\square$

Alternative proof of Theorem 11.4.1 It suffices to prove that, for any finite subset $\{t_1, \ldots, t_n\} \subset T$ and bounded continuous function F on R^n,

$$E\left(F\left(G_{t_1} + f(t_1), \ldots, G_{t_n} + f(t_n)\right)\right) \tag{11.122}$$
$$= e^{-\|f\|^2/2} E\left(e^{G(f)} F\left(G_{t_1}, \ldots, G_{t_n}\right)\right).$$

Assume first that $f(t) = \sum_{i=1}^n a_i \Gamma(t_i, t)$ for $a_i \in R^1$. Define the $n \times n$ matrix

$$C_{i,j} = \{C(n)\}_{i,j} = \Gamma(t_i, t_j). \tag{11.123}$$

Without loss of generality we can assume that C^{-1} exists. It follows from (4.10) that

$$E\left(F\left(G_{t_1} + f(t_1), \ldots, G_{t_n} + f(t_n)\right)\right) \tag{11.124}$$
$$= \frac{1}{(2\pi)^{n/2}\sqrt{\det C}} \int F\left(z_1 + f(t_1), \ldots, z_n + f(t_n)\right)$$
$$e^{-\sum_{i,j=1}^n C_{i,j}^{-1} z_i z_j / 2} \prod_{i=1}^n dz_i$$
$$= \frac{1}{(2\pi)^{n/2}\sqrt{\det C}} \int F\left(z_1, \ldots, z_n\right)$$
$$e^{-\sum_{i,j=1}^n C_{i,j}^{-1} (z_i - f(t_i))(z_j - f(t_j))/2} \prod_{i=1}^n dz_i.$$

Using the fact that $f(t_j) = \sum_{i=1}^n C_{j,i} a_i$, we have

$$\sum_{i,j=1}^n C_{i,j}^{-1} (z_i - f(t_i))(z_j - f(t_j)) \tag{11.125}$$
$$= \sum_{i,j=1}^n C_{i,j}^{-1} z_i z_j - 2 \sum_{i=1}^n z_i a_i + \sum_{i,j=1}^n C_{i,j} a_i a_j.$$

Note that by (5.76),

$$\|f\|^2 = (f, f) = \sum_{i,j=1}^n a_i a_j (f_i, f_j) = \sum_{i,j=1}^n C_{i,j} a_i a_j. \tag{11.126}$$

Combining the last three displays and using (4.10) once more, we have

$$E\left(F\left(G_{t_1} + f(t_1), \ldots, G_{t_n} + f(t_n)\right)\right) \tag{11.127}$$
$$= \frac{e^{-\|f\|^2/2}}{(2\pi)^{n/2}\sqrt{\det C}} \int F\left(z_1, \ldots, z_n\right)$$

$$e^{\sum_{i=1}^{n} z_i a_i} e^{-\sum_{i,j=1}^{n} C_{i,j}^{-1} z_i z_j / 2} \prod_{i=1}^{n} dz_i$$

$$= e^{-\|f\|^2/2} E\left(F\left(G_{t_1}, \ldots, G_{t_n}\right) e^{\sum_{i=1}^{n} a_i G_{t_i}}\right).$$

Since

$$G(f) = G\left(\sum_{i=1}^{n} a_i \Gamma(t_i, t)\right) = \sum_{i=1}^{n} a_i G_{t_i}, \qquad (11.128)$$

we see that (11.127) is (11.122) in the case when $f(t) = \sum_{i=1}^{n} a_i \Gamma(t_i, t)$. Furthermore, since n is arbitrary, we actually have (11.122) for any $f(t)$ in the span of $\{\Gamma(t, s) \, ; \, s \in T\}$.

To get (11.122) for general $f \in H(\Gamma)$, note that we can find a sequence $\{f_k\}$ in the the span of $\{\Gamma(t, s) \, ; \, s \in T\}$ such that $f_k \to f$ in $H(\Gamma)$. By (11.114), $f_k(t_j) \to f(t_j)$ for $j = 1, \ldots, n$. Also, since Θ_P is an isometry, $G(f_k) \to G(f)$ in L^2, and since $\{G(f_k)\}$ and $G(f)$ are Gaussian, this implies that $\exp(G(f_k)) \to \exp(G(f))$ in L^2. Of course, $\|f_k\| \to \|f\|$. Thus (11.122) for f follows from (11.122) for the f_k, which we have proved. $\qquad \square$

11.5 A probability estimate for fractional Brownian motion

We can now give the analogue of Lemma 11.2.1, which enables us to find a version of Theorem 11.2.6 for the local times of symmetric stable processes.

Lemma 11.5.1 *For fractional Brownian motion of index $0 < \gamma \leq 1$ we have that, for all $\epsilon > 0$ sufficiently small,*

$$P\left(\sup_{|x| \leq 1} \eta(x) < \lambda\right) \leq d\lambda^{2(1-\epsilon)/\gamma} \qquad (11.129)$$

for some constant $d < \infty$.

Although this estimate also applies to two-sided Brownian motion, it is not strong enough to yield Theorem 11.2.6. In fact, for Brownian motion we get $2\lambda^2/\pi$ on the right-hand side of (11.129). This follows from Lemma 11.2.1 and the independence of Brownian motion on R_+ and $(-\infty, 0)$.

Proof We use the Cameron–Martin Formula. It is clear, since d is arbitrary, that we only need to prove (11.129) for all $0 < \lambda \leq \lambda_0$ for some λ_0 sufficiently small. Let $f \in C_0^\infty$ be an even function supported on $[-1, 1]$ that is strictly decreasing on $[0, 1]$ with $f(0) = 1$ and $f(1) = 0$. For $0 < h \le 1$, set $f^{(h)}(x) = hf(x/h^{2/\gamma})$ and $\bar{f}^{(h)}(x) = h - f^{(h)}(x)$ so that in particular $\bar{f}^{(h)}(0) = 0$. We have

$$P\left(\sup_{|x| \le 1} \eta(x) < h\right) = P\left(\eta(x) - \bar{f}^{(h)}(x) < f^{(h)}(x); |x| \le 1\right). \quad (11.130)$$

Let $\mathcal{K}_\gamma$ denote the reproducing kernel Hilbert space for η. By the Cameron–Martin Formula (Theorem 11.4.1),

$$P\left(\eta(x) - \bar{f}^{(h)}(x) < f^{(h)}(x); |x| \le 1\right) \quad (11.131)$$

$$= E\left(1_{[\eta(x) - \bar{f}^{(h)}(x) < f^{(h)}(x); |x| \le 1]}\right)$$

$$= e^{-\|\bar{f}^{(h)}\|^2_{\mathcal{K}_\gamma}/2} E\left(e^{-\eta(\bar{f}^{(h)})} 1_{[\eta(x) < f^{(h)}(x); |x| \le 1]}\right),$$

where $\eta(\bar{f}^{(h)})$ is a mean zero Gaussian random variable with variance $\|\bar{f}^{(h)}\|^2_{\mathcal{K}_\gamma}$. Then, by Hölder's inequality,

$$P\left(\eta(x) - \bar{f}^{(h)}(x) < f^{(h)}(x); |x| \le 1\right) \quad (11.132)$$

$$\le e^{(p-1)\|\bar{f}^{(h)}\|^2_{\mathcal{K}_\gamma}/2} \left(P\left(\eta(x) < f^{(h)}(x); |x| \le 1\right)\right)^{1/q},$$

where $1/p + 1/q = 1$. We show in Lemma 11.5.2 that

$$\|\bar{f}^{(h)}\|_{\mathcal{K}_\gamma} = \|\bar{f}^{(1)}\|_{\mathcal{K}_\gamma} < \infty \quad (11.133)$$

for all $h > 0$. Therefore, by taking $q = 1/(1 - \epsilon)$, we see that to obtain (11.129) we need only show that

$$P\left(\eta(x) < f^{(h)}(x); |x| \le 1\right) \le d h^{2/\gamma} \quad (11.134)$$

for some constant $d < \infty$. Consider the event $\{\eta(x) < f^{(h)}(x); |x| \le 1\}$. For this to occur we must have that $\eta(x) < 0$ outside of $(-h^{2/\gamma}, h^{2/\gamma})$, the support of $f^{(h)}(x)$. In addition we have $\eta(0) = 0$. Consequently,

$$P\left(\eta(x) < f^{(h)}(x); |x| \le 1\right) \quad (11.135)$$

$$\le P\left(\sup_{|y| < h^{2/\gamma}} \eta(y) > \sup_{h^{2/\gamma} \le |y| \le 1} \eta(y)\right)$$

$$\le P\left(\sup_{|y| < h^{2/\gamma}} \eta(y) - \eta(-1) > \sup_{h^{2/\gamma} \le |y| \le 1} \eta(y) - \eta(-1)\right)$$

$$= P\left(\sup_{|y-1| < h^{2/\gamma}} \eta(y) > \sup_{h^{2/\gamma} \le |y-1| \le 1} \eta(y)\right)$$

$$= P \left(\sup_{1-h^{2/\gamma} < y < 1+h^{2/\gamma}} \eta(y) > \sup_{0 \le y \le 1-h^{2/\gamma}, 1+h^{2/\gamma} \le y \le 2} \eta(y) \right),$$

where the next to last equality follows from the fact that η has stationary increments.

Set

$$F(y) = P \left(\sup_{0 \le x < y} \eta(x) > \sup_{y \le x \le 2} \eta(x) \right) \qquad y \in [0, 2] \qquad (11.136)$$

and note that the last term in (11.135) is equal to

$$F \left(1 + h^{2/\gamma} \right) - F \left(1 - h^{2/\gamma} \right). \qquad (11.137)$$

Since $F(y)$ is nondecreasing in y, we have that $F(y)$ is differentiable almost everywhere on $[0, 2]$. Pick $3/2 < y_0 < 2$ such that $F(y)$ is differentiable at y_0 and set $\psi(y_0) = F'(y_0)$. Note that $0 \le \psi(y_0) < \infty$. We then have that, for some $\delta_0 > 0$,

$$F((1+\delta)y_0) - F((1-\delta)y_0) \le (\psi(y_0) + 1)4\delta \qquad \delta \le \delta_0. \qquad (11.138)$$

By (11.5) we have that for any $0 \le z \le 2$ and $k > 0$,

$$F(z) = P \left(\sup_{0 \le x < kz} \eta(x) > \sup_{kz \le x \le 2k} \eta(x) \right). \qquad (11.139)$$

Taking $k = y_0$ we see that

$$F(1+\delta) - F(1-\delta) \qquad (11.140)$$

$$= P \left(\sup_{(1-\delta)y_0 \le x < (1+\delta)y_0} \eta(x) > \sup_{0 \le x < (1-\delta)y_0, (1+\delta)y_0 \le x \le 2y_0} \eta(x) \right)$$

$$\le P \left(\sup_{(1-\delta)y_0 \le x < (1+\delta)y_0} \eta(x) > \sup_{0 \le x < (1-\delta)y_0, (1+\delta)y_0 \le x \le 2} \eta(x) \right)$$

$$= F((1+\delta)y_0) - F((1-\delta)y_0).$$

Using (11.140) and (11.138), we see that (11.137) is bounded by $dh^{2/\gamma}$ for some constant $d > 0$. Thus (11.134) follows from (11.135). $\qquad \square$

Lemma 11.5.2 *Let f be an even function in C_0^∞ and let $f^{(h)}(x) = hf(x/h^{2/(\beta-1)})$, $0 < h \le 1$. Then*

$$\| f^{(h)} - f^{(h)}(0) \|_{\mathcal{K}_{\beta-1}} = \| f - f(0) \|_{\mathcal{K}_{\beta-1}} < \infty. \qquad (11.141)$$

Proof Note that $\widehat{f^{(h)}}(\lambda) = h^{1+2/(\beta-1)} \widehat{f}(\lambda h^{2/(\beta-1)})$, so that, by (11.109),

$$
\begin{aligned}
\| f^{(h)} - f^{(h)}(0) \|^2_{\mathcal{K}_{\beta-1}} &= \frac{1}{2\pi} h^{2+4/(\beta-1)} \int |\lambda|^\beta \, |\widehat{f}(\lambda h^{2/(\beta-1)})|^2 \, d\lambda \\
&= \frac{1}{2\pi} \frac{h^{2+4/(\beta-1)}}{h^{(1+\beta)2/(\beta-1)}} \int |\lambda|^\beta \, |\widehat{f}(\lambda)|^2 \, d\lambda \\
&= \| f - f(0) \|^2_{\mathcal{K}_{\beta-1}} < \infty.
\end{aligned}
\tag{11.142}
$$

$\square$

Corollary 11.5.3 *For fractional Brownian motion of index $0 < \gamma \leq 1$, we have that for any constants $c, \epsilon > 0$,*

$$
P\left(\sup_{|x| \leq \lambda^{1/\gamma}} \eta(x) < c\lambda \right) \leq d\lambda^{(1-\epsilon)/\gamma}
\tag{11.143}
$$

for some constant $d < \infty$.

Proof By changing the constant d in (11.129) we can write it as

$$
P\left(\sup_{|x| \leq 1} \eta(x) < c\lambda \right) \leq d\lambda^{2(1-\epsilon)/\gamma}.
\tag{11.144}
$$

Furthermore, by (11.5),

$$
P\left(\sup_{|x| \leq \lambda^{1/\gamma}} \eta(x) < c\lambda \right) = P\left(\sup_{|x| \leq 1} \eta(x) < c\lambda^{1/2} \right).
\tag{11.145}
$$

Thus we get (11.143). $\square$

Remark 11.5.4 Fractional Brownian motion was defined in (11.3). Consider the broader class $\eta_C = \{ \eta_C(s), s \in R^1 \}$ of mean zero Gaussian processes with stationary increments and

$$
E\left(|\eta_C(s) - \eta_C(t)|^2 \right) = C|s - t|^{\beta-1},
\tag{11.146}
$$

where $\beta \in (0, 2]$ and $C > 0$. The scaling property (11.5) holds for all η_C. Furthermore, since d in (11.129) is arbitrary, it is clear that if (11.129) holds for some $\eta_{C'}$, then it holds for all η_C, where, of course, the constant d depends on C. This also applies to (11.143).

11.6 Most visited sites of symmetric stable processes

We dealt with this question for Brownian motion in Section 11.2. Therefore, although some of the results in this section apply to Brownian motion, in this section we are really only interested in symmetric stable processes with index $1 < \beta < 2$.

We begin with a probability estimate for the local times of symmetric stable processes.

Lemma 11.6.1 *Let $X = \{X_t \,;\, t \geq 0\}$ be a symmetric stable process of index $1 < \beta \leq 2$ in R^1 and denote its local times by $L = \{L_t^x, (x, t) \in R^1 \times R_+\}$. Let $\tau(\cdot)$ denote the inverse local time of $L_{\cdot}^0$. Then, for any $\epsilon > 0$,*

$$P^0 \left(\sup_{|x| \leq \lambda^{1/(\beta-1)}} L_{\tau(1)}^x \leq 1 + \lambda \right) \leq d\lambda^{(1-\epsilon)/(\beta-1)} \tag{11.147}$$

for some $d < \infty$.

Proof By (11.5), the scaling property of fractional Brownian motion, we have

$$h_\lambda := P_\eta \left(\sup_{|x| \leq \lambda^{1/(\beta-1)}} \eta^2(x) \leq 2\lambda \right) \tag{11.148}$$

$$= P_\eta \left(\sup_{|x| \leq 1} \eta^2(x) \leq 2 \right)$$

so that $h_\lambda = h_1 > 0$. Consequently,

$$P^0 \left(\sup_{|x| \leq \lambda^{1/(\beta-1)}} L_{\tau(1)}^x \leq 1 + \lambda \right) \tag{11.149}$$

$$= \frac{1}{h_1} P^0 \times P_\eta \left(\sup_{|x| \leq \lambda^{1/(\beta-1)}} L_{\tau(1)}^x \leq 1 + \lambda \text{ and } \sup_{|x| \leq \lambda^{1/(\beta-1)}} \eta^2(x) \leq 2\lambda \right)$$

$$\leq \frac{1}{h_1} P^0 \times P_\eta \left(\sup_{|x| \leq \lambda^{1/(\beta-1)}} L_{\tau(1)}^x + \frac{1}{2}\eta^2(x) \leq 1 + 3\lambda \right).$$

Therefore, by Theorem 8.2.2, the Generalized Second Ray–Knight Theorem,

$$P^0 \left(\sup_{|x| \leq \lambda^{1/(\beta-1)}} L_{\tau(1)}^x \leq 1 + \lambda \right) \tag{11.150}$$

$$\leq \frac{1}{h_1} P_\eta \left(\sup_{|x| \leq \lambda^{1/(\beta-1)}} \frac{\left(\eta_x + \sqrt{2}\right)^2}{2} \leq 1 + 3\lambda \right)$$

$$\leq \frac{1}{h_1} P_\eta \left(\sup_{|x| \leq \lambda^{1/(\beta-1)}} \frac{\eta^2(x)}{2} + \eta(x) \leq 3\lambda \right)$$

$$\leq \frac{1}{h_1} P_\eta \left(\sup_{|x| \leq \lambda^{1/(\beta-1)}} \eta(x) \leq 3\lambda \right).$$

The inequality (11.147) now follows from Lemma 11.5.3. □

Lemma 11.6.2 *Let $X = \{X_t \,;\, t \geq 0\}$ be a symmetric stable process of index $1 < \beta \leq 2$ in R^1 and denote its local times by $L = \{L_t^x, (x, t) \in R^1 \times R_+\}$. Let $\tau(\cdot)$ denote the inverse local time of L^0. Then, for any $\epsilon > 0$ and all $v > 0$,*

$$P^0 \left(\sup_{|s| \leq (v\lambda)^{1/(\beta-1)}} L_{\tau(v)}^s - v \leq \lambda v \right) \leq d\lambda^{(1-\epsilon)/(\beta-1)} \qquad (11.151)$$

for some constant $d < \infty$. Therefore, in particular,

$$P^0 \left(L_{\tau(v)}^* - v \leq \lambda v \right) \leq d\lambda^{(1-\epsilon)/(\beta-1)}, \qquad (11.152)$$

where $L^* := \sup_x L^x$ is the maximal local time.

Proof By (11.6),

$$L_{\tau(1)}^x \overset{law}{=} \frac{1}{v} L_{\tau(v)}^{xv^{1/(\beta-1)}} \qquad (11.153)$$

Using this in (11.147) we get (11.151). □

We can now give a version of Lemma 11.2.5 for the symmetric stable process of index $1 < \beta < 2$.

Lemma 11.6.3 *Let $X = \{X_t \,;\, t \geq 0\}$ be a symmetric stable process of index $1 < \beta < 2$ in R^1 and denote its local times by $L = \{L_t^x, (x, t) \in R^1 \times R_+\}$. Let $\tau(\cdot)$ denote the inverse local time of L^0. Then, for any $b > \frac{\beta-1}{2-\beta}$,*

$$\lim_{r \to \infty} \frac{(\log r)^b (\sup_x L_{\tau(r)}^x - r)}{r} = \infty \qquad (11.154)$$

and

$$\lim_{r \to 0} \frac{(\log 1/r)^b (\sup_x L_{\tau(r)}^x - r)}{r} = \infty. \qquad (11.155)$$

The proof is essentially the same as the proof of Theorem 11.2.5, but since it is short and since this is an important result, we repeat the proof with the necessary modifications.

Proof Let $v = v_n = e^{n^a}$ for $0 < a < 1$ and $\lambda = n^{-(\beta-1)(1+2\epsilon)}$ in (11.152). It follows from the Borel–Cantelli Lemma that

$$L^*_{\tau(v_n)} - v_n \geq \frac{v_n}{n^{(\beta-1)(1+2\epsilon)}} \tag{11.156}$$

for all $n \geq n_0(\omega)$ on a set of probability one, where ω denotes the paths of the stable process. Let $v_n \leq r \leq v_{n+1}$. Then, restricted to this set,

$$L^*_{\tau(r)} - r \geq -(r - v_n) + \frac{v_n}{n^{(\beta-1)(1+2\epsilon)}} \tag{11.157}$$

for all $n \geq n_0(\omega)$ and

$$\frac{L^*_{\tau(r)} - r}{r} \geq -\frac{v_{n+1} - v_n}{v_n} + \frac{v_n}{v_{n+1}} \frac{1}{n^{(\beta-1)(1+2\epsilon)}} \tag{11.158}$$

for all $n \geq n_0(\omega)$. Then by a simple calculation we see that

$$\frac{L^*_{\tau(r)} - r}{r} \geq -\frac{2}{n^{1-a}} + \frac{(1-\delta)}{n^{(\beta-1)(1+2\epsilon)}} \tag{11.159}$$

for all $\delta > 0$ and for all $n \geq n_0(\omega)$. It is easy to see that if $\beta - 1 < 1 - a$, then, for ϵ small enough, $(\beta - 1)(1 + 2\epsilon) < 1 - a$. Thus

$$\frac{L^*_{\tau(r)} - r}{r} \geq \frac{(1-\delta)}{n^{(\beta-1)(1+2\epsilon)}} \tag{11.160}$$

for all $n \geq n'_0(\omega)$ and $v_n \leq r \leq v_{n+1}$ on a set of probability one. For r in this range, $n \sim (\log r)^{1/a}$. Since we can take a arbitrarily close to $2 - \beta$, we get (11.154).

We get (11.155) from (11.158) by taking $v_n = e^{-n^a}$ for $0 < a < 1$ and repeating the rest of the above argument. □

We can now prove the "most visited sites" theorem for symmetric stable processes of index $1 < \beta < 2$.

Theorem 11.6.4 *Let $X = \{X_t ; t \geq 0\}$ be a symmetric stable process of index $1 < \beta < 2$ in R^1 and denote its local times by $L = \{L^x_t, (x,t) \in R^1 \times R_+\}$. Let V_t be given by (11.2). Then, for any $\gamma > \beta/(2-\beta)(\beta-1)$, we have*

$$\lim_{t \to \infty} \frac{(\log t)^\gamma}{t^{1/\beta}} V_t = \infty \qquad P^0 \text{ a.s.} \tag{11.161}$$

and

$$\lim_{t \to 0} \frac{|\log t|^\gamma}{t^{1/\beta}} V_t = \infty \qquad P^0 \text{ a.s.} \tag{11.162}$$

In particular, (11.161) tells us that $\lim_{t\to\infty} V_t = \infty$, that is, the process $\{V_t, t \in R_+\}$ is transient. Moreover, we can make a similar comment about (11.162) as we made for (11.79) following Theorem 11.2.6.

Proof Here again the proof is a very minor modification of the proof of Theorem 11.2.6 for Brownian motion. Let $t \in [\tau(r^-), \tau(r)]$ for r large. By (11.154) we have that

$$\sup_x L_t^x > r + r/(\log r)^b \qquad (11.163)$$

for all $b > (\beta - 1)/(2 - \beta)$.

On the other hand, let h_a be as defined in (11.10) with $a = 2(1 + \epsilon)b$ for some $\epsilon > 0$. It follows from (11.12) with $\gamma = b$ and the fact that $L_{\tau(r)}^0 = r$, that

$$\sup_{0 \leq |x| \leq h_a(L_t^0)} L_t^x \leq \sup_{0 \leq |x| \leq h_a(r)} L_{\tau(r)}^x \qquad (11.164)$$

$$< \quad r + r/(\log r)^b$$

for all r sufficiently large.

Comparing (11.163) and (11.164), we see that for all t sufficiently large,

$$V_t > h_a(L_t^0). \qquad (11.165)$$

Using Lemma 11.1.3 and the fact that h_a is increasing, this shows that

$$V_t > h_a\left(\frac{t^{1/\bar\beta}}{(\log t)^{(1+\epsilon)}}\right) \qquad (11.166)$$

$$\geq \frac{t^{1/\beta}}{\left((\log t)^{(2b+1)(1+\epsilon)}\right)^{1/(\beta-1)}}.$$

Since we can take b arbitrarily close to $(\beta - 1)/(2 - \beta)$ and ϵ arbitrarily close to zero, we get (11.161).

The proof of (11.162) is essentially the same as the proof of (11.79) in Theorem 11.2.6. $\qquad \square$

11.7 Notes and references

The first investigation of the most visited sites question for Lévy processes was by Bass and Griffin (1985), who considered Brownian motion. They obtained (11.78) for all $\gamma > 11$. They also showed that

$$\liminf_{t\to\infty} \frac{(\log t)^\gamma}{t^{1/2}} V_t = 0 \qquad P^0 \quad \text{a.s.} \qquad (11.167)$$

for all $\gamma < 1$. Many people think that (11.78) should hold for all $\gamma > 1$ but this has remained an open problem for the last 20 years.

Bass, Eisenbaum and Shi (2000) obtained (11.154) for all $b > \frac{9}{\beta-1}$. This is better than our estimate for b when β is near 2 but not as good when β is near 1. We use a different method of proof in Theorem 11.6.3 than the one in Bass, Eisenbaum and Shi (2000). In that work, a clever use of Slepian's Lemma allowed them to get a version of the critical Lemma 11.5.1, by considering the amount of time Brownian motion spends within a cone. We think it takes us far too afield to develop all the prerequisites needed to proof this result. (In Bass, Eisenbaum and Shi (2000), they referred to Bañuelos and Smits (1997).) Instead, we use a rather sharp direct estimate for fractional Brownian motion by Molchan (1999). Molchan's estimate uses the Cameron–Martin Formula, which involves reproducing kernel Hilbert spaces, thus our Sections 11.3 and 11.4. For other results about the most visited sites of symmetric stable processes, see Eisenbaum (1997) and Eisenbaum and Khoshnevisan (2002).

Both Bass, Eisenbaum and Shi (2000) and Molchan (1999) restrict their attention to symmetric stable process and hence can use the scaling property in Lemma 11.1.1. Furthermore, in applying the Ray–Knight Theorem or the Generalized Second Ray–Knight Theorem both in Bass, Eisenbaum and Shi (2000) and in our presentation in this chapter, the scaling property of fractional Brownian motion, (11.5), is used often. This is analytically convenient but may not be necessary. In Marcus (2001), the approach of Bass, Eisenbaum and Shi (2000) is extended to Lévy processes with regularly varying characteristic exponents. Extending the approach of this chapter beyond symmetric stable processes remains to be done.

The most visited sites problem is only one of the many questions that one can ask about the behavior of local times $\{L_t^x, (x, t) \in R^1 \times R_+\}$ as $t \to \infty$. We do not pursue these questions in this chapter because our guiding principle is to concentrate on results that employ the isomorphism theorems of Chapter 8. That is except for (11.40), which we mention as a complement of Lemma 11.1.3. A good starting point for considering these questions are Donsker and Varadhan (1977), Lacey (1990), Marcus and Rosen (1994a), Marcus and Rosen (1994b), Bertoin (1995), Bertoin and Caballero (1995), and Blackburn (2000).

The proofs in Section 11.3 extend easily to all Gaussian processes associated with Lévy processes and even to all Gaussian processes with spectral densities. As an example consider the Gaussian processes associated with recurrent Lévy processes killed the first time they hit zero. Let X be a recurrent Lévy process with Lévy exponent ψ. Recall that, by Theorem 4.2.4, the 0-potential density of X killed the first time it hits zero is

$$u_{T_0}(x, y) = \phi(x) + \phi(y) - \phi(x - y), \tag{11.168}$$

where

$$\phi(x) := \frac{1}{\pi} \int_0^\infty \frac{1 - \cos \lambda x}{\psi(\lambda)} \, d\lambda \qquad x \in R. \tag{11.169}$$

Consider the complex Hilbert space $L_\psi^2 = L_C^2(R^1, (\psi(\lambda)/2\pi) \, d\lambda)$ and denote its norm by $\| \cdot \|_\psi$, that is, $g \in L_\psi^2$ is a measurable function with

$$\|g\|_\psi = \left(\frac{1}{2\pi} \int |g(\lambda)|^2 \, \psi(\lambda) \, d\lambda \right)^{1/2} < \infty. \tag{11.170}$$

For $g \in L_\psi^2$, let

$$Ig(x) := \frac{1}{2\pi} \int (1 - e^{-ix\lambda}) g(\lambda) \, d\lambda. \tag{11.171}$$

It follows from the Cauchy-Schwarz inequality that Ig exists and is uniformly continuous.

The proof of the following theorem is a simple generalization of the proof of Theorem 11.3.1.

Lemma 11.7.1 *The linear space*

$$H_\psi = \{Ig : g \in L_\psi^2\} \tag{11.172}$$

equipped with the norm

$$\|Ig\|_{H_\psi} := \|g\|_\psi \tag{11.173}$$

is a separable complex Hilbert space of continuous functions. Furthermore,

$$u_{T_0}(t, \cdot) \in H_\psi \qquad \forall t \in R^1 \tag{11.174}$$

and

$$(f(\cdot), u_{T_0}(t, \cdot))_{H_\psi} = f(t) \qquad \forall f \in H_\psi, \tag{11.175}$$

where $(\cdot, \cdot)_{H_\psi}$ denotes the inner product on H_ψ.

Let $\mathcal{K}_\psi$ denote the reproducing kernel Hilbert space for the covariance u_{T_0} and let $\| \cdot \|_{\mathcal{K}_\psi}$ denote the norm of $\mathcal{K}_\psi$. Then

$$\mathcal{K}_\psi = \{f \in H_\psi \mid f = \bar{f}\} \tag{11.176}$$

and, for $f \in \mathcal{K}_\psi$,

$$\|f\|_{\mathcal{K}_\psi} = \|f\|_{H_\psi}. \qquad (11.177)$$

In a similar way one can consider the reproducing kernel Hilbert spaces for Gaussian processes associated with Lévy processes killed at the end of an independent exponential time.

12

Local times of diffusions

12.1 Ray's Theorem for diffusions

In Subsection 8.2.1 we show that we can generalize the Second Ray–Knight Theorem for recurrent diffusions so that the process can start at an arbitrary point in the state space. We do the same thing here in the transient case. The Second Ray–Knight Theorem in the transient case is given in Theorem 8.2.3. As one can see from the proof, it is actually a theorem about the local times of the h-transform of a process. (We pass from the h-transform to the original process in (8.61).) In Theorem 8.4.1, in particular in (8.127), we make this more explicit. In this section, for diffusions, we extend Theorem 8.4.1 to an h-transform process starting at x and terminating at its last exit from y for $y \neq x$. This is how Ray's original result was given, although our result appears quite different from his.

Let X be a transient regular diffusion on an interval $I \subset R^1$, as described in Section 4.3, with symmetric 0-potential density $u(r, s)$. Recall that $u(r, s)$ has the form given in (4.111),

$$u(r, s) = \begin{cases} p(r)q(s) & r \leq s \\ p(s)q(r) & s < r, \end{cases} \qquad (12.1)$$

where p and q are positive continuous functions, with p strictly increasing and q strictly decreasing.

For $h_y(r) = u(r, y)/u(y, y)$, let X^{h_y} denote the h_y-transform of X. (X^{h_y} is an alternative notation for $\widetilde{X} = (\Omega_{h_y}, \mathcal{F}^{h_y}, \mathcal{F}_t^{h_y}, X_t, \theta_t, P^{x/h_y})$ in Theorem 3.9.2 that makes more explicit the dependence on the function h_y.)

Let $\overline{L} = \{\overline{L}_t^y \, ; \, (y, t) \in I \times R_+\}$ denote the local time for X^{h_y} normalized so that

$$E^{x,y}\left(\overline{L}_\infty^r\right) = \frac{u(x,r)h_y(r)}{h_y(x)} = \frac{u(x,r)u(r,y)}{u(x,y)} \tag{12.2}$$

(recall that we use $P^{x,y}$ to denote the law of X^{h_y}, starting at x; see also Remark 3.9.3).

Let $G = \{G_r, r \in I\}$ be a mean zero Gaussian process with covariance $u(r,s)$ and let $G_{r,z}$ denote the projection of G_r on the orthogonal complement of G_z, that is,

$$G_{r,z} = G_r - \frac{E(G_r G_z)}{E(G_z G_z)}G_z = G_r - \frac{u(r,z)}{u(z,z)}G_z. \tag{12.3}$$

In the next lemma we collect some facts about $G_{r,z}$ that are used throughout this chapter.

Lemma 12.1.1

(1) If $y < z$, then $G_{z,y} = G_z - \dfrac{q(z)}{q(y)}G_y$, and it is independent of $\mathcal{G}_{\leq y}$, the σ-algebra generated by $\{G_r, r \leq y\}$.

(2) If $y < z$, then $G_{y,z} = G_y - \dfrac{p(y)}{p(z)}G_z$, and it is independent of $\mathcal{G}_{\geq z}$, the σ-algebra generated by $\{G_r, r \geq z\}$.

(3) If $t < s < r$, then

$$G_{r,t} = G_{r,s} + \frac{q(r)}{q(s)}G_{s,t}, \tag{12.4}$$

and $G_{r,s}$ and $G_{s,t}$ are independent.

Proof (1) If $y < z$, then $\dfrac{u(y,z)}{u(y,y)} = \dfrac{q(z)}{q(y)}$. Since $G_{z,y}$ is the projection of G_z on the orthogonal complement of G_y, it is orthogonal to G_y. Indeed, a simple calculation using (12.1) shows that $G_{z,y}$ is orthogonal to each G_r, $r \leq y$. Consequently it is independent of $\mathcal{G}_{\leq y}$.

(2) The proof is similar to the proof of (1).

(3) If $t < s < r$, then, by (1)

$$\begin{aligned} G_{r,t} &= G_{r,s} + \frac{q(r)}{q(s)}G_s - \frac{q(r)}{q(t)}G_t \tag{12.5}\\ &= G_{r,s} + \frac{q(r)}{q(s)}\left(G_s - \frac{q(s)}{q(t)}G_t\right) = G_{r,s} + \frac{q(r)}{q(s)}G_{s,t}. \end{aligned}$$

Since $G_{s,t} \in \mathcal{G}_{\leq s}$, it is independent of $G_{r,s}$ by (1). $\qquad \square$

Theorem 12.1.2 (Ray's Theorem for Diffusions) *Let* $\overline{L} = \{\overline{L}_t^x, (x,t)$
$\in I \times R_+\}$ *denote the local times of* X^{h_v} *normalized as in (12.2). Let* $\overline{G}$
be an independent copy of G. *Let* $x \leq y$. *Then, under* $P^{x,y} \times P_{G,\overline{G}}$ *or*
$P^{y,x} \times P_{G,\overline{G}}$,

$$\left\{ \overline{L}_\infty^r + \left(\frac{G_{r,x}^2}{2} + \frac{\overline{G}_{r,x}^2}{2} \right) 1_{\{r \leq x\}} + \left(\frac{G_{r,y}^2}{2} + \frac{\overline{G}_{r,y}^2}{2} \right) 1_{\{r \geq y\}} : r \in I \right\}$$

$$\stackrel{law}{=} \left\{ \frac{G_r^2}{2} + \frac{\overline{G}_r^2}{2} : r \in I \right\}, \tag{12.6}$$

where $I \subset R^1$.

Proof This proof is an easy application of Lemma 3.10.4. Using it and
Lemma 5.2.1, all we need to do is verify that

$$v(r,s) := u(r,s) - \frac{h_y(r)u(s,x)}{h_y(x)} = u(r,s) - \frac{u(r,y)u(s,x)}{u(x,y)} \tag{12.7}$$

is the covariance of the Gaussian process $G_{r,x}1_{\{r \leq x\}} + \overline{G}_{r,y}1_{\{r \geq y\}}$ when
$x \leq y$, and similarly with x and y interchanged when $y \leq x$. (It is not
clear a priori that $v(r,s)$ is the covariance of any Gaussian process, or
even positive definite. This follows from the proof.)

Suppose $x \leq y$. Then, using (12.1), we see that for $x \leq r \leq y$,

$$v(r,r) = p(r)q(r) - \frac{p(r)q(y)p(x)q(r)}{p(x)q(y)} = 0. \tag{12.8}$$

We leave it to the reader to check that for $r < x \leq s \leq y$ and for
$x \leq r \leq y < s$, $v(r,s) = v(s,r) = 0$.

Suppose $r, s < x$. Then

$$\begin{aligned} v(r,s) &= u(r,s) - \frac{p(r)u(s,x)}{p(x)} = u(r,s) - \frac{u(r,x)u(s,x)}{u(x,x)} \\ &= EG_{r,x}G_{s,x}. \end{aligned} \tag{12.9}$$

Similarly when $r, s > y$, $v(r,s) = EG_{r,y}G_{s,y}$. Finally, we again leave it
to the reader to check that, when $r < x < y < s$, $v(r,s) = v(s,r) = 0$,
which is consistent with the fact that $G_{r,x}1_{\{r \leq x\}}$ and $G_{s,y}1_{\{s \geq y\}}$ are
independent. (By Lemma 12.1.1 (2), $G_{s,y}1_{\{s \geq y\}} \in \mathcal{G}_{\geq x}$.)

The same arguments work when $y \leq x$. $\qquad \square$

We now give an analog of Theorem 8.2.6 for transient diffusions.

Corollary 12.1.3 *Under the same hypotheses as Theorem 12.1.2, under*

$P^{y,0} \times P_{G,\overline{G}}, \, y > 0,$

$$\left\{ \overline{L}_{T_0}^r + \left(\frac{G_{r,y}^2}{2} + \frac{\overline{G}_{r,y}^2}{2} \right) 1_{\{r \geq y\}} : r \in I \right\} \tag{12.10}$$

$$\overset{law}{=} \left\{ \frac{G_{r,0}^2}{2} + \frac{\overline{G}_{r,0}^2}{2} : r \in I \right\},$$

where $I \subset R_+$.

Proof By the additivity property of local times, $\{\overline{L}_\infty^r = \overline{L}_{T_0}^r + \overline{L}_\infty^r \circ \theta_{T_0}\}$ and by the strong Markov property, $\overline{L}_{T_0}^r$ and $\overline{L}'^r_\infty := \overline{L}_\infty^r \circ \theta_{T_0}$ are independent. Thus, by (12.6), under $P^{y,0} \times P^{0,0} \times P_{G,\overline{G}}$ (noting that $I \subset R_+$),

$$\left\{ \overline{L}_{T_0}^r + \overline{L}'^r_\infty + \left(\frac{G_{r,y}^2}{2} + \frac{\overline{G}_{r,y}^2}{2} \right) 1_{\{r \geq y\}} : r \in I \right\} \overset{law}{=} \left\{ \frac{G_r^2}{2} + \frac{\overline{G}_r^2}{2} : r \in I \right\}. \tag{12.11}$$

Also, by (12.6), under $P^{0,0} \times P_{G,\overline{G}}$,

$$\left\{ \overline{L}'^r_\infty + \left(\frac{G_{r,0}^2}{2} + \frac{\overline{G}_{r,0}^2}{2} \right) : r \in I \right\} \overset{law}{=} \left\{ \frac{G_r^2}{2} + \frac{\overline{G}_r^2}{2} : r \in I \right\}. \tag{12.12}$$

Substituting the left-hand side of (12.12) into the right-hand side of (12.11) and canceling $\overline{L}'^r_\infty$, we get (12.10). $\qquad\square$

Theorem 8.2.6 and Corollary 12.1.3 have the same form. On their right-hand sides we have the Gaussian process with covariance u_{T_0} and on their left-hand sides we have the orthogonal compliment of the projection of this process onto its value at y.

Remark 12.1.4 We come to a very interesting question. Is it possible that Theorem 8.2.6 holds for processes other than diffusions? The proof of the theorem holds the answer, which is no, at least in whatever cases we can think of. It is very easy to see this. In order for the theorem to hold we need that $v(r, s)$ in (12.7) is the covariance of a Gaussian process. Unfortunately this function is not even symmetric in general. Consider the canonical symmetric p-stable process killed at the end of an exponential time with mean α. In this case,

$$u(r, s) = \frac{1}{\pi} \int_0^\infty \frac{\cos \lambda(r - s)}{\alpha + \lambda^p} \, d\lambda \tag{12.13}$$

as we show in (4.84). (For $v(r, s)$ to be symmetric, we would need $u(r, y)u(s, x) = u(s, y)u(r, x)$.)

One obvious way to make $v(r,s)$ symmetric is to take $x = y$, that is, to make the process start and stop at the same point. The results obtained by doing this already appear in Section 8.2.

Example 12.1.5 Theorem 12.1.2 applied to Brownian motion killed at the end of an independent exponential time with mean $1/2$ gives Theorem 2.8.1. As a somewhat more esoteric application of Theorem 12.1.2 we give an interesting modification of Theorem 8.2.6. We consider standard Brownian motion $B = \{B(r), r \in R_+\}$ starting at $x > 0$ and killed the first time it hits 0. But now we use the h-transform $h_y(\cdot)$ to condition this process to hit $y > x$ and die at y, so that the process never does hit 0. The 0-potential of B for $x, y \geq 0$ is $2(x \wedge y)$); see (2.139). So G in Theorem 12.1.2 is $\sqrt{2}B$. It follows from Theorem 12.1.2 that, under $P^{x,y} \times P_{B,\overline{B}}$,

$$\{\overline{L}^r_\infty + ((B_r - \frac{r}{x}B_x)^2 + (\overline{B}_r - \frac{r}{x}\overline{B}_x)^2))1_{\{r \leq x\}} \tag{12.14}$$

$$+ (B^2_{r-y} + \overline{B}^2_{r-y})1_{\{r \geq y\}} : r \in R_+\} \overset{law}{=} \{B^2_r + \overline{B}^2_r : r \in R_+\},$$

where $\overline{L}^\cdot_\infty$ is the total accumulated local time of the h_y-transform of the killed Brownian motion. Note the two independent Brownian bridges between 0 and x.

12.2 Eisenbaum's version of Ray's Theorem

Ray's Theorem has been the subject of many investigations. It has been reformulated and reproved in different ways. References are given in Section 12.6. In all of these papers the law of $\{L^r_\infty ; r \in I\}$ is described piecewise, in three separate regions: $r \leq x$, $x \leq r \leq y$, and $r \geq y$, conditioned to agree at the endpoints.

The version of Ray's Theorem closest to Theorem 12.1.2 is in Eisenbaum (1994). We derive her result from Theorem 12.1.2. Let $Z = \{Z_t(x); (x,t) \in R^1 \times R^+\}$ be a zero-dimensional squared Bessel process starting at x (see Section 14.2), and let $\bar{Z} = \{\bar{Z}_t(x); (x,t) \in R^1 \times R^+\}$ be an independent copy of Z. In addition, let B_t be a standard two-dimensional Brownian motion, also called a planar Brownian motion, independent of Z and $\bar{Z}$.

For p and q as in (12.1), set $\tau(r) = p(r)/q(r)$ and $\phi(r) = q(r)/p(r)$, and note that $\tau(r)$ is increasing and $\phi(r)$ is decreasing.

Theorem 12.2.1 *Let* $\{\overline{L}_\infty^r ; r \in I\}$ *be as in Theorem 12.1.2. Then*

$$\{\overline{L}_\infty^r ; r \in I\} \overset{law}{=} \{\Psi_r ; r \in I\}, \tag{12.15}$$

where

$$\Psi_r = \tfrac{1}{2}q^2(r)|B_{\tau(r)}|^2 \qquad x \leq r \leq y, \tag{12.16}$$

$$\Psi_r = \tfrac{1}{2}q^2(r)Z_{\tau(r)-\tau(y)}(|B_{\tau(y)}|^2) \qquad r \geq y, \tag{12.17}$$

$$\Psi_r = \tfrac{1}{2}p^2(r)\bar{Z}_{\phi(r)-\phi(x)}(\phi^2(x)|B_{\tau(x)}|^2) \qquad r \leq x. \tag{12.18}$$

Note that $|B_t|^2$ is a two-dimensional squared Bessel process starting at 0 (a $BESQ^2(0)$ process in the notation of Section 14.2).

Proof To begin, we motivate the choice of $\Psi(r)$ in (12.16)–(12.18). Let W_t denote a Brownian motion and let G be a mean zero Gaussian process with covariance $u(r, s)$. By checking covariances it is easy to verify that

$$\{G_r ; r \in I\} \overset{law}{=} \{q(r)W_{\tau(r)} : r \in I\} \overset{law}{=} \{p(r)W_{\phi(r)} : r \in I\}. \tag{12.19}$$

Then using Lemma 12.1.1 (1) and (2), we obtain

$$\{G_{r,y} ; r \geq y\} \overset{law}{=} \{q(r)(W_{\tau(r)} - W_{\tau(y)}) : r \geq y\}, \tag{12.20}$$

$$\{G_{r,x} ; r \leq x\} \overset{law}{=} \{p(r)(W_{\phi(r)} - W_{\phi(x)}) : r \leq x\}. \tag{12.21}$$

As we show in the proof of Theorem 12.1.2, $\{G_{r,x}; r \leq x\}$ and $\{G_{r,y}; r \geq y\}$ are independent.

Let $\bar{B} = \{\bar{B}_t, t \in R_+\}$ and $\widehat{B} = \{\widehat{B}_t, t \in R_+\}$ be two planar Brownian motions independent of each other. It follows from (12.19)–(12.21) that (12.6) is equivalent to

$$\{\overline{L}_\infty^r + \tfrac{1}{2}p^2(r)|\bar{B}_{\phi(r)-\phi(x)}|^2 I_{\{r \leq x\}} \tag{12.22}$$
$$+ \tfrac{1}{2}q^2(r)|\widehat{B}_{\tau(r)-\tau(y)}|^2 I_{\{r \geq y\}} : r \in I\}$$
$$\overset{law}{=} \{ \tfrac{1}{2}q^2(r)|\widehat{B}_{\tau(r)}|^2 : r \in I\}$$
$$\overset{law}{=} \{ \tfrac{1}{2}p^2(r)|\bar{B}_{\phi(r)}|^2 : r \in I\},$$

where $\overline{L} = \{\overline{L}_\infty^r, r \in I\}$ is independent of $\bar{B}$ and $\widehat{B}$. Using this we see that

$$\overline{L}_\infty^r \overset{law}{=} \tfrac{1}{2}q^2(r)|\widehat{B}_{\tau(r)}|^2 \qquad x \leq r \leq y$$

$$\overline{L}_\infty^r + \tfrac{1}{2}q^2(r)|\widehat{B}_{\tau(r)-\tau(y)}|^2 I_{\{r \geq y\}} \overset{law}{=} \tfrac{1}{2}q^2(r)|\widehat{B}_{\tau(r)}|^2 \qquad r \geq y$$

$$\overline{L}_\infty^r + \tfrac{1}{2}p^2(r)|\bar{B}_{\phi(r)-\phi(x)}|^2 I_{\{r \leq x\}} \overset{law}{=} \tfrac{1}{2}p^2(r)|\bar{B}_{\phi(r)}|^2 \qquad r \leq x. \tag{12.23}$$

The last two equalities can be written as

$$\overline{L}^r_\infty + \tfrac{1}{2}q^2(r)|\widehat{B}_{\tau(r)-\tau(y)}|^2 I_{\{r\geq y\}} \tag{12.24}$$

$$\overset{law}{=} \tfrac{1}{2}q^2(r)|\widehat{B}_{\tau(y)} + \widehat{B}_{\tau(r)-\tau(y)}|^2 \qquad r \geq y$$

and

$$\overline{L}^r_\infty + \tfrac{1}{2}p^2(r)|\bar{B}_{\phi(r)-\phi(x)}|^2 I_{\{r\leq x\}} \tag{12.25}$$

$$\overset{law}{=} \tfrac{1}{2}p^2(r)|\bar{B}_{\phi(x)} + \bar{B}_{\phi(r)-\phi(x)}|^2 \qquad r \leq x.$$

By Remark 14.2.3, and using the fact that $|B_{\phi(x)}|^2 \overset{law}{=} \phi^2(x)|B_{\tau(x)}|^2$ in (12.25), we see that $\overline{L}^r_\infty$ is equal in law to Ψ_r, separately, in each of the three regions (12.16)–(12.18). What is not yet clear is that (12.15) holds globally when we take Z, $\bar{Z}$, and B in (12.16)–(12.18) to be independent.

To complete the proof we show that (12.22) also holds with $\overline{L}^r_\infty$ replaced by Ψ_r, since, for example, by taking Laplace transforms, we can show that for all finite sets $D \subset R^1$, $\{\overline{L}^r_\infty ; r \in D\} \overset{law}{=} \{\Psi_r ; r \in D\}$ and thus obtain (12.15).

Let $B = \{B_t, t \in R_+\}$ be a planar Brownian motion independent of $\bar{B}$ and $\widehat{B}$. Let $Z_{n,t}(x)$ denote an n-th order squared Bessel process starting at x. It follows from (12.18) and Remarks 14.2.1 and 14.2.3 that for $r \leq x$,

$$\Psi_r + \tfrac{1}{2}p^2(r)|\bar{B}_{\phi(r)-\phi(x)}|^2 \tag{12.26}$$

$$= \tfrac{1}{2}p^2(r)\left(Z_{0,\phi(r)-\phi(x)}(\phi^2(x)|B_{\tau(x)}|^2) + Z_{2,\phi(r)-\phi(x)}(0)\right)$$

$$= \tfrac{1}{2}p^2(r)\left(Z_{2,\phi(r)-\phi(x)}(\phi^2(x)|B_{\tau(x)}|^2)\right)$$

$$= \tfrac{1}{2}p^2(r)|\bar{B}_{\phi(r)-\phi(x)} + \phi(x)B_{\tau(x)}|^2.$$

A similar argument, using (12.17), shows that for $r \geq y$,

$$\Psi_r + \tfrac{1}{2}q^2(r)|\widehat{B}_{\tau(r)-\tau(y)}|^2 = \tfrac{1}{2}q^2(r)|\widehat{B}_{\tau(r)-\tau(y)} + B_{\tau(y)}|^2. \tag{12.27}$$

Therefore, by (12.26), (12.27), and (12.16), we see that for $r \in I$,

$$\{\Psi_r + \tfrac{1}{2}p^2(r)|\bar{B}_{\phi(r)-\phi(x)}|^2 I_{\{r\leq x\}} + \tfrac{1}{2}q^2(r)|\widehat{B}_{\tau(r)-\tau(y)}|^2 1_{\{r\geq y\}}\}$$

$$\overset{law}{=} \{ \tfrac{1}{2}p^2(r)|\bar{B}_{\phi(r)-\phi(x)} + \phi(x)B_{\tau(x)}|^2 1_{\{r\leq x\}} \tag{12.28}$$

$$+ \tfrac{1}{2}q^2(r)|B_{\tau(r)}|^2 1_{\{x<r<y\}} + \tfrac{1}{2}q^2(r)|\widehat{B}_{\tau(r)-\tau(y)} + B_{\tau(y)}|^2 1_{\{r\geq y\}}\}.$$

Since this last process is clearly Markovian, it suffices to show that it agrees in law with $\tfrac{1}{2}q^2(r)|B_{\tau(r)}|^2$ separately in the regions $\{r \leq x\}$, $\{x \leq r \leq y\}$, and $\{r \geq y\}$. This is obvious for the latter two regions. As for the region $\{r \leq x\}$, we again use the fact that $q(r)B_{\tau(r)} \overset{law}{=} p(r)B_{\phi(r)}$

to see that

$$\{\tfrac{1}{2}p^2(r)|\bar{B}_{\phi(r)-\phi(x)} + \phi(x)B_{\tau(x)}|^2 : r \le x\}$$
$$\overset{law}{=} \{\tfrac{1}{2}p^2(r)|B_{\phi(r)}|^2 : r \le x\} \tag{12.29}$$
$$\overset{law}{=} \{\tfrac{1}{2}q^2(r)|B_{\tau(r)}|^2 : r \le x\}.$$

$\square$

12.3 Ray's original theorem

The designation of Theorem 12.1.2 as Ray's Theorem is honorific since Ray's description of the local times of diffusions X^{h_y} is quite different from the one in Theorem 12.1.2. We give a relatively simple derivation of the theorem D. Ray proved in Ray (1963), using some of the ideas that go into the proof of Theorem 12.1.2. We use the same notation as in Theorem 12.1.2. In addition, we introduce the following random variables:

$$J := \inf_{0<t<\zeta} X_t^{h_y} \quad \text{and} \quad S := \sup_{0<t<\zeta} X_t^{h_y}, \tag{12.30}$$

where ζ is the lifetime of X^{h_y}. Recall that X^{h_y} starts at x and $x \le y$, so $J \le x$. Also, using the facts that X^{h_y} has continuous paths, $\zeta < \infty$, and $\lim_{t\uparrow\zeta} X^{h_y} = y$, it is clear that J and S are finite random variables and $S \ge y$.

It is useful to have a better description of J and S before we state Ray's original theorem. We write $u(r,s)$, the 0-potential of X as in (12.1), and, as in Theorem 12.2.1, we set $\tau(r) = p(r)/q(r)$.

Lemma 12.3.1 *For* $x \le y$,

$$P^{x,y}(J \le z) = \frac{\tau(z)}{\tau(x)} \wedge 1 := F(z) \tag{12.31}$$

and

$$P^{x,y}(S \ge z) = \frac{\tau(y)}{\tau(z)} \wedge 1 := H(z). \tag{12.32}$$

Proof By the strong Markov property,

$$E^{x,y}(\overline{L}_\infty^z) = P^{x,y}(T_z < \infty)E^{z,y}(\overline{L}_\infty^z). \tag{12.33}$$

Therefore, by (12.2),

$$P^{x,y}(T_z < \infty) = \frac{u(x,z)u(z,y)}{u(x,y)u(z,z)}, \tag{12.34}$$

and using (12.1), for $z \leq x$, we get

$$P^{x,y}(T_z < \infty) = \frac{\tau(z)}{\tau(x)} \wedge 1. \qquad (12.35)$$

Similarly for $z \geq y$,

$$P^{x,y}(T_z < \infty) = \frac{u(x,z)u(z,y)}{u(x,y)u(z,z)} = \frac{\tau(y)}{\tau(z)} \wedge 1. \qquad (12.36)$$

Using the continuity of X^{h_y} it is easy to see that for $z \leq x$,

$$P^{x,y}(J \leq z) = P^{x,y}(T_z < \infty). \qquad (12.37)$$

This proves (12.31). Similarly, for $z \geq y$, $P^{x,y}(S \geq z) = P^{x,y}(T_z < \infty)$, which gives (12.32). $\qquad \square$

J and H are real-valued random variables. We use P_J and P_H to denote their probability measures.

Theorem 12.3.2 (Ray's Theorem, original version) *Let $\overline{L} = \{\overline{L}_t^x, (x,t) \in I \times R_+\}$ denote the local times of X^{h_y} and $G = \{G_r, r \in R^1\}$ be as in Theorem 12.1.2. Let $G^{(i)} = \{G_r^{(i)}, r \in R^1\}$, $i = 1, \ldots, 4$ be four independent copies of G and let $P_{\mathcal{G}_j} := P_{G^{(1)},\ldots,G^{(j)}}$. Let $x < y$. Then*

$$\left\{ J, L_\infty^r \, ; \, r \leq x, P^{x,y} \right\} \overset{law}{=} \left\{ J, \Lambda_r \, ; \, r \leq x, P_J \times P_{\mathcal{G}_4} \right\} \qquad (12.38)$$

$$\left\{ L_\infty^r \, ; \, x \leq r \leq y, P^{x,y} \right\} \overset{law}{=} \left\{ \frac{1}{2}\left(\left(G_r^{(1)}\right)^2 + \left(G_r^{(2)}\right)^2 \right) \, ; \, x \leq r \leq y, P_{\mathcal{G}_2} \right\}, \qquad (12.39)$$

and

$$\left\{ S, L_\infty^r \, ; \, r \geq y, P^{x,y} \right\} \overset{law}{=} \left\{ S, \Gamma_r \, ; \, r \geq y, P_S \times P_{\mathcal{G}_4} \right\}, \qquad (12.40)$$

where

$$\Lambda_r = \tfrac{1}{2} 1_{\{r > J\}} \sum_{i=1}^4 \left(G_{r,J}^{(i)}\right)^2 \qquad (12.41)$$

and

$$\Gamma_r = \tfrac{1}{2} 1_{\{r < S\}} \sum_{i=1}^4 \left(G_{r,S}^{(i)}\right)^2 \qquad (12.42)$$

(here, $G_{r,J}^{(i)}$ and $G_{r,S}^{(i)}$ are defined as in (12.3), with G replaced by $G^{(i)}$).

Throughout this book we have been able to exploit relationships involving the moment generating functions of Gaussian processes. We continue to explore these relationships for Gaussian processes with co-variances of the form (12.1). (We know by Lemma 5.1.9 that these are Gaussian Markov processes.)

Lemma 12.3.3 *Let* $t < s \leq r_j \leq \cdots r_2 \leq r_1$. *Consider* $\{G_{r_i,s}, 1 \leq i \leq j\}$ *(see Lemma 12.1.1), and let* Σ_s *denote its covariance matrix and* $\widetilde{\Sigma}_s := (\Sigma_s^{-1} - \Lambda)^{-1}$, *and similarly for* $\{G_{r_i,t}, 1 \leq i \leq j\}$. *Let* $\tau(s) = p(s)/q(s)$. *Then*

$$\frac{\det(I - \Sigma_s \Lambda)}{\det(I - \Sigma_t \Lambda)} = \frac{1}{1 - 2K(q, \Lambda, \widetilde{\Sigma}_s)(\tau(s) - \tau(t))}, \qquad (12.43)$$

where $K(q, \Lambda, \widetilde{\Sigma}_s) = (1/2)(q\Lambda q^t + q\Lambda\widetilde{\Sigma}_s\Lambda q^t)$ *for* Λ *as given in Lemma 5.2.1, with* $n = j$ *and* $q = (q(r_1), \ldots, q(r_j))$.

Furthermore, consider $\{G_{r_i}, 1 \leq i \leq j\}$ *and let* Σ *denote its covariance matrix. Then*

$$\tau(s)\det(I - \Sigma_t \Lambda) - \tau(t)\det(I - \Sigma_s \Lambda) = (\tau(s) - \tau(t))\det(I - \Sigma \Lambda). \qquad (12.44)$$

Proof For each $i \leq j$, by Lemma 12.1.1 (3), we can write

$$\begin{aligned} G_{r_i,t} &= G_{r_i,s} + \frac{q(r_i)}{q(s)}G_{s,t} && (12.45) \\ &= G_{r_i,s} + q(r_i)\left(\frac{1}{q(s)}G_s - \frac{1}{q(t)}G_t\right) \\ &:= G_{r_i,s} + \rho_{r_i,s,t}, \end{aligned}$$

where the two processes $\{G_{r_i,s}, i = 1, \ldots, j\}$ and $\{\rho_{r_i,s,t}, i = 1, \ldots, j\}$ are independent. Also, note that

$$E_G\left(\left(\frac{1}{q(s)}G_s - \frac{1}{q(t)}G_t\right)^2\right) = \frac{u(s,s)}{q^2(s)} + \frac{u(t,t)}{q^2(t)} - 2\frac{u(t,s)}{q(t)q(s)} = \tau(s) - \tau(t). \qquad (12.46)$$

It follows from (5.43) that

$$(\det(I - \Sigma_t \Lambda))^{-1/2} = (\det(I - \Sigma_s \Lambda))^{-1/2} \qquad (12.47)$$

$$E_G \exp\left(K(q, \Lambda, \widetilde{\Sigma}_s)(\frac{1}{q(s)}G_s - \frac{1}{q(t)}G_t)^2\right).$$

The left-hand side is obtained by using (5.43) on $\sum_{i=1}^{j} G_{r_i,t}^2/2$ and the right-hand side by using (5.43) on $\sum_{i=1}^{j}(G_{r_i,s} + \rho_{r_i,s,t})^2/2$ and using the independence of $\{G_{r_i,s}, i = 1, \ldots, j\}$ and $\{\rho_{r_i,s,t}, i = 1, \ldots, j\}$. Equation (12.43) now follows from (12.46) and (12.47).

To obtain (12.44) we first consider $\det(I - \Sigma_s \Lambda)$. The entries of $I - \Sigma_s \Lambda$ are of the form

$$\delta_{l,m} - (u(r_l, r_m) - \tau(s)q(r_l)q(r_m))\lambda_m \qquad 1 \le l, m \le j. \qquad (12.48)$$

For each $l = 2, \ldots, m$, multiply the first row of $\det(I - \Sigma_s \Lambda)$ by $q(r_l)/q(r_1)$ and subtract it from the l-th row. The resulting matrix has no terms in $\tau(s)$ in rows 2 through j. This implies that $\det(I - \Sigma_s \Lambda)$ is of the form $A + \tau(s)B$, where neither A nor B contains terms in $\tau(s)$. It is also clear that $A = \det(I - \Sigma \Lambda)$ (to see this, think of what we get if $\tau(s) = 0$). Thus we have shown that

$$\det(I - \Sigma_s \Lambda) = \det(I - \Sigma \Lambda) + \tau(s)B, \qquad (12.49)$$

where B is not a function of $\tau(s)$. Exactly the same argument shows that

$$\det(I - \Sigma_t \Lambda) = \det(I - \Sigma \Lambda) + \tau(t)B \qquad (12.50)$$

since the only difference in the entries of $\det(I - \Sigma_s \Lambda)$ and $\det(I - \Sigma_t \Lambda)$ is the terms $\tau(s)$ and $\tau(t)$. Equation (12.44) follows from (12.49) and (12.50). $\qquad \square$

Lemma 12.3.4 *Let $t < s \le r_j \le \cdots r_2 \le r_1 \le x$. Then, with Σ_s and Σ_t as defined in Lemma 12.3.3,*

$$\int_t^s \left(E \exp\left(\frac{1}{2} \sum_{i=1}^j \lambda_i G_{r_i,r}^2 \right) \right)^4 dF(r) = \frac{F(s) - F(t)}{\det(I - \Sigma_s \Lambda) \det(I - \Sigma_t \Lambda)}, \qquad (12.51)$$

where $F(r)$ is given in (12.31). In particular, when $t = -\infty$,

$$\int_{-\infty}^s \left(E \exp\left(\frac{1}{2} \sum_{i=1}^j \lambda_i G_{r_i,r}^2 \right) \right)^4 dF(r) = \frac{F(s)}{\det(I - \Sigma_s \Lambda) \det(I - \Sigma \Lambda)}, \qquad (12.52)$$

where Σ is the covariance matrix of $\{G_{r_i}\}_{i=1}^j$.

Proof Let $K = K(q, \Lambda, \widetilde{\Sigma}_s)$. Setting $t = r$ in (12.43) and then integrating, it follows that

$$(\det(I - \Sigma_s \Lambda))^2 \int_t^s \left(\frac{1}{\det(I - \Sigma_r \Lambda)} \right)^2 dF(r) \qquad (12.53)$$

$$= \int_t^s \left(\frac{1}{1 - 2K(\tau(s) - \tau(r))} \right)^2 dF(r).$$

Set $v = \tau(r)/\tau(s)$. The right-hand side of (12.53) is equal to

$$\frac{\tau(s)}{\tau(x)} \int_{\tau(t)/\tau(s)}^{1} \left(\frac{1}{1 - 2K\tau(s)(1-v)} \right)^2 dv \qquad (12.54)$$

$$= \frac{\tau(s)}{\tau(x)} \frac{1}{2K\tau(s)} \left(\frac{1}{1 - 2K(\tau(s) - \tau(t))} - 1 \right)$$

$$= \frac{(\tau(s) - \tau(t))/\tau(x)}{1 - 2K(\tau(s) - \tau(t))}$$

$$= (F(s) - F(t)) \frac{\det(I - \Sigma_s \Lambda)}{\det(I - \Sigma_t \Lambda)},$$

where, for the last equality, we use (12.43). Combining (12.53)–(12.54) we get

$$\int_t^s \left(\frac{1}{\det(I - \Sigma_r \Lambda)} \right)^2 dF(r) = \frac{F(s) - F(t)}{\det(I - \Sigma_s \Lambda) \det(I - \Sigma_t \Lambda)}, \qquad (12.55)$$

which is (12.51).

We obtain (12.52) by taking the limit on both sides, as $t \downarrow -\infty$. $\square$

The case $r_1 \leq t < s \leq x$ is a degenerate form of (12.51) in which there are no squares of Gaussian processes present. In this case, both sides of (12.51) are equal to $F(s) - F(t)$.

Proof of Theorem 12.3.2 Equation (12.39) follows immediately from (12.6). We proceed to the proof of (12.38). Let $-\infty = r_{n+1} < r_n \leq \cdots \leq r_{j+1} \leq t < s \leq r_j \leq \cdots r_2 \leq r_1 \leq x$. It follows from Lemma 12.3.4 that

$$E^{x,y} \left(\left(\left(E_G \exp\left(\frac{1}{2} \sum_{i=1}^n \lambda_i I_{\{r_i > J\}} G_{r_i, J}^2 \right) \right)^4, t < J \leq r_j \right) \right.$$

$$= \frac{F(r_j) - F(t)}{\det(I - \Sigma_{r_j} \Lambda) \det(I - \Sigma_t \Lambda)} \qquad (12.56)$$

for Σ. as in Lemma 12.3.4.

We now consider the local time process of X^{h_y}. It follows from the Markov property that

$$E^{x,y} \left(\exp\left(\sum_{i=1}^j \lambda_i \overline{L}_\infty^{r_i} \right), T_t < \infty \right) \qquad (12.57)$$

$$= E^{x,y} \left(\exp\left(\sum_{i=1}^j \lambda_i \overline{L}_{T_t}^{r_i} \right), T_t < \infty \right) E^{t,y} \exp\left(\sum_{i=1}^j \lambda_i \overline{L}_\infty^{r_i} \right)$$

$$= E^{x,y}\left(\exp\left(\sum_{i=1}^{j}\lambda_i\overline{L}_{T_t}^{r_i}\right), T_t < \infty\right)\det(I - \Sigma\Lambda),$$

where, in the last equality, we use Theorem 12.1.2 with x, y replaced by t, y and the fact that $r_1, \ldots, r_j$ are between t and y. Using (3.209) with the element 0 replaced by y, so that $\{\mathcal{L} > 0\} = \{T_y < \infty\}$, we see that

$$E^{x,y}\left(\exp\left(\sum_{i=1}^{j}\lambda_i\overline{L}_{T_t}^{r_i}\right), T_t < \infty\right) \tag{12.58}$$

$$= \frac{1}{h_y(x)}E^x\left(\exp\left(\sum_{i=1}^{j}\lambda_i\overline{L}_{T_t}^{r_i}\right)1_{\{T_t<\infty\}}\circ k_\mathcal{L}, T_y < \infty\right)$$

$$= \frac{1}{h_y(x)}E^x\left(\exp\left(\sum_{i=1}^{j}\lambda_i\overline{L}_{T_t}^{r_i}\right), T_t < \infty, T_y \circ \theta_{T_t} < \infty\right)$$

$$= \frac{1}{h_y(x)}E^x\left(\exp\left(\sum_{i=1}^{j}\lambda_i\overline{L}_{T_t}^{r_i}\right), T_t < \infty\right)P^t(T_y < \infty)$$

$$= \frac{h_t(x)}{h_y(x)}E^{x,t}\left(\exp\left(\sum_{i=1}^{j}\lambda_i\overline{L}_{T_t}^{r_i}\right)\right)P^t(T_y < \infty),$$

where, for the last line, we use (3.209) with the element 0 replaced by t.

By Corollary 12.1.3 with $y, 0$ replaced by x, t and the fact that $r_1, \ldots, r_j$ are between t and x,

$$E^{x,t}\left(\exp\left(\sum_{i=1}^{j}\lambda_i\overline{L}_{T_t}^{r_i}\right)\right) = \frac{1}{\det(I - \Sigma_t\Lambda)}. \tag{12.59}$$

Also, by (3.109),

$$\frac{h_t(x)}{h_y(x)}P^t(T_y < \infty) = \frac{u(x,t)}{u(t,t)}\frac{u(y,y)}{u(x,y)}\frac{u(t,y)}{u(y,y)} = \frac{\tau(t)}{\tau(x)} = F(t). \tag{12.60}$$

Combining (12.57)–(12.60) we obtain

$$E^{x,y}\left(\exp\left(\sum_{i=1}^{j}\lambda_i\overline{L}_{\infty}^{r_i}\right), T_t < \infty\right) = \frac{1}{\det(I - \Sigma_t\Lambda)}F(t)\frac{1}{\det(I - \Sigma\Lambda)}. \tag{12.61}$$

It follows from (12.61) and (12.44) that

$$E^{x,y}\left\{\exp\left(\sum_{i=1}^{n}\lambda_i L_{\infty}^{r_i}\right), t \leq J < r_j\right\} \tag{12.62}$$

$$= E^{x,y}\left\{\exp\left(\sum_{i=1}^{j}\lambda_i L_{\infty}^{r_i}\right), t \leq J < r_j\right\}$$

$$= \frac{1}{\det(I - \Sigma\Lambda)} \left(\frac{F(r_j)}{\det(I - \Sigma_{r_j}\Lambda)} - \frac{F(t)}{\det(I - \Sigma_t\Lambda)} \right)$$

$$= \frac{F(r_j) - F(t)}{\det(I - \Sigma_{r_j}\Lambda)\det(I - \Sigma_t\Lambda)}.$$

Equation (12.38) follows from (12.56) and (12.62).

The proof of (12.40) is an exact replication of the proof of (12.38). We consider $y \leq r_1 \leq \cdots \leq r_j \leq s < t$ and set

$$\begin{aligned} G_{r_i,t} &= G_{r_i,s} + \left(\frac{u(r_i,s)}{u(s,s)}G_s - \frac{u(r_i,t)}{u(t,t)}G_t \right) \qquad (12.63)\\ &= G_{r_i,s} + p(r_i)(\frac{1}{p(s)}G_s - \frac{1}{p(t)}G_t). \end{aligned}$$

We also note that $P^{x,y}(S \leq z) = I(z)$, where $I(z) = (1 - \tau(y)/\tau(z)) \vee 0$. We proceed to obtain analogies of Lemmas 12.3.3 and 12.3.4 and the proof of (12.38) (essentially, $p(\cdot)$ replaces $q(\cdot)$, $1/\tau(\cdot)$ replaces $\tau(\cdot)$, and $H(\cdot)$ replaces $F(\cdot)$). □

Example 12.3.5 Ray's Theorem for Brownian motion killed at the end of an independent exponential time with mean $1/2$ is given in Theorem 2.8.1. Ray's original theorem allows us also to describe the local time in terms of fourth-order Bessel processes. We point out on page 58 that the associated Gaussian process in this case, which is an Ornstein–Uhlenbeck process, can be represented by $e^{-x}B_{e^{2x}}$ $x \in R^1$, where B is Brownian motion. Consequently, for example, for $r \geq J$, $G_r = e^{-r}(B(e^{2r}) - B(e^{2J}))$. This shows that for $r \geq J$, $\Lambda_r = (1/2)e^{2r}Z_{4,e^{2r}}(e^{2J})$, where $Z_{4,\cdot}(\cdot)$ is defined on page 536.

12.4 Markov property of local times of diffusions

We present a very interesting property of the local times of transient regular diffusions that can be understood by realizing that the diffusions themselves are rescaled time changed Brownian motion.

Theorem 12.4.1 *The total accumulated local time process* $\{\overline{L}^r_\infty, r \in I\}$ *in Theorem 12.1.2 is a Markov process under* $P^{x,y}$.

This theorem is almost immediately obvious from Theorem 12.2.1 since, in each of the three regions considered, the local time is equal in law to a Bessel process, which itself is a Markov process. We make this more precise in the next proof.

Proof of Theorem 12.4.1 using Theorem 12.2.1 It suffices to show that

$$E\left(H(\Psi_s)\big|\,\mathcal{F}_r^\Psi\right) = E\left(H(\Psi_s)\big|\,\Psi_r\right) \qquad (12.64)$$

for each $r < s$ and any measurable function H, where $\mathcal{F}_t^\Psi$ denotes the σ-algebra generated by the stochastic process Ψ_u, $u \le t$.

This is not difficult; for example, take $y \le r < s$. For each $a \in R^1$, the continuous process $Z = \{Z_t(a)\,;\, t \in R_+\}$ induces a probability measure $P_Z^a(\cdot)$ on $C(R^1, R_+)$. Let X_t denote the canonical coordinate process on $C(R^1, R_+)$. Then, under $\{P_Z^a(\cdot)\,;\, a \in R^1\}$, X_t is a time-homogeneous strong Markov process. Using (12.17), independence, and the strong Markov property, we have

$$
\begin{aligned}
E\left(H(\Psi_s)\big|\,\mathcal{F}_r^\Psi\right) &= E\left(H\left(\frac{1}{2}q^2(s)Z_{\tau(s)-\tau(y)}\left(|B_{\tau(y)}|^2\right)\right)\Big|\,\mathcal{F}_r^\Psi\right) \\
&= E_Z^{Z_{\tau(r)-\tau(y)}\left(|B_{\tau(y)}|^2\right)}\left(H\left(\frac{1}{2}q^2(s)X_{\tau(s)-\tau(r)}\right)\right),
\end{aligned}
$$

which implies (12.64) when $y \le r < s$. (We introduce X and $P_Z^a(\cdot)$ to make it clear that we are not taking expectations with respect to B.)

In a similar manner we can check that (12.64) holds for all $r < s$. $\square$

The problem with the above proof, from the perspective of this book, is that we do not really cover Bessel processes or use stochastic differential equations. We give another proof of Theorem 12.4.1, based on some simple computations involving the isomorphisms in Theorem 12.1.2, which follows easily from the next lemma.

Let G be a mean zero Gaussian process with covariance $u(r,s)$ as in (12.1). For $z \le w$ we define

$$
\begin{aligned}
c &= \frac{u(w,z)}{u(z,z)} = \frac{q(w)}{q(z)} \qquad (12.65) \\
\sigma^2 &= u(w,w) - \frac{u^2(w,z)}{u(z,z)} = EG_{w,z}^2 \\
K = K_\lambda &= \frac{1}{1 - \lambda\sigma^2} = \left(E\left(\exp\left(\frac{\lambda}{2}G_{w,z}^2\right)\right)\right)^2.
\end{aligned}
$$

Lemma 12.4.2 *Let $\{\overline{L}_\infty^x, x \in I\}$ be as in Theorem 12.1.2. Let $x_1 < \ldots < x_n = z \le w$ and let H be a bounded continuous function on R^n. Let $H_n(\overline{L}_\infty) := H(\overline{L}_\infty^{x_1}, \ldots, \overline{L}_\infty^{x_n})$.*

For all $\lambda \in C$ sufficiently small the following hold:

If $x \leq y \leq z \leq w$,

$$E^{x,y}\left(\exp\left(\lambda \overline{L}_\infty^w\right) H_n(\overline{L}_\infty^\cdot)\right) = E^{x,y}\left(\exp\left(c^2\lambda K_\lambda \overline{L}_\infty^z\right) H_n(\overline{L}_\infty^\cdot)\right).$$
$$(12.66)$$

If $x \leq z \leq w \leq y$,

$$E^{x,y}\left(\exp\left(\lambda \overline{L}_\infty^w\right) H_n(\overline{L}_\infty^\cdot)\right) = K_\lambda E^{x,y}\left(\exp\left(c^2\lambda K_\lambda \overline{L}_\infty^z\right) H_n(\overline{L}_\infty^\cdot)\right).$$
$$(12.67)$$

If $z \leq w \leq x \leq y$,

$$E^{x,y}\left(\exp\left(\lambda \overline{L}_\infty^w\right) H_n(\overline{L}_\infty^\cdot)1_{\{\overline{L}_\infty^z \neq 0\}}\right) \qquad (12.68)$$
$$= K_\lambda^2 E^{x,y}\left(\exp\left(c^2\lambda K_\lambda \overline{L}_\infty^z\right) H_n(\overline{L}_\infty^\cdot)1_{\{\overline{L}_\infty^z \neq 0\}}\right)$$

and

$$E^{x,y}\left(\exp\left(\lambda \overline{L}_\infty^w\right) H_n(\overline{L}_\infty^\cdot)1_{\{\overline{L}_\infty^z = 0\}}\right) \qquad (12.69)$$
$$= \left(\frac{1 - F(w) + (F(w) - F(z))K_\lambda}{1 - F(z)}\right) E^{x,y}\left(H_n(\overline{L}_\infty^\cdot)1_{\{\overline{L}_\infty^z = 0\}}\right).$$

We prove this lemma after we show how it immediately implies Theorem 12.4.1.

Proof of Theorem 12.4.1 using Theorem 12.1.2 Fix $x < y$ and let $z \leq w$. To establish the Markov property we show that for every function $f \in \mathcal{S}$, the Schwarz space of rapidly decreasing functions on R_+, we can find a bounded continuous function $\bar{f}$ possibly depending on z and w, such that for any finite sequence of points $x_1 < \ldots < x_n = z \leq w$,

$$E^{x,y}\left(f(\overline{L}_\infty^w)H_n(\overline{L}_\infty^\cdot)\right) = E^{x,y}\left(\bar{f}(\overline{L}_\infty^z)H_n(\overline{L}_\infty^\cdot)\right). \qquad (12.70)$$

This implies that

$$E^{x,y}\left(f(\overline{L}_\infty^w) \mid \sigma(\overline{L}_\infty^r; r \leq z)\right) = E^{x,y}\left(f(\overline{L}_\infty^w) \mid \sigma(\overline{L}_\infty^z)\right) \qquad (12.71)$$

for every function $f \in \mathcal{S}$ and hence for all bounded measurable functions f, by the conditional Dominated Convergence Theorem. This establishes the Markov property.

Refer to Lemma 12.4.2. It is easy to see that both sides of equations (12.66)–(12.69) are analytic in λ in the region Re $\lambda < \delta$ for some $\delta > 0$. Therefore, they hold for λ purely imaginary. This gives us (12.70) for $f(x) = e^{ipx}$ for any real p and also an explicit formula for $\bar{f}$. This leads easily to (12.70) for any $f \in \mathcal{S}$ and completes the proof of Theorem 12.4.1 using Theorem 12.1.2. $\qquad \square$

In preparation for the proof of Lemma 12.4.2, we make the following definitions. Let $\bar{\lambda} = (\lambda_1, \ldots, \lambda_n)$ and

$$B_G \stackrel{def}{=} \exp\left(\frac{1}{2}\sum_{i=1}^{n}\lambda_i G_{x_i}^2\right), \tag{12.72}$$

$$B_{G,x} \stackrel{def}{=} \exp\left(\frac{1}{2}\sum_{i=1}^{n}\lambda_i G_{x_i,x}^2 1_{\{x_i \leq x\}}\right), \tag{12.73}$$

$$B_{G,y} \stackrel{def}{=} \exp\left(\frac{1}{2}\sum_{i=1}^{n}\lambda_i G_{x_i,y}^2 1_{\{x_i \geq y\}}\right), \tag{12.74}$$

$$\Pi_{\overline{L}} = \Pi_{\overline{L}}(\bar{\lambda}) \stackrel{def}{=} \exp\left(\sum_{i=1}^{n}\lambda_i L_\infty^{x_i}\right). \tag{12.75}$$

We next note the following simple lemma.

Lemma 12.4.3 *Let* $x_1, \ldots, x_n < z \leq w$. *Then*

$$E\left(B_G \exp\left(\frac{\lambda}{2}G_w^2\right)\right) = K^{1/2} E\left(B_G \exp\left(\frac{\lambda}{2}c^2 K G_z^2\right)\right). \tag{12.76}$$

Proof We write

$$G_w = G_{w,z} + c\,G_z. \tag{12.77}$$

Since $G_{w,z}$ is independent of G_{x_i}, $i = 1, \ldots, n$ by Lemma 12.1.1 (1) and $EG_{w,z}^2 = \sigma^2$, we see from Lemma 5.2.1 that

$$E\left(B_G \exp\left(\frac{\lambda}{2}G_w^2\right)\right) = K^{1/2} E\left(B_G \exp\left(\frac{\lambda c^2 G_z^2}{2} + \frac{\lambda^2 c^2 \sigma^2 G_z^2}{2(1-\lambda\sigma^2)}\right)\right), \tag{12.78}$$

from which we get (12.76). □

Proof of Lemma 12.4.2 We begin with (12.67). Using the same analyticity arguments as in the proof of Theorem 12.4.1, we see that in order to prove (12.67), it suffices to show that for all $\lambda, \lambda_1, \ldots, \lambda_n \in \mathcal{C}$ sufficiently small,

$$E^{x,y}\left(\Pi_{\overline{L}}(\bar{\lambda}) \exp\left(\lambda \overline{L}_\infty^w\right)\right) = K_\lambda E^{x,y}\left(\Pi_{\overline{L}}(\bar{\lambda}) \exp\left(c^2\lambda K_\lambda \overline{L}_\infty^z\right)\right). \tag{12.79}$$

Since $x \leq z \leq w \leq y$, $1_{\{x_i \geq y\}}$, $1_{\{w \geq y\}}$, and $1_{\{w \leq x\}}$ are all 0. Using this, Theorem 12.1.2, Lemma 12.4.3, and Theorem 12.1.2 again, we get

$$E^{x,y} E_{G,\bar{G}}\left(\Pi_{\overline{L}} B_{G,x} B_{\bar{G},x} \exp\left(\lambda \overline{L}_\infty^w\right)\right) \tag{12.80}$$

$$= E_{G,\bar{G}}\left(B_G B_{\bar{G}} \exp\left(\lambda\left(\frac{G_w^2}{2} + \frac{\bar{G}_w^2}{2}\right)\right)\right)$$

$$= K E_{G,\bar{G}} \left(B_G B_{\bar{G}} \exp \left(\lambda c^2 K \left(\frac{G_z^2}{2} + \frac{\bar{G}_z^2}{2} \right) \right) \right)$$

$$= K E^{x,y} E_{G,\bar{G}} \left(\Pi_{\underline{L}} B_{G,x} B_{\bar{G},x} \exp \left(\lambda c^2 K \overline{L}_\infty^z \right) \right),$$

from which we get (12.79), by canceling $E_{G,\bar{G}} \left(B_{G,x} B_{\bar{G},x} \right)$ from the first and last terms in (12.80), and consequently (12.67). (Actually, Lemma 12.4.3 is proved for moment generating functions. It is easy to see that it is also valid when $\lambda, \lambda_1, \ldots, \lambda_n$ are complex, as long as $|\mathrm{Re}\,\lambda|$ is sufficiently small. In the remainder of this proof we use other lemmas proved for real $\lambda, \lambda_1, \ldots, \lambda_n$ in this way.)

We now prove (12.66). It follows from Theorem 12.1.2, Lemma 12.4.3, and the fact that $B_{G,x}$ and $B_{G,y}$ are independent that

$$E^{x,y} E_{G,\bar{G}} \left(\Pi_{\underline{L}} B_{G,x} B_{\bar{G},x} B_{G,y} B_{\bar{G},y} \exp \left(\lambda \left(L_\infty^w + \frac{G_{w,y}^2}{2} + \frac{\bar{G}_{w,y}^2}{2} \right) \right) \right)$$

$$= E_{G,\bar{G}} \left(B_G B_{\bar{G}} \exp \left(\lambda \left(\frac{G_w^2}{2} + \frac{\bar{G}_w^2}{2} \right) \right) \right) \qquad (12.81)$$

$$= K E_{G,\bar{G}} \left(B_G B_{\bar{G}} \exp \left(\lambda c^2 K \left(\frac{G_z^2}{2} + \frac{\bar{G}_z^2}{2} \right) \right) \right)$$

$$= K E^{x,y} \left(\Pi_{\underline{L}} \exp \left(\lambda c^2 K L_\infty^z \right) \right)$$
$$\left(E_G \left(\exp \left(\lambda c^2 K \frac{G_{z,y}^2}{2} \right) B_{G,y} \right) \right)^2 \left(E_G B_{G,x} \right)^2,$$

where, at the last step, we use Theorem 12.1.2 again. Also, obviously,

$$E^{x,y} E_{G,\bar{G}} \left(\Pi_{\underline{L}} B_{G,x} B_{\bar{G},x} B_{G,y} B_{\bar{G},y} \exp \left(\lambda \left(L_\infty^w + \frac{G_{w,y}^2}{2} + \frac{\bar{G}_{w,y}^2}{2} \right) \right) \right)$$

$$= E^{x,y} \left(\Pi_{\underline{L}} \exp \left(\lambda L_\infty^w \right) \right) \left(E_G B_{G,x} \right)^2 \left(E_G \left(\exp \left(\lambda \frac{G_{w,y}^2}{2} \right) B_{G,y} \right) \right)^2.$$
$$(12.82)$$

We claim that

$$E_G \left(B_{G,y} \exp \left(\lambda \frac{G_{w,y}^2}{2} \right) \right) = K^{1/2} E_G \left(B_{G,y} \exp \left(\lambda c^2 K \frac{G_{z,y}^2}{2} \right) \right).$$
$$(12.83)$$

Substituting this in the right-hand side of (12.82) and comparing it to the last line of (12.81) gives (12.66).

To establish (12.83), note that by Lemma 12.1.1,

$$G_{w,y} = G_{w,z} + c\,G_{z,y} \qquad (12.84)$$

and that $G_{w,z}$ is independent of both $G_{z,y}$ and $B_{G,y} \in \mathcal{G}_{\leq z}$. Using (12.84) in place of (12.77) and continuing with the proof of Lemma 12.4.3, we get (12.83).

To obtain (12.68) we use the additivity of local time, $L_\infty = L_{T_z} + L_\infty^\cdot \circ \theta_{T_z}$, and the strong Markov property to obtain

$$E^{x,y} \left(\Pi_{\overline{L}} \exp \left(\lambda L_\infty^w \right) 1_{\{T_z < \infty\}} \right)$$
$$= E^{x,y} \left(\exp \left(\lambda L_{T_z}^w \right) 1_{\{T_z < \infty\}} \right) E^{z,y} \left(\Pi_{\overline{L}} \exp \left(\lambda L_\infty^w \right) \right) \text{(12.85)}$$

since, in the first passage from x to z, $\Pi_{\overline{L}} = 1$. It follows from (12.58)–(12.60) and Corollary 12.1.3 that

$$E^{x,y} \left(\exp \left(\lambda L_{T_z}^w \right) 1_{\{T_z < \infty\}} \right) = E^{x,y} \left(\exp \left(\lambda L_{T_z}^w \right) \right) F(z) \text{(12.86)}$$
$$= K F(z).$$

Furthermore, since $z \leq w \leq y$, we can use (12.67) to get

$$E^{z,y} \left(\Pi_{\overline{L}} \exp \left(\lambda L_\infty^w \right) \right) = K E^{z,y} \left(\Pi_{\overline{L}} \exp \left(c^2 \lambda K L_\infty^z \right) \right). \text{(12.87)}$$

Using (12.85)–(12.87) we get

$$E^{x,y} \left(\Pi_{\overline{L}} \exp \left(\lambda L_\infty^w \right) 1_{\{T_z < \infty\}} \right) = K^2 F(z) E^{z,y} \left(\Pi_{\overline{L}} \exp \left(c^2 \lambda K L_\infty^z \right) \right). \text{(12.88)}$$

Also, by (12.85) and (12.86) with w replaced by z and λ replaced by $c^2 \lambda K$, we see that

$$E^{x,y} \left(\Pi_{\overline{L}} \exp \left(c^2 \lambda K L_\infty^z \right) 1_{\{T_z < \infty\}} \right) \text{(12.89)}$$
$$= E^{x,z} \left(\exp \left(c^2 \lambda K L_{T_z}^z \right) \right) F(z) E^{z,y} \left(\Pi_{\overline{L}} \exp \left(c^2 \lambda K L_\infty^z \right) \right).$$

Recognizing that the first expectation to the right of the equality sign in (12.89) is equal to 1 (since $L_{T_z}^z = 0$), we can substitute (12.89) into (12.88) to get (12.68).

To obtain (12.69) we note that

$$E^{x,y} \left(\exp(\lambda L_\infty^w) | L_\infty^{x_i}, i = 1, \dots, n; L_\infty^z = 0 \right)$$
$$= E^{x,y} \left(\exp \left(\lambda L_\infty^w \right) | z \leq J \right) \text{(12.90)}$$
$$= \frac{P^{x,y}(w \leq J) + E^{x,y}(\exp(\lambda L_\infty^w), z \leq J \leq w)}{P^{x,y}(z \leq J)}.$$

By (12.31), $P^{x,y}(z \leq J) = 1 - F(z)$ and by (12.62),

$$E^{x,y}(\exp(\lambda L_\infty^w), z \leq J \leq w) = \frac{F(w) - F(z)}{\det(I - \Sigma_w \Lambda) \det(I - \Sigma_z \Lambda)}$$
$$= (F(w) - F(z)) K \text{(12.91)}$$

since $\Sigma_w = 0$ and $\det(I - \Sigma_z \Lambda) = 1/K$. Consequently,

$$E^{x,y}\left(\exp(\lambda L_\infty^w)|L_\infty^{x_i}, i = 1, \ldots, n; L_\infty^z = 0\right) \qquad (12.92)$$
$$= \left(\frac{1 - F(w) + (F(w) - F(z))K_\lambda}{1 - F(z)}\right).$$

Multiplying by $H_n(\overline{L}_\infty^{\cdot})1_{\{\overline{L}_\infty^z = 0\}}$ and taking the expectation, we get (12.69). □

12.5 Local limit laws for *h*-transforms of diffusions

We obtain a version of Theorem 9.5.25 for the first hitting time of 0 a diffusion that is conditioned to die at its last exit from 0. Let X be a transient regular diffusion as in Section 12.1, with 0-potential density $u(r,s)$ and local times $\{L_t^r; (r,t) \in R^1 \times R_+\}$. Recall that

$$u_{T_0}(r,s) := E^r(L_{T_0}^s) = u(r,s) - \frac{u(r,0)u(s,0)}{u(0,0)}. \qquad (12.93)$$

Theorem 12.5.1 *Let X^{h_0} be the h_0-transform of X and let $\{\overline{L}_t^r; (r,t) \in R^1 \times R_+\}$ denote its local times. For $y > 0$ we have*

$$\limsup_{x \downarrow 0} \frac{\overline{L}_{T_0}^x}{u_{T_0}(x,x)\log\log(1/u_{T_0}(x,x))} = 1 \qquad P^{y,0} \quad a.s. \qquad (12.94)$$

Proof Let $G_{r,0}$ denote the Gaussian process with covariance $u_{T_0}(r,s)$. This theorem follows from Corollary 12.1.3 and the fact that

$$\limsup_{t \downarrow 0} \frac{G_{r,0}^2 + \overline{G}_{r,0}^2}{2u_{T_0}(r,r)\log\log(1/r)} = 1 \qquad a.s., \qquad (12.95)$$

where $\overline{G}_{r,0}^2$ is an independent copy of $G_{r,0}^2$. Here we used the fact that the term $\left(\frac{G_{r,y}^2}{2} + \frac{\overline{G}_{r,y}^2}{2}\right)1_{\{r \geq y\}}$ in (12.10) contributes nothing when we take the limsup at 0. It is easy to see that (12.95) holds. By (12.20),

$$\{G_{r,0}; r \geq 0\} \stackrel{law}{=} \{q(r)\left(B\left(\tau(r) - \tau(0)\right)\right); r \geq 0\}. \qquad (12.96)$$

Using Lemma 9.5.24 and the fact that $u_{T_0}(r,r) = q^2(r)\left(\tau(r) - \tau(0)\right)$, we get (12.95). □

Example 12.5.2 For standard Brownian motion, $u_{T_0}(x,x) = 2x$; see Lemma 2.5.1. Thus, for Brownian motion, (2.217) follows from Theorem 9.5.25. If we apply Theorem 12.5.1 to the h_0-transform of Brownian motion killed at the end of an independent exponential time with mean $1/2$, we get $u_{T_0}(x,x) = 1 - e^{-2x}$; see Section 2.8 and use (12.93). Thus we get the same result in (12.94) that we get in (9.130) for standard Brownian motion.

12.6 Notes and references

Most of the results in this chapter are taken from Marcus and Rosen (2003). They are mostly consequences of Theorem 12.1.2, which we refer to as Ray's Theorem. Ray's original version of this result, which is given in Theorem 12.3.2, has been studied by Williams (1974), Sheppard (1985), Biane and Yor (1988), and Eisenbaum (1994). In some of these works the description of $\{\overline{L}_\infty^r \, ; \, r \in I\}$ looks quite different from the one given by Ray, as does ours in Theorem 12.1.2.

We show in Section 12.4 that Theorem 12.4.1, which states that the total accumulated local time of a transient diffusion under $P^{x,y}$ is a Markov process in the spatial variable, is an immediate consequence of Theorem 12.2.1. In Ray (1963) and Sheppard (1985), this property is proved by computing the explicit conditional expectations that define the Markov property. In Theorem 12.4.2, we use Theorem 12.1.2 to simplify the computations in these papers. (To see that our (12.69) is the same as Sheppard (1985, (2.19)) note that $q^2(x)(\tau(x) - \tau(z)) = u(x,x) - (u^2(x,z)/u(z,z))$.)

See Walsh (1983) for a proof of Theorem 12.4.1 using excursion theory.

13

Associated Gaussian processes

In Section 7.4 we develop and exploit some special properties of Gaussian processes that are associated with Borel right processes. In this chapter we consider the question of characterizing associated Gaussian processes. In order to present these results in their proper generality, we must leave the familiar framework of Borel right processes and consider local Borel right process, which are introduced in the final sections of Chapter 4. The reader should note that this is the first place in this book, after Chapter 4, that we mention local Borel right processes. We remind the reader that Borel right processes are local Borel right processes, and for compact state spaces, there is no difference between local Borel right processes and Borel right processes.

Let S be a locally compact space with a countable base. A Gaussian process $\{G_x \, ; \, x \in S\}$ is said to be associated with a strongly symmetric transient local Borel right process X on S, with reference measure m, if the covariance $\Gamma = \Gamma(x,y) = E(G_x G_y)$ is the 0-potential density of X for all $x, y \in S$. Not all Gaussian processes are associated. It is remarkable that some very elementary observations about the 0-potential density of a strongly symmetric transient Borel right process show what is special about associated Gaussian processes.

One obvious condition is that $\Gamma(x,y) \geq 0$ for all $x, y \in S$, since the 0-potential density of a strongly symmetric transient Borel right process is nonnegative (see Remark (3.3.5)). Also, since Γ is the 0-potential density of X, it follows from (3.109) that

$$\Gamma(x,y) = P^x(T_y < \infty)\Gamma(y,y). \tag{13.1}$$

Obviously this holds with x and y interchanged. Consequently,

$$\Gamma(x,y) \leq \Gamma(x,x) \wedge \Gamma(y,y). \tag{13.2}$$

Furthermore, since X is transient, it follows from Lemma 3.6.12 that we cannot have both $P^x(T_y < \infty) = 1$ and $P^y(T_x < \infty) = 1$. This means that

$$\Gamma(x, y) < \Gamma(x, x) \vee \Gamma(y, y). \tag{13.3}$$

Therefore, the 2×2 covariance matrix

$$\overline{\Gamma} = \begin{pmatrix} \Gamma(x, x) & \Gamma(x, y) \\ \Gamma(x, y) & \Gamma(y, y) \end{pmatrix} \tag{13.4}$$

is invertible and

$$\overline{\Gamma}^{-1} = \frac{1}{\det\left(\overline{\Gamma}\right)} \begin{pmatrix} \Gamma(y, y) & -\Gamma(x, y) \\ -\Gamma(x, y) & \Gamma(x, x) \end{pmatrix}. \tag{13.5}$$

Thus we see that the off-diagonal terms in $\overline{\Gamma}^{-1}$ are negative. Furthermore, by (13.2), the row sums of $\overline{\Gamma}^{-1}$ are positive, and, as we just noted, to show that Γ is invertible, at least one of the row sums is strictly positive.

Generalizations of these conditions on the covariance matrix of the associated Gaussian process to any p points in the state space of X lead to necessary and sufficient conditions for a Gaussian process to be associated with a local Borel right process.

Here is another precursor of the material covered in this chapter. Let g be a mean zero normal random variable. It is easy to see that g^2 is infinitely divisible. This is because the Laplace transform of g^2 is $(1 + 2\lambda E g^2)^{-1/2}$, and it is easy to check that $(1 + 2\lambda E g^2)^{-1/(2n)}$ is a completely monotone function of λ. We show in Section 13.2 that all mean zero Gaussian vectors in R^2 also have infinitely divisible squares and that this is no longer true in R^p, for $p \geq 3$. Let $G = \{G(x), x \in S\}$ be a mean zero Gaussian process. There is an intimate connection between G having infinitely divisible squares and G being an associated Gaussian process.

13.1 Associated Gaussian processes

Let $A = \{a_{i,j}\}_{1 \leq i,j \leq n}$ be an $n \times n$ matrix. If $a_{i,j} \geq 0$ for all i, j, we say that A is a positive matrix and write $A \geq 0$. If $a_{i,j} \leq 0$ for all $i \neq j$, we say that A has negative off-diagonal elements. If $\sum_{j=1}^{n} a_{i,j} \geq 0$ for all $i = 1, \ldots, n$, we say that A has positive row sums.

Theorem 13.1.1 *Let S be a locally compact space with a countable base. Let $G = \{G_x \, ; \, x \in S\}$ be a Gaussian process with continuous covariance $\Gamma(x, y)$. G is associated with a strongly symmetric transient* *local Borel right process X on S if and only if, for every finite collection $x_1, \ldots, x_n \in S$, the matrix $\Gamma = \{\Gamma(x_i, x_j)\}_{1 \leq i,j \leq n}$ is invertible and Γ^{-1} has negative off-diagonal elements and positive row sums.*

Theorem 13.1.1 is an immediate consequence of the following theorem:

Theorem 13.1.2 *Let S be a locally compact space with a countable base. Let $\Gamma(x, y)$ be a continuous symmetric function on $S \times S$. The following are equivalent:*

(1) *$\Gamma(x, y)$ is the 0-potential density of a strongly symmetric transient local Borel right process X on S.*

(2) *For every finite subset $S' \subseteq S$, $\Gamma(x, y)$ restricted to $S' \times S'$ is the 0-potential density of a strongly symmetric transient Borel right process X on S'.*

(3) *For every finite collection $x_1, \ldots, x_n \in S$, the $n \times n$ matrix $\Gamma = \{\Gamma(x_i, x_j)\}_{1 \leq i,j \leq n}$ is invertible and Γ^{-1} has negative off-diagonal elements and positive row sums.*

Proof (1) $\Rightarrow$ (3) We define the following stopping time:

$$\sigma = \inf\{t \geq 0 \,|\, X_t \in \{x_1, \ldots, x_p\} \cap \{X_0\}^c\} \tag{13.6}$$

(note that σ may be infinite). Let $\{L_t^x \,;\, (x, t) \in S \times R^1\}$ be the local times of X. Since $\Gamma(x_i, x_j)$ is the 0-potential density of X, we can normalize the local time so that

$$\Gamma(x_i, x_j) = E^{x_i}\left(L_\infty^{x_j}\right). \tag{13.7}$$

This result is proved in (3.108) for Borel right processes. We leave to the reader the verification that it also holds for local Borel right processes. The key point is that whenever we use the strong Markov property in the steps leading to (3.108), we can just as well use the local strong Markov property, Lemma 4.8.5, which is satisfied by local Borel right processes..

Using (13.7) and the local strong Markov property (Lemma 4.8.5), we see that

$$\begin{aligned}
\Gamma(x_i, x_j) &= E^{x_i}\left(L_\sigma^{x_j}\right) + E^{x_i}\left(E^{X_\sigma}(L_\infty^{x_j}); \sigma < \infty\right) \tag{13.8} \\
&= b_{i,j} + \sum_{k=1}^{p} h_{i,k}\Gamma(x_k, x_j),
\end{aligned}$$

where $b_{i,j} = E^{x_i}(L_\sigma^{x_j})$ and $h_{i,k} = P^{x_i}(X_\sigma = x_k)$.

Let $B = \{b_{i,j}\}_{1 \leq i,j \leq p}$ and $H = \{h_{i,j}\}_{1 \leq i,j \leq p}$. We can write (13.8) as

$$\Gamma = B + H\Gamma \tag{13.9}$$

so that $(I - H)\Gamma = B$. Moreover, B is a diagonal matrix with all its diagonal elements strictly positive. This follows because, starting from $X_0 = x_i$, $\sigma > 0$, which implies that each $b_{i,i} > 0$. On the other hand, the process is killed the first time it hits any $x_j \neq x_i$. Thus, starting from x_i, $L_\sigma^{x_j} = 0$, $j \neq i$.

Since B is invertible, both $(I - H)$ and Γ are invertible and

$$\Gamma^{-1} = B^{-1}(I - H). \qquad (13.10)$$

It is clear that $H \geq 0$. It follows from this that Γ^{-1} has negative off diagonal elements. Furthermore,

$$\sum_{j=1}^{p} h_{i,j} = P^{x_i}(\sigma < \infty) \leq 1 \qquad \forall\, i = 1 \dots, p, \qquad (13.11)$$

from which it follows that Γ^{-1} has positive row sums.

(3) $\Rightarrow$ (2) To begin, we show that an $n \times n$ matrix that has negative off-diagonal elements and positive row sums can be used to construct a strongly symmetric Borel right process on $S = \{1, 2, \dots, n\}$. Let $A = \{a_{i,j}\}_{1 \leq i,j \leq n}$ be such a matrix. Let

$$P_t(i,j) = \{e^{-tA}\}_{i,j}, \qquad (13.12)$$

that is,

$$P_t(i,j) = \sum_{k=0}^{\infty} \frac{(-t)^k A_{i,j}^k}{k!} \qquad (13.13)$$

with $A^0 := I$. We write this compactly as $P_t = e^{-tA}$. We show that $\{P_t\,;\, t \geq 0\}$ is a sub-Markov semigroup on $(S, \mathcal{B})$.

The semigroup property is clear. Furthermore, since A has negative off-diagonal elements, if we take c to be the largest diagonal element of A, then $B := cI - A \geq 0$. Hence

$$P_t = e^{-tA} = e^{-tc}e^{tB} \geq 0. \qquad (13.14)$$

To complete the proof that P_t is a sub-Markov semigroup on S, we must show that $P_t(x, S) \leq 1$ for all $x \in S$. This comes down to showing that

$$\sum_{j=1}^{n} P_t(i,j) = \sum_{j=1}^{n}\{e^{-tA}\}_{i,j} \leq 1 \qquad \forall\, i \in S. \qquad (13.15)$$

Let $h_i(t) = \sum_{j=1}^{n}\{e^{-tA}\}_{i,j}$. Then

$$h_i'(t) = -\sum_{j=1}^{n}\{e^{-tA}A\}_{i,j} = -\sum_{j=1}^{n}\sum_{p=1}^{n}\{e^{-tA}\}_{i,p}A_{p,j} \qquad (13.16)$$

$$= -\sum_{p=1}^{n} \{e^{-tA}\}_{i,p} \sum_{j=1}^{n} A_{p,j}.$$

Note that $\sum_{j=1}^{n} A_{p,j} \geq 0$, since A has positive row sums. It follows from (13.14) and (13.16) that $h_i'(t) \leq 0$. Since $h_i(0) = 1$, this gives us (13.15).

It is easy to see that $\{P_t; t \geq 0\}$ is a strongly continuous contraction semigroup on $C(S)$. Here we are identifying $C(S)$ with R^n. Properties (2) and (3) on page 122 follow from (13.14)–(13.15) and (13.13), respectively. Therefore, we can use Theorem 4.1.1 to construct a Feller process X on S with transition semigroup $\{P_t; t \geq 0\}$. If A is symmetric, then clearly $\{P_t; t \geq 0\}$ is also symmetric. In this case, X is a strongly symmetric Borel right process. Here we use counting measure on S as the reference measure.

We now show that if A is invertible, then X is transient and A^{-1} is the 0-potential of X. For any $\alpha > 0$,

$$U^\alpha = \int_0^\infty e^{-\alpha t} P_t \, dt = \int_0^\infty e^{-\alpha t} e^{-tA} \, dt = (\alpha + A)^{-1}. \qquad (13.17)$$

To see this, note that since $\{P_t; t \geq 0\}$ is a contraction semigroup, the integrals converge. Therefore,

$$(\alpha + A) \int_0^\infty e^{-\alpha t} e^{-tA} \, dt = \int_0^\infty (\alpha + A) e^{-t(\alpha+A)} \, dt \qquad (13.18)$$

$$= -\int_0^\infty \frac{d}{dt} \left(e^{-t(\alpha+A)} \right) dt = I.$$

Taking the limit in (13.17) as $\alpha \to 0$, we see that $U^0 = A^{-1}$. Therefore, X is transient. (Strictly speaking, in Lemma 3.6.11 we show that a process is transient if the 0-potential density exists. But note that on a finite state space, with the counting measure as the reference measure, the 0-potential and the 0-potential density are the same.) (3) $\Rightarrow$ (2) now follows on identifying $x_1, \ldots, x_n$ with $\{1, 2, \ldots, n\}$ and the matrix Γ^{-1} with the matrix A.

(2) $\Rightarrow$ (1) This is the crux of the theorem and the reason that we must consider local Borel right processes rather than Borel right processes. The proof of this implication is Theorem 4.10.1. $\qquad \square$

In preparation for Section 13.3, we continue to explore some properties of Borel right processes with finite-dimensional state spaces.

Let $\{P_t, ; t \geq 0\}$ be a right continuous semigroup of $n \times n$ matrices with $P_t \geq 0$ and $P_0 = I$. (Note that when $n = 1$, this is equivalent to having a function $f(t) \geq 0$ satisfying $f(t+s) = f(t)f(s)$ for all $s, t \geq 0$.)

As in the case $n = 1$, one can verify that $P_t = e^{-tA}$ for some $n \times n$ matrix A. The matrix $-A$ is called the generator of the semigroup $\{P_t, ; t \geq 0\}$. If X is a Borel right processes with semigroup $\{P_t, ; t \geq 0\}$, we also say that $-A$ is the generator of X. We formulate our results in terms of A rather than the generator $-A$, because our focus is on potentials, and as we see in the proof of Theorem 13.1.1, if A is invertible, A^{-1} is the 0-potential of X.

Let $A = \{a_{i,j}\}_{1 \leq i,j \leq n}$ be an $n \times n$ matrix. If $\sum_{j=1}^{n} a_{i,j} = 0$ for all $i = 1, \ldots, n$, we say that A has zero row sums.

Remark 13.1.3 Note that a matrix A with positive row sums that is invertible cannot have zero row sums, that is, at least one of its row sums is strictly positive. This is obvious because if all the row sums of A are equal to 0, $A\mathbf{1}^t = 0$, where $\mathbf{1}$ denotes the vector in R^n with all its elements equal to 1. This is impossible if A is invertible. Therefore the matrix Γ^{-1} in Theorems 13.1.1 and 13.1.2 has at least one row sum that is strictly positive. Consequently, since the off-diagonal elements of Γ^{-1} are negative, there is at least one row of Γ^{-1} in which the diagonal element is greater than the sum of the absolute values of the off-diagonal elements. Such a matrix is sometimes said to be "weakly diagonally dominant."

(3) $\Rightarrow$ (2) of Theorem 13.1.2 also implies that Γ is positive and positive definite, since U^0 is a zero potential. Furthermore, since it is invertible, it must be strictly positive definite. Thus we see that an invertible matrix (Γ^{-1}) with negative off-diagonal elements and positive row sums is weakly diagonally dominant and has a positive, strictly positive definite inverse.

Lemma 13.1.4 *Let A be an $n \times n$ matrix. $P_t = e^{-tA}$ is the semigroup of a Borel right process X on $S = \{1, 2, \ldots, n\}$ if and only if A has negative off-diagonal elements and positive row sums. In this case, the α-potential*

$$U^\alpha = (\alpha + A)^{-1} \quad \forall \, \alpha > 0. \tag{13.19}$$

Furthermore, A is symmetric if and only if X is symmetric.

If, in addition, A has zero row sums, then X is recurrent.

A is invertible if and only if X is transient, in which case $U^0 = A^{-1}$. (In this case, at least one of the row sums of A is strictly positive.)

Proof Suppose that A has negative off-diagonal elements and positive row sums. The fact that $P_t = e^{-tA}$ is the semigroup of a Borel right process X on $S = \{1, 2, \ldots, n\}$ is contained in the proof of $(3) \Rightarrow (2)$ of Theorem 13.1.2.

Assume now that $P_t = e^{-tA}$ is the semigroup of a Borel right process X on $\{1, 2, \ldots, n\}$. Since

$$P_t(i, j) = \{e^{-tA}\}_{i,j} = 1_{\{i=j\}} - A_{i,j}t + O(t^2) \qquad \text{as } t \to 0, \quad (13.20)$$

we immediately see from the positivity of $P_t(i, j)$ that A must have negative off-diagonal elements. Summing over j we see that

$$P_t(i, S) = 1 - \sum_{j=1}^{n} A_{i,j}t + O(t^2) \qquad \text{as } t \to 0, \quad (13.21)$$

and since $P_t(i, S) \leq 1$, we must have $\sum_{j=1}^{n} A_{i,j} \geq 0$ for each $i = 1, \ldots, n$. Thus A has positive row sums.

The representation of U^α in (13.19) is given in (13.17), and, of course, the statement about symmetry is immediate.

Suppose that A also has zero row sums. Set $h_i(t) = \sum_{j=1}^{n} \{e^{-tA}\}_{i,j}$. We see from (13.16) that $h_i'(t) \equiv 0$. Consequently, $P_t(i, S) = P_0(i, S) = 1$ for all $i \in S$. Let U be the 0-potential of X. Then

$$\sum_{j=1}^{n} U(i, j) = U(i, S) = \int_0^\infty P_t(i, S)\, dt = \infty \qquad \forall i \in S. \quad (13.22)$$

Using (3.68) we see that $U(i, i) = \infty$ for all i, which implies that X is recurrent.

Finally, we know that X is transient if and only if U^0 exists. Taking the limit as $\alpha \to 0$ in (13.19), we see that this is equivalent to A being invertible and equal to U^0. $\qquad \square$

Let A be a symmetric $n \times n$ matrix that has negative off-diagonal elements and positive row sums. By Lemma 13.1.4 there exists symmetric Borel right process X on $S = \{1, 2, \ldots, n\}$ with semigroup $P_t = e^{-tA}$. We show here how the elements of A are related to the probabilistic structure of X. Define the stopping time

$$\sigma = \inf\{t \geq 0 \mid X_t \in S \cap \{X_0\}^c\} \quad (13.23)$$

(note that σ may be infinite). Since $\sigma = s + \sigma \circ \theta_s$ on $\{\sigma > s\}$, it follows from the strong Markov property that

$$\begin{aligned} P^i(\sigma > t + s) &= P^i(\sigma \circ \theta_s > t, \sigma > s) \quad (13.24) \\ &= P^i(P^{X_s}(\sigma > t); \sigma > s) = P^i(\sigma > t)P^i(\sigma > s), \end{aligned}$$

where the last equality uses the fact that if $\sigma > s$, then X_s is still at its initial position. Since $P^i(\sigma > s)$ is right continuous, we see that σ is an exponential random variable. Therefore, $P^i(\sigma > s) = e^{-q_i s}$ for some $q_i \in R_+$. Let

$$h_{i,j} = P^i(X_\sigma = j) \tag{13.25}$$

and note that $h_{i,i} = 0$ for all i.

Lemma 13.1.5

$$q_i = A_{i,i} \geq 0 \tag{13.26}$$

and, for $i \neq j$

$$h_{i,j} = \begin{cases} 0 & A_{i,i} = 0 \\ -\dfrac{A_{i,j}}{A_{i,i}} & A_{i,i} > 0. \end{cases} \tag{13.27}$$

Proof In the next paragraph we show that as $t \to 0$,

$$P^i(X_t = j) = q_i h_{i,j} t + O(t^2) \qquad i \neq j \tag{13.28}$$

and

$$P^i(X_t = i) = 1 - q_i t + O(t^2). \tag{13.29}$$

Using these and (13.20) we obtain (13.26) and (13.27),

We now obtain (13.28) and (13.29). When $q_i = 0$ this is obvious, so we can assume that $\sigma < \infty$ almost surely. Let $\sigma' = \sigma \circ \theta_\sigma$. By the Markov property, conditional on X_σ, we see that σ and σ' are independent exponential random variables. We have

$$P^i(X_t = j) = P^i(\sigma > t; X_t = j) + P^i(\sigma \leq t < \sigma + \sigma'; X_t = j)$$
$$+ P^i(\sigma + \sigma' \leq t; X_t = j). \tag{13.30}$$

Clearly,

$$P^i(\sigma > t; X_t = j) = e^{-q_i t} 1_{\{i=j\}} \tag{13.31}$$

and

$$P^i(\sigma + \sigma' \leq t; X_t = j) \leq P^i(\sigma + \sigma' \leq t) = O(t^2). \tag{13.32}$$

Also, when $i \neq j$

$$P^i(\sigma \leq t < \sigma + \sigma'; X_t = j) \tag{13.33}$$
$$= P^i(\sigma \leq t < \sigma + \sigma'; X_\sigma = j)$$
$$= q_i q_j \int_{\{r \leq t < r+s\}} e^{-q_i r} e^{-q_j s} \, dr \, ds P^i(X_\sigma = j)$$
$$= \frac{q_i}{q_i - q_j} \left(e^{-q_j t} - e^{-q_i t} \right) P_{i,j} = q_i h_{i,j} t + O(t^2).$$

Using (13.31)–(13.33) in (13.30), we get (13.28) and (13.29).   □

Lemma 13.1.6 *Let X be a symmetric Borel right process on $S = \{1, 2, \ldots, n\}$ with semigroup $P_t = e^{-tA}$ (A is an $n \times n$ symmetric matrix with elements $\{A_{i,j}\}_{1 \leq i,j \leq n}$). Let X' be the symmetric Borel right process on $S' = \{1, 2, \ldots, n-1\}$ obtained by killing X the first time it hits $\{n\}$. Then X' has semigroup $P'_t = e^{-tA'}$, where A' is the symmetric $(n-1) \times (n-1)$ matrix with elements $\{A_{i,j}\}_{1 \leq i,j \leq n-1}$.*

Proof That X' is a symmetric Borel right process follows from Theorem 4.5.2. We show below that (13.28)–(13.29) hold when $P^i(X_t = j)$ is replaced by $P^i(X'_t = j)$ for $i, j \in S'$. Therefore, by Lemma 13.1.5, $P'_t = e^{-tA'}$ is the semigroup for X'.

Note that

$$P^i(X'_t = j) = P^i(X_t = j, T_n > t). \tag{13.34}$$

Therefore,

$$\begin{aligned}
P^i(X'_t = j) \;=\; & P^i(\sigma > t;\, X_t = j,\, T_n > t) \tag{13.35}\\
& + P^i(\sigma \leq t < \sigma + \sigma';\, X_t = j,\, T_n > t)\\
& + P^i(\sigma + \sigma' \leq t;\, X_t = j,\, T_n > t).
\end{aligned}$$

It is easy to see that for $i, j \neq n$,

$$P^i(\sigma > t;\, X_t = j,\, T_n > t) = P^i(\sigma > t;\, X_t = j) = e^{-q_i t} 1_{\{i=j\}} \tag{13.36}$$

and

$$P^i(\sigma + \sigma' \leq t;\, X_t = j,\, T_n > t) \leq P^i(\sigma + \sigma' \leq t) = O(t^2). \tag{13.37}$$

Also, the condition $\sigma \leq t < \sigma + \sigma'$ means that there is precisely one jump up to time t. Therefore, as long as $i, j \neq n$, we have

$$P^i(\sigma \leq t < \sigma + \sigma';\, X_t = j,\, T_n > t) = P^i(\sigma \leq t < \sigma + \sigma';\, X_t = j). \tag{13.38}$$

Comparing (13.35)–(13.38) with (13.30)–(13.32), we see that (13.28)–(13.29) hold when $P^i(X_t = j)$ is replaced by $P^i(X'_t = j)$ for $i, j \in S'$. □

In the next example we use Theorem 13.1.2 to obtain a nice smoothness condition for the increments variance of an associated Gaussian process that has stationary increments.

Example 13.1.7 Let $G := \{G(x), x \in S\}$ be an associated Gaussian process with covariance matrix $u(x, y) = u(x - y)$. Let $\sigma^2(x - y) :=$

$E(G(x) - G(y))^2$. Consider $\overline{G} := (G(x_1), G(x_2), G(x_3))$. By Theorem 13.1.2, the inverse of the covariance matrix of $\overline{G}$ has negative entries off the diagonal. Considering the $(1,3)$ entry of the inverse, we get

$$u(x_1, x_2)u(x_2, x_3) \leq u(x_2, x_2)u(x_1, x_3). \tag{13.39}$$

Since $\sigma^2(x) = 2(u(0) - u(x))$, this gives

$$\sigma^2(x_3 - x_1) \leq \sigma^2(x_2 - x_1) + \sigma^2(x_3 - x_2) - \frac{\sigma^2(x_3 - x_2)\sigma^2(x_2 - x_1)}{2}. \tag{13.40}$$

Set $y = x_3 - x_1$ and $x = x_2 - x_1$; we get

$$\begin{aligned} \sigma^2(y) - \sigma^2(x) &\leq \sigma^2(y - x) - \frac{\sigma^2(y - x)\sigma^2(x)}{2u(0)} \tag{13.41} \\ &= \frac{\sigma^2(y - x)u(x)}{u(0)}. \end{aligned}$$

In particular,

$$|\sigma^2(y) - \sigma^2(x)| \leq \sigma^2(y - x). \tag{13.42}$$

This is stronger than the result in Lemma 7.4.2 and can be used to simplify some arguments in Section 7.4. (To appreciate the significance of this result, see (7.254).)

Clearly this observation applies to the α-potential density of Lévy processes (see (4.84)) and compliments (7.232), which applies to Lévy processes killed the first time they hit zero (see Lemma 4.2.4).

13.2 Gaussian precesses with infinitely divisible squares

In this section we show that associated Gaussian processes have infinitely divisible squares. In Section 13.3 we will show that, properly formulated, this property characterizes associated Gaussian processes.

Let $G = (G_1, \dots, G_p)$ be an R^p-valued Gaussian random variable. G is said to have infinitely divisible squares if $G^2 := (G_1^2, \dots, G_p^2)$ is infinitely divisible, that is, for any n we can find an R^p-valued random vector Z_n such that

$$G^2 \overset{law}{=} Z_{n,1} + \dots + Z_{n,n}, \tag{13.43}$$

where $\{Z_{n,j}\}, j = 1, \dots, n$ are independent identically distributed copies of Z_n.

The Gaussian process $G = \{G_x, x \in S\}$ is said to have infinitely divisible squares if, for every finite collection $x_1, \ldots, x_p \in S$, the R^p-valued random variable $(G_{x_1}, \ldots, G_{x_p})$ has infinitely divisible squares. We also express this by saying that G^2 is infinitely divisible.

Let $A = \{a_{i,j}\}_{1 \leq i,j \leq n}$ be an $n \times n$ matrix. We call A a positive matrix and write $A \geq 0$ if $a_{i,j} \geq 0$ for all i, j.

The matrix A is said to be an M-matrix if

(1) $a_{i,j} \leq 0$ for all $i \neq j$.
(2) A is nonsingular and $A^{-1} \geq 0$.

A diagonal matrix is called a signature matrix if its diagonal entries are either one or minus one.

The following theorem characterizes Gaussian processes with infinitely divisible squares.

Theorem 13.2.1 *Let* $G = (G_{x_1}, \ldots, G_{x_p})$ *be a mean zero Gaussian random variable with strictly positive definite covariance matrix* $\Gamma = \{\Gamma_{i,j}\} = \{E(G(x_i)G(x_j))\}$. *Then* G^2 *is infinitely divisible if and only if there exists a signature matrix* $\mathcal{N}$ *such that* $\mathcal{N}\Gamma^{-1}\mathcal{N}$ *is an* M-*matrix.*

Proof For a $p \times p$ matrix A set $|A| = \det A$. It follows from Lemma 5.2.1 that, for $s_1, \ldots, s_p \in [0, 1)$,

$$E \exp \left(-\frac{1}{2} \sum_{i=1}^{p} a(1 - s_i) G^2(x_i) \right) = \frac{1}{|I + \Gamma a(I - S)|^{1/2}}$$
$$:= P(\mathbf{s}) \qquad (13.44)$$

for all $a > 0$, where $\mathbf{s} = (s_1, \ldots, s_p)$ and S is a diagonal matrix with diagonal entries $s_1, \ldots, s_p$.

Assume that G^2 is infinitely divisible. Let Y_n be an R^p-valued random vector for which

$$G^2 \overset{law}{=} Y_{n,1} + \cdots + Y_{n,n}, \qquad (13.45)$$

where the $Y_{n,j}$, $j = 1, \ldots, n$, are independent identically distributed copies of Y_n. Clearly we can take $Y_n \geq 0$. It follows from (13.44) that for each $1 \leq j \leq n$,

$$E \exp \left(-\frac{1}{2} \sum_{i=1}^{p} a(1 - s_i) Y_{n,j}(x_i) \right) = P^{1/n}(\mathbf{s}). \qquad (13.46)$$

Note that all the terms in the power series expansion of $P^{1/n}(\mathbf{s})$ are positive. This follows since

$$\exp\left(-\frac{1}{2}a(1-s_i)Y_{n,j}(x_i)\right) = \exp\left(-\frac{aY_{n,j}(x_i)}{2}\right)\sum_{k=0}^{\infty}\frac{(as_iY_{n,j})^k}{2^k k!}.$$

$$(13.47)$$

Obviously $P(\mathbf{s}) > 0$, so we can take

$$\log P(\mathbf{s}) = \lim_{n\to\infty}\left(n\left(P^{1/n}(\mathbf{s})-1\right)\right). \tag{13.48}$$

We show immediately below that $\log P(\mathbf{s})$ has a power series expansion. From (13.48) and the other observations above we see that, except for the constant term, all the terms in this expansion must be greater than or equal to zero.

Let $Q = I - (I + a\Gamma)^{-1}$. Note that if λ is an eigenvalue of Γ, then $\frac{a\lambda}{1+a\lambda}$ is an eigenvalue of Q, so that, in particular, all the eigenvalues of Q are in $[0,1)$. Then

$$\begin{aligned}
P^2(\mathbf{s}) &= |I + a\Gamma - a\Gamma S|^{-1} && (13.49) \\
&= |(I-Q)^{-1} - ((I-Q)^{-1} - I)S|^{-1} \\
&= |I-Q||I-QS|^{-1}.
\end{aligned}$$

Therefore,

$$\begin{aligned}
2\log P(\mathbf{s}) &= \log|I-Q| - \log|I-QS| && (13.50) \\
&= \log|I-Q| + \sum_{n=1}^{\infty}\frac{\text{trace}\{(QS)^n\}}{n}.
\end{aligned}$$

To understand the last term, let $\lambda_1,\dots,\lambda_p$ be the eigenvalues of the symmetric matrix $S^{1/2}QS^{1/2}$. Considering the conditions on S and the fact that the eigenvalues of the symmetric matrix Q are all in $[0,1)$, we see that $\|S^{1/2}QS^{1/2}\| \le \|Q\| \le 1$ and $(u, S^{1/2}QS^{1/2}u) = (S^{1/2}u, QS^{1/2}u) \ge 0$ for any u. It follows that all of the eigenvalues of $S^{1/2}QS^{1/2}$ are also in $[0,1)$. Therefore,

$$\begin{aligned}
|I-QS| &= |I - S^{1/2}QS^{1/2}| && (13.51) \\
&= (1-\lambda_1)\cdots(1-\lambda_p),
\end{aligned}$$

so that

$$\log|I-QS| = -\sum_{i=1}^{p}\sum_{n=1}^{\infty}\frac{\lambda_i^n}{n}. \tag{13.52}$$

Also,

$$\text{trace}\{(QS)^n\} = \text{trace}\{(S^{1/2}QS^{12})^n\} \tag{13.53}$$

$$= \sum_{i=1}^{p} \lambda_i^n.$$

Thus we get (13.50) and it is clear that $\log P(\mathbf{s})$ has a convergent power series.

Let $\{i_1, \ldots, i_k\}$ be any subset of $\{1, \ldots, p\}$ and denote the elements of Q by $q_{j,k}$. We now show that for all $k \geq 2$,

$$q_{i_1,i_2} q_{i_2,i_3} \cdots, q_{i_{k-1},i_k} q_{i_k,i_1} \geq 0. \tag{13.54}$$

This is trivial for $k = 2$ since the term in (13.54) is simply q_{i_1,i_2}^2 (recall that Q is symmetric). Assume (13.54) holds for $k = 3, \ldots, m - 1$. We show that it also holds when $k = m$.

We first consider the case in which $q_{i_j,i_n} = 0$ when $(i_j, i_n) \neq (i_j, i_{j+1})$ or (i_m, i_1). In this case, the expansion of $\text{trace}\{(QS)^m\}$ is considerably simplified and we see that $q_{i_1,i_2} q_{i_2,i_3} \cdots, q_{i_{k-1},i_k} q_{i_m,i_1}$ is the coefficient of $s_{i_1} \cdots s_{i_m}$ in $\text{trace}\{(QS)^m\}/m$. Thus it is nonnegative because all the terms in the power series for $\log P(\{\mathbf{s}\})$, other than the constant term, are greater than or equal to zero.

It remains to consider the cases in which there exists an (i_j, i_n) with $q_{i_j,i_n} \neq 0$, where either $2 \leq j < m - 1$ and $j + 1 < n \leq m$ or $j = 1$ and $2 < n < m$. When this occurs we can write the product in (13.54), with $k = m$, as

$$q_{i_1,i_2} q_{i_2,i_3} \cdots q_{i_{j-1},i_j} \ q_{i_j,i_n} \ q_{i_n,i_{n+1}} \cdots q_{i_m,i_1} \tag{13.55}$$
$$\times q_{i_j,i_{j+1}} q_{i_{j+1},i_{j+2}} \cdots q_{i_{n-1},i_n} q_{i_n,i_j} \times q_{i_j,i_n}^{-2}.$$

(Note that we use q_{i_j,i_n} as a bridge between q_{i_{j-1},i_j} and $q_{i_n,i_{n+1}}$ and place the removed terms on the second line in (13.55).) Thus we create two terms of the type (13.54), each containing at most $m - 1$ terms. Therefore, by the induction hypothesis, they both must be greater than or equal to zero.

We can write

$$Q = I - \left(a^{-1}I + \Gamma\right)^{-1} \left(a^{-1}I\right). \tag{13.56}$$

For an invertible matrix A, we use $A^{i,j}$ to denote $\{A^{-1}\}_{i,j}$. Also, let $W = \Gamma^{-1}$ and set $w_{i,j} = \{W\}_{i,j}$. For $i \neq j$ we have

$$q_{i,j} = -\left(a^{-1}I + \Gamma\right)^{i,j} a^{-1}. \tag{13.57}$$

Consequently,

$$\lim_{a \to \infty} a q_{i,j} = -w_{i,j} \tag{13.58}$$

Considering (13.54), we see that for all $k \geq 2$

$$(-1)^k w_{i_1,i_2} w_{i_2,i_3} \cdots, w_{i_{k-1},i_k} w_{i_k,i_1} \geq 0. \qquad (13.59)$$

We use (13.59) to show that there exists a signature matrix $\mathcal{N}$ such that $\mathcal{N}W\mathcal{N}$ is an M-matrix. This is trivial when W has order $p = 1$. Suppose it holds when W has order $p-1$. Now let W have order p. Let U be the $(p-1) \times (p-1)$ matrix with entries $U_{i,j} = W_{i,j}$, $i,j = 1,\ldots,p-1$. By the induction hypothesis, there exists a signature matrix E such that EUE is an M-matrix. To simplify the notation, we assume, without loss of generality, that U itself is an M-matrix.

Let $\{1,\ldots,p-1\}$ be partitioned into sets $G_1,\ldots,G_m$. Let $U[G_i, G_i]$ denote the submatrix of U formed by rows and columns indexed by G_i. We choose the $\{G_i\}$ so that each $U[G_i, G_i]$ is irreducible (see page 600). Note that U is the direct sum of $U[G_i, G_i]$, $\{1,\ldots,m\}$.

We claim that for all pairs $j,l \in G_i$, for each $1 \leq i \leq m$, $w_{j,p}w_{l,p} \geq 0$. By Lemma 14.9.3 there exists an r and different integers $k_1,\ldots,k_r$ in $\{1,\ldots,p-1\}$ such that $k_1 = l$, $k_r = j$ and the elements $w_{k_1,k_2}, w_{k_2,k_3},\ldots,$ w_{k_{r-1},k_r} are all nonzero. This means they are all less than zero since U is an M-matrix. Now consider the cycle product

$$w_{j,p} w_{p,l} w_{k_1,k_2} w_{k_2,k_3} \cdots w_{k_{r-1},k_r}. \qquad (13.60)$$

If $w_{j,p}w_{l,p} < 0$, one and only one of the terms in the product (13.60) is greater than zero, while the remaining r terms are less than zero. This contradicts (13.59), which implies that $(-1)^{r+1}$ times the product (13.60) is nonnegative. This proves our claim.

For $i \in \{1,\ldots,m\}$ we say that G_i is type 1 if $w_{j,p} \geq 0$ for all $j \in G_i$, and type 2 if $w_{j,p} \leq 0$ for all $j \in G_i$ and $w_{j,p} < 0$ for at least one $j \in G_i$. We define a $p \times p$ signature matrix D with diagonal $\{d_1,\ldots,d_p\}$ as

$$d_j = \begin{cases} 1 & \text{if } j \in G_i \text{ and } G_i \text{ is type 1} \\ -1 & \text{if } j \in G_i \text{ and } G_i \text{ is type 2} \\ -1 & \text{if } j = p. \end{cases}$$

We now show that matrix DWD is an M-matrix. Using the fact that U is an M-matrix, it is easy to verify that $\{DWD\}_{j,k} \leq 0$ for $j \neq k$. Also, since U is an M-matrix and $D^2 = I$, it follows from Lemmas 14.9.1 and 14.9.4 that the first $p-1$ principle minors of DWD are strictly positive. Furthermore, since Γ is strictly positive definite, so is W. Therefore, $\det(DWD) > 0$ also. Consequently, by Lemmas 14.9.1

and 14.9.4 again, DWD is an M-matrix. This completes the "only if" part of this theorem.

We step out of the proof for a while to make some observations about Laplace transforms. Let $\phi(\lambda)$ be the Laplace transform of a positive real-valued random variable X, that is,

$$\phi(\lambda) = \int_0^\infty e^{-\lambda x} \, dF(x) \qquad \lambda > 0, \tag{13.61}$$

where F is the probability distribution function of X. Let $s \in [0, 1)$ and consider $\phi(a(1-s))$, with $a > 0$. Its power series, expanded about $s = 0$, is

$$\phi(a(1 - s)) = \sum_{n=0}^\infty \frac{(-a)^n \phi^{(n)}(a)}{n!} s^n. \tag{13.62}$$

It follows from (13.61) that ϕ is completely monotone (see, e.g., Feller (1971, XIII.4)). Consequently, the coefficients of the series in (13.62) are all positive. It is easy to check that $\phi(a(1 - e^{-\lambda/a}))$ is the Laplace transform of a discrete probability measure that puts mass $(-a)^n \phi^{(n)}(a)/n!$ on the point n/a, $n = 0, 1, \ldots$. Furthermore,

$$\lim_{a \to \infty} \phi(a(1 - e^{-\lambda/a})) = \phi(\lambda). \tag{13.63}$$

Indeed, we can use these ideas in reverse to construct infinitely divisible Laplace transforms.

Lemma 13.2.2 *Let $\psi : (R_+)^n \to (R_+)^n$ be a continuous function. Let $\mathbf{s} \in (R_+)^n$ and suppose that, for all $a > 0$ sufficiently large, $\log \psi(a(1 - s_1), \ldots, a(1 - s_n))$ has a power series expansion at $\mathbf{s} = \mathbf{0}$ with all its coefficients positive, except for the constant term. Then ψ is the Laplace transform of an infinitely divisible random variable in $(R_+)^n$.*

Proof We give the proof when $n = 1$. Exactly the same argument applies to random variables in $(R_+)^n$, only the notation required to mimic the proof is more complicated. Let $\sum_{n=0}^\infty b_n(a)s^n$ denote the power series expansion of $\log \psi(a(1 - s))$ at $s = 0$. Then

$$e^{\log \psi(a(1-s))} = e^{b_0(a)} e^{\sum_{n=1}^\infty b_n(a)s^n} \tag{13.64}$$

has a power series expansion about zero with all its coefficients positive. Therefore, $\psi(a(1 - e^{-\lambda/a}))$ is the Laplace transform of a discrete probability measure on the points k/a, $k = 0, 1, \ldots$. It follows from the

Extended Continuity Theorem (see Feller (1971, XIII.1, Theorem 2a)) that

$$\psi(\lambda) = \lim_{a \to \infty} e^{\log \psi(a(1 - e^{-\lambda/a}))} \tag{13.65}$$

is a Laplace transform. The same procedure applied to $(\log \psi(a(1 - s)))/n$ shows that $\psi^{1/n}(\lambda)$ is a Laplace transform. $\qquad\square$

Proof of Theorem 13.2.1 continued By Lemma 5.2.1, the Laplace transform of G^2 is $|I + \Gamma\Lambda|^{-1/2}$. We want to show that this is the Laplace transform of an infinitely divisible random variable. Consistent with (13.44), set $P_a(\mathbf{s}) = |I + \Gamma a(I - S)|^{-1/2}$. Let $\{\lambda_i\}$ denote the diagonal elements of Λ and let $s_i = \exp(-\lambda_i/a)$. (Recall that $\mathbf{s} = (s_1, \ldots, s_p)$ and S is a diagonal matrix with diagonal entries $s_1, \ldots, s_p$.) Then

$$\lim_{a \to \infty} P_a(\mathbf{s}) = |I + \Gamma\Lambda|^{-1/2}. \tag{13.66}$$

By Lemma 13.2.2 we see that, to show that G^2 is infinitely divisible, it suffices to show that $\log P_a(\mathbf{s})$ has a power series expansion at zero, with all its coefficients positive, except for the constant term. By the hypothesis there exists a signature matrix D such that DWD is an M-matrix. Therefore, by Lemma 14.9.4,

$$DWD = \lambda I - B, \tag{13.67}$$

where $B \geq 0$ and λ is greater than the absolute value of any eigenvalue of B.

Using (13.67) we have

$$
\begin{aligned}
|I + \Gamma a(I - S)| &= |\Gamma(W + a(I - S))| \tag{13.68} \\
&= |\Gamma||\lambda I - B + aI - aS| \\
&= |\Gamma|(\lambda + a)\left|I - \frac{1}{\lambda + a}(B + aS)\right|.
\end{aligned}
$$

Therefore, by (13.50),

$$2\log P_a(\mathbf{s}) = -\log|\Gamma| - \log|\lambda + a| + \sum_{n=1}^{\infty} \frac{\text{trace}\{(B + aS)^n\}}{n(\lambda + a)^n}. \tag{13.69}$$

Since $B \geq 0$, the coefficients of the terms involving any combination of $s_1, \ldots, s_p$ are all positive. $\qquad\square$

Remark 13.2.3 Let $G = (G_{x_1}, \ldots, G_{x_p})$ be a Gaussian vector with covariance matrix Γ. Let $\mathcal{N}$ be a signature matrix with diagonal elements $s(x_i)$. If $\mathcal{N}\Gamma^{-1}\mathcal{N}$ is an M-matrix, $\mathcal{N}\Gamma\mathcal{N} \geq 0$. Therefore, $\widetilde{G} := (s(x_1)G_{x_1}, \ldots, s(x_p)G_{x_p})$ has a positive covariance. Since $G^2 = \widetilde{G}^2$, when considering Gaussian vectors with infinitely divisible squares we can always assume that the covariance of the vector is positive.

There is nothing mysterious about the signature matrix in Theorem 13.2.1. It simply accounts for the fact that if G has an infinitely divisible square, then so does $\widetilde{G}$, for any choice of $s(x_i) = \pm 1$, $i = 1, \ldots, p$. By considering different signs for the $s(x_i)$, one can see which configurations of covariance matrices with strictly negative terms are possible for Gaussian vectors with infinitely divisible squares. For example, when $p = 3$, the covariance matrix has either no strictly negative terms or exactly four strictly negative terms.

We now give a remarkable property of associated Gaussian processes.

Corollary 13.2.4 *Let S be a locally compact space with a countable base. Let $G = \{G_x \, ; \, x \in S\}$ be a Gaussian process with continuous covariance that is associated with a strongly symmetric transient Borel right process X on S. Then G has infinitely divisible squares.*

Proof By Theorem 13.1.2, for every finite set $x_1, \ldots, x_n \in S$, $\Gamma = \{\Gamma(x_i, x_j)\}_{1 \leq i,j \leq p}$ is invertible and Γ^{-1} has negative off-diagonal elements. Also, since $\Gamma(x, y)$ is the 0-potential of a strongly symmetric Borel right process, $\Gamma \geq 0$. This follows from (3.65); see also Remark 3.3.5. Therefore, Γ is an M-matrix. Consequently, to apply Theorem 13.2.1 with $\mathcal{N} = I$ to show that G^2 is infinitely divisible, it only remains for us to show that Γ is also strictly positive definite.

It is a simple fact that an invertible symmetric positive definite matrix is strictly positive definite, since such a matrix has positive eigenvalues. Furthermore, the eigenvalues must be strictly positive since, if zero were an eigenvalue, the matrix would not be invertible. □

In Remark 13.2.3 we discussed the essential role that the signature matrices play in considering whether Gaussian vectors have infinitely divisible squares. Nevertheless, in Corollary 13.2.4 we can take the signature matrix to be the identity. This is generally the case for Gaussian processes with a continuous covariance defined on a reasonably nice space.

A topological space S is said to be pathwise connected if, for any $x, y \in S$, we can find a continuous function $r : [0, 1] \mapsto S$ with $r(0) = x$, $r(1) = y$.

Corollary 13.2.5 *Let S be a pathwise connected topological space. Let $G = \{G_x \, ; \, x \in S\}$ be a Gaussian process with continuous covariance*

$\Gamma = \{\Gamma(x,y), x, y \in S\}$ *with the property that, for any* $x_1, \ldots, x_p$, *the covariance matrix* $\overline{\Gamma} = \{\overline{\Gamma}\}_{i,j} = \Gamma(x_i, x_j)$ *is strictly positive definite. If G has infinitely divisible squares, then* $\overline{\Gamma}^{-1}$ *is an M-matrix. This holds for all* $x_1, \ldots, x_p$ *and* $1 \le p < \infty$.

Proof We first show that

$$\Gamma(x,y) \ge 0 \tag{13.70}$$

for all $x, y \in S$. Pick $x, y \in S$ and a continuous function $r : [0,1] \mapsto S$ with $r(0) = x$, $r(1) = y$. By considering the Gaussian process $\{G_{r(t)} ; t \in [0,1]\}$, it suffices to prove (13.70) with $S = [0,1]$.

We show below that for some function $s : [0,1] \mapsto \{-1, 1\}$,

$$s(x)s(y)\Gamma(x,y) = |\Gamma(x,y)| \qquad \forall x, y \in [0,1]. \tag{13.71}$$

We show here that this implies (13.70). Since $\overline{\Gamma}$ is strictly positive definite for all $x_1, \ldots, x_p \in S$, it follows that $\Gamma(x,x) > 0$ for each $x \in [0,1]$. Therefore, by continuity, $\Gamma(x,y) > 0$ for each $x \in [0,1]$ and all y in some neighborhood of x. Using the continuity of $\Gamma(x,y)$ again, we see that (13.71) implies that $s(y)$ is continuous. Consequently, $s(y)$ is constant on $[0,1]$. Using this in (13.71), we get (13.70).

To construct the function s, let $D_n = \{k2^{-n} ; 0 \le k \le 2^n\}$. By Theorem 13.2.1 and the first three sentences of Remark 13.2.3, we can find a function $s_n : D_n \mapsto \{-1, 1\}$ with

$$s_n(x)s_n(y)\Gamma(x,y) = |\Gamma(x,y)| \qquad \forall x, y \in D_n. \tag{13.72}$$

For any $x \in [0,1]$, let $r_n(x)$ be the sum of the first n terms in the dyadic expansion of x. Clearly, $r_n(x) \in D_n$. Then, using (13.72) and the continuity of $\Gamma(x,y)$, we see that for any $x, y \in [0,1]$ with $\Gamma(x,y) \ne 0$,

$$h(x,y) := \lim_{n \to \infty} s_n(r_n(x))s_n(r_n(y)) \tag{13.73}$$

exists and

$$h(x,y)\Gamma(x,y) = |\Gamma(x,y)|. \tag{13.74}$$

We point out above that $\Gamma(x,x) > 0$ for each $x \in [0,1]$. Therefore, it follows from continuity of Γ that there exists a $\delta > 0$ such that $\Gamma(x,y) > 0$ for all $x, y \in [0,1]$ with $|x - y| \le \delta$. For any $x, y \in [0,1]$ we can choose a finite sequence $x = z_0, z_1, \ldots, z_p = y$ with $|z_j - z_{j-1}| \le \delta$, $1 \le j \le p$. Then, by (13.73), for all $x, y \in [0,1]$,

$$h(x,y) := \lim_{n \to \infty} s_n(r_n(x))s_n(r_n(y)) = \lim_{n \to \infty} \prod_{j=1}^{p} s_n(r_n(z_{j-1}))s_n(r_n(z_j)) \tag{13.75}$$

exists (since $(s_n(z))^2 = 1$ for any z). Since Γ is continuous, we see that (13.74) holds for all $x, y \in [0,1]$.

It follows from the definition of h that $h(x,y) = h(y,x) \in \{-1,1\}$ and

$$h(x,y) = h(x,z)h(z,y) \qquad \forall\, x,y,z \in [0,1]. \qquad (13.76)$$

Pick $z_0 \in [0,1]$ and set $s(x) = h(x, z_0)$. This gives us (13.71) and hence (13.70).

Now choose any $x_1, \ldots, x_p \in S$. By Theorem 13.2.1 there exists a signature matrix $\mathcal{N}$ such that $\mathcal{N}\overline{\Gamma}^{-1}\mathcal{N}$ is an M-matrix. In particular, this implies that $\mathcal{N}\overline{\Gamma}\mathcal{N} \geq 0$. By (13.70) we have that $\mathcal{N}_i\mathcal{N}_j = 1$ whenever $g(x_i, x_j) > 0$. Suppose $g(x_i, x_j) = 0$. If there exists a sequence $i = r(1), r(2), \ldots, r(k) = j$ with $r : [1, \ldots, k] \mapsto [1, \ldots, p]$ such that $g(x_{r(l)}, x_{r(l+1)}) > 0$, then we again obtain $\mathcal{N}_i\mathcal{N}_j = 1$. If there is no such sequence, then $\overline{\Gamma}$ is reducible. We can write it as the direct sum of irreducible matrices and use the argument just given on each of them. Thus $\overline{\Gamma}^{-1}$ is an M-matrix. $\qquad \square$

Example 13.2.6 The matrix Γ^{-1} in Theorem 13.1.2 (3) is an M-matrix. This follows from the proof of (3) $\Rightarrow$ (2) of Theorem 13.1.2 in which we show that when Γ^{-1} is invertible, Γ is a 0-potential. As we point out in the proof of Corollary 13.2.4, this implies that $\Gamma \geq 0$.

To understand Section 13.3 it is important to note that an M-matrix need not have positive row sums. This can be seen from the following M-matrix A:

$$A = \begin{pmatrix} 1 & -\frac{3}{4} & -\frac{1}{3} \\ -\frac{3}{4} & 1 & -\frac{1}{4} \\ -\frac{1}{3} & -\frac{1}{4} & 1 \end{pmatrix} \qquad A^{-1} = \frac{36}{5} \begin{pmatrix} \frac{15}{16} & \frac{5}{6} & \frac{25}{48} \\ \frac{5}{6} & \frac{8}{9} & \frac{1}{2} \\ \frac{25}{48} & \frac{1}{2} & \frac{7}{16} \end{pmatrix}. \qquad (13.77)$$

Example 13.2.7 Let $g = \{(g_1, g_2)\}$ be a mean zero Gaussian vector. Let Γ be the covariance matrix of G. Then Γ and Γ^{-1} have the forms

$$\Gamma = \begin{pmatrix} a & c \\ c & b \end{pmatrix} \qquad \text{and} \qquad \Gamma^{-1} = \frac{1}{ab - c^2} \begin{pmatrix} b & -c \\ -c & a \end{pmatrix}$$

when $ab > c^2$ (We always have $ab \geq c^2$. If $ab = c^2$, $g_2 = \frac{c}{a}g_1$. Thus G^2 is one-dimensional and thus infinitely divisible, as we pointed out in the introduction to this chapter.) Whether $c \geq 0$ or $c < 0$, it is easy to see that we can find a signature matrix $\mathcal{N}$ such that $\mathcal{N}\Gamma^{-1}\mathcal{N}$ is an M-matrix. (We take the diagonal components of $\mathcal{N}$ equal when $c \geq 0$

and to have opposite signs when $c < 0$.) Thus all mean zero Gaussian vectors in R^2 have infinitely divisible squares.

The squares of Gaussian vectors in R^3 need not be infinitely divisible. It is easy to find strictly positive definite matrices in R^3 with all off-diagonal entries less than zero. We point out in Remark 13.2.3 that Gaussian vectors with such covariance matrices do not have infinitely divisible squares. The same argument applies to R^p for $p > 3$. We take up this point in greater detail in Example 13.3.4.

13.3 Infinitely divisible squares and associated processes

In Corollary 13.2.4 we showed that if S is a locally compact space with a countable base and $G = \{G_x \,;\, x \in S\}$ is a Gaussian process with continuous covariance that is associated with a strongly symmetric transient local Borel right process on S, then G has infinitely divisible squares. We know that the converse is false because a Gaussian vector with a positive covariance matrix has infinitely divisible squares if the inverse of its covariance matrix is an M-matrix, whereas it is an associated process only if its inverse is an M-matrix with positive row sums. We show in Example 13.2.6 that M-matrices need not have positive row sums. Therefore, to get an equivalence between Gaussian processes with infinitely divisible squares and associated Gaussian processes, we must have more than simply the condition that the Gaussian process has infinitely divisible squares. We do this in the next theorem.

Theorem 13.3.1 *Let S be a locally compact space with a countable base. Let $G = \{G_x \,;\, x \in S\}$ be a Gaussian process with strictly positive definite continuous covariance $\Gamma(x, y)$. The following are equivalent:*

(1) G is associated with a strongly symmetric transient local Borel right process X on S.

(2) $\{(G_x + c)^2 \,;\, x \in S\}$ is infinitely divisible for all $c \in R^1$.

(3) $\{(G_x + b\xi)^2 \,;\, x \in S \cup \{\delta\}\}$, $G(\delta) \equiv 0$, where $\delta \notin S$, is infinitely divisible, for some $b \neq 0$. Here, ξ is a standard normal random variable independent of G. Furthermore, if this holds for some $b \neq 0$, it holds for all $b \in R^1$.

Proof It follows from Theorem 13.1.2 that (1) holds if and only if it holds for every finite subset $S' \subseteq S$. Since infinite divisibility only involves finite-dimensional distributions, it suffices to prove this theorem for finite sets S. Without loss off generality we take $S = \{1, 2, \ldots, n\}$. $(1) \Rightarrow (2)$ We show below that, given any strongly symmetric transient

Borel right process X on a finite set S, we can find a strongly symmetric recurrent Borel right process Y on $S \cup \{0\}$ with $P^x(T_0 < \infty) > 0$ for all $x \in S$ such that X is the process obtained by killing Y the first time it hits 0. Let L_t^x denote the local time of Y. It follows from Theorem 8.2.2 that under $P^0 \times P_G$,

$$\left\{ L_{\tau(t)}^x + \frac{1}{2}G_x^2; \, x \in S \right\} \overset{law}{=} \left\{ \frac{1}{2}\left(G_x + \sqrt{2t} \right)^2; \, x \in S \right\} \qquad (13.78)$$

for all $t \in R_+$.

We know from Corollary 13.2.4 that $\{G_x^2, x \in S\}$ is infinitely divisible. Also, it follows from the additivity of local time that for any integer m,

$$L_{\tau(t)}^x = \sum_{k=1}^{m} L_{\tau(t/m)}^x \circ \theta_{\tau(t(k-1)/m)}. \qquad (13.79)$$

Using the strong Markov property we see that under P^0,

$$\{L_{\tau(t/m)}^x \circ \theta_{\tau(t(k-1)/m)}, x \in S\} \overset{law}{=} \{L_{\tau(t/m)}^x, x \in S\} \qquad (13.80)$$

and the m processes $\{L_{\tau(t/m)}^x \circ \theta_{\tau(t(k-1)/m)}, x \in S\}$, $1 \le k \le m$, are independent. Thus $\{L_{\tau(t)}^x, x \in S\}$ is also infinitely divisible. Combining these facts with (13.78) we have (2) of this theorem for $c \ge 0$. However since $\left(G_x + \sqrt{2t} \right)^2 \overset{law}{=} \left(G_x - \sqrt{2t} \right)^2$, it holds for all $c \in R^1$.

To simplify the notation we replace the state 0 by $n+1$. To complete the proof of (1) $\Rightarrow$ (2) we show that for any strongly symmetric transient Borel right process X on S, we can find a strongly symmetric recurrent Borel right process $\overline{X}$ on $\overline{S} = \{1, 2, \ldots, n+1\}$ such that X is the process obtained by killing $\overline{X}$ the first time it hits $n+1$ and $P^x(T_{n+1} < \infty) > 0$ for all $x \in S$.

To see this, recall that by Lemma 13.1.4 the semigroup for X is of the form $Q_t = e^{-tA}$ with A an invertible symmetric $n \times n$ matrix that has negative off-diagonal elements and positive row sums. We define the symmetric $(n+1) \times (n+1)$ matrix $\overline{A}$ by setting

$$
\begin{aligned}
\overline{A}_{i,j} &= A_{i,j} & 1 \le i,j \le n \\
\overline{A}_{i,n+1} &= \overline{A}_{n+1,i} = -\sum_{j=1}^{n} A_{i,j} & 1 \le i \le n \\
\overline{A}_{n,n} &= \sum_{i=1}^{n}\sum_{j=1}^{n} A_{i,j}.
\end{aligned}
$$

It follows that $\overline{A}$ has negative off-diagonal elements and zero row sums. Therefore, by Lemma 13.1.4, $P_t = e^{-t\overline{A}}$ is the semigroup of a recurrent strongly symmetric Borel right process $\overline{X}$ on $\overline{S}$, and by Lemma 13.1.6, X is the process obtained by killing $\overline{X}$ the first time it hits $n+1$.

Let P^x denote the probability of $\overline{X}$ starting at some point $x \in S$. We now show that $P^x(T_{n+1} < \infty) > 0$ for all $x \in S$. Assume, to the contrary, that $P^x(T_{n+1} < \infty) = 0$ for one or more points $x \in S$. By relabeling these points, if necessary, there exists a $1 \le k \le n$ such that $P^i(T_{n+1} < \infty) = 0$ for $1 \le i \le k$ and $P^i(T_{n+1} < \infty) > 0$ for $k+1 \le i \le n$. Therefore, by the strong Markov property, $P^i(T_j < \infty) = 0$ for all $1 \le i \le k$ and $k+1 \le j \le n$. Equivalently,

$$P_t(i,j) = 0 \qquad \forall 1 \le i \le k, \ k+1 \le j \le n. \tag{13.81}$$

It follows then from (13.27), (13.28), and the definition of $\overline{A}$ that $A_{i,j} = 0$ for all $1 \le i \le k$ and $k+1 \le j \le n$.

Also, since $P^i(T_{n+1} < \infty) = 0$ for $1 \le i \le k$, $P_t(i, n+1) = 0$, so that by the argument just given $\overline{A}_{i,n+1} = 0$ for all $1 \le i \le k$. Therefore, by the definition of $\overline{A}$, the first k row sums in A are equal to zero. Let $v \in R^n$ be defined by $v_i = 1$ for all $1 \le i \le k$ and $v_i = 0$ for all $k+1 \le i \le n$. Since the first k row sums in A are equal to zero, we see that $Av = 0$. Therefore A is not invertible. This contradicts the assumption that A is an M-matrix. Therefore $P^x(T_{n+1} < \infty) > 0$ for all $x \in S$.

$(2) \Rightarrow (3)$ Let $\Gamma = \{\Gamma(i,j)\}_{1 \le i,j \le n}$. Let Λ denote the diagonal matrix with diagonal entries $\lambda_1, \ldots, \lambda_n \in R_+$ and set $G_0 \equiv 0$.

By Lemma 5.2.1, for some function $F(\Lambda, \lambda_0, \Gamma)$,

$$E\left(e^{-\sum_{i=0}^n \lambda_i (G_i + \sqrt{n}c)^2}\right) = \frac{1}{\det(I + \Gamma\Lambda)} \exp\left(nc^2 F(\Lambda, \lambda_0, \Gamma)\right) \tag{13.82}$$

and

$$E\left(e^{-\sum_{i=0}^n \lambda_i (G_i + \xi)^2}\right) = \frac{1}{\det(I + \Gamma\Lambda)} E_\xi \exp\left(\xi^2 F(\Lambda, \lambda_0, \Gamma)\right). \tag{13.83}$$

Let $\widetilde{\psi}(\Lambda, \lambda_0, \Gamma)$ denote the left-hand side of (13.82). Then we have

$$\widetilde{\psi}^{1/n}(\Lambda, \lambda_0, \Gamma) = \frac{1}{(\det(I + \Gamma\Lambda))^{1/n}} \exp\left(c^2 F(\Lambda, \lambda_0, \Gamma)\right). \tag{13.84}$$

By (2), $((G_1 + \sqrt{n}c)^2, \ldots, (G_n + \sqrt{n}c)^2)$ is infinitely divisible. Then clearly $((\sqrt{n}c)^2, (G_1 + \sqrt{n}c)^2, \ldots, (G_n + \sqrt{n}c)^2)$ is also infinitely divisible. Therefore, $\widetilde{\psi}^{1/n}(\Lambda, \lambda_0, \Gamma)$ is the Laplace transform of a random variable for every n. We take the limit in (13.84) as $n \to \infty$. Since $(\det(I+\Gamma\Lambda))^{-1/n} \to 1$, it follows from the continuity theorem for Laplace transforms that $\exp\left(c^2 F(\Lambda, \lambda_0, \Gamma)\right)$ is a Laplace transform for any c.

Integrating, we see that $E_X \exp\left(X F(\Lambda, \lambda_0, \Gamma)\right)$ is a Laplace transform for any nonnegative random variable X. Since ξ^2 is infinitely divisible, for any m we can write $\xi^2 \overset{law}{=} X_1 + \cdots + X_m$, where $X_1, \ldots, X_m$ are independent identically distributed nonnegative random variables.

This shows that for any $c \neq 0$, $E_\xi \exp\left(c^2 \xi^2 F(\Lambda, \lambda_0, \Gamma)\right)$ is the Laplace transform of an infinitely divisible random variable. By (2) with $c = 0$, we have that $(\det(I + \Gamma\Lambda))^{-1}$ is the Laplace transform of an infinitely divisible random variable. Therefore, it follows from (13.83) that $(c^2\xi^2, (G_1 + c\xi)^2, \ldots, (G_n + c\xi)^2)$ is infinitely divisible for all $c \neq 0$. As we just pointed out, it follows from (2) that this is infinitely divisible when $c = 0$.

Let $G_{b\xi} = \{G_x + b\xi \, ; \, x \in S \cup \{\delta\}\}$. Before going on to the proof of (3) $\Rightarrow$ (1) we explore the relationship between the covariance matrices of G and $G_{b\xi}$ on S and $S \cup \{\delta\}$. Here we take $\delta = n + 1$.

Lemma 13.3.2 *Let Γ be a symmetric strictly positive definite $n \times n$ matrix and let $b \in R^1$, $b \neq 0$. Consider the $(n+1) \times (n+1)$ symmetric matrix $\overline{\Gamma}$ defined by $\overline{\Gamma}_{i,j} = \Gamma_{i,j} + b^2$, $i, j = 1, \ldots n$ and $\overline{\Gamma}_{n+1,j} = b^2$, $j = 1, \ldots n + 1$. Then $\overline{\Gamma}^{-1}$ exists and the following are equivalent:*

(1) Γ^{-1} has negative off-diagonal elements and positive row sums.

(2) $\overline{\Gamma}^{-1}$ has negative off-diagonal elements.

Proof   Since Γ is strictly positive definite, it is invertible. We first note that

$$\overline{\Gamma}^{i,j} = \Gamma^{i,j} \qquad\qquad i, j = 1, \ldots, n$$

$$\overline{\Gamma}^{n+1,j} = -\sum_{i=1}^{n} \Gamma^{i,j} \qquad j = 1, \ldots, n \qquad\qquad (13.85)$$

$$\overline{\Gamma}^{n+1,n+1} = \frac{1}{b^2} + \sum_{i,j=1}^{n} \Gamma^{i,j},$$

where, for an invertible matrix A, we use $A^{i,j}$ to denote $\{A^{-1}\}_{i,j}$. To prove (13.85), we simply go through the elementary steps of taking the inverse of $\overline{\Gamma}$. We begin with the array

$$
\begin{array}{ccccc|ccc}
\Gamma_{1,1} + b^2 & \cdots\cdots & \Gamma_{1,n} + b^2 & b^2 & & 1 & \cdots\cdots & 0 \quad 0 \\
\vdots & \ddots & \vdots & \vdots & & \vdots & \ddots & \vdots \quad \vdots \\
\Gamma_{n,1} + b^2 & \cdots\cdots & \Gamma_{n,n} + b^2 & b^2 & & 0 & \cdots\cdots & 1 \quad 0 \\
b^2 & \cdots\cdots & b^2 & b^2 & & 0 & \cdots\cdots & 0 \quad 1
\end{array}
$$

Next we subtract the last row from each of the other rows and then divide the last row by b^2 to get

$$
\begin{array}{ccccc|ccccc}
\Gamma_{1,1} & \cdots\cdots & \Gamma_{1,n} & 0 & & 1 & \cdots\cdots & 0 & -1 \\
\vdots & \ddots & \vdots & \vdots & & \vdots & \ddots & \vdots & \vdots \\
\Gamma_{n,1} & \cdots\cdots & \Gamma_{n,n} & 0 & & 0 & \cdots\cdots & 1 & -1 \\
1 & \cdots\cdots & 1 & 1 & & 0 & \cdots\cdots & 0 & 1/b^2
\end{array}
$$

This shows that $\det(\overline{\Gamma}) = b^2 \det(\Gamma)$ and consequently Γ is invertible if and only if $\overline{\Gamma}$ is invertible.

We now work with the first n rows to get the inverse of Γ so that the array looks like

$$
\begin{array}{ccccc|cccc}
1 & \cdots\cdots & 0 & 0 & & \Gamma^{1,1} & \cdots\cdots & \Gamma^{1,n} & a_1 \\
\vdots & \ddots & \vdots & \vdots & & \vdots & \ddots & \vdots & \vdots \\
0 & \cdots\cdots & 1 & 0 & & \Gamma^{n,1} & \cdots\cdots & \Gamma^{n,n} & a_n \\
1 & \cdots\cdots & 1 & 1 & & 0 & \cdots\cdots & 0 & 1/b^2
\end{array}
$$

At this stage we do not know the a_j, $j = 1, \ldots, n$.

Finally, we subtract each of the first n rows from the last row to obtain

$$
\begin{array}{ccccc|cccc}
1 & \cdots\cdots & 0 & 0 & & \Gamma^{1,1} & \cdots\cdots & \Gamma^{1,n} & a_1 \\
\vdots & \ddots & \vdots & \vdots & & \vdots & \ddots & \vdots & \vdots \\
0 & \cdots\cdots & 1 & 0 & & \Gamma^{n,1} & \cdots\cdots & \Gamma^{n,n} & a_n \\
0 & \cdots\cdots & 0 & 1 & & -\sum_{i=1}^{n}\Gamma^{i,1} & \cdots\cdots & -\sum_{i=1}^{n}\Gamma^{i,n} & a_{n+1}
\end{array}
$$

where

$$
a_{n+1} = (1/b^2 - \sum_{j=1}^{n} a_j). \tag{13.86}
$$

Since the inverse matrix is symmetric, we see that $a_j = -\sum_{i=1}^{n}\Gamma^{i,j}$, $j = 1, \ldots, n$. This verifies (13.85).

Note the following interesting property of the row sums:

$$
\begin{aligned}
\sum_{j=1}^{n+1} \overline{\Gamma}^{i,j} &= 0 \qquad j = 1, \ldots, n \\[1mm]
\sum_{j=1}^{n+1} \overline{\Gamma}^{n+1,j} &= \frac{1}{b^2}.
\end{aligned} \tag{13.87}
$$

This shows, in particular, that when $\overline{\Gamma}$ is invertible the row sums of $\overline{\Gamma}^{-1}$ are positive.

We can now complete the proof. When Γ^{-1} has negative off-diagonal elements and positive row sums, it is obvious from the first two lines of

(13.85) that $\overline{\Gamma}^{-1}$ has negative off-diagonal elements. It is equally obvious from the same two lines that (2) $\Rightarrow$ (1).

Proof of Theorem 13.3.1 continued (3) $\Rightarrow$ (1) We show that (1) follows if (3) holds for some $b \neq 0$. Let $\overline{\Gamma}$ be the covariance matrix of $\{G_x + b\xi, x \in S \cup \{n+1\}\}$ with $G_{n+1} \equiv 0$. Consequently, $\overline{\Gamma}$ is a symmetric positive definite matrix. By hypothesis, Γ is strictly positive definite and therefore invertible. Therefore, by Lemma 13.3.2, $\overline{\Gamma}$ is invertible and, since it is positive definite, it must be strictly positive definite. Consequently, by Theorem 13.2.1, there exists a signature matrix $\mathcal{N}$ such that $\mathcal{N}\,\overline{\Gamma}^{-1}\,\mathcal{N}$ is an M-matrix. In particular, this implies that $\mathcal{N}\,\overline{\Gamma}\,\mathcal{N} \geq 0$. Since $\overline{\Gamma}_{i,n+1} = b^2 > 0$ for all $1 \leq i \leq n+1$, we must have $\mathcal{N}_i \mathcal{N}_{n+1} = 1$ for all $1 \leq i \leq n+1$, and therefore $\mathcal{N}_i \mathcal{N}_j = 1$ for all $1 \leq i,j \leq n+1$. Without loss of generality we can take $\mathcal{N} = I$, the identity matrix. Thus we see that $\overline{\Gamma}^{-1}$ is an M-matrix, which implies, in particular, that it has negative off-diagonal elements. Therefore, by Lemma 13.3.2, Γ^{-1} has negative off-diagonal elements and positive row sums. Statement (1) now follows from Theorem 13.1.1. □

Theorem 13.3.3 *Let S be a locally compact space with a countable base. Let $G = \{G_x \,;\, x \in S\}$ be a Gaussian process with continuous covariance $\widetilde{\Gamma}(x,y)$. Assume that for every finite collection $x_1, \ldots, x_p \in S$, the matrix $\widetilde{\Gamma} = \{\widetilde{\Gamma}(x_i, x_j)\}_{1 \leq i,j \leq p}$ is positive and strictly positive definite. Assume that $\widetilde{\Gamma}(0, x) > 0$ for some element $0 \in S$ and all $x \in S$. If G^2 is infinitely divisible, then*

$$\widetilde{\Gamma}(x,y) = \frac{\widetilde{\Gamma}(x,0)}{\widetilde{\Gamma}(0,0)} u(x,y) \frac{\widetilde{\Gamma}(y,0)}{\widetilde{\Gamma}(0,0)} \qquad \forall x, y \in S, \qquad (13.88)$$

where $u(x,y)$ is the 0-potential density of a strongly symmetric transient local Borel right process X on S.

Proof Let $h(x) = \dfrac{\widetilde{\Gamma}(x,0)}{\widetilde{\Gamma}(0,0)}$. By Theorem 13.1.1, it suffices to show that, for every finite subset $S' = \{x_1, \ldots, x_p\} \subseteq S$, the matrix

$$\overline{\Gamma} = \left\{ \frac{\widetilde{\Gamma}(x_i, x_j)}{h(x_i) h(x_j)} \right\}_{1 \leq i,j \leq p} \qquad (13.89)$$

is invertible and $\overline{\Gamma}^{-1}$ has negative off-diagonal elements and positive row sums. The fact that $\overline{\Gamma}$ is invertible follows from the hypothesis that $\widetilde{\Gamma}$ is strictly positive definite.

Suppose that $0 \notin S'$. Then we add it as an element x_{p+1} and consider the matrix defined in (13.89) for $1 \leq i, j \leq p+1$ and denote it by $\overline{\Gamma}'$. Suppose we show that $\overline{\Gamma}'$ is invertible and that $(\overline{\Gamma}')^{-1}$ has negative off-diagonal elements and positive row sums. Then it follows from Lemma 13.1.4 that $\overline{\Gamma}'$ is the 0-potential of a transient strongly symmetric Borel right process on $S' \cup \{x_{p+1}\}$. It then follows from Theorem 13.1.1 that $\overline{\Gamma}^{-1}$ has negative off-diagonal elements and positive row sums. Consequently, we may as well assume that $0 \in S'$.

It is convenient to relabel the elements of S' so that $x_p = 0$. For $x \in S'$, write $G_x = \eta_x + h(x)G_{x_p}$ so that $\eta_{x_p} = 0$ and $\{\eta_{x_j}; j = 1, \ldots, p-1\}$ is independent of G_{x_p}. By hypothesis, $\{\eta_x + h(x)G_{x_p}, x \in S'\} = \{G_x, x \in S'\}$ has infinitely divisible squares. Therefore,

$$\left(\frac{\eta}{h} + G_{x_p}\right) := \left\{ \frac{\eta_x}{h(x)} + G_{x_p}, x \in S' \right\} \qquad (13.90)$$

has infinitely divisible squares. Let $S'' = \{x_1, \ldots, x_{p-1}\}$ and $\left(\frac{\eta}{h}\right) := \left\{ \frac{\eta_x}{h(x)}, x \in S'' \right\}$.

We now use the notation of Lemma 13.3.2. Let Γ denote the covariance matrix of $\left(\frac{\eta}{h}\right)$ and set $b = EG_{x_1}^2$. The matrix $\overline{\Gamma}$ of that lemma is precisely the matrix $\overline{\Gamma}$ in (13.89), which is the covariance matrix of $\left(\frac{\eta}{h} + G_{x_1}\right)$. By Lemma 13.2.1 and the assumption of positivity, $\overline{\Gamma}^{-1}$ is an M-matrix. In particular it has negative off-diagonal elements. Furthermore, as we point out following (13.87), it has positive row sums. $\square$

Example 13.3.4 Consider the M-matrix A in (13.77) that does not have positive row sums. To avoid confusion in the rest of this paragraph, relabel this matrix Γ^{-1}. Clearly Γ is strictly positive definite (the determinants of its principle minors are positive). Let $\eta = (\eta_1, \eta_2, \eta_3)$ be a mean zero Gaussian vector with covariance matrix Γ. Let ξ be a mean zero real-valued normal random variable independent of η. Then η has infinitely divisible squares, but $\eta_\xi := (\eta_1 + \xi, \eta_2 + \xi, \eta_3 + \xi, \xi)$ does not, whatever the value of $E\xi^2 > 0$. To see this, let $\overline{\Gamma}$ be the covariance matrix of η_ξ. The proof of (3) $\Rightarrow$ (1) of Theorem 13.3.1 shows that if η_ξ^2 is infinitely divisible, $\overline{\Gamma}^{-1}$ is an M-matrix. Therefore, in particular, it has negative off-diagonal elements. Consequently, by Lemma 13.3.2, Γ^{-1} has positive row sums. This is a contradiction.

On the other hand, one can do the arithmetic to check that $\widetilde{\eta} =$

$(\eta_1 + \xi, \eta_2 + \xi, \eta_3 + \xi)$ does have infinitely divisible squares for any value of $E\xi^2$.

We know that if $(\eta_1 + \xi, \eta_2 + \xi, \eta_3 + \xi, \xi)$ has infinitely divisible squares then not only does (η_1, η_2, η_3) have infinitely divisible squares, but the inverse of its covariance matrix also has positive row sums. This is true whatever the value of $E\xi^2$, as long as it is not zero. The example just given shows that even if $(\eta_1 + \xi, \eta_2 + \xi, \eta_3 + \xi)$ has infinitely divisible squares for ξ of all variances including zero, it is not necessarily true that the inverse of the covariance matrix of (η_1, η_2, η_3) has positive row sums.

We consider a related question. When $(\eta_1 + \xi, \eta_2 + \xi, \eta_3 + \xi)$ has infinitely divisible squares for some normal random variable ξ independent of (η_1, η_2, η_3), does this imply that (η_1, η_2, η_3) has infinitely divisible squares? In the next example we show that this is not necessarily true. Suppose that the covariance matrix of $(\eta_1 + \xi, \eta_2 + \xi, \eta_3 + \xi)$ is given by

$$\begin{pmatrix} 1+b & b-a & b-a \\ b-a & 1+b & b-a \\ b-a & b-a & 1+b \end{pmatrix},$$

where $E\xi^2 = b$ and $0 < a < 1/2$. By computing the determinants of its principle minors, we see that it is strictly positive definite. The inverse of this matrix has all its off-diagonal terms equal to $-(b-a)(1+a)$. Thus it is an M-matrix if and only if $b \geq a$. In particular, when $b = 0$, it is the inverse of the covariance matrix of (η_1, η_2, η_3).

This leads to the following interesting example. Let

$$\begin{aligned} \eta_1 &= \xi_1 + c\xi_2 + c\xi_3 \qquad\qquad (13.91) \\ \eta_2 &= c\xi_1 + \xi_2 + c\xi_3 \\ \eta_3 &= c\xi_1 + c\xi_2 + \xi_3, \end{aligned}$$

where ξ_i, $i = 1, \ldots, 3$ are independent normal random variables with mean zero and variance one. Then (η_1, η_2, η_3) has infinitely divisible squares if and only if $c \geq 0$. In fact, one can check that this example extends to n vectors, $(\eta_1, \ldots, \eta_n)$, defined similarly, when $|c|$ is sufficiently small, depending on n. (One term in the minors determining the off-diagonal terms of the inverse of the covariance matrix dominates, and these terms have the opposite sign of c.)

13.4 Additional results about M-matrices

The matrices Γ^{-1} and $\overline{\Gamma}^{-1}$ in Lemma 13.3.2 are both M-matrices. More significantly, so are the matrices in Lemma 13.1.2 (3), as we point out in Remark 13.1.3, although we do not refer to them as such. The proof of this fact in Lemma 13.1.2 uses Markov chains. It is desirable to have a proof using simple linear algebra. We do this in the next lemma, which also explores further the relationship between Γ^{-1} and $\overline{\Gamma}^{-1}$.

Lemma 13.4.1 *Let Γ^{-1} and $\overline{\Gamma}^{-1}$ be as given in Lemma 13.3.2, but assume only that Γ^{-1} is symmetric and invertible. Then $\overline{\Gamma}^{-1}$ is also invertible and the following are equivalent:*

(1) Γ^{-1} has negative off-diagonal elements and positive row sums.
(2) Γ^{-1} is an M-matrix with positive row sums.
(3) $\overline{\Gamma}^{-1}$ is an M-matrix with positive row sums.
(4) $\overline{\Gamma}^{-1}$ is an M-matrix.
(5) $\overline{\Gamma}^{-1}$ has negative off-diagonal elements.

Recall that we showed in Remark 13.1.3 that an invertible matrix with positive row sums has at least one of its row sums strictly greater than zero.

Proof The only way we use the fact that Γ is strictly positive definite in the proof of Lemma 13.3.2 is to show that Γ is invertible. Since this is now a hypothesis, we can use this conclusion of Lemma 13.3.2 as well as two items mentioned in its proof. These are that Γ is invertible if and only if $\overline{\Gamma}$ is invertible and that $\overline{\Gamma}^{-1}$ always has positive row sums. (Thus (3) $\iff$ (4) is trivial.)

Furthermore, since Γ is invertible, if it is positive definite, it must be strictly positive definite. We next show that Γ is strictly positive definite if and only if $\overline{\Gamma}$ is strictly positive definite, whatever the value of $b \neq 0$.

Let $\overline{u} = (u_1, \ldots, u_n, u_{n+1}) = (u, u_{n+1})$ with $u = (u_1, \ldots, u_n)$. We have

$$(\overline{u}, \overline{\Gamma}\overline{u}) = (u, \Gamma u) + b^2 \left(\sum_{i=1}^{n+1} u_i \right)^2. \tag{13.92}$$

When Γ is strictly positive definite, if $(u, \Gamma u) = 0$, then $u = 0$. Therefore, if $\overline{u} \neq 0$, $\sum_{i=1}^{n+1} u_i = u_{n+1} \neq 0$, so that $b^2 (\sum_{i=1}^{n+1} u_i)^2 > 0$. This shows that $\overline{\Gamma}$ is strictly positive definite.

On the other hand, if Γ is not strictly positive definite, then $(u, \Gamma u) \leq 0$ for some $u = (u_1, \ldots, u_n) \neq 0$. Let $\overline{u} = (u, u_{n+1})$ with $u_{n+1} =$

$-\sum_{i=1}^{n} u_i$. We see from (13.92) that $(\overline{u}, \overline{\Gamma}\overline{u}) \leq 0$ with $\overline{u} \neq 0$. Thus, $\overline{\Gamma}$ is not strictly positive definite.

(1) $\Rightarrow$ (2) Suppose that (1) holds. Then, by Lemma 13.3.2, $\overline{\Gamma}^{-1}$ has negative off-diagonal elements for all $b \neq 0$. It is obvious that $\overline{\Gamma} \geq 0$ for all $|b|$ sufficiently large. By definition, for these values of $|b|$, $\overline{\Gamma}^{-1}$ is an M-matrix. By Lemma 14.9.4, for these values of $|b|$, $\overline{\Gamma}^{-1}$ is strictly positive definite. Therefore, as we just showed, Γ^{-1} is strictly positive definite. So, by Lemma 14.9.4 again, Γ^{-1} is an M-matrix.

(2) $\Rightarrow$ (3) By Lemma 13.3.2, $\overline{\Gamma}^{-1}$ has negative off-diagonal elements. Also, $\Gamma \geq 0$, since it is an M-matrix. Therefore, $\overline{\Gamma} \geq 0$, and hence is also an M-matrix;

(3) $\Longleftrightarrow$ (4) and (4) $\Rightarrow$ (5) are trivial. (5) $\Longleftrightarrow$ (1) is given in Lemma 13.3.2. □

13.5 Notes and references

This chapter is based on Eisenbaum (2003), Eisenbaum (2005), and Eisenbaum and Kaspi (2006). We introduce local Borel right processes to make the equivalences in Theorems 13.1.2 and 13.3.1 more concrete when the state space of the Markov process is locally compact. The inequality in (13.42) has been observed several times; see Marcus and Rosen (1992b) and the references therein.

Theorem 13.2.1 combines the work of Griffiths (1984) and Bapat (1989). This problem has a long history and was originally posed as characterizing the infinite divisibility of the multivariate gamma distribution; see Eisenbaum and Kaspi (2006) for further references. Many of the arguments in Griffiths (1984) and Bapat (1989) are special cases of more general results found in the 1974 text Berman and Plemmons (1994) (the reference given is to the 1994 reprint); see in particular Theorems 1.3.20 and 2.2.1 and Section 3 of Chapter 6. Lemma 14.9.1 is taken from Horn and Johnson (1999).

Lemmas 13.3.2 and 13.4.1 are new observations that shed light on the relationship between Gaussian processes with infinitely divisible squares and associated Gaussian processes, and simplify the proof of Theorem 13.3.1.

14

Appendix

14.1 Kolmogorov's Theorem for path continuity

Kolmogorov's Theorem gives a simple condition for Hölder continn a complete separable metric space. Since we only use it in Chapter 2 for processes on R^d, we simplify matters and consider only processes on $[0,1]^d$. This result is interesting from a historical perspective since it contains the germs of the much deeper continuity conditions obtained in Chapter 5 (actually, in Chapter 5, we only consider Gaussian processes, but the methods developed have a far larger scope, as is shown in Ledoux and Talagrand (1991)).

Let D_m be the set of d-dimensional vectors in $[0,1]^d$ with components of the form $i/2^m$ for some integer $i \in [0, 2^m]$. Let D denote the set of dyadic numbers in $[0,1]^d$, that is, $D = \cup_{m=0}^{\infty} D_m$.

Theorem 14.1.1 *Let $X = \{X_t, t \in [0,1]^d\}$ be a stochastic process satisfying*

$$E\left(|X_t - X_s|^h\right) \le c|t - s|^{d+r} \qquad \forall\, s, t \in [0,1]^d \qquad (14.1)$$

for constants $c, r > 0$ and $h \ge 1$. Then, for any $\alpha < r/h$,

$$|X_t - X_s| \le C(\omega)|t - s|^\alpha \qquad \forall\, s, t \in D \qquad (14.2)$$

for some random variable $C(\omega) < \infty$ almost surely.

Proof Let N_m be the set of nearest neighbors in D_m. This is the set of pairs $s, t \in D_m$ with $|s - t| = 2^{-m}$. Using (14.1) and the fact that $|N_m| \le 2d2^{dm}$, we have

$$E\left(\sup_{(s,t)\in N_m} |X_t - X_s|^h\right) \le \sum_{(s,t)\in N_m} E\left(|X_t - X_s|^h\right) \le c2d2^{-mr}.$$

$$(14.3)$$

For any $s \in D$, let s_m be the vector in D_m with $s_m \leq s$ that is closest to s. Then, for any m, we have that either $s_m = s_{m+1}$ or else the pair $\{s_m, s_{m+1}\} \in N_{m+1}$. Now let $s, t \in D$ with $|s - t| \leq 2^{-m}$. Then either $s_m = t_m$ or $\{s_m, t_m\} \in N_m$. In either case,

$$X_t - X_s = \sum_{i=m}^{\infty} (X_{t_{i+1}} - X_{t_i}) + X_{t_m} - X_{s_m} + \sum_{i=m}^{\infty} (X_{s_{i+1}} - X_{s_i}). \quad (14.4)$$

Using (14.3) we see that

$$\left\| \sup_{\substack{s,t \in D \\ |s-t| \leq 2^{-m}}} |X_t - X_s| \right\|_h \leq 3 \sum_{i=m}^{\infty} (cd2^{-ir})^{1/h} \leq C2^{-mr/h}. \quad (14.5)$$

Fix $\alpha < r/h$. Since for any pair $s, t \in D$ with $s \neq t$ we can find some m such that $2^{-m-1} \leq |s - t| \leq 2^{-m}$, we have

$$\left\| \sup_{\substack{s,t \in D \\ s \neq t}} |X_t - X_s|/|s - t|^{\alpha} \right\|_h \leq 2 \sum_{m=0}^{\infty} C2^{\alpha(m+1)}2^{-mr/h} < \infty. \quad (14.6)$$

This gives (14.2). $\qquad\qquad\qquad\qquad\qquad\qquad\qquad\qquad\qquad\qquad\square$

14.2 Bessel processes

In this book, Bessel processes are mainly used to give classical versions of some results that we obtain by other methods. Therefore, we only give a brief description of some of their properties. In this section we assume that the reader has some familiarity with stochastic calculus and stochastic differential equations. For an excellent introduction to Bessel processes, see Revuz and Yor (1991, Chapter XI). Consider the stochastic differential equation

$$Z_t = x + 2 \int_0^t \sqrt{Z_s}\, dB_s + \delta t \quad (14.7)$$

for $x \in R^1$ and $\delta \geq 0$, where B_s is a Brownian motion. This equation has a unique positive strong solution $\{Z_t,\ t \in R_+\}$ called the square of the δ-dimensional Bessel process started at x and denoted by $BESQ^{\delta}(x)$. We also refer to this process as a δ-th ordered squared Bessel process.

The motivation for the designation $BESQ^{\delta}(x)$ comes from the case in which δ is an integer. Let

$$W_t = (W_t^{(1)} + a_1, \ldots, W_t^{(n)} + a_n), \quad (14.8)$$

where $(W_t^{(1)}, \ldots, W_t^{(n)})$ is an n-dimensional Brownian motion, that is,

$W_t^{(1)}, \ldots, W_t^{(n)}$ are n independent Brownian motions. It follows from Ito's formula that

$$|W_t|^2 = |W_0|^2 + 2 \left(\sum_{i=1}^{n} \int_0^t W_s^{(i)} \, dW_s^{(i)} \right) + n \, t. \qquad (14.9)$$

This equation can be written in the form of (14.7) by taking $Z_t = |W_t|^2$ and

$$B_t = \sum_{i=1}^{n} \int_0^t \frac{W_s^{(i)}}{|W_s|} \, dW_s^{(i)} \qquad (14.10)$$

and noting that B_t is a Brownian motion. To prove this last assertion, note that $|W_t| > 0$ for almost all t and $< B, B >_t = t$ (this last term is the quadratic variation of B up to time t).

Remark 14.2.1 In the above notation, let $\widetilde{W}_t = (W_t^{(1)}, \ldots, W_t^{(n)})$. Writing $\widetilde{W}_t = \widetilde{W}_{t-s} + \widetilde{W}_s$, we see by (14.9) that for $t \geq s$,

$$|\widetilde{W}_t|^2 = |\widetilde{W}_s|^2 + |\widetilde{W}_{t-s}|^2. \qquad (14.11)$$

Let $Z_{n,t}(|W(0)|) := |W_t|^2$ for W_t in (14.8). It follows from (14.9) and (14.11) that

$$Z_{n,t}(0) \overset{law}{=} Z_{n,t-s}(|W(s)|^2). \qquad (14.12)$$

The following additivity property of squared Bessel processes is used in the proof of Theorem 2.6.3.

Theorem 14.2.2 *Let Z and Z' be independent stochastic processes with Z a $BESQ^\delta(x)$ and Z' a $BESQ^{\delta'}(x')$, $\delta, \delta' \geq 0$. Then $Z + Z'$ is a $BESQ^{\delta+\delta'}(x + x')$.*

Proof Z satisfies (14.7) and Z' satisfies

$$Z_t' = x' + 2 \int_0^t \sqrt{Z_s'} \, dB_s' + \delta' t, \qquad (14.13)$$

where B' is a Brownian motion independent of B. Therefore, $X_t = Z_t + Z_t'$ satisfies

$$X_t = x + x' + 2 \int_0^t (\sqrt{Z_s} \, dB_s + \sqrt{Z_s'} \, dB_s') + (\delta + \delta') t. \qquad (14.14)$$

Set

$$W_t = \int_0^t \frac{\sqrt{Z_s} \, dB_s + \sqrt{Z_s'} \, dB_s'}{\sqrt{X_s}} \qquad (14.15)$$

and note that $< W, W >_t = t$. Therefore, W_t is a linear Brownian motion and the integral in (14.14) can be written as $\int_0^t \sqrt{X_s}\, dW_s$. □

Remark 14.2.3 The assertion in the second paragraph of Theorem 2.6.3 follows immediately from (2.184) and Theorem 14.2.2. Note that in the notation of Remark 14.2.1, for $r \geq x$, $B_{r-x}^2 + \bar{B}_{r-x}^2 \overset{law}{=} Z_{2,r-x}(0)$ and $B_r^2 + \bar{B}_r^2 \overset{law}{=} Z_{2,r-x}(Y_x)$, where

$$Y_x = B_x^2 + \bar{B}_x^2 \overset{law}{=} Z_{2,x}(0). \tag{14.16}$$

Therefore, by (2.184) and Theorem 14.2.2, for $r \geq x$, $L_{T_0}^r$ is $BESQ^0(Y_x)$. Given this, it is clear that, conditioned on B_x, $\{L_{T_0}^r, r \geq x\}$ is independent of $\{B_r, 0 \leq r \leq x\}$.

14.3 Analytic sets and the Projection Theorem

We prove the Projection Theorem, which has a critical role in the proof of Theorem 3.2.6.

Theorem 14.3.1 (The Projection Theorem) *Let* $(\Omega, \mathcal{F}, P)$ *be a complete probability space, and let* π *denote the projection from* $R^1 \times \Omega$ *to* Ω. *Let* $\mathcal{B}$ *denote the Borel* σ-*algebra for* R^1 *and let* $\mathcal{B} \times \mathcal{F}$ *denote the product* σ-*algebra. Then* $\pi(\mathcal{B} \times \mathcal{F}) \subseteq \mathcal{F}$.

The proof of the Projection Theorem follows from a sequence of lemmas and theorems including Lusin's Theorem on analytic sets. We begin by defining these sets. Let $\mathcal{O}$ be a set and $\mathcal{G}$ a collection of subsets of $\mathcal{O}$ that contain the empty set. A Souslin scheme $\{A.\}$ over $\mathcal{G}$ is a collection consisting of a set $A_{n_1,\ldots,n_k} \in \mathcal{G}$ for each $k \in \mathbb{N}$ and $(n_1, \ldots, n_k) \in \mathbb{N}^k$. Let $\sigma = (\sigma(1), \sigma(2), \ldots) \in \mathbb{N}^{\mathbb{N}}$. We set $\sigma|n = (\sigma(1), \sigma(2), \ldots, \sigma(n))$. The kernel K of the Souslin scheme $\{A.\}$ is given by

$$K = \bigcup_\sigma \bigcap_n A_{\sigma|n}.$$

A set is called analytic over $\mathcal{G}$ if it is the kernel of some Souslin scheme over $\mathcal{G}$. The collection of all analytic sets over $\mathcal{G}$ is denoted $\mathcal{A}(\mathcal{G})$. Trivially, $\mathcal{G} \subseteq \mathcal{A}(\mathcal{G})$.

We first show that $\mathcal{A}(\mathcal{G})$ is closed under countable unions. Let $K^{(i)}$ be the kernel of the Souslin scheme $\{A^{(i)}\}$ and $\psi(n) = (\psi_1(n), \psi_2(n))$ be a bijection between $\mathbb{N}$ and $\mathbb{N}^2$. Then $\cup_i K^{(i)}$ is the kernel of the Souslin scheme given by $A_{n_1,\ldots,n_k} = A_{\psi_2(n_1),n_2,\ldots,n_k}^{(\psi_1(n_1))}$. We next show that $\mathcal{A}(\mathcal{F})$ is closed under countable intersections. Let $\psi(n) = (\psi_1(n), \psi_2(n))$ be a bijection between $\mathbb{N}$ and $\mathbb{N}^2$ such that $\psi_2(n) < \psi_2(n')$ whenever $n < n'$ and $\psi_1(n) = \psi_1(n')$. For example, we can take

$$\psi\left(\sum_{j=1}^{n} j + k\right) = (k, n - k + 1) \qquad 1 \le k \le n.$$

Then, if $K^{(i)}$ is the kernel of the Souslin scheme $\{A^{(i)}\}$, $\cap_i K^{(i)}$ is the kernel of the Souslin scheme given by $A_{n_1,\ldots,n_k} = A^{(\psi_1(k))}_{<\psi_2(n_1),\ldots,\psi_2(n_k)\,|\,\psi_1(k)>}$, where $< \psi_2(n_1), \ldots, \psi_2(n_k)\,|\,\psi_1(k) >$ is the ordered collection of those $\psi_2(n_j)$, $j \le k$ such that $\psi_1(j) = \psi_1(k)$.

Lemma 14.3.2 *If $G^c \in \mathcal{A}(\mathcal{G})$ for every $G \in \mathcal{G}$, then $\sigma(\mathcal{G}) \subseteq \mathcal{A}(\mathcal{G})$, where $\sigma(\mathcal{G})$ is the σ-algebra generated by $\mathcal{G}$.*

Proof Let $\widetilde{\mathcal{A}}(\mathcal{G}) = \{K \in \mathcal{A}(\mathcal{G}) \,|\, K^c \in \mathcal{A}(\mathcal{G})\}$. Note that $\widetilde{\mathcal{A}}(\mathcal{G})$ is closed under countable unions and intersections. To see this, suppose that $K_n \in \widetilde{\mathcal{A}}(\mathcal{G})$ and $\cup K_n = K$. Then it follows from the fact that $\mathcal{A}(\mathcal{G})$ is closed under countable unions and intersections and $\widetilde{\mathcal{A}}(\mathcal{G}) \subseteq \mathcal{A}(\mathcal{G})$ that $K \in \mathcal{A}(\mathcal{G})$ and also that $K^c = \cap K_n^c \in \mathcal{A}(\mathcal{G})$. Thus $K \in \widetilde{\mathcal{A}}(\mathcal{G})$. A similar argument works for intersections. Obviously $\widetilde{\mathcal{A}}(\mathcal{G})$ is closed under complementation. Thus $\widetilde{\mathcal{A}}(\mathcal{G})$ is a σ-algebra. Since $\mathcal{G} \in \mathcal{A}(\mathcal{G})$, it follows by the hypothesis that $\mathcal{G} \subseteq \widetilde{\mathcal{A}}(\mathcal{G})$ and the lemma follows. □

Let $(\Omega, \mathcal{F}, P)$ be a probability space. For any $B \subseteq \Omega$ let

$$P^*(B) = \inf_{\substack{F \supseteq B \\ F \in \mathcal{F}}} P(F). \tag{14.17}$$

It is easy to check that $P^*(B) \le P^*(C)$ if $B \subseteq C$, and that $P^*(\cup_n B_n) \le \sum_n P^*(B_n)$ for any sequence $B_1, B_2, \ldots$ of sets in Ω. Thus P^* is an outer measure on Ω.

We observe that $P^*(F) = P(F)$ for all $F \in \mathcal{F}$. Furthermore, for any $B \subseteq \Omega$, we can find a $\widetilde{B} \in \mathcal{F}$ with $\widetilde{B} \supseteq B$ such that $P^*(B) = P(\widetilde{B})$. To see this, let $F_1, F_2, \ldots$ be a minimizing sequence in (14.17). By taking intersections it is easily seen that we can assume that the sequence $F_1, F_2, \ldots$ is decreasing, and then we can take $\widetilde{B} = \cap_n F_n$.

We also observe that

$$P^*(\cup_n B_n) = \lim_{n \to \infty} P^*(B_n) \tag{14.18}$$

for any increasing sequence $B_1, B_2, \ldots$ of sets in Ω. To see this, note that

$$P^*(\cup_n B_n) \le P(\cap_n \cup_{k \ge n} \widetilde{B}_k) \le \liminf_{n \to \infty} P(\widetilde{B}_n) = \lim_{n \to \infty} P^*(B_n). \tag{14.19}$$

Finally, we note that if $(\Omega, \mathcal{F}, P)$ is a complete probability space and if $P^*(B) = 0$ for some $B \subseteq \Omega$, then $B \in \mathcal{F}$. This follows since $\widetilde{B} \supseteq B$ and $P(\widetilde{B}) = 0$.

Theorem 14.3.3 (Lusin) *Let $(\Omega, \mathcal{F}, P)$ be a complete probability space. Then $\mathcal{A}(\mathcal{F}) = \mathcal{F}$.*

Proof Let K be the kernel of a Souslin scheme $\{A_{\cdot}\}$ on $\mathcal{F}$. Since P^* is an outer measure, $P^*(\widetilde{K}) \leq P^*(K) + P^*(\widetilde{K} - K)$. We show below that

$$P^*(\widetilde{K}) \geq P^*(K) + P^*(\widetilde{K} - K), \qquad (14.20)$$

which implies that $P^*(\widetilde{K} - K) = 0$. By the observation immediately preceding this theorem, we have $\widetilde{K} - K \in \mathcal{F}$. Since $K = \widetilde{K} \cap (\widetilde{K} - K)^c$, we also have $K \in \mathcal{F}$, which proves the theorem.

Thus it remains to obtain (14.20). To do this we first introduce some notation. For $(h_1, h_2, \ldots, h_m) \in \mathbb{N}^m$ set

$$K_{h_1, h_2, \ldots, h_m} = \bigcup_{n_1 \leq h_1, \cdots, n_m \leq h_m} \bigcap_{j=1}^{m} A_{n_1, n_2, \ldots, n_j}$$

and

$$K^{h_1, h_2, \ldots, h_m} = \bigcup_{\{n_i\}: n_1 \leq h_1, \cdots, n_m \leq h_m} \bigcap_{k=1}^{\infty} A_{n_1, n_2, \ldots, n_k},$$

where the sets $A_{\cdot, \ldots, \cdot}$ are members of the Souslin scheme $\{A_{\cdot}\}$.

Fix $\epsilon > 0$ and let $h \in \mathbb{N}$. Since $K^h \uparrow K$ and $K^{h_1, h_2, \ldots, h_m, h} \uparrow K^{h_1, h_2, \ldots, h_m}$ as $h \to \infty$, we can inductively define a sequence of integers $h_1, h_2, \ldots$ such that, for each m,

$$P^*(\widetilde{K} \cap K^{h_1, h_2, \ldots, h_m}) \geq P^*(K) - \epsilon. \qquad (14.21)$$

Then, since $K_{h_1, h_2, \ldots, h_m} \supset K^{h_1, h_2, \ldots, h_m}$ and $K_{h_1, h_2, \ldots, h_m} \in \mathcal{F}$, we have

$$\begin{aligned} P^*(\widetilde{K}) &= P^*(\widetilde{K} \cap K_{h_1, h_2, \ldots, h_m}) + P^*(\widetilde{K} \cap (K_{h_1, h_2, \ldots, h_m})^c) \quad (14.22) \\ &\geq P^*(K) + P^*(\widetilde{K} \cap (K_{h_1, h_2, \ldots, h_m})^c) - \epsilon. \end{aligned}$$

Note that $K_{h_1, h_2, \ldots, h_m}$ is decreasing in m. We show immediately below that the limit

$$\cap_{m=1}^{\infty} K_{h_1, h_2, \ldots, h_m} \subseteq K. \qquad (14.23)$$

Therefore, $(K_{h_1, h_2, \ldots, h_m})^c$ is increasing in m and the limit contains K^c. Since $\epsilon > 0$, we get (14.20).

To prove (14.23), let $x \in \cap_{m=1}^{\infty} K_{h_1, h_2, \ldots, h_m}$. We say that x is (r, k)–representable if $x \in A_r \cap_{m=2}^{k} A_{r, n_2, \ldots, n_m}$ for some $n_i \leq h_i$, $i = 2, \ldots, k$. Let

$$\mathcal{Q}_r = \{k \in \mathbb{N} \mid x \text{ is } (r, k)\text{–representable}\}.$$

Clearly, since $x \in \cap_{m=1}^{\infty} K_{h_1, h_2, \ldots, h_m}$, we have $\cup_{r \leq h_1} \mathcal{Q}_r = \mathbb{N}$, so that we can find some $l_1 \leq h_1$ such that $|\mathcal{Q}_{l_1}| = \infty$. Thus $\mathcal{Q}_{l_1} = \mathbb{N}$. We then say that x is (l_1, s, k)–representable if $x \in A_{l_1, s} \cap_{m=3}^{k} A_{l_1, s, n_3, \ldots, n_m}$ for some $n_i \leq h_i$; $i = 3, \ldots, k$ and set

$$\mathcal{Q}_{l_1, s} = \{k \in \mathbb{N} \mid x \text{ is } (l_1, s, k)\text{–representable}\}.$$

As before, $\mathcal{Q}_{l_1} = \mathbb{N}$ implies that $\cup_{s \leq h_2} \mathcal{Q}_{l_1, s} = \mathbb{N}$, so that we can find some $l_2 \leq h_2$ such that $|\mathcal{Q}_{l_1, l_2}| = \infty$, so that in fact $\mathcal{Q}_{l_1, l_2} = \mathbb{N}$.

Proceeding inductively, we obtain a sequence $l_i \leq h_i$, $i = 1, 2, \ldots$ with $x \in \cap_{m=1}^{\infty} A_{l_1, l_2, \ldots, l_m} \subseteq K$. This completes the proof of (14.23). $\quad\square$

The Projection Theorem follows from Lusin's Theorem and the following theorem:

Theorem 14.3.4 *Let $(\Omega, \mathcal{F}, P)$ be a complete probability space, and let π denote the projection from $R^1 \times \Omega$ to Ω. Let $\mathcal{B}$ denote the Borel σ-algebra of R^1 and let $\mathcal{B} \times \mathcal{F}$ denote the product σ-algebra. Then $\pi(\mathcal{B} \times \mathcal{F}) \subseteq \mathcal{A}(\mathcal{F})$.*

Proof Let $\mathcal{K}$ denote the collection consisting of all compact subsets of R^1 together with the empty set. Recall that, as above, $\mathcal{K} \subseteq \mathcal{A}(\mathcal{K})$ and $\mathcal{A}(\mathcal{K})$ is closed under countable unions. Let $B \in \mathcal{K}$. Then, since the complement of any compact set in R^1 is a countable union of compact sets, we have that $B^c \in \mathcal{A}(\mathcal{K})$. It follows from this that if $C \in \mathcal{K} \times \mathcal{F}$, then $C^c \in \mathcal{A}(\mathcal{K} \times \mathcal{F})$. Therefore, by Lemma 14.3.2, $\sigma(\mathcal{K} \times \mathcal{F}) \subseteq \mathcal{A}(\mathcal{K} \times \mathcal{F})$ and since $\mathcal{K}$ generates $\mathcal{B}$, we have

$$\mathcal{B} \times \mathcal{F} \subseteq \mathcal{A}(\mathcal{K} \times \mathcal{F}). \tag{14.24}$$

Let $H \in \mathcal{B} \times \mathcal{F}$. Then, by (14.24), H is the kernel of a Souslin scheme $K_{n_1, \ldots, n_k} \times F_{n_1, \ldots, n_k}$ over $\mathcal{K} \times \mathcal{F}$. $\pi(H)$ is then the kernel of a Souslin scheme $G_{n_1, \ldots, n_k}$ over $\mathcal{F}$, where $G_{n_1, \ldots, n_k} = F_{n_1, \ldots, n_k}$ if $\cap_{j=1}^{k} K_{n_1, \ldots, n_j} \neq \emptyset$, and $G_{n_1, \ldots, n_k} = \emptyset$ otherwise. Here we use the fact that if $\cap_{j=1}^{k} K_{n_1, \ldots, n_j} \neq \emptyset$ for all k, then $\cap_{j=1}^{\infty} K_{n_1, \ldots, n_j} \neq \emptyset$, which holds because the sets $K_{n_1, \ldots, n_j}$ are compact. $\quad\square$

14.4 Hille–Yosida Theorem

Let $(B, \|\cdot\|)$ be a Banach space. When dealing with operators from B to B, we also use $\|\cdot\|$ to denote the operator norm. The material in this section is used in Section 4.1, in which one can find the definition of a strongly continuous contraction semigroup and contraction resolvent. If $\{R_\lambda, \lambda > 0\}$ is a contraction resolvent and in addition

$$\lim_{\lambda \to \infty} \lambda R_\lambda f = f \qquad \forall f \in B, \tag{14.25}$$

$\{R_\lambda, \lambda > 0\}$ is a called a strongly continuous contraction resolvent.

Theorem 14.4.1 (Hille–Yosida Theorem) *Let $\{R_\lambda; \lambda > 0\}$ be a strongly continuous contraction resolvent on B. Then we can find a strongly continuous contraction semigroup $\{P_t; t \geq 0\}$ on B with*

$$R_\lambda = \int_0^\infty e^{-\lambda t} P_t \, dt. \tag{14.26}$$

Here the integral is the Bochner integral.

Proof Let us define

$$P_{t,\lambda} = e^{-\lambda t} \sum_{n=0}^\infty (\lambda t)^n (\lambda R_\lambda)^n / n!, \tag{14.27}$$

where $(\lambda R_\lambda)^0 = I$. (We can express this as $P_{t,\lambda} = \exp\left(-\lambda t(1 - \lambda R_\lambda)\right)$. Using this representation, it is easy to verify, at least heuristically, the many relationships given below.)

Using the fact that $\|\lambda R_\lambda\| \leq 1$, it is easy to check that

$$\|P_{t,\lambda}\| \leq 1. \tag{14.28}$$

Furthermore, using the identity $\exp(a + b) = \sum_{n,m=0}^\infty a^n b^m / n! m!$, one can show that $\{P_{t,\lambda}; t \geq 0\}$ is a semigroup and that

$$\lim_{t \to 0} \frac{P_{t,\lambda} - I}{t} = \lambda(\lambda R_\lambda - I). \tag{14.29}$$

It follows from the resolvent equation that R_λ, R_μ, and, consequently, $P_{t,\lambda}$ and $P_{t,\mu}$ commute. Therefore, for any $n > 1$,

$$P_{t,\lambda} - P_{t,\mu} = \sum_{k=1}^n P_{\frac{(k-1)t}{n},\lambda} P_{\frac{(n-k)t}{n},\mu} \left\{ P_{\frac{t}{n},\lambda} - P_{\frac{t}{n},\mu} \right\} \tag{14.30}$$

(this is merely a telescoping sum). Using (14.28), we now see that

$$\|P_{t,\lambda} f - P_{t,\mu} f\| \leq n \| \left\{ P_{\frac{t}{n},\lambda} f - f \right\} - \left\{ P_{\frac{t}{n},\mu} f - f \right\} \|, \tag{14.31}$$

so that, by (14.29), we have

$$\|P_{t,\lambda}f - P_{t,\mu}f\| \le t\|\lambda(\lambda R_\lambda - I)f - \mu(\mu R_\mu - I)f\|. \tag{14.32}$$

Let $\mathcal{R} :=$ the range of R_λ. It follows from the resolvent equation that $\mathcal{R}$ is the same for all $\lambda > 0$. Therefore, using the fact that $\lim_{\lambda\to\infty} \lambda R_\lambda f = f$, we see that $\mathcal{R}$ is dense in B. Furthermore, R_λ is one–one, since if $R_\lambda h = 0$, then from the resolvent equation we see that $R_\mu h = 0$ for all μ, which implies that $h = 0$ since $\lim_{\mu\to\infty} \mu R_\mu h = h$. Hence we can define R_λ^{-1} on $\mathcal{R}$ and check by the resolvent equation that

$$G := \lambda - R_\lambda^{-1} = \mu - R_\mu^{-1} \quad \forall \lambda, \mu > 0. \tag{14.33}$$

Note that if $f \in \mathcal{R}$, we can write

$$\lambda(\lambda R_\lambda - I)f = \lambda R_\lambda(\lambda - R_\lambda^{-1})f = \lambda R_\lambda Gf. \tag{14.34}$$

Thus for $f \in \mathcal{R}$ we see that $\lim_{\lambda\to\infty} \lambda(\lambda R_\lambda - I)f = Gf$, so that by (14.32) we see that

$$P_t f := \lim_{\lambda\to\infty} P_{t,\lambda}f \tag{14.35}$$

converges locally uniformly in t. Since (14.32) implies that $P_{t,\lambda}f$ is continuous, we have that $P_t f$ is continuous in t for each $f \in \mathcal{R}$. However, using (14.28) again together with the density of $\mathcal{R}$, we see that (14.35) holds for all $f \in B$ and that $\|P_t\| \le 1$ and $P_t f$ is continuous in t. In other words, $\{P_t; t \ge 0\}$ is a strongly continuous contraction semigroup.

It remains to show (14.26). To see this, we note that by a direct calculation using (14.27),

$$\int_0^\infty e^{-\lambda t} P_{t,\mu}\, dt = ((\lambda + \mu)I - \mu^2 R_\mu)^{-1}. \tag{14.36}$$

Let $\alpha = \dfrac{\mu\lambda}{\lambda + \mu}$. Using the resolvent equation for R_μ and R_α, it is easy to show that

$$((\lambda + \mu)I - \mu^2 R_\mu)^{-1} = \frac{\mu^2}{(\lambda + \mu)^2} R_\alpha + \frac{1}{(\lambda + \mu)}I. \tag{14.37}$$

It follows from the resolvent equation that

$$R_\alpha\left(I - \frac{\lambda^2}{\lambda + \mu} R_\lambda\right) = R_\lambda. \tag{14.38}$$

Taking the limit of the left-hand side as $u \to \infty$, we see that $R_\alpha = R_\lambda$. Therefore, taking the limit on the right-hand side of (14.37), we obtain (14.26). $\qquad\square$

The next theorem shows that the hypotheses of Theorem 4.1.3 are sufficient for the Hille–Yosida Theorem.

Theorem 14.4.2 *Let S be a locally compact space with a countable base and let $C_0(S)$ denote the Banach space of continuous functions on S that vanish at infinity, with the uniform norm. Let $\{R_\lambda; \lambda > 0\}$ be a contraction resolvent on $C_0(S)$ that, in addition, satisfies*

$$\lim_{\lambda \to \infty} \lambda R_\lambda f(x) = f(x) \qquad \forall f \in C_0(S) \text{ and } x \in S. \tag{14.39}$$

Then, $\{R_\lambda; \lambda > 0\}$ is a strongly continuous contraction resolvent on $C_0(S)$.

Proof We show that the pointwise convergence of (14.39) implies the strong convergence

$$\lim_{\lambda \to \infty} \lambda R_\lambda f = f \qquad \forall f \in C_0(S). \tag{14.40}$$

As in the previous proof, the resolvent equation implies that $\mathcal{R}$ the range of R_μ is the same for all $\mu > 0$. Furthermore, rearranging the resolvent equation, we see that for any $g \in C_0(S)$,

$$(\lambda R_\lambda - I)R_\mu g = \frac{\mu}{\lambda - \mu} R_\mu g - \frac{\lambda}{\lambda - \mu} R_\lambda g. \tag{14.41}$$

In operator terminology, (4.9) states that $\|\lambda R_\lambda\| \leq 1$. Therefore, it follows from (14.41) that, for any $f \in \mathcal{R}$, $\lim_{\lambda \to \infty} \lambda R_\lambda f = f$. Using the property that $\|\lambda R_\lambda\| \leq 1$ once again, we see that $\lim_{\lambda \to \infty} \lambda R_\lambda f = f$ for any $f \in \overline{\mathcal{R}}$, the closure of $\mathcal{R}$ in $C_0(S)$.

It only remains to show that $\overline{\mathcal{R}} = C_0(S)$. Suppose it does not. Then, by the Hahn-Banach Theorem, there is a nontrivial continuous linear functional h on $C_0(S)$ that vanishes on $\overline{\mathcal{R}}$ and a fortiori on $\mathcal{R}$. Representing h by a finite signed measure ν on S, we therefore have $\int R_\lambda f(x) \, d\nu(x) = 0$ for all $\lambda > 0$ and $f \in C_0(S)$. Then, by (14.39), the fact that $\|\lambda R_\lambda\| \leq 1$, and the Dominated Convergence Theorem, we have that for all $f \in C_0(S)$,

$$h(f) = \int f(x) \, d\nu(x) = \lim_{\lambda \to \infty} \int \lambda R_\lambda f(x) \, d\nu(x) = 0. \tag{14.42}$$

This contradicts the fact that h is not trivial. Thus $\overline{\mathcal{R}} = C_0(S)$. $\qquad\square$

14.5 Stone–Weierstrass Theorems

A real vector space $\mathcal{H} \subseteq C(S)$ is called a vector lattice if it is closed under $\vee$ and $\wedge$. $\mathcal{H}$ is said to separate the points of S if, for any $x, y \in S$, we can find an $f \in \mathcal{H}$ with $f(x) \neq f(y)$.

Theorem 14.5.1 (Stone–Weierstrass) *Let S be a compact space and let $\mathcal{H} \subseteq C(S)$ be a vector lattice, containing the constants, which separates the points of S. Then $\mathcal{H}$ is dense in $C(S)$ in the uniform topology.*

This is a classical theorem; see, for example, Royden (1988, Chapter 9, Proposition 30).

Theorem 14.5.2 *Let S be a locally compact space and let $\mathcal{H} \subseteq C(S)$ be a vector lattice, containing the constants, such that $\mathcal{H} \cap C_0(S)$ separates the points of S. Then $\mathcal{H} \cap C_0(S)$ is dense in $C_0(S)$ in the uniform topology.*

Proof In order to apply the Stone–Weierstrass Theorem (Theorem 14.5.1), which requires a compact space, we consider S_Δ, the one-point compactification of S, where Δ is the point at infinity. Let $C(S, \Delta)$ denote the continuous functions on S that have a continuous extension to S_Δ. It is clear that $C(S, \Delta)$ includes $C_0(S)$ and the functions that are constant on S. Under the hypothesis that $\mathcal{H} \cap C_0(S)$ separates the points of S, we see that the continuous extension $\mathcal{L}$ of $\mathcal{H} \cap C(S, \Delta)$ to S_Δ separates the points on S_Δ. Since $\mathcal{L}$ is a vector lattice that also contains the constants, it follows from Theorem 14.5.1 that $\mathcal{L}$ is dense in $(C(S_\Delta), \|\cdot\|_\infty)$. Let $f \in C_0(S)$, which we consider as a function in $C(S_\Delta)$ by setting $f(\Delta) = 0$. Fix $\epsilon > 0$. By what we have just shown, we can find $g \in \mathcal{L}$ such that

$$\sup_{x \in S_\Delta} |g(x) - f(x)| \leq \epsilon. \tag{14.43}$$

Since $f(\Delta) = 0$, this shows that $|g(\Delta)| \leq \epsilon$. Then, by (14.43) we have

$$\sup_{x \in S_\Delta} |(g(x) - g(\Delta)) - f(x)| \leq 2\epsilon. \tag{14.44}$$

Let $\widetilde{g}$ denote the restrictions of g to S. We have $\widetilde{g} - g(\Delta) \in C_0(S)$. Also, since both $\widetilde{g}$ and the constant function $g(\Delta)$ are in $\mathcal{H}$, we have $\widetilde{g} - g(\Delta) \in \mathcal{H} \cap C_0(S)$. Since ϵ can be arbitrarily close to 0, we see from (14.44) that $\mathcal{H} \cap C_0(S)$ is dense in $C_0(S)$ in the uniform norm. $\quad\square$

14.6 Sums of independent symmetric random variables

Let B be a real separable Banach space, $\{X_n, n \geq 1\}$ a sequence of independent symmetric B-valued random variables, and $S_n = \sum_{j=1}^n X_j$. The next lemma is a generalization of a well-known inequality of Lévy for real-valued random variables. Its proof is essentially the same as in the real valued case (see, e.g., Ledoux and Talagrand (1991, Proposition 2.3)).

Lemma 14.6.1 (Lévy's inequality) *Let N be a measurable seminorm on B and suppose that S_n converges almost surely to a limit S. Then*

$$P\left(\sup_{n \geq 1} N(S_n) > \lambda\right) \leq 2P(N(S) > \lambda). \tag{14.45}$$

We now show that, for sums of independent symmetric random variables, many different types of convergence are equivalent.

Theorem 14.6.2 (Lévy–Ito–Nisio Theorem) *The following statements are equivalent:*

(1) $\{S_n\}$ converges almost surely to a B-valued random variable S.

(2) $\{S_n\}$ converges in probability to a B-valued random variable S.

(3) Let μ_n denote the probability distribution of S_n. Then there exists a probability measure μ on B such that $\mu_n \xrightarrow{dist} \mu$.

(4) For $f \in B^*$,

$$\int e^{i(f,x)} \, d\mu_n \to \int e^{i(f,x)} \, d\mu \tag{14.46}$$

where (f, x) denotes the evaluation of f at x.

Proof By the Banach–Mazur Theorem (see Wojtaszczyk (1991, Chapter II, B, Theorem 4)), B is isometrically isomorphic to a closed subspace of $C[0, 1]$, the space of continuous functions on $[0, 1]$ with the sup norm, that is, for $f \in C[0, 1]$, $\|f\| = \sup_{0 \leq t \leq 1} |f(t)|$. We represent B in this way.

The proof that (1) implies (2) is exactly the same as for real-valued random variables.

To show that (2) implies (3), we must show that the finite-dimensional distributions of $\{\mu_n\}$ converge to the corresponding finite-dimensional distributions of μ and that $\{\mu_n\}$ is a tight family on $C[0, 1]$. Tightness is equivalent to the following: Given $\epsilon > 0$, there exist $a > 0$ and $\delta > 0$ such that

$$\sup_n \mu_n(\{x : |x(0)| > a\}) < \epsilon \qquad (14.47)$$

and

$$\sup_n \mu_n(\{x : \|x\|_\delta > \epsilon\}) < \epsilon, \qquad (14.48)$$

where $\|x\|_\delta := \sup_{|s-t| \leq \delta;\, s,t \in [0,1]} |x(s) - x(t)|$.

Suppose that (2) holds and let μ be the probability distribution of S. Given $t_1, \ldots, t_j$ in $[0,1]$, the vectors $S_n(t_1), \ldots, S_n(t_j)$ converge in probability to $S(t_1), \ldots, S(t_j)$. Thus the finite-dimensional distributions converge and, clearly, (14.47) holds. Also,

$$\begin{aligned}
\mu_n(\{x : \|x\|_\delta > \epsilon\}) &= P(\|S_n\|_\delta > \epsilon) & (14.49)\\
&\leq P(\|S_n - S\|_\delta > \epsilon//2) + P(\|S\|_\delta > \epsilon/2)\\
&\leq P(\|S_n - S\| > \epsilon/4) + P(\|S\|_\delta > \epsilon/2)
\end{aligned}$$

since $\|x\|_\delta \leq 2\|x\|$. Since S_n converges to S in probability, given $\epsilon > 0$ we can find an n_0 such that for $n \geq n_0$, the first probability in the last line of (14.49) is less than $\epsilon/2$. Furthermore, since continuous functions on compact sets are uniformly continuous, we can take δ sufficiently small so that

$$\sup_{1 \leq j < n_0} \mu_j(\{x : \|x\|_\delta > \epsilon\}) < \epsilon/2 \qquad \text{and} \qquad P(\|S\|_\delta > \epsilon/2) < \epsilon/2.$$
$$(14.50)$$

Therefore (2) implies (3). Statement (3) implies (4) by the definition of weak convergence of measures.

We now show that (4) implies (3) implies (2) implies (1). Suppose that (4) holds. By considering characteristic functions we see, in particular, that given $t \in [0,1]$, the real-valued random variables $S_n(t)$ converge in distribution. Therefore, by Lévy's Theorem, in the real-valued case

$$S_n(t) \to S(t) \qquad \text{a.s.,} \qquad (14.51)$$

where $S(t)$ denotes the limit random variable (here we use the fact that we are considering sums of independent random variables; see, e.g., Chung (1974, Theorem 9.5.5)). Consequently, for $t_1, \ldots, t_j$ in $[0,1]$ and A a measurable set in $\mathcal{B}(R_j)$,

$$P((S(t_1), \ldots, S(t_j)) \in A) = \mu(\{x : (x(t_1), \ldots, x(t_j)) \in A\}). \quad (14.52)$$

μ is a probability measure on $C[0,1]$. By (14.52) and Billingsley (1968, Theorem 9.2), there is a separable version of the process $\{S(t), t \in [0,1]\}$

that is continuous almost surely and for which (14.51) and (14.52) continue to hold. Therefore we can consider S as a $C[0,1]$-valued random variable. Furthermore, since the finite-dimensional distributions of S_n and $S - S_n$ are independent and symmetric, S_n and $S - S_n$ are independent, symmetric $C[0,1]$-valued random variables. Applying Lemma 14.6.1 to the series with two terms, S_n and $S - S_n$, we see that

$$P(\|S_n\|_\delta > \epsilon) \le 2P(\|S\|_\delta > \epsilon). \tag{14.53}$$

This shows that $\{\mu_n\}$ is tight; (3) follows.

Assume (3). This implies that $\{S_n\}$ is a tight family and, as we have already shown, that (4) holds, where S is a random variable in $C[0,1]$. Therefore, $\{S_n - S\}$ is also a tight family. Let $t_1, \dots, t_j \in [0,1]$ be such that each $t \in [0,1]$ satisfies $|t - t_k| < \delta$ for some $1 \le k \le j$. By (4), given $\epsilon > 0$, we can find an n so that

$$P\left(\max_{1 \le k \le j} |S_n(t_k) - S(t_k)| > \epsilon/2 \right) < \epsilon/2. \tag{14.54}$$

Also, since $\{S_n - S\}$ is a tight family,

$$\sup_{n \ge 1} P\left(\|S_n - S\|_\delta > \epsilon/2\right) < \epsilon/2. \tag{14.55}$$

Noting that

$$\|S_n - S\| \le \|S_n - S\|_\delta + \max_{1 \le k \le j} |S_n(t_k) - S(t_k)| \tag{14.56}$$

and using (14.54) and (14.55), we get (2).

By Lemma 14.6.1,

$$P\left(\max_{1 \le p \le r} \|X_m + \cdots + X_{m+p}\| > \epsilon \right) \le 2P(\|X_m + \cdots + X_{m+r}\| > \epsilon). \tag{14.57}$$

The same argument that was used for real-valued random variables shows that (2) implies (1). □

Remark 14.6.3 Statements (1), (2), and (3) in Theorem 14.6.2 hold without the assumption of symmetry. However, it is shown in Ito and Nisio (1968a) (see also Jain and Marcus (1978, II, Remark 3.5)) that (4) does not necessarily imply (3) without the assumption of symmetry.

We have the following important corollary of Theorem 14.6.2.

Corollary 14.6.4 *Let* $X = \{X(t), t \in T\}$ *be a process of class* $\mathcal{S}$*. Assume further that* X *has a version with continuous sample paths. Then* *the series in (5.94) converges uniformly on T almost surely (and thus is a concrete version of X).*

Proof Since X has a version with continuous paths, there is a measure μ on $C(T)$, the space of continuous functions on T, that has the same finite distributions as X. Also, $X_j(t) := \phi_j(t)\xi_j$ are independent symmetric $C(T)$-valued random variables. For fixed $t \in T$, by hypothesis, $S_n(t)$ converges to $S(t)$ almost surely. This is all we used to show that (4) implies (1) in the proof of Theorem 14.6.2. $\square$

14.7 Regularly varying functions

Regularly varying functions are introduced in Section 7.2. In results involving these functions we use well-known integrability theorems. We state them here without proof. Essentially these results say that regularly varying functions of index p integrate like pure p-th powers. The only exception is the very interesting case of regularly varying functions of index minus one.

For details and proofs we refer the reader to Bingham, Goldie and Teugels (1987), Feller (1971) and Pitman (1968) and to the seminal paper, Karamata (1930).

Theorem 14.7.1 *If L is slowly varying at infinity and is locally bounded on $[M, \infty)$ for some $0 < M < \infty$, then for all $p < 1$,*

$$\int_M^x \frac{L(u)}{u^p}\, du \sim \frac{x^{1-p}L(x)}{1-p} \qquad as\ x \to \infty. \tag{14.58}$$

If L is slowly varying at infinity and $p > 1$, then $\int_M^\infty (L(u)/u^p)\, du < \infty$ for some $0 < M < \infty$ and

$$\int_x^\infty \frac{L(u)}{u^p}\, du \sim \frac{L(x)}{(p-1)x^{p-1}} \qquad as\ x \to \infty. \tag{14.59}$$

If $L(u) \equiv 1$ in (14.58) or (14.59), the two sides of these relationships are obvious. Therefore, one can think of these relationships as showing that the slowly varying functions vary so slowly with respect to powers that they do not affect the rate of growth of the integrals of powers, except for regularly varying functions of index minus one, which is not included in the above theorem. We treat this case next.

Theorem 14.7.2 *If L is slowly varying at infinity and is in $L^1([a, b])$ for*

*all intervals $[a, b] \subset [M, \infty)$ for some $0 < M < \infty$, then $\int_M^x (L(u)/u) \, du$
is slowly varying and*

$$\frac{1}{L(x)} \int_M^x \frac{L(u)}{u} \, du \to \infty \qquad \text{as } x \to \infty. \qquad (14.60)$$

*If L is slowly varying at infinity and $\int_M^\infty (L(u)/u) \, du < \infty$ for some
$0 < M < \infty$, then $\int_x^\infty (L(u)/u) \, du$ is slowly varying and*

$$\frac{1}{L(x)} \int_x^\infty \frac{L(u)}{u} \, du \to \infty \qquad \text{as } x \to \infty. \qquad (14.61)$$

If $L(x)$ is slowly varying at zero, $\widetilde{L}(1/x) = L(x)$ is slowly varying at
infinity. Substituting $\widetilde{L}(1/x)$ into the above results gives comparable
results as $x \to 0$. We give them for the convenience of the reader.

Theorem 14.7.3 *If L is slowly varying at zero and is bounded on $[a, \delta]$,
for some $\delta > 0$ and all $0 < a < \delta$, then for all $p > 1$,*

$$\int_x^\delta \frac{L(u)}{u^p} \, du \sim \frac{L(x)}{(p-1)x^{p-1}} \qquad \text{as } x \to 0. \qquad (14.62)$$

*If L is slowly varying at zero and $p < 1$, then $\int_0^\delta (L(u)/u^p) \, du < \infty$
for some $\delta > 0$ and*

$$\int_0^x \frac{L(u)}{u^p} \, du \sim \frac{x^{1-p}L(x)}{1-p} \qquad \text{as } x \to 0. \qquad (14.63)$$

Theorem 14.7.4 *If L is slowly varying at zero and is bounded on $[a, \delta]$
for some $\delta > 0$ and all $0 < a < \delta$, then $\int_x^\delta (L(u)/u) \, du$ is slowly varying
and*

$$\frac{1}{L(x)} \int_x^\delta \frac{L(u)}{u} \, du \to \infty \qquad \text{as } x \to 0. \qquad (14.64)$$

*If L is slowly varying at zero and $\int_0^\delta (L(u)/u) \, du < \infty$ for some $\delta > 0$,
then $\int_0^x (L(u)/u) \, du$ is slowly varying and*

$$\frac{1}{L(x)} \int_0^x \frac{L(u)}{u} \, du \to \infty \qquad \text{as } x \to 0. \qquad (14.65)$$

The next result is the Monotone Density Theorem.

Theorem 14.7.5 *Let*

$$F(x) = \int_0^x f(x) \, dx \qquad x \in [0, M] \qquad (14.66)$$

for $M > 0$, where f is integrable on $[0, M]$. If $F(x) = x^p L(x)$ as $x \downarrow 0$, *where $p \geq 0$ and L is slowly varying at 0, and if f is monotone on $[0, \delta]$ for some $\delta > 0$, then*

$$f(x) \sim px^{p-1}L(x). \tag{14.67}$$

It follows from Theorem 14.7.5 that functions that are concave on $[0, \delta]$ for some $\delta > 0$ and are regularly varying at 0 are in fact normalized regularly varying functions; (see (7.123)).

Let U be nondecreasing on R_+ with $U(0) = 0$. Let

$$\widehat{U}(s) := \int_0^\infty e^{-sx}\, dU(x) \tag{14.68}$$

for $s > \sigma$, for some $\sigma \geq 0$ for which the integral is finite.

The following is from Feller (1971, Theorem 3, Chapter XIII, Section 5).

Theorem 14.7.6 *Assume that $\widehat{U}(s) < \infty$ for all s sufficiently large and let L be a slowly varying function at infinity. For all $c \geq 0$ and $p \geq 0$, the following are equivalent:*

$$U(x) \quad \sim \quad cx^p L(1/x)/\Gamma(1+p) \qquad as \quad x \to 0+, \tag{14.69}$$

$$\widehat{U}(s) \quad \sim \quad cs^{-p}L(s) \qquad as \quad s \to \infty. \tag{14.70}$$

When $c = 0$, (14.69) is interpreted as $U(x) = o(x^p L(1/x))$ and similarly for (14.70).

14.8 Some useful inequalities

Lemma 14.8.1 *Let $f(x)$ be a positive decreasing convex function on $[a, \infty]$ that is differentiable at a. Then*

$$\int_a^\infty f(x)\, dx \geq -\frac{f^2(a)}{2f'(a)}. \tag{14.71}$$

Proof The tangent line to $f(x)$ at $x = a$ is given by

$$y = f(a) + f'(a)(x - a) \tag{14.72}$$

(since it has slope $f'(a)$ and contains $(a, f(a))$. $f(x) \geq f(a) + f'(a)(x-a)$ by convexity. The line (14.72) intercepts the x-axis at $b = (af'(a) - f(a))/f'(a)$. The right-hand side of (14.71) is $\int_a^b (f(a) + f'(a)(x - a))dx$. (To evaluate this integral easily note that it is the area of a right triangle, with the sides that meet at the right angle having lengths $f(a)$ and $b - a = -f(a)/f'(a)$.) $\qquad\square$

Lemma 14.8.2 (Paley–Zygmund Lemma) *For a positive random variable X and $0 < \lambda < 1$,*

$$P\left(X \geq \lambda EX\right) \geq (1-\lambda)^2 \frac{(EX)^2}{EX^2}. \tag{14.73}$$

Proof This follows immediately from the following two elementary inequalities:

$$EXI_{[X \geq \lambda EX]} \leq \left(EX^2 P(X \geq \lambda EX)\right)^{1/2} \tag{14.74}$$

and

$$EXI_{[X \geq \lambda EX]} = EX - EXI_{[X < \lambda EX]} \geq (1-\lambda)EX. \tag{14.75}$$

$\square$

Lemma 14.8.3 (Boas's Lemma) *Let $a_j \geq 0$, $j \geq 1, \ldots$ and $g_n^2 = \sum_{j=n}^{\infty} a_j^2$. Then*

$$\sum_{n=1}^{\infty} a_n \leq 2 \sum_{n=1}^{\infty} \frac{g_n}{\sqrt{n}}. \tag{14.76}$$

If, in addition, the $\{a_j\}$ are nonincreasing,

$$\sum_{n=1}^{\infty} \frac{g_n}{\sqrt{n}} \leq 2 \sum_{n=1}^{\infty} a_n. \tag{14.77}$$

Proof Inequality (14.76) is a simple consequence of the Schwarz inequality,

$$\sum_{n=1}^{\infty} a_n = \sum_{n=1}^{\infty} \sum_{k=n}^{\infty} \frac{a_k}{k} \leq \sum_{n=1}^{\infty} g_n \left(\sum_{k=n}^{\infty} \frac{1}{k^2}\right)^{1/2} \leq 2 \sum_{n=1}^{\infty} \frac{g_n}{\sqrt{n}}. \tag{14.78}$$

To obtain (14.77) we note that for all $m \geq n$,

$$a_m^2 = \left(\left(\sum_{k=n}^{m} a_k^2\right)^{1/2} - \left(\sum_{k=n}^{m-1} a_k^2\right)^{1/2}\right)\left(\left(\sum_{k=n}^{m} a_k^2\right)^{1/2} + \left(\sum_{k=n}^{m-1} a_k^2\right)^{1/2}\right)$$

$$\geq 2\left(\left(\sum_{k=n}^{m} a_k^2\right)^{1/2} - \left(\sum_{k=n}^{m-1} a_k^2\right)^{1/2}\right)a_m(m-n)^{1/2}. \tag{14.79}$$

Consequently,

$$\sum_{m=n+1}^{\infty} \frac{a_m}{2(m-n)^{1/2}} \geq \sum_{m=n+1}^{\infty} \left(\left(\sum_{k=n}^{m} a_k^2\right)^{1/2} - \left(\sum_{k=n}^{m-1} a_k^2\right)^{1/2}\right)$$

$$= g_n - a_n \tag{14.80}$$

or, equivalently,

$$g_n \leq a_n + \frac{1}{2} \sum_{m=n+1}^{\infty} \frac{a_m}{(m-n)^{1/2}}. \tag{14.81}$$

Multiplying this by $n^{-1/2}$ and summing on n gives

$$\sum_{n=1}^{\infty} \frac{g_n}{\sqrt{n}} \leq a_1 + \sum_{m=1}^{\infty} a_{m+1} \left(\frac{1}{(m+1)^{1/2}} + \frac{1}{2} \sum_{n=1}^{m} \frac{1}{(n(m+1-n))^{1/2}} \right), \tag{14.82}$$

where we also use a change of the order of summation. Inequality (14.77) follows by noting that

$$\sup_{m \geq 1} \left(\frac{1}{(m+1)^{1/2}} + \frac{1}{2} \sum_{n=1}^{m} \frac{1}{(n(m+1-n))^{1/2}} \right) \leq 2. \tag{14.83}$$

$\square$

14.9 Some linear algebra

We collect some results about symmetric matrices that are used in this book.

Lemma 14.9.1 *A symmetric matrix is strictly positive definite if and only if all its principle minors are strictly positive.*

Proof Let A be an $n \times n$ symmetric matrix with all of its principle minors strictly positive. We show by induction on n that A is strictly positive definite. This is clear if $n = 1$, so assume that $n \geq 2$ and we have established this result for all symmetric $(n-1) \times (n-1)$ matrices. Let $u_1, \ldots, u_n$ be eigenvectors of A that are an orthonormal basis for R^n. Let $\rho_1, \ldots, \rho_n$ denote their corresponding eigenvalues. It is clear that none of the eigenvalues are zero because $\det A = \prod_{j=1}^{n} \rho_j > 0$. Note that

$$\left(\left(\sum_{j=1}^{n} b_j u_j \right), A \left(\sum_{j=1}^{n} b_j u_j \right) \right) = \sum_{j=1}^{n} \rho_j b_j^2. \tag{14.84}$$

Thus, to show that A is strictly positive definite, we need to show that all the eigenvalues $\rho_1, \ldots, \rho_n$ are strictly greater than zero. Since $\det A > 0$, we see that it is not possible for A to have exactly one strictly negative eigenvalue. Assume that A has at least two strictly negative eigenvalues.

Relabeling the eigenvalues, if necessary, we can assume that ρ_1 and ρ_2 are strictly less than zero. Then, by (14.84), for any real number r,

$$((u_1 + ru_2), A(u_1 + ru_2)) = \rho_1 + r^2 u_2^2 \rho_2 < 0. \tag{14.85}$$

Choose r so that the n-th component of $v = u_1 + ru_2$ is equal to zero. Consider v as an element of R^{n-1}. Let A_{n-1} denote the $(n-1) \times (n-1)$ matrix $\{A\}_{1 \le i,j \le n-1}$. Then (14.85) states that $(v, A_{n-1}v) < 0$, which contradicts the induction hypothesis. This shows that all the eigenvalues of A are strictly greater than zero, and hence A is strictly positive definite.

For the other direction, let A_p denote the $p \times p$ matrix $\{A_{i,j}\}_{1 \le i,j \le p}$, $1 \le p \le n$. Since A is strictly positive definite, so is A_p, for all $p = 1, \ldots, n$. This implies that all the eigenvalues of A_p are strictly positive. Consequently, $\det A_p > 0$, for all $p = 1, \ldots, n$. □

Lemma 14.9.2 *Let $B = \{b_{i,j}\}_{1 \le i,j \le p}$ be a symmetric matrix with $B \ge 0$ and let $\rho(B)$ denote the largest eigenvalue of B. Then we can find an eigenvector $x = (x_1, \ldots, x_p)$ with eigenvalue $\rho(B)$ such that $x_i \ge 0$ for all $1 \le i \le p$, with strict inequality for at least one i.*

Proof Let $u_1, \ldots, u_p$ be eigenvectors of B that form an orthonormal basis for R^p. Let $\rho_1, \ldots, \rho_p$ denote their corresponding eigenvalues. Note that any $y \in R^p$, with $|y| = 1$, can be written as $y = \sum_{j=1} a_j u_j$ with $\sum_{j=1} a_j^2 = 1$. Thus

$$(By, y) = \sum_{j=1} \rho_j a_j^2 \le \rho(B). \tag{14.86}$$

Let $x = (x_1, \ldots, x_p)$, $|x| = 1$, be an eigenvector of B with eigenvalue $\rho(B)$. Then $(Bx, x) = \rho(B)$, that is,

$$(Bx, x) = \sum_{j,k=1}^{p} b_{i,j} x_i x_j = \rho(B). \tag{14.87}$$

Let $\bar{x} = (|x_1|, \ldots, |x_p|)$ and note that, trivially, $|\bar{x}|$ is also equal to 1. Since, by hypothesis, all the $b_{i,j} \ge 0$, (14.87) clearly implies that

$$(B\bar{x}, \bar{x}) = \sum_{j,k=1}^{p} b_{i,j} |x_i| |x_j| \ge \rho(B). \tag{14.88}$$

Therefore, by (14.86), $\bar{x}$ is an eigenvector of B with eigenvalue $\rho(B)$. □

A $p \times p$ matrix $A = \{a_{i,j}\}_{1 \leq i,j \leq p}$ is said to be reducible if there exists a permutation matrix P such that

$$PAP^t = \begin{pmatrix} B & 0 \\ C & D \end{pmatrix}, \tag{14.89}$$

where B and D are square matrices. (A $p \times p$ matrix P is a permutation matrix if exactly one entry in each row and column is equal to 1 and all the other entries are 0. Note that the effect of left multiplication by P in (14.89) is to permute rows of A, whereas the effect of right multiplication by P^t in (14.89) is to perform the same permutation on the columns of A.)

A matrix that is not reducible is called irreducible. If A is a symmetric, reducible matrix, one can apply a sequence of these interchanges to bring A to the form in which it can be expressed as a direct sum of irreducible matrices. When this is done PAP^t appears as a string of irreducible square matrices aligned along its diagonal, with 0 as the entries in the rows and columns outside these matrices.

There is a nice interpretation of irreducible matrices as graphs that is particularly simple for symmetric matrices. For a $p \times p$ matrix $A = \{a_{i,j}\}_{1 \leq i,j \leq p}$, consider a graph with p nodes. If $a_{j,p} \neq 0$, we say that one can go from node j to node p in the graph. If A is symmetric, then, clearly, one can also go from node p to node j. We can break the nodes up into disjoint subsets of nodes that communicate with each other but not with nodes in any other subset. Then we can reorder the indices of these nodes as

$$\{i_{1,1}, \ldots, i_{1,n_1}, i_{2,1} \ldots, i_{2,n_2}, \ldots, i_{k,1}, \ldots, i_{k,n_k}\}, \tag{14.90}$$

where $n_1 + \cdots + n_k = p$ and where the elements in each set $i_{j,1} \ldots, i_{j,n_j}$ communicate with each other. The permutation of $\{1, \ldots, p\}$ into the sequence in (14.90) is the same as the permutation that defines the permutation matrix P that makes PAP^t irreducible.

Lemma 14.9.3 *Let $A = \{a_{i,j}\}_{1 \leq i,j \leq p}$ be a symmetric irreducible matrix, $A \neq 0$. Then, for all $l, j \in \{1, \ldots, p\}$, there exists an r and different integers $k_1, \ldots, k_r$ in $\{1, \ldots, p\}$ such that $k_1 = l$, $k_r = j$ and*

$$a_{k_1,k_2}, a_{k_2,k_3}, \ldots, a_{k_{r-1},k_r} \tag{14.91}$$

are all nonzero.

Proof Consider the graph $\mathcal{G}$ with vertices $\{i \mid 1 \leq i \leq p\}$ and with a edge between i, j if and only if $a_{i,j} \neq 0$. The fact that A is irreducible means that $\mathcal{G}$ is connected, that is, that each element of $\mathcal{G}$ can be reached from any other element of $\mathcal{G}$. Choose some path from i to j and remove any loops it may contain. This gives the different integers $k_1, \ldots, k_r$ in the sequence of elements of A that connect i to j in (14.91). □

$A = \{a_{i,j}\}_{1 \leq i,j \leq p}$ is an M-matrix if $a_{i,j} \leq 0$, $i \neq j$, A is nonsingular, and $A^{-1} \geq 0$; see page 561.

Lemma 14.9.4 *Let $A = \{a_{i,j}\}_{1 \leq i,j \leq p}$ be a symmetric matrix with $a_{i,j} \leq 0$, $i \neq j$. Then the following are equivalent:*

(1) A is an M-matrix.

(2) For all $\lambda > 0$ sufficiently large, we can write $A = \lambda I - B$, where $B \geq 0$ and λ is greater than the absolute value of any eigenvalue of B.

(3) A is strictly positive definite.

Proof (1) $\Rightarrow$ (2) By choosing λ sufficiently large we can always write $A = \lambda I - B$, where $B \geq 0$ and is strictly positive definite. This is because the larger we take λ, the larger the diagonal elements of B must be. Since the off-diagonal elements of B remain unchanged, by taking λ sufficiently large, B has strictly positive principle minors. Thus it is strictly positive definite, and, in particular, all the eigenvalues of B are positive. We write

$$A = \lambda \left(I - \frac{1}{\lambda} B \right) := \lambda (I - T). \tag{14.92}$$

By the hypothesis, $(I - T)^{-1}$ exists and $(I - T)^{-1} \geq 0$. Let $\rho(T)$ denote the largest eigenvalue of T. Since T is symmetric and $T \geq 0$, it follows from Lemma 14.9.2 that there exists a vector $x = (x_1, \ldots, x_p)$ with $x_i \geq 0$ for all $1 \leq i \leq p$ and with strict equality for at least one i, such that $Tx = \rho(T)x$. The fact that $(I - T)^{-1}$ exists means that $\rho(T) \neq 1$. Thus

$$(I - T)x = (1 - \rho(T))x, \tag{14.93}$$

and since $(I - T)^{-1}$ exists,

$$(I - T)^{-1}x = \frac{1}{(1 - \rho(T))} x. \tag{14.94}$$

Since $(I - T)^{-1} \geq 0$, we see that $\rho(T) < 1$. Thus λ is greater than the absolute value of any eigenvalue of B.

(2) $\Rightarrow$ (3) Let x be an eigenvector of A with eigenvalue ξ. Then, by (14.92), $\lambda - \xi$ is an eigenvalue of B. Since B is symmetric, $\lambda - \xi$ is real.

Therefore, the hypothesis that $\lambda > |\lambda - \xi|$ implies that $\xi > 0$. Thus all the eigenvalues of A are strictly greater than zero, and therefore A is strictly positive definite.

$(3) \Rightarrow (1)$ Since A is strictly positive definite, writing A as in (14.92) shows that λ is greater than any of the eigenvalues of B. Note that

$$\left(I - \frac{1}{\lambda}B\right)\left(\sum_{j=0}^{k}\left(\frac{1}{\lambda}B\right)^{j}\right) = I - \left(\frac{1}{\lambda}B\right)^{k+1}. \tag{14.95}$$

Therefore, taking the limit as $k \to \infty$, we see that

$$A^{-1} = \frac{1}{\lambda}\sum_{k=0}^{\infty}\left(\frac{1}{\lambda}B\right)^{k}, \tag{14.96}$$

where the sum converges in the operator norm. Since $B \geq 0$, we see that $A^{-1} \geq 0$. Thus A is an M-matrix. $\qquad\square$

References

The numbers following each entry refer to the pages on which the entry is cited in this book.

Adler, R. J. (1991). *An Introduction to Continuity, Extrema, and Related Topics for General Gaussian processes*. Hayward, CA: Institute of Mathematical Statistics. [241]

Adler, R. J. and Pyke, R. (1993). Uniform quadratic variation for Gaussian processes. *Stochastic Process. Appl.*, *48*, 191–210. [495]

Ahlfors, L. (1966). *Complex Analysis* (second ed.). New York: McGraw Hill. [142]

Bakry, D. and Ledoux, M. (1996). Lévy–Gromov's isoperimetric inequality for an infinite dimensional diffusion generator. *Invent. Math.*, *123*, 259–281. [242]

Bañuelos, R. and Smits, R. G. (1997). Brownian motion in cones. *Prob. Theory Related Fields*, *108*, 299–319. [527]

Bapat, R. (1989). Infinite divisibility of multivariate gamma distributions and *M*–matrices. *Sankhya*, *51*, 73–78. [6, 579]

Barlow, M. (1985). Continuity of local times for Lévy processes. *Z. Wahrscheinlichkeitstheorie und Verw. Gebiete*, *69*, 23–35. [2, 120]

Barlow, M. (1988). Necessary and sufficient conditions for the continuity of local time of Lévy processes. *Ann. Probab.*, *16*, 1389–1427. [2, 120, 455]

Barlow, M. and Hawkes, J. (1985). Applications de l'entropie métrique à la continuité des temps locaux des processus de Lévy. *C. R. Acad. Sc. Paris*, *301*, 237–239. [2, 120]

Bass, R. F., Eisenbaum, N., and Shi, Z. (2000). The most visited sites of symmetric stable processes. *Prob. Theory Related Fields*, *116*, 391–404. [5, 527]

Bass, R. F. and Griffin, P. (1985). The most visited site of Brownian motion and simple random walk. *Z. Wahrscheinlichkeitstheorie und Verw. Gebiete*, *70*, 417–436. [60, 526]

Belyaev, Y. K. (1961). Continuity and Hölder's conditions for sample functions of stationary Gaussian processes. In *Fourth Berkeley Sympos. Math. Statist. and Prob.*, volume 2 (pp. 23–33). Berkeley: University of California Press. [241]

Berman, A. and Plemmons, R. J. (1994). *Nonnegative Matrices in the Mathematical Sciences*. Philadelphia: Classics in Applied Mathematics, SIAM. [579]

Bertoin, J. (1995). Some applications of subordinators to local times of Markov processes. *Forum Math.*, *7*, 629–644. [527]

Bertoin, J. (1996). *Lévy Processes*. Cambridge: Cambridge University Press. [142, 188]

Bertoin, J. and Caballero, M. (1995). On the rate of growth of subordinators with slowly varying Laplace exponents. In *Sem. de Prob. XXIX*, volume 1613 of *Lecture Notes Math* (pp. 125–132). Berlin: Springer–Verlag. [527]

Biane, P. and Yor, M. (1988). Sur la loi des temps locaux browniens en un temps exponential. In *Séminaire de Probabilités XXII*, volume 1321 of *Lecture Notes Math* (pp. 454–466). Berlin: Springer-Verlag. [61, 550]

Billingsley, P. (1968). *Convergence of Probability Measures*. New York: Wiley. [592]

Bingham, N., Goldie, C., and Teugels, J. (1987). *Regular Variation*. Cambridge: Cambridge University Press. [304, 438, 594]

Blackburn, R. (2000). Large deviations of local times of Lévy processes. *J. Theor. Probab.*, *18*, 825–842. [527]

Blumenthal, R. and Getoor, R. (1968). *Markov Processes and Potential Theory*. New York: Academic Press. [61, 119, 120]

Bobkov, S. (1996). A functional form of the isoperimetric inequality for the Gaussian measure. *J. Funct. Anal.*, *135*, 39–49. [242]

Borell, C. (1975). The Brunn-Minkowski inequality in Gauss space. *Invent. Math.*, *30*, 207–216. [241]

Bouleau, N. and Yor, M. (1981). Sur la variation quadratique de temps locaux de certaines semi-martingales. *C. R. Acad. Sc. Paris*, *292*, 491–492. [496]

Boylan, E. (1964). Local times for a class of Markov processes. *Ill. J. Math.*, *8*, 19–39. [120]

Brydges, D., Fröhlich, J., and Spencer, T. (1982). The random walk representation of classical spin systems and correlation inequilties. *Comm. Math. Phys.*, *83*, 123–150. [1]

Chung, K. L. (1967). *Markov Chains with Stationary Transition Probabilities* (second ed.). New York: Springer-Verlag. [455]

Chung, K. L. (1974). *A Course in Probability Theory* (second ed.). San Diego: Academic Press. [592]

Chung, K. L. (1982). *Lectures from Markov Processes to Brownian Motion*. New York: Springer-Verlag. [61, 119]

Chung, K. L., Erdös, P., and Sirao, T. (1959). On the Lipchitz's condition for Brownian motion. *J. Math. Soc. Japan*, *11*, 263–274. [361]

de la Vega, F. W. (1973). On almost sure convergence of quadratic Brownian variation. *Ann. Probab.*, *2*, 551–552. [456, 495]

Dellacherie, C., Maisonneuve, B., and Meyer, P.-A. (1992). *Probabilitiés et Potential, Chapitres XVII à XXIV*. Paris: Hermann. [119]

Dellacherie, C. and Meyer, P.-A. (1978). *Probabilities and Potential*. Paris, Amsterdam: Hermann, North Holland. [175]

Dellacherie, C. and Meyer, P.-A. (1980). *Probabilities and Potential B, Theory of Martingales*. Amsterdam: North Holland. [73, 180]

Dellacherie, C. and Meyer, P.-A. (1987). *Probabilitiés et Potential, Chapitres XII à XVI*. Paris: Hermann. [119, 188]

Donoghue, W. (1969). *Distributions and Fourier Transforms*. New York: Academic Press. [236]

Donsker, M. and Varadhan, S. R. S. (1977). On laws of the iterated logarithm for local times. *Comm. Pure Appl. Math.*, *30*, 707–753. [527]

Doob, J. (1953). *Stochastic Processes.* New York: Willey. [237]

Dudley, R. M. (1967). The sizes of compact subsets of Hilbert space and the continuity of Gaussian process. *J. Funct. Anal*, *1*, 290–330. [280]

Dudley, R. M. (1973). Sample functions of the Gaussian process. *Ann. Probab.*, *1*, 66–103. [5, 280, 281, 456, 495]

Dudley, R. M. (1989). *Real Analysis and Probability.* Belmont, CA: Wadsworth. [215]

Dudley, R. M. (1999). *Uniform Central Limit Theorems.* Cambridge: Cambridge University Press. [4, 241]

Dynkin, E. B. (1983). Local times and quantum fields. In *Seminar on Stochastic Processes*, volume 7 of *Progress in Probability* (pp. 64–84). Boston: Birkhäuser. [1, 394]

Dynkin, E. B. (1984). Gaussian and non-Gaussian random fields associated with Markov processes. *J. Fcnl. Anal.*, *55*, 344–376. [1, 394]

Ehrhard, A. (1983). Symétrisation dans l'espace de Gauss. *Math. Scand.*, *53*, 281–301. [241]

Eisenbaum, N. (1994). Dynkin's isomorphism theorem and the Ray–Knight theorems. *Prob. Theory Related Fields*, *99*, 321–335. [61, 534, 550]

Eisenbaum, N. (1995). Une version sans conditionnement du theoreme d'isomorphisme de Dynkin. In *Séminaire de Probabilités XXIX*, volume 1613 of *Lecture Notes Math* (pp. 266–289). Berlin: Springer-Verlag. [1, 394]

Eisenbaum, N. (1997). On the most visited sites by a symmetric stable process. *Prob. Theory Related Fields*, *107*, 527–535. [527]

Eisenbaum, N. (2003). On the infinite divisibility of squared Gaussian processes. *Prob. Theory Related Fields*, *125*, 381–392. [6, 579]

Eisenbaum, N. (2005). A connection between Gaussian processes and Markov processes. *Elec. J. Probab.*, *10*, 202–215. [6, 579]

Eisenbaum, N. and Kaspi, H. (2006). A characterization of the infintely divisible squared Gaussian process. *Ann. Probab.*, *34*. [6, 579]

Eisenbaum, N., Kaspi, H., Marcus, M. B., Rosen, J., and Shi, Z. (2000). A Ray–Knight theorem for symmetric Markov processes. *Ann. Probab.*, *28*, 1781–1796. [1, 61, 395]

Eisenbaum, N. and Khoshnevisan, D. (2002). On the most visited sites of symmetric Markov process. *Stoch. Proc. Appl.*, *101*, 241–256. [527]

Eisenbaum, N. and Shi, Z. (1999). Measuring the rarely visited sites of Brownian motion. *J. Theor. Probab.*, *12*, 595–613. [60]

Feller, W. (1971). *An Introduction to Probability Theory and its Applications, Vol. II.* New York: Wiley. [91, 141, 166, 382, 565, 566, 594, 596]

Fernique, X. (1970). Intégrabilité des vecteurs Gaussiens. *Comptes Rendus Acad. Sci. Paris Sér A*, *270*, 1698–1699. [242]

Fernique, X. (1971). Regularité de processus Gaussiens. *Invent. Math.*, *12*, 1304–320. [242]

Fernique, X. (1974). Des résultats nouveaux sur les processus Gaussiens. *Comptes Rendus Acad. Sci. Paris Sér A–B*, *278*, A363–A365. [242, 280]

Fernique, X. (1975). Regularité des trajectoires des fonctions aléatoires Gaussiennes. In *École d,Été de Probabilités de Saint–Flour, IV–1974*, volume 480 of *Lecture Notes Math* (pp. 1–96). Berlin: Springer Verlag. [241, 280]

Fernique, X. (1997). *Fonctions Aléatoires Gaussiennes Vecteurs Aléatoires Gaussiennes.* Montreal: CRM. [4, 241, 242, 280, 281, 361]

Fitzsimmons, P. and Pitman, J. (1999). Kac's moment formula for aditive functionals of a Markov process. *Stochastic Process. Appl.*, *79*, 117–134. [61, 120]

Fukushima, M., Oshima, Y., and Takeda, M. (1994). *Dirichlet Forms and Symmetric Markov Processes*. New York: Walter de Gruyter. [119]

Garsia, A. M., Rodemich, E., and Rumsey, Jr., H. (1970). A real variable lemma and the continuity of paths of some Gaussian processes. *Indiana Math. J.*, *20*, 565–578. [280, 361]

Getoor, R. (1975). *Markov Processes: Ray Processes and Right Processes*. New York: Volume 440 of Lecture Notes in Mathematics, Springer-Verlag. [119, 188]

Getoor, R. and Kesten, H. (1972). Continuity of local times of Markov processes. *Comp. Math.*, *24*, 277–303. [120]

Giné, E. and Kline, R. (1975). On quadratic variation of processes with Gaussian increments. *Ann. Probab.*, *3*, 716–721. [495]

Griffiths, R. C. (1984). Characterizations of infinitely divisible multivariate gamma distributions. *Jour. Multivar. Anal.*, *15*, 12–20. [6, 579]

Hawkes, J. (1985). Local times as stationary processes. In *From Local Times to Global Geometry, Control and Physics*, volume 150 of *Pitman Research notes in Math.* (pp.1). New York: Longman. [120]

Horn, R. A. and Johnson, C. R. (1999). *Matrix Analysis*. Cambridge: Cambridge Univ. Press. [579]

Hu, Y. and Shi, Z. (1998). The limits of Sinai's simple random walk in random environment. *Ann. Probab.*, *26*, 1477–1521. [60]

Ibragamov, I. and Linnik, Y. (1971). *Independent and Stationary Sequences of Random Variables*. Groningen, Netherlands: Wolters–Noordhoff. [142]

Ito, K. and McKean, H. (1974). *Diffusion Processes and Their Sample Paths*. New York: Springer-Verlag. [59]

Ito, K. and Nisio, M. (1968a). On the convergence of sums of independent Banach space valued random variables. *Osaka J. Math.*, *5*, 35–48. [593]

Ito, K. and Nisio, M. (1968b). On the oscillation function of a Gaussian process. *Math. Scand.*, *22*, 209–233. [213, 241]

Jain, N. and Kallianpur, G. (1972). Oscillation function of a multiparameter Gaussian process. *Nagoya Math. J.*, *47*, 15–28. [213, 241]

Jain, N. and Marcus, M. B. (1978). Continuity of subgaussian processes. In *Probability on Banach Spaces*, volume 4 of *Advances in Probability* (pp. 81–196). New York: Marcel Dekker. [241, 242, 280, 281, 593]

Jain, N. and Monrad, D. (1983). Gaussian measures in b_p. *Ann. Probab.*, *11*, 46–57. [495]

Kahane, J.-P. (1985). *Some Random Series of Functions*. Cambridge: Cambridge University Press. [229]

Kallenberg, O. (2001). *Foundations of Modern Probability* (second ed.). New York: Springer-Verlag. [188]

Karamata, J. (1930). Sur un mode de croissance régulière des fonctions. *Mathematica (Cluj)*, *4*, 38–53. [594]

Katzenelson, Y. (1968). *Harmonic Analysis*. New York: Wiley. [336]

Kawada, T. and Kono, N. (1973). On the variation of Gaussian processes. In *Proceedings of the Second Japan-USSR Symposium on Probability Theory*, volume 330 of *Lecture Notes Math* (pp. 175–192). Berlin: Springer-Verlag. [495]

Khoshnevisan, D. (2002). *Multiparameter Processes*. New York: Springer-Verlag. [188]

Knight, F. (1969). Random walks and a sojourn density process of Brownian motion. *Trans. Amer. Math. Soc.*, *143*, 173–185. [49, 58, 61]

Knight, F. (1981). *Essentials of Brownian motion and diffusion.* Providence, RI: AMS. [61]

Kolmogorov, A. N. (1951). On the differentiability of the transition probabilities in homogeneous Markov processes with a denumerable number of states. *Učhen. Zap. MGY*, *148*, 53–59. [441]

Kono, N. (1969). Oscillation of sample functions in stationary Gaussian processes. *Osaka J. Math.*, *6*, 1–12. [495]

Kono, N. (1970). On the modulous of continuity of sample functions of Gaussian processes. *J. Math. Kyoto Univ.*, *10*, 493–536. [361]

Lacey, M. (1990). Large deviations for the maximal local time of stable Lévy processes. *Ann. Probab.*, *18*, 1669–1674. [527]

Landau, H. J. and Shepp, L. A. (1971). On the supremum of a Gaussian process. *Sankhyā* Ser. A., *32*, 369–378. [242]

Le Jan, Y. (1988). On the Fock space representation of functionals of the occupation field and their renormalization. *J. Fcnl. Anal.*, *80*, 88–108. [119]

Ledoux, M. (1998). A short proof of the Gaussian isoperimetric inequality. In *High Dimensional Probability*, volume 43 of *Progress in Probability* (pp. 229–232). Boston: Birkhäuser. [241, 242]

Ledoux, M. and Talagrand, M. (1991). *Probability in Banach Spaces.* New York: Springer-Verlag. [4, 241, 242, 280, 580, 591]

Lévy, P. (1940). Le mouvement brownian plan. *Amer. J. Math.*, *62*, 487–550. [456]

Lévy, P. (1948). *Processus stochastiques et mouvement brownien.* Paris: Gauthier-Villars. [61]

Li, W. and Shao, Q. (2002). A normal comparison inequality and its applications. *Prob. Theory Related Fields*, *122*, 494–508. [242]

Lifshits, M. A. (1995). *Gaussian Random Functions.* Dordrecht: Kluwer. [241]

Marcus, M. B. (1968). Hölder conditions for Gaussian processes with stationary increments. *Trans. Amer. Math. Soc.*, *134*, 29–52. [361]

Marcus, M. B. (1970). Hölder conditions for continuous Gaussian processes. *Osaka J. Math.*, *7*, 483–494. [361]

Marcus, M. B. (1972). Gaussian lacunary series and the modulus of continuity for Gaussian processes. *Z. Wahrscheinlichkeitstheorie und Verw. Gebiete*, *22*, 301–322. [361]

Marcus, M. B. (1991). Rate of growth of local times of strongly symmetric Markov processes. In *Proceedings Seminar on Stochastic Processes, 1990*, volume 24 (pp. 253–885). Boston: Birkhauser. [455]

Marcus, M. B. (2001). The most visited sites of certain Lévy processes. *J. Theor. Probab.*, *14*, 867–886. [527]

Marcus, M. B. and Pisier, G. (1981). *Random Fourier Series with Applications to Harmonic Analysis.* Princeton, NJ: Princeton University Press. [241, 280, 281]

Marcus, M. B. and Rosen, J. (1992a). Moduli of continuity of local times of strongly symmetric Markov processes via Gaussian processes. *J. Theor. Probab.*, *5*, 791–825. [5, 361, 454]

Marcus, M. B. and Rosen, J. (1992b). Moment generating functions for the local times of symmetric Markov processes and random walks. In *Probability in Banach Spaces 8, Proceedings of the Eighth International Conference, Progress in Probability series*, volume 30 (pp. 364–376). Boston: Birkhauser. [579]

Marcus, M. B. and Rosen, J. (1992c). p-variation of the local times of symmetric stable processes and of Gaussian processes with stationary increments. *Ann. Probab.*, *20*, 1685–1713. [495, 496]

Marcus, M. B. and Rosen, J. (1992d). Sample path properties of the local times of strongly symmetric Markov processes via Gaussian processes. *Ann. Probab.*, *20*, 1603–1684. [1, 2, 5, 120, 152, 188, 241, 394, 395, 454, 455]

Marcus, M. B. and Rosen, J. (1993). ϕ-variation of the local times of symmetric Lévy processes and stationary Gaussian processes. In *Seminar on Stochastic Processes, 1992*, volume 33 of *Progress in Probability* (pp. 209–220). Boston: Birkhauser. [455, 457, 495]

Marcus, M. B. and Rosen, J. (1994b). Laws of the iterated logarithm for the local times of recurrent random walks on Z^2 and of Lévy processes and recurrent random walks in the domain of attraction of Cauchy random variables. *Ann. Inst. H. Poincaré Prob. Stat.*, *30*, 467–499. [527]

Marcus, M. B. and Rosen, J. (1994a). Laws of the iterated logarithm for the local times of symmetric Lévy processes and recurrent random walks. *Ann. Probab.*, *22*, 626–658. [527]

Marcus, M. B. and Rosen, J. (1999). *Renormalized Self-intersection Local Times and Wick Power Chaos Processes*, volume 675. Providence: Memoirs of the AMS. [455]

Marcus, M. B. and Rosen, J. (2001). Gaussian processes and the local times of symmetric Lévy processes. In *Lévy Processes–Theory and Applications*, volume 1 of *Math* (pp. 67–89). Boston: Birkhauser. [395]

Marcus, M. B. and Rosen, J. (2003). New perspectives on Ray's theorem for the local times of diffusions. *Ann. Probab.*, *31*, 882–913. [550]

Marcus, M. B. and Shepp, L. A. (1972). Sample behavior of Gaussian processes. In *Proceedings of the Sixth Berkeley Symposium Math. Statist. Prob.*, volume 2 (pp. 423–441). Berkeley: University of California Press. [242, 361]

McKean, H. P. (1962). A Hölder condition for Brownian local time. *J. Math. Kyoto Univ.*, *1*, 195–201. [120]

McShane, E. J. and Botts, T. A. (1959). *Real Analysis*. Princeton, NJ: Van Nostrand. [210]

Meyer, P.-A. (1966). Sur les lois de certaines fonctionnelles multiplicative. *Publ. Inst. Statist. Univ. Paris*, *15*, 295–310. [120]

Millar, P. W. and Tran, L. T. (1974). Unbounded local times. *Z. Wahrscheinlichkeitstheorie und Verw. Gebiete*, *30*, 87–92. [120]

Molchan, G. (1999). Maximum of fractional Brownian motion: Probabilities of small values. *Comm. Math. Phys.*, *205*, 97–111. [5, 527]

Nisio, M. (1967). On the extreme values of Gaussian processes. *Osaka J. Math.*, *4*, 313–326. [361]

Perkins, E. (1982). Local time is a semimartingale. *Z. Wahrscheinlichkeitstheorie und Verw. Gebiete*, *60*, 79–117. [59, 496]

Pitman, E. J. G. (1968). On the behavior of the characteristic function of a probability distribution in the neighbourhood of the origin. *J. Australian Math. Soc. Series A*, *8*, 422–443. [594]

Port, S. C. and Stone, C. J. (1978). *Brownian Motion and Classical Potential Theory*. New York: Academic Press. [61]

Preston, C. (1971). Banach spaces arising from some integral inequalities. *Indiana Math. J.*, *20*, 997–1015. [280]

Preston, C. (1972). Continuity properties of some Gaussian processes. *Ann. math. Statist.*, *43*, 285–292. [280]

Ray, D. (1963). Sojourn times of a diffusion processs. *Ill. J. Math.*, *7*, 615–630. [5, 49, 58, 59, 61, 120, 144, 188, 537, 550]

Reuter, G. E. H. (1969). Remarks on a Markov chain example of Kolmogorov. *Z. Wahrscheinlichkeitstheorie und Verw. Gebiete*, *13*, 315–320. [455]

Revuz, D. and Yor, M. (1991). *Continuous Martingales and Brownian Motion.* New York: Springer-Verlag. [56, 61, 123, 125, 173, 581]

Rogers, L. C. G. and Williams, D. (2000a). *Diffusions, Markov Processes, and Martingales. Volume Two: Foundations.* Cambridge: Cambridge University Press. [144, 175]

Rogers, L. C. G. and Williams, D. (2000b). *Diffusions, Markov Processes, and Martingales. Volume One: Foundations.* Cambridge: Cambridge University Press. [61, 86, 101, 119, 124, 150, 366]

Rosen, J. (1986). Tanaka's formula for multiple intersections of planar Brownian motion. *Stochastic Process. Appl.*, *23*, 131–141. [61]

Rosen, J. (1991). Second order limit laws for the local times of stable processes. In *Séminaire de Probabilités XXV*, volume 1485 of *Lecture Notes Math* (pp. 407–424). Berlin: Springer-Verlag. [496]

Rosen, J. (1993). *p*-variation of the local times of stable processes and intersection local time. In *Seminar on Stochastic Processes, 1991*, volume 33 of *Progress in Probability* (pp. 157–168). Boston: Birkhauser. [496]

Royden, H. L. (1988). *Real Analysis, third edition.* New York: Macmillan. [590]

Sato, K. (1999). *Lévy Processes and Infinitely Divisible Distributions.* Cambridge: Cambridge University Press. [212]

Sharpe, M. (1988). *General Theory of Markov Processes.* New York: Academic Press. [119]

Sheppard, P. (1985). On the Ray–Knight property of local times. *J. London Math. Soc.*, *31*, 377–384. [61, 550]

Sirao, T. and Watanabe, H. (1970). On the upper and lower class for stationary Gaussian processes. *Trans. Amer. Math. Soc.*, *147*, 301–331. [361]

Slepian, D. (1962). The one-sided barrier problem for Gaussian noise. *Bell System Tech. J.*, *41*, 463–501. [242]

Stroock, D. W. (1993). *Probability Theory: An Analytic View.* Cambridge: Cambridge University Press. [188]

Sudakov, V. N. (1973). A remark on the criterion of continuity of Gaussain sample functions. In *Proceedings Second Japan-USSR Symposium on Probabability Theory Kyoto, 1972*, volume 330 of *Lecture Notes in Math.*, (pp. 444–454)., Berlin. Springer-Verlag. [242]

Sudakov, V. N. and Tsirelson, B. S. (1978). Extremal properties of half–spaces for spherically invariant measures. *J. Soviet. Math.*, *9*, 9–18. Translated from Zap. Nauch. Sem. L.O.M.I. 41, 14–24, 1974. [241]

Talagrand, M. (1987). Continuity of Gaussian processes. *Acta Math.*, *159*, 99–149. [280]

Talagrand, M. (1992). A simple proof of the majorizing measure theorem. *Geom. Funct. Anal.*, *2*, 118–125. [242, 280]

Talagrand, M. (2005). *The Generic Chaining.* Berlin: Springer-Verlag. [281]

Taylor, S. J. (1972). Exact asymptotic estimates of Brownian path variation. *Duke. Math. J.*, *39*, 219–242. [5, 456, 495]

Trotter, H. (1958). A property of Brownian motion paths. *Ill. J. Math.*, *2*, 425–433. [61, 120]

van der Hofstad, R., den Hollander, F., and Konig, W. (1997). Central limit theorem for the Edwards model. *Ann. Probab.*, *25*, 573–597. [61]

Walsh, J. (1978). A diffusion with a discontinuous local time. *Asterisque, 52,* 37–46. [455]

Walsh, J. (1983). Stochastic integration with respect to local time. In *Seminar on Stochastic Processes, 1982*, volume 16 of *Progress in Probability* (pp. 237–302). Boston: Birkhauser. [496, 550]

Williams, D. (1974). Path decomposition and continuity of local time for one-dimensional diffusion, I. *Proc. London Math. Soc, 28*, 738–768. [61, 550]

Wittman, R. (1986). Natural densities for Markov transition probabilities. *Prob. Theory Related Fields, 73,* 1–10. [78]

Wojtaszczyk, P. (1991). *Banach Spaces for Analysts.* Cambridge: Cambridge University Press. [591]

Index of notation

Author index

Subject index